T0140119

More information about this series at http://www.springer.com/series/417

Springer Texts in Statistics

Mervyn G. Marasinghe • Kenneth J. Koehler

Statistical Data Analysis Using SAS

Intermediate Statistical Methods

Second Edition

Mervyn G. Marasinghe
Department of Statistics
Iowa State University
Ames, IA, USA

Kenneth J. Koehler
Department of Statistics
Iowa State University
Ames, IA, USA

Additional material to this book can be downloaded from http://extras.springer.com.

ISSN 1431-875X ISSN 2197-4136 (electronic)
Springer Texts in Statistics
ISBN 978-3-319-69238-8 ISBN 978-3-319-69239-5 (eBook)
https://doi.org/10.1007/978-3-319-69239-5

Library of Congress Control Number: 2017959325

Printed on acid-free paper

This Springer imprint is published by the registered company Springer International Publishing AG part of Springer Nature.
The registered company address is: Gewerbestrasse 11, 6330 Cham, Switzerland

To the memory of my father
and gratefully to my mother
and to my loving family Sumi, Kevin, and Neal

— M.G.M.

To my incredible wife, Michelle,
and our children and their families

— K.J.K.

Preface

One of the hazards of writing a book based on a software system is that the release of a newer version of the software on which the book is based may supersede the appearance of the book in print. This happened to the authors with the publication of the earlier edition of this book. However, with a large and well-developed software system like SAS, this is not really an issue, particularly for the beginning user. Because of its complexity and the availability of a variety of analytical tools, the task of learning SAS and then mastering it for everyday use for data analysis has become a long-term project. That is what we found with the earlier edition. Although it was based on SAS Version 9.1, we find that the earlier version is still in use today particularly as a reference and also by international SAS users to whom a later version of SAS may not be available. The new edition is based on the current version of SAS, Version 9.4, although it was released almost 4 years ago.

As discussed in the preface of the first edition, the aim of this book is to teach how to use the SAS software system for statistical analysis of data. While the book is intended to be used as a textbook in a second course in statistical methods taught primarily to advanced undergraduates in statistics and graduate students in many other disciplines that involve the use of statistics for data analysis, it would be a valuable source of information for researchers in the academic setting as well as professionals in the industry and business that use the SAS system in their work. In particular, data analysis has become an important tool in the general area of data science now being offered as a separate area of study.

The style of presentation of material in the revised book is the same as before: introduction of a brief theoretical and/or methodological description of each topic under discussion including the statistical model used if applicable and presentation of a problem as an application, followed by a SAS analysis of the data provided and a discussion of the results.

The primary reason for planning this revision is the fact that SAS has made a large number of changes beginning with SAS Version 9.2, as well as the introduction of a new system of statistical graphics that essentially replaced the SAS/GRAPH system that existed prior to that version. This necessitated modifications to most of the SAS programs used in the book as well as the rewriting of an entire chapter. The second reason was the incorporation of the ODS system for managing the tabular and graphical output produced from SAS procedures. Not only did this require the reproduction of all output presented in the older version of the textbook, it also required adding additional textual material explaining these changes and the new commands that were required to use the new facility.

This book is intended for use as the textbook in a second course in applied statistics that covers topics in multiple regression and analysis of variance at an intermediate level. Generally, students enrolled in such courses are primarily graduate majors or advanced undergraduate students from a variety of disciplines. These students typically have taken an introductory-level statistical methods course that requires the use of a software system such as SAS for performing statistical analysis. Thus, students are expected to have an understanding of basic concepts of statistical inference such as estimation and hypothesis testing when they begin on a course based on this book.

While the same approach that was used in the first edition is continued, we have rewritten material in almost every chapter; added new examples; completely replaced a chapter; added a new chapter based on SAS procedures for the analysis of nonlinear and generalized linear models; updated all SAS output, including graphics, that appears in the previous version; added more exercise problems to several chapters; and included completely new material on SAS templates in the appendix. These changes necessitated the book to be lengthened by about 200 pages.

We started with a more gentle introductory example but proceed quickly to present more advance material and techniques, especially concerning the SAS data step. Important features such as data step programming, pointers, and line-hold specifiers are described in detail. Chapter 3 which originally contained descriptions of how to use the SAS/GRAPH package was completely rewritten to describe new Statistical Graphics (SG) procedures that are based on ODS Graphics.

The basic theory of statistical methods covered in the text is discussed briefly and then is extended beyond the elementary level. Particular attention has been given to topics that are usually not included in introductory courses. These include discussion of models involving random effects, covariance analysis, variable subset selection methods in regression methods, categorical data analysis, graphical tools for residual diagnostics, and the analysis of nonlinear and generalized linear models. We provide just sufficient information to facilitate the use of these techniques without burgeoning theoretical details. A thorough knowledge of advanced theoretical material such as the theory of the linear model or the theory of maximum likelihood estimation is neither assumed nor required to assimilate the material presented.

SAS programs and SAS program outputs are used extensively to supplement the description of the analysis methods. Example data sets are taken from the areas of biological and physical sciences and engineering. Exercises are included in each chapter. Most exercises involve constructing SAS programs for the analysis of given observational or experimental data. Complete text files of all SAS examples used in the book can be downloaded from the Springer website for this book. Text versions of all data sets used in examples and exercises are also available from the website. Statistical tables are not reprinted in the book.

The first author has taught a one-semester course based on material from this book for many years. The coverage depends on the preparation and maturity level of students enrolled in a particular semester. In a class mainly composed of graduate students from disciplines other than statistics, with adequate knowledge of statistical methods and the use of SAS, the instructor may select more advanced topics for coverage and skip most of the introductory material. Otherwise, in a mixed class of undergraduate and graduate students with little experience using SAS, the coverage is usually 5 weeks of introduction to SAS, 5 weeks on regression and graphics, and 5 weeks of ANOVA applications. This amounts to approximately 60% of the material in the textbook. The structure of sections in the chapters facilitates this kind of selective coverage.

The first author wishes to thank Professor Kenneth J. Koehler, the former chair of the Department of Statistics at Iowa State University, for agreeing to be a coauthor of this book and also to write Chap. 7. He has taught several courses based on the material for that chapter, and some of the examples are taken from his consulting projects.

Mervyn G. Marasinghe
Associate Professor Emeritus
Department of Statistics
Iowa State University, Ames, IA 50011, USA

Kenneth J. Koehler
Professor
Department of Statistics
Iowa State University, Ames, IA 50011, USA

Contents

1

Introduction to the SAS Language

1.1 Introduction

The SAS system is a computer package program for performing statistical analysis of data. The system incorporates data manipulation and input/output capabilities as well as an extensive collection of procedures for statistical analysis of data. The SAS system achieves its versatility by providing users with the ability to write their own program statements to manipulate data as well as call up SAS routines called *procedures* for performing major statistical analysis on specified *data sets*. The user-written program statements usually perform data modifications such as transforming values of existing variables, creating new variables using values of existing variables, or selecting subsets of observations. The statements and the syntax available to perform these manipulations are quite extensive so that these comprise an entire programming language. Once data sets have thus been prepared, they are used as input to statistical procedures that performs the desired analysis of the data. SAS will perform any statistical analysis that the user correctly specifies using appropriate SAS procedure statements.

When SAS programs are run under the SAS windowing environment, the source code is entered in the SAS *Program Editor* window and submitted for execution. A *Log* window which shows the details of execution of the SAS code and an *Output* window which shows the results are also parts of this system. Traditionally, results of a SAS procedure were displayed in the output window in the `listing` format using monospace fonts with which users of SAS in its previous versions are more familiar. SAS provides the user the ability to manage where (the *destination*) and in what format the output is produced and displayed, via the SAS Output Delivery System (ODS). For example, output from executing a SAS procedure may be directed to a `pdf` or an `html` formatted file, the content to be included in the output selected and

M. G. Marasinghe, K. J. Koehler, *Statistical Data Analysis Using SAS*,
Springer Texts in Statistics, https://doi.org/10.1007/978-3-319-69239-5_1

formatted by the user to produce a desired appearance (called an *ODS style*). Thus ODS allows the user the flexibility in presenting the output from SAS procedures in a style of user's own choice. Beginning with SAS Version 9.3, instead of routing the output to a `listing` destination in the output window, SAS windowing system is set up by default to use an `html` destination and for the resulting html file to be automatically displayed using an internal browser. The user may modify these default settings by selecting `Tools` ➡ `Options` ➡ `Preferences` from the main menu system on the SAS window. Figure 1.1 shows the default settings under the `Results` tab of the `Preferences` window.

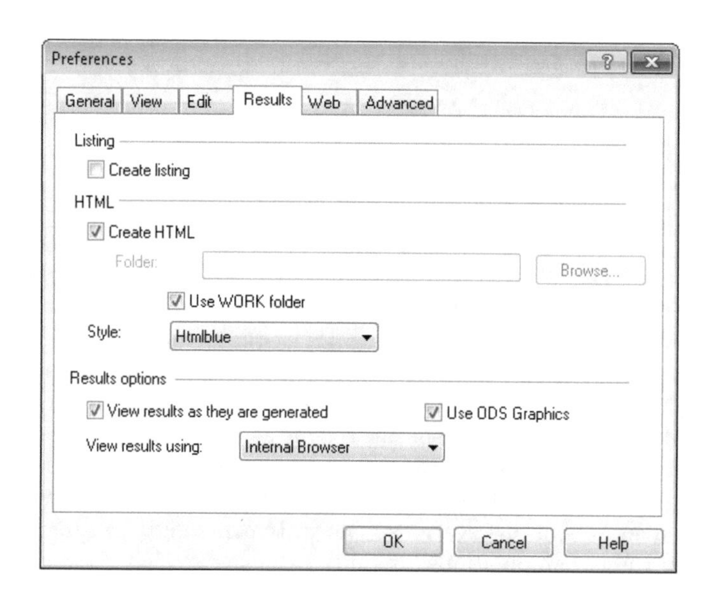

Fig. 1.1. Screenshot of the results tab on the preferences dialog box

Note the check boxes that are selected on this dialog. Thus the creation of `html` output is enabled by default, while the creation of the `listing` output is not. Also note that the *style* selected (from a drop-down list) is `Htmlblue`, the default style associated with the `html` destination. An ODS style is a description of the appearance and structure of tables and graphs in the ODS output and how these are integrated in the output and is specified using a *style template*. The `Htmlblue` style is an all-color style that is designed to integrate tables and statistical graphics and present these as a single entity. Note that the `Use ODS Graphics` box is checked meaning that the creation of ODS Graphics, the functionality of automatically creating statistical graphics, is also enabled. This is equivalent to including a `ODS Graphics On` statement within the SAS program, whenever ODS Graphics are to be produced by default or as a result of a user request initiated from a procedure that supports ODS Graphics. The following example illustrates the default ODS output produced by SAS.

```
data biology;
input Id Sex $ Age Year Height Weight;
BMI=703*Weight/Height**2;
datalines;
7389   M  24  4  69.2  132.5
3945   F  19  2  58.5  112.0
4721   F  20  2  65.3   98.6
1835   F  24  4  62.8  102.5
9541   M  21  3  72.5  152.3
2957   M  22  3  67.3  145.8
2158   F  21  2  59.8  104.5
4296   F  25  3  62.5  132.5
4824   M  23  4  74.5  184.4
5736   M  22  3  69.1  149.5
8765   F  19  1  67.3  130.5
5734   F  18  1  64.3  110.2
4529   F  19  2  68.3  127.4
8341   F  20  3  66.5  132.6
4672   M  21  3  72.2  150.7
4823   M  22  4  68.8  128.5
5639   M  21  3  67.6  133.6
6547   M  24  2  69.5  155.4
8472   M  21  2  76.5  205.1
6327   M  20  1  70.2  135.4
8472   F  20  4  66.8  142.6
4875   M  20  1  74.2  160.4
;

proc means data=biology mean std min max maxdec=3;
  class Sex;
  var BMI;
  title "Biology class: BMI Statistics by Gender";
run;
```

Fig. 1.2. Illustrating ODS output

An Introductory SAS Program

The SAS code displayed in Fig. 1.2 is used here to give the reader a quick introduction to a complete SAS program. The raw data consists of values for several variables measured on students enrolled in an elementary biology class at a college during a particular semester. In this program an `input` statement reads raw data from data lines embedded in the program (called *instream data*) and creates a SAS data set named `biology`.

The *list input style* used in this program scans the data lines to access values for each of the variables named in the input statement. Notice that the data values are aligned in columns but also are separated by (at least) one blank. The "$" symbol used in the input statement indicates that the variable named `Sex` contains character values. The SAS expression `703*Weight/Height**2` calculates a new value using the values of the two variables `Weight` and `Height` obtained from the current data line being processed and assigns it to a (newly created) variable named `BMI` representing the body mass index of the individual (the conversion factor 703 is required as the two variables Weight and Height were not recorded in metric units as needed by the definition of body mass index). Once the SAS data set is created and saved in a temporary folder, the SAS procedure named MEANS is used to produce an analysis containing some statistics for the new variable BMI separately for the females and males in the class. Figure 1.3 displays a reproduction of the default html output displayed by the Results Viewer in SAS and illustrates the *Htmlblue* style.

Biology class: BMI Statistics by Gender

The MEANS Procedure

Analysis Variable : BMI					
Sex	**N Obs**	**Mean**	**Std Dev**	**Minimum**	**Maximum**
F	10	20.366	2.341	16.256	23.846
M	12	21.236	1.775	19.085	24.638

Fig. 1.3. ODS output

In most of the SAS examples used in this book, the pdf-formatted ODS version of the resulting output will be used to display the output. An ODS statement (not shown in all SAS programs) will be used to direct the output produced to a pdf destination. Note carefully that since the destination is different from html, the output produced is in a different style than Htmlblue; that is, the output is formatted for printing rather than for being displayed in a browser window.

An alternative way of running SAS programs for producing ODS-formatted output is to use the SAS Enterprise Guide (SAS/EG). SAS/EG is a point-and-click interface for managing data, performing a statistical analysis, and generating reports. Behind the scenes, SAS/EG generates SAS programs that are submitted to SAS, and the results returned back to SAS/EG. Since the focus of this book is SAS programming, general instructions on how to use SAS/EG is not discussed here. However, SAS/EG includes a full programming interface that uses a color-coded, syntax-checking SAS language editor that can be used to write, edit, and submit SAS programs and is available to SAS programmers as an alternative to using the SAS windowing environment. Further, the output in SAS/EG is automatically produced in ODS format, and the user can select options for the output to be directed to a destination such as a pdf or an html file.

Most statistical analysis does not require knowledge of the considerable number of features available in the SAS system. However, even a simple analysis will involve the use of some of the extensive capabilities of the language. Thus, to be able to write SAS programs effectively, it is necessary to learn at least a few SAS statement structures and how they work. The following SAS program contains features that are common to many SAS programs.

SAS Example A1

The data to be analyzed in this program consist of gross income, tax, age, and state of individuals in a group of people. The only analysis required is to obtain a SAS *listing* of all observations in the data set. The statements necessary to accomplish this task are given in the program for SAS Example A1 shown in Fig. 1.4.

```
data first ; 2
input (Income Tax Age State)(@4 2*5.2 2. $2.);
datalines ; 1
123546750346535IA
234765480895645IA
348578650595431IA
345786780576541NB
543567511268532IA
231785870678528NB
356985650756543NB
765745630789525IA
865345670256823NB
786567340897534NB
895651120504545IA
785650750654529NB
458595650456834IA
345678560912728NB
346685960675138IA
546825750562527IA
;
proc print ; 3
title 'SAS Listing of Tax data';
run;
```

Fig. 1.4. SAS Example A1: program

In this program those lines that end with a semicolon can be identified as *SAS statements*. The statements that follow the `data first;` statement up to and including the semicolon appearing by itself in a line signaling the end of the lines of data, cause a *SAS data set* to be created. Names for the SAS variables to be created in the data set and the location of their values on each line of data are specified in the `input` statement. The raw data are embedded in the *input stream* (i.e., physically inserted within the SAS program) preceded by a `datalines;` statement 1. The `proc print;` performs the requested analysis of the SAS data set created, namely, to print a listing of the entire SAS data set.

As observed in the SAS Example A1, SAS programs are usually made up of *two kinds of statements*:

- Statements that lead to the creation of SAS data sets
- Statements that lead to the analysis of SAS data sets

The occurrence of a group of statements used for creating a SAS data set (called a *SAS data step*) can be recognized because it begins with a `data` statement **2**, and a group of statements used for analyzing a SAS data set (called a *SAS proc step*) can be recognized because it begins with a `proc` statement **3**. There may be several of each kind of these steps in a SAS program that logically defines a data analysis task.

SAS interprets and executes these steps in their order of appearance in a program. Therefore, the user must make sure that there is a logical progression in the operations carried out. Thus, a `proc` step must follow the `data` step that creates the SAS data set to be analyzed by that `proc` step. Although statements in a data step are executed sequentially, in order that computations are carried out on the data values as expected, statements within the step must also satisfy this requirement, in general, except for certain *declarative* or *nonexecutable* statements. For example, an `input` statement that defines variables must precede executable SAS statements, such as SAS programming statements, that references those variable names.

One very important characteristic of the execution of a SAS data step is that the statements in a data step are executed and an observation written to the output SAS data set, repeatedly for every line of data input in cyclic fashion, until every data line is processed. A detailed discussion of *data step processing* is given in Sect. 1.6.

The first statement following the `data` statement **2** in the data step usually (but not always) is an `input` statement, especially when raw data are being accessed. The input statement used here is a moderately complex example of *a formatted input* statement, described in detail in Sect. 1.4. The *symbols* and *informats* used to read the data values for the variables `Income, Tax, Age,` and `State` from the data lines in SAS Example A1 and their effects are itemized as follows:

- `@4` causes SAS to begin reading each data line *at* column 4.
- `2*5.2` reads data values for `Income` and `Tax` from columns 4–8 and 9–13, respectively, using the informat `5.2` twice, that is, two decimal places are assumed for each value.
- `2.` reads the data value for `Age` from columns 14 and 15 as a whole number (i.e., a number without a fraction portion) using the informat `2.`
- `$2.` reads the data value for `State` from columns 16 and 17 as a character string of length 2, using the informat `$2.`

A semicolon symbol ";" appearing by itself in the first column in a data line signals the end of the lines of raw data supplied instream in the current data step. On its encounter, SAS proceeds to complete the creation of the SAS data set named `first` by closing the file. The `proc print;` **3** that follows the data step signals the beginning of a proc step. The SAS data set processed in this proc step is, by default, the data set created immediately preceding it (in this program the SAS data set `first` was the only one created). Again, by default, all variables and observations in the SAS data set will be processed in this proc step.

The output from execution of the SAS program consists of two parts: the *SAS Log* (see Fig. 1.5), which is a running commentary on the results of ex-

```
2          data first ;
3          input (Income Tax Age State)(@4 2*5.2 2. $2.);
4          datalines;

NOTE: The data set WORK.FIRST has 16 observations and 4 variables.
NOTE: DATA statement used (Total process time): 4
      real time            0.01 seconds
      cpu time             0.01 seconds


21         ;
22         proc print ;
23         title 'SAS Listing of Tax data';
24         run;

NOTE: There were 16 observations read from the data set WORK.FIRST.
NOTE: The PROCEDURE PRINT printed page 1.
NOTE: PROCEDURE PRINT used (Total process time):
      real time            0.03 seconds
      cpu time             0.03 seconds
```

Fig. 1.5. SAS Example A1: log

ecuting each *step* of the entire program, and the *SAS Output* (see Fig. 1.6), which is the output produced as a result of the statistical analysis. In interactive mode under the SAS windowing environment, SAS will display these in separate windows called the *log* and *output* windows. When the results of a program executed in the batch mode are printed, the SAS log and the SAS output will begin on new pages.

SAS Listing of Tax data

Obs	Income	Tax	Age	State
1	546.75	34.65	35	IA
2	765.48	89.56	45	IA
3	578.65	59.54	31	IA
4	786.78	57.65	41	NB
5	567.51	126.85	32	IA
6	785.87	67.85	28	NB
7	985.65	75.65	43	NB
8	745.63	78.95	25	IA
9	345.67	25.68	23	NB
10	567.34	89.75	34	NB
11	651.12	50.45	45	IA
12	650.75	65.45	29	NB
13	595.65	45.68	34	IA
14	678.56	91.27	28	NB
15	685.96	67.51	38	IA
16	825.75	56.25	27	IA

Fig. 1.6. SAS Example A1: pdf-formatted output

The *SAS log* contains error messages and warnings and provides other useful information via `NOTES` **4**. For example, the first NOTE in Fig. 1.5 indicates that a work file containing the SAS data set created is saved in a system folder and is named `WORK.FIRST`. This file is a *temporary* file because it will be discarded at the end of the current SAS session.

The printed output produced by the `proc print;` statement appears in Fig. 1.6. It contains a listing of data for all 16 observations and 4 variables in the data set. By default, *variable names* are used in the SAS output to identify the data values for each variable, and an observation number is automatically generated that identifies each observation. Note also that the data values are also automatically *formatted* for printing using default format specifications. For example, values of both the `income` and `Tax` variables are printed correct to two decimal places, those of the variable `Age` as whole numbers and those of the variable `State` as a string of two characters. These are default formats because it was not specified in the program how these values must appear in the output.

1.2 Basic Language: A Summary of Rules and Syntax

Data Values

Data values are classified as either *character* values or *numeric* values. A character value may consist of as many as 32,767 characters. It may include letters, numbers, blanks, and special characters. Some examples of character values are

MIG7, D'Arcy, 5678, South Dakota

A standard numeric value is a number with or without a decimal point that may be preceded by a plus or minus sign but may not contain commas. Some examples are

71, 0.0038, –4., 8214.7221, 8.546E–2

Data values that are not one of these standard types (such as dates with slashes or numbers with embedded commas) may be accessed using special *informats*, which converts them to an internal value. These are stored then in SAS data sets as character or numeric values as appropriate.

SAS Data Sets

SAS data sets consist of *data values* arranged in a rectangular array as displayed in Fig. 1.7. Data values in a column represents a *variable* and those in a row comprise an *observation*. In addition to the data values, *attributes* associated with each variable, such as the name and type of a variable, are also kept in the *data descriptor* part of the SAS data set. Internally, SAS data sets have a special organization that is different from that of data sets created

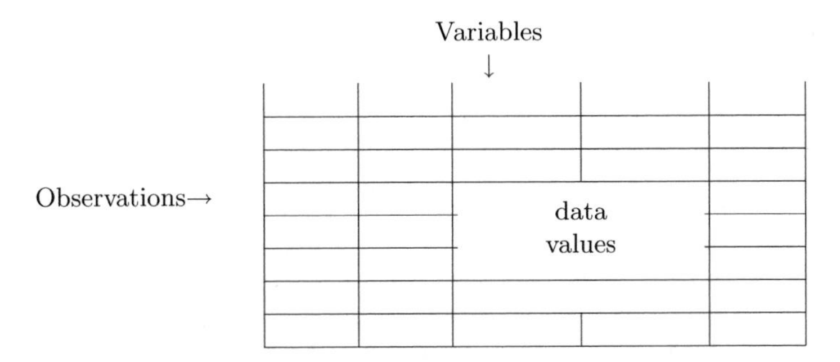

Fig. 1.7. Structure of a SAS data set

using simple editing (e.g., ASCII or flat files). SAS data sets are ordinarily created in a SAS data step and may be stored as *temporary* or *permanent* files. SAS procedures can access data only from SAS data sets. Some procedures are also capable of creating SAS data sets to save information computed as results of an analysis.

Variables

Each column of data values in a SAS data set represents a SAS variable. Variables are of two types: *numeric* or *character*. Values of a numeric variable must be numeric data values, and those of a character variable must be character data values. A character variable can include values that are numbers, but they are treated like any other sequence of characters. SAS cannot perform arithmetic operations on values of a character variable. Certain character strings such as dates are usually converted and stored in a data set numeric values using *informats* when those values are read from external data.

SAS variables have several *attributes* associated with them. The *name* of the variable and its *type* are two examples of variable attributes. The other attributes of a SAS variable include *length* (in bytes), *relative position* in the data set, *informat*, *format*, and *label*. In addition to data values, attribute information of SAS variables is also saved in a SAS data set (as part of the descriptor information).

Observations

An observation is a group of data values that represent different measurements on the same individual. "Individual" here can mean a person, an experimental animal, a geographic region, a particular year, and so forth. Each row of data values in a SAS data set may represent an observation. However, it is possible for each observation in a SAS data set to be formed using data values obtained from several input data lines.

SAS Names

SAS users select names for many elements in a SAS program, including variables, SAS data sets, statement labels, etc. Many SAS names can be up to 32 characters long; others are limited to a length of 8 characters. The first character in a SAS name must be an alphabetic character. Embedded blanks are not allowed. Characters after the first can be alphabetic (upper or lowercase), numeric, or the underscore character. SAS is not case sensitive, except inside of quoted strings. However, SAS will remember the case of variable names used when it displays them later, so it might be useful to capitalize the first letter in variable names. Names beginning with the underscore character are reserved for special system variables. Some examples of variable names are `H22A,` `RepNo,` and `Yield`.

SAS Variable Lists

A list of SAS variables consists of the names of the variables separated by one or more blanks. For example,

```
H22A  RepNo  Yield
```

A user may `define` or `reference` a sequence of variable names in SAS statements by using an abbreviated list of the form

```
charsxx-charsyy
```

where “chars” is a set of characters and the “xx” and “yy” indicate a sequence of numbers. Thus, the list of indexed variables `Q2` through `Q9` may appear in a SAS statement as

```
Q2  Q3  Q4  Q5  Q6  Q7  Q8  Q9
```

or equivalently as `Q2-Q9`.

Using this form in an `input` statement implies that a variable corresponding to each intermediate number in the sequence will be created in the SAS data set and values for them therefore must be available in the lines of data. For example, `Var1-Var4` implies that `Var2` and `Var3` are also to be defined as SAS variables.

Any subset of variables already in a SAS data set may be *referenced*, whether the variable names are numbered sequentially or not, by giving the first and last names in the subset separated by two dashes (e.g., `Id--Grade`). To be able to do this, the user must make sure that the list of variables referenced appears consecutively in the SAS data set. The lists `Id-numeric-Grade` and `Id-character-Grade`, respectively, refer to the subsets of numeric and character variables in the specified range.

SAS Statements

In every SAS documentation describing syntax of particular SAS statements, the general form of the statement is given. In these descriptions, words in boldface letters are *SAS keywords*. Keywords must be used exactly as they appear in the description. SAS keywords may not be used as SAS names. Words in lowercase letters specified in the general form of a SAS statement describe the information a user must provide in those positions.

For example, the general form of the `drop` statement is specified as

`DROP` *variable-list;*

To use this statement, the keyword *drop* must be followed by the names of the variables that are to be omitted from a SAS data set. The variable-list may be one or more variable names (or it may be in any form of a SAS variable list); for example,

`drop X Y2 Age;` or `drop Q1-Q9;`

The individual statement descriptions indicate what information is optional, usually by enclosing them in angled brackets < >; several choices are indicated by the term <`options`>. Some examples are

`OUTPUT` <*data-set-name(s)*>;

`FILENAME` *fileref* <*device-type*><*options*>
<*operating-environment-options*>;

`PROC MEANS` <option(s)> <statistic-keyword(s)>;
`VAR` variable(s) </`WEIGHT`=weight-variable>) ;
`CLASS` variable(s) </option(s >) ;

Syntax of SAS Statements

Some general rules for writing SAS statements are as follows:

- SAS statements can begin and end in any column.
- SAS statements end with a semicolon.
- More than one SAS statement can appear on a line.
- SAS statements can begin anywhere on one line and continue onto any number of lines.
- Items in SAS statements should be separated from neighboring items by one or more blanks. If items in a statement are connected by special symbols such as +, −, /, *, or =, blanks are unnecessary. For example, in the statement `X=Y;` no blanks are needed. However, the statement could also be written in one of the forms `X = Y;` or `X= Y;` or `X =Y;`, all of which are acceptable.

Statements beginning with an asterisk (*) are treated as comments. Multiple comments may be enclosed within of a /* and a */ used at the beginning of a new line. In general, SAS statements are used for data step programming or in the proc step for specifying information to a SAS procedure. Other statements are global in scope and can be used anywhere in a SAS program.

Missing Values

A missing value indicates that no data value is stored for the variable in the current observation. Once SAS determines a value to be missing in the current observation, the value of the variable for that observation will be set to the SAS missing value indicator.

When inputting data, a missing numeric value in the data line can be represented by blanks or a single period, depending on how the values on a data line are input (i.e., what type of *input* statement is used; see below). A missing character value in SAS data is represented by a blank character. SAS also uses this representation when printing missing values of SAS variables.

SAS variables can be assigned a missing value by using statements such as `Score=.` for numeric variables or `Name=' '` for a character variable. Similarly, missing value can be used in comparison operations. For example, to check whether a value of a numeric variable, say `Age`, is missing for a particular observation and then to remove the entire observation from the data set, the following SAS *programming statement* may be used:

```
if Age=. then delete;
```

When a missing value is used in an arithmetic calculation, SAS sets the result of that calculation to a missing value. This is called missing value propagation. Several operations, such as dividing by a zero or numerical calculations that result in overflow, automatically generate a missing value. In comparison operations a numeric missing value is considered smaller than all numbers, and a character missing value is smaller than any printable character value.

A *special missing value* can be used to differentiate among different categories of missing value by using the letters A–Z or an underscore. For example, if a user wants to represent a special type of missing value by the letter A, then the special missing value symbol .A is used to represent the missing value both in the data line and in conditional and/or assignment statements. For example, to process such a missing value a statement such as

```
if Score=.A then Score=0;
```

may be used.

SAS Programming Statements

SAS programming statements are *executable* statements used in `data` step programming and are discussed in Sect. 1.5. Other SAS statements such as the *drop* statement discussed earlier are declarative (i.e., they are used to assign various attributes to variables) and thus are nonexecutable statements.

These include *data, datalines, array, label, length, format, informat, by,* and *where* statements.

1.3 Creating SAS Data Sets

Creating a SAS data set suitable for subsequent analysis in a `proc` step involves the following three actions by the user:

a. Use the `data` statement to indicate the beginning of the `data` step and, optionally, name the data set.
b. Use one of the statements `input` or *set*, to specify the location of the information to be included in the data set.
c. Optionally, modify the data before inclusion in the data set by means of user-written `data` step programming statements. Some of the statements that could be used to do this are described in Sect. 1.5.

```
data first ; 1
input (Income Tax Age State)(@4 2*5.2 2. $2.);
datalines;
123546750346535IA
234765480895645IA
348578650595431IA
345786780576541NB
543567511268532IA
231785870678528NB
356985650756543NB
765745630789525IA
865345670256823NB
786567340897534NB
895651120504545IA
785650750654529NB
458595650456834IA
345678560912728NB
346685960675138IA
546825750562527IA
;
data second; 2
set first;
if Age<35 & State='IA';
run;
proc print; 3
title 'Selected observations from the Tax data set';
run;
```

Fig. 1.8. SAS Example A2: program

Note also that the statements *set*, *merge*, `update`, or `modify` statements may also follow a data statement for creating a new SAS data set using various methods of combining SAS data sets such as concatenating, interleaving, merging, updating, and modifying. Some examples of these methods will be provided in Chap. 2. The basic use of the *input* and the *set* statements for creating and modifying SAS data sets are discussed in this chapter. In this section, the SAS `data step` is used for the creation of SAS data sets and is illustrated by means of some examples. These examples are also used to introduce some variations in the use of several related SAS statements.

SAS Example A2

In the program for SAS Example A2, shown in Fig. 1.8, two SAS data sets are created in separate `data` steps. The first data set (named `first` **1**) uses data included instream preceded by a `datalines;` statement, as in SAS Example A1. The second data set (named `second` **2**) is created by extracting a subset of observations from the existing SAS data set`first`. This is done in the second step of the SAS program.

```
1     data first ;
2     input (Income Tax Age State)(@4 2*5.2 2. $2.);
3     datalines;

NOTE: The data set WORK.FIRST 4 has 16 observations and 4 variables.
NOTE: DATA statement used (Total process time):
      real time           0.29 seconds
      cpu time            0.01 seconds

20    ;
21    data second;
22    set first;
23    if Age<35 & State='IA';
24    run;

NOTE: There were 16 observations read from the data set WORK.FIRST.
NOTE: The data set WORK.SECOND 5 has 5 observations and 4 variables.
NOTE: DATA statement used (Total process time):
      real time           0.01 seconds
      cpu time            0.01 seconds

25    proc print;
NOTE: Writing HTML Body file: sashtml.htm
26    title 'Selected observations from the Tax data set';
27    run;

NOTE: There were 5 observations read from the data set WORK.SECOND.
NOTE: PROCEDURE PRINT used (Total process time):
      real time           0.98 seconds
      cpu time            0.20 seconds
```

Fig. 1.9. SAS Example A2: log

In the second data step, a subset of observations from the SAS data set `first` are used to create the new SAS data set named `second`. The observations that form this subset are those that satisfy the condition(s) in the `if` data modification statement that follows the `set` statement. The input data for this `data step` are already available in the SAS data set `first` which is named in the `set` statement. Note that the if statement used here is of the form `if (expression);`, where the *expression* is a SAS logical expression. As will be discussed in detail in a later section, such expressions may have one of two possible values: TRUE or FALSE. In this form of the if statement, the resulting action is to write the current observation to the output SAS data set if the expression evaluates to a TRUE value. The `if` statement, when present, must follow the `set` statement. (As a rule, SAS programming statements follow the `input` or the `set` statement in data steps.) Clearly, two *data* steps and one *proc* step **3** can be identified in this SAS program.

The SAS log obtained from executing the SAS Example A2 program is reproduced in Fig. 1.9. Note carefully that this indicates the creation of two temporary data sets: `WORK.FIRST` **4** and `WORK.SECOND` **5**. The output from executing the SAS Example A2 program, shown in Fig. 1.10, displays the listing of the observations in the SAS data set named `second` because the `proc print;` step, by default, processes the most recently created SAS data set. It can be verified that these constitute the subset of the observations in the SAS data set named `first` for which the values for the variable `Age` are less than 35 and those for `State` are equal to the character string IA. By executing this program, an ODS-formatted output is also obtained and is displayed in Fig. 1.10. In many of the examples in the rest of this chapter, the output displayed has been produced in the ODS format.

Selected observations from the Tax data set

Obs	Income	Tax	Age	State
1	578.65	59.54	31	IA
2	567.51	126.85	32	IA
3	745.63	78.95	25	IA
4	595.65	45.68	34	IA
5	825.75	56.25	27	IA

Fig. 1.10. SAS Example A2: pdf-formatted output

SAS Example A3

The SAS Example A3 program, shown in Fig. 1.11, illustrates how the `proc step` in SAS Example A2 can be modified to obtain the listing of the same subset of observations without the creation of a new SAS data set. This is achieved by the use of the `where` statement in the `proc step`. The `where` statement **1** is an example of a *procedure information statement* described in Sect. 1.8.

```
data first ;
input (Income Tax Age State)(@4 2*5.2 2. $2.);
datalines;
123546750346535IA
234765480895645IA
348578650595431IA
345786780576541NB
543567511268532IA
231785870678528NB
356985650756543NB
765745630789525IA
865345670256823NB
786567340897534NB
895651120504545IA
785650750654529NB
458595650456834IA
345678560912728NB
346685960675138IA
546825750562527IA
;
proc print;
where Age<35 & State='IA'; 1
title 'Selected observations from the Tax data set';
run;
```

Fig. 1.11. SAS Example A3: program

1.4 The INPUT Statement

The *input statement* describes the arrangement of data values in each data line. SAS uses the information supplied in the input statement to produce observations in a SAS data set being created by reading in data values for each of the variables listed in the input statement. There are several methods to input values for variables to form a data set; three of these are summarized below.

List Input

When the data values are separated from one another by one or more blanks, a user may describe the data line to SAS with

INPUT *variable_name_list*;

In this style of data input, the data value for the next variable is read beginning from the first non-blank column that occurs in the data line following the previous value. The variable names are those chosen to be assigned to the variables that are to be created in the new SAS data set. These names follow the rules for valid SAS names. Examples of the use of `list input` are

```
input Age Weight Height;

input Score1-Score10;
```

SAS assigns the first value in each data line to the first variable, the second value to the second variable, and so on. Note that the second statement is a convenient shortened form to read data values into a sequence of ten variables named `Score1, Score2,...,Score10`, respectively.

List input can be used for reading data values for either numeric or character variables. To describe character variables with *list input*, the $ symbol is entered following each character variable name in the list of variables in the input statement. For example, when

```
input State $ Pop Income;
```

is used, SAS infers that the variable `State` will contain character values and Pop and `Income` will contain numeric values. SAS allocates character variables described in this way a maximum length of eight characters (bytes) by default. If a value read from a data line has fewer than eight characters, then it is filled on the right with blanks up to eight characters total. If a value is longer than eight characters, it is truncated on the right to eight characters. Character variables expected to contain values of length more than eight characters can be read using an *informat* in the formatted input method discussed below.

If SAS does not find a value for the next variable on the current data line when using list input, it will move to the next data line and continue to scan for a value. For this reason, when using the list input method, if there are any missing data values, they must be indicated on the data line by entering a period (the SAS missing value indicator as described previously) separated from other data values by at least one blank on either side of the period, instead of leaving it blank.

Formatted Input

For many instream data sets, or those accessed from recording media such as disks or CDs, `list input` may be inappropriate. This is because, in order to save space, the data values contiguous to one another may have been prepared with no spaces or, other characters such as commas, separating them. In such cases, `SAS informats` must be used to input the data.

In general, `informats` can be used to read data lines present in almost any form. They provide information to SAS such as how many columns are occupied by a data value, how to read the data value, and how to store the data value in the SAS data set. The two most commonly used `informats` are those available for the purpose of inputting numeric and character data values.

To read a data value from a data line, the user must specify in which column the data value begins, how many columns to scan, whether the data value is numeric or character, and where, if needed, a decimal point should be placed in the case of a numeric value.

If the data values are in specific columns in the data line (but do not necessarily begin in column 1), to indicate the column to begin reading a data value, the character "@" followed by the column number, placed before the name of a variable, may be used. For example,

```
input @26 Store @45 Sales;
```

tells SAS that a value for the variable `Store` is to be read beginning in column 26 and a value for `Sales` beginning in column 45. Here it is assumed that the values in each data line are separated by blanks (as when using the list input style); otherwise, informats are required to read these values, as described below. When the data values appear in consecutive columns, the use of "@" symbol is not necessary to indicate the position to begin accessing the next value, because the next value is read beginning at the column number immediately following the columns from which the previous value was accessed.

For a numeric variable, the `informat` "`w.`" specifies that the next `w` columns beginning at the current column be read as the variable's value. The `w` must be a positive integer. For example,

```
input @25 Weight 3.;
```

tells SAS to move to column 25 and read the next three columns (i.e., columns 25, 26, and 27) and store the numeric value (in floating point form) as the value for the variable `Weight` in the current observation.

The `informat` "`w.d`" tells SAS to read the variable's value as above and then insert a decimal point before the last `d` digits. For example,

```
input @10 Price 6.2;
```

tells SAS to begin at column 10 and to read the next six columns as a value of `Price`, inserting a decimal point before the last two digits. If a data value already has a decimal point entered, SAS leaves it in place, overriding the specification given in the informat. In the latter case, the `w` in "`w.d`" must also count a column for the decimal point.

For a character variable, the `informat` "`$w.`" tells SAS to begin in the current column and to read the next `w` columns as a character value. Leading and trailing blanks are removed. For example,

```
input @30 Name $45.;
```

tells SAS to read columns 30–74 as a value of the character variable `Name`. To retain leading and trailing blanks if they appear in the data line, a user may use the `$CHARw. informat` instead of `$w`. Some examples below illustrate the use of `informats` in practice. Suppose a data line contains

```
0001IA005040891349
```

where 0001 is the I.D. number of a survey response, IA is the state in which the respondent resides, 5.04 is the number of tons of fertilizer sold in February 1985, 0.89 is the percentage of sales to members, and 1349 is the number of members for this responding farmers' cooperative. Let `Id, State, Fert,`

`Percent`, and `Members` be the names assigned by the user to the corresponding variables. An appropriate `input` statement would be

```
input Id 4.  State $2.  Fert 5.2  Percent 3.2  Members 4.;
```

It is important to note that an "@" symbol is not necessary here to read any of these data values because data values are read beginning in column 1, data values appear consecutively in the data line, and the fields do not contain any blank columns. Thus an "@" symbol is not needed for skipping to any position at the beginning or in the interior of the line of data. Thus SAS automatically accesses the data value for the next variable beginning from the column following the last value.

Suppose, instead, that the data line has the following appearance:

```
0001xxxxIA00504x089xxxxxx1349
```

where the `x`'s represent columns of data that are not of interest for the current analysis; these columns may or may not be blanks. Instead of reading these columns, it is possible to skip over to the appropriate column using the "@" symbol or the "+" symbol. For example, after reading a value for `Id`, the value for `State` is read beginning in column 9, using "@9," and after reading values for `State` and `Fert` using appropriate informats, one column is skipped using "+1." The input statement thus could be of the form

```
input Id 4. @9 State $2. Fert 5.2 +1 Percent 3.2 @26 Members 4.;
```

Symbols, such as "@"and "+" that could be used on input statements are called *pointer control* symbols. The use of the *pointer* and *pointer controls* in reading data from an input data line is described in detail in Sect. 1.7.

Finally, the variable names and informats (including pointer controls) that occur on an input statement can be grouped into two separate *lists* enclosed in parentheses. For example, the above statement could also be written as

```
input (Id State Fert Percent Members)(4. @9 $2 5.2 +1 3.2 @26 4.);
```

Here, each informat or pointer control-informat combination is associated with a variable name in the list sequentially. If the informat list is shorter than the number of variables present, then the entire informat list is reapplied to the remaining variables as required.

Column INPUT

Column input is another alternative to list input when the data values are not separated by blanks or other separators, but the user prefers not to use informats. In this case, the values must occupy the same columns on all data lines, a requirement that is also necessary for using formatted input. However, in the input statement, the variable name is followed by the range of columns that the data value occupies in the data line, instead of an informat. The column numbers are specified in the form `begin-end` and are optionally followed by an integer preceded by a decimal point to indicate the number of decimal places to be assumed for the data value. For inputting character strings, the "$" symbol must follow the variable name but before the column specification. Blanks occurring both *before* and *after* the data value are ignored. For example, if the data line has the appearance

```
0001IA  5.04 891349
```

then it could be read, using column input as

```
input Id 1-4 State $ 5-6 Fert 7-12 Percent 13-15 .2 Members 16-19;
```

This reads the value for Id from columns 1 through 4 as an integer and the value for State as a character string from the next two columns. The value for Fert is read as the value exactly as it appears in columns 7 through 12, i.e., as a number with a fractional part. The .2 following 13–15 indicates where the decimal point must be assumed when reading the value for Percent. The value for Percent will thus be read as 0.89 and the value for Members as 1349 from the above data line.

Combining INPUT Styles

An input statement may contain a combination of the above styles of input. For example, as in the previous example, if the data line has the appearance

```
0001IA  5.04 891349
```

then it could be read, using a combination of column, formatted, and list input styles as

```
input Id 1-4 State $2. Fert Percent 2.2 Members 16-19;
```

Here, column input is used to read the value for Id, formatted input to read the value for State, and switches to list input style to read the value for Fert. As mentioned above (and discussed later in Sect. 1.7), this causes the *pointer* to move to column 14 after reading the value for Fert (as it is the next non-blank column). Thus, when using an informat to read the value for Percent, the width of field w must be 2 instead of 3 (i.e., no leading blank). Consequently, the informat 2.2 is used instead of 3.2, as was used in the previous example. Then the value for Members is read using column input again. Thus, a knowledge of how the *pointer* is handled by the three styles of input is necessary to combine them correctly in a single statement. Additionally, the *: modifier* may be used with informats for reading data values of varying widths, as will be illustrated in SAS Example A8 (see Fig. 1.23).

1.5 SAS Data Step Programming Statements and Their Uses

SAS allows the user to perform various kinds of modification to the variables and observations in the data set as it is being created in the data step. The use of the `if Age<35 & State='IA';` statement to obtain a subset of observations in SAS Example A2 is an example of a typical SAS *programming statement*. SAS programming statements are generally used to modify the data during the process of creating a new SAS data set, either from raw data or from data already available in a SAS data set; hence, they must follow an `input` or a `set`, statement. The syntax and usage of several statements available for SAS *data step programming* are discussed below.

Assignment Statements

Assignment statements are used to create new variables and change the values of existing ones. The general form of the `assignment` statement is

`variable_name=` *expression;*

New variables can be created by combining one or more existing variables in an *arithmetic expression*. This may involve combining *arithmetic operators*, *SAS functions*, and other arithmetic expressions enclosed in *parentheses* and assigning the value of that expression to a new variable name. For example, in the SAS data step in Example 1.5.1,

Example 1.5.1

```
data sample;
input(X1-X7) (@5 3*5.1 4*6.2);
Y1 = X1+X2**2;
Y2 = abs(X3)
Y3 = sqrt(X4+4.0*X5**2)-X6;
X7 = 3.14156*log(X7);
datalines;
    .
    .
    .
;
```

three new variables `Y1, Y2,` and `Y3` are created. The value of `Y1` for each observation in the data set, for example, will be the sum of the value of `X1` and the square of the value of `X2` in that observation. The `variable_name` in an assignment statement may be a new variable to be added to the data set and assigned the value of the expression; or it may be a variable already present in the data set, in which case the original value of the variable is replaced by the value resulting from evaluating the expression. Thus, in the above data step, each value of the variable X7 that is input will be replaced by the natural logarithm of the original value of X7 multiplied by 3.14156.

Arithmetic expressions are normally evaluated beginning from the left and proceeding to the right, but applying the Rules 1, 2, and 3, given in Fig. 1.12, may change the order of evaluation. The result of an arithmetic expression containing a missing value is a missing value. The SAS system incorporates a large number of mathematical functions that can be used in the expressions, as shown in the above example. Some examples of the commonly used mathematical functions are `abs`, `log`, and `sqrt`.

SAS Functions

A SAS function is internal code that returns a value that is determined using the current values of user-specified arguments. The general form of a function call is

function-name(argument1,argument2, . . .);

Some examples of function calls are

`mean` (Flavor, Texture, Looks)
`mdy` (Month, Day, Year)
`substr` (Item, 3, 5)

Respectively, in each of the above calls, the `mean` function calculates the average of values of the variables `Flavor`, `Texture`, and `Looks`, the `mdy` function forms a SAS date value using numerical values of `Month`, `Day`, and `Year`, and the `substr` function extracts a substring of length 5 from the character string in the variable `Item`, beginning at character position 3. In general, functions are available for performing mathematical, numerical, probability, and combinatorial operations, computing descriptive statistics including percentiles, manipulating SAS dates and time values, converting state and zip codes, extracting and matching character strings, and performing many other tasks including complex financial calculations.

Arithmetic expressions are evaluated according to a set of rules called *precedence rules*. These rules, summarized in Fig. 1.12, specify the order of evaluation of entities within an expression. It is good programming practice to follow these rules when writing expressions. Some details on the use of the operators in Fig. 1.12 are listed below:

- An infix operator applies to the operands on each side of it. Infix operators $+$, $-$, $*$, $/$ perform the standard arithmetic operations of addition, subtraction, multiplication, and division, respectively. For example, $X+Y$ forms the sum of the values of variables X and Y.
- Infix operators include all comparison, logical, and concatenation operators (i.e., those listed in Groups IV to VII).

`Rule 1.` Expressions within parenthesis are evaluated first.

`Rule 2.` An operator in a higher ranking group below has higher priority and therefore is evaluated before an operator in lower ranking group.

Group I	**, +(prefix), −(prefix), (NOT), >< (MIN), <> (MAX)
Group II	*, /
Group III	+(infix), −(infix)
Group IV	||
Group V	<, <=, =, =, >=, >, >
Group VI	&(AND)
Group VII	|(OR)

`Rule 3.` Operators with the same priority (same group) are evaluated from left to right of the expression (except for Group I operators, which are evaluated right to left).

Fig. 1.12. Order of evaluating expressions

- As a prefix operator, the plus (+) sign or the minus sign (−) can be used to change the sign of a variable, constant, function, or a parenthetical expression. Thus $-(X*Y)$ negates the value of the result of the computation $X*Y$.
- The infix operator ** performs exponentiation, i.e., X**2 raises the value of X to the power of 2. Because Group I operators are evaluated from right to left, the expression $X = -A**2$ is evaluated as $X = -(A**2)$.
- The concatenation operator (||) concatenates character values. For example, `Auto` =‘Chevy’||‘Camaro’ produces the string ‘Chevy Camaro’ as the value of the variable `Auto`.
- The operators in Group V are comparison operators used in logical expressions as described in the next paragraph.
- Depending on the characters available on your keyboard, the symbol for NOT may be one of the not sign (¬), tilde (˜), or caret (ˆ), and the symbol (||) may be represented by (¦¦) or (!!).
- The logical `AND` operator (&) or the `OR` operator (|) is used to form complex expressions by combining several logical expressions. The broken vertical bar (¦) or exclamation mark (!) may be used for the NOT operator.

The assignment statements used in Example 1.5.1 contain only arithmetic expressions. However, variable names may be combined using comparison operators to form *logical expressions* as described in the paragraph below. Both arithmetic and logical expressions may be combined using *logical operators* such as the `and` operator (&) or the `or` operator (|) to form more complex expressions.

Conditional Execution

As in any programming language, several constructs for altering the normal top-down flow of a program are available in SAS. The `if-then` and `else` statements allow the execution of SAS programming statements that depend on the value of an *expression*. The syntax of the statements are

IF *expression* THEN *statement;*
< ELSE *statement;* >

The *expression*, in many cases, is a *logical expression* that evaluates to a *one* if the expression is TRUE or a *zero* if the expression is FALSE. A *logical expression* consists of numerical or character comparisons made using *comparison operators*. These may be combined using *logical operators* such as the `and` operator (&) or the `or` operator (|) to form more complex logical expressions. The *statement* in the above syntax is any executable SAS statement; however, several SAS statements enclosed in a `do-end` group may be used in place of a single SAS statement.

The following examples illustrate typical uses of `if-then/else` statements.

Example 1.5.2

```
if Score < 80 then Weight=.67;
else Weight=.75;
```

In this example, the expression `Score < 80` evaluates to a *one* if the current value of the variable `Score` is less than 80, and in this case, the assignment statement `Weight=.67` will be executed; otherwise, the expression evaluates to a *zero* and the statement `Weight=.75` will be executed. The following statement illustrates a more advanced method for obtaining the same result using the numerical values of the comparisons `Score < 80` and `Score >= 80`:

```
Weight=(Score < 80) *.67 + (Score >= 80) *.75;
```

It becomes clear that this statement will evaluate to the required value depending on the value of the variable `Score` by assigning numerical values 0 or 1 as the resulting values of the parenthesized expressions.

Example 1.5.3

```
if State= 'CA' | State= 'OR' then Region='Pacific Coast';
```

This is an example of the use of an `if-then` statement without the use of an `else` statement. The expression here is a *logical expression* that will evaluate to a *one* if at least one of the comparisons `State= 'CA'` or `State= 'OR'` is true or to a *zero* otherwise. Thus, the current value of the SAS variable `Region` will be set to the character string `'Pacific Coast'` if the current value of the SAS variable `State` is either `'CA'` or `'OR'`. If this is not so, then the current value of `Region` will be determined by `if-then` statements appearing later in the SAS data step, or otherwise will be left blank.

Example 1.5.4

```
if Income= . then delete;
```

The special SAS program statement, `delete`, stops the current data line from being processed further. This observation is not written to the SAS data set being created, and control returns to the beginning of the data step to process the next line of data. In this example, if the current value of the variable `Income` is found to be a SAS missing value, then the observation is not written into the data set as a new observation. The result is that no observation is created from the data line being processed.

In SAS Example A2 (see program in Fig. 1.8), the *subsetting if* statement used was of the form

IF *expression*;

This statement is equivalent to the statement

IF not *expression* THEN delete;

The result is that if the computed value of the expression is FALSE, then the current observation is not written to the output SAS data set. On the other hand, it will be written to the output SAS data set if the expression evaluates to TRUE.

Example 1.5.5

```
if 6.5<=Rate<=7.5 then go to useit;
          ⋮
  ··· SAS program statements ···
  ··· to calculate new rate ···
          ⋮
useit: Cost= Hours*Rate;
          ⋮
```

Sometimes it may be required to avoid executing (or jump over) a few SAS program statements depending on the value of an expression. For this purpose, SAS program statements could be *labeled* using the `label:` notation. In the above example, `useit:` is the label that identifies the SAS statement `Cost= Hours*Rate;` if the expression `if 6.5<=Rate<=7.5` evaluates to TRUE, then control transfers to this statement. Note that the `if 6.5<=Rate<=7.5` statement is a condensed version of the equivalent statement `Rate>=6.5 & Rate<=7.5`, which will evaluate to a *one* only if both of the comparisons `Rate>=6.5` AND `Rate<=7.5` are true or to a *zero* otherwise.

Example 1.5.6

```
if Score < 80 then do;
    Weight=.67;
    Rate=5.70;
    end;
else do;
    Weight=.75;
    Rate=6.50;
    end;
```

A `do-end` group can be used to extend the conditional evaluation of single SAS statements to conditionally executing groups of SAS statements. The above example is a straightforward extension of Example 1.5.2.

SAS Example A4

The extended example shown in Fig. 1.13 illustrates how consecutive `if-then/else` statements can be used to create values for a new variable, as well as how they may be avoided using a convenient transformation.

In the SAS Example A4 program, there are three different data steps, and they create SAS data sets named `group1`, `group2`, and `group3`, respectively. In the first data step **1**, data are read using list input with the statement `input Age @@;`. The `@@` pointer control symbol causes the `input` statement to be repeatedly executed for the data line. Thus, the data set named `group1` will have 14 observations, each with a single value for the variable `Age`.

In the second data step **2**, the SAS data set `group2` will be formed using the observations from `group1` as input, with a new variable named `AgeGroup` being created. The variable `AgeGroup` will be assigned a value for each observation as determined by the value of `Age` in the current observation, by executing the series of `if-then/else` statements. Thus, for example, `AgeGroup` will be assigned a value of zero, since the value of `Age` is 1 in the first observation read.

In the third data step **3**, the SAS data set `group3` will be formed using the observations in `group1` as in the previous step. However, the values for the new variable `AgeGroup` this time are determined simply by executing the

```
data group1; 1
   input Age @@;
datalines;
1 3 7 9 12 17  21 26 30 32 36 42 45 51
;

data group2; 2
set group1;
   if 0<=Age<10 then AgeGroup=0;
   else if 10<=Age<20 then AgeGroup=10;
   else if 20<=Age<30 then AgeGroup=20;
   else if 30<=Age<40 then AgeGroup=30;
   else if 40<=Age<50 then AgeGroup=40;
   else if Age >=50 then AgeGroup=50;
run;

proc print;run;

data group3; 3
set group1;
AgeGroup=int(Age/10)*10;
run;

proc print; run;
```

Fig. 1.13. SAS Example A4: program

arithmetic expression `int`(Age/10) $*$ 10 that converts the value of `Age` to the required values of `AgeGroup`, by a simple mathematical calculation. Note that the `int` function is a SAS function that truncates the result of execution of a numerical expression to the lower integer value.

```
                         The SAS System                         1

                     Obs    Age    AgeGroup

                       1      1        0
                       2      3        0
                       3      7        0
                       4      9        0
                       5     12       10
                       6     17       10
                       7     21       20
                       8     26       20
                       9     30       30
                      10     32       30
                      11     36       30
                      12     42       40
                      13     45       40
                      14     51       50
```

Fig. 1.14. SAS Example A4: listing output

The two `proc print;` statements constitute two proc steps that list two of these data sets `group2` and `group3`, which are identical in content. One of the two data sets is displayed in Fig. 1.14.

Repetitive Computation

Repetitive computation is achieved through the use of *do loops* or *for loops*, respectively, in commonly known low-level programming languages such as Fortran or C. In the SAS data step language, several forms of `do` statements, in addition to the `do-end` groups discussed earlier are available. The statements iterative `do`, `do while`, and `do until` are very flexible, allow a variety of uses, and can be combined. The use of iterative `do` loops in the data step is illustrated in Examples 1.5.7–1.5.10.

Example 1.5.7

```
data scores;
input Quiz1-Quiz5 Test1-Test3;
array scores {8} Quiz1-Quiz5 Test1-Test3;
do I= 1 to 8;
    if scores{I}= . then scores{I}= 0;
end;
datalines;
        ⋮
```

An *iterative do loop*, in general, is used to perform the same operation on a sequence of variables. This requires the sequence of variables to be defined as *elements of an array*, using the `array` statement. This statement, being nonexecutable, may appear anywhere in the data step, but in practice, it is inserted immediately after the variables are defined (usually in the `input` statement). The array definition allows the user to reference a set of variables using the corresponding array elements. This is achieved by the use of *subscripts*.

In Example 1.5.7, the variables `Quiz1,...,Quiz5, Test1,...,Test3` are defined as elements of the array named `scores`, and they are referenced in the do loop as `scores{1},...,scores{8}`, respectively, where the values $1,\ldots,8$ are called the *subscripts*. Within the do loop, the subscripts are assigned by using an *index variable*, here named `I`, that is used as a *counting variable* in the `do` statement. During the execution of the loop (i.e., statements enclosed within the `do` through the `end` statements), the variable `I` takes the values $1,\ldots,8$, sequentially. The task performed by the do loop in Example 1.5.7 is to convert a missing value, read from any data line for any of the above eight variables, to a zero in the corresponding observation written to the data set created.

Example 1.5.8

```
data load;
input D1-D7;
array day {7} D1-D7;
array hour {7} H1-H7;
    do I= 1 to 7;
    if day{I}= 999 then day{I}=.;
    hour{I}= day{I}*12;
end;
datalines;
        ⋮
```

Variables defined in two different arrays may be processed in a single do loop if the two arrays are of the same length. In this example, two arrays, `day` and `hour`, are defined—the first consisting of the variables `D1`–`D7` and the second consisting of a new set of variables `H1`–`H7`. In the do loop, first the value of each of the variables `D1`–`D7` is converted to a missing value if the current value of that variable is 999. Then the current value of each of the variables `H1`–`H7` is set to 12 times the value of each of the corresponding variables `D1`–`D7`, respectively. Note carefully that the second array statement assigns an array name to a set variables `H1`–`H7` yet to be used in the data step.

Example 1.5.9

```
data index;
do A= 1 to 4;
    do B= 3,6,9;
        C=(A-1)*10+B;
        output;
    end;
end;
proc print data=index;
title 'Creating indices';
run;
```

In this SAS program, a *nested do loop* is illustrated using an example where the counting variables `A` and `B` of the do statements are manipulated to create the values of a new variable `C`. This technique is often used for generating factor levels of combinations of factors or interactions in factorial experiments. The `output` statement inside the loop forces a new observation containing current values of the variables `A`, `B`, and `C` to be written to the data set, each pass through the loop. Thus at the end of the processing of the loop, the SAS data set `index` will contain 12 observations corresponding to all 12 combinations of the 4 values of `A` and the 3 values of `B`. The printed listing of this data set is

```
                Creating indices

        Obs     A      B       C

          1     1      3       3
          2     1      6       6
          3     1      9       9
          4     2      3      13
          5     2      6      16
          6     2      9      19
          7     3      3      23
          8     3      6      26
          9     3      9      29
         10     4      3      33
         11     4      6      36
         12     4      9      39
```

The `do while` statement repeatedly executes statements in a do loop repetitively as long as a condition, checked before each iteration, evaluates to TRUE. The `do until` statement executes statements similarly but checks the condition at the end of the loop. The SAS program below calculates the time in months needed to pay off a loan of $5000 that accrues interest at annual rate of 12% if paid off at $500 dollar monthly installments:

Example 1.5.10

```
data loan;
Balance=5000;
   do while (Balance>500);
        Balance+Balance*0.01;
        Balance+ -500;
        Month+1;
        output;
   end;
proc print data=loan;
title 'Loan Amortization';
run;
```

Note that the statements within the loop that involve the + sign are a special type of assignment statements called *sum statements* and have the general form

$$variable + expression;$$

This statement adds the value of *expression* on the right side of the plus sign to the current value of the *variable* which must be of numeric type. This variable automatically *retains* its current value until it is updated during the execution of the loop. If the expression evaluates to a missing value, it is treated as zero. If the variable is not assigned an initial value, it is automatically initialized to zero before the DO loop begins (note the variable `Month` in this example). The printed listing of data set loan is

```
Loan Amortization

Obs     Balance    Month

  1     4550.00      1
  2     4095.50      2
  3     3636.46      3
  4     3172.82      4
  5     2704.55      5
  6     2231.59      6
  7     1753.91      7
  8     1271.45      8
  9      784.16      9
 10      292.00     10
```

The `do until` statement can be used in a similar manner; which version is preferred depends on the application.

1.6 Data Step Processing

A basic understanding of the operations in the SAS data step is necessary to effectively use the capabilities, such as data step programming, available in the data step. The discussion here is kept to a minimum technical level by making use of illustrations and examples. When SAS begins execution of a data step, the statements are first syntax checked and compiled into machine code. At this stage, SAS has sufficient information to create the following:

- an *input buffer*, an area in the memory where the current line of data can be temporarily stored
- a *program data vector* (PDV), an area in the memory where SAS builds an observation to be written to a SAS data set

The PDV is a temporary *placeholder* for *a single value* of each of the variables in the list of variables recognized by SAS to exist at this stage. These locations are all initialized to SAS missing values when the data step processing begins. If some of these variables are not assigned a value either by accessing a value from the input buffer or as a result of a calculation by executing a SAS programming statement, they will remain as missing values until the end of the data step processing. At the discretion of the user, some or all of the variables in the PDV may form the *observation* written to the SAS data set at the end of the data step. The SAS data set is a file in which each observation is written as a separate record and thus will contain the entire set of observations the user opts to include in the data set. On the other hand, the PDV contains only those values of the variables obtained from the *current data line* (or new values calculated using them) at any point in the execution of the data step.

The basic SAS data step begins at the *data* statement. Values for the variables in the PDV are initialized to SAS missing values, a line of data is read into the *input buffer*, and data values transferred into the PDV from the input buffer, replacing the missing values in the PDV. The pointer control

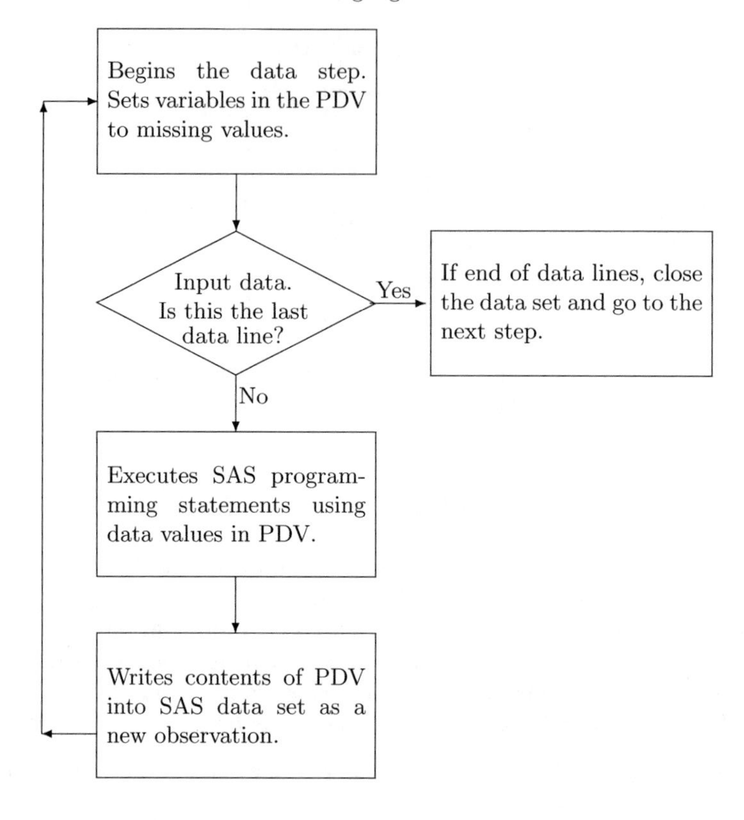

Fig. 1.15. Flow of operations in a data step

symbols and informats in the `input` statement facilitate the conversion of the columns in the input buffer into data values for the variables in the PDV. SAS programming statements are then executed using the *current values* of the variables in the PDV, values in the PDV are then output as a new observation in the SAS data set, and control returns back to the beginning of the data step. Recall that, as explained above, the values of variables in the PDV are reset to missing values at this stage. This is an *iteration* of a SAS data step and the automatically generated SAS variable _N_ keeps track of the current iteration number. The user may make use of this variable in any programming statement in the data step.

In this description it has been assumed that the data step is operating under its default behavior. It is possible for the user to alter the flow of operations described above by various actions, implemented via the inclusion of one or more *executable* SAS programming statements at different points in the data step. For example, if an `output` statement is inserted among the SAS programming statements in the data step, instead of waiting to write an observation to the SAS data set at the end of an iteration of the SAS data step,

SAS will write the current values of the variables as a new observation at the point the `output` statement is encountered. The user may also use a `retain` statement (see Sect. 1.7.4) to keep selected variables from being initialized to a missing value. The flow of operations in a data step is summarized in the chart shown in Fig. 1.15.

SAS Example A5

```
data four;
input X1-X3;
X3= 3*X3-X1**2; 1
X4=sqrt(X2); 2
drop X1 X2; 3
datalines;
3 4 5
-2 9 3
. 16  8
-3 1 4
;
proc print data=four;
title 'Flow of operations in a data step';
run;
```

Fig. 1.16. SAS Example A5: program

A simple example is used to illustrate the flow of operations in a data step described above. Consider the data step in the SAS program shown in Fig. 1.16. This data step creates the SAS data set named `four`. Four data lines with data values for three variables `X1, X2,` and `X3` are read instream. The values of variable `X3` are transformed **1**, and a new variable `X4` **2** is created. Further, only variables `X3` and `X4` are written to the data set.

At the beginning of each iteration of the data step, variables `X1, X2, X3,` and `X4` are initialized to missing values in the PDV because SAS has detected their presence in the data step during the compile stage. The data step execution proceeds as follows:

- The first line of data is transferred to the input buffer.
- The values 3, 4, and 5 are accessed using list input from the input buffer and transferred to the PDV as the new values of variables `X1, X2,` and `X3`; the value for `X4` still remains a missing value.
- A value of 6 is computed by substituting the values of `X1=3` and `X3=5` in the expression $3 * X3 - X1 * *2$. This replaces the current value 5 of the variable `X3` in the PDV.
- The square root of 4, the value of `X2` in the PDV, replaces the missing value of the variable `X4`.
- SAS ascertains that the end of the data step has been reached and writes the observation into the data set (named `four`) using the current values of the variables in the PDV. Variables `X1` and `X2` are excluded from the data set because they appear in the `drop` statement **3**.

- Next, SAS goes back to the beginning of the data step (the `input` statement) and reinitializes the PDV to missing values, and the next line of data is transferred to the input buffer.

The appearance of the PDV (just before writing the first observation to the SAS data set) is

X1	X2	X3	X4	_N_	_ERROR_
3	4	6	2	1	0

The sequence of operations described above continues until end-of-file is detected (i.e., end of the data is encountered) by the `input` statement. SAS then closes the data set and proceeds to the next step. The `drop` statement is a nonexecutable SAS statement and thus may appear anywhere in the data step. It results in the variables listed in the statement being marked so that those variables will be omitted from the observations written to the SAS data set.

```
                  Flow of operations in a data step                      1

                           Obs      X3      X4

                            1        6       2
                            2        5       3
                            3        .       4
                            4        3       1
```

Fig. 1.17. SAS Example A5: output

```
2     data four;
3     input X1-X3;
4     X3= 3*X3-X1**2;
5     X4=sqrt(X2);
6     drop X1 X2;
7     datalines;

NOTE: Missing values were generated as a result of performing an
      operation on missing values.
      Each place is given by: (Number of times) at (Line):(Column).
      1 at 4:9    1 at 4:12
NOTE: The data set WORK.FOUR has 4 observations and 2 variables.
NOTE: DATA statement used (Total process time):
      real time           0.00 seconds
      cpu time            0.01 seconds

11    ;
12    proc print data=four;
13    title 'Flow of operations in a data step';
14    run;

NOTE: There were 4 observations read from the data set WORK.FOUR.
NOTE: PROCEDURE PRINT used (Total process time):
      real time           0.00 seconds
      cpu time            0.01 seconds
```

Fig. 1.18. SAS Example A5: log

A listing of the data set created in the data step described above is produced as the output from the next step in the program and is shown in Fig. 1.17. The SAS Log is shown in Fig. 1.18.

SAS Example A6

The SAS program shown in Fig. 1.19 uses the **array**, **do**, and **output** statements to create a SAS data set that is markedly different in appearance from the instream data set used to create it. This technique, called *transposing*, is useful for preparing SAS data sets for analysis of data obtained from statistically designed experiments (e.g., factorial experiments).

The data for this example, displayed in Fig. 1.19, are scores received by students for five quizzes. The name of the student and the five scores are entered with at least one blank as a separator so that the data can be read using list input. It is important to note that when using list input, missing values must be indicated by a period. If a blank is entered as the missing value, the input statement will mistakenly read the next data value available as the value for the variable for which a value is actually missing in the current line of data.

```
data quizzes;
input Name $ Quiz1-Quiz5;
array qz {5} Quiz1-Quiz5; 1
drop Quiz1-Quiz5;
do Test= 1 to 5;
  if qz{Test}=. then qz{Test}= 0 ;
  Score = qz{Test};
  output; 2
end;
datalines;
Smith  8  7  9  .  3
Jones  4  5 10  8  4
;
proc print data=quizzes;
run;
```

Fig. 1.19. SAS Example A6: program

The five variable names for the quiz scores `Quiz1,...,Quiz5` are declared in an array named `qz` 1 and then used in the `do` loop with a counter variable named `Test`. At the beginning of the data step, variables `Name, Quiz1,...,Quiz5, Test,` and `Score` are all initialized to missing values in the PDV (program data vector). Thus, the appearance of the PDV at the beginning of the data step is

```
Name Test Quiz1 Quiz2 Quiz3 Quiz4 Quiz5  Score   _N_  _ERROR_
       .     .     .     .     .     .      .     1      0
```

The first line of data `Smith 8 7 9 . 3` is transferred to the input buffer. The `input` statement reads these values from the input buffer using the list input style and assigns them as new values of variables `Quiz1,...,Quiz5` in the PDV. Thus, the appearance of the PDV at this stage is

```
Name   Test Quiz1 Quiz2 Quiz3 Quiz4 Quiz5 Score  _N_  _ERROR_
Smith    .     8     7     9     .     3     .   1      0
```

The statements in the `do` loop are executed with the counter variable `Test` taking values 1 through 5, incremented by +1. With the value of `Test` set to 1, `if qz{Test}=. then qz{Test}= 0;` determines whether the value for `Quiz1` in the PDV is a missing value and, if so, replaces it with a zero. Here `qz{Test}=.` is false so SAS proceeds to execute the next statement.

```
                                                    1
         Obs    Name    Test    Score

           1    Smith     1        8
           2    Smith     2        7
           3    Smith     3        9
           4    Smith     4        0
           5    Smith     5        3
           6    Jones     1        4
           7    Jones     2        5
           8    Jones     3       10
           9    Jones     4        8
          10    Jones     5        4
```

Fig. 1.20. SAS Example A6: listing output

The next statement `Score= qz{Test};` causes the value of `Score` to be set to the value of `Quiz1` since `qz{1}` refers to the variable `Quiz1`. Thus, the appearance of the PDV at this point is

```
Name   Test Quiz1 Quiz2 Quiz3 Quiz4 Quiz5 Score  _N_  _ERROR_
Smith    1     8     7     9     .     3     8   1      0
```

A new observation containing current values of the variables `Name`, `Test`, and `Score` is written to the SAS data set `quizzes` at this time because the `output` statement **2** is encountered. The observation written to the SAS data set is

```
Obs    Name    Test    Score

  1    Smith     1       8
```

because the variables `Quiz1,...,Quiz5` are not included in the SAS data set as they are named in a `drop` statement.

The above steps are repeated for each pass through the *do loop* for the set of data values already in the PDV (i.e., without reading in a new data line). Thus, for each line of data input, five observations are written to the SAS data set, each observation in the data set corresponding to a quiz score for a student. The printed listing of this data set will thus have the appearance shown in Fig. 1.20.

SAS Example A7

Figure 1.21 displays an example of construction of a SAS data set intended to be used as input to a SAS analysis of variance procedure. This program uses nested do loops. The data comes from an experiment that involves two factors: `Amount` with levels `0.9, 0.8, 0.7, 0.6` and `Concentration` with levels `1%, 1.5%, 2%, 2.5%, 3%`. The data values, consisting of reaction times measured for each combination of `Amount` and `Concentration`, are available as a table, with the columns corresponding to the levels of `Concentration` and the rows to the levels of `Amount`. Since data from factorial experiments are typically tabulated in this fashion, entering the data with each row in the table as a line of data is convenient.

```
data reaction;
length Conc $4;
do Amount =.9 to .6 by -.1; ❷
   do Conc = '1%' , '1.5%' , '2%' , '2.5%' , '3%' ;
   input Time @@; ❶
   output; ❸
   end;
end;
datalines;
10.9 11.5 9.8 12.7 10.6
 9.2 10.3 9.0 10.6  9.4
 8.7  9.7 8.2  9.4  8.5
 7.2  8.6 7.5  9.7  7.7
;
proc print;
title 'Reaction times for biological substrate';
run;
```

Fig. 1.21. SAS Example A7: program

The data are entered instream, as shown in Fig. 1.21. The `input time @@;` ❶ statement inside two nested do loops with index variables `Conc` and `Amount` is used to read the data values one at a time. Note carefully that `Conc` runs through the set of *character* values `1%, 1.5%, 2%, 2.5%,` and `3%` for each value of `Amount`, and that `Amount` runs through the values `0.9, 0.8, 0.7,` and `0.6`, in that order ❷. The `@@` pointer control symbol ❶ causes the `input` statement to read values from the same line of data, until the end of that data line is reached (i.e., until 5 values are input). This enables values for the variable `Time` to be read, one at a time, from each line of data. (See Sect. 1.7 for more about the use the `@@` pointer control.)

Reaction times for biological substrate

Obs	Conc	Amount	Time
1	1%	0.9	10.9
2	1.5%	0.9	11.5
3	2%	0.9	9.8
4	2.5%	0.9	12.7
5	3%	0.9	10.6
6	1%	0.8	9.2
7	1.5%	0.8	10.3
8	2%	0.8	9.0
9	2.5%	0.8	10.6
10	3%	0.8	9.4
11	1%	0.7	8.7
12	1.5%	0.7	9.7
13	2%	0.7	8.2
14	2.5%	0.7	9.4
15	3%	0.7	8.5
16	1%	0.6	7.2
17	1.5%	0.6	8.6
18	2%	0.6	7.5
19	2.5%	0.6	9.7
20	3%	0.6	7.7

Fig. 1.22. SAS Example A7: pdf-formatted output

The `output` statement **3** will cause an observation containing the current values in the PDV for the variables `Conc`, `Amount`, and `time` to be written to the SAS data set named `reaction`. This will be repeated for all combinations of the index variables `Conc` and `Amount`; that is, 20 observations will be written, one for each combination of values for these variables. The ODS-formatted listing of this data set, shown in Fig. 1.22, displays the values of these variables for each observation.

1.7 More on INPUT Statement

In this section, the column pointer controls `@` and `+`, line-hold specifiers `trailing @` and `trailing @@`, and the line pointer control `#n` are discussed.

1.7.1 Use of Pointer Controls

The SAS `input` statement uses a *pointer* to track the position in the input buffer where it begins reading a data value for each of the variables in the PDV. At the start of the execution of the input statement, the pointer is positioned at the beginning of the input buffer (position one) and then moves along the buffer as each successive informat or pointer control in the input statement is encountered. As the pointer moves along the input buffer, the input statement reads data values from the input buffer beginning at the current pointer position and converts them to values for successive variables in the PDV. This conversion is done using the informats supplied by the user in the input statement (or using a default informat if one is not supplied, as in the case of list input). For example, the following `input` statement

```
input Id 4. @9 State $2. Fert 5.2 +1 Percent 3.2 @26 Members 4.;
```

was used in Sect. 1.4 to read the data line

```
0001xxxxIA00504x089xxxxxx1349
```

Suppose that the data line has been moved to the input buffer and that the pointer is positioned at the beginning of the buffer as follows:

```
0001xxxxIA00504x089xxxxxx1349
↑
```

The SAS numeric informat `4.` reads the value `0001` for the variable `Id` (and inserts it in the PDV), and the pointer is repositioned at column 5 of the input buffer:

```
0001xxxxIA00504x089xxxxxx1349
    ↑
```

The pointer control `@9` then causes the pointer to move to column 9:

```
0001xxxxIA00504x089xxxxxx1349
        ↑
```

The SAS character informat `$2` reads the value `IA` as the value for `State` and inserts it in the PDV, and the pointer moves over those two columns to column 11 of the input buffer:

```
0001xxxxIA00504x089xxxxxx1349
          ↑
```

Next, the SAS numeric informat `5.2` is used to read the value `00504` as the current value 5.04 of the variable `Fert` in the PDV and the pointer moves over five columns:

```
0001xxxxIA00504x089xxxxxx1349
               ↑
```

The pointer control `+1` moves the pointer one more column:

```
0001xxxxIA00504x089xxxxxx1349
                ↑
```

Three columns are read using the informat `3.2` to obtain the current value of the variable `Percent` in the PDV, and the pointer moves a further three columns:

```
0001xxxxIA00504x089xxxxxx1349
                   ↑
```

The pointer control `@26` moves the pointer to column 26:

```
0001xxxxIA00504x089xxxxxx1349
                         ↑
```

Notice that it would have been more convenient to use `+6` to move the pointer to column 26 than determining the required number, 26, of columns to move the pointer to the current position from position one. At this stage the value `1349` for the variable `Members` is read from the input buffer, using the informat `4.`, and inserted in the PDV. The pointer moves to column 31:

```
0001xxxxIA00504x089xxxxxx1349
                              ↑
```

At this point, SAS recognizes that the end of the input statement has been reached and the execution of the SAS programming statements using the values in the PDV begins.

Note that the pointer will move *backward* along the input buffer if the `@` pointer control is used with a value that points to a position to the left of the current pointer position. There are other variations of the use of the pointer controls `@` and `+` available. For example, the form `@`*`numeric-variable`* or `@`*`numeric-expression`* can be used to read subsequent data by positioning the pointer at the value of a *numeric-variable* or a *numeric-expression*, respectively. Thus, moving of the pointer can be made dependent on a value of a variable or an expression. The pointer control `+` can also be used to move the pointer backward. For example, `+(-3)` or `+num`, where the value of `num` is set to -3, moves the pointer back three columns from the current position.

1.7.2 The `trailing @` Line-Hold Specifier

So far, it is understood that when the end of the input statement is reached (the SAS pointer is positioned at the end of the input buffer), SAS proceeds to execute the programming statements that follow using the values in the PDV. In addition, the input buffer will be replaced with the next line of data when this occurs.

Sometimes it may become necessary to execute SAS programming statements after reading only some of the data values from the input buffer. One situation of this kind occurs when reading the rest of the data values depends on the value(s) of variable(s) read so far. Obviously, it is necessary to use a second `input` statement to read the rest of the data values from the input buffer. However, this is not possible because, ordinarily, completing the execution of an `input` statement will cause the input buffer to be replaced with the next line of data. Thus, the values yet to be read from the previous line of data will become unavailable.

```
data garden;
input Store : $13. Count  @; 1
    do i=1 to Count;
          input Item : $10. Price : 5.2 @; 2
          output;
       end;
drop i;
datalines;
JMart 4 rake 1250 sprinkler 875 bench 12000 chair 3525
Woodsons 3 edging 750 planter 1365 basket 870
Home'nGarden 5 sweeper 1185 gloves 350 shears 2100
         spade 3450 trimmer 7640
;
proc print data=garden;
title 'Gardening materials purchased Spring 2004 ';
run;
```

Fig. 1.23. SAS Example A8: program

The use of an `@` symbol appearing by itself as the last item on an input statement (i.e., just before the semicolon), called a `trailing @`, is one solution to this problem. The `trailing @` forces SAS to hold the pointer at the current position on the input buffer and allows SAS to execute another `input` statement before the contents of the current input buffer are replaced.

SAS Example A8

Figure 1.23 shows an example where the `trailing @` is used twice to hold the same data line (in the input buffer). First, it is used on the statement `input`

`Store : $13. Count @;` ❶ to hold the data line after reading values for the variables `Store` and `Count`. The variable `Count` contains the number of pairs of values for the variables `Item` and `Price` to be read from the same data line. These pairs of values are read using the statement `input Item : $10. Price : 5.2 @;`. This statement appears within a do loop that executes a number of times equal to the value of the `Count` variable, read previously from the same data line.

The second use of `trailing @` in this example occurs in the `input Item : $10. Price : 5.2 @;` ❷. It holds the line after each pair of values for `Item` and `Price` is read, leaving the pointer at the correct position for the next execution of the same `input` statement. After a pair of values for `Item` and `Price` is read, the `output` statement causes the values in the PDV for the variables `Store, Count, Item,` and `Price` to be written as an observation in the SAS data set named `garden` created in this data step.

Gardening materials purchased Spring 2004

Obs	Store	Count	Item	Price
1	JMart	4	rake	12.50
2	JMart	4	sprinkler	8.75
3	JMart	4	bench	120.00
4	JMart	4	chair	35.25
5	Woodsons	3	edging	7.50
6	Woodsons	3	planter	13.65
7	Woodsons	3	basket	8.70
8	Home'nGarden	5	sweeper	11.85
9	Home'nGarden	5	gloves	3.50
10	Home'nGarden	5	shears	21.00
11	Home'nGarden	5	spade	34.50
12	Home'nGarden	5	trimmer	76.40

Fig. 1.24. SAS Example A8: pdf-formatted output

Note that the data values for the last observation in the input stream continues on to a second data line. These data values are processed correctly because a value of 5 read for the variable `Count` results in the statement `input Item : $10. Price : 5.2 @;` being repeatedly executed five times. This causes SAS to encounter the end of a data line after reading the third pair of values `Item` and `Price` from the input buffer and thus move the second data line into the input buffer. The next two pairs of values for Item and Price are then read using the same input statement.

Additionally, this program illustrates the use of the : *modifier* with both character and numeric informats in the list input style, for reading data values of varying widths. First, `Store : $13.` allows the reading of character strings shorter than the specified width of 13 columns of the data value to be read ❶. The : modifier causes `$13.` to recognize the first blank encountered in the data field as a delimiter, as is the case when using list input with simply a single $ symbol without specifying a length. Thus, data values of shorter lengths than 13 characters are read correctly as the values of Store. Second, `Price : 5.2`❷ allows the reading of numeric data values of varying widths delimited by blanks using the numeric informat `5.2`. The ODS-formatted listing of the data set produced by the SAS Example A8 is displayed in Fig. 1.24.

1.7.3 The `trailing @@` Line-Hold Specifier

It was stated in Sect. 1.7.2 that SAS assumes implicitly that the processing of a data line is over when the end of the input statement is reached and automatically goes to read a new line of data. There, `trailing @` pointer control was used to hold the current data line for further processing.

Another situation in which it is necessary that SAS does not assume that processing of a data line is complete when the end of an input statement is reached occurs when information for multiple observations are to be read from the same line of data. The trailing @@ causes the `input` statement to be repeatedly executed for the same data line, and each time the `input` statement is executed, a new iteration of the data step is also executed (as if another data line has been moved into the buffer).

Example 1.7.1

```
data sat;
input Name $ Verbal Math @@;
Total= Verbal + Math;
datalines;
Sue 610 560 John 720 640 Mary 580 590
Jim 650 760 Bernard 690 670 Gary 570 680 Kathy 720 780
Sherry 640 720
;
proc print;
run;
```

In Example 1.7.1, several sets of data values, consisting of the values for the variables Name, Verbal, and Math, are entered into several data lines in the input stream. Three values at a time are read from each line using list input from the input buffer and transferred to the PDV, with the pointer maintaining its current position, while the three values are being processed. The program statement `total= Verbal + Math;` is then executed, and an observation is written to the data set, as SAS has reached the end of the data step. The above actions describe a single iteration of the data step.

Instead of returning to read a new data line, the next set of values will now be read from the input buffer beginning from the current position of the pointer. The presence of the **trailing @@** caused the data line to be *held* in the input buffer. Note, however, that using a **trailing @** instead of the **trailing @@** will not work in this case. This is because the input buffer would have been reinitialized to missing values at the end of each iteration of the data step, thus wiping out the data values that are yet to be transferred to the PDV. The printed output from `proc print;` in Example 1.7.1 is shown as follows:

```
Obs    Name       Verbal    Math    Total

 1     Sue          610      560     1170
 2     John         720      640     1360
 3     Mary         580      590     1170
 4     Jim          650      760     1410
 5     Bernard      690      670     1360
 6     Gary         570      680     1250
 7     Kathy        720      780     1500
 8     Sherry       640      720     1360
```

1.7.4 Use of RETAIN Statement

Recall that each time SAS returns to the top of the data step, every variable value in the PDV is initialized to missing values. The `retain` statement is a declarative statement that causes the value of each variable listed in the statement to be retained in the PDV from one iteration of the data step to the next. The general form of the `retain` statement is

RETAIN *variable-list* < *(initial-values)* >;

By default, the *initial-values* assigned to the variables in the list are missing values; however, the retain statement allows the user to specify the values to be used for initializing as well. Note that it is redundant to use this statement in a data step where data are accessed from an existing SAS data set to create a new SAS data set (as when using statements such as `set` or `merge` to be discussed in Chap. 2) because the values of variables are automatically *retained* from one iteration to the next in such a data step. The SAS Example A9 program displayed in Fig. 1.25 illustrates the use of the `retain` statement.

SAS Example A9

The `retain` statement is most useful when multiple types of data lines are to be processed in a data step. It is necessary to retain data values read in one type of a data line in the PDV, so that they can be combined with information read from other types of data lines to form a single observation to be output to the SAS data set. Usually, values retained in the PDV remain there until they are overwritten by new values read from a data line of the same type.

```
data ledger;
retain Store Region Month;
input Type $1. @;
if Type='S' then input @3 Store 4. Region ~10. Month : $8. ; ❷
else do;
     input @4 Date ddmmyy8. Sales 7.2; ❸
     output;
     end;
drop Type;
datalines;
S 0021 Southeast  March ❶
   10/05/04 134510
   12/05/04  23675
   21/05/04  96860
   28/05/04 265036
S 0173 Northwest  January
   15/05/04  67200
   18/05/04 158325
   29/05/04 127950
   30/05/04  45845
   02/06/04 304730
;
proc print data=ledger;
 Id Store;
 format Store z4. Date ddmmyy8. Sales dollar10.2 ;
 title 'Sales Analysis for Martin & Co.';
run;
```

Fig. 1.25. SAS Example A9: program

In the SAS Example A9 program, there are two kinds of data lines: one kind, identified by an "S" entered in the first column ❶, specifies values for the variables `Store, Region,` and `Month`, and the other containing a blank in the first column specifies values for the variables `Date` and `Sales`. Notice that in the second `input` statement ❷, the : modifier is used with the $8. informat to read a value for the variable `Month`, and the informat `ddmmyy8.` is used in the third `input` statement ❸ to read a date value.

As seen from the ODS-formatted listing of the output data set displayed in Fig. 1.26, the values for `Date` and `Sales` have been combined with those of `Store, Region,` and `Month` to form each individual observation in the data set. It is important to recognize that the values for `Date` and `Sales` are read from a new data line using the statement `input @4 Date ddmmyy8. Sales 7.2;`, following which an observation is written to the SAS data set

Sales analysis for Martin & Co.

Store	Region	Month	Date	Sales
0021	Southeast	March	10/05/04	$1,345.10
0021	Southeast	March	12/05/04	$236.75
0021	Southeast	March	21/05/04	$968.60
0021	Southeast	March	28/05/04	$2,650.36
0173	Northwest	January	15/05/04	$672.00
0173	Northwest	January	18/05/04	$1,583.25
0173	Northwest	January	29/05/04	$1,279.50
0173	Northwest	January	30/05/04	$458.45
0173	Northwest	January	02/06/04	$3,047.30

Fig. 1.26. SAS Example A9: pdf-formatted output

(named `ledger`). Then SAS returns to the top of the data step and variables in the PDV are all initialized to missing values *except* for `Store`, `Region`, and `Month`. The values for these variables in the PDV remain the same as those that were previously read from the last type "S" data line. A new set of values for `Date` and `Sales` are read from the next data line, unless the next data line is of type "S." Note that the data lines are required to be arranged precisely in the sequence they appear in the SAS program for the example to work as described.

1.7.5 The Use of Line Pointer Controls

The pointer controls discussed in Sect. 1.7.1 are called *column pointer controls* because they facilitate the movement of the pointer along the columns of a data line (in the input buffer). The pointer control `#n` moves the pointer to the first column of the nth data line in the input buffer. This implies that it is possible for the input buffer to contain multiple data lines. The largest value of `n` that is used in an input statement is used by SAS to determine how many data lines will be read into the input buffer at a time. The user may specifically state the number of lines to be read using the `n=` option in the `infile` statement. Once several lines are in the input buffer, `#n` or one of its other forms `#numeric-variable` or `#numeric-expression` may be used in the input statement, to move the pointer among these lines of data to read data values into the PDV. The pointer may move either forward or backward among these lines depending on the value of `n`, the `numeric-variable`, or the `numeric-expression`.

Example 1.7.2

```
data survey;
input #1 Id 1-4 Gender $ Bdate ddmmyy6. (Race Marital Educ)
        (1.) #2 @5 (Q1-Q20) (1.) ;
format Bdate ddmmyy8. ;
datalines;
3241 F 100287012
    13431043321110310022
5673 M 211178124
    11031002231134310433
4702 M 170780025
    31134310433211103100
2496 F 030979013
    22311542102231152111
6543 M 090885124
    03100343104332111031
;
proc print;
run;
```

Data lines continued onto several lines may also occur due to reasons different from the situation described in Sect. 1.7.4. An observation may constitute information entered on several data lines simply because the data values are too numerous to be entered on a single line. Such data may arise as a result of surveys, longitudinal studies in which many variables are measured over time on each experimental subject, or experiments involving repeated measures. In Example 1.7.2, the data set is the result of a questionnaire in which demographic data are entered in the first data line, and the responses to 20 questions are recorded in the second data line as a series of single-digit numbers corresponding to responses made by the subject (identified by the I.D. number on the first line). The printed output from `proc print;` in Example 1.7.2 is as follows:

The SAS System 1

Obs	Id	Gender	Bdate	Race	Marital	Educ	Q1	Q2	Q3	Q4	Q5	Q6	Q7	Q8	Q9	Q10	Q11	Q12	Q13	Q14	Q15	Q16	Q17	Q18	Q19	Q20
1	3241	F	10/02/87	0	1	2	1	3	4	3	1	0	4	3	3	2	1	1	1	0	3	1	0	0	2	2
2	5673	M	21/11/78	1	2	4	1	1	0	3	1	0	0	2	2	3	1	1	3	4	3	1	0	4	3	3
3	4702	M	17/07/80	0	2	5	3	1	1	3	4	3	1	0	4	3	3	2	1	1	1	0	3	1	0	0
4	2496	F	03/09/79	0	1	3	2	2	3	1	1	5	4	2	1	0	2	2	3	1	1	5	2	1	1	1
5	6543	M	09/08/85	1	2	4	0	3	1	0	0	3	4	3	1	0	4	3	3	2	1	1	1	0	3	1

Note that the informat `1.` is applied repeatedly to each of the variables in the list `(Race Marital Educ)` to read the responses to these variables, which are single-digit numbers entered in adjoining columns in the first line of data. Similarly, the informat `1.` is applied to each variable in the list `(Q1-Q20)`.

1.8 Using SAS Procedures

The Proc Step

It has previously been noted that the group of SAS statements used to invoke a procedure for performing a desired statistical analysis of a SAS data set is designated as a `proc step` and that the group begins with a statement of the form

`PROC` *procedure_name;*

One may use options and parameters in the `proc` statement to provide additional information to the procedure. Some procedures also allow optional *procedure information statements*, which usually follow the `proc` statement, to be included in the proc step. Thus the most general form of a proc step is

`PROC` *proc_name options_list;*
  <procedure information statements;>
  < variable attribute statements;>

If the user only requires that

- the most recently created data set will be analyzed,
- all variables in the data set are to be processed, and
- the entire data set is to be processed instead of subsets of observations,

then most SAS procedures can be invoked by using a simple `proc` statement as in the SAS Example A1 program. For example, in the code

```
data new;
input X Y Z;
datalines;
     ⋮
;
proc print;
run;
```

`proc print` will produce a listing of the data values in the entire SAS data set created in the `data step` immediately preceding the `proc print` statement.

Specifying Options in the PROC *Statement*

On the other hand, if a user intends to analyze a data set that is not the most recently created one or if a user wishes to specify additional information to the procedure, these may be specified as `options` in the `proc` statement. For example,

```
proc print data=one;
```

specifies that the `print` procedure uses the data from the data set named `one`, irrespective of whether it is the most recently created SAS data set in the current job. Thus, the data set named `one` may have been created in `any` one of the several data steps preceding the current `proc step`.

```
proc corr kendall;
```

provides an example of using a keyword option to specify a type of computation to be performed. Here the keyword `kendall` specifies that Kendall's tau-b correlation coefficients be computed when procedure `corr` is executed, instead of the Pearson correlations that would have been computed by default.

Procedure Information Statements

Certain statements may be optionally included in a `proc` step to provide additional information to be used by the procedure in its execution. Some statements of this type are `var, by, output,` and `title`. For example, the requirement that the analysis is to be performed only on some of the variables in the data set can be specified by using the procedure information statement `var`. In the example

```
proc means data=Store mean std;
var Bolts Nuts Screws;
```

the `var` statement requires that procedure `means` computes the mean and standard deviation only of the variables named `Bolts, Nuts,` and `Screws` in the data set named `Store`. These variables are thus identified as the *analysis variables.*

A special procedure information statement, the `by` statement, allows many SAS procedures to process subsets of a specified data set based on the values of the variable (or variables) listed in the `by` statement. This in effect means that SAS executes the procedure repeatedly on each subset of data separately. The form of the `by` statement is

`BY` *variables_list*;

When a `by` statement appears, the SAS procedure expects the data set to be arranged in the order of values of the variable(s) listed in the `by` statement. The essential requirement is that those observations with identical values for each of these variables occur together in the input data set. This is required so that these observations can be analyzed by the procedure as subsets called `by groups`. See Example 1.8.1 for an illustration.

The formation of by groups is most conveniently achieved by using the SAS procedure named SORT to rearrange the data set prior to analyzing it. When used in the `proc sort` step, a `by` statement specifies the key variables to be used for performing the sort. In the SORT procedure, observations are first arranged in the increasing order of the values of the first variable specified in the `by` statement. Within each of the resulting groups, observations are then arranged in the increasing order of the values of the second variable specified and so on.

For numeric sort keys, the signed value of a variable is used to determine the ordering with SAS missing value assigned the lowest rank. For character variables, ordering is determined using the ASCII sequence in UNIX and Windows operating environments. The main features of the ASCII sequence are that digits are sorted before uppercase letters, and uppercase letters are sorted before lowercase letters. The blank is the smallest displayable character. Thus, the string 'South' is larger than the string 'North' but is smaller than the string 'Southern'.

By default, `proc sort` will replace the input SAS data set with the data set rearranged as requested in the `by` statement. However, if needed, the sorted output can be written to a new SAS data set using an `out=` option on the `proc sort` statement to name the new data set.

Example 1.8.1 As an example, suppose that it is required to analyze the variables in a data set by *Gender* and *Income category* where the respondents were assigned to one of, say, 3 Income categories 1, 2, or 3. Once the SAS data set is created, a sequence of SAS statements comparable to those shown below may be used to obtain a required analysis. First, `proc sort` uses the SAS data set as input, rearranges it as specified in the `by` statement, and replaces the input data set with the modified data set. The `proc print` uses this data set to produce a listing of the rearranged data.

```
        ⋮
proc sort;
by Gender Income;
proc print;
by Gender;
        ⋮
```

The output from `proc print` will result in a listing of observations that are grouped by `Gender` first, and within each group arranged in the increasing values of `Income`, as shown in Output 1.

Output 1

```
                Gender = F

listing of observations in ascending order of
INCOME values and a value of F for Gender

                Gender = M

listing of observations in ascending order of
Income values and a value of M for Gender
```

If, in addition, it is required to find the mean and variance of the variables in the data set by both Gender and Income class, the following statements may be added:

```
proc means mean var;
by Gender Income;
var Age Food Rent;
       ⋮
```

producing the table of means and variances for the variables `Age`, `Rent`, and `Food` for each subgroup defined by values of `Gender` and `Income`, as shown in Output 2. The vertical dots represent the statistics (means and variances) calculated using data values for variables Age, Rent, and Food, etc. in each of these subgroups.

Output 2

```
Gender = F Income= 1
           ⋮
Gender = F Income= 2
           ⋮
Gender = F Income= 3
           ⋮
Gender = M Income= 1
           ⋮
Gender = M Income= 2
           ⋮
Gender = M Income= 3
           ⋮
```

The `where` statement used in SAS Example A3 (see Fig. 1.11) is another example of a procedure information statement.

Variable Attribute Statements

These statements allow SAS users to specify the `format`, `informat`, `label`, and `length` of the variables in a `proc step`. Such specifications are associated with the variables only during the execution of a `proc step` if specified in that `proc step`. On the other hand, these statements may also be used in a `data step` to specify attributes of SAS variables, in which case they would be permanently associated with the variables in the data set created by the `data step`. Thus, these attributes will be available subsequently to any SAS procedure for use within a `proc step`. The `format` and `label` statements are two variable attribute statements frequently used in `proc` steps.

The FORMAT Statement

The `format` statement may be used both in the data step and the proc step to specify *formats*. SAS `formats` are used for converting data values stored in a SAS data set to the form desired in displayed output. They provide information to SAS such as how many character positions are to be used by a data value, in what form the data value must appear in the displayed output, and what additional symbols, such as decimal points, commas, dollar signs, etc., must appear in the printed form of the data value. For example, a date value stored internally as a binary value may be displayed in one of several date formats provided by SAS for displaying dates. The two most commonly used `formats` are those available for the purpose of displaying numeric and character data values. The general form of the statement is

FORMAT var-1=*format*...<var-n=*format*>;

The LABEL Statement

The `label` statement is also used in both the data and the proc steps to give more descriptive *labels* than the variable names to identify the data values (or statistics computed on the data values) in the output. The general form of the statement is

LABEL var-1=*label-1*...<var-n=*label-n*>;

where *label-1*, ··· *label-n* are character strings enclosed in single quotation marks (or double quotation marks if the label includes single quotation marks). Any number of variables may be associated with labels in a single LABEL statement.

The LENGTH Statement

The `length` statement is used to assign a length (in bytes) to one or more variables. It is most useful when character variables containing strings of length more than eight characters, the default length allocated to character variables stored in a data set, need to be defined. The general form of the statement when used for this purpose is

LENGTH *variable(s)* <$ > *length*;

where the $ specifies that the preceding variables are character variables and *length* is a number in the range 1 to 32,767, specifying how many bytes are required to be allocated.

SAS Example A10

The SAS Example A10 program, displayed in Fig. 1.27, illustrates several features of the `data` and `proc` steps discussed so far. The two different `proc print` steps provide listings of the same SAS data set; the first step is a simple invocation of the procedure, whereas in the second step, several procedure information and variable attribute statements are used to produce more complete annotation.

In the `input` statement **1**, the : modifier is used to read the value for the variable `Region` and the date informat `monyy5.` to read the `date` value. The `format` statement **2** used in this data step specifies formats for the variables `Month` and `Revenue`. Thus, these formats were used in the first `proc print` for printing the values of these variables, as illustrated in Output 1 shown in Fig. 1.28.

The statement `by Region State Month;` **3** was used with `proc sort`; thus, as expected, in the listing produced by the first `proc print` step (Output 1), observations appear arranged in the increasing order of `Month` values within groups with the same `Region` and `State` values. These appear in the increasing order of `State` values within groups with the same `Region` values and, finally, in the increasing order of values of the `Region` variable. However, these `by` groups are not clearly separated in the output from the first `proc print` step. This is because a `by` statement is not used in the first proc step for identifying these as separate groups.

In the second `proc print` step, the statement `by Region State` **4** is used, causing separate listing for each by group as defined by identical values for `State` within groups with the same `Region` values, as seen in Fig. 1.29. The `format` **5** statement in this proc step provides a format for printing values of the variable `Expenses`. In addition, the statements `sum Revenue Expenses;` **6** and `sumby Region;` in the second `proc print` step illustrate how totals are calculated for each of the numeric variables `Revenue` and `Expenses` and are displayed for each group defined by the same value for the by variable `Region`, respectively.

```
data sales;
input Region : $8. State $2. +1 Month monyy5. HeadCnt Revenue
          Expenses; 1
format Month monyy5. Revenue dollar12.2; 2
label Region="Sales Region" HeadCnt="Sales Personnel";
datalines;
SOUTHERN FL JAN78 10 10000 8000
SOUTHERN FL FEB78 10 11000 8500
SOUTHERN FL MAR78 9 13500 9800
SOUTHERN GA JAN78 5 8000 2000
SOUTHERN GA FEB78 7 6000 1200
PLAINS NM MAR78 2 500 1350
NORTHERN MA MAR78 3 1000 1500
NORTHERN NY FEB78 4 2000 4000
NORTHERN NY MAR78 5 5000 6000
EASTERN NC JAN78 12 20000 9000
EASTERN NC FEB78 12 21000 8990
EASTERN NC MAR78 12 20500 9750
EASTERN VA JAN78 10 15000 7500
EASTERN VA FEB78 10 15500 7800
EASTERN VA MAR78 11 16600 8200
CENTRAL OH JAN78 13 21000 12000
CENTRAL OH FEB78 14 22000 13000
CENTRAL OH MAR78 14 22500 13200
CENTRAL MI JAN78 10 10000 8000
CENTRAL MI FEB78 9 11000 8200
CENTRAL MI MAR78 10 12000 8900
CENTRAL IL JAN78 4 6000 2000
CENTRAL IL FEB78 4 6100 2000
CENTRAL IL MAR78 4 6050 2100
;
proc sort;
by Region State Month; 3
run;
proc print;
run;
proc print label;
 by Region State ; 4
 format Expenses dollar10.2 ; 5
      label State= State Month= Month Revenue="Sales Revenue"
     Expenses="Overhead Expenses";
 id Region State;
 sum Revenue Expenses; 6
 sumby Region;
 title " Sales Report by State and Region";
run;
```

Fig. 1.27. SAS Example A10: program

Although labels for the variables `Region` and `HeadCnt` were also specified in a `label` statement in the data step, they do not appear in the printed output from the first `proc print`. This is because attributes available in a data set will not be used by some proc steps, unless they are specifically requested by the use of an option. The *label* option in `proc print label;` in the second `proc print` step is an example. The output from this step, displayed in Output 2, shows the labels for `Region` and `HeadCnt` being used (see Fig. 1.29). Note also that a format for printing values of `Expenses` was not available in the data set, so the default format is used in the first `proc print`. In the second `proc print`, a format specifically for printing values of `Expenses` was included.

Sales Report by State and Region

Obs	Region	State	Month	Headcnt	Revenue	Expenses
1	CENTRAL	IL	JAN78	4	$6,000.00	2000
2	CENTRAL	IL	FEB78	4	$6,100.00	2000
3	CENTRAL	IL	MAR78	4	$6,050.00	2100
4	CENTRAL	MI	JAN78	10	$10,000.00	8000
5	CENTRAL	MI	FEB78	9	$11,000.00	8200
6	CENTRAL	MI	MAR78	10	$12,000.00	8900
7	CENTRAL	OH	JAN78	13	$21,000.00	12000
8	CENTRAL	OH	FEB78	14	$22,000.00	13000
9	CENTRAL	OH	MAR78	14	$22,500.00	13200
10	EASTERN	NC	JAN78	12	$20,000.00	9000
11	EASTERN	NC	FEB78	12	$21,000.00	8990
12	EASTERN	NC	MAR78	12	$20,500.00	9750
13	EASTERN	VA	JAN78	10	$15,000.00	7500
14	EASTERN	VA	FEB78	10	$15,500.00	7800
15	EASTERN	VA	MAR78	11	$16,600.00	8200
16	NORTHERN	MA	MAR78	3	$1,000.00	1500
17	NORTHERN	NY	FEB78	4	$2,000.00	4000
18	NORTHERN	NY	MAR78	5	$5,000.00	6000
19	PLAINS	NM	MAR78	2	$500.00	1350
20	SOUTHERN	FL	JAN78	10	$10,000.00	8000
21	SOUTHERN	FL	FEB78	10	$11,000.00	8500
22	SOUTHERN	FL	MAR78	9	$13,500.00	9800
23	SOUTHERN	GA	JAN78	5	$8,000.00	2000
24	SOUTHERN	GA	FEB78	7	$6,000.00	1200

Fig. 1.28. SAS Example A10: output from the first proc print step

SAS Example A11

As alluded to in the introduction, predefined *style definitions* determine the appearance of tables and graphs from SAS procedures. A style definition is a complete description of all the attributes to use when creating a specific output. For tables, attributes specify features such as background color, font size and color of table contents, table border, etc. Attributes are collected into groups called *style elements*, which describe the attributes of distinct part of the table, e.g., the table header. Thus the entire appearance of a table is controlled by the selected style definition, and that may be changed by choosing a different style definition. For example, the appearance of output tables and graphics are quite different in *HTMLBlue*, *Statistical*, and *Journal* styles. An existing style definition may be modified by overriding parts of the style template of a specific style. A more elaborate description of style attributes and style elements and how they may be modified is presented in Appendix A.

Sales Report by State and Region

Sales Region	State	Month	Sales Personnel	Sales Revenue	Overhead Expenses
CENTRAL	IL	JAN78	4	$6,000.00	$2,000.00
		FEB78	4	$6,100.00	$2,000.00
		MAR78	4	$6,050.00	$2,100.00

Sales Region	State	Month	Sales Personnel	Sales Revenue	Overhead Expenses
CENTRAL	MI	JAN78	10	$10,000.00	$8,000.00
		FEB78	9	$11,000.00	$8,200.00
		MAR78	10	$12,000.00	$8,900.00

⋮ ⋮ ⋮ ⋮

Sales Region	State	Month	Sales Personnel	Sales Revenue	Overhead Expenses
EASTERN	NC	JAN78	12	$20,000.00	$9,000.00
		FEB78	12	$21,000.00	$8,990.00
		MAR78	12	$20,500.00	$9,750.00

⋮ ⋮ ⋮ ⋮

Sales Region	State	Month	Sales Personnel	Sales Revenue	Overhead Expenses
NORTHERN	NY	FEB78	4	$2,000.00	$4,000.00
		MAR78	5	$5,000.00	$6,000.00
NORTHERN				**$8,000.00**	**$11,500.00**

⋮ ⋮ ⋮ ⋮

Sales Region	State	Month	Sales Personnel	Sales Revenue	Overhead Expenses
SOUTHERN	FL	JAN78	10	$10,000.00	$8,000.00
		FEB78	10	$11,000.00	$8,500.00
		MAR78	9	$13,500.00	$9,800.00

Sales Region	State	Month	Sales Personnel	Sales Revenue	Overhead Expenses
SOUTHERN	GA	JAN78	5	$8,000.00	$2,000.00
		FEB78	7	$6,000.00	$1,200.00
SOUTHERN				**$48,500.00**	**$29,500.00**
				$282,250.00	**$162990.00**

Fig. 1.29. SAS Example A10: output from the second proc print step

Style elements can be modified by specifying the `style=` options in the procedure statement or as options for the statement that produce various parts of the output table, e.g., the `id` or `sum` statements. In the next example, the `style=` option is used in the `proc print` statement to override style attributes that were defined for the style definition in use (e.g., *HTMLBlue* style used in SAS Example A10). The simplified syntax of a style specification used here is

STYLE <(*location(s)*)> = [*style-attribute-name-1*=*style-attribute-value-1*
<*style-attribute-name-2*=*style-attribute-value-2* ...>]

For the PRINT procedure, there are nine defined locations: `table`, `obs`, `data`, `obsheader`, `header`, `bylabel`, `total`, `grandtotal`, and `N`. These refer to various parts of the table that the style specification applies to, e.g., `header` refers to header of columns (other than OBS and ID), `data` refers to values in columns (other than OBS and ID), etc. More details can be found under the syntax description of the `proc print` statement.

```
proc print label sumlabel='Regional Total' grandtotal_label="All Sales Total"
      style(bysumline)=[background=skyblue foreground=linen]
      style(grandtotal)=[foreground=darklblue background=cornflowerblue]
      style(header)=[font_style=italic background=lightcyan];;
 by Region State ;
 format Expenses dollar10.2 ;
 label State= State Month= Month Revenue="Sales Revenue"
       Expenses="Overhead Expenses";
 id Region State;
 sum Revenue Expenses;
 sumby Region;
 title " Sales Report by State and Region";
run;
```

Fig. 1.30. SAS Example A11: using `style=` options

SAS Example A11 (see Fig. 1.30) illustrates the use of some `style=` options to modify the appearance of the output table from SAS Example A10, resulting in the new output table shown in Fig. 1.31.

By examining the proc step in Fig. 1.30, it can be easily observed that the locations of the table modified were `bysumline`, `grandtotal`, and `header`, and style attributes changed were background color, foreground color, and font style. The keyword *background* is synonymous with `backgroundcolor`, and the keyword *foreground* is synonymous with *color*, which specifies the color of the text, in case of a table element. (Note that style attributes tables are available as part of the description of the TEMPLATE procedure which will be briefly discussed in Appendix A. Definitions for a specific table can be found by examining the corresponding template supplied by SAS using the TEMPLATE procedure or by browsing the templates through the SAS Results window as explained in the appendix.)

Sales Report by State and Region

Region	State	*Month*	*Headcnt*	*Sales Revenue*	*Overhead Expenses*
CENTRAL	IL	JAN78	4	$6,000.00	$2,000.00
		FEB78	4	$6,100.00	$2,000.00
		MAR78	4	$6,050.00	$2,100.00

Region	State	*Month*	*Headcnt*	*Sales Revenue*	*Overhead Expenses*
CENTRAL	MI	JAN78	10	$10,000.00	$8,000.00
		FEB78	9	$11,000.00	$8,200.00
		MAR78	10	$12,000.00	$8,900.00

Region	State	*Month*	*Headcnt*	*Sales Revenue*	*Overhead Expenses*
CENTRAL	OH	JAN78	13	$21,000.00	$12,000.00
		FEB78	14	$22,000.00	$13,000.00
		MAR78	14	$22,500.00	$13,200.00
Regional Total				$116,650.00	$69,400.00

⋮ ⋮ ⋮ ⋮

Region	State	*Month*	*Headcnt*	*Sales Revenue*	*Overhead Expenses*
SOUTHERN	FL	JAN78	10	$10,000.00	$8,000.00
		FEB78	10	$11,000.00	$8,500.00
		MAR78	9	$13,500.00	$9,800.00

Region	State	*Month*	*Headcnt*	*Sales Revenue*	*Overhead Expenses*
SOUTHERN	GA	JAN78	5	$8,000.00	$2,000.00
		FEB78	7	$6,000.00	$1,200.00
Regional Total				$48,500.00	$29,500.00
All Sales Total				$282,250.00	$162990.00

Fig. 1.31. SAS Example A11: edited output from the modified proc step

1.9 Exercises

1.1 Each of the following cases shows the observations written to the SAS data set (variable names and corresponding data values) when the given lines of data are read by the given INPUT statement:

a. `input Id Gender $ Age GPA SAT;`
`1906␣F␣19␣3.25␣725`
`2045␣M␣.␣2.95␣␣690`
`3117␣␣M␣␣24␣␣␣3.72␣␣␣793`

b. `input (Q1-Q4) (3.);`
`432␣16798␣␣2`
`57␣␣␣36␣␣␣84`

c. `input Id $1-4 @8 Pulse 3. +2 Weight 4.1 Runtime;`
`C37A236␣87␣92517␣9.5`
`D45B␣54␣␣␣␣␣1423␣8.6␣␣`
`␣␣␣␣158␣69␣8␣965␣7.8`

d. `input Name :$11. Id 4. Visit : mmddyy8.;`
`Wilson␣3974␣11/25/47`
`Worthington␣1598␣7/16/86`

NOTE: The symbol ␣ denotes one space.

1.2 Sketch the output resulting from executing the following SAS program. Describe in your own words the flow of operations in the data step in creating this data set.

```
data two;
input Score @@;
if Score > 70 then do;
   Adjust = "y";
   Index = Score -70;
   end;
else if Score < 70 then do;
   Adjust = "n";
   end;
datalines;
67 69 70 72 75 .
;
proc print data=two; run;
```

1.3 The following data lines are input in a data step:

```
21   50.2 17 47.5 54 32.1
12.  54.3
23.
45.6
```

What would be the contents of the SAS data set if the input statement used was each of the following? Write a brief explanation of what takes place in each data step.

a. `input Id  Score1;`
b. `input Id  Score1 @@;`
c. `input Id;`
d. `input Id  Score1  Score2;`

1.4 The program data vector in a SAS data step has variables with values as follows:

```
Code = 'VLC'
Size = 'M'
V1 = 2
V2 = 3
V3 = 7
V4 = .
```

Determine the results of the following SAS expressions:

a. `(V1 + V2 - V3)/3`
b. `V3 - V2/V1`
c. `V1*V2 - V3`
d. `V2*V3/V1`
e. `V1**2 + V2**2`
f. `Code = 'VLC'`
g. `Code = 'VLC' & size = 'M'`
h. `Code = 'VLC'|size = 'M'`
i. `Code = 'VLC' & V4^=.`
j. `(V3=.) + (V2=3)`
k. `V1 + V2 + V3 ^= 12`
l. `Code = 'VLC' | (Size = 'M' & V1 = 3)`
m. `3 < V2 < 5`

`Hint:` Recall that logical expressions evaluate to numeric values 1 (for 'TRUE') or 0 (for 'FALSE').

1.5 Show the values for the variable `Miles` that will be stored in the SAS data set `distance`:

```
data distance;
input Miles 5.2;
datalines;
1
12
123
1234
12345
1.
 12.
```

```
    12.3
    1234.5
;
```

1.6 Display the printed output produced by executing the following SAS program. Show what is in the program data vector at the point the first observation is to be written to the SAS data set.

```
data b21;
input Y1 Y2 @@;
Y3=Y2**2-5.0;
Y4=sqrt(Y1)/2+1;
drop Y1 Y2;
datalines;
4 -3 0 2 9 . 16 5 1 12
;
proc print data=b21;
run;
```

1.7 Display the printed output produced from executing the following SAS program. Show what is in the program data vector immediately after processing the first line of data.

```
data carmart;
input Dept $ Id $ P82 P83 P84;
Drop P82 P83 P84;
Year=1982; Sales=P82; output;
Year=1983; Sales=P83; output;
Year=1984; Sales=P84; output;
datalines;
parts    176  3500  2500   800
parts    217  2644  3500  3000
tools    124  5672  6100  7400
tools     45  1253  4698  9345
repairs   26  9050  5450  8425
repairs  142
;
proc print; run;
```

1.8 Sketch the printed output produced by executing the following SAS program. Display the contents of the program data vector immediately after processing the first line of data (just before it is written to the SAS data set).

```
data compete;
input Red Blue Grey Green White;
array grade8(5) Red Blue Grey Green White;
```

```
drop Team;
do Team=1 to 5;
   if grade8(Team)=. then grade8(Team)=0;
   grade8(Team)= grade8(Team)*10;
   Total + grade8(Team);
end;
datalines;
4  6  0  1  .
3  2  8  9  12
5  .  4  7  6
7  5  10  4  5
;
proc print; run;
```

1.9 Write a SAS data step to create a data set named `corn` with variables `Variety` and `Yield` using input data lines entered with varying numbers of pairs of values for the two variables as shown in the following:

```
A 24.2  B 31.5  B 32.0  C 43.9
C 45.2  A 21.8
B 36.1  A 27.2  C 34.6
```

1.10 Consider the following SAS data step:

```
data result;
input Type C1 C2 ;
datalines;
5 0 2
7 3 1
. 0 0
;
proc print;
```

Display appearance of the output from the `print` if each of the following sets of statements, respectively, appeared between `input` and `datalines;` when the program is executed:

```
a. Index = (2*C1) + C2;
b. if Type <= 6 then do;
            Index = (2*C1) + C2;
            output;
            end;
c. if Type <= 6 then do;
            Index = (2*C1) + C2;
            end;
      else delete;
d. if Type > 6 then delete;
```

```
e. if Type > 6 then delete;
       Index = (2*C1) + C2;
```

1.11 Study the following program:

```
data tests;
input Name $ Score1 Score2 Score3 Team $ ;
datalines;
Peter 12 42 86 red
Michael 14 29 72 blue
Susan 15 27 94 green
;
run;
proc print; run;
```

a. Sketch the printed output produced from executing this program.

b. What would be the printed output if the `input` statement is changed to the following:

```
input Name $ Score1 Score2 Score3;
```

c. What would you do to modify the above program if the data value for the variable `Score2` was missing for Michael?

d. Would the above `input` statement still work if the data lines were of the form given below. Explain why or why not.

```
Peter
12 42 86
red
Michael
...
```

e. Use the SAS function `sum()` in a single SAS assignment statement to create a new variable called `total`. Where would you insert this statement in the above program?

1.12 You have five plots randomly assigned to fertilizer A and five to fertilizer B a yield variable is measured on each plot. One would, for example, like to structure the SAS data set to look as the following:

Fert	Yield
A	67
A	66
A	64
A	62
A	68
B	70
B	74
B	78
B	77
B	80

a. Write an `input` statement to read the data values entered exactly as shown above, i.e., with one or more spaces between the Fert type and the Yield, with data values on separate lines for each pair.
b. Suppose you arrange your data values on two data lines like this:

```
A 67 A 66 A 64 A 62 A 68
B 70 B 74 B 78 B 77 B 80
```

Write an `input` statement for this arrangement.
c. Instead, if the five yield values for the plots assigned Fert A are placed on the first line and the five yield values for those assigned Fert B on the next data line like this:

```
A 67 66 64 62 68
B 70 74 78 77 80
```

Write a `data` step to read these data. Make sure the data set contains a `Fert` variable as well as a `Yield` variable.
d. Modify part (c) program so that the five plots in each group has a plot number from 1 to 5.

1.13 A research project at a college department has collected data on athletes. A subset of the data is given below. We will construct a single SAS program to do the tasks described in the itemized parts below.

IdNo	Age	Race	Systolic blood pressure	Diastolic blood pressure	Heart rate
4101	18	W	130	80	60
4102	18	W	140	90	70
4103	19	B	120	70	64
4104	17	B	150	90	76
4105	18	B	124	86	72
4106	19	W	145	94	70
4107	23	B	125	78	68
4108	21	W	140	85	74
4109	18	W	150	82	65
4110	20	W	145	95	75

Write SAS code to accomplish all of the tasks described below in a single SAS program. Put appropriate titles on each listing produced (i.e., use appropriate `title` statements in the proc steps) for the purpose of identifying parts of the output clearly. Execute the complete program.

a. Write SAS statements necessary to create a SAS data set named `athlete`. Name your variables as `IdNo, Age, Race, SPB, DBP`, and `HR`, respectively. Include a *label* statement for the purpose of describing the variables `SPB, DBP`, and `HR`. Enter the data instream, leaving a blank between fields and use the *list input style* to read the data in. [`This is the first data step in your program`]
b. *Average* blood pressure is defined as a weighted average of systolic blood pressure and diastolic blood pressure. Since heart spends more

time in its relaxed state (*diastole*), the diastolic pressure is weighted two-thirds, and the systolic blood pressure is weighted one-third. Add a SAS programming statement to the data step to create a new variable named `ABP` which contains values of average blood pressure computed for each athlete. Label this variable also. [`This modifies the first data step.`]

c. Add a PROC step to obtain a SAS listing of the data set `athlete`. [`This would be the first proc step in your program`]
d. Add SAS statements to create a new data set named `project` containing a subset of observations from the above data set. This subset will consist of only those athletes with a value greater than or equal to 100 for *average blood pressure* and a heart rate greater than 70 and provide a SAS listing of this data set. Omit the observation number from this listing; instead identify the athletes in the output by their ID numbers. [`You will add a second data step and a second proc step to do this part.`]
e. Obtain the same listing as in part (d), but without creating a new SAS data set to do it. Instead use the SAS statement `where` within the `proc print` step to select the subset of observations to be processed. [`This would require a third proc step.`]

1.14 The admissions office of a college has collected data on prospective undergraduate students. A subset of the data is given blow. We will construct a single SAS program to do the tasks outlined in the steps below:

Id	Age	Gender	High school GPA	College entrance Exam score
2101	18	M	3.7	650
2102	18	F	2.4	490
2103	19	M	3.3	580
2104	17	F	3.5	630
2105	18	F	3.1	610
2106	19	M	2.8	530
2107	23	M	3.2	590
2108	21	M	3.4	620
2109	18	F	3.1	630
2110	19	F	2.7	540
2111	22	M	3.1	580
2112	20	M	3.2	610
2113	18	F	3.6	640

SAS code to accomplish all of the parts below must be in a `single` SAS program. Put different titles on each report produced.

a. Write SAS statements necessary to create a SAS data set named `admit`. Name your variables as `Id, Age, Gender, GPA`, and `SAT`, respectively. Enter the data *instream*, leaving a blank between fields and entering a period to denote the decimal point. Use the *list input* style to read the data into SAS. [`This is the first data step in your program`]

b. A rating index for each applicant is to be computed (to be used for scholarship awards) using the following formula:

$$\text{rating} = \text{gpa} + 3 \times (\text{entrance exam score} \div 500)$$

Include a SAS programming statement in the above data step to add a new variable named `Rating` which contains values of the above index computed for each student. [`This would modify your first data step`]

c. Add a PROC step to your program to obtain a SAS Report (i.e., listing) of the data set `admit`. [`This would be your first proc step`]

d. Students who have a rating index of over 7 will be considered for academic scholarships. Create a new data set named `schols` using a subset of observations from the the SAS data set `admit`. The data set `schols` must contain only those applicants with a rating index greater than or equal to 7.0. Obtain a SAS report of the new data set. Suppress the observation number from this listing, instead identifying the students in the output by their ID number. [`This would add a second data step and a second proc step.`]

e. Obtain exactly the same listing as in part (d), without creating a new SAS data set to do it. Instead, use the SAS data set `admit` in a new `proc print` step and use the SAS statement `where` in this procedure step to select the subset of observations to be processed. [`This would be the third proc step in the SAS program.`]

1.15 Ms. Anderson wants to use a SAS program to compute the total score, assign letter grades, and compute summary statistics for her college Stat 101 class. A maximum of 50 points each could be earned for the quizzes, 100 points each for the midterm exams and the labs, and 200 points for the final exam. Data for the entire semester are available in the text file `stat101.txt` and a subset of the data is shown below:

Id	Major	Year	Quiz	Exam1	Exam2	Lab	Final
5109	Psych	4	50	93	93	98	162
7391	Econ	4	49	95	98	97	175
4720	Math	4	39	63	84	95	95
4587	Stat	3	46	92	96	88	150
⋮	⋮	⋮	⋮	⋮	⋮	⋮	⋮
3907	C E	4	44	80	99	99	134
4013	Econ	2	48	86	87	96	165
4456	Acct	4	36	83	88	91	154
7324	Psych	3	42	78	98	95	102
0746	Chem	4	48	84	84	97	154

Note: `Year` = the year in school, `Quiz` = the total for 5 quizzes, `Lab` = the total for ten labs

Do not run the SAS program until all of the steps described below in parts (a) to (e) have been completed.

a. Write SAS statements necessary to create a SAS data set named `stat101`. Name your variables as `Id, Major, Year, Quiz, Exam1, Exam2, Lab`, and `Final`, respectively. Enter the data instream, with at least one blank between data values and use the *list input style* to read the data in (You may cut and paste the data from the data file into your program). [`This is the first data step in your program`]
b. Write SAS statement(s) to be added to the above data step to create
   i. a new numeric variable `Score` containing the value of the course percentage, based on weighting the points obtained for the quizzes by 10%, each of the two midterms by 20%, the lab total by 10%, and the final by 40% computed for each student.
   ii. a new character variable `Grade` containing letter grades A, B, C, D, and F, using 90, 80, 70, 60 percent cutoffs, respectively. You may use the variable `Score` you created in part (i), in the SAS statements needed for this part.

   [`These would modify the first data step.`]
c. Add a `proc` step and provide a SAS listing of the data set `stat101`. [`This would be the first proc step in your program.`]
d. Students who are juniors and seniors and obtained A's from this class will qualify for applying to a research internship next summer. Create a SAS data set named `intern` containing only those juniors and seniors earning a letter grade A, using the observations from the data set `stat101`. Provide a SAS listing of the new data set that show only the variables `Id`, `Major`, `Year`, and `Score`. Include a statement to suppress the observation number from this listing, instead identifying the students in the output by their ID number. [`It would require adding a second data step and a second proc step to your program.`]
e. Obtain exactly the same listing as in part (d), without creating a new SAS data set to do it. Instead use the SAS statement `where` within a `proc print` procedure step to select the subset of observations to be processed. [`This would require adding a third proc step.`]

1.16 A local high school collects data on student performance in grades 9 through 12. In grades 9 and 10, data were collected for Science and English only, while for grades 11 and 12, Math scores were also recorded. Unfortunately, the data so collected were recorded as described below resulting in two completely different data layouts. Write a SAS program to create a temporary SAS data set called `perform` which contains observations for all four years of high school by accessing raw data (may be, containing 100s of data lines) as described below. Turn in your SAS program.

```
Sample Data Lines:

                    1         2         3
  Columns:123456789012345678901234567890
          ------------------------------
          0962432736578
          118091315736792
          0945712817859
          125916294847689
          1076543057182
          112479329697883
```

`Data Description:` (note the two types of data lines depending on grade)

Field	Variable description	Columns	Type
1	Grade	1–2	Char (09 or 10)
2	Student Id	2–6	Char
3	GPA	7–9	Numeric (with 2 decimals)
4	Science	10–11	Numeric (whole number)
5	English	12–13	Numeric (whole number)
1	Grade	1–2	Char (11 or 12)
2	Student Id	2–6	Char
3	GPA	7–9	Numeric (with 2 decimals)
4	Science	10–11	Numeric (whole number)
5	Math	12–13	Numeric (whole number)
6	English	14–15	Numeric (whole number)

2

More on SAS Programming and Some Applications

Although several approaches are possible for introducing the SAS language, in presenting the material in Chap. 1 in this book, the authors have consciously avoided a cookbook approach. The earlier students encountered the concepts of pointers and program data vectors, for example, the better their understanding of the fundamentals of the SAS data step. Without a basic understanding of the flow of operations in the data step, they will be ill equipped to use the data step effectively. From experience, it has been observed that this technique is more effective in getting the students to a higher point in the learning curve much earlier than using a cookbook approach. Having mastered the material in Chap. 1, readers will be ready to examine the use of SAS for data analysis in greater detail. In this chapter, several SAS procedures are used to illustrate the use of statements common to many SAS procedures as well as some that are specific to each procedure. To begin Chap. 2, some useful SAS statements available in both data and proc steps not encountered previously are introduced and discussed in detail.

2.1 More on the DATA and PROC Steps

In the previous chapter, the presentation of many important aspects of both the data and proc steps was deferred in order to keep the material presented in those sections, to some degree, more accessible. Some of these topics are covered in detail in this section. Many readers who are already familiar with SAS may have observed that in previous examples, the raw data were input as instream data when the content of the data sets called for the data to be read from external text files. This choice was made because it has been the experience of the authors that the introduction of the `infile` statement, the primary tool for accessing data from external files, at an early stage impedes

M. G. Marasinghe, K. J. Koehler, *Statistical Data Analysis Using SAS*,
Springer Texts in Statistics, https://doi.org/10.1007/978-3-319-69239-5_2

the beginning SAS user from understanding the basic data step operations. This is due to the fact it makes it more difficult to follow the data step processing clearly when the raw data lines are not directly visible to the user. Also, it is common for some beginners to confuse the external data file with the SAS data set being created. Thus, the decision to primarily use instream data for creating SAS data sets in Chap. 1 was made.

2.1.1 Reading Data from Files

The `infile` statement is primarily used to specify the external file containing the raw data, but it includes options to allow the user more control during the process of transferring data values from the raw data file into a SAS data set. For example, the user may use an option available in the `infile` statement to change the "delimiter," used by the *list input style* for reading data with the input statement, from a blank space to another character such as a comma. Another option allows the user to be given control when end-of-file is reached when reading external data so that other actions may be initiated before closing the new SAS data set.

The INFILE Statement

In SAS examples presented in Chap. 1, the data lines were inserted instream preceded by a `datalines` statement to identify the beginning of the data lines (see SAS Example A1 program in Fig. 1.4 in Chap. 1). The `infile` statement is an executable statement required to access data from an external file. In a SAS data step, it must obviously be present before the input statement because the execution of input statement requires the knowledge of the source of the raw data. The general form of the `infile` statement is

INFILE *file-spec* *<options>* ;

where *file-spec* represents a file specification. In the Windows environment, the file specification is easiest to be given directly as a path name to a file inserted within quotes; for example,

```
infile "C:\users\user_name\Documents\...\demogr.txt";
```

However, this may become cumbersome if some options are also to be included in the `infile` statement. Thus, it may be convenient to use a *fileref*.

The FILENAME Statement

The nonexecutable `filename` statement associates the physical name and location of an external file with a *fileref*, which is an alias for the file. The fileref is then available for use within the current SAS program. Under the Windows environment, a fileref is synonymous with the path name of the file. Text files previously saved in a folder can be given a fileref by including a `filename` statement in the SAS program. The following statements assign the fileref `mydata` to the raw data file named `demogr.txt` and uses it in an infile statement:

```
filename mydata "C:\users\user_name\Documents\...\demogr.txt";
infile mydata;
```

When a datalines statement is used to process instream data, SAS automatically assumes an infile statement with the file specification datalines; thus the infile statement is not required, unless the user wants to use one or more of the options available on the infile statement. In that case, the user must include an infile statement even if the data are included instream. An example of the use of an option while reading instream data is

```
filename datalines eof=last;
```

The above option specifies that once the last data line has been processed, the data step is to be continued by transferring control to the SAS statement labeled last instead of the default action of closing the SAS data set and terminating the data step. For instance, if the last observation has not yet been written to the SAS data set when end-of-file is encountered (for some reason, such as the last data line being incomplete), this allows the user to define how that situation should be handled.

Example 2.1.1 This is a simple modification of SAS Example A1 displayed in Fig. 1.4 in Chap. 1 to read the raw data set from a file instead from data entered instream. First, suppose that the data set is available as a text file prepared by entering the data lines into a simple text editor such as Notepad (if a word processor is used to enter the data, the user must make sure that the file is saved as a simple text file). Assume the file is named, say, wages.txt and is saved in a folder under the Windows environment. The SAS Example A1 program must be modified to access the data from this file as follows:

```
data first ;
infile "C:\users\user_name\Documents\...\wages.txt";
input (Income Tax Age State)(@4 2*5.2 2. $2.);
run;
proc print ;
title 'SAS Listing of Tax data';
run;
```

Here the file specification is a quoted string giving the path to the file containing the raw data.

Some Infile Statement Options

There are several infile statement options that may be useful for managing the conversion of information in data lines to an observation, such as the eof= option discussed earlier or the n= option discussed in Sect. 1.7.5. They are too numerous to be discussed in detail in this book; however, a few are sufficiently important to be briefly mentioned here. The delimiter= or the dlm= option allows the user to change the default value of the separator of data values when using the list input style to read data, from a space to the character specified. To read data separated by commas, use

```
infile datalines delimiter=',';
```

For example, this option can be used when reading data from a file of type csv:

```
infile "C:\users\user_name\Documents\...\sales.csv" dlm=',';
```

If any of the data values in the input data contain an embedded comma, this option will not work; instead, the option dsd must be used:

```
infile datalines dsd;
```

With this option in force, a missing value is assumed if two consecutive commas are detected. When using a list input style, if a line contains fewer data values than the number of variables listed in the input statement, use the missover option to prevent SAS from moving the input pointer to the next line in order to read the values not available in the current line:

```
infile datalines missover;
```

The missover option sets the remaining input statement variables to missing values. The option flowover is the default. The default causes the pointer to move to the next input data line if the current input line is not complete. Options such as firstobs= and obs= allow the user to access a specified number of data lines beginning from a specified line of data in the external data set. For example, the following processes data lines 20 through 50:

```
infile datalines firstobs=20 obs=50;
```

If firstobs= is omitted, SAS will access the first 50 data lines. The n= option specifies the number of lines that the pointer can move to in the input buffer using the # pointer control in a single execution of the input statement. The default value is 1. See Sect. 1.7.5 for more details.

2.1.2 Combining SAS Data Sets

When several data sets are created using multiple sources, they must be combined before a meaningful statistical analysis can be performed. Depending on the structure and the format of the input data sets and those required of the output data set, a variety of methods are available in SAS to form a combined data set. The SAS data step statements set, merge, and update are the primary tools available for combining data sets in a SAS data step. In SAS Example A2 (see Fig. 1.8 in Chap. 1 for the program), the set statement was used to illustrate how a SAS data set containing a subset of another SAS data set may be created as follows:

```
data second;
set first;
if Age<35 & State='IA';
run;
```

Here an `if` statement was used to select those observations that satisfy the logical condition specified, thus creating a subset of the original data set named `first`. An if statement used in this context is called a *subsetting if.*

SAS Example B1

In this section, an example is used to illustrate the use of the `set` statement to combine two SAS data sets by appending observations from one data set to those of the other. This process is called *concatenation* and allows the combination of several data sets. It is usually practicable when the data sets contain data from similar studies. This implies that the input data sets are expected to contain exactly the same variables (i.e., variables with identical names). It is possible that a few variables are different among some data sets due to decisions taken during the data collection process. If one or more of the data sets contain variables that are not common to all, the combined data set will contain those variables, but with missing values in the observations formed from the data sets that do not contain those variables.

```
data third ;
input W 1-2 X 3-5 Y 6;
datalines;
211023
312034
413045
;
data fourth;
input X Y Z;
datalines;
14 5 7862
15 6 6517
16 7 8173
;
data fifth;
set third fourth;
run;
proc print;
title 'Combining SAS data sets end-to-end ';
run;
```

Fig. 2.1. SAS Example B1: program

Three SAS data sets named `third`, `fourth`, and `fifth` are created in the SAS Example B1 program (see Fig. 2.1), the first two using external data and the other by combining the two SAS data sets previously created. The first data step uses the column input style to create the data set `third`, and the second uses the list input style to create the data set `fourth`, both containing `three` observations and `three` variables, respectively (see the abbreviated SAS log in Fig. 2.2). By observing the program, it can be determined that the variable `Z` is not present in the data set `third` and the variable `W` is not present in the data set `fourth`, whereas the variables `X` and `Y` are common to both data sets. The data set `fifth` is created using the data step:

```
data fifth;
set third fourth;
```

The data set `fifth` is formed by concatenating the observations in the two data sets `third` and `fourth` and so will contain *six* observations and the *four* variables `W, X, Y`, and `Z`. In the simplest use of the set statement illustrated here, SAS reads observations from the first data set in the list, `third`, transfers data values to the PDV, and then writes them sequentially to the new data set `fifth`. For example, the PDV following reading the first observation from `third` is

```
 W    X   Y   Z   _N_  _ERROR_
21  102   3   .    1      0
```

Although only the variables `W, X,` and `Y` are in data set `third`, SAS has detected the presence of the variable `Z` in the data step during the parsing stage. Thus, a slot for `Z` is created in the PDV and is initialized to a missing

```
1    data third ;
2    input W 1-2 X 3-5 Y 6;
3    datalines;

NOTE: The data set WORK.THIRD has 3 observations and 3 variables.

7    ;
8    data fourth;
9    input X Y Z;
10   datalines;

NOTE: The data set WORK.FOURTH has 3 observations and 3 variables.

14   ;
15   data fifth;
16   set third fourth;
17   run;

NOTE: There were 3 observations read from the data set WORK.THIRD.
NOTE: There were 3 observations read from the data set WORK.FOURTH.
NOTE: The data set WORK.FIFTH has 6 observations and 4 variables.

18   proc print;
19   title 'Combining SAS data sets end-to-end ';
20   run;

NOTE: There were 6 observations read from the data set WORK.FIFTH.
```

Fig. 2.2. SAS Example B1: log

value. When the observation is written to the output data set `fifth`, it will contain the values for the *four* variables `W, X, Y,` and `Z` as given above. Once the data in the first data set listed is exhausted, SAS begins reading data from the next data set listed in the set statement, `fourth`, and transfers the data values to the PDV. The PDV following reading the first observation from the data set `fourth` is

```
W   X   Y     Z   _N_  _ERROR_
.  14   5  7862    1       0
```

with `W` initialized to a missing value. Again, an observation containing values for the variables `W, X, Y,` and `Z` is written to the output data set `fifth`. This process continues until data set `fourth` reaches end-of-file. Then the data step comes to an end and SAS closes the output data set `fifth` and exits. The number of observations in the new data set is the total number of observations in the two input data sets, and the order of appearance is all observations from the first data set listed in the set statement, followed by all observations from the second data set listed, with missing values inserted appropriately for `Z` and `W`, respectively. The output from `proc print` (shown in Fig. 2.3) displays a listing of the data set `fifth`.

Combining SAS data sets end-to-end

Obs	W	X	Y	Z
1	21	102	3	.
2	31	203	4	.
3	41	304	5	.
4	.	14	5	7862
5	.	15	6	6517
6	.	16	7	8173

Fig. 2.3. SAS Example B1: output

The SET Statement

The general form of the `set` statement is

SET <*SAS-data-set(s)* <(*data-set-option(s)*)>> <options> ;

where *data-set-options* are those options that may be specified in parentheses after a SAS data set name, whether it is an input data set (as in a `set` statement) or an output data set (as in an `input` statement). More commonly used options such as `firstobs=, obs=,` or `where=` specify observations to be selected; those such as `drop=, keep=,` and `rename=` have variable names as arguments. When a set statement is used, it is more efficient to use an option to access only those variables required:

```
data fifth;
set third(keep=X Y) fourth(drop=Z);
run;
```

This will result in a SAS data set named `fifth` without any missing values:

Obs	X	Y
1	102	3
2	203	4
3	304	5
4	14	5
5	15	6
6	16	7

If the variable `Z` in the SAS data set `fourth` is renamed to be `W`, it will also result in a SAS data set with no missing values (albeit one different from the above):

```
data fourth;
set third fourth(rename=(Z=W));
run;
```

This would be an option if variables measuring the same characteristic or trait have been assigned different names in the two data sets. By renaming `Z` to be `W`, a variable that already exists in the data set `third`, the user is in fact recognizing this fact. The resulting data set is

Obs	W	X	Y
1	21	102	3
2	31	203	4
3	41	304	5
4	7862	14	5
5	6517	15	6
6	8173	16	7

In the SAS Example A2 program (see Fig. 1.8), the data set option `where=` could have been used to select the required subset of observations; thus,

```
data second;
set first(where=(Age<35 & State='IA'));
run;
```

There are several options that are unique to the `set` statement; among them are those that enable accessing observations nonsequentially according to a value given in the `key=` option or according to the observation number in the `point=` option.

Programming statements other than the *subsetting if*, such as assignment statements, may be used following the `set` statement, just as one would use following an `input` statement. In particular, one could use the `output` statement to create multiple observations in the output data set from a single observation in the input data set, similar to its use in SAS Example A7 (see Fig. 1.21). The use of a `by` statement following the set statement allows the creation of new observations by interleaving observations from several data sets. The observations in the output data set are arranged by the values of the `by` variable(s), in the order of the data sets in which they occur. Consider the two data sets AAA and BBB containing information for identical subjects:

```
  Data set AAA                    Data set BBB
 Id        Height               Id        Weight
 111         65                 111          145
 222         70                 222          156
 333         58                 333          148
 444         71                 444          166
 555         69                 555          175
 777         70                 666          136
```

The following SAS data step results in the formation of an interleaved SAS data set:

```
data CCC;
  set AAA BBB;
  by id;
run;
```

The resulting data set CCC (a listing is shown below) has 12 observations, which is the total number of observations from both data sets. The new data set contains all variables from both data sets. The values of variables found in one data set but not in the other are set to a missing value, and the observations are arranged in the order of the values of the variable `id`. In particular, note that the observation with `id` equal to 666 occurs before that with the `id` equal to 777 in the output data set, although the second observation came from the data set AAA listed first in the `set` statement. Note that observations in each of the original data sets were already arranged in the increasing order of the values of `id`. Thus, it is required for interleaving to ensure that the observations are sorted or grouped in each input data set by the variable or variables that are in the `by` statement.

```
 id     Height     Weight

111      65           .
111       .         145
222      70           .
222       .         156
```

```
333    58      .
333     .    148
444    71      .
444     .    166
555    69      .
555     .    175
666     .    136
777    70      .
```

Instead of the `set` statement, a `merge` statement may be more appropriate for combining these two data sets:

```
data CCC;
  merge AAA BBB;
  by id;
run;
```

A listing of the resulting data set is

```
Obs    id    Height    Weight

 1    111      65        145
 2    222      70        156
 3    333      58        148
 4    444      71        166
 5    555      69        175
 6    666       .        136
 7    777      70          .
```

Note that missing values are generated for the variables `Height` and `Weight` for those observations with no common `id` values in the two data sets.

2.1.3 Saving and Retrieving Permanent SAS Data Sets

In SAS examples discussed so far in this chapter, raw data, input either in-stream or from text files, were used to create temporary SAS data sets. As discussed in Sect. 1.2, SAS data sets contain not only the rectangular array of data but also other information such as variable attributes. In practice, the creation of a SAS data set requires substantial effort so that a user may want to save it permanently for future analysis using SAS procedures for performing different statistical applications. The availability of a carefully constructed permanent SAS data set allows the user to bypass the data set creation step at least for the duration of a research project. In addition, SAS data sets have become a convenient vehicle for transfer of large data sets to other users.

Two SAS examples are used in this subsection to illustrate how to use raw data to create a permanent data set and how to retrieve data for analysis from a previously saved SAS data set. The concept of a SAS *library* is easily understood in the context of running SAS programs under the Windows environment. Recall that the complete path name of a file was used with the `filename` statement to associate a fileref with the physical name and location of a file. Similarly, the `libname` statement associates the physical name and location of an external folder (directory) with a *libref*, which is an alias for the complete path name of the folder (directory). The following statement assigns the libref `mylib` to the folder named `projectA`:

```
libname mylib "C:\users\user_name\Documents\...\projectA\";
```

To save a SAS data set in a folder given in a libref as above, the user must specify a *two-level* SAS data set name, where the first level is the libref and the second level is the actual data set name. A two-level SAS data set name, in general, is a name that contains two parts separated by a period of the form `libref.membername` and is used to refer to *members* of a library libref. The *membername* is the name of a SAS data set when the members stored in the library are SAS data sets. Under the Windows operating system, a library is synonymous with a folder (or a directory). Thus, SAS data sets can be saved in a folder directly by executing statements in SAS programs giving two-level names to the data sets to be saved.

For example, `mylib.survey` refers to a SAS data set named `survey` to be saved in the above folder `projectA`. The libref defined in a SAS program is available for use only within the current SAS program. Many SAS data sets may be saved in the same folder (as members of the library) by using the libref `mylib` as the first-level name as many times as needed in the same program. A different name may be used as a libref to associate the same library in another SAS program, thus allowing the user to access previously stored members or add new members to the library.

SAS Example B2

```
libname mylib1 "C:\users\user_name\Documents\...\Class\";
data mylib1.first;
input X1-X5;
datalines;
1 2 3 4 5
2 3 4 5 6
6 5 4 3 2
1 2 1 2 1
7 2 55 5 5
;
run;
```

Fig. 2.4. SAS Example B2: program

The SAS Example B2 program (see Fig. 2.4) is a simple example illustrating the use of the `libname` statement and two-level names to create and access permanent SAS data sets. Instream raw data lines are used to create a SAS

```
1    libname mylib1 "C:\users\user_name\Documents\...\Class\" ;
NOTE: Libref MYLIB1 was successfully assigned as follows:
      Engine:        V9
      Physical Name: C:\users\user_name\Documents\...\Class
2    data mylib1.first;
3    input X1-X5;
4    datalines;

NOTE: The data set MYLIB1.FIRST has 5 observations and 5 variables.

10   ;
```

Fig. 2.5. SAS Example B2: log page

data set using the two-level name `mylib1.first`. The first part of the two-level name `mylib1` refers to a folder in a disk mounted on a zip drive. Thus, the SAS data set named `first` created in the data step is saved as a permanent file in the specified folder. Thus, the data set `first` will be a member of this library. The SAS log reproduced in Fig. 2.5 indicates this fact by listing the two-level name MYLIB1.FIRST and identifying the name of the folder as `C:\users\user_name\Documents\...\Class`. The actual physical name of the file saved is `first.sas7bdat`, as can be verified by manually obtaining a listing of the `Class` folder (see Fig. 2.6). Obviously, if a single-level name, say `first`, was used instead in the data statement, the SAS data set would have been temporarily saved in the WORK folder (and the SAS data set thus created referred to as WORK.FIRST in the log page).

Name	Date modified	Type	Size
demogr.txt	10/2/2005 2:32 PM	Text Document	3 KB
first.sas7bdat	6/26/2015 4:57 PM	SAS7BDAT File	128 KB
fuel.txt	11/3/2004 11:45 AM	Text Document	2 KB
fueldat.sas7bdat	9/18/2012 10:13 AM	SAS7BDAT File	9 KB
oranges.sas7bdat	9/12/2013 5:52 PM	SAS7BDAT File	5 KB
oranges_out.rtf	8/28/2014 9:47 AM	Rich Text Format	35 KB

Fig. 2.6. Screenshot of `Class` folder listing

SAS Example B3

By including a `libname` statement of the form shown in the SAS program shown in Fig. 2.4 (possibly with a different libref, but the same physical path name of the folder), one or more SAS data sets stored permanently in a library can be accessed for further processing in another SAS program to be executed subsequently.

```
libname mydef1 "C:\users\user_name\Documents\...\Class\";

proc print data=mydef1.first; run;

proc means data=mydef1.first; run;

data mydef1.second;
input Y1-Y3;
datalines;
31 34 38
43 45 47
10 11 12
908 97 96
;

proc means data=mydef1.second mean std var range maxdec=3;
run;
```

Fig. 2.7. SAS Example B3: program

In the SAS Example B3 program (see Fig. 2.7), the SAS data set `first` is accessed from the library for processing using this method. The following `libname` statement in this program associates the libref `mydef1` with the same library where the data set `first` was saved when the SAS program in Fig. 2.4 was executed:

```
libname mydef1 "C:\users\user_name\Documents\...\Class\";
```

This allows the two-level name `mydef1.first` to be used as shorthand for accessing the SAS data set `first` from the library to be analyzed using the SAS procedure `proc print` by naming it in the `data=` option. The listing resulting from this statement is shown on page 1 of the output produced by the SAS Example B3 program, displayed in Fig. 2.8.

The SAS System

Obs	X1	X2	X3	X4	X5
1	1	2	3	4	5
2	2	3	4	5	6
3	6	5	4	3	2
4	1	2	1	2	1
5	7	2	55	5	5

The SAS System

The MEANS Procedure

Variable	N	Mean	Std Dev	Minimum	Maximum
X1	5	3.4000000	2.8809721	1.0000000	7.0000000
X2	5	2.8000000	1.3038405	2.0000000	5.0000000
X3	5	13.4000000	23.2873356	1.0000000	55.0000000
X4	5	3.8000000	1.3038405	2.0000000	5.0000000
X5	5	3.8000000	2.1679483	1.0000000	6.0000000

Fig. 2.8. SAS Example B3: output pages 1 and 2

The statement `proc means data=mydef1.first;` produces the statistical analysis shown on page 2 of the output from the SAS Example B3 program displayed in Fig. 2.8. Again, `proc means` accesses the SAS data set `first` from the same library and computes the default statistics for all variables in the data set as shown on page 2. There is nothing to preclude the user from adding new SAS data sets to the same library, in the same program or in separate SAS programs. The data step shown in Fig. 2.7 reads instream data using list input as usual and creates the SAS data set named `second` and then saves it permanently in the library identified by the libref `mydef1`.

The SAS System

The MEANS Procedure

Variable	Mean	Std Dev	Variance	Range
Y1	248.000	440.211	193786.000	898.000
Y2	46.750	36.372	1322.917	86.000
Y3	48.250	35.122	1233.583	84.000

Fig. 2.9. SAS Example B3: output page 3

The data set `second` just created may be accessed from the library in the same SAS program if required for more processing. For example, in the last proc step shown in Fig. 2.7, `proc means` is used to produce selected descriptive statistics using options on the proc statement. The additional option `maxdec=3` limits the values of the statistics output to three decimals. The output from this proc step appears in Fig. 2.9.

2.1.4 User-Defined Informats and Formats

Before discussing the use of the `format` procedure for creating user-defined informats and formats, a review of these two variable attributes is informative. Several simple informats such as `$10.` or `5.2` and more complex informats such as `dollar10.2` or `monyy5.` were used in several examples in Chap. 1. `Informats` determine how raw data values are read and converted to a number or a character string to be stored in memory locations. An informat contains information of the type of data (character or numeric) to be read; the length it occupies in the data field; how to handle leading, trailing, or embedded blanks and zeros; where to place the decimal point; and so forth. For example, the informat `ddmmyy8.` converts the date value 19/10/07 entered in a data line into the binary number 17458 to be stored as a value of a SAS variable. Similarly, `formats` convert data values from internal form into a form the user wants them to appear in printed output. For example, the format `dollar15.2` prints the value of `Cost`=2317438.3921, which is (say) the result of the product of the values of `Quantity`=2346.678 and `Price`=987.54, as $2,317,438.39.

SAS system contains a large number of predefined informats and formats to handle many types of data conversion. However, it is not possible to provide informats or formats for every conceivable need; for example, an informat might be needed to convert the character strings "YES" and "NO" to be stored as the numeric values one or zero, respectively, or a numeric value stored internally as a one or a two to indicate gender will be best converted to be output as character strings "Female" or "Male," respectively. FORMAT is a SAS procedure that allows the user to define informats or formats to do these kinds of specialized conversion. In this section, the emphasis will be on the use of `proc format` for creating output formats, although the importance of user-defined informats cannot be overstated. In practical terms, the primary use of user-constructed informats is for data validation, and some examples of this type of application appear below. The general structure of a `proc format` step (with three procedure information statements to be illustrated below) is

```
PROC FORMAT <option(s)>;
  INVALUE <$>name <(informat-option(s))> value-range-set(s);
  PICTURE name <(format-option(s))> value-range-set-1
                                  <(picture-1-option(s))>
             <...value-range-set-n <(picture-n-option(s))> >;
  VALUE <$>name <(format-option(s))> value-range-set(s);
```

In the above description, the phrase *value-range-set* refers to an assignment type specification that defines a one-to-one or a many-to-one relationship between value or values to be converted to another value. The specification and action of a `value-range-set` depend on the context of its usage.

In the case of an `invalue` statement, *value-range-set* is of the form

```
value or range = informatted-value|[existing-informat]
```

where `value` is a value such as 1 or "NB," `range` is a list of values usually specified in the form 100–999 or "A"–"Z." The words `low` and `high` may be used to define the end points of any range (numeric or character), implying that the specified range covers the entire range of values below the upper end point or above the lower end point, respectively. For example, the range `100-high` covers every value greater than 100. The less than ($<$) symbol may be to exclude values from ranges. For example, the range `10-<20` is equivalent to $10 \leq \text{value} < 20$, and the range `10<-20` is equivalent to $10 < \text{value} \leq 20$. The `informatted value` (on the right-hand side of the equal sign) specifies the internal value that the raw data value that is equal to the value (or in the range of values) on the left is to be converted.

In the case of a `value` statement, the *value-range-set* is of the form

```
value or range = 'formatted-value'|[existing-format].
```

The definition of value and range is the same as for the `invalue` statement. The '`formatted-value`' specifies a character string to which the value (or the range of values) that appears on the left side of the equal sign is to be converted for printing. The '`formatted-value`' is a character string, regardless of whether the format created is a character or numeric format.

User-defined formats may be used to avoid creating category variables in the data step when it is not really necessary to do so. The idea is to use the formatted value of the actual variable value to represent a category. This may be a little more efficient when the category variable created may be used only once. The following example illustrates the use of this technique:

Example 2.1.2

```
proc format;
  value af low-<10='0'
           10-<20='10'
       20-<30='20'
       30-<40='30'
       40-<50='40'
       50-high='50';
run;

data group1;
input Age @@;
datalines;
1 3 7 9 12 17  21 26 30 32 36 42 45 51
;
proc print data=group1;
format Age af.;
run;
```

Here the user-defined format `af.` converts all values for the `Age` variable to be character strings when they are output (as displayed below). It is important to note that a new variable is not created for this purpose. Also the formatted variable may be used as a category variable in other procedures where such variables are used in the analyses, such as `proc freq` and `proc anova`, without actually creating a new variable from another numerical variable.

Obs	Age
1	0
2	0
3	0
4	0
5	10
6	10
7	20
8	20

```
 9   30
10   30
11   30
12   40
13   40
14   50
```

A discussion of several possible options on the `value` and `invalue` statements are omitted here but can be found under the description of `proc format`. For example, the option `fuzz=` allows the user to specify a fuzz factor for matching values to a range. If a value does not exactly match or falls in a range but comes within the fuzz factor, then the format or the informat will consider it to be a match or in the range. This facility is useful especially when the raw data contains fractions that need to be rounded up or down to be exactly in a prespecified range. For example, the value 99.9 may be considered in the range 100–200 if a fuzz factor of 0.1 has been specified (`fuzz=.1`) and values below 100 are not considered to be in the conversion range.

SAS Example B4

```
proc format;
    invalue $st 'IA'='Iowa'
                'NB'='Nebraska';
run;
data first;
length State $ 12;
infile "C:\users\user_name\Documents\...\Class\wages.txt";
input (Income Tax Age State)(@4 2*5.2 2. $st2.);
run;

proc print noobs;
format Income Tax dollar8.2  State $12. ;
var Income Tax Age State;
title 'SAS Listing of Tax data';
run;
```

Fig. 2.10. SAS Example B4: program

In the SAS Example B4 program displayed in Fig. 2.10, the two-character state codes used in the raw data set of SAS Example A1 (see Fig. 1.4) are converted to values that are longer character strings identifying the name of the respective state. Since there are several SAS functions (e.g., `stnamel()`) available for state name conversions, this informat (named $st) is created only as an illustration.

In the proc format step, an `invalue` statement is used to define the required conversion. Note that only values expected to be in the data for the character variable `State` are used in this definition. If other state values are to be converted, then they must be included in the format definition. Since character-type variables are assigned lengths of 2 bytes by default, a `length` statement (that must appear before the input statement) specifies the length of the State variable to be 12 bytes. Thus, a format with a width of at least 12 positions is needed to print values of the State variable. As seen in the SAS program, the format $12. is associated with the State variable, and the resulting output is shown in Fig. 2.11.

If the only state codes allowed in the data set are "NB" and "IA," the invalue statement may be modified to flag any other state code used as an error as follows:

```
invalue $st 'IA'='Iowa' 'NB'='Nebraska' other='Invalid St.';
```

In the above, the word other is a SAS keyword that will match any value that is not the strings "NB" or "IA." Thus, if this informat is used to input values for a character variable with values other than "NB" or "IA," the respective observations will contain the string "Invalid St." as the value of that variable. This is an example of the use of an informat for data validation, an important step in data analysis.

SAS Listing of Tax data

Income	Tax	Age	State
$546.75	$34.65	35	Iowa
$765.48	$89.56	45	Iowa
$578.65	$59.54	31	Iowa
$786.78	$57.65	41	Nebraska
$567.51	$126.85	32	Iowa
$785.87	$67.85	28	Nebraska
$985.65	$75.65	43	Nebraska
$745.63	$78.95	25	Iowa
$345.67	$25.68	23	Nebraska
$567.34	$89.75	34	Nebraska
$651.12	$50.45	45	Iowa
$650.75	$65.45	29	Nebraska
$595.65	$45.68	34	Iowa
$678.56	$91.27	28	Nebraska
$685.96	$67.51	38	Iowa
$825.75	$56.25	27	Iowa

Fig. 2.11. SAS Example B4: output

It is important to note that user-defined informats read only `character` values, although these can be converted to either character or numeric values. In the above example, if the state FIPS codes were input as numbers (19 for IA and 31 for NB), they must still be accessed as character data by the informat. So the appropriate `invalue` statement is

```
invalue $st '19'='Iowa' '31'='Nebraska';
```

If the expansion of the state code was required only for the printed output, then it would have been sufficient to create a format (as opposed to an informat) for this purpose. The above program is modified as shown in Fig. 2.12 (the output from this SAS program is not shown).

```
proc format;
    value $st 'IA'='Iowa'
              'NB'='Nebraska';
run;
data first;
infile "C:\users\user_name\Documents\...\Class\wages.txt";
input (Income Tax Age State)(@4 2*5.2 2. $2.);
run;
proc print noobs;
format Income Tax dollar8.2  State $st10. ;
var Income Tax Age State;
title 'SAS Listing of Tax data';
run;
```

Fig. 2.12. SAS Example B4: modified program

A `value` statement is used to define a format for printing values of the variable `State`. Note that the new format $st is used in a `format` statement to specify the values of the variable `State`. Note carefully that the values to be stored in `State` were read using the informat $2. and hence will be one of the strings "IA" or "NB." The conversion takes place when they are output using the format $st10., where these values will be printed as "Iowa" and "Nebraska," respectively, using ten print positions aligned to the left.

Note the difference between the FORMAT procedure and the `format` statement carefully. As in the above example, `proc format` is used to create user-defined formats or informats. The `format` (or the `informat`) statement associates a currently existing format (or informat) with one or more variables. Either standard SAS or user-defined formats or informats can be associated with variables this way. For example, the statement

```
format Income Tax dollar8.2  State $st10.;
```

associates the SAS format `dollar8.2` with the variables `Income` and `Tax`, whereas the user-defined format `$st10.` is associated with the variable `State`. `Proc format` stores user-defined informats and formats as entries in *SAS cat-*

alogs (specially structured files), either temporarily in the WORK library or permanently in a user-specified library.

Finally, the following example illustrates how an existing SAS informat or a format can be used as an informatted or a formatted value in `value-range-set` definitions in invalue or value statements, respectively. Recall that the definitions of `value-range-sets` in these two statements were

```
value or range = informatted-value|[existing-informat]

value or range = 'formatted-value'|[existing-format].
```

Instead of an informatted or a formatted value, the user can specify an existing SAS informat or a format placed inside box brackets that will be used for the conversion of the value or the range on the left hand side of the `value-range-set` definition.

Example 2.1.3

```
proc format;
  invalue ff 0-high=[4.2]
                 -1 = .;
run;

data ex212;
input A 2.0 B ff4.2 ;
datalines;
10 205
20 216
30 237
40 257
50 -1
60 469
;
```

The user-defined numeric informat (named `ff`) converts all positive data values using the SAS informat `4.2`. When the data value is −1, it is converted to the SAS missing value for a numeric variable, i.e., a period. A scenario for the need to use such an informat may arise if the raw data set has been prepared where a −1 has been entered instead of using SAS missing values or spaces to indicate missing data values where the actual data values are all positive numbers. The output data set is

Obs	A	B
1	10	2.05
2	20	2.16
3	30	2.37

```
4    40    2.57
5    50     .
6    60    4.69
```

2.1.5 Creating SAS Data Sets in Procedure Steps

Several SAS procedures used for statistical analysis have the capability to let the user specify which statistics, calculated by the procedure, are to be saved in newly created SAS data sets. In some procedures, these data sets are organized in special structures that allow them to be read by another SAS procedures for further analysis by specifying the `type=` attribute of the data set. For example, `proc corr` creates a data set with the attribute `type=corr` containing a correlation matrix, which can be directly input to a procedure such as `proc reg` as an input data set. If the required analysis performed by `proc reg` is solely based on the correlation matrix, then much of the overhead spent on recomputing the correlation matrix can be avoided.

In this subsection, the discussion is limited to a description of the use of the `output` statement in several SAS procedures that compute an extensive number of statistics for variable values. In most of these procedure steps, a `class` statement specifies classification variables in the data set that are discrete-valued variables that identify groups, classes, or categories of observations in the data set. They may be numeric or character-valued and may be observed ordinal- or nominal-valued variables or user-constructed variables. In practice, continuous-valued variables may be used to define new grouping variables that can then be used in `class` statements. An example would be the creation of a variable defining income groups with values, say, 1, 2, and 3, or “Low,” “Medium,” and “High,” using the values of the continuous-valued variable `Income` to form the groups. A `var` statement (i.e., the variables statement) identifies the analysis variables (that must all be of numeric type). The statistics are computed on the values of analysis variables for subsets of observations defined by the classification variables.

In the SAS Example B5 program, `proc means` is used to introduce the basic use of the output statement. A simplified general form of the `output` statement used in this example is

```
OUTPUT <OUT=SAS-data-set> <output-statistic-specification(s)>;
```

where an *output-statistic-specification* is of the general form

```
statistic-keyword<(variable-list)>=<name(s)>
```

where *statistic-keyword* specifies the statistic to be calculated and is stored as a value of a variable in the output data set. Some of available statistic keywords are `n, mean, median, var, cv, std, stderr, max, min, range, cv, skewness, kurtosis, q1, q3, qrange, p1, p5, p10, p90,` `p95, p99, t,` and `probt`. The optional *variable-list* specifies the names of one or more analysis variables on whose values the specified statistic is to be computed. If this list is omitted, the specified statistic is computed for all the analysis variables. The optional *name(s)* specifies one or more names for the variables in the output data set that will contain the analysis variable statistics in the same sequence that the analysis variables are listed in the `var` statement. The first name contains the statistic for the first analysis variable, the second name contains the statistic for the second analysis variable, and so on. If the names are omitted, the analysis variable names are used to name the variables in the output data set.

```
data biology;
input Id Sex $ Age Year Height Weight;
datalines;
7389   M  24  4  69.2  132.5
3945   F  19  2  58.5  112.0
4721   F  20  2  65.3   98.6
1835   F  24  4  62.8  102.5
9541   M  21  3  72.5  152.3
2957   M  22  3  67.3  145.8
2158   F  21  2  59.8  104.5
4296   F  25  3  62.5  132.5
4824   M  23  4  74.5  184.4
5736   M  22  3  69.1  149.5
8765   F  19  1  67.3  130.5
5734   F  18  1  64.3  110.2
4529   F  19  2  68.3  127.4
8341   F  20  3  66.5  132.6
4672   M  21  3  72.2  150.7
4823   M  22  4  68.8  128.5
5639   M  21  3  67.6  133.6
6547   M  24  2  69.5  155.4
8472   M  21  2  76.5  205.1
6327   M  20  1  70.2  135.4
8472   F  20  4  66.8  142.6
4875   M  20  1  74.2  160.4
;

proc means data=biology order=data;
  class Year Sex;
  var Height Weight;
  output out=stats mean=Av_Ht Av_Wt stderr=SE_Ht SE_Wt;
run;

proc print data=stats;
  title "Biology Class Data Set: Output Statement";
run;
```

Fig. 2.13. SAS Example B5: program

SAS Example B5

The SAS Example B5 program (see Fig. 2.13) illustrates the use of `proc means` to calculate and print statistics for an input data set named `biology` (used previously in Fig. 1.2) and, in addition, save the statistics *in a new SAS data set* created in the proc step. The simple data set to be analyzed includes a numeric variable `Year` (indicating class in college) and a character variable `Sex` that will be used as classification variables and two numeric analysis variables `Height` and `Weight`. Ordinarily, `proc means` produces printed output of five default statistics (n (sample size), mean, standard deviation, minimum, and maximum) calculated for every variable in the `var` statement list, for subsets of observations formed by all combinations of the levels of the `class` variables. The option `maxdec=3` used on the proc statement limits the number of decimal places output when printing all calculated statistics. Page 1 of the SAS output (see Fig. 2.14) displays the printed output in its standard format. As described above, the five default statistics are computed for the variables `Height` and `Weight` for groups observations defined by the levels "F" and "M," respectively, of the `Sex` variable within each value 1, 2, 3, or 4, of the `Year` variable, respectively.

The MEANS Procedure

Year	Sex	N Obs	Variable	N	Mean	Std Dev	Minimum	Maximum
1	F	2	Height	2	65.800	2.121	64.300	67.300
			Weight	2	120.350	14.354	110.200	130.500
	M	2	Height	2	72.200	2.828	70.200	74.200
			Weight	2	147.900	17.678	135.400	160.400
2	F	4	Height	4	62.975	4.614	58.500	68.300
			Weight	4	110.625	12.455	98.600	127.400
	M	2	Height	2	73.000	4.950	69.500	76.500
			Weight	2	180.250	35.143	155.400	205.100
3	F	2	Height	2	64.500	2.828	62.500	66.500
			Weight	2	132.550	0.071	132.500	132.600
	M	5	Height	5	69.740	2.481	67.300	72.500
			Weight	5	146.380	7.535	133.600	152.300
4	F	2	Height	2	64.800	2.828	62.800	66.800
			Weight	2	122.550	28.355	102.500	142.600
	M	3	Height	3	70.833	3.182	68.800	74.500
			Weight	3	148.467	31.183	128.500	184.400

Fig. 2.14. SAS Example B5: output page 1

The *output-statistic-specifications* used in the `output` statement has the basic form of `statistic=names`. They are `mean=Av_Ht Av_Wt` and `stderr= SE_Ht SE_Wt`. Since the `var` statement used in the proc step is `var Height Weight;`, the above specifications request that the means and standard errors of the variables `Height` and `Weight` are to be computed and stored in the new variables `Av_Ht`, `Av_Wt`, `SE_Ht`, and `SE_Wt`, respectively. Page 2 of the SAS output (see Fig. 2.15) displays a listing of the SAS data set named `stats` produced in the proc means step. It can be observed that there are 15 observations displaying values for the variables Year, Sex, _TYPE_, _FREQ_, Av_Ht, Av_Wt, SE_Ht, and SE_Wt. The value of the _TYPE_ variable (0, 1, 2, or 3) indicates which combinations of the class variables are used to define the subgroups of observations used for computing the statistics. For example, there is exactly one observation (Observation 1) with the value _TYPE_=0. In this observation, the variables Year and Sex are set to respective missing values, indicating that both of these variables are ignored in determining the sample used to compute the statistics shown for this observation; that is, the subgroup for their computation is the entire data set as evidenced by the value of _FREQ_=22.

Biology Class Data Set: Output Statement

Obs	Year	Sex	_TYPE_	_FREQ_	Av_Ht	Av_Wt	SE_Ht	SE_Wt
1	.		0	22	67.8955	137.591	0.97773	5.4643
2	.	F	1	10	64.2100	119.340	1.03489	4.8887
3	.	M	1	12	70.9667	152.800	0.85390	6.4766
4	1		2	4	69.0000	134.125	2.11069	10.3181
5	2		2	6	66.3167	133.833	2.72255	16.4964
6	3		2	7	68.2429	142.429	1.30783	3.4515
7	4		2	5	68.4200	138.100	1.89642	13.3320
8	1	F	3	2	65.8000	120.350	1.50000	10.1500
9	1	M	3	2	72.2000	147.900	2.00000	12.5000
10	2	F	3	4	62.9750	110.625	2.30701	6.2277
11	2	M	3	2	73.0000	180.250	3.50000	24.8500
12	3	F	3	2	64.5000	132.550	2.00000	0.0500
13	3	M	3	5	69.7400	146.380	1.10932	3.3698
14	4	F	3	2	64.8000	122.550	2.00000	20.0500
15	4	M	3	3	70.8333	148.467	1.83697	18.0037

Fig. 2.15. SAS Example B5: output page 2

Similarly, for _TYPE_=1, there are two subgroups formed for each of the values of the Sex variable ignoring the Year variable. Note carefully that Sex variable appears rightmost in the variable list in the class statement; hence, its levels form _TYPE_=1 groups. Observations 2 and 3 list statistics computed based on these groups of observations and note sample sizes given by _FREQ_=10 and _FREQ_=12, respectively.

There are four observations with _TYPE_=2, and these statistics are based on the groups of observations that correspond to each level of Year ignoring the levels of Sex. Observations with _TYPE_=3 correspond to subgroups defined by all combinations of levels of `Year` and the levels of `Sex`. Thus, the complete set is formed by 1+2+4+8 = 15; thus, 15 observations are included in `stats`.

Other forms of the *output-statistic-specifications* used in the `output` statement can be used to alter the appearance of the SAS data set created. Several procedure information statements and proc statement options are available for controlling the contents of this data set. The following examples of the `ways` and `types` statements illustrate some of these choices. These two statements allow the user to select the subset of observations to be included in the output data set as defined by the _TYPE_ variable discussed earlier. The `ways` statement uses integers to indicate number of class variables chosen to form the combinations; for example, one may specify two to request that subgroups are to be formed by combining all possible pairs of class variables in the class variable list. The `types` statement, on the other hand, allows the user to specify class variables and how they are to be combined directly.

Including the statement `ways 1;` in the proc step in the above example produces the output data set shown in Fig. 2.16. This requests that subgroups are to be defined by the levels of the class variables taken one at a time. Here,

Biology Class Data Set: Output Statement

Obs	Year	Sex	_TYPE_	_FREQ_	Av_Ht	Av_Wt	SE_Ht	SE_Wt
1	.	F	1	10	64.2100	119.340	1.03489	4.8887
2	.	M	1	12	70.9667	152.800	0.85390	6.4766
3	1		2	4	69.0000	134.125	2.11069	10.3181
4	2		2	6	66.3167	133.833	2.72255	16.4964
5	3		2	7	68.2429	142.429	1.30783	3.4515
6	4		2	5	68.4200	138.100	1.89642	13.3320

Fig. 2.16. SAS Example B5: result using the WAYS statement

Biology Class Data Set: Output Statement

Obs	Year	Sex	_TYPE_	_FREQ_	Av_Ht	Av_Wt	SE_Ht	SE_Wt
1	1	F	3	2	65.8000	120.350	1.50000	10.1500
2	1	M	3	2	72.2000	147.900	2.00000	12.5000
3	2	F	3	4	62.9750	110.625	2.30701	6.2277
4	2	M	3	2	73.0000	180.250	3.50000	24.8500
5	3	F	3	2	64.5000	132.550	2.00000	0.0500
6	3	M	3	5	69.7400	146.380	1.10932	3.3698
7	4	F	3	2	64.8000	122.550	2.00000	20.0500
8	4	M	3	3	70.8333	148.467	1.83697	18.0037

Fig. 2.17. SAS Example B5: result from using TYPES statement

two sets of statistics are produced for the two levels of `Sex` and four sets for the four levels of `Year`. The printed output (not shown) is similarly structured; two tables of statistics are produced for each class variable separately. Other possibilities in this example are `ways 0;` when no subgroups are formed, meaning statistics are computed for the entire data set, and `ways 2;` when subgroups are formed for all eight combinations of the two class variables.

Including the statement `types Year Sex;` in the proc step produces the same data set shown in Fig. 2.16 and the corresponding printed output (not shown). If, instead, `types Year;` is used, then only those statistics for the subgroups defined by the four levels of `Year` will be calculated. The statement `types Year*Sex;` produces statistics for subgroups formed for the eight combinations of the two class variables `Year` and `Sex` as shown in Fig. 2.17.

2.2 SAS Procedures for Descriptive Statistics

While `proc means` used in previous examples is classified as a Base SAS procedure, UNIVARIATE procedure is a SAS procedure classified as a Base statistical procedure. As an introduction to how the ODS system may be used to extract tables and graphs from these procedures, this section begins with a simple example.

SAS Example B6

The SAS Example B6 program (see Fig. 2.18) illustrates the use of `proc univariate` to calculate and print statistics and graphics for the input data set named `biology` used previously in several SAS examples. An `infile` statement is used to access the data from a text file from a folder. In the proc univariate step, several tables of staistics and a normal probability plot are produced for the `Height` variable. The `ods select` statement may be used to identify tables and statistics to be output. Procedures assign a name to each table and graph that it creates. These names can be found, respectively, in tables labeled `ODS Table Names` and `ODS Graph Names` under the details section of the procedure description. Note carefully that except for default tables or graphs produced by a procedure, some tables are only produced by including the required options on the proc statement, and many graphs are produced only by including specific statements in the proc step. Many of these options and statements available with `proc univariate` will be illustrated in other examples that follow. In SAS Example B6, the `normaltest` option produces a table of results of tests for normality, and the `probplot` statement produces a normal probability plot of the variable `Height` in the biology data set. Thus the normal probability plot is output to the specified destination in addition to tables of basic statistical measures, quantiles, and tests for normality. These are displayed in Figs. 2.19 and 2.20, respectively.

```
data biology;
infile "C:\users\user_name\Documents\...\Class\biology.txt";
input Id Sex $ Age Year Height Weight;
run;

ods select BasicMeasures Quantiles TestsForNormality ProbPlot;

proc univariate data=biology normaltest;
  var Height;
  probplot Height;
  title 'Biology class: Analysis of Height Distribution';
run;
```

Fig. 2.18. SAS Example B6: program

The top table in Fig. 2.20 shows descriptive statistics computed by default for the variable `Height`. The proc statement option `normaltest` in this program produces results of several statistical tests for normality. For example, the p-value for the Shapiro–Wilk test, 0.9231, in Fig. 2.20 indicates that the null hypothesis that the data is a random sample from a normal distribution will not be rejected. Marked departures from a straight line indicates that the sample distribution deviates from a normal distribution. In such a case, points in a normal probability plot show some identifiable curvature pattern. Most

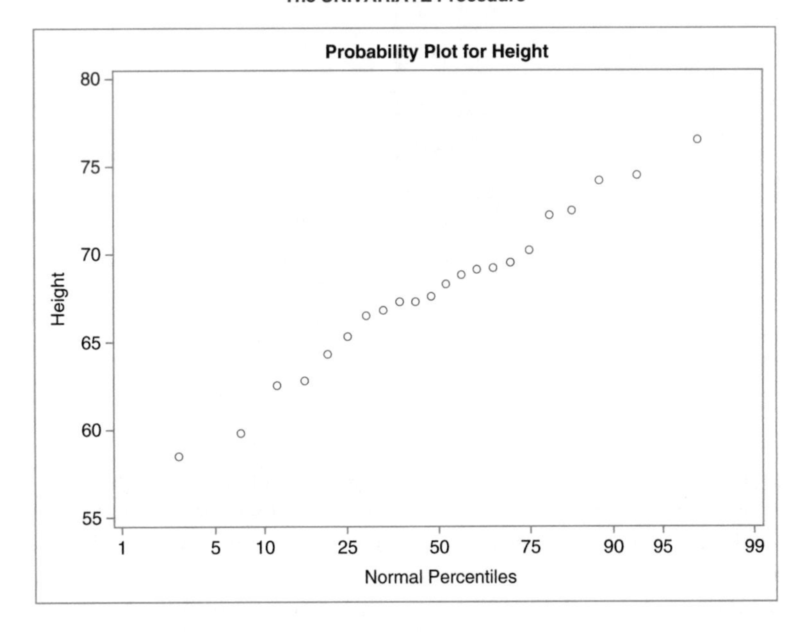

Fig. 2.19. SAS Example B6: normal probability plot of `Height`

Biology class: Analysis of Height Distribution

The UNIVARIATE Procedure
Variable: Height

Basic Statistical Measures			
Location		Variability	
Mean	67.89545	Std Deviation	4.58595
Median	67.95000	Variance	21.03093
Mode	67.30000	Range	18.00000
		Interquartile Range	4.90000

Tests for Normality				
Test	Statistic		p Value	
Shapiro-Wilk	W	0.980466	Pr < W	0.9231
Kolmogorov-Smirnov	D	0.107727	Pr > D	>0.1500
Cramer-von Mises	W-Sq	0.034689	Pr > W-Sq	>0.2500
Anderson-Darling	A-Sq	0.206227	Pr > A-Sq	>0.2500

Quantiles (Definition 5)	
Level	Quantile
100% Max	76.50
99%	76.50
95%	74.50
90%	74.20
75% Q3	70.20
50% Median	67.95
25% Q1	65.30
10%	62.50
5%	59.80
1%	58.50
0% Min	58.50

Fig. 2.20. SAS Example B6: selected statistical tables

commonly, a bowl-shaped pattern of the points indicates a right-skewed distribution, a mound-shaped pattern indicates a left-skewed distribution, and an S-shaped pattern indicates a short-tailed (i.e., heavy-tailed) distribution, relative to the shape of a normal distribution. Note that one must make sure that normal percentiles appear on the x-axis when identifying these patterns, as done in SAS procedures that produce these plots. If one cannot identify a specific pattern clearly, then it can be concluded that no evidence is provided by the plot to suspect the plausibility of the normality of the data. The normal probability plot produced by inclusion of the `probplot` statement (see Fig. 2.19) affirms that the population distribution is normal agreeing with the result of the Shapiro–Wilk test as it does not imply a departure from a straight line.

Note that the bottom table in Fig. 2.20 shows percentiles calculated using `definition 5` by default.

SAS Example B7

The data set shown in Table B.1 of Appendix B appeared in Weisberg (1985) and was extracted from the *American Almanac* and the *World Almanac* for 1974. It lists the values of fuel consumption for each of the 48 contiguous states, in addition to several other measured variables. As a prelude to the use of several SAS procedures for analysis, a SAS data set that contains user-generated labels, formats, grouping variables (ordinal or nominal variables with values that identify groups of observations belonging to different classes or strata), etc. is created and stored in a library. This data set is then accessed repeatedly in several SAS proc steps.

The SAS data set `fueldat` is created in the SAS Example B7 program shown in Fig. 2.21. The following actions are taken in the data step of this program. The data are input from a text file using an `infile` statement. The SAS data set is saved in a folder using the two-level name `mylib.fueldat` to be accessed in other SAS programs later. Mnemonic variable names are used, but `label` statements are included to provide more descriptive labeling as necessary. In the same data step, five new variables are created as follows:

a. A numeric variable `Percent` that will contain the percent of population with driving licenses in each state
b. A numeric variable `Fuel` that measures the per capita motor fuel consumption in gallons in each state
c. An ordinal variable called `IncomGrp` assigned the value 1, 2, or 3 according to whether the per capita Income (in thousands of dollars) is less than or equal to 3.8, greater than 3.8 and less than or equal to 4.4, or over 4.4, respectively
d. A nominal variable called `TaxGrp` with a value of "`Low` " when the fuel tax is less than 8 cents and a value of "`High`" otherwise
e. A character variable named `State` containing the state name in uppercase and lowercase, for example, Kansas

A `format` statement ensures that values of the variable created in (a) are printed rounded to one decimal place and those of the variable created in (b) are printed as whole numbers (i.e., appropriate print formats are associated with `Percent` and `Fuel` variables). A `drop` statement is used to exclude variables `Fuelc` and `St` from the data set created. Printed output from the program, a partial listing of the data set, is not reproduced here.

```
libname mylib "C:\users\user_name\Documents\...\Class\stat479\";
data mylib.fueldat;
filename fueldd 'C:\Documents and Settings\...\fuel.txt';
input (St Pop Tax Numlic Income Roads Fuelc)
      ( $2.  5. 2.1     5.    4.3   5.3    5.);
label Pop='Population(in thousands)'
      Tax='Motor Fuel Tax Rate(in cents/gallon)'
      Numlic='No. of Licensed Drivers'
      Income='Per capita Income(in thsnds.)'
      Roads='Miles of Primary Highways(in thsnds.)' ;
Percent=100*Numlic/Pop;
Fuel=1000*Fuelc/Pop;

if Income=<3.8 then IncomGrp=1;
else if 3.8<Income=<4.4 then IncomGrp=2;
else IncomGrp=3;
if Tax<8.0 then TaxGrp='Low ';
else TaxGrp='High';

label Percent='% of Population with Driving Licenses'
      Fuel='Fuel Consumption (in gallons/person)'
      IncomGrp='Income Level'
      TaxGrp='Fuel Tax Level'
      State='State' ;

format Percent 4.1 Fuel 7. ;
State=stnamel(St);
drop Fuelc St;
run;

options orientation=landscape;
proc print data=mylib.fueldat(obs=20) label;
title 'Subset of the Fuel Data Set' ;
run;
```

Fig. 2.21. SAS Example B7: program

2.2.1 The UNIVARIATE Procedure

Although there are several SAS procedures that produce descriptive statistics, `proc univariate` is best suited for studying the empirical distributions of variables in a data set. It produces a variety of descriptive statistics such as moments and percentiles and optionally creates output SAS data sets containing selected sample statistics. In addition, `proc univariate` can be used to produce high-resolution graphics such as histograms with overlayed kernel density estimates, quantile-quantile plots, and probability plots supplemented with goodness-of-fit statistics for a variety of distributions. A discussion of the statements that produce high-resolution graphics is deferred until Chap. 3. In this subsection, a brief discussion of several statements available for calculating sample statistics and saving those in a SAS data set is presented. This is followed by an illustrative example. The general structure of a `proc univariate` step (that includes five of the procedure information statements to be illustrated) is

```
PROC UNIVARIATE < options > ;
  BY variables ;
  CLASS variable-1 <(v-options)> < variable-2 <(v-options)> >
              ...< / KEYLEVEL= value1 | ( value1 value2 ) >;
  VAR variables ;
  ID variables ;
  OUTPUT < OUT=SAS-data-set >
     < keyword1=names...keywordk=names > < percentile-options >;
  HISTOGRAM <variables> < / options>;
  PROBPLOT <variables> < / options>;
  QQPLOT <variables> < / options>;
  INSET keyword-list </ options>;
```

A large number of `proc` statement options are available for `proc univariate`. Although some of these are standard options such as the `data=` option for naming the data set to be analyzed or the `noprint` for suppressing printed output, others are more specialized. Some of these special `proc` statement options are summarized below:

Some PROC Statement Options

alpha= option specifies an α for calculating $(1-\alpha)100\%$ confidence intervals, the default being 0.05

cibasic < (< type= ><alpha= >) > option calculates $(1-\alpha)100\%$ confidence intervals for the mean, standard deviation, and variance assuming that the data are normally distributed. Optionally, `type` may be set equal to one of the keywords `lower, upper,` or `two sided`. The defaults are `type=twosided` and `alpha` value set in the above `alpha=` proc option or the default value of 0.05.

mu0= option is used to list value(s) (μ_0) stipulated in the hypotheses for tests concerning population means corresponding to the variables listed in the `var` statement. The tests performed are the Student's t-test, the sign test, and the Wilcoxon signed rank test.

normaltest option requests tests for normality. Computed test statistics and p-values for the Shapiro–Wilk test (for sample sizes less than or equal to 2000), the Kolmogorov–Smirnov test, the Anderson–Darling test, and the Cramér–von Mises test are output.

pctldef= gives the user the option of selecting one of five methods (labeled 1, 2, 3, 4, or 5) that `proc univariate` uses for calculating sample percentiles. These methods depend on the sample size n and the percentile p and are described in the documentation. The default method is 5.

plots option requests a panel of plots that contains a horizontal histogram, a box plot, and a normal probability plot.

trim=values < (<type= ><alpha= >) > option requests the computation of trimmed means for each variable in the `var` list, where the *values* list is the numbers k or the fractions p of observations to be *trimmed* from both ends of the n observations ordered smallest to largest. If p are specified, then the numbers trimmed equal np rounded up to the nearest integer, respectively. Confidence intervals for the population means are also calculated based on the trimmed means and estimates of their standard errors; the options `type=` and `alpha=` may be used to change their default settings, as described for the `cibasic` proc option earlier.

vardef= specifies the divisor to be used in the calculation of variance and standard deviation. The default value for the divisor is `df` when the degrees of freedom $n - 1$ will be used. Other possible values that may be specified are `n`, `wdf`, and `Weight` or `wgt`, respectively, when the sample size n, the weighted degrees of freedom $\sum w_i - 1$, or the sum of weights $\sum w_i$ (where w_i are the weights specified in `Weight` statement) will be used.

Some CLASS Statement Options

The variables list in the `class` statement specifies groups into which the observations in a data set are classified into for the purpose of calculating statistics. The values of these variables can be numeric or character and are called *levels*. For the purpose of displaying output from such an analysis (e.g., tables), procedures such as `univariate` must be provided with a way to determine in what order the statistics calculated for each level of a class variable are to be displayed. In many procedures, the `order=` option is available as a `proc` statement option to be used for this purpose. In `proc univariate`, this option is available as one of the *v-options* in the `class` statement, the levels of each class variable may be separately ordered.

The `class` statement allows the *v-options* `missing` and `order=` to be specified, enclosed in parentheses, for each of the variables in the *class variable list*. For example, using the `order=` option for each variable allows the user to specify the display order of the levels of each of the class variables separately. The default setting for the `order=` option is `internal`, which specifies that the internal unformatted (character or numeric) value of a variable be used for this purpose. The other available choices are `data`, in which case the levels will be displayed in the order they appeared in the input data, `formatted`, which requests that the levels be ordered by their formatted values, and `freq`, which requests that levels be listed in the decreasing order of frequency of observations for each level.

SAS Example B8

Several variables in the SAS data set created in SAS Example B7 on fuel consumption data are analyzed using `proc univariate` in the SAS Example B8 program (see Fig. 2.22) to illustrate the use of the procedure options and

```
libname mylib "C:\users\user_name\Documents\...\stat479\";
proc univariate data=mylib.fueldat plots normaltest cibasic mu0=4 500 trim=2;
  var  Income Fuel;
  id State;
  title 'Use of Proc Univariate to Compute Statistics:2';
run;

proc univariate data=mylib.fueldat noprint;
  var Fuel Percent;
  output out=stats pctlpts=33.3 66.7 pctlpre=Fuel Lic;
  title 'Calculation of User Specified Percentile Points';
run;

proc print data=stats;
run;
```

Fig. 2.22. SAS Example B8: program

statements discussed in this section. The previously saved SAS data set named `fueldat` is accessed using the two-level name `mylib.fueldat`.

Use of Proc Univariate to Compute Statistics:2

The UNIVARIATE Procedure
Variable: Income (Per capita Income(in thsnds.))

Basic Confidence Limits Assuming Normality			
Parameter	**Estimate**	**95% Confidence Limits**	
Mean	4.24183	4.07527	4.40840
Std Deviation	0.57362	0.47752	0.71851
Variance	0.32904	0.22803	0.51626

Tests for Location: Mu0=4				
Test	**Statistic**		**p Value**	
Student's t	t	2.920853	Pr > |t|	0.0053
Sign	M	7	Pr >= |M|	0.0595
Signed Rank	S	248.5	Pr >= |S|	0.0093

Tests for Normality				
Test	**Statistic**		**p Value**	
Shapiro-Wilk	W	0.975229	Pr < W	0.3988
Kolmogorov-Smirnov	D	0.080296	Pr > D	>0.1500
Cramer-von Mises	W-Sq	0.058391	Pr > W-Sq	>0.2500
Anderson-Darling	A-Sq	0.375093	Pr > A-Sq	>0.2500

Trimmed Means								
Percent Trimmed in Tail	**Number Trimmed in Tail**	**Trimmed Mean**	**Std Error Trimmed Mean**	**95% Confidence Limits**		**DF**	**t for H0: Mu0=4.00**	**Pr > |t|**
4.17	2	4.239795	0.087055	4.064233	4.415358	43	2.754533	0.0086

Fig. 2.23. SAS Example B8: confidence intervals and tests for the `Income` variable

The first proc step analyzes the distribution of the variables `Income` and `Fuel` and includes the procedure options `cibasic`, `mu0=4 500`, and `trim=2`. By default, the `cibasic` option produces 95% confidence intervals for the population mean μ, the population standard deviation σ, and the population variance σ^2 for each of the variables, calculated under the normality assumption for the data. These are shown in Fig. 2.23 for the `Income` variable and in Fig. 2.24 for the `Fuel` variable. Recall that the data values for `Income` are per capita income figures in thousands of dollars and those for `Fuel` are per capita fuel consumption in gallons.

Note that the percentiles and extreme values are also part of the output but shown here except the extreme values for the Fuel variable (see Fig. 2.27). The `id State;` statement results in these values being identified by the corresponding state name.

Use of Proc Univariate to Compute Statistics:2

The UNIVARIATE Procedure
Variable: Fuel (Fuel Consumption (in gallons/person))

Basic Confidence Limits Assuming Normality			
Parameter	Estimate	95% Confidence Limits	
Mean	576.74280	544.25372	609.23188
Std Deviation	111.88866	93.14382	140.14953
Variance	12519	8676	19642

Tests for Location: Mu0=500				
Test	Statistic		p Value	
Student's t	t	4.751954	Pr > |t|	<.0001
Sign	M	13	Pr >= |M|	0.0002
Signed Rank	S	415	Pr >= |S|	<.0001

Tests for Normality				
Test	Statistic		p Value	
Shapiro-Wilk	W	0.933466	Pr < W	0.0092
Kolmogorov-Smirnov	D	0.133491	Pr > D	0.0313
Cramer-von Mises	W-Sq	0.116579	Pr > W-Sq	0.0687
Anderson-Darling	A-Sq	0.802331	Pr > A-Sq	0.0368

Trimmed Means								
Percent Trimmed in Tail	Number Trimmed in Tail	Trimmed Mean	Std Error Trimmed Mean	95% Confidence Limits		DF	t for H0: Mu0=500.00	Pr > |t|
4.17	2	570.3739	14.65103	540.8273	599.9205	43	4.803342	<.0001

Fig. 2.24. SAS Example B8: confidence intervals and tests for the `Fuel` variable

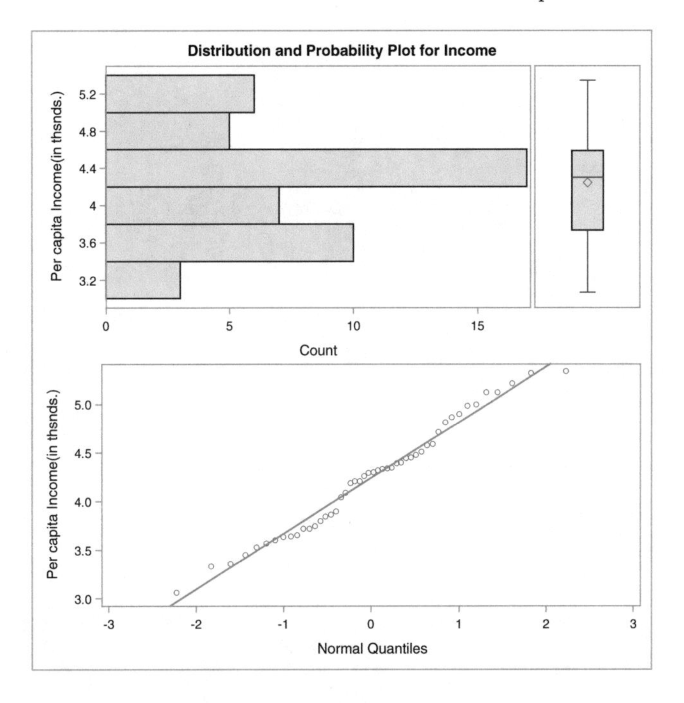

Fig. 2.25. SAS Example B8: panel of basic summary plots for `Income`

From Fig. 2.23, for the `Income` variable, the confidence interval reported for μ is (4.07527, 4.40840) and the p-value for the two-sided t-test of $H_0 : \mu = 4$ versus $H_a : \mu \neq 4$ is reported as 0.0053. The p-value for the corresponding one-sided test is, of course, $0.0053/2 = 0.0026$. The tests and confidence intervals are not reproduced here for the `Fuel` variable, but they are found in Fig. 2.24 and can be similarly interpreted. The estimate, confidence interval, and associated t-test for the mean μ under trimming for the `Income` variable appear in the bottom table titled Trimmed Means in Fig. 2.23. The option `trim=2` requested that the `trimmed mean` be computed after the *two smallest* and the *two largest* observations are deleted from the sample, which is equivalent to approximately 4% trimming from the tails of the distribution of the `Income` variable. Associated confidence intervals and a t-test for the population mean μ are computed based on the standard error of the trimmed mean. For a symmetric distribution, the trimmed mean is an unbiased estimate of the population mean. The results under trimming here indicate that the estimates and test statistics are not significantly different from those statistics calculated from the complete sample (see Figs. 2.23 and 2.24). This is also an indicator of the symmetry of the population distribution of the `Income` variable. Similar interpretations of Trimmed Means follow for the `Fuel` variable as well.

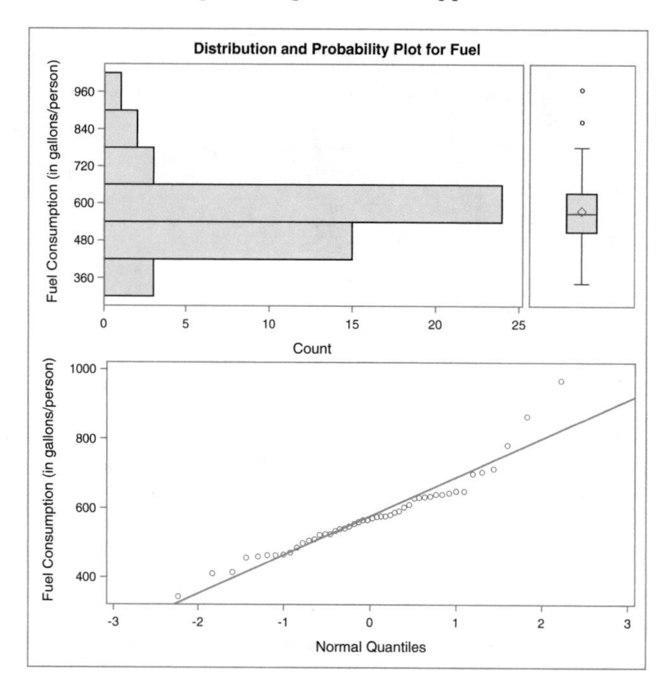

Fig. 2.26. SAS Example B8: panel of basic summary plots for `Fuel`

The normal probability plots produced for the two variables, `Income` and `Fuel`, respectively, are shown in Figs. 2.25 and 2.26, and they support the findings of the respective tests of normality of the two variables. For example, the histogram and the box plot (see Fig. 2.26) show a highly positively skewed population distribution with two extreme values for the `Fuel` variable, which were identified as those that correspond to the states of South Dakota and Wyoming in Fig. 2.27.

The final proc step requests the calculation of the 33.3 and 66.7 percentiles of the `Fuel` (Fuel Consumption (in gallons/person)) and `Percent` (% of Pop-

Extreme Observations					
Lowest			Highest		
Value	State	Obs	Value	State	Obs
343.679	New York	7	703.755	Montana	38
410.124	Rhode Island	5	713.608	North Dakota	18
414.031	Massachusetts	4	781.784	Nevada	45
456.846	Connecticut	6	865.285	South Dakota	19
459.854	West Virginia	25	968.116	Wyoming	40

Fig. 2.27. SAS Example B8: extreme values for `Fuel`

ulation with Driving Licenses) variables. In this example, the printed output is suppressed (as a result of the `noprint` proc option); however, the output is written to a new SAS data set named `stats`. The percentiles that cannot be directly requested via the usage of a keyword such as `p1`, `p10`, or `p90` are calculated by the use of the pair of keywords `pctlpts=` and `pctlpre=` used concurrently. For example, the use of `pctlpts=33.3 66.7 pctlpre=fuel lic` generates the 33.3 and 66.7 percentiles for the two analysis variables (i.e., variables in the `var` statement), respectively, and adds them to the new data set as values of new variables named `fuel33_3`, `fuel66_7`, `lic33_3`, and `lic66_7`. The `proc print data=stats;` statement produces the output of these values shown in Fig. 2.28.

Calculation of User Specified Percentile Points

Obs	Fuel33_3	Lic33_3	Fuel66_7	Lic66_7
1	524.994	54.4421	609.991	58.0087

Fig. 2.28. SAS Example B8: calculating user-specified percentiles

2.2.2 The FREQ Procedure

The FREQ procedure in SAS computes many statistics and measures related to the analysis of categorical data. The discussion in this subsection is primarily intended to illustrate the use of statements and options to generate these statistics rather than a presentation of statistical methodology involved in the analysis of categorical data. It is recommended that the prospective user of `proc freq` consult references cited to learn more about techniques available for hypothesis testing and measuring association among categorical variables. Moreover, the type of inference depends on many factors such as sampling strategy; thus, a knowledge of how the data are collected is also necessary for making relevant conclusions.

A chi-square goodness-of-fit test can be used to test several types of hypothesis using frequency counts. For example, using a one-way frequency table with k classes, one could compute a chi-square statistic to test whether the counts conform to sampling from a multinomial population with specified probabilities. In this case, the null hypothesis of interest is

$$H_0 : p_i = p_{i0}, \quad i = 1, 2, \ldots, k$$

where the p_{i0}'s are postulated values of the multinomial probabilities. The Pearson chi-square statistic is given by

$$\chi^2 = \sum_{i=1}^{k} \frac{(f_i - e_i)^2}{e_i}$$

where f_i is the observed frequency count in class i and e_i is the expected frequency calculated under the null hypothesis (i.e., $e_i = p_{i0}N$ where N is the total number of responses). Another application of a chi-square test is for testing *homogeneity* of several multinomial populations. In this case, random samples are taken from each population and then classified by a categorical variable. The populations are usually defined by levels of variables such as gender, age group, state, etc., and the levels of the categorical variable form the k categories of the multinomial populations.

For example, suppose samples are drawn from two populations (say, males and females or persons below and above the age of 40) and they are grouped into three categories (say, according to three levels of support for a certain local bond issue). Suppose that the multinomial probabilities for each population are as given in the following table:

	Groups		
Populations	p_{11}	p_{12}	p_{13}
	p_{21}	p_{22}	p_{23}

Then the null hypothesis of homogeneity of populations (i.e., whether random samples were drawn from the same multinomial population) is given by

$$H_0 : p_{11} = p_{21},\ p_{12} = p_{22},\ p_{13} = p_{23}$$

Note carefully that the sampling procedure here is different from the process used in the construction of a *contingency table*. In the above situation, random samples are drawn from two different populations, and then each sample is classified into three different groups. Contingency tables are constructed by multiple classification of a single random sample. Observations in a sample may be cross-classified by variables with ordinal or nominal data values defining *categorical variables*. These variables may be covariates already present in the data set (e.g., gender, marital status, or region), thus forming natural subsets or strata of the data, or may be generated from other quantitative variables such as `Population` or `Income`. For example, observations in a sample may be categorized into three income groups (say "low," "middle," or "high") by creating a new variable, say `IncomGrp`, and assigning the above character strings as its values according to whether the value of the income variable is below $30,000, between $30,000 and $70,000, or above $70,000, respectively.

A chi-square statistic can be computed for a two-way $r \times c$ contingency table to test whether the two categorical variables are independent; that is, the null hypothesis tested is of the form

$$H_0 : p_{ij} = p_{i.}p_{.j},\quad i = 1, 2, \ldots, r,\ j = 1, 2, \ldots, c$$

where the p_{ij} are the probabilities considering that the entire sample is from a multinomial population and $p_{i.}$ and $p_{.j}$, called the *marginal* probabilities, are probabilities for multinomial populations defined by each categorical variable. The chi-square statistic for the test of this hypothesis is given by

$$\chi^2 = \sum_i \sum_j \frac{(f_{ij} - e_{ij})^2}{e_{ij}}$$

where f_{ij} is the observed frequency in the ijth cell and $e_{ij} = f_{i.}f_{.j}/N$, where $f_{i.}$ and $f_{.j}$ are the observed row and column marginal frequencies, respectively.

When the row and column variables are independent, the above statistic has an asymptotic chi-square distribution with $(r-1)(c-1)$ degrees of freedom. Instead of χ^2, the likelihood ratio chi-square statistic, usually denoted by G^2, that has the same asymptotic null distribution may be computed. If the row and columns are ordinal variables, the Mantel-Haenszel chi-square statistic tests the alternative hypothesis that there is a linear association between them. Fisher's exact test is another test of association between the row and column variables that does not depend on asymptotic theory. It is thus suitable for small sample sizes and for sparse tables. One may compute measures of association between variables that may or may not depend on the chi-square test of independence. Some of these are illustrated in SAS Example B8.

One would use `proc freq` to analyze count data using one-way frequency tables or two-way or higher-order contingency tables. In addition to chi-square statistics for testing whether two categorical variables are independent, for a two-way contingency table, `proc freq` also computes measures of association that estimate the strength of association between the pair of variables. In this subsection, a brief discussion of several statements available in `proc freq` followed by an illustrative example is presented. The general structure of a `proc freq` step is

```
PROC FREQ < options > ;
  BY variables ;
  TABLES requests < / options > ;
  EXACT statistic-options < / computation-options > ;
  TEST options ;
  OUTPUT < OUT=SAS-data-set > options ;
```

The primary statement in `proc freq` is the `tables` statement for requesting tables having different structures with options for selecting statistics to be included in those tables. Since the computation of frequencies requires that the variables used in the `tables` statement are necessarily discrete valued (containing either nominal or ordinal data) such as category-, classification-, or grouping-type variables, statements such as `var` or `class` are not available in `proc freq`.

The syntax of the tables statement allows the user to request one-way tables just by listing the variables in the tables statement and two-way tables by two variable names combined with an asterisk between them. Thus, the statement `tables Region;` produces one-way tables with frequency counts of observations for each *level* of `Region`, and the statement `tables TaxGrp*Region;` produces a two-way cross-tabulation with the *levels* of the variable `TaxGrp` as the rows of the table and the *levels* of `Region` as the columns. A combination of a level of `TaxGrp` and a level of `Region` forms a *cell* in the table. In this case, frequencies of observations are tallied for every possible combination of the two variables and entered in the respective cells; they are called *cell frequencies*. Multiway combinations of variables such as $P*Q*R*S$ are possible in which case two-way tables are produced for every combination of levels of each of the variables P and Q. The cells in each two-way table are formed by combinations of a level of variables R and S. The proc statement option `page` may be used to force these tables to be output on different pages. The `tables` statement syntax also allows variations such as `tables` $Q*(R\ S)$, which is equivalent to the specification `tables` $Q*R\ Q*S$, or `tables` $(P\ Q)*(R\ S)$, which is equivalent to `tables` $P*R\ Q*R\ P*S\ Q*S$.

By default, one-way frequency tables contain the statistics frequency, cumulative frequency, percentage frequency, and cumulative percentage computed for each level of the variable and two-way or multi-way tables may include cell frequency, cell percentage of the total frequency, cell percentage of row frequency, and cell percentage of column frequency computed for each cell. Many options are available with the `tables` statement to control the statistics that are calculated and output by `proc freq`. While some of these are simple options for suppressing the statistics computed by default, others request additional statistics such as goodness-of-fit statistics and measures of association to be computed. An abbreviated description of these options is provided below:

Some TABLES Statement Options

nocol suppresses printing the column percentage for each cell.

nocum suppresses printing the cumulative frequencies and cumulative percentages in one-way frequency tables and in list format.

norow suppresses printing the row percentage for each cell.

nopercent suppresses printing the percentage, row percentage, and column percentage in two-way tables or percentages and cumulative percentages in one-way tables and in list format.

noprint suppresses printing the frequency table but displays other statistics.

list prints multiway tables in list format.

binomial requests binomial proportion, confidence limits, and test for one-way tables.

testf= specifies expected frequencies for a one-way table chi-square test.

testp= specifies expected proportions for a one-way table chi-square test.

chisq requests chi-square tests and measures of association based on chi-square.

cellchi2 prints each cell's contribution to the total Pearson chi-square statistic.

deviation prints the deviation of the cell frequency from the expected value for each cell.

expected prints the expected cell frequency for each cell under the null.

fisher requests Fisher's exact test for tables larger than 2×2.

cmh requests all Cochran-Mantel-Haenszel statistics.

measures requests measures of association and their asymptotic standard errors.

cl requests confidence limits for the measures statistics.

alpha= sets the confidence level for confidence limits.

agree requests tests and measures of classification agreement.

all requests tests and measures of association produced by the chisq, measures, and cmh options.

Options such as `binomial`, `testf=`, and `testp=` are used for specifying either the postulated probabilities (p_{i0}'s) where $H_0 : p_i = p_{i0},\ i = 1, 2, \ldots, k$ or the expected frequencies in a sample of size n classified in a one-way frequency table for performing a chi-square goodness-of-fit test. An application is provided as an exercise at the end of this chapter.

SAS Example B9

In the following contrived example of a two-way table, suppose that subjects are classified according to levels of two variables `A` and `B`. The column variable `A` has three categories, say a_1, a_2, and a_3, and the row variable `B` has three categories, say b_1, b_2, and b_3. Most often, the row variable is called the *dependent variable* if the categories of the variable are recognized as possible *outcomes* or *responses*. An example would be where factor `B` is `Marital Status` and factor A is a response to a question with three possibilities. Consider the table of frequencies:

	b_1	b_2	b_3	Total
a_1	8	16	31	55
a_2	9	18	74	101
a_3	34	23	17	74
Total	51	57	122	230

In this case, the column variable is called the independent variable with categories being *classes*, *groups*, or *strata*. The designation of whether the two types of variable are assigned to rows or columns is usually a matter of choice.

In the above setup, subjects from each of the column categories (independent variable) can be viewed as being classified into one of the row categories (dependent variable). The choice of the independent and dependent variables does not affect the statistical analysis of the data except when part of the inference is measuring the predictability of a response category given that an object belongs to a certain group or class. Otherwise, when variables cannot be clearly identified as independent and dependent variables, statistics and measures unaffected by an arbitrary designation are preferred.

In the SAS Example B9 program (see Fig. 2.29), the cell frequencies are directly input to `proc freq` instead of raw data (which are not available in this example). The use of the `weight` statement allows SAS to construct the two-way cross-tabulation using the cell counts. In the first part of the output (see Fig. 2.30), only the statistics observed count f_i, the expected frequency e_i, and its contribution to the total chi-square statistic are displayed in each cell (i.e., percentage of the total frequency, percentage of row frequency, and percentage of column frequency are suppressed using `tables` statement options given earlier). It is clear that the cells (2,1), (2,3), (3,1), and (3,3) provide the largest contributions to the total chi-square statistic of 52.4. It is observed that at the lowest level of B, the response is smaller than expected for group 2 of A and larger than expected for group 3 of A. This pattern is reversed at the highest level of B.

```
data exb9;
input A $ B $  Count @@;
datalines;
a1 b1  8 a1 b2 16 a1 b3 31
a2 b1  9 a2 b2 18 a2 b3 74
a3 b1 34 a3 b2 23 a3 b3 17
;
proc freq data=exb9;
weight Count;
tables A*B/chisq expected cellchi2 nocol nopercent norow measures;
title "Example B9: Illustration of Tables Options";
run;
```

Fig. 2.29. SAS Example B9: program

For two-way frequency tables, the chi-square test of independence is a test of *general association*, where the null hypothesis is that the row and column variables are independent (no association) and the alternative hypothesis is that an association exists between the two variables with the type of association unspecified. The chi-square statistic and the likelihood ratio statistic are both suitable for testing this hypothesis. `proc freq` computes these statistics in response to `chisq` option in the `tables` statement. For large sample sizes and if the null hypothesis is true, these test statistics have approximately a chi-square distribution. (For small samples, the user may request that Fisher's exact test be computed by specifying the `exact` option in the `tables` statement.) In Fig. 2.30, the p-values of both the chi-square statistic and the likelihood ratio statistic are smaller than, say, 0.01. Thus, the null hypothesis that the two variables are independent is rejected, leading to the conclusion that there is some type of association between these two variables.

Example B9: Illustration of Tables Options

The FREQ Procedure

Frequency Expected Cell Chi-Square

Table of A by B				
	B			
A	b1	b2	b3	Total
a1	8 12.196 1.4434	16 13.63 0.4119	31 29.174 0.1143	55
a2	9 22.396 8.0124	18 25.03 1.9747	74 53.574 7.7878	101
a3	34 16.409 18.859	23 18.339 1.1846	17 39.252 12.615	74
Total	51	57	122	230

Statistics for Table of A by B

Statistic	DF	Value	Prob
Chi-Square	4	52.4031	<.0001
Likelihood Ratio Chi-Square	4	53.1793	<.0001
Mantel-Haenszel Chi-Square	1	25.0216	<.0001
Phi Coefficient		0.4773	
Contingency Coefficient		0.4308	
Cramer's V		0.3375	

Fig. 2.30. SAS Example B9: `A` × `B` chi-square test

Beyond the use of the chi-square statistic for testing the hypothesis of no association between two variables, researchers also use the magnitude of the statistic (and thus the p-value itself) as a measure of the strength of any association that may exist especially when the null hypothesis is rejected. This is because the tables with the larger chi-square value generally provide stronger evidence for association between the two variables. However, the problem with this interpretation is that the value of the chi-square statistic is sensitive to the sample size. That is, tables with larger sample sizes but similar frequency distribution may actually result in significantly larger chi-square values. One approach to solve this problem is to adjust the chi-square statistic for either the sample size or the dimension of the table or for both. `Proc freq` provides three measures `phi coefficient` ϕ, `contingency coefficient` C, and `Cramer's` V that are suitable for measuring the strength of the dependency between nominal variables but are also applicable for ordinal variables.. For this data, these are displayed in the second table of Fig. 2.30.

Phi Coefficient, ϕ The ϕ statistic is a measure that adjusts only for the sample size n and is defined as

$$\phi = \sqrt{\frac{\chi^2}{n}}$$

This adjustment makes sure that tables that have widely differing values of the chi-square statistic because they are based on different sample sizes result in similar values for the phi coefficient, which reflect the true association between the variables. The range for ϕ is $0 < \phi < \min\{\sqrt{r-1}, \sqrt{c-1}\}$. For the above data set, Fig. 2.30 displays a value of 0.48 for ϕ. Since, for a 3×3 table, the upper limit of the statistic is 1.4, the strength of association between the variables A and B is shown to be rather moderate to weak.

Contingency Coefficient, C The C coefficient is a direct transformation of the ϕ statistic and is defined as

$$C = \sqrt{\frac{\chi^2}{n + \chi^2}}$$

The advantage of C is that it is constrained to be in the range of 0–1. Thus the value of C is zero if there is no association between the two variables but has a value that is less than 1 even with perfect dependence. Its value is dependent on the size of the table with a maximum value of $\sqrt{(r-1)/r}$ for an $r \times r$ table. For a 3×3 table, this value is 0.816. Thus, the value of 0.43 of C appears to indicate a strength midway between no association and a perfect association.

Cramer's V This measure adjusts for both the sample size and the dimension of the table and is defined as

$$V = \sqrt{\frac{\chi^2}{nt}}$$

where $t = \min\{\sqrt{r-1}, \sqrt{c-1}\}$. Cramer's V is a normed measure, so its value is between 0 and 1; thus, a value of 0.34 is approximately in the bottom third of the scale. This indicates a slightly weaker association than that the contingency coefficient shows but similar to that indicated by ϕ.

The above three measures of association are all derived from the Pearson chi-square statistic. There are many other measures of association between two categorical variables that `proc freq` calculates. Some of these statistics are briefly discussed here. Many of these statistical measures also require the assignment of a dependent variable and an independent variable, as the goal is to predict a rank (category) of an individual on the dependent variable given that the individual belongs to a certain category in the independent variable.

Statistics for Table of A by B

Statistic	Value	ASE
Gamma	-0.4375	0.0828
Kendall's Tau-b	-0.2981	0.0586
Stuart's Tau-c	-0.2804	0.0555
Somers' D C|R	-0.2891	0.0575
Somers' D R|C	-0.3074	0.0601
Pearson Correlation	-0.3306	0.0626
Spearman Correlation	-0.3354	0.0650
Lambda Asymmetric C|R	0.1574	0.0607
Lambda Asymmetric R|C	0.2326	0.0622
Lambda Symmetric	0.1983	0.0540
Uncertainty Coefficient C|R	0.1138	0.0293
Uncertainty Coefficient R|C	0.1082	0.0281
Uncertainty Coefficient Symmetric	0.1109	0.0286

Sample Size = 230

Fig. 2.31. SAS Example B9: `A` $\times$ `B` measures of association

For calculating the following measures for the two variables under consideration, pairs of observations are first classified as concordant or discordant. A pair is concordant if the observation with the larger value for variable one also has the larger value for variable two, and it is discordant if the observation with the larger value for variable one has the smaller value for variable two. Thus, the pair of observations (12, 2.7) and (15, 3.1) are concordant and the pair (12, 2.7) and (10, 3.1) are discordant.

Gamma γ, Kendall's tau-b, Somers' D

These measures are considered when both variables are *ordinal* and the first two are called *symmetrical* because their values are not affected by the selection of the variable for predicting the other variable. `Gamma` is a normed measure of association based on the numbers of concordant and discordant pairs. If there are no discordant pairs, `gamma` is $+1$ and perfect positive association exists between the two variables, and if there are no concordant pairs, `gamma` is -1 and perfect negative association exists between the two variables. Values in between -1 and $+1$ measure the strength of negative or positive association. If the numbers of discordant and concordant pairs are equal, `gamma` is zero, and the rank of the independent variable cannot be used to predict the rank of the dependent variable. In the SAS Example B9 output (see Fig. 2.31), `gamma`$= -0.4375$ with an estimated asymptotic standard error (ASE) of 0.0828, indicating a moderate to weak negative association between the two variables.

`Kendall's tau-b` is the ratio of the difference between the number of concordant and discordant pairs to the total number of pairs. It is scaled to be between -1 and $+1$ when there are no ties, but not otherwise. An asymmetric version of tau-b, the ordinal measure `Somers' D`, on the other hand adjusts for ties by counting pairs where ties occur only on the independent variable so that the value of the statistic lies between -1 and $+1$ when such ties occur. Usually, two values of this statistic are computed: one when the row variable is considered the independent variable (Somers' D $C|R$) and one when the column is considered the independent variable (Somers' D $R|C$). The values differ because of the way ties are counted. In SAS Example B9, Somers' D $R|C = -0.3074$, showing a moderate negative association.

Proportional Reduction in Error (PRE) Measures

When one of the variables is to be considered an independent variable for predicting the other (dependent) variable and if the measure of association depends on this choice, it is called an *asymmetrical measure.* The extent to which the error in prediction is reduced by using the value of the independent variable for prediction compared to ignoring this knowledge underlies the definition and interpretation of several measures of association, called *pre* measures. If the independent variable is ignored, an individual can be allocated into a category or class according to the observed proportions of the dependent variable. If the independent variable is used, the allocation of the individual is done on the basis of the observed proportions of the dependent variable *for each category of the independent variable.* Then the proportional reduction in error *pre* is defined as

$$\text{PRE} = \frac{e1 - e2}{e1}$$

where $e1$ = errors of prediction made when the independent variable is ignored and $e2$ = errors of prediction made when the prediction is based using the independent variable.

For example, in the cross-tabulation used in SAS Example B9, considering the row variable A as the dependent variable, an individual may be assigned to category $a2$ based just on the row frequencies (which are 55, 101, and 74, respectively). The number of errors in assigning individuals to category $a2$ ignoring the column variable B is then $230 - 101 = 129$. However, if the assigning of an individual to a category of variable A is made based on the cell frequencies (i.e., using the classification based on levels of variable B), the number of errors made will be $(51-34)+(57-23)+122-74) = 17+34+48 = 99$. Thus PRE is $(129 - 99)/129 = 0.2326$ or 23.26%.

The nominal asymmetric measure `lambda`, $\lambda(R|C)$, is interpreted as the proportional improvement in predicting the dependent (row) variable given the independent (column) variable. Asymmetric lambda has the range $0 \leq \lambda(R|C) \leq 1$, although values around 0.3 are considered high. The measure $\lambda(C|R)$ may be interpreted similarly. In SAS Example B9, if information about variable B is used to predict A, the *proportional reduction in error* in the prediction according to $\lambda(R|C)$ is 23.26% compared to not using that information, exactly the value calculated above. `Stuart's tau c` makes an adjustment for table size in addition to a correction for ties. Tau-c is appropriate only when both variables lie on an ordinal scale. Tau-c also is in the range $-1 \leq \tau_c \leq 1$.

Pearson's Correlation Coefficient, r^2, and Spearman's Rank-Order Correlation Coefficient, ρ

`Pearson's correlation coefficient` and `Spearman's rank-order correlation coefficient` are also appropriate for ordinal variables. Pearson's correlation coefficient describes the strength of the linear association between the row and column variables. This statistic may also be interpreted as a proportional reduction in error *pre*. It is computed using the row and column scores specified by the `scores=` option in the `tables` statement. By default, the row or column scores are the integers 1, 2,... for character variables and the actual variable values for numeric variables. Consult SAS documentation for other options. Spearman's correlation coefficient is computed with rank scores.

Similar to Pearson's correlation coefficient, the value of Spearman's ρ lies in the range -1 and $+1$, with these values indicating perfect negative or positive association, respectively. For example, if the ranks of one variable agrees perfectly with the ranks of the other variable, $\rho = +1$. Just as with Pearson's correlation coefficient, it is possible to conduct tests based on the t-statistic:

$$t = \hat{\rho}\sqrt{\frac{n-2}{1-\hat{\rho}^2}}$$

where $\hat{\rho}$ is the sample rank-correlation coefficient, for testing hypotheses about the population rank-correlation coefficient ρ for sample sizes larger than 10.

SAS Example B10

```
libname mylib "C:\users\user_name\Documents\...\stat479\"

data fueldat2;
set mylib.fueldat;

if Fuel=<525 then FuelGrp=1;
else if 525<Fuel=<610 then FuelGrp=2;
else FuelGrp=3;

if Percent=<54 then LicGrp=1;
else if 54<Percent=<58 then LicGrp=2;
else LicGrp=3;

label LicGrp='% Driving Licenses'
      FuelGrp='Fuel Consumption';
run;

proc format;
   value lg 1='below 54%'
            2='54 to 58%'
            3='above 58%' ;
   value fg  1 = 'Low Fuel Use'
             2 = 'Medium Fuel Use'
             3 = 'High Fuel Use';
   value ing 1 = 'Low Income'
             2 = 'Middle Income'
             3 = 'High Income';
run;

proc freq data=fueldat2;
  tables FuelGrp*(TaxGrp IncomGrp)/chisq expected
              cellchi2 nocol nopercent norow;
  tables FuelGrp*LicGrp/chisq expected nocol nopercent norow measures;
  tables TaxGrp*FuelGrp/list ;
  format FuelGrp fg. LicGrp lg. IncomGrp ing. ;
  title 'Output from Proc Freq';
run;
```

Fig. 2.32. SAS Example B10: program

The analysis of the SAS data set on fuel consumption created in SAS Example B7 is continued in SAS Example B10 (see Fig. 2.32 for the program) using `proc freq` to illustrate the statistics resulting from some of the `tables` statement options discussed in this section. A new SAS data set is created by supplementing the original data set with two category variables, `FuelGrp` and `LicGrp`, each with three levels, in the data step. The 33.3 and 66.7 percentiles of the `Fuel` and `Percent` variables calculated in SAS Example B8 (see Fig. 2.28) aid in the determination of cutoff values for creating the corresponding category variables. Thus, the category variables `FuelGrp` and `LicGrp` will have three levels each. In addition, `proc format` described in Sect. 2.1.4 facilitates the creation of output formats to convert the ordinal levels of the three categorical variables FuelGrp, IncomGrp, and LicGrp to more descriptive strings when they are printed.

Output from Proc Freq

The FREQ Procedure

Frequency Expected Cell Chi-Square

Table of FuelGrp by TaxGrp			
FuelGrp(Fuel Consumption)	TaxGrp(Fuel Tax Level)		
	High	Low	Total
Low Fuel Use	10 7.3333 0.9697	6 8.6667 0.8205	16
Medium Fuel Use	10 7.7917 0.6259	7 9.2083 0.5296	17
High Fuel Use	2 6.875 3.4568	13 8.125 2.925	15
Total	22	26	48

Statistics for Table of FuelGrp by TaxGrp

Statistic	DF	Value	Prob
Chi-Square	2	9.3275	0.0094
Likelihood Ratio Chi-Square	2	10.2233	0.0060
Mantel-Haenszel Chi-Square	1	7.2412	0.0071
Phi Coefficient		0.4408	
Contingency Coefficient		0.4034	
Cramer's V		0.4408	

Sample Size = 48

Fig. 2.33. SAS Example B10: fuel group $\times$ tax group cross-tabulation

The proc step generates three contingency tables for combinations of the variable FuelGrp with each of TaxGrp,IncomGrp, and LicGrp. The output is shown in Figs. 2.33, 2.34, and 2.35. The FuelGrp by TaxGrp table is a 3×2 cross-tabulation where the levels of FuelGrp are ordered by their internal (unformatted) values of 1, 2, and 3. However, internal values of the two levels of TaxGrp are the strings "Low " and "High"; thus, they are ordered by their alphanumeric values. The p-values of both the chi-square statistic and the likelihood ratio statistic are smaller than, say, 0.01. Thus, the null hypothesis that the two variables are independent is rejected.

To study the strength of association between the two variables FuelGrp and TaxGrp, the statistics in Fig. 2.33 are used here. The three measures phi coefficient ϕ, contingency coefficient C, and Cramer's V displayed next in the output are suitable statistics for measuring the strength of the dependency between nominal variables and are also applicable for ordinal variables, as in this example. The value of C is zero if there is no association between the two variables but has a value that is less than 1 even with perfect dependence. Its value is dependent on the size of the table with a maximum value of $\sqrt{(r-1)/r}$ for an $r \times r$ table. For a 3×3 table, this value is 0.816. Thus, the value of 0.40 for C appears to indicate a strength of about 50% of a perfect association.

Output from Proc Freq

The FREQ Procedure

Frequency Expected Cell Chi-Square

Table of FuelGrp by IncomGrp				
	IncomGrp(Income Level)			
FuelGrp(Fuel Consumption)	Low Income	Middle Income	High Income	Total
Low Fuel Use	1 4.3333 2.5641	3 6 1.5	12 5.6667 7.0784	16
Medium Fuel Use	8 4.6042 2.5046	7 6.375 0.0613	2 6.0208 2.6852	17
High Fuel Use	4 4.0625 0.001	8 5.625 1.0028	3 5.3125 1.0066	15
Total	13	18	17	48

Statistics for Table of FuelGrp by IncomGrp

Statistic	DF	Value	Prob
Chi-Square	4	18.4040	0.0010
Likelihood Ratio Chi-Square	4	18.7393	0.0009
Mantel-Haenszel Chi-Square	1	7.2622	0.0070
Phi Coefficient		0.6192	
Contingency Coefficient		0.5265	
Cramer's V		0.4378	
WARNING: 33% of the cells have expected counts less than 5. Chi-Square may not be a valid test.			

Sample Size = 48

Fig. 2.34. SAS Example B10: fuel group $\times$ income group cross-tabulation

The other statistics in Fig. 2.33 also lead to similar conclusions. Cramer's V is a normed measure, so its value is between 0 and 1; thus, a value of 0.44 is about in the middle of the scale. The range for ϕ is $0 < \phi < \min\{\sqrt{r-1}, \sqrt{c-1}\}$. Thus, for this table, the maximum is 1, so again a strength of association similar to the above measures is indicated. The cross-tabulation `FuelGrp` by `IncomGrp` shown in Fig. 2.34 is a 3×3 table. Again, the chi-square and the likelihood ratio statistic are both significant at the 0.01 level, indicating dependency. A study of the values for the three measures discussed above, shown in Fig. 2.34, indicates a slightly stronger association between the variables `FuelGrp` and `IncomGrp`.

Output from Proc Freq

The FREQ Procedure

Frequency Expected				
Table of FuelGrp by LicGrp				
	LicGrp(%% Driving Licenses)			
FuelGrp(Fuel Consumption)	**below 54%**	**54 to 58%**	**above 58%**	**Total**
Low Fuel Use	6 4.6667	9 6	1 5.3333	16
Medium Fuel Use	7 4.9583	5 6.375	5 5.6667	17
High Fuel Use	1 4.375	4 5.625	10 5	15
Total	14	18	16	48

Statistics for Table of FuelGrp by LicGrp

Statistic	DF	Value	Prob
Chi-Square	4	14.6905	0.0054
Likelihood Ratio Chi-Square	4	16.2966	0.0026
Mantel-Haenszel Chi-Square	1	9.9989	0.0016
Phi Coefficient		0.5532	
Contingency Coefficient		0.4841	
Cramer's V		0.3912	
WARNING: 33% of the cells have expected counts less than 5. Chi-Square may not be a valid test.			

Fig. 2.35. SAS Example B10: fuel group × licenses group cross-tabulation

The hypothesis of independence between the two variables `FuelGrp` and `LicGrp` is rejected at 0.05 (the likelihood ratio statistic has a p-value of 0.0026; see Fig. 2.35). This is one situation where Fisher's exact test may be performed instead of the chi-square test because conditions for that test are clearly not met due to several small cell frequencies. In this example, the inclusion of the `exact` option produces the output in Fig. 2.36. The p-value is smaller than 0.05 so the conclusion is that the independence hypothesis is rejected at $\alpha = 0.05$.

Fisher's Exact Test	
Table Probability (P)	<.0001
Pr <= P	0.0046

Fig. 2.36. SAS Example B10: Fisher's exact test for fuel × licenses

If it is concluded that there is association between the two variables, several statistics are available for evaluating the strength of such association. The output resulting from including the `measures` option in a `tables` statement is shown in Fig. 2.37. For interpretating `gamma` and `tau-b`, consider `LicGrp` as the independent variable that is used to predict the dependent variable `FuelGrp` and that both variables are ordinal. The value of 0.5641 for `gamma` indicates a positive association between the two variables. This implies that the ordering of the ranks of states for these two variables is positively correlated. Further, if the category of percentage of licenses is used to predict the category of fuel use, the proportional reduction in error PRE compared to randomly assigning a state to a fuel use category is 56%. The ordinal measures of association `Kendall's` τ_b and `Somers' D` $R|C$ both have a value of 0.40 again confirming the positive association between these two variables. Rather than being concerned with the ranking of pairs of observations on the two variables (concordancy or discordancy) as discussed previously, `Spearman's` ρ measures the strength of the relationship between the overall ranks of each observation (or subject) on the two variables.

Statistics for Table of FuelGrp by LicGrp

Statistic	Value	ASE
Gamma	0.5641	0.1272
Kendall's Tau-b	0.4024	0.0983
Stuart's Tau-c	0.4010	0.0988
Somers' D C|R	0.4016	0.0979
Somers' D R|C	0.4031	0.0989
Pearson Correlation	0.4612	0.1049
Spearman Correlation	0.4640	0.1090
Lambda Asymmetric C|R	0.2667	0.1456
Lambda Asymmetric R|C	0.2903	0.1463
Lambda Symmetric	0.2787	0.1335
Uncertainty Coefficient C|R	0.1553	0.0656
Uncertainty Coefficient R|C	0.1547	0.0654
Uncertainty Coefficient Symmetric	0.1550	0.0655

Sample Size = 48

Fig. 2.37. SAS Example B10: measures of association—fuel and population

Pearson's correlation coefficient and Spearman's ρ are both close to 0.5 indicating moderate positive association, supporting conclusions made with previous statistics. If the percentage of licenses category is used to predict the fuel use category of a state, the nominal measure `asymmetric lambda`, $\lambda(R|C)$, can be used to obtain the proportional reduction in error PRE. Here the value is 0.2903, so that that proportional reduction in error is 29% compared to prediction not based on the number of licenses issued.

Including `cl` along with `measures` as `tables` statement options will lead to the computation of asymptotic confidence intervals for the measures of association discussed earlier. The default confidence coefficient is 0.05, which may

Gamma	
Gamma	0.5641
ASE	0.1272
95% Lower Conf Limit	0.3148
95% Upper Conf Limit	0.8134

Spearman Correlation Coefficient	
Correlation	0.4640
ASE	0.1090
95% Lower Conf Limit	0.2503
95% Upper Conf Limit	0.6776

Test of H0: Gamma = 0	
ASE under H0	0.1390
Z	4.0587
One-sided Pr > Z	<.0001
Two-sided Pr > IZI	<.0001

Test of H0: Correlation = 0	
ASE under H0	0.1098
Z	4.2268
One-sided Pr > Z	<.0001
Two-sided Pr > IZI	<.0001

Fig. 2.38. Result of TEST statement in PROC FREQ

be changed by including an optional `alpha=` option with the required value. For some of the measures, adding a `test` statement will produce an asymptotic test of whether the measure is equal to zero as well as an asymptotic confidence interval. These association measures are gamma, Kendall's τ_b, Stuart's τ_c, Somers' D, and Pearson's and Spearman's ρ. Figure 2.38 shows the output resulting from including the statement `test gamma scorr;`. In both cases, the null hypotheses are rejected at reasonable α values. Finally, Fig. 2.39 illustrates how the `list` option may be used to format a cross-tabulation as a one-way table.

TaxGrp	FuelGrp	Frequency	Percent	Cumulative Frequency	Cumulative Percent
High	Low Fuel Use	10	20.83	10	20.83
High	Medium Fuel Use	10	20.83	20	41.67
High	High Fuel Use	2	4.17	22	45.83
Low	Low Fuel Use	6	12.50	28	58.33
Low	Medium Fuel Use	7	14.58	35	72.92
Low	High Fuel Use	13	27.08	48	100.00

Fig. 2.39. SAS Example B10: tax group × fuel group cross-tabulation as a list

2.3 Some Useful Base SAS Procedures

There are many Base SAS procedures that calculate a variety of statistics as well as others that perform utility functions. Some of these, such as `proc print`, `proc means`, `proc sort`, and `proc format`, were previously discussed or used in SAS Example programs. Some others such as `proc rank` and `proc corr` will be used in examples to follow in later chapters.

In Sect. 2.2, the `plots` option used in `proc univariate` produced a histogram (or a stem-and-leaf plot), a box plot, and a normal probability plot in high-resolution as an ODS Graphics panel. In addition, easy-to-use statistical graphics procedures such as SGPLOT and SGCHART are available for producing high-resolution graphics. Although graphs created by SAS statistical graphics procedures are preferable for use in presentations or publications, low-resolution graphics produced by some SAS procedures also play a role, for example, in routine exploratory data analysis or as diagnostic tools. Base SAS procedures PLOT and CHART are two procedures available for specifically producing this type of graphics. However, a discussion of these two procedures is omitted from this section. Instead, two useful Base SAS procedures for presentation of data summaries will be introduced and their use illustrated through SAS Example programs.

2.3.1 The TABULATE Procedure

The TABULATE procedure is an extremely versatile procedure for producing display-quality tables containing descriptive statistics. Using an extremely simple and flexible system of syntax, the user is able to customize the appearance of the tables incorporating labeling and formatting as well as generate tabular reports that contain many of the same descriptive statistics that are computed by several other statistical procedures. More recently incorporated innovations allow *style elements* (e.g., colors) to be specified to enhance the appearance of tables output in HTML and RTF formats using ODS graphics. The statements available in `proc tabulate` and options that can be specified are too numerous to be described in detail in this text. A brief description useful for understanding the example presented below follows. A general structure of an abbreviated `proc tabulate` step (that omits several important statements) is

```
PROC TABULATE <options>;
  CLASS variable(s) ;
  VAR variable(s) ;
  BY variable(s) ;
  TABLE expression, expression, ... < / options >;
  KEYLABEL keyword='text' ... ;
```

Options available for the proc statement are `data=`, `out=`, `missing`, `order=`, `formchar< (position(s)) >='formatting-character(s)'`, `noseps`, alpha=, vardef=, pctldef=, format= , and style=. Since the reader has encountered many of these options before, only those that are relevant to proc tabulate are described here. The formchar= option is not used for ODS output so is not discussed. A value specified for the format= option is any valid SAS or user-defined format for printing each cell value in the table, the default being best12.2. The style= specifies the style element (or style elements) (for the Output Delivery System) to use for each cell of the table. For example, style=[background=gray] specifies that the background color for data cells be of the color gray. At a secondary level, style elements can also be specified in *dimension expressions* (as described below) to control the appearance of table elements such as analysis variable name headings, class variable name headings, class variable level value headings, data cells, keyword headings, and page dimension text.

The table statement is the primary statement in proc tabulate. The main components of table statements are *dimension expressions*. A dimension expression specifies combinations of classification variables to define categories for which the statistics displayed in the cells of the table are calculated. The expression may also combine an analysis variable on which the specified statistic is to be computed. By default, the statistic calculated is the sum of the values of the analysis variable for each category; alternatively, the user may combine the name of a statistic to be computed into the expression. A simplified form of the table statement is

```
table expression-1, expression-2, expression-3 </ options> ;
```

where *expression-1*, *expression-2*, and *expression-3*, respectively, define the appearance of the page, row, and column components of these combinations. If only two-dimensional expressions are present, the left expression specifies the rows, and the right expression defines the columns. If only a single expression is present, it defines the columns of the table. A dimension expression consists of combinations of the following elements separated by asterisks, blanks, or parentheses:

- Classification variable(s) (variables in the class statement)
- Analysis variables (variables in the var statement)
- Statistics (n, mean, std, min, max, etc.)
- Format specifications (e.g., f=7.2)
- Nested dimension expressions in parentheses

As an example, suppose that the statements

```
class Region PopGrp TaxGrp;
var Fuel Income;
```

are present in a proc tabulate step and that the expressions below are used to define the rows. Some examples of possible dimension expressions are

Region defines rows for the different values of the nominal variable Region.

Region PopGrp defines rows by stacking the values of Region and PopGrp (an operation called *concatenation*, represented in the expression as a blank or a space).

Region*PopGrp defines rows as all combinations of the values of Region and PopGrp (an operation called *crossing*, represented in the expression as an "*").

Region*Fuel defines rows for the different values of Region and statistics (the sum, by default) are calculated for the analysis variable Fuel

Region*PopGrp*TaxGrp defines rows of all combinations of values of Region, PopGrp, and TaxGrp (i.e., all three variables are crossed with each other).

Region PopGrp*TaxGrp defines rows by concatenating different values of Region with all combinations of PopGrp crossed with TaxGrp.

(Region PopGrp)*TaxGrp defines rows by concatenating values of Region crossed with TaxGrp and PopGrp crossed with TaxGrp.

Region*mean*Fuel defines rows by values of Region and the means are calculated for the analysis variable Fuel.

Region*(mean stderr)*Fuel*f=6.2 defines rows by values of Region and the means and standard deviations are calculated for each region for the variable Fuel and appear side by side (i.e., the two statistics are concatenated). They are output using the specified format.

The column specification of mean*Fuel causes the mean of the fuel consumption variable to be computed and appear as a single column. Thus the table statement table Region, mean*Fuel; will result in a table with a single column of fuel consumption means computed for all regions shown as rows. The statement table Region*PopGrp, mean*Fuel; will similarly result in a single column of fuel consumption means but for all combinations of the classification variables Region and PopGrp tabulated as rows. The statement table Region, PopGrp*mean*Fuel; will, on the other hand, produce a two-way table of means for the variable Fuel (i.e., the same values as in the previous table) now tabulated with regions shown as rows and values of PopGrp in columns. If no statistic keyword appears in the dimension expressions in the above table statements, the sum (or the frequencies, if no analysis variable is present) will be calculated by default. SAS documentation on proc tabulate provides many examples of various combinations used to define dimension expressions; thus, an extensive discussion is not provided here.

SAS Example B11

The SAS data set created in SAS Example B6 on fuel consumption data is used again in the SAS Example B11 program (see Fig. 2.40) to illustrate the use of proc tabulate for producing tables of statistics. Recall that IncomGrp and TaxGrp are two category variables created previously and included in the data set. In addition, another grouping variable LicGrp is also created using the values of the variable Percent, which contains the percentage with driving licenses in the population. From among several analysis variables present in

```
libname mylib "C:\users\user_name\Documents\...\stat479\";

data fueldat3;
set mylib.fueldat;

if Percent=<54 then LicGrp=1;
else if 54<Percent=<58 then LicGrp=2;
else LicGrp=3;
run;

proc format;
   value ing 1 = 'Low Income'
             2 = 'Middle Income'
             3 = 'High Income';
    value lg 1='below 54%'
             2='54 to 58%'
             3='above 58%' ;
run;
proc tabulate data=fueldat3;
   var Fuel;
   class IncomGrp TaxGrp LicGrp;
   format IncomGrp ing. LicGrp lg.;
   table IncomGrp*TaxGrp,Fuel*(n mean stderr);
   table TaxGrp*LicGrp='Percent Driver Licenses',Fuel*(n='Sample Size'*f=4.0
         (mean='Sample Mean' stderr='Standard Error of the Mean')*f=8.1);
   title 'Illustrating PROC TABULATE': Simple Example;
run;
```

Fig. 2.40. SAS Example B11: program

the data set, `Fuel` (per capita fuel consumption) is selected for the computation of statistics to be tabulated.

In the SAS program, the `class` statement lists the three category variables, and the `var` statement lists the analysis variable. A simple `table` statement is first used to produce a two-way table that tabulates the statistics sample size (n), the mean, and the standard error of the mean (stderr) for the variable `Fuel`. The first dimension expression `IncomGrp*TaxGrp` defines the rows of the table, with the rows representing combinations of the levels of `IncomGrp` and `TaxGrp`. The second dimension `Fuel*(n mean stderr)` defines the columns to be the sample size, the mean, and the standard error of the mean computed for the variable `Fuel`. As can be observed, previously defined labels and formats are used for the variables and their levels. The default format of `best12.2` is used for printing all cell values. This table is shown in Fig. 2.41.

The second `table` statement also produces a two-way table with `TaxGrp* LicGrp` defining the row and the same statistics computed for the `Fuel` variable defining the columns. However, `format` specifications are added to control the formatting of the cell values in the table (e.g., `n*f=4.0`). Also, `label` parameters are added to assign more elaborate labels for class variable names (e.g., `LicGrp='Percent Driver Licenses'`) and statistic keyword headings (e.g., `mean='Sample Mean'`). These produce the table shown in Fig. 2.42.

Illustrating PROC TABULATE

		Fuel Consumption (in gallons/person)		
		N	Mean	StdErr
Income Level	Fuel Tax Level			
Low Income	High	7	561.51	18.72
	Low	6	626.14	27.27
Middle Income	High	7	547.93	26.69
	Low	11	649.03	35.21
High Income	High	8	463.14	20.02
	Low	9	590.71	49.73

Fig. 2.41. Output from PROC TABULATE: simple example (Table 1)

SAS Example B12

This example is a simple modification of SAS Example B11 in order to illustrate how `style=` options may be used to modify the appearance of selected parts of the output table. TABULATE and other Base SAS report writing procedures that are compatible with the ODS system use *table templates* to produce output tables. These templates specify the appearance of various predefined parts of the output using one or more style elements that, by default, are determined by the style template currently in use. For

Illustrating PROC TABULATE

		Fuel Consumption (in gallons/person)		
		Sample Size	Sample Mean	Standard Error of the Mean
Fuel Tax Level	Percent Driver Licenses			
High	below 54%	8	495.0	26.2
	54 to 58%	10	512.2	19.3
	above 58%	4	597.1	27.5
Low	below 54%	6	552.6	42.3
	54 to 58%	8	591.4	25.7
	above 58%	12	680.5	37.5

Fig. 2.42. Output from PROC TABULATE: simple example (Table 2)

```
proc tabulate data=fueldat3;
   var Fuel / style=[background=bisque];
   class IncomGrp TaxGrp LicGrp/style=[background=lightgreen color=red];
   classlev IncomGrp TaxGrp LicGrp/style=[background=yellow];
   format IncomGrp ing. LicGrp lg.;
   keyword all/ style=[background=linen color=blue];
   keylabel all='Total';
   table IncomGrp*TaxGrp,Fuel*(n mean stderr)*([style=[background=lightcyan]]);
   table TaxGrp*LicGrp='Percent Driver Licenses' all,Fuel*(n='Sample Size'*f=4.0
         (mean='Sample Mean'
          stderr='Standard Error of the Mean')*f=8.1)*([style=[background=lavender]])/
          box=[label='Fuel Use by Tax and %Licenses' style=[backgroundcolor=coral]];
   title 'Illustrating PROC TABULATE: Adding Labels, Formats and Styles';
run;
```

Fig. 2.43. SAS Example B12: proc tabulate Step: adding labels, formats and styles

example, most tables in this chapter are produced using the *HTMLBlue* style template which is the default style for HTML destination as mentioned previously.

A *style* template is an ODS template that defines the visual aspects (colors, fonts, lines, markers, and so on) of SAS output. Style templates consist of *style elements*, each of which is a named collection of *style attributes* that describe the appearance of distinct regions of the output table. Examples of regions of an output table are header, footer, row header, cells, etc. Style elements that correspond to each of these are identified by a reserved name, and their style attributes can be found by examining the style templates (e.g., that for *HTMLBlue*) supplied by SAS using the TEMPLATE procedure.

For example, the name of the style element that describes the characteristics of the row headings region is `Rowheader`, and `fontstyle=` and `backgroundcolor=` are two of its attributes. The style element name for the cells region is `Data`, and `fontstyle=` and `backgroundcolor=` also are two of its attributes.

Values for style attributes are preassigned for each *style element* within a style template. For example, the default value under the *HTMLBlue* style for `fontstyle=` is `roman`, and that `backgroundcolor=` is `cxedf2f9`. These may have different values assigned to them in other style templates.

When a certain style is in effect (say, e.g., *HTMLBlue*), the default appearance of the table is thus determined by the content of the corresponding style template. While the default style templates may be modified to create one's own customized template, the discussion here is limited to an explanation of how the `style=` options can be used within a procedure step to alter the appearance of a table.

The `style=` option used with specific procedure statements modifies the appearance of the area of the table affected by that particular statement. This allows the user to make changes to a particular region of the output table without affecting other parts of the table. In SAS Example B12 program (see Fig. 2.43), visual properties of several parts of the HTML output table are modified using `style=` options used with several different statements in the proc step. This way the user can override the default settings in the default style being used at the time the procedure is executed.

A `style=` option used with a statement modifies a specific area of a table. Information on the areas of a table that can be modified with different statements can be found in the description for TABULATE procedure. For example, a `style=` option in the `proc tabulate` statement changes the specific default style attributes of all cells in the table. However, style= option may be used in a dimension expression to modify style attributes of an individual table. In Fig. 2.43, different background colors are specified by using style= options in the column expression. Specifying style= options in the `class`, `classlev`, `var`, and `keyword` statements override the attributes of the class variable name headings, class-level value headings, analysis variable name headings, and keyword headings, respectively. Note that `style=` options specified in `table` statements override the same specification in any of the above statements. For example, this allows the user to specify different cell attributes for tables as seen in Fig. 2.43.

Note that the keyword statement refers to statistic keywords or the universal keyword `all` used to refer to averaging over all of the categories for class variables in the same parenthetical group or dimension. See the use of `all` in the row expression for the second table in Fig. 2.43.

The `box=` option in the table statement modifies the appearance of the empty box above the row titles. It is important to carefully note that if no `style=` options are used (as in SAS Example B11), all style attributes used are those specified in the style template in effect. Recall that the default

Illustrating PROC TABULATE: Adding Labels, Formats and Styles

		Fuel Consumption (in gallons/person)		
		N	Mean	StdErr
Income Level	Fuel Tax Level			
Low Income	High	7	561.51	18.72
	Low	6	626.14	27.27
Middle Income	High	7	547.93	26.69
	Low	11	649.03	35.21
High Income	High	8	463.14	20.02
	Low	9	590.71	49.73

Fig. 2.44. Output from PROC TABULATE: adding labels, formats and styles (Table 1)

Illustrating PROC TABULATE: Adding Labels, Formats and Styles

Fuel Use by Tax and %Licenses		Fuel Consumption (in gallons/person)		
		Sample Size	Sample Mean	Standard Error of the Mean
Fuel Tax Level	Percent Driver Licenses			
High	below 54%	8	495.0	26.2
	54 to 58%	10	512.2	19.3
	above 58%	4	597.1	27.5
Low	below 54%	6	552.6	42.3
	54 to 58%	8	591.4	25.7
	above 58%	12	680.5	37.5
Total		48	576.7	16.1

Fig. 2.45. Output from PROC TABULATE: adding labels, formats and styles (Table 2)

style being used in these examples is *HTMLblue*. The output resulting from executing the SAS Example B12 is shown in Figs. 2.44 and 2.45.

2.3.2 The REPORT Procedure

REPORT is a powerful, highly versatile, and flexible SAS procedure for report writing. It makes available capabilities of the PRINT, MEANS, and TABULATE procedures in a single procedure that enables the user to create a variety of reports and tables. However, its flexibility implies that it might require some effort to master. The basic introduction provided here is designed to encourage progressing to more advanced applications and uses statements that can be used only with Windows applications with ODS destinations.

The REPORT procedure can be used to present data in tables (*a detail report*) with calculated summary lines printed for subsets of data as requested by the user much like the PRINT procedure, to produce summary reports much like the MEANS procedure, or to arrange these summaries in display tables (*a summary report*) designed by the user as in the TABULATE procedure. While REPORT can be used to produce acceptable reports using its default behavior, just as with TABULATE, its power lies in that the reports it produces are highly customizable. It is this aspect of `proc report` that makes it more difficult to master as well. The statements available in `proc report` and options that can be specified are too numerous and complex to be described in its entirety in this section. Instead, several selected statements and some key options that illustrate their primary uses will be introduced in the examples that follow. The general structure of a `proc report` step (that excludes several important statements) is

```
PROC REPORT <options>;
  COLUMN column-specification(s);
  DEFINE column/ column usege and attributes ;
  BREAK location break-variable </ option(s)>;
  RBREAK location </ option(s)>;
  COMPUTE column;
      compute column statements;
      ENDCOMP;
```

Options available with the proc statement are too numerous for each of them to be covered in detail. A few of the important options are briefly discussed below. The options `data=, out-, pctldef=, vardef=, split=, ps=, ls=` and `missing` have been introduced in discussions on several other SAS Base procedures and so are not duplicated here. Discussions on several other keyword options that affect only LISTING output are also be omitted. Options such as `bypageno=, center|nocenter, completecols|nocompletecols, completerows|nocompleterows, showall` are mostly self-explanatory and could be looked up when needed. A more important option that is useful when reports are intended for an ODS destination is

`style` <(*location(s)*)>=<*style-override(s)*>

which specifies one (or more) style overrides to be used for different parts of the report. Most Base SAS procedures that support ODS use one or more templates to produce output tables. These table templates include separate templates for table elements such as columns, headers, and footers. A style template for the entire table has been created using the TEMPLATE procedure, but one can use the `style=` option in the proc statement or in specific statements within the procedure to override various style elements to modify the appearance of the table. Style elements that are usually overridden this way are background color, foreground color, font faces, font sizes, font styles, etc. This approach was illustrated previously in Chap. 1 in SAS Example A11 (see Fig. 1.30) in a `proc print` step and in SAS Example B12 (see Fig. 2.43) in a `proc tabulate` step in this chapter.

The syntax for a `proc report` step is significantly different from that for most SAS Base procedures. Most notably, the VAR and CLASS statements are not used in `proc report`; however, the WEIGHT and FREQ statements as well as BY and WHERE statements are still available. The COLUMN statement lists the variables that are used to generate the report and usually precedes other statements that reference them. A DEFINE statement is used for each variable listed in the COLUMN statement to specify attributes such as *usage type* of the column, the summary statistic to be computed, and the format to be used for printing its value. Although there are default values for all of these attributes for each column, it is a good idea to include a DEFINE statement for each column even if some of the default attributes are acceptable.

Usage type for a column variable is specified as one of *display, group, order, analysis, computed, order,* or *across*, where

display displays the values of the variable with each row representing an observation in the input data set.

group consolidates multiple observations in the input data set into one row.

analysis specifies a numeric variable whose values are used to calculate a statistic for all the observations in the input data set.

computed a new variable created by the user whose value is a computed value.

order specifies a variable that is to be used to order the *detail rows* according to the ascending, formatted values.

across a variable that creates a column for each value of the variable; usually this column variable is a category or a nominal variable.

By default, `proc report` will produce a *detail* report containing detail rows each displaying a line corresponding to an observation in the input data set unless `define` statement(s) are used to *consolidate* rows in the data set into groups identified by values of variables (columns) in the data set. Obviously, grouping is done using category, class, or user-created variables, and consolidation is achieved by summarizing across the set of observations belonging to groups defined by those variables. The report will thus consist of *summary rows* with each row being represented by statistics computed on multiple observations in a group as described. By default, the summary statistic used for consolidation is the *sum*; however, keywords as those used in SAS procedures such as MEANS or UNIVARIATE (e.g., `mean`, `var`, or `stderr`) may be used to specify the statistics to be computed in a summary report thus produced. One important thing to note is that a summary report may contain detail rows of observations that are not *consolidated* by all category variables in the data set. This implies that to form consolidated rows containing statistics for grouped observations, a define statement must exist for every category variable declared in the `column` statement.

Variables in the SAS data set named `fueldat` concerning fuel use by 48 contiguous states in the United States, used in the SAS Example B11 program (see Fig. 2.40), are used here to illustrate the use of some `proc report` statements. Each observation in the data set consists of values measured on several variables such as `Income` and `NumLic`, for a state. Recall that `IncomGrp` and `TaxGrp` are two category variables created previously and included in the data set, with three income groups (1, 2, and 3) and two tax groups (low and high), respectively. From among several analysis variables present in the data set, `Income`, `NumLic`, and `Fuel` are selected for the computation of statistics.

Consider the following two statements in a proc report step:

```
column IncomGrp TaxGrp Income NumLic Fuel;

define IncomGrp/group order=internal;
```

Here the variables that will appear in the resulting report are listed in the column statement. However, only one of the category variables is defined as a group type; thus, consolidation will not take place. Instead a detail report is produced with rows for each state, the observations being grouped by values IncomGrp. If, in addition, the following statement

```
define TaxGrp/group;
```

is also included, a summary table is produced for the six combinations of IncomGrp and TaxGrp with each cell of the table representing the sum (by default) of the values of the numeric variables Income, NumLic, and Fuel. Thus consolidation will take place. The user can specify that the statistics sample size, mean, and variance are calculated for each combination of IncomGrp and TaxGrp by modifying the column statement, thus

```
column IncomGrp TaxGrp (Income NumLic Fuel),(n mean var);
```

Note that in the above statement, the *comma* operator is used to nest or stack columns by associating keywords for statistics with variable names. Parenthesis can be used to associate a set of these statistics to a combined set of variables as shown above. Thus the three statistics sample size, mean, and variance are nested within each variable (i.e., computed values of the statistics will be stacked below each of the variables) in the report. Consolidation may be specifically requested in the detail report by replacing define TaxGrp/group; with the following statement:

```
break after TaxGrp/summarize;
```

The break statement produces a default summary row after the last row for each TaxGrp. Thus consolidation is forced to occur but the detail rows are still retained in the report. The following statement produces a summary across the entire report:

```
rbreak after TaxGrp/summarize;
```

Earlier, statistics were computed using keywords. Instead, a define statement can be used specify a *statistic* to be computed for each variable. The following three statements (along with the define statements for IncomGrp and TaxGrp used above) will calculate different statistics for each of the variables Income, NumLic and Fuel. Among other options, a format= and a text string to be used as the column heading are used here to describe the calculated columns:

```
define Income/analysis mean format=7.2 "Average Fuel Use";
```

```
define NumLic/analysis max format=4. "Maximum Drivers";

define Fuel/analysis stderr format=7.2 "Std. Err. of Fuel Mean";
```

In SAS Examples below, three programs with increasing complexity are presented to illustrate the techniques discussed above.

SAS Example B13

```
libname mylib "C:\users\user_name\Documents\...\stat479\";

data fueldat3;
set mylib.fueldat;

proc format;
   value ing 1 = 'Low Income'
             2 = 'Middle Income'
             3 = 'High Income';
run;

proc report data=fueldat3;
   column IncomGrp TaxGrp (Income NumLic Fuel),(N Mean StdErr);
   define IncomGrp/ group order=internal format=ing.;
   define TaxGrp/group;
   title 'Illustrating PROC REPORT': Simple Example;
run;
```

Fig. 2.46. SAS Example B13: program

The SAS data set created in SAS Example B6 on fuel consumption data is used again in the SAS Example B13 program (see Fig. 2.46). Recall that `IncomGrp` and `TaxGrp` are two category variables created previously and included in the data set. In addition, another grouping variable `LicGrp` was also created using the values of the variable `Percent`, which contains the percentage of persons with driving licenses in the population.

The column statement names the category variables `IncomGrp, TaxGrp` for identifying groups of observations and the analysis variables `Income, NumLic, Fuel` on which the specified statistics (sample size, mean, and the standard error of the mean) are to be computed. The define statements identify the two category variables as *group* types, which request that observations are to be consolidated for combinations of values of these two variables. This will cause the production of a summary table where the above statistics will be calculated for each `IncomGrp`$\times$`TaxGrp` combination.

This description may be more easily understood by studying the output report shown in Fig. 2.47 and figuring out how the layout of the report (e.g., rows and columns) is produced and how the statistics shown are computed as a result of the statements in this proc step.

Note that mixed case letters were used in the proc step (e.g., Mean) for naming the list of statistics to be computed because of the fact that SAS

Illustrating PROC REPORT: Simple Example

		Per capita Income(in thsnds.)			No. of Licensed Drivers			Fuel Consumption (in gallons person)		
Income Level	Fuel Tax Level	N	Mean	StdErr	N	Mean	StdErr	N	Mean	StdErr
Low Income	High	7	3.5095714	0.0810825	7	1440.5714	301.93936	7	561.50722	18.720052
	Low	6	3.5748333	0.0744363	6	1080.5	292.71656	6	626.13707	27.269403
Middle Income	High	7	4.2191429	0.0725053	7	1660	554.84369	7	547.92835	26.689706
	Low	11	4.1449091	0.0630903	11	2128.1818	515.78029	11	649.0267	35.208534
High Income	High	8	4.8955	0.1281738	8	3223.125	983.75951	8	463.13744	20.018598
	Low	9	4.8111111	0.0897089	9	3999.7778	1261.7806	9	590.70997	49.734505

Fig. 2.47. Output from PROC REPORT: introductory example

ignores case when parsing these keywords; thus, the output column headers would appear more legible. In Fig. 2.47, the labels `N`, `Mean`, and `StdErr` are used as column titles, for the three different variables.

A better alternative to control the appearance of the report is to create an *alias* for each statistic computed for an analysis variable. Aliases are different names assigned to the same variable, and the user can create as many aliases as are needed. Aliases are specified in the COLUMN statement when a single analysis variable is to appear in more than one way in the report. This generally occurs when several statistics are to be computed on the same variable, as in SAS Example B13. Once aliases are created, a DEFINE statement is used to specify how the values of each of the alias variables is to be computed and also provide details (such as a format and a label) of the appearance of the alias variable in the report. The use of aliases is illustrated in SAS Example B14.

SAS Example B14

SAS Example B14 program (see Fig. 2.48) uses the same data set as in the previous example but replaces `TaxGrp` with another grouping variable `LicGrp`, created using the values of the variable `Percent`, that contains the percentage with driving licenses in the population. The last four variables in the COLUMN statement are *aliases*: `NumLicMin` and `NumLicMax` are aliased with `Numlic`, and `FuelMin` and `FuelMax` are aliased with `Fuel` using the "=" symbol in each case. Using aliases not only enables calculating different statistics for the same variable but also allows defining formatting, column titles, etc. for each alias individually. In this program, four DEFINE statements are used for each of the four aliases used.

The last four `define` statements are used to calculate the minima and maxima of the `Numlic` and `Fuel` variables using the *analysis* option

```
libname mylib "C:\users\user_name\Documents\...\stat479\";

data fueldat3;
set mylib.fueldat;

if Percent=<54 then LicGrp=1;
else if 54<Percent=<58 then LicGrp=2;
else LicGrp=3;
run;

proc format;
   value ing 1 = 'Low Income'
             2 = 'Middle Income'
             3 = 'High Income';
    value lg 1='below 54%'
             2='54 to 58%'
             3='above 58%' ;
run;

proc report data=fueldat3;
   column IncomGrp LicGrp,(Numlic=NumLicMin Numlic=NumLicMax Fuel=FuelMin Fuel=FuelMax);
   define IncomGrp/ group order=internal format=ing.;
   define LicGrp/across order=internal format=lg.;
   define NumLicMin/analysis min  format=5. 'Minimum No.of Licenses';
   define NumLicMax/analysis max format=5. 'Maximum No.of Licenses';
   define FuelMin/analysis min  format=4.'Minimum Fuel Use';
   define FuelMax/analysis max  format=4. 'Maximum Fuel Use';
   title 'Illustrating PROC REPORT: Aliases and Define Statement';
run;
```

Fig. 2.48. SAS Example B14: illustrating the use of aliasing

alongwith the appropriate statistic keyword (here min and max), along with appropriate formats and column headers. Although a statement such as `column IncomGrp LicGrp,(Numlic Fuel),(min max)`allows the user to

Illustrating PROC REPORT: Aliases and Define Statement

	Percent of Licensed Drivers			
	below 54%			
Income Level	Minimum No.of Licenses	Maximum No.of Licenses	Minimum Fuel Use	Maximum Fuel Use
Low Income	341	2088	487	714
Middle Income	2463	2804	547	580
High Income	2073	8278	344	471

Percent of Licensed Drivers							
54 to 58%				above 58%			
Minimum No.of Licenses	Maximum No.of Licenses	Minimum Fuel Use	Maximum Fuel Use	Minimum No.of Licenses	Maximum No.of Licenses	Minimum Fuel Use	Maximum Fuel Use
600	2835	566	699	501	501	648	648
441	6595	410	640	232	2368	561	968
982	5948	457	525	340	12130	524	865

Fig. 2.49. Output from PROC REPORT: aliasing

compute these statistics without using aliases, the use of aliases provides a way of formatting each variable separately.

Another feature illustrated in SAS Example B14 is the use of the across option available with the DEFINE statement in proc report to define a column item as an *across* variable. In this example LicGrp is defined as an across variable so that, while the values of IncGrp appear in rows, the values of LicGrp appear across the page as columns ordered from left to right, according to the internal values of LicGrp, as seen in Fig. 2.49. Since the statistics calculated for Numlic and Fuel, defined as analysis variables, are nested within the across variable LicGrp (observe the COLUMN statement carefully), these appear stacked below each value of LicGrp.

SAS Example B15

This example (see SAS program in Fig. 2.50) uses the raw data set used in SAS Examples A10 and A11 (see Figs. 1.27 and 1.30) accessed from an external text file and uses proc report to produce a report similar to the ones created in those examples. In addition, this example is also used to illustrate the use of break and rbreak statements and compute groups to enhance the report produced. Other than Region, State, Month, Headcnt, Revenue, and Expenses, all variables available from the SAS data set sales, a new variable named Profit appears in the report and is a calculated variable using the compute block:

```
compute Profit;
   Profit = Revenue.sum-Expenses.sum;
endcomp;
```

Clearly this compute block is associated with the computed variable Profit as opposed to the compute block that appears later in this proc step which is associated with a location. Note also that while Headcnt, Revenue, and Expenses are all defined as *analysis* variables, the new variable Profit is defined as a *computed* variable using appropriate define statement in the proc report step. In the compute block, the value of Profit variable is calculated as the difference between Revenue and Expenses; however, these variables are referenced in the block using their *compound names.* The convention is that when constructing a report that involves sharing a column with a statistic, a compound name must be used in a compute block.

A compound name has the form *variable-name.statistic* and identifies both the variable and the name of the statistic that was used when the analysis variables were defined. For example, the variable (or column) Revenue was defined as an analysis variable with sum as the statistic to be computed; thus the compound name assigned to the variable is Revenue.sum. Within the compute block, the value of Revenue.sum depends on which part of the report is involved in the computation: in detail rows, the value is the revenue for each observation in the input data; in the region summary lines, the value is the total sales for all the states in a region; and in the report summary line, it is the total revenue for all regions.

```
data sales;
infile "C:\users\user_name\Documents\...\stat479\data\sales.txt";
length Region $ 12;
input Region : $8. State $2. +1 Month monyy5. Headcnt Revenue
        Expenses;
run;
proc report data=sales split="*";
   column Region State Month Headcnt Revenue Expenses Profit;
   define Region/group order=internal "Sales*Region";
   define State/group order=internal;
   define Month/group order=internal format=monyy5. left;
   define Headcnt/analysis sum center "Sales*Personnel";
   define Revenue/analysis sum format=dollar12.2 "Sales*Revenue";
   define Expenses/analysis sum  format=dollar12.2 "Overhead*Expenses";
   define Profit /computed format=dollar12.2 "Monthly*Profit"
                                         style(column)=[backgroundcolor=linen];
   compute Profit;
     Profit = Revenue.sum-Expenses.sum;
   endcomp;
   break after Region/summarize style=[font_style=italic color=cornflowerblue];
   rbreak after/summarize style=[font_weight=bold color=darkcyan];
   compute after;
     Region = "All Regions";
   endcomp;
   title "Sales Analysis using Proc Report";
run;
```

Fig. 2.50. SAS Example B15: illustrating the use of define statement

While used here for a simple computation using an assignment statement, compute blocks can be used to perform complex calculations that may involve many SAS data step programming devices such as arrays, IF-THEN-ELSE structures, and various DO-END loops.

The `break` statement in the SAS Example B14 produces a default summary row after the last row for each `Region`. So consolidation takes place after displaying rows for each region value (of course, the regions are ordered by their internal value. The `rbreak` statement produces a summary at the end because the keyword *after* is used as the location. With both the break and rbreak statements, style options are used to enhance the output as shown in Fig. 2.51. Lastly, a compute block is included with the location designated as *after* to be executed at the end of the report (as a specific target such as a break variable is omitted). Here the compute block just assigns a value to the `Region` column in the summary line produced as result of the `rbreak` statement

Sales Analysis using Proc Report

Sales Region	State	Month	Sales Personnel	Sales Revenue	Overhead Expenses	Monthly Profit
CENTRAL	IL	JAN78	4	$6,000.00	$2,000.00	$4,000.00
		FEB78	4	$6,100.00	$2,000.00	$4,100.00
		MAR78	4	$6,050.00	$2,100.00	$3,950.00
	MI	JAN78	10	$10,000.00	$8,000.00	$2,000.00
		FEB78	9	$11,000.00	$8,200.00	$2,800.00
		MAR78	10	$12,000.00	$8,900.00	$3,100.00
	OH	JAN78	13	$21,000.00	$12,000.00	$9,000.00
		FEB78	14	$22,000.00	$13,000.00	$9,000.00
		MAR78	14	$22,500.00	$13,200.00	$9,300.00
CENTRAL			*82*	*$116,650.00*	*$69,400.00*	*$47,250.00*
EASTERN	NC	JAN78	12	$20,000.00	$9,000.00	$11,000.00
		FEB78	12	$21,000.00	$8,990.00	$12,010.00
		MAR78	12	$20,500.00	$9,750.00	$10,750.00
	VA	JAN78	10	$15,000.00	$7,500.00	$7,500.00
		FEB78	10	$15,500.00	$7,800.00	$7,700.00
		MAR78	11	$16,600.00	$8,200.00	$8,400.00
EASTERN			*67*	*$108,600.00*	*$51,240.00*	*$57,360.00*
NORTHERN	MA	MAR78	3	$1,000.00	$1,500.00	$-500.00
	NY	FEB78	4	$2,000.00	$4,000.00	$-2,000.00
		MAR78	5	$5,000.00	$6,000.00	$-1,000.00
NORTHERN			*12*	*$8,000.00*	*$11,500.00*	*$-3,500.00*
PLAINS	NM	MAR78	2	$500.00	$1,350.00	$-850.00
PLAINS			*2*	*$500.00*	*$1,350.00*	*$-850.00*
SOUTHERN	FL	JAN78	10	$10,000.00	$8,000.00	$2,000.00
		FEB78	10	$11,000.00	$8,500.00	$2,500.00
		MAR78	9	$13,500.00	$9,800.00	$3,700.00
	GA	JAN78	5	$8,000.00	$2,000.00	$6,000.00
		FEB78	7	$6,000.00	$1,200.00	$4,800.00
SOUTHERN			*41*	*$48,500.00*	*$29,500.00*	*$19,000.00*
All Regions			204	$282,250.00	$162,990.00	$119,260.00

Fig. 2.51. Output from PROC REPORT: break, rbreak, and compute statements

2.4 Exercises

2.1 The following SAS program inputs information about hospitals in cities in different states, such as the size of the state and the cities (i.e., population), median income and median housing costs, the number of admissions, and the number of beds, using three different data types 1, 2, and 3. Answer the questions below:

```
data hospital;
retain Form_Id State Stpop CitySize Income Housing Admit Beds;
input Form_Id $3. @5 Type $1. @ ;
if Type ='1' then input @8 State $2. Stpop 10.;
if Type ='2' then do;
       input @8 CityPop 8.  Income 5. Housing 6.;
         if CityPop<60000 then CitySize='Small';
         else CitySize='Large';
       end;
if Type ='3' then do;
  input @8  Admit 6. Beds 4.;
  output;
  end;
drop Type CityPop;
datalines;
v12 1  IA   1708232
v12 2      53620 5240 14236
v12 3     5126 178
v12 3     3364 134
v12 3     4857 184
v12 1  KS   1575899
v12 2      86610 4879 18154
v12 3     3916 156
v12 3     5527 182
v12 3    12139 351
v12 3     8257 238
v12 2      36574 3754 12739
v12 3     3465 112
v12 3     4576 142
;
proc print;
run;
proc means data=hospital noprint;
  class State CitySize;
  var Admit Beds;
  output out=stats mean= Av_Adms Av_Beds std=S_Adms S_Beds ;
run;
proc print data=stats;
run;
```

a. Show the contents of the PDV immediately after processing the first line of data.
b. Show the contents of the PDV immediately after processing the second line of data.
c. Show the contents of the PDV immediately after processing the third line of data.
d. Show the contents of the PDV immediately after processing the fifth line of data (before the observation is output to the SAS data set).
e. Display the first observation written to the SAS data set.
f. Run this program and turn-in the output only.
g. Examine the output produced by the second proc print step. Describe the statistics printed in each line of this output, i.e., explain what the computed numbers are in each line. Make sure that, for each value of the _TYPE_ variable, you identify the group of observations used to compute the statistics that appear in that line.

2.2 Organizations interested in making sure that accused persons have a trial of their peers often compare the distributions of jurors by age, education, and other socioeconomic variables. One such study provided the following information on the education of 1000 jurors:

Education	Elementary	Secondary	College credits	College degree
No. of jurors	278	523	98	101

The national percentages of the population in these educational levels are 39.2%, 40.5%, 9.1%, and 11.2%, respectively. Is there significant evidence of a difference between education distribution of jurors and the national education distribution? Use a SAS program to perform a chi-square goodness-of-fit at $\alpha = 0.05$ to answer this question.

2.3 Ott and Longnecker (2001) present an example in which the number of cell clumps per algae species was fitted to a Poisson distribution. A lake sample was analyzed to determine the number of clumps of cells per microscope field. The data are summarized below for 150 fields examined. Here, x_i denotes the number of cell clumps per field, and n_i denotes the frequency of occurrence of fields of each cell clump count.

x_i	0	1	2	3	4	5	6	≤ 7
n_i	6	23	29	31	27	13	8	13

Write a SAS program to perform a chi-square goodness-of-fit at $\alpha = 0.05$ to test the hypothesis that the observed counts were drawn from a Poisson probability distribution.

2.4 A political organization in a certain state, interested in making sure that they reach persons of every racial category in their advertising, obtains a random sample of 200 from their lists. The following counts are obtained from the sample using the available demographic information:

Race	White	African American	Hispanic	Asian	Other
No. in sample	142	28	16	10	4

Experts believe that the statewide percentages of the population in these racial categories are 70.1%, 12.4%,10.7%, 5.3%, and 1.5%, respectively. Is there significant evidence that the observed proportions in the sample differ significantly from these hypothesized proportions? Use a SAS program to perform a chi-square goodness-of-fit at $\alpha = 0.05$ to answer this question.

2.5 Devore (1982) discussed an example in which it is examined whether the phenotypes produced from a dihybrid cross of tall cut-leaf tomatoes with dwarf potato-leaf tomatoes obey the Mendelian laws of inheritance. There are four categories corresponding to the four possible phenotypes, tall cut-leaf, tall potato-leaf, dwarf cut-leaf, and dwarf potato-leaf, with respective expected probabilities p_1, p_2, p_3, and p_4. The null hypothesis of interest is

$$H_0 : p_1 = \frac{9}{16}, \; p_2 = \frac{3}{16}, \; p_3 = \frac{3}{16}, \; p_4 = \frac{1}{16}$$

Write a SAS program to perform a chi-square goodness-of-fit at $\alpha = 0.05$ of this hypothesis given that the observed counts in each category in a sample of size 1611 are 926, 288, 293, and 104, respectively.

2.6 The number of noxious weeds found in 1/4 ounce samples of *Phleum pratense* (meadow grass) is recorded below, where x_i denotes the number of noxious weeds and n_i denotes the number of samples with x_i noxious weeds out of a total of 98 samples:

x_i	0	1	2	3	4	5	6	7	≥ 8
n_i	3	17	26	16	18	9	3	5	1

Write a SAS program to perform a chi-square goodness-of-fit at $\alpha = 0.05$ to test the hypothesis that the observed counts were drawn from a Poisson probability distribution. Hint: For this data the Poisson mean λ is estimated as

$$\hat{\lambda} = \frac{\sum n_i x_i}{n} = 295/98 \approx 3.02.$$

To fit the data to a Poisson distribution, the probabilities for a Poisson distribution with mean = 3.02 are needed. These are calculated as follows (using one of many Poisson probabilities available via the Internet):

x_i	0	1	2	3	4	5	6	7	≥ 8
$P(X = x_i)$	0.049	0.147	0.223	0.224	0.169	0.102	0.051	0.022	0.013

2.7 The following data, taken from Rice (1987), represent the incidence of tuberculosis in relation to blood groups in a sample of Eskimos. Is there any association of the disease and blood group?

Severity	O	A	AB	B
Moderate-advanced	7	5	3	13
Minimal	27	32	8	18
Not present	55	50	7	24

Write a SAS program with a `proc freq` step for performing a chi-square test using $\alpha = 0.05$ to answer the above question.

2.8 In a study of the relationship between hair color and eye color among 592 students in a statistics course, the following data were obtained. Use the data to compute a chi-square test of independence using `proc freq`. Use nominal measures of association, the contingency coefficient C and Cramer's V, to comment on the strength of association if present.

Eye color	Hair color			
	Black	Brown	Red	Blond
Green	5	29	14	16
Hazel	15	54	14	10
Blue	20	84	17	94
Brown	68	119	26	7

2.9 An example in Schlotzhauer and Littell (1997) presented data from an experiment conducted by an epidemiologist who classified the disease severity of dairy cows (none, low, high) by analyzing blood samples for the presence of a bacterial disease. The size of the herd that each cow belonged to was classified as large, medium, or small. One of the aims of this study was to determine if disease severity was affected by herd size.

Severity	Herd size		
	Large	Medium	Small
None	11	88	136
Low	18	4	19
High	9	5	9

Use a SAS program to analyze these data using `proc freq`. Is there any association between two variables? Use the various measures to interpret any association present considering that the two variables are ordinal and that the experimenter is planning to use herd size for predicting disease severity. Obtain confidence intervals for the measures that you discuss.

2.10 Write a SAS program containing an `infile` statement to access the data set used in SAS Example B5 (see Fig. 2.13) from a text file. Use `proc tabulate` to obtain a table laid out as follows. The rows of the table consist of the combinations of year in school and gender, with the levels of gender appearing within each level of year. The column must present mean and standard deviation of the two variables height and weight. Use formats and labels to enhance your table. How can you add a *single* column containing the sample size formatted to print as a four-digit integer?

2.11 The data for this problem consist of measurements made on a group of people participating in a physical fitness program. The name of the file is `fitness.txt` and is described in Table B.4 of Appendix B. Use the following input statement to read these data:

```
input Id 2. WtLoss 2.1 Height 2. Weight 3. Intake 4. Aero 2.
      BodyFat 3.1 RunTime 3.1  RstPulse 2. Oxygen 3.1 Age 2.
                                                    Gender $1.;
```

To access this data set, include an appropriate *infile* statement in your SAS program. Input the data and create a SAS data set named `fitness`. Use SAS statements in the data step to do the following:

a. Exclude observations with missing values for `Aero` or `BodyFat` from this data set.
b. Provide more descriptive labeling as necessary using a *label* statement.
c. Exclude variables `Height` and `WtLoss` from the data set `fitness`.
d. Create four new variables as follows:
   i. A numeric variable `BurnRate` that measures the rate at which calories are burned, computed as Monthly Weight Loss times Food Intake (Cal/day) per each pound of Body Weight.
   ii. A numeric variable `BMI` containing values of Body Mass Index, the ratio of weight to height squared (in kg per meter^2), a standard measure of obesity.
   iii. A numeric category variable `WtGrp` (Weight Group) that takes values 1, 2, or 3 accordingly as the subject's weight is $\leq$ 140 lbs, over 140 but $\leq$ 165 lbs, or above 165 lbs, respectively.
   iv. A numeric category variable `AgeGrp` (age group) that takes values "A," "B," or "C" accordingly as the subject's age is $\leq$ 25, between 25 and $\leq$ 45, or above 45, respectively.
e. Include *format* statements to ensure that values for the variable created in A. above appear rounded to one decimal place and those of the variable created in B. appear rounded to a whole number in the printed output (i.e., associate appropriate formats with `BurnRate` and `BMI` variables using a *format* statement).

2.12 Those participants of the physical fitness program discussed in Exercise 2.11 with BMI exceeding 25 or BurnRate less than 15 were selected for an advanced aerobics program. Add a `proc print` step to the SAS program in Exercise 2.11 to obtain a SAS listing (i.e., a SAS report) that contains only the variables `BodyFat`, `RstPulse`, `BurnRate`, and `BMI` of those participants. The data must appear grouped into observations for each weight group and age group combination. You must not create additional SAS data sets to do this and make sure that all variables are labeled and data values are appropriately formatted.

2.13 Add a `proc means` step containing appropriate *class* and *output* statements, to the same SAS program created in Exercise 2.12, to create a SAS data set named `stats1`. The data set `mystats` must contain sample means and standard errors of the means of the variables `BMI`, `RunTime`, and `Oxygen` for each of the nine subgroups defined by combinations of values of the category variables `WtGrp` and `AgeGrp`. Use the *types* statement to ensure that the statistics are calculated only for *combinations of levels* of `WtGrp` and `AgeGrp`. Suppress the printed output produced in the `proc means` step. Print the data set `mystats`. Label all new variables on this output.

Exercises 2.14–2.20 concern the demographic data set on countries obtained from Ott et al. (1987) (see Table B.5 of Appendix B). To access this data set from a text file, include an appropriate `infile` statement in your SAS program. A `filename` statement may also be included if you prefer. Use the following input statement to read these data:

```
input Country $20. Birthrat Deathrat Infmort Lifeexp
                   Popurban Percgnp Levtech Civillib;
```

2.14 Write a SAS program to create a SAS data set named `world` using the data in Table B.5. Label variables as appropriate. Create category variables as described below:

Variable	Groupings	Category variable
Infmort	$< 24 = 1$ (low) 24–73 = 2 (moderate) $\geq 74 = 3$ (high)	`Infgrp`
Levtech	$< 24 = 1$ (low) $\geq 24 = 2$ (high)	`Techgrp`
Degree of civil liberties	$1, 2 = 1$ (low degree of denial) $3, 4, 5 = 2$ (moderate degree of denial) $6, 7 = 3$ (high degree of denial)	`Civilgrp`

Use a `libname` statement and a two-level name to save the SAS data set permanently in a folder in your computer, so that you will be able to access it for analyses later.

2.15 In a SAS program, use `proc univariate` to compute 33.3 and 66.7 percentiles of the variables `Birthrat`, `Deathrat`, and `Popurban` available in the SAS data set `world`. Access the SAS data set saved previously in Exercise 2.14. Use the `output` statement to save these statistics in a temporary SAS data set named, say, `stats`. Obtain a listing of this data set.

2.16 Use the printed output from the analysis performed in Exercise 2.15 to determine good cutoff values for creating additional category variables `Birthgrp`, `Deathgrp`, and `Popgrp` (each with three categories) corresponding to these variables. In a data step of a new SAS program, access the SAS data set `world` saved previously. Add statements to the data step to create the category variables described in Exercice 2.7, name the resulting SAS data set `world2`, and save the new data set in the same folder.

2.17 In a new SAS program, use the SAS data set `world2` saved in Exercise 2.16 in a `proc univariate` step to compute descriptive statistics, extreme values, percentiles, and low-resolution plots for the variables `Lifeexp`, `Percgnp`, and `Popurban`. Use an appropriate option to calculate

t-tests for the hypotheses that population means for each of these variables exceed 70 years, below \$3000, and above 60%, respectively. Also, include an option for calculating 95% confidence intervals for these parameters. Interpret the results of the t-tests using the p-values printed. Comment on the shape of the distribution of each of these variables using the printed output produced. Do the Shapiro–Wilk tests for normality conducted for each variable above support your conclusions?

2.18 Add a `proc means` step containing appropriate `class` and `output` statements, to the same SAS program used in Exercise 2.17, to create a SAS data set named `stats1`. The data set `stats1` must contain sample means, standard errors of the means, and maximum and minimum values of the variables `Birthrat`, `Deathrat`, and `Infmort` calculated separately for each of the nine groups defined by combinations of levels of the category variables `Birthgrp` and `Popgrp`. Use the `types` statement to ensure that statistics are calculated only for *combinations of levels* of `Birthgrp` and `Popgrp`. Suppress printed output in `proc means`. Also, add a `proc format` step to your SAS program to define user formats for use when printing all category variables. Obtain a listing of the data set `stats1`. Label all new variables on output.

2.19 In a new SAS program, use the SAS data set `world2` saved in Exercise 2.16 in a `proc freq` step to do the following:

a. Obtain two-way frequency tables (in the cross-tabulation format) for the variable `Techgrp` with `Infgrp`, `Techgrp` with `Civilgrp`, and `Popgrp` with `Infgrp`. Compute chi-square statistics, cell χ^2, and cell expected values but no column, row, or cell percentages.
b. Obtain a two-way frequency table (in the list format) for `Infgrp` and `Civilgrp`.
c. Use the chi-square statistic to test hypotheses of independence between pairs of variables considered in part (a) and state your results.
d. Use the contingency coefficient C to comment on the strength of association for the pair of variables `techgrp` and `civilgrp`.
e. Use the values of gamma, Kendall's tau-b, and Spearman's correlation coefficient to comment on the association between `Techgrp` and `Infgrp`.
f. Use appropriate measures to evaluate the strength of association between `popgrp` and `infgrp`. Considering that `popgrp` is an independent variable useful for predicting `infgrp` for each country, interpret the appropriate measures of association. Explain.

2.20 Write a SAS program to use the data set (named `world2`) saved in Exercise 2.16, in a `proc tabulate` step to print a tabulation giving the sample size, sample mean, sample standard deviation, and the standard error of the mean of the variables `popurban` for subgroups of observations defined by the combination of values of `Techgrp` and `Deathgrp`. The row analysis consists of combinations of `Techgrp` and `Deathgrp`, and the statistics for `Popurban` must appear on the columns. Print the sample size without decimals, the sample mean with four decimal places, and the other two statistics with two decimal places each. Also, use appropriate text strings to label all statistic keyword headings (e.g., print Standard Deviation for `std`).

3

Introduction to SAS Graphics

3.1 Introduction

As discussed in Chap. 1, SAS ODS (Output Delivery System) manages procedure output for display in a variety of formats, such as html, pdf, rtf, etc. (called *destinations*). This allows the flexibility in organizing the output from SAS procedures and thus enables the user to present the output in a more desirable manner. Traditionally, under the SAS windowing environment, results of a SAS procedure were displayed in the output window in the SAS *listing* format. As output from SAS procedures, ODS creates objects that have basically three components: data component, table definition (order of columns, rows, etc.), and an output destination (.html, .rtf, etc.). Examples of currently available ODS destinations are

LISTING: produces traditional SAS monospace typeset output
PS or PDF: output that is formatted for a high-resolution printer
HTML: output that is formatted in various markup languages such as HTML
RTF: output that is formatted for use with Microsoft Word

The default destination in the SAS windowing environment beginning with SAS 9.3 is HTML (a change from traditional LISTING destination used in previous versions). Under the windowing environment, the *Results* tab in the *Preferences* window (use the sequence `Tools`→ `Options`→ `Preferences` to get to the *Preferences* window) may be used to change back to the LISTING destination as the default if so desired. In the past, SAS users used SAS/GRAPH (and other procedures that produced high-resolution graphics as output) to produce high-quality graphics, henceforth, referred to as traditional SAS Graphics in this book. Beginning with SAS 9.3, SAS procedures will produce graphs under *ODS Graphics* automatically using default settings, in the same way tables containing statistics output from those procedures are produced.

M. G. Marasinghe, K. J. Koehler, *Statistical Data Analysis Using SAS*,
Springer Texts in Statistics, https://doi.org/10.1007/978-3-319-69239-5_3

Template-Based Graphics (ODS Graphics)

Template-based graphics produced by ODS Graphics (different from traditional SAS Graphics) is currently the default for producing graphs in most SAS procedures. In addition, ODS Graphics is enabled by default at the SAS start-up; thus no action is necessary by the user to enable graphical output produced by various procedures to appear in ODS Graphics format in the output. In previous versions of SAS, notably SAS 9.2 under the SAS windowing environment, ODS GRAPHICS ON statement was required to enable ODS Graphics (and ODS GRAPHICS OFF statement to disable or turn off ODS Graphics). In SAS 9.4, ODS GRAPHICS ON is the default. Thus all graphics produced by SAS procedures are ODS Graphics by default.

With ODS Graphics, *styles* and *templates* control the appearance of tables and graphics output from procedures in Base SAS as well as many other statistical procedures. Since under ODS Graphics, default templates and styles for graphs produced by various procedures are provided by SAS software, the user can create statistical graphics that are consistent in appearance across procedures. SAS 9.4 uses the default style called *HTMLBlue* available for the HTML destination. This new style is an all-color style that is designed to integrate tables and modern statistical graphics, viewable as a single entity.

In the previous edition of this book, the main emphasis was on producing traditional SAS Graphics using SAS/GRAPH procedures and SAS-/GRAPH-related statements in other SAS statistical procedures. The availability of ODS Graphics as a part of SAS/Base in SAS 9.4 makes it possible to create statistical graphics without having access to SAS/GRAPH. Since ODS Graphics in SAS 9.4 produces high-quality statistical graphics with minimal syntax in addition to the fact that it allows graphs and tables to be integrated in the same ODS output destinations, many of the SAS programs and output presented in this edition will use ODS Graphics. If so desired, using the Graph Template Language, one can modify a default template so that the changes are in effect each time the user runs a procedure to create the graph. The user may also make changes to graphs using the ODS Graphics Editor, a point-and-click interface.

In SAS Version 9.4, ODS Statistical Graphics output is also produced by SAS procedures such as UNIVARIATE when statements that produce these graphics are included in the proc step. For example, the HISTOGRAM statement used in a `proc univariate` step will produce ODS Graphics output in HTML by default in SAS 9.4; this output is automatically displayed in an internal viewer. Several examples of these procedures will be illustrated in later sections. The user may request that this graphical and other output be sent to a destination such as a pdf or an rtf file by enclosing the procedure step within appropriate ODS statements.

Prior to SAS 9.2, the plots produced by proc univariate were extremely basic by default. Producing more elaborate graphical output required the specification of colors, fonts, and other graphical elements via SAS/GRAPH statements and plot statement options. Beginning with SAS 9.2, the default appearance of the graphs is governed by the ODS style in operation, which produces attractive graphical output with consistent appearance across procedures using standard templates. See the SAS program in Fig. 3.4 and the resulting graphical output for an example.

It is to be noted that the default style for the pdf destination is quite different from *HTMLBlue*, the default style for HTML destination. In this chapter and elsewhere, the pdf output for publication was created using a style that mimics the appearance of the output produced by the *HTMLBlue* style.

ODS Statistical Graphics Procedures

As discussed in the above paragraph, beginning with SAS Version 9.4, ODS Graphics output is produced by SAS procedures such as UNIVARIATE when statements that produce these graphics are included in the proc step. This output is called ODS Statistical Graphics (to distinguish them from statistical graphics produced using traditional SAS/GRAPH procs or statements). For example, the HISTOGRAM statement used in a proc univariate step will produce ODS Graphics output in HTML, and in SAS 9.4, this output can be automatically displayed in an internal viewer. The user may request that this graphical and other output (such as tables produced by the procedure) be sent to a destination such as a *pdf* or an *rtf* file by enclosing the procedure step within appropriate ODS statements.

SAS Example C1

In addition, SAS 9.4 also makes available several procedures under Base SAS for producing ODS Statistical Graphics using raw data or output from other procedures. These procedures are called statistical graphics procedures. For example, the SAS program shown in Fig. 3.1 illustrates the use of the SGPLOT procedure for obtaining a simple regression plot enhanced with confidence and prediction bands, using the biology data set used in the SAS examples in Sect. 1.1: Here the data is accessed from the text file *biology.txt* available in a folder using an `infile` statement **1**. The SAS data set created is then

```
libname lib9  "C:\users\user_name\Documents\...\Class\"; 2
data lib9.biology; 3
infile "C:\users\user_name\Documents\...\Class\biology.txt"; 1
input Id Sex $ Age Year Height Weight;
run;

title "Regression of Weight on Height: Biology class data";
proc sgplot data=lib9.biology;
  reg x=Height y=Weight/CLM CLI;
run;
```

Fig. 3.1. Illustrating ODS output

saved in the library (folder) specified in the `libname` statement 2 using the two-level name *lib9.biology* 3 in the `data` statement.

The SAS data set *biology* is then accessed in the `proc sgplot` step to create the graphical output reproduced in Fig. 3.2. Note that here the graph shown is exactly as produced using the *HTMLBlue* style:

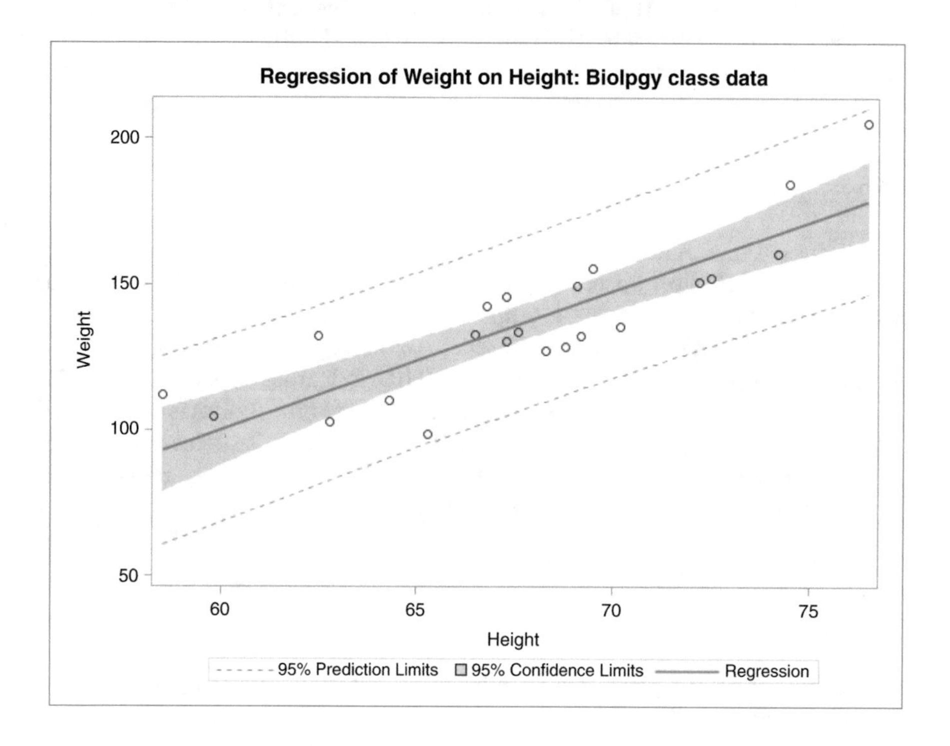

Fig. 3.2. Results of SAS Example C1

Traditional SAS Graphics via SAS/GRAPH

Traditional graphics can be produced using SAS/GRAPH procedures and statements (if SAS/GRAPH is licensed in your installation). Traditional graphics are saved in graphics catalogs. Their appearance is controlled by the SAS/GRAPH GOPTIONS, AXIS, and SYMBOL statements (as described in SAS/GRAPH: Reference) and numerous other specialized plot statement options.

The SAS/GRAPH statements and procedure options for controlling graph appearance continue to be available for producing traditional graphics. However, the NOGSTYLE system option must be specified to prevent the prevailing ODS style from affecting the appearance of traditional graphs. For example, this enables existing proc univariate or proc reg programs to produce customized graphs that appear as they did under previous SAS releases.

On the other hand, the appearance of ODS Graphics output controlled by the prevailing ODS style is not affected by SAS/GRAPH statements nor plot statement options that govern the appearance of traditional graphics. For example, the *CAXIS=* option used to specify the color of graph axes in traditional graphics is ignored when producing ODS Graphics output.

3.2 Template-Based Graphics (SAS/ODS Graphics)

As outlined above, SAS 9.4, many procedures will create template-based graphics and, by default, direct them to an HTML destination. The user may change the destination, say, to an RTF destination using an ODS statement. As noted previously, the default style in effect depends on the destination; however, the user may select a different style appropriate for the specific destination or modify an existing style template to create a customized style.

SAS Example C2

This example illustrates template-based statistical graphics produced by a Base SAS procedure. The SAS program shown in Fig. 3.3 accesses the SAS data set *biology* from a library (where it was previously saved in SAS Example C1), conducts a distribution analysis using the UNIVARIATE procedure, and delivers the output to an *rtf* destination **1**. The output from this SAS program is not displayed here but consists of the usual output that includes tables of computed statistics such as basic statistical measures and supplemented by a table containing results of several tests for normality, as requested:

```
libname libc "C:\users\user_name\Documents\...\Class\";

ods rtf file="C:\users\user_name\Documents\...\Class\c2out.rtf"; 1

proc univariate data=libc.biology Normal;
  var Height;
  title "Biology class: Analysis of Height Distribution";
run;
ods rtf close;
```

Fig. 3.3. SAS Example C2: illustrating ODS output

The user may use the ODS SELECT statement to include only a specified subset of the tables or graphs in the ODS destination. For example in a `proc reg` step, using the statement *ods select ParameterEstimates;* selects the table of parameter estimates to be in the output destination. Every output table and graph from a SAS procedure has an associated name and label. These table names are listed in the individual procedure documentation chapter or in the individual procedure section of the SAS online Help system. One can also use the SAS Results window (accessed from the SAS Explorer) to view the names of the tables that are created during a SAS session.

The UNIVARIATE procedure was introduced in Sect. 2.2.1. A discussion of procedure statements available in proc univariate for producing statistical graphics was not included in that introduction. Currently there are several statements that produce useful statistical graphics in the UNIVARIATE procedure. These are

```
CDFPLOT <variables> < / options> ;
HISTOGRAM <variables> < / options> ;
PPPLOT <variables> < / options> ;
PROBPLOT <variables> < / options> ;
QQPLOT <variables> < / options> ;
```

Some of these statements will be illustrated in various SAS Examples in this book, and some relevant statement options will be discussed whenever the statement is introduced. The HISTOGRAM statement is used to generate histograms and optionally superimpose them with estimated parametric or nonparametric density estimates. When the user requests that a parametric density function is to be fitted to the data, the user must specify the name of the distribution selected from a list of available distributions. The user may also specify a value for each of the parameters of the distribution. For example, if a normal density with parameters $\mu = 5$, $\sigma = 1$ is to be fitted, the required option is of the form `normal(mu=5 sigma=1)`. The user may set these parameters equal to the value `est` to specify that those parameters are to be set equal to their maximum likelihood estimates calculated from the data. Otherwise, if the parameter values are to be estimated from data, the specifications of parameter values may be omitted altogether, i.e., the `normal` option may be used by itself without any sub-options, which is equivalent to specifying `normal(mu=est sigma=est)`.

Lists of values for parameters may be specified to superimpose multiple fitted curves from the same distribution family on a histogram, the values being used sequentially according the position in the list. In this case, the two curves are identified on the plot using different colors; the `color=` sub-option allows the user to select the different color values.

SAS Example C3

In the following modified version of SAS Example C2, displayed in Fig. 3.4, portions of the output produced are selected to be included in the *rtf* destination using ODS SELECT. Note that a `histogram` statement (with the option `normal`) is present thus resulting in a histogram created using ODS Graphics. Note that the user selected the tables of basic statistical measures and the table containing the results of the tests for normality along with the histogram produced by incorporating the names *BasicMeasures*, *TestsForNormality*, and

```
libname libc "C:\users\user_name\Documents\...\Class\";

ods select BasicMeasures TestsForNormality Histogram; 1

proc univariate data=libc.biology Normal;
  var Height;
  histogram Height/midpoints=60 to 78 by 3 normal;
  title "Biology class: Analysis of Height Distribution";
run;
```

Fig. 3.4. SAS Example C3: illustrating ODS SELECT

Histogram in the `ods select` statement **1**. The name *Histogram* is an example of an ODS Graph Name. These are listed in a table in the `Details` section of procedure descriptions under the heading titled `ODS Graphics` and is commonly associated with the name of the statement that produces the plot.

Biology class: Analysis of Height Distribution

The UNIVARIATE Procedure
Variable: Height

Basic Statistical Measures			
Location		Variability	
Mean	67.89545	Std Deviation	4.58595
Median	67.95000	Variance	21.03093
Mode	67.30000	Range	18.00000
		Interquartile Range	4.90000

Tests for Normality				
Test	Statistic		p Value	
Shapiro-Wilk	W	0.980466	Pr < W	0.9231
Kolmogorov-Smirnov	D	0.107727	Pr > D	>0.1500
Cramer-von Mises	W-Sq	0.034689	Pr > W-Sq	>0.2500
Anderson-Darling	A-Sq	0.206227	Pr > A-Sq	>0.2500

Biology class: Analysis of Height Distribution

The UNIVARIATE Procedure

Distribution of Height

Percent
40
30
20
10
0
60 63 66 69 72 75 78
Height
Curve —— Normal(Mu=67.895 Sigma=4.5859)

Fig. 3.5. Results of using ODS SELECT

The output produced from SAS Example C3, sent to the default `html` destination and displayed by the SAS Results Viewer, is displayed in Fig. 3.5. It is important to note that this output is displayed using the default HTML-Blue style. The style may be changed to a different one by using the menu sequence `Tools`→ `Options`→ `Preferences...` and then using the `Results` tab to select the preferred style or by naming it in the `style=` option on the ODS statement. As usual, this output may be directed to other destinations such as a pdf file by including appropriate ODS statements.

SAS Example C4

This example shown in Fig. 3.6 uses the SAS data set named *biology* again as input. However, a data step is used to create the new variable BMI (as discussed in Chap. 1) and include it in a temporary data set (called *new* here, for simplicity) **1**. This data set is used in the SAS/STAT procedure TTEST to perform a two-sample t-test for testing whether the population means of the BMI variable are the same for the two male and female groups. The op-

```
libname libc "C:\users\user_name\Documents\...\Class\";

data new; 1
set libc.biology;
BMI=703*Weight/Height**2;
run;

proc ttest data=new cochran ci=equal; 2
  class Sex;
  var BMI;
  title "Biology class: Two Sample T-test";
run;
```

Fig. 3.6. SAS Example C4: statistical graphics from PROC TTEST

tion `cochran` is included for testing the homogeneity of variances for the two groups as well as the keyword option `ci=equal` for computing 95% confidence interval calculated based on the assumption that the population variances are indeed the same **2**. The output displayed in Figs. 3.7, 3.8, and 3.9 includes ODS Statistical Graphics produced by default. These are histograms superimposed by two smoothers, side-by-side box plots, and normal probability plots, respectively, for the two groups respectively.

There are many procedures in SAS/STAT package and others that support ODS Statistical Graphics. Commonly used procedures such as ANOVA, GLM, REG, and MIXED are a few examples. These procedures automatically produce some relevant graphics that are usually associated with the statistical techniques considered in the same manner they produce tables of relevant statistics. The TTEST procedure is an example of such a procedure. Some of these graphical tools will be discussed throughout the rest of this book when the relevant procedures are introduced.

Biology class: Two Sample T-test

The TTEST Procedure

Variable: BMI

Sex	N	Mean	Std Dev	Std Err	Minimum	Maximum
F	10	20.3660	2.3415	0.7404	16.2557	23.8458
M	12	21.2358	1.7746	0.5123	19.0845	24.6376
Diff (1-2)		-0.8698	2.0492	0.8774		

Sex	Method	Mean	95% CL Mean		Std Dev	95% CL Std Dev	
F		20.3660	18.6910	22.0410	2.3415	1.6105	4.2746
M		21.2358	20.1083	22.3633	1.7746	1.2571	3.0130
Diff (1-2)	Pooled	-0.8698	-2.7001	0.9604	2.0492	1.5677	2.9592
Diff (1-2)	Satterthwaite	-0.8698	-2.7732	1.0335			

Method	Variances	DF	t Value	Pr > |t|
Pooled	Equal	20	-0.99	0.3333
Satterthwaite	Unequal	16.571	-0.97	0.3479
Cochran	Unequal	.	-0.97	0.3578

Equality of Variances				
Method	Num DF	Den DF	F Value	Pr > F
Folded F	9	11	1.74	0.3818

Fig. 3.7. SAS Example C4: output from PROC TTEST (Part 1)

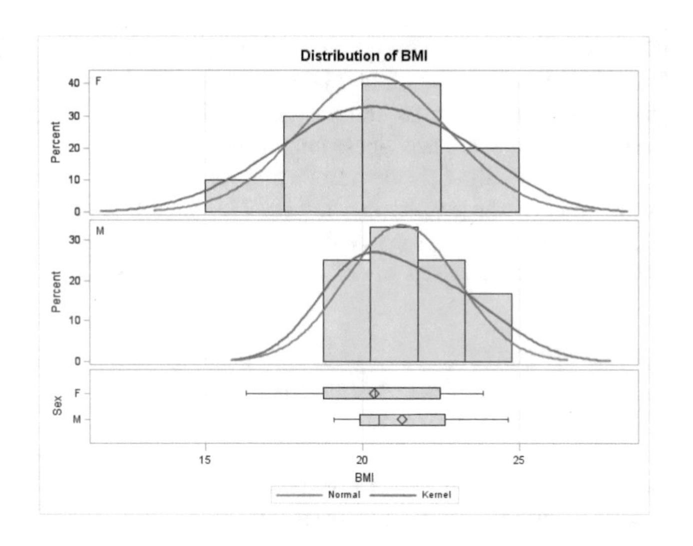

Fig. 3.8. SAS Example C4: output from PROC TTEST (Part 2)

3.3 SAS Statistical Graphics Procedures

As illustrated in SAS Example C1 in Sect. 3.1, Base SAS now includes several procedures for creating single-cell or multicell plots or panels of plots using simple syntax that are useful for creating many of the plots required for sta-

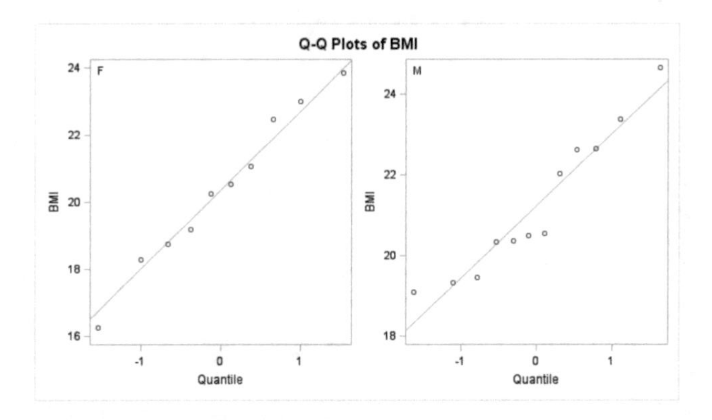

Fig. 3.9. SAS Example C4: output from PROC TTEST (Part 3)

tistical analysis. Several of these procedures will be illustrated in this section via examples. A subset of the statements and options required precedes a brief description of each procedure.

3.3.1 The SGPLOT Procedure

SGPLOT is an ODS Statistical Graphics procedure available for creating basic plots such as scatter plots, line plots, histograms, and bubble plots. However, its strength lies in the ability to enhance some of these with overlays of regression lines, confidence ellipses, loess smoothers, normal and kernel density estimates, penalized B-spline curves, etc., using easy-to-use options. Additionally, the appearance of the graph may be enhanced by adding features such as legends and reference lines using an extensive set of statements and options. The following is an abbreviated set of statements available for use in a`proc sgplot` step:

```
PROC SGPLOT < option(s)> ;
  DENSITY response-variable </option(s)>;
  DOT category-variable </option(s)>;
  ELLIPSE X= numeric-variable Y= numeric-variable </option(s)>;
  HBAR category-variable < /option(s) >
  HBOX response-variable </option(s)>;
  HISTOGRAM response-variable < /option(s)>
  HLINE category-variable < /option(s)>
  INSET "text-string-1" <... "text-string-n"> | (label-list);
  KEYLEGEND <"name-1" ... "name-n"> </option(s)>;
  LOESS X= numeric-variable Y= numeric-variable </option(s)>;
  REFLINE value(s) </option(s)>;
  REG X= numeric-variable Y= numeric-variable </option(s)>;
```

```
SCATTER X= variable Y= variable </option(s)>;
SERIES X= variable Y= variable </option(s)>;
STEP X= variable Y= variable </option(s)>;
VBAR category-variable < /option(s)>
VBOX response-variable </option(s)>;
VLINE category-variable < /option(s)>
XAXIS <option(s)>;
YAXIS <option(s)>;
```

A few of the statements available in `proc sgplot` will be illustrated via examples below. The first statement discussed is the `scatter` statement. The syntax allows the user to specify an X variable and a Y variable to be plotted on the horizontal and the vertical axes, respectively. Basically, this statement produces *scatter plots*, ordered pairs (x, y) plotted as points that visually display the relationship between the two variables such as trends in the data or the occurrence of interesting clusters.

Some SCATTER Statement Options

datalabel uses the Y values to label the points.

datalabel= uses the values of a variable to label the points.

group= specifies a variable that is used to group the data.

markerattrs= specifies appearance of markers in the plot (as a style element or using sub-options `color=`, `size=`, `symbol=`).

markerchar= specifies a variable whose values replace the marker symbols in the plot.

markercharattrs= specifies the appearance of markers in the plot when the `markerchar=` option is used.

name= specifies a name for the plot.

Marker attributes referenced above may be specified using a *style element* or by using sub-options. By default, the values of the sub-options are preset by the specific style element of the current style in operation. To change these settings, use the sub-options `color=`, `size=`, and `symbol=`. To set the `color` sub-option, use the same names from the SAS/GRAPH color naming schemes, set `size=` using units of measurement from Table 3.3 (default unit is pixels), and to specify a symbol, use one given in Table 3.1.

Table 3.1. List of marker symbols

ArrowDown	Ibeam	TriangleLeft	HomeDownFilled
Asterisk	Plus	TriangleRight	SquareFilled
Circle	Square	Union	StarFilled
Diamond	Star	X	TriangleFilled
GreaterThan	Tack	Y	TriangleDownFilled
LessThan	Tilde	Z	TriangleLeftFilled
Hash	Triangle	CircleFilled	TriangleRightFilled
HomeDown	TriangleDown	DiamondFilled	

For non-grouped data (i.e., when `group=` option is not specified), the attributes of symbols used may be modified using the `markerattrs=` option in the `scatter` statement; `markerattrs=` (`color=`blue `size=`2 mm `symbol=`CircleFilled) is an example. For grouped data, attributes used for representing different groups cycle through a sequence of default colors and marker symbols depending on the style in effect. These default values may be changed using `proc template` to modify the style to alter the settings for each of the attributes the user wants to be different for representing each of the groups in the graph. This procedure will be discussed in detail in Appendix A.

Table 3.2. Line patterns

Pattern	Number
Solid	1
ShortDash	2
MediumDash	4
LongDash	5
MediumDashShortDash	8
DashDashDot	14
DashDotDot	15
Dash	20
LongDashShortDash	26
Dot	34
ThinDot	35
ShortDashDot	41
MediumDashDotDot	42

For more details, see the description of the `scatter` statement under SGPLOT procedure. The `ellipse` statement may be used along with the `scatter` statement to create bivariate confidence or prediction regions for a specified level $100(1-\alpha)\%$. These intervals are calculated under the assumption that pairs of data values (x, y) have a bivariate normal distribution. A *95% prediction ellipse* is generated by default. By definition, the probability of a point falling within this region is 0.95. On the other hand, one would have 95% confidence that the bivariate mean of the distribution lies within a *95% confidence ellipse*. It is possible to draw multiple ellipses by including more than one ellipse statement specifying different types or alpha levels.

Some ELLIPSE Statement Options

alpha= specifies the confidence level for the ellipse.

clip the ellipse will be clipped because the axes are determined without the ellipse.

legendlabel="text-string" specifies a label that identifies the ellipse in the legend.

lineattrs= specifies appearance of plotted lines (as a style element or using sub-options `color=`, `pattern=`, `thickness=`).

name= specifies a name for the plot.

outline|nooutline specifies whether the outlines for the ellipse are visible.

type= mean | predicted specifies the type of ellipse. `mean` specifies a confidence ellipse for the population mean. `predicted` specifies a prediction ellipse for a new observation. Default is `predicted`.

Default legends for each ellipse, such as "95% Prediction Ellipse," are automatically generated; however, the user may specify one using the `legendlabel=` option. The line attributes for the outline of the region may be specified using a style element or by using sub-options. By default, the values of the sub-options are set by the specific style element of the current style in operation. To change these settings, use sub-options `color=`, `pattern=`, and `thickness=`. Set the `color` sub-option as described under the options for the `scatter` statement, set `pattern=` using line types from Table 3.2 (default is set by the current style), and specify the thickness using units of measurement in Table 3.3.

Table 3.3. Units of measurement

Unit	Description
cm	Centimeters
in	Inches
mm	Millimeters
pct or %	Percentage
pt	Point size, calculated at 72 dpi
px	Pixels

SAS Example C5

This example (see SAS program in Fig. 3.10) illustrates the `scatter` statement used in the SAS statistical graphic procedure SGPLOT. In the first `proc sgplot` step, the `scatter` statement produces a simple scatter plot **1**; the marker attributes are style elements of the current style in operation (e.g., plot symbols are empty blue circles) by default. The `ellipse` statement is used to overlay a 95% prediction curve **2**, which is drawn in red as specified in the `lineattrs=` option. In the next `proc sgplot` step, the option `group=Sex` identifies the points in the scatter plot by the two gender groups. Thus the group variable is necessarily a classification or a category variable. The `keylegend=` statement is used to position the default legend inside the plot region **3**. Figures 3.11, 3.12, and 3.13 show the results of these three scatter statements.

```
libname libc "C:\users\user_name\Documents\...\Class\";

title "Plot of Height vs. Weight with Prediction Ellipse ";
proc sgplot data=libc.biology;
  scatter x=Height y=Weight; 1
  ellipse x=Height y=Weight/lineattrs=(color=red); 2
run;

title "Plot of Height vs. Weight with Age as Data Labels";
proc sgplot data=libc.biology;
 scatter x=Height y=Weight/datalabel=Age
        markerattrs= (color=magenta size=2 mm symbol=Asterisk);
run;

title "Height vs. Weight grouped by Gender";
proc sgplot data=libc.biology;
  scatter x=Height y=Weight/group=Sex;
  keylegend / location=inside position=bottomright; 3
run;
```

Fig. 3.10. Introducing ODS Statistical Graphics procedures: SGPLOT

The `histogram` statement in `proc sgplot` produces a histogram of a continuous variable determining number of bins (classes) and binwidths (class interval widths) automatically. Options available for use with the statement allow the user to specify values for the start of the first bin, number of bins, and the binwidth. The vertical axis represents the frequency of observations that fall into each of the bins that could be displayed as a count or as a percentage or proportion of the total number of observations. Percentage frequency is plotted on the vertical axis by default; the `scale=` option can be used to change the default setting of *percent.*

Some HISTOGRAM Statement Options

binstart= specifies the X coordinate of the first bin.

binwidth= specifies the bin width.

boundary=lower|upper specifies how boundary values are assigned to bins (default=upper).

fill|nofill specifies whether the area fill is visible.

fillattrs= specifies appearance of the area fill (as a style element or using sub-option `color=`).

legendlabel="text-string" specifies a label that identifies the histogram in the legend.

name="text-string" specifies a name for the plot.

nbins= specifies the number of bins.

outline|nooutline specifies whether the outlines of the bars are displayed.

scale=count|percent|proportion specifies the scale of the vertical axis. Default is percent

The `density` statement in `proc sgplot` allows the user to overlay a density plot fitted to the data. One type of a density plot is obtained by fitting a normal distribution to the data. The user has the option of specifying val-

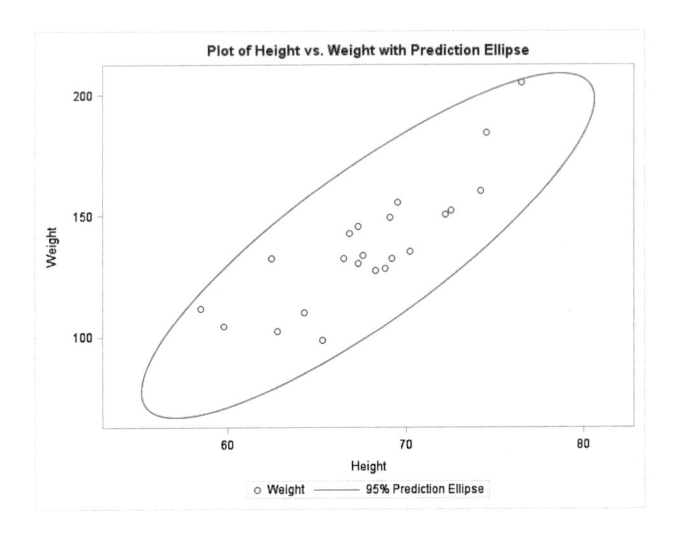

Fig. 3.11. Output from SGPLOT: SCATTER statement

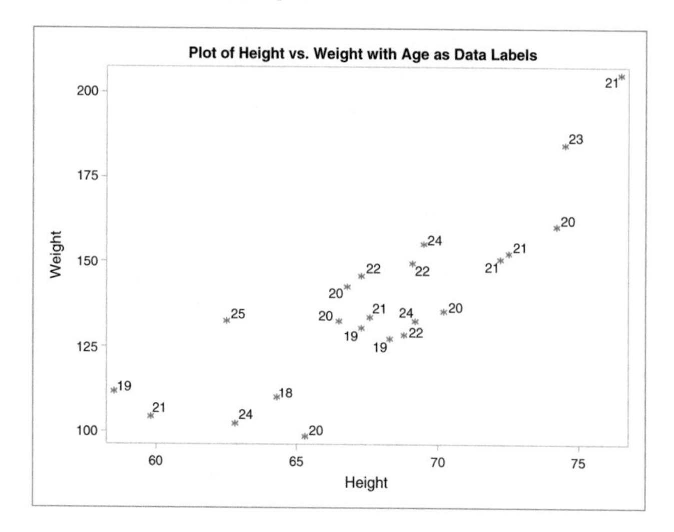

Fig. 3.12. Output from SGPLOT: SCATTER statement

ues for the two parameters μ and/or σ using sub-options `mu=` and `sigma=` (see *normal-opts* in the statement description). The `mu=` and `sigma=` specify values for the parameters μ and σ of the normal distribution. If values for either parameter are not specified, they will be estimated using the data.

The other type of density plot available is a nonparametric kernel density estimate. The user may specify a standardized bandwidth `c=` and a kernel function `weight=` (see *kernel-opts* in the statement description). The standardized bandwidth is a value between 0 and 100 and controls the level of

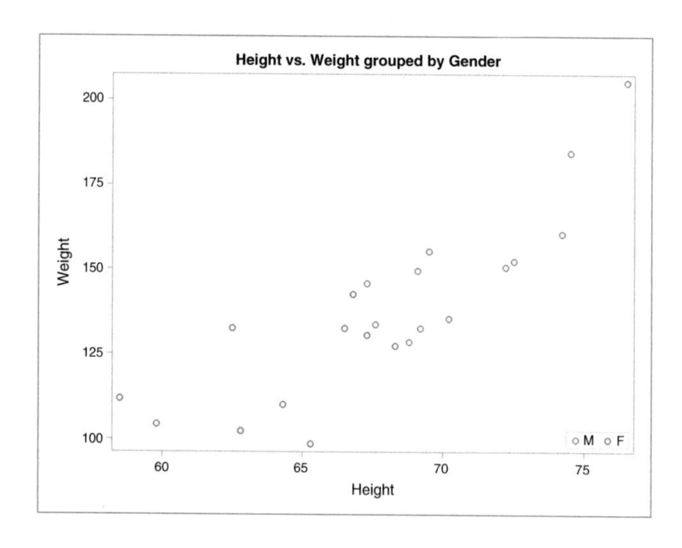

Fig. 3.13. Output from SGPLOT: SCATTER statement

smoothing: too small a value will show little smoothing showing spikes by attempting to fit every detail or a too large a value will perform oversmoothing hiding most of the structure in the data. An optimal bandwidth is one that results in a density estimate that is close to the true density. The `weight=` sub-option accepts three possible kernel functions normal, quadratic, or triangular as its value, with normal being the default.

Some DENSITY Statement Options

legendlabel="text-string" specifies a label that identifies the ellipse in the legend.

lineattrs= specifies appearance of plotted lines (as a style element or using sub-options `color=`,`pattern=`,`thickness=`.

name="text-string" specifies a name for the plot.

scale=count|density|percent|proportion specifies the scale of the vertical axis. Default is density

type= normal < (*normal-opts*)> | kernel < (*kernel-opts*)> specifies the type of distribution curve that is used for the density plot. Default is normal.

SAS Example C6

In an experiment to investigate whether the addition of an antibiotic to the diet of chicks promotes growth over a standard diet described in Ott and Longnecker (2001), an animal scientist rears and feeds 100 chicks in the same environment, with individual feeders for each chick. The weight gain for the 100 chicks are recorded after an 8-week period. The construction of a frequency table suggests class intervals of width 0.1 units beginning with a midpoint at the smallest data value of 3.6.

The SAS Example C6 program, shown in Fig. 3.14, uses the `histogram` statement (with no options specified) in the ODS Statistical Graphics procedure `proc sgplot` to construct a histogram. The first `density` statement specifies that a normal density curve be superimposed on the histogram **1**. By default, the procedure fits a normal density to the data using the sample mean and sample standard deviation estimated from the data. The second density statement will superimpose a kernel density plot using default values for the bandwitdth and kernel (weight) function **2**. The output is shown in Fig. 3.15.

It is possible to experiment with several options available with the `histogram` statement to control the appearance of the histogram; specifically, the options `binstart=`, `nbins=`, and `binwidth=` are useful for this purpose. In SAS Example C6 program the statements

```
data chicks;
input Wtgain @@;
label Wtgain ='Weight gain (in gms) after 8-weeks';
datalines;
3.7 4.2 4.4 4.4 4.3 4.2 4.4 4.8 4.9 4.4
4.2 3.8 4.2 4.4 4.6 3.9 4.1 4.5 4.8 3.9
4.7 4.2 4.2 4.8 4.5 3.6 4.1 4.3 3.9 4.2
4.0 4.2 4.0 4.5 4.4 4.1 4.0 4.0 3.8 4.6
4.9 3.8 4.3 4.3 3.9 3.8 4.7 3.9 4.0 4.2
4.3 4.7 4.1 4.0 4.6 4.4 4.6 4.4 4.9 4.4
4.0 3.9 4.5 4.3 3.8 4.1 4.3 4.2 4.5 4.4
4.2 4.7 3.8 4.5 4.0 4.2 4.1 4.0 4.7 4.1
4.7 4.1 4.8 4.1 4.3 4.7 4.2 4.1 4.4 4.8
4.1 4.9 4.3 4.4 4.4 4.3 4.6 4.5 4.6 4.0
;

proc sgplot data=chicks;
  title "Weight Gain Distribution";
  histogram Wtgain;
  density Wtgain; 1
  density Wtgain / type=kernel; 2
  keylegend / location=inside position=topright;
run;
```

Fig. 3.14. Histogram with SGPLOT procedure

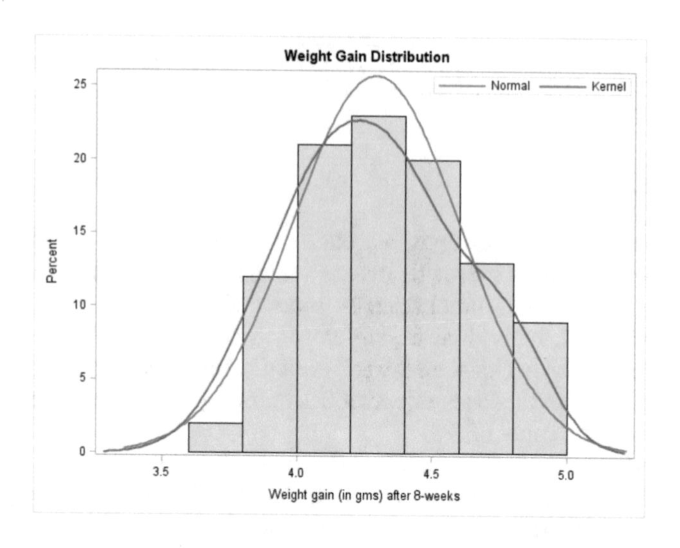

Fig. 3.15. Output from SAS Example C6

```
histogram Wtgain/binstart=3.5 nbins=7;
histogram Wtgain/binstart=3.75 binwidth=.25;
```

produce histograms (not shown) that are slightly different from the default one shown in Fig. 3.15.

Some VBOX Statement Options

boxwidth= specifies the width of the box, as a value between 0.0 (0% of the available width) and 1.0 (100% of the available width). Default is 0.4.

category= specifies the category variable for the plot. A box plot is created for each distinct value of the category variable.

connect=mean|median|q1|q3|min|max specifies that a connect line joins a statistic from box to box.

connectattrs= specifies appearance of the connecting lines (as a style element or using sub-options `color=`, `pattern=`, and `thickness=`).

datalabel <= variable> adds data labels for the outlier markers. If you specified a variable, then the values for that variable are used for the data labels. If you did not specify a variable, then the values of the analysis variable are used.

datalabelattrs= specifies appearance of labels (as a style element or using sub-options `color=`, `family=` "*font-family*", `size=`, `style=italic|normal`, `weight=bold|normal`) .

fill|nofill specifies whether the boxes are filled with color.

fillattrs= specifies appearance of the fill of the boxes (as a style element or using sub-option `color=`.

group= specifies a variable that is used to group the data.

legendlabel="text-string" specifies a label that identifies the ellipse in the legend.

lineattrs= specifies appearance of the box outlines (as a style element or using sub-options `color=`, `pattern=`, and `thickness=`).

meanattrs= specifies appearance of the box outlines (as a style element or using sub-options `color=`, `pattern=`, and `thickness=`).

medianattrs= specifies appearance of the box outlines (as a style element or using sub-options `color=`, `pattern=`, and `thickness=`).

name= specifies a name for the plot.

outlierattrs= specifies appearance of the box outlines (as a style element or using sub-options `color=`, `pattern=`, and `thickness=`).

outline|nooutline specifies whether the outlines for the bars are displayed.

percentile=1|2|3|4|5 specifies a method for computing the percentiles for the plot. Default is 5.

type= mean | predicted specifies the type of ellipse. `mean` specifies a confidence ellipse for the population mean. `predicted` specifies a prediction ellipse for a new observation. Default is `predicted`.

whiskerattrs= specifies appearance of the box outlines (as a style element or using sub-options `color=`, `pattern=`, and `thickness=`).

SAS Example C7

The data in Table B.6 taken from Koopmans (1987) give hydrocarbon (HC) emissions at idling speed, in parts per million (ppm), for automobiles of various years of manufacture. The data were extracted via random sampling from that of an extensive study of the pollution control existing in automobiles in current service in Albuquerque, New Mexico.

```
data emissions;
input period @;
hc=1;
do until (hc<0); [1]
      input hc @;
      if (hc<0) then return;
      output;
          end;
label period = 'Year of Manufacture'
      hc = 'Hydrocarbon Emissions (ppm)';
datalines;
1 2351 1293 541 1058 411 570 800 630 905 347 -1 [2]
2 620 940 350 700 1150 2000 823 1058 423 270 900 405 780 -1
3 1088 388 111 558 294 211 460 470 353 71 241 2999 199 188 353 117 -1
4 141 359 247 940 882 494 306 200 100 300 190 140 880 200 223 188 435 940 241 223 -1
5 140 160 20 20 223 60 20 95 360 70 220 400 217 58 235 1880 200 175 85 -1
;

proc format; [3]
 value pp 1='Pre-1963'
          2='1963-1967'
          3='1968-1969'
          4='1970-1971'
          5='1972-1974';
run;

proc sgplot data=emissions;
  title "Hydrocarbon Emissions Distribution by Period";
  vbox hc / category=period datalabel; [4]
  format period pp.;
run;
```

Fig. 3.16. Histogram with SGPLOT procedure

The SAS Example C7 program, shown in Fig. 3.16, uses the statistical graphics procedure SGPLOT to construct side-by-side vertical box plots for the five periods. Note that since the number of data values for each period is different, a new technique is employed to input the data. An artificially created data value for the variable `hc` of -1 is first inserted at the end of the set of data values for each period (note that this is not required at the end of every data line) **2**. A `do until` loop is used to read the data values for `hc` until a -1 is encountered, holding the data line after reading each value and using an `output;` statement to write an observation with a pair of values for `period` and `hc`. The loop is restarted after the processing of all hc values for a period is completed. A new value for the variable `period` is read after completion of each `do until` loop **1**.

A *schematic box plot* is the default style of the box plots produced by the `vbox` statement, where the whiskers are drawn to the *lower and upper adjacent values* from the edges of the box. Lower and upper adjacent values are, respectively, the smallest observed value inside the lower fence and the largest observed value inside the upper fence. Options to specify other styles of box plots and various other modifications are available. A box plot is created for each value of the category variable specified in the `category=` option in the vbox statement. A `datalabel` option is used to label the symbols plotted. This causes the observations plotted outside the upper and lower fences to be identified using the values of the variable `hc` **3**. A format statement specifies a user-defined format (used in the `proc format` step **4**) for labeling the values of the variable `period`.

The output shown in Fig. 3.17 is used in the following to compare and comment on features such as shape, location, dispersion (spread), and outliers, if any, of the distributions of HC emission in the five periods.

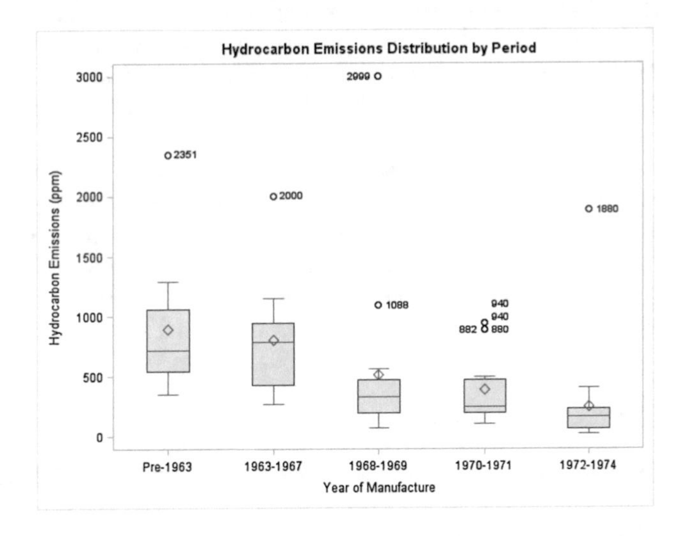

Fig. 3.17. Output from SAS Example C7

Side-by-side box plots are one of the most useful methods available for visually comparing features of sample distributions. They enable the comparison of the location, spread, and shape of the distributions by examining the relative positions of the median and the mean, the heights of the boxes which measure the interquartile ranges (IQRs), and the relative placing of the medians (and the means) between the ends of the boxes (i.e., the quartiles), the relative lengths of the whiskers, and the presence of outside values at either end of the whiskers. By observing the presence of trends in these characteristics, experimenters will be able to compare distributions across populations defined over time, location, treatments, or predefined experimental or observational groups. As an example, some observations regarding the empirical distributions of hydrocarbons over the five periods of study that may be made using Fig. 3.17 and useful conclusions that may be drawn from these observations are itemized as follows:

a. There is a decreasing trend in the magnitudes of location and spread of HC levels over years of manufacture except in the first two periods. The location, as measured by the median, and spread, as measured by the IQR and the lengths of the whiskers, appear to follow the same pattern. During the first two periods, there appears to be no significant shift of both location and spread.
b. The observed shapes of the distribution of HC levels have some similarity over the five periods. For example, the sample mean is larger than the median in all five samples and the upper (or right) whisker is longer in all but one. There is also at least one outside value on the upper side. Thus, all of the evidence indicates right-skewed distributions for all five periods.
c. There is an apparent change in HC emission coincident with the establishment of federal emission control standards in 1967–1968 (period 3). This is clearly observed as both the location and spread of the data decrease significantly following period 2.
d. Statistical analysis of data (say, using the one-way ANOVA model) may not be straightforward because assumptions such as normality and homogeneity of variances across periods may not hold. The graphical analysis shows that these two assumptions may not be plausible for this data; particularly, there is an obvious heterogeneity of variances as observed by the differences in the spread of the data as measured by the heights of the boxes.

Some VLINE Statement Options

`alpha=` specifies the confidence level for the confidence limits. Default is 0.05.

`categoryorder=respasc|respdesc` specifies the order in which the response values are arranged. By default, the plot is sorted in ascending order of the category values.

`clusterwidth=` specifies the cluster width as a fraction of the midpoint spacing. Default is 0.8

curvelabel <="text-string"> adds a label to the line plot.

curvelabelattrs= specifies the appearance of the labels in the plot when you use the CURVELABEL= option (as a style element or using sub-options `color=`, `family=` "*font-family*", `size=`, `style=italic|normal`, and `weight=bold|normal`).

curvelabelloc=outside|inside specifies whether the curve label is placed inside the plot axes (INSIDE) or outside of the plot axes (OUTSIDE), which is the default.

curvelabelpos=auto|end|max|min|start specifies the location of the curve label. Default is end.

datalabel= uses the values of a variable to label the points.

datalabelattrs= specifies appearance of labels (as a style element or using sub-options).

datalabelpos=data|bottom|top specifies the location of the data label. Default is data.

group= specifies a category variable to divide the values into groups.

groupdisplay=cluster|overlay specifies how to display grouped bars. Default is overlay.

grouporder=data|reversedata|ascending|descending specifies the ordering of lines within a group. Default is ascending.

legendlabel="text-string" specifies a label that identifies the line plot in the legend.

lineattrs= specifies the appearance of the lines in the line plot (as a style element or using sub-options).

markers adds markers to the plot.

markerattrs= specifies the appearance of the markers in the plot (as a style element or using sub-options).

missing a missing value is taken as a valid category and creates a bar for it.

name="text-string" specifies a name for the plot.

nostatlabel removes the statistic name from the axis and legend labels.

response= specifies a numeric response variable for the plot.

stat=freq|mean|median|percent|sum specifies the statistic for the vertical axis. Default is sum when `response=` is used; else default is freq.

x2axis assigns the category variable to the secondary (top) horizontal axis.

y2axis assigns the response variable to the secondary (right) vertical axis.

SAS Example C8

In profile plots of means or interaction plots, the levels of one experimental factor (say, A) are plotted on the horizontal axis. The vertical axis represents the sample means of responses resulting from the application of combinations of the levels of A and the levels of a second factor (say, B) to experimental units, independently. This plot is sometimes called the *profile plot of means*, because the sample means corresponding to the same levels of factor B are joined by line segments, thus showing the pattern of variation of the average response across the levels of factor A at each level of B.

Table 3.4. Mice survival time data (Box et al. 1978)

Poison	Drug			
	A	B	C	D
I	0.31	0.82	0.43	0.45
	0.45	1.10	0.45	0.71
	0.46	0.88	0.63	0.66
	0.43	0.72	0.76	0.62
II	0.36	0.92	0.44	0.56
	0.29	0.61	0.35	1.02
	0.40	0.49	0.31	0.71
	0.23	1.24	0.40	0.38
III	0.22	0.30	0.23	0.30
	0.21	0.37	0.25	0.36
	0.18	0.38	0.24	0.31
	0.23	0.29	0.22	0.33

This plot is also known as the *interaction plot* because it is useful for interpreting significant interaction that may be present between the two factors A and B and may help in explaining the basis for such interaction. This type of a plot is also useful in *profile analysis* of independent samples in multivariate data analysis, where means of several variables, such as responses to test scores measured on different subjects, are compared across independent groups of subjects such as classrooms.

Consider the survival times of groups of four mice randomly allocated to each combination of three poisons and four drugs shown in Table 3.4. The experiment described in Box et al. (1978) was an investigation to compare several antitoxins to combat the effects of certain toxic agents. The SAS Example C8 program shown in Fig. 3.18 draws an interaction plot of this data. The data are entered so that each line of data includes responses to the four

Table 3.5. Data arranged for input to `proc means`

Poison	Drug	Time
1	1	0.31
1	2	0.82
1	3	0.43
1	4	0.45
1	1	0.45
1	2	1.10
⋮	⋮	⋮
3	3	0.24
3	4	0.31
3	1	0.23
3	2	0.29
3	3	0.22
3	4	0.33

levels of `Drug` for each level of `Poison`. Four data lines represent the replicates for each level of `poison` so that there are 12 data lines in the input stream. The `trailing @` symbol in the `input` statements and an `output` in a `do` loop **1** enable a SAS data set to be prepared in the form required to be analyzed by procedures such as `means` or `glm`. As shown in Table 3.5, the data set named `survival` will have 48 observations, each corresponding to a response to a combination of a level of `Drug` and a level of `Poison`.

As demonstrated in SAS Example C8, the `vline` statement in SGPLOT is ideal for constructing an interaction plot. The cell means need not be calculated in advance as they are automatically computed using the `stat=` option. The `vline` statement is used when the *vertical* axis is used to represent the response variable (or a statistic computed from the response). In this example, using the `stat=mean` **2** option, means of the response variable `Time` are calculated for each value of the category variable `Poison` which is plotted on the horizontal axis. However, since the category variable `Drug` is used as a grouping variable (see the `group=` option **2**), the means are computed and

```
data survival;
input  Poison 1. @; 1
   do Drug=1 to 4;
      input Time 3.2 @;
      output;
   end;
datalines;
1 31 82 43 45
1 45110 45 71
1 46 88 63 66
1 43 72 76 62
2 36 92 44 56
2 29 61 35102
2 40 49 31 71
2 23124 40 38
3 22 30 23 30
3 21 37 25 36
3 18 38 24 31
3 23 29 22 33
;
title1 "Analysis of Survival Time data";
title2 "Interaction Plot of Cell Means";

proc sgplot data=survival;
vline Poison/response=Time stat=mean group=Drug markers; 2
run;
```

Fig. 3.18. SAS Example C8: interaction plot with SGPLOT procedure

plotted separately for each value of `Drug`. These line plots representing each unique value of `Drug` variable are automatically identified visually using different colors for both the plot symbols and line segments selected by default (see Fig. 3.19). The statistical interpretation of this interaction plot will be presented in Chap. 5.

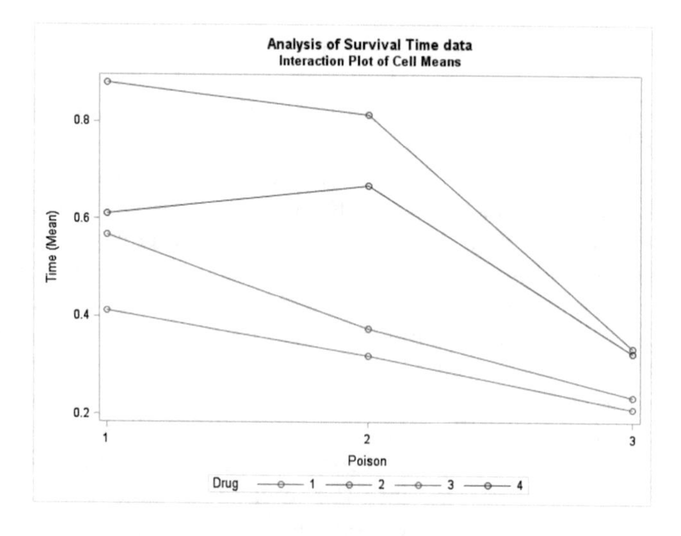

Fig. 3.19. Output from SAS Example C8

3.3.2 The SGPANEL Procedure

Essentially, the SGPANEL procedure creates the same plots as the SGPLOT procedure but they appear in separate panels. The `panelby` statement names the category variable that classifies the data used for plots that appear in different panels. Other than the `panelby` statement (that is required), most of the statements are common to both procedures. In SGPANEL, the `colaxis` and the `rowaxis` statements replace the `xaxis` and the `yaxis` statements used in SGPLOT, respectively.

```
PROC SGPANEL < option(s)>;
  PANELBY variable(s) </option(s)>;
  COLAXIS <option(s)>;
  DENSITY response-variable </option(s)>;
  DOT category-variable </option(s)>;
  HBAR category-variable </option(s)>
  HBOX response-variable </option(s)>;
  HISTOGRAM response-variable </option(s)>
  HLINE category-variable </option(s)>
  KEYLEGEND <"name(s)"> </option(s)>;
  HLINE variable </option(s)>;
  LOESS X= numeric-variable Y= numeric-variable </option(s)>;
  NEEDLE X= variable Y= numeric-variable </option(s)>;
  PBSPLINE X= numeric-variable Y= numeric-variable </option(s)>;
  REFLINE value(s) </option(s)>;
  REG X= numeric-variable Y= numeric-variable </option(s)>;
  ROWAXIS <option(s)>;
  SCATTER X= variable Y= variable </option(s)>;
  SERIES X= variable Y= variable </option(s)>;
  STEP X= variable Y= variable </option(s)>;
  VBAR category-variable </option(s)>
  VBOX response-variable </option(s)>;
  VLINE category-variable </option(s);
```

Some PANELBY Statement Options

border|noborder specifies whether borders are to be drawn around each cell in the panel display. Default depends on style in effect.

colheaderpos=top|bottom|both specifies the location of the column headings in the panel. Default is top.

columns=n specifies the number of columns in the panel.

layout=lattice|panel|columnlattice|rowlattice specifies the type of layout that is used for the panel. Default is panel which arranges cells in rows and columns.

missing a missing value is taken as a valid category and creates a cell for it.

novarname removes the variable names from the cell headings of a panel layout, or from the row and column headings of a lattice layout.

onepanel places the entire panel in a single output image.

rowheaderpos=left|right|both specifies the location of the row headings in the panel. Default is right.

rows=*n* specifies the number of rows in the panel.

spacing=*n* specifies the number of pixels between the rows and columns in the panel.

sparse creates empty cells in the panel for combinations of classifications with no observations.

start=topleft|bottomleft specifies whether the first cell in the panel is placed at the upper left or the lower left corner.

uniscale=column|row|all scales the shared axes in the panel to be identical.

SAS Example C9

As in several previous SAS programs, the SAS Example C9 shown in Fig. 3.20 accesses the SAS data set named `fueldat` from a library (a folder, in this case) using the two-level name `mylib.fueldat`. The program consists of two proc steps, illustrating the use of the `hbar` and `vbar` statements, respectively, to obtain bar charts that present the same information in two different ways: first, the menu sequence `Tools`→ `Options`→ `Preferences...` and then using the `Results` tab to change the ODS `style` from the default of *HTMLBlue* to *Science*, before execution of the program in the SAS windowing environment. Switching to a new ODS style changes the overall appearance of the graphs and will emphasize the difference in the two plots. This could also be accomplished using the `style=` option in the ODS destination statement (e.g., `style=`science).

The `proc sgplot` step contains a `hbar` statement **1** that produces a horizontal bar chart of the means of the `Roads` variable (1971 miles of highway, in thousands) for each income group as defined earlier. The `group=` option is used with the `hbar` statement so that each bar is subdivided into two fuel tax groups Low and High. The `keylegend` statement controls the positioning of the legend within the graph. The resulting graph is shown in Fig. 3.21. The second step uses `proc sgpanel` to present the same graphical analysis as a vertical barchart, the difference being instead of grouping by the fuel tax category variable, a **panelby** statement displays the average fuel usage for each income group in separate panels **2**. The resulting graph is shown in Fig. 3.22.

```
libname mylib "C:\users\user_name\Documents\...\Class\";

proc format;
   value ing 1 = 'Low Income'
             2 = 'Middle Income'
             3 = 'High Income';
run;

title 'Horizontal Bar Chart of Miles of Primary Highways';
proc sgplot data=mylib.fueldat;
  hbar Incomgrp/response=Roads stat=mean group=Taxgrp; 1
  keylegend / title="Fuel Tax" location=outside position=bottom;
  format Incomgrp  ing.;
run;

title 'Vertical Bar Chart Fuel Use by Income Group for each Fuel Tax Group';
proc sgpanel data=mylib.fueldat;
  panelby Taxgrp; 2
  vbar Incomgrp/response=Fuel stat=mean;
  format Incomgrp  ing.;
run;
```

Fig. 3.20. SAS Example C9: bar charts with SGPLOT and SGPANEL procedures

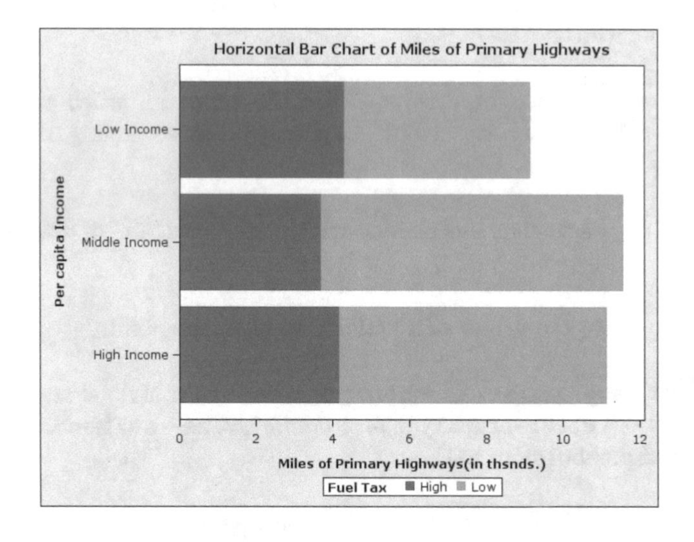

Fig. 3.21. Output from SAS Example C9: horizontal bar chart

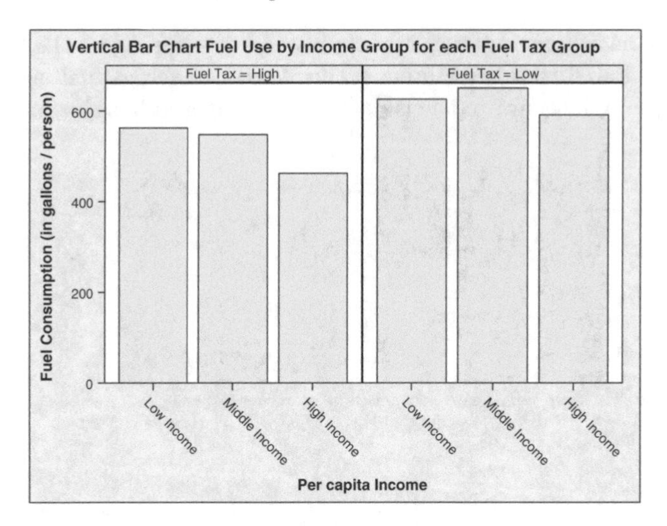

Fig. 3.22. Output from SAS Example C9: vertical bar charts in row lattices

Some VBAR Statement Options

alpha= specifies the confidence level for the confidence limits.

barwidth= specifies the width of the bars as a fraction of the maximum available width.

categoryorder=respasc|respdesc specifies the order in which the response values are arranged. By default, plot ordered by ascending order of category values.

clusterwidth= specifies the cluster width as a fraction of the maximum width.

datalabel= uses the values of a variable to label the points.

datalabelattrs= specifies appearance of labels (as a style element or using sub-options `color=`, `family=` "*font-family*", `size=`, `style=italic|normal`, and `weight=bold|normal`).

dataskin=none|crisp|gloss|matte|pressed|sheen specifies a special effect to be used on all filled bars. Default is none.

fill|nofill specifies whether the boxes are filled with color.

fillattrs= specifies appearance of the fill of the boxes (as a style element or using sub-option `color=`.

group= specifies a category variable to divide the values into groups.

groupdisplay=stack|cluster specifies how to display grouped bars. Default is stack.

grouporder=data|reversedata|ascending|descending specifies the ordering of bars within a group. Default is ascending.

legendlabel="text-string" specifies a label that identifies the bar chart in the legend.

missing a missing value is taken as a valid category and creates a bar for it.

name="text-string" specifies a name for the plot.

nostatlabel removes the statistic name from the axis and legend labels.

outline|nooutline specifies whether the outlines for the bars are displayed.

response= specifies a numeric response variable for the plot.

stat=freq|mean|median|percent|sum specifies the statistic for the vertical axis. Default is sum when `response=` is used; else default is freq

SAS Example C10

In this example the demographic data set on 60 countries (the data are displayed in Table B.5) is used to obtain histograms classified by a category variable created by grouping data in the data step. First, the SAS data set named `world` is created using raw data read from a text file instead of data entered instream (see Fig. 3.23). The same data set was used in an exercise in Chap. 2. Although there are many useful options available with the `infile` statement, for the purpose of reading a text file, all that is needed is to supply the path name of the data file as a quoted string:

```
infile  "C:\users\user_name\Documents\...\Class\demogr.txt" ;
```

An ordinal variable called `TechGrp` assigned the value 1, 2, 3, or 4 according to whether the Level of Technology variable had values that were less than 13, greater than or equal to 13 and less than 25, greater than or equal to 25 and less than 60, or over 60, respectively. These cutoff values were somewhat arbitrarily chosen to divide the 60 countries into four groups according to their level of technological advancement. A proc format step is used to define a numeric format `tg.` for printing values of this category variable.

```
data world;
infile  "C:\users\user_name\Documents\...\Class\demogr.txt";
input Country $20. Birthrate Deathrate InfMort LifeExp PopUrban
              PercGnp LevTech CivilLib;

if LevTech<13 then TechGrp=1;
else if 13<= LevTech<25 then TechGrp=2;
else if 25<= LevTech<60 then TechGrp=3;
else if LevTech>= 60 then TechGrp=4;

label LifeExp="Life Expectancy in yrs."
         TechGrp="Level of Technology";
run;
proc format;
     value tg 1='Low'
              2='Moderate'
              3='High'
              4='Advanced' ;
run;

proc sgpanel data=world;
  panelby TechGrp;
  histogram LifeExp/binstart=45 binwidth=5 nbins=8;
  format TechGrp tg.;
run;
```

Fig. 3.23. SAS Example C10: multiple histograms with SGPANEL procedure

A few of the statements available in the SGPANEL procedure will be illustrated via examples below. This procedure requires a `panelby` statement to specify one or more category variables that defines how the the panel is divided into cells of individual graphs. This statement must appear before any other statements that are required later to initiate plots. Options such as `rows=`, `columns=` , or `layout=` available with the `panelby` statement enable the user to control how the cells are organized within the panel. In SAS Example C10, the statement `panelby TechGrp;` produces a panel layout with four cells (corresponding to the four values of TechGrp) that are arranged automatically as a 2×2 matrix in the increasing order of the values of the category variable `TechGrp`.

The next statement in the step is the `histogram` statement to specify the graph that will appear in each cell. The options available with the histogram statement are the same as those described previously for `proc sgplot` (see page 160 for some histogram statement options). In this example, the options `binstart=`, `binwidth=`, and `nbins=` are used to specify exactly how each of the histograms is to be constructed for the `LifeExp` variable in the data set containing the life expectancies for each country. The graphical output from this SGPANEL step is shown in Fig. 3.24.

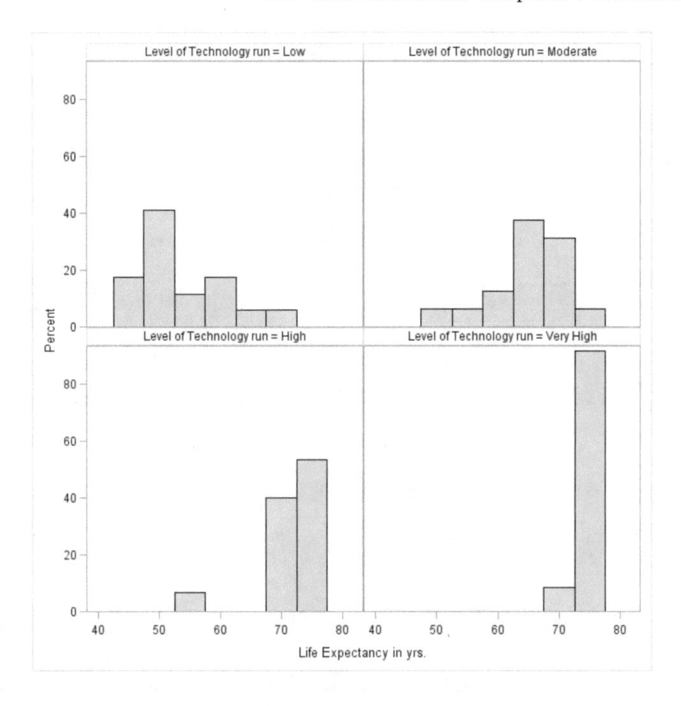

Fig. 3.24. Output from SAS Example C10: histograms in panel cells

Some DOT Statement Options

alpha= specifies the confidence level for the confidence limits.

categoryorder=respasc|respdesc specifies the order in which the response values are arranged. By default, plot ordered by ascending order of category values.

clusterwidth= specifies the width of the group clusters as a fraction of the midpoint spacing.

datalabel= uses the values of a variable to label the points.

datalabelattrs= specifies appearance of labels (as a style element or using sub-options `color=`, `family=` "*font-family*", `size=`, `style=italic|normal`, and `weight=bold|normal`) .

discreteoffset= numeric value that specifies an amount of offset of all dots.

group= specifies a category variable to divide the values into groups.

groupdisplay=cluster|overlay specifies how to display grouped dots. Default is overlay.

legendlabel="text-string" specifies a label that identifies the dot plot in the legend.

limitattrs= specifies appearance of the limit lines (as a style element or using sub-options `color=`, `pattern=`, and `thickness=`).

limits=upper|lower|both specifies which limit lines to display.

limitstat=clm|stddev|stderr specifies the statistic for the limit lines. Default is clm.

markerattrs= specifies appearance of markers in the plot (as a style element or using sub-options `color=`, `size=`, and `symbol=`).

missing a missing value is taken as a valid category and creates a cell for it.

name="text-string" specifies a name for the plot.

nostatlabel removes the statistic name from the axis and legend labels.

numstd=*n* specifies the number of standard units for the limit lines, when you specify `limitstat= stddev` or `limitstat=stderr.`

response= specifies a numeric response variable for the plot.

stat=freq|mean|median|percent|sum specifies the statistic for the vertical axis. Default is sum when `response=` is used; else default is freq.

SAS Example C11

When the values of a quantitative variable are identified by values of a nominal variable with a large range of values or a category variable with a large number of levels are required to be summarized graphically, the dot chart is an extremely useful tool. In these situations, not only are dot plots more compact than, say horizontal bar charts, they are easily modified if the category variable is divided into groups or when the quantitative variable is itself classified into several groups. A detailed early discussion of dot charts is found in Cleveland (1985). In this example (see SAS program in Fig. 3.25), the quantitative

```
libname mylib "C:\users\user_name\Documents\...\Class\";

proc sgpanel data=mylib.fueldat;
  title "Dot Plot of Fuel Use by Fuel Tax";
  panelby Taxgrp/layout=rowlattice;
  dot State/response=Fuel categoryorder=respasc;
  rowaxis valueattrs=(color=maroon size=5 px);
run;
```

Fig. 3.25. SAS Example C11: dot plots with SGPANEL procedure

variable `Fuel` from the fuel data set used earlier in Chap. 2 is used to display per capita fuel use by states. The use of the SGPANEL procedure here instead of SGPLOT allows the user to split the states according to the levels of the fuel tax category variable created earlier. Thus the dot plots (shown in Fig. 3.26) appear in two separate panels laid out in a single column (the *row lattice* structure is illustrated here in contrast to the panel arrangement).

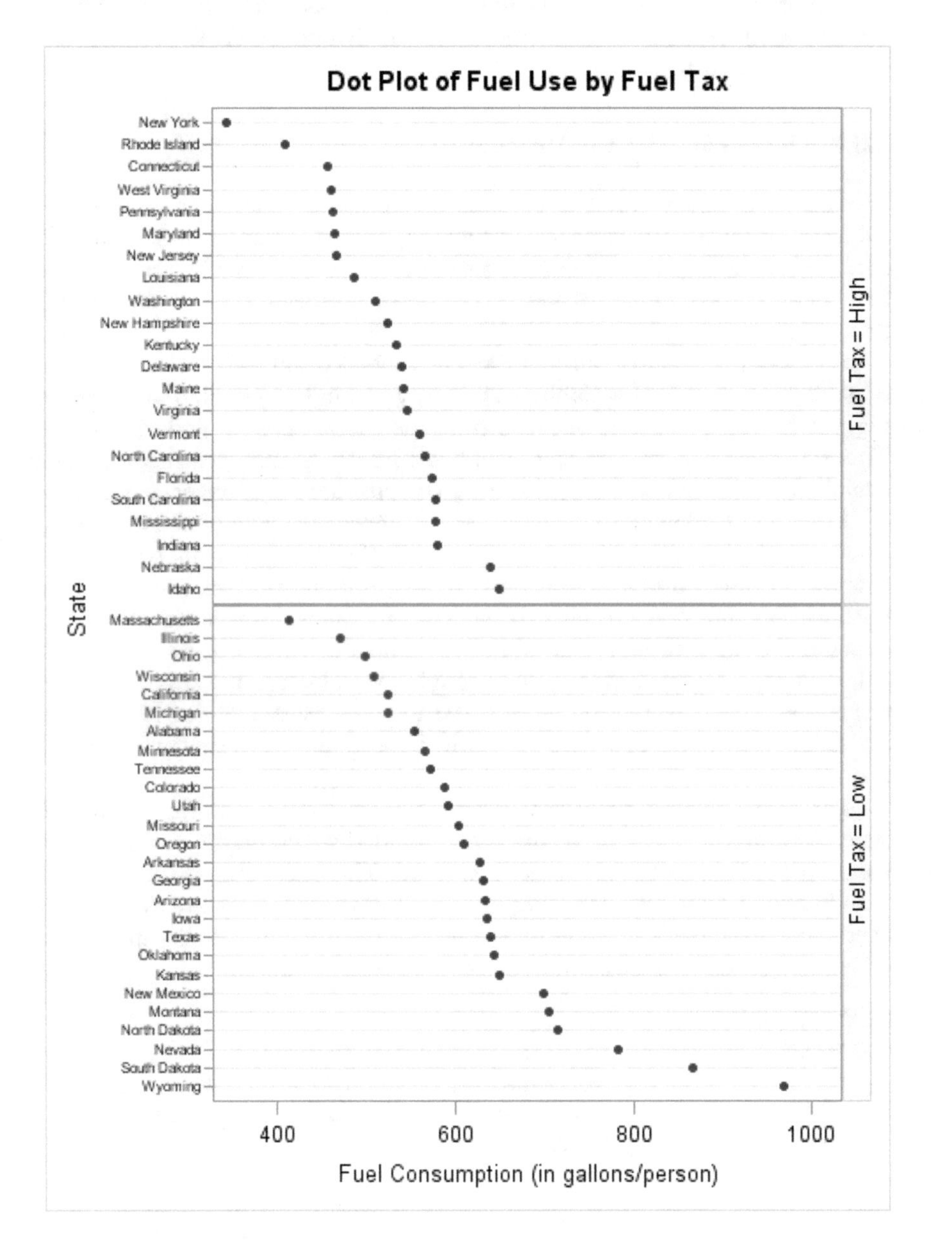

Fig. 3.26. Output from SAS Example C11: dot plots in panels

The categoryorder=respasc option is used to order the state values alphabetically in ascending order. However, the values of the category variable (State in this example) may not be printed at all the tick marks on the row axis (here the vertical axis) because the default size of the characters is too large for the space available for placing the tick marks. Since, for a dot plot to be useful, one needs to be able to identify all markers, the size of characters used in printing the value labels may be modified using the valueattrs= parameter available in a rowaxis statement. In Example C11, the size of value labels is changed to 5 pixels.

3.3.3 The SGSCATTER Procedure

When more than two variables are measured on observational or experimental subjects (or units), the relationships among several variables may need to be analyzed simultaneously. One possible approach is to obtain bivariate scatter plots of all pairs of variables and display them arranged in a two-dimensional array of plots.

For example, if three variables are present, six pairwise scatter plots are possible, since a pair of variables can be graphed in two possible ways (choosing one as the x variable and the other as the y variable and vice versa). These six plots can be displayed as a 3×3 matrix of plots, each row-column combination of the matrix representing the position for placement of a plot.

Although displaying only the lower (or upper) triangle of the matrix appears sufficient, most software programs display the entire matrix since it enables the user to observe patterns or associations that may exist among the variables that may not have dependency relationships that may exist, say, in regression applications. An additional numeric variable (usually an ordinal variable, category variable, or a grouping variable created from the values of another variable) may be represented in these scatter plots either by using different symbols or colors (or both) to examine whether clusters of the observations exist as may be defined by such categories.

The SGSCATTER procedure creates scatter plots for multiple combinations of variables that may be arranged as side-by-side panels or a matrix of panels depending on the plot statement used. Three plot statements available with SGSCATTER are plot, compare, and matrix. Options are available for overlaying fit plots and ellipses on the scatter plots, changing plot appearance including marker attributes, and controlling legends and labels:

```
PROC SGSCATTER <options>;
  COMPARE X=variable |(variable-1...variable-n)
          Y=variable |(variable-1 ...variable-n) </options>;
  MATRIX variable-1 variable-2 < ...variable-n > </options>;
  PLOT plot-request(s) </options>;
```

Some MATRIX Statement Options

datalabel= specifies a variable that is used to create data labels for each point in the plot.

datalabelpos= specifies the location of the data label with respect to the plot. Position can be one of the following values: bottom, bottomleft, bottomright, left, center, right, top, topleft, and topright.

ellipse <=(options)> adds a confidence or prediction ellipse to each cell that contains a scatter plot.

group= specifies a variable that is used to group the data.

legend=(options) specifies the appearance of the legend for the scatter plot.

markerattrs= specifies appearance of markers in the plot (as a style element or using sub-options `color=`, `size=`, and `symbol=`).

nolegend= removes the legend from the graph.

```
data world;
infile  "C:\users\user_name\Documents\...\Class\demogr.txt";
input Country $20. Birthrate Deathrate InfMort LifeExp PopUrban
              PercGnp LevTech CivilLib;
label Birthrate='Crude Birth Rate'
      Deathrate='Crude Death Rate'
      InfMort='Infant Mortality Rate'
      LifeExp='Life Expectancy in yrs.'
      PopUrban='Percent of Urban Population'
      PercGnp='Per capita GNP in U.S. dollars'
      LevTech='Level of Technology'
      CivilLib='Degree of Denial of Civil Liberties';

if LevTech<13 then TechGrp=1;
else if 13<= LevTech<25 then TechGrp=2;
else if 25<= LevTech<60 then TechGrp=3;
else if LevTech>= 60 then TechGrp=4;

label TechGrp='Level of Technology';
run;
proc format;
     value tg 1='Low'
              2='Moderate'
              3='High'
              4='Very High' ;
run;

proc sgscatter data=world;
  title "Scatterplot Matrix of Demographic Variables";
  matrix PopUrban PercGnp InfMort LifeExp
        /group=TechGrp;
  format TechGrp tg.;
run;
```

Fig. 3.27. SAS Example C12: scatter plot matrix of world demographic data

SAS Example C12

In a multiple regression situation, scatter plot matrices are especially useful for establishing types of relationship that individual independent variables (regressors, explanatory variables) may have with the dependent variable (response). This may help in the model building stage, by suggesting the form the variables may enter the model (e.g., by suggesting possible transformations, etc. that linearize relationships). The scatter plot matrix may also help in studying pairwise collinearities that may exist among the independent variables, information useful in model selection procedures.

The data set with demographic variables measured for 60 countries used in SAS Example C10 is used again in SAS Example C12 program shown in Fig. 3.27. Note that labels have been added to the program so that the graphical output is enhanced. Suppose that a scatter plot matrix of the four variables named `PopUrban`, `PercGnp`, `InfMort`, and `LifeExp` is required. With *proc sgscatter*, all that is required to accomplish this is to use a `matrix` statement, list the four variables as the required arguments, and include any other desired options following a backslash. In this example, the `group=TechGrp` option causes the data values to be grouped by the values of the category variable `TechGrp`. The scatter plot matrix is reproduced in Fig. 3.28. From this plot, it can be observed, as one might expect, that there is a strongly negative linear relationship between infant mortality rate and life expectancy. There also appears to be linear relationships between each of these variables with percent of urban population that, in each case, is not particularly strong. These two variables also appear to have different nonlinear relationships with Per Capita GNP.

Attribute Map Data Sets

It may be desirable to use different visual attributes of elements of a graph rather than the default settings when the `group=` option is used in a plot statement such as in the `matrix` statement in `proc sgscatter`, for example. The easiest way to accomplish this is by creating an attribute map data set. This data set will contain *variables* that correspond to each of the attributes for which one desires to assign new values and variables named `Id` and `Value` that contains the group variable name and its possible values, respectively.

Table 3.6. Attribute map for TechGrp

Id	Value	MarkerColor	MarkerSymbol	MarkerSize
TechGrp	Low	darkcyan	"CircleFilled"	8
TechGrp	Moderate	blueviolet	"CircleFilled"	8
TechGrp	High	crimson	"CircleFilled"	8
TechGrp	Very High	green	"CircleFilled"	8

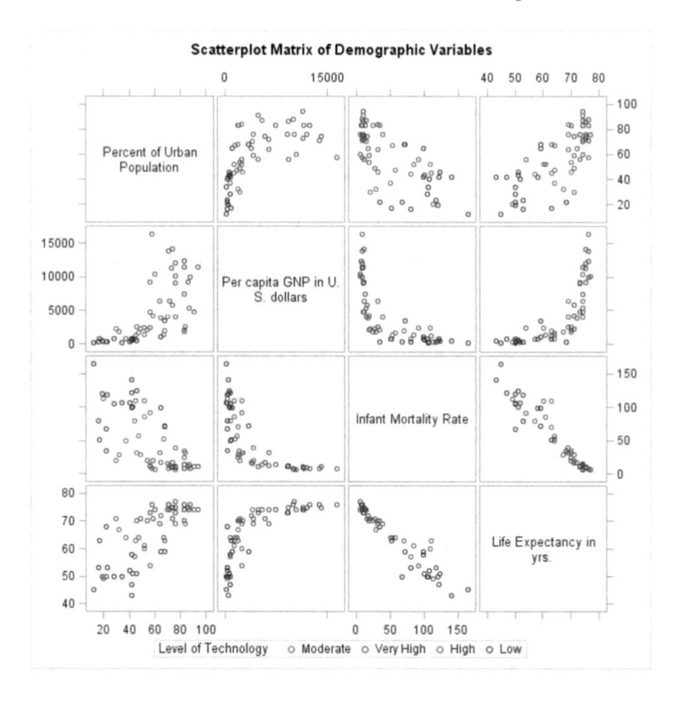

Fig. 3.28. Output from SAS Example C12: scatter plot matrix of world demographic data

The number of observations of this data set will be exactly the same as the number of possible values of the group variable. Suppose that in SAS Example C12, it is desired to change the plot symbols to filled circles, and the size of the symbols to 8 pixels, but use four different colors that correspond to the four different values of the `TechGrp` variable, as the color attribute of the plot symbols. The attribute map data set required to achieve this objective is shown in Table 3.6.

The names of the variables other than `Id` and `Value` are established *Style Attribute* names associated with ODS graphics, available from Style Attribute tables in the ODS User's Guide. Some common attributes include *Color*, *Font*, *FillColor*, *FillPattern*, *LineColor*, *LineStyle*, *LineThickness*, *MarkerSize*, *MarkerSymbol*, and *TextColor*. The attribute map data set shown in Table 3.6 specifies values for MarkerColor, MarkerSymbol, and MarkerSize attributes. In the SAS program shown in Fig. 3.29, the `dattrmap=` option in the proc sgscatter statement references the data set, named `mymap` and created in the data step shown that is designed to create the attribute map data set as shown in Table 3.6. The `attrid=` references the value of the Id variable, here `TechGrp`. Note carefully that as the value of the variable `TechGrp` changes, style attributes such as the marker color assumes different values, while style attributes such as the marker symbol stays unchanged.

The scatter plot matrix output from this program is not reproduced here. In addition to SGSCATTER, attribute maps may be used with procedures SGPLOT and SGPANEL. A discrete attribute map data set can contain more than one attribute map. This capability enables the user to apply different attribute maps to several group variables in a graph.

```
**** insert SAS code that creates the data set world as in SAS Example C12 *****

data mymap;
length MarkerColor MarkerSymbol $ 12;
input Id $ Value $9. MarkerColor $ ;
MarkerSymbol="CircleFilled";
MarkerSize=8;
datalines;
TechGrp Low      darkcyan
TechGrp Moderate blueviolet
TechGrp High     crimson
TechGrp Very High green
;

proc sgscatter data=world dattrmap=mymap;
  title "Scatterplot Matrix of Demographic Variables";
  matrix PopUrban PercGnp InfMort LifeExp
        /group=TechGrp attrid=TechGrp;
  format TechGrp tg.;
run;
```

Fig. 3.29. SAS Example C12: scatter plot matrix of world demographic data

3.4 ODS Graphics from Other SAS Procedures

In Sect. 3.1, it was stated that ODS Graphics output is produced automatically by many SAS procedures when statements that create these graphics are included in the proc step. This output is called ODS Graphics (to distinguish them from statistical graphics produced using traditional SAS/GRAPH procs or statements). For example, the HISTOGRAM statement used in a `proc univariate` step will produce ODS Graphics output in *HTMLBlue* style, and in SAS 9.4, this output can be automatically displayed in an internal viewer or directed to an ODS destination that accommodates graphics. ODS Graphics produces graphs in standard image file formats, and the consistent appearance and individual layout of these graphs are controlled by ODS *styles* and *templates*, respectively, as stated previously. Since the default templates for ODS Graphics are provided by SAS software, detailed setting of parameter values is not necessary to produce these graphs. These are produced as automatically as tables of statistical analysis output by these procedures. Among the commonly used SAS procedures that produce ODS Graphics are the Base SAS procedures CORR, FREQ, and UNIVARIATE and the SAS/STAT procedures such as REG, GLM, CLUSTER, CALIS, MIXED, LOGISTIC, PRINCOMP, TTEST, PHREG, MDS, and KDE among many others. The following two examples are used to illustrate ODS Graphics in this section. Several other examples of ODS Graphics will be found in other chapters where these procedures are used.

SAS Example C13

The Base SAS procedure UNIVARIATE is used to construct a histogram and a normal probability plot. The `histogram` and the `probplot` statements are used here in their simplest forms. All quantities required to make each of the plots are computed internally.

```
libname mylib "C:\users\user_name\Documents\...\Class\";

proc univariate data=mylib.fueldat noprint;
  var Income Fuel;
  histogram Income;
  probplot Fuel/normal(mu=est sigma=est) odstitle=title;
  title 'Use of Proc Univariate for Creating ODS Graphics:1';
run;
```

Fig. 3.30. SAS Example C13: program

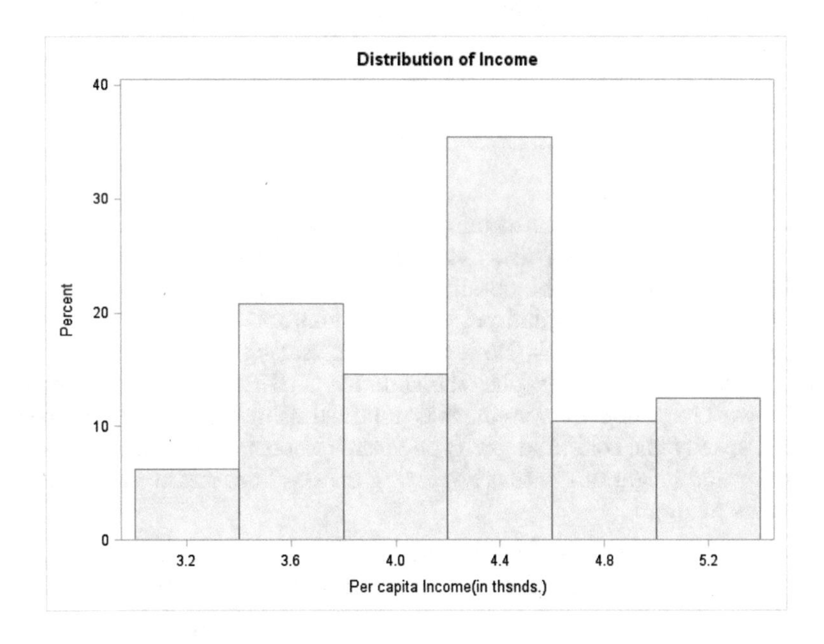

Fig. 3.31. SAS Example C13: histogram

SAS Example C13 shown in Fig. 3.30 accesses the SAS data set named `fueldat` from a library using the two-level name `mylib.fueldat`. As usual, the appearance of the graphical output from this program is controlled by default style elements for each graph under ODS Graphics (see Figs. 3.31 and 3.32). Note that the histogram has a title generated automatically under the default style; however, the text in the title statement supplied in the program is used as the title for the normal probability plot as a result of including the `odstitle=` option in the `probplot` statement.

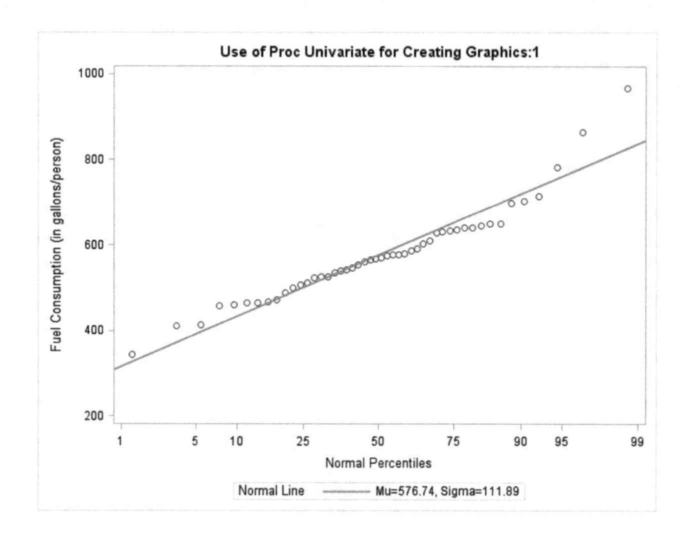

Fig. 3.32. SAS Example C13: normal probability plot

If ODS Graphics is turned off (either by using a ODS GRAPHICS OFF statement or interactively, using the `Preferences` dialog), a graph based on traditional graphics may be obtained. Traditional graphics are controlled by SAS/GRAPH statements and procedure options and are not discussed in the current edition of this book. However, for illustrating the differences between the appearance of these graphs, the code for SAS Example C13 is modified as follows. The `probplot` statement is modified using traditional graphics options to specify the color and line type of the reference line, color of axis lines, and color and height of the text appearing on axis lines and to add minor tick marks, as follows:

```
probplot Fuel/normal(mu=est sigma=est color=red l=2)
               caxis=blue ctext= red height= 2 vm=9 pctlminor;
```

In addition the symbol statement (as one may use with SAS/GRAPH)

```
symbol1 c=steelblue v=dot i=none;
```

may also be added to enhance the appearance of the plot symbols. The modified graph obtained via traditional SAS Graphics is shown in Fig. 3.33. It is immediately noticeable that this graph is different from Fig. 3.32 which was

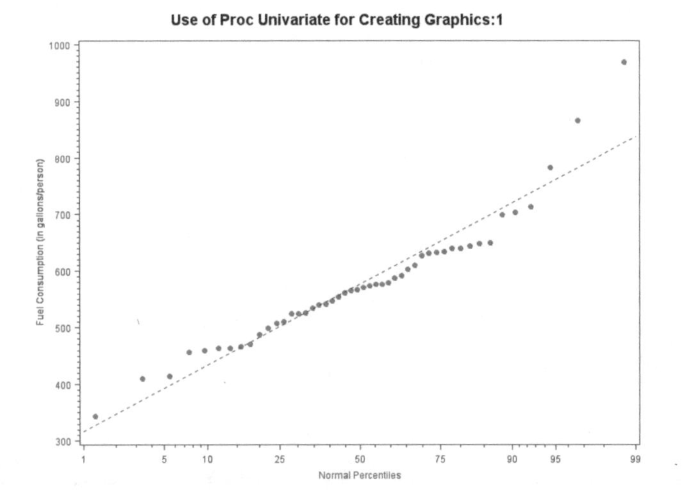

Fig. 3.33. SAS Example C13: normal probability plot (modified using traditional SAS Graphics options)

produced via ODS Graphics. The graph in Fig. 3.33 has the appearance of a graph produced by *traditional* SAS Graphics. The use of these options enables SAS to use traditional graphics options if ODS Graphics is turned off, and these options will be ignored if the program is executed with ODS Graphics in effect.

```
libname mylib "C:\users\user_name\Documents\...\Class\";

proc univariate data=mylib.fueldat noprint;
  var Fuel;
  class Incomgrp;
  histogram Fuel/nrows=3;
  title 'Use of Proc Univariate for Creating Graphics:2';
run;
```

Fig. 3.34. SAS Example C14: program

SAS Example C14

It would be useful if histograms for different levels of a category variable can be constructed and displayed in panels. The options in the `class` and `histogram` statements under the UNIVARIATE procedure enable the creation of histograms for different groups. Using the same SAS data set `fueldat` as in SAS Examples C9 and C13, SAS Example C14 displayed in Fig. 3.34 produces histograms of fuel consumption for three different income groups. The category variable `Incomgrp` is specified in the `class` statement. The three histograms are displayed in three single-cell plot panels when `nrows=3` is specified in the histogram statement. The graph in Fig. 3.35 shows the histograms produced by this proc univariate step.

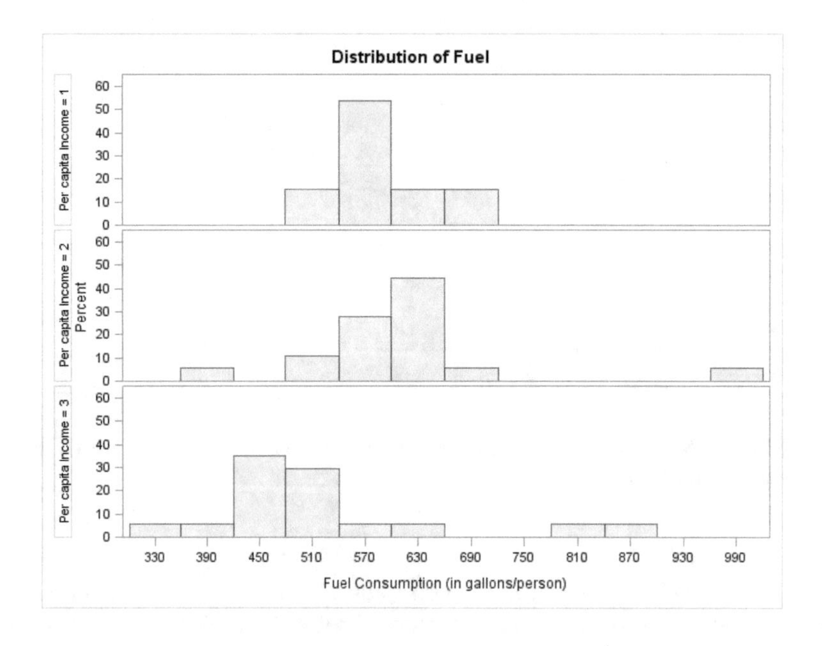

Fig. 3.35. SAS Example C14: histograms by income groups

SAS Example C15

A fitted normal density curve and computed statistics can be overlaid into the ODS Graphics produced by the UNIVARIATE procedure using the options under the histogram and inset statements. By using ODS SELECT statement, SAS Example C15 shown in Fig. 3.36 outputs specific objects into the ODS destination, including parameter estimates, goodness of fit test, quantiles of the fitted normal distribution, and histograms.

```
libname mylib "C:\users\user_name\Documents\...\Class\";

ods select ParameterEstimates GoodnessOfFit FitQuantiles Histogram;
proc univariate data=mylib.fueldat noprint;
  var Income;
  histogram Income/normal(percents=20 30 40 50 60 70 80);
  inset n mean std  normal(ad adpval) / pos = ne format = 6.3;
  title 'Use of Proc Univariate for Creating Graphics:3';
run;
```

Fig. 3.36. SAS Example C15: program

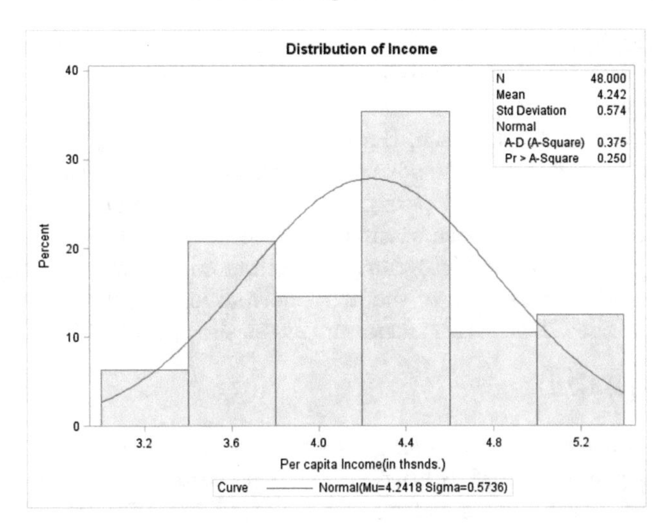

Fig. 3.37. SAS Example C15: histograms with normal fit overlay

Figure 3.37 shows the histogram produced by the `histogram` statement in SAS Program C15, which is essentially the same as that produced in SAS Example C13.

Use of Proc Univariate for Creating Graphics:3

The UNIVARIATE Procedure
Fitted Normal Distribution for Income (Per capita Income(in thsnds.))

Parameters for Normal Distribution		
Parameter	Symbol	Estimate
Mean	Mu	4.241833
Std Dev	Sigma	0.573624

Goodness-of-Fit Tests for Normal Distribution				
Test	Statistic		p Value	
Kolmogorov-Smirnov	D	0.08029628	Pr > D	>0.150
Cramer-von Mises	W-Sq	0.05839137	Pr > W-Sq	>0.250
Anderson-Darling	A-Sq	0.37509336	Pr > A-Sq	>0.250

Quantiles for Normal Distribution		
	Quantile	
Percent	Observed	Estimated
20.0	3.65600	3.75906
30.0	3.84600	3.94102
40.0	4.18800	4.09651
50.0	4.29800	4.24183
60.0	4.34500	4.38716
70.0	4.47600	4.54264
80.0	4.81700	4.72461

Fig. 3.38. SAS Example C15: selected tables and graphics from proc univariate

However, by adding the normal option after the backslash in the histogram statement, a normal density curve with the fitted mean and standard deviation is overlaid on the histogram. The `inset` statement augments the plot with selected statistics, in this case, the number of observations, the fitted mean, and standard deviation of the normal density, as well as the Anderson–Darling statistic and the associated p-value for the fitted normal distribution. The position and format of these statistics are decided by the options `pos=` and `format=`. Except for the histogram, the estimated parameters, the goodness of fit test, and the quantiles of the fitted normal distribution are produced as the result of the ODS SELECT statement, as shown in Fig. 3.38.

SAS Example C16

As will be illustrated in many examples throughout this book, SAS statistical procedures generate ODS Statistical Graphics automatically as part of the output (just as tables of statistics are produced as part of the results). The GLM procedure is used on the hydrocarbon emissions data set used previously

```
.... insert data step and the oroc format step from SAS Example C7 here.....

proc glm data=emissions;
  title "Hydrocarbon Emissions Distribution by Period";
  class period;
  model hc=period;
  format period pp.;
run;
```

Fig. 3.39. SAS Example C16: program

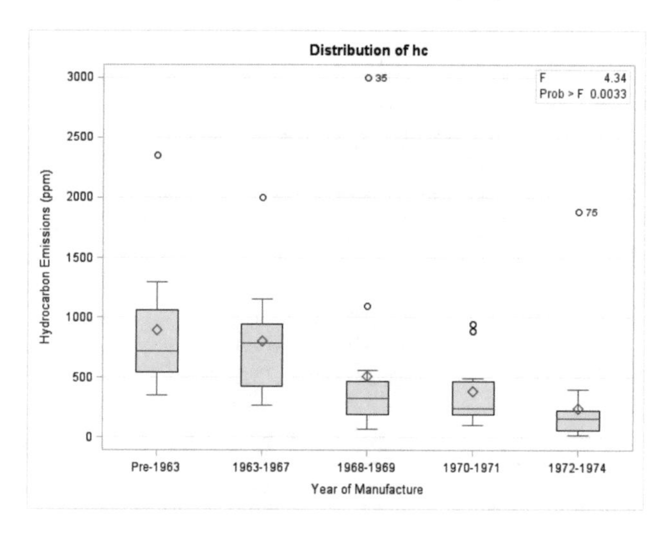

Fig. 3.40. SAS Example C16: graphics from proc glm

in SAS Example C7 (see Figs. 3.16 and 3.17). Since the statistical tables' output are not relevant to this example, they are not reproduced here (Fig. 3.39). The set of side-by-side box plots displayed in Fig. 3.40 is a commonly used graphical tool for examining homogeneity of variances in analysis of variance setting.

3.5 Exercises

3.1 Carry out an analysis similar to SAS Example C3 for the `Weight` variable in the biology data set using `proc univariate` in a SAS program. Obtain a histogram using midpoints that you select and an overlaid normal curve fitted to the data. Also use an ODS SELECT statement to create tables and graphics to be included in the output as shown in that example. You may read the data from the text file `biology.txt` or create and save a SAS data set beforehand to use in this SAS program.

3.2 Groundwater quality is affected by the geological formations in which it is found. A study of chlorine concentration was carried out to determine whether differences in quality existed for water tables on the east and west sides of the Rio Grande River in New Mexico. The data from wells on each side in (milliequivalents), reported in Koopmans (1987)), are

West Side: 0.58, 0.38, 0.32, 0.55, 0.56, 0.62, 0.61, 0.63, 0.52, 0.53, 0.49, 0.37, 0.40, 0.62, 0.44, 0.18, 0.24, 0.21

East Side: 0.34, 0.24, 1.03, 0.68, 0.29, 1.14, 0.34, 0.46, 0.53, 0.40, 0.33, 0.37, 0.40, 0.55, 0.76, 0.37, 0.40, 0.45, 0.30, 0.46, 0.12, 0.39, 0.65

3.8 Use `proc ttest` (as in SAS Example C4) to obtain histograms with overlaid normal fits and side-by-side box plots of chlorine concentrations for the two sides of the river. Use information provided by the box plots and Q–Q plots to describe and compare (location, dispersion, and shape of) chlorine concentration distributions on the two sides of the river. Use the output tables to comment on the equality of variances of the two populations. Perform a t-test of $\mu_{west} = \mu_{east}$ based on an appropriate p-value (based on pooled t-test or the t-test based on Satterthwaite approximation) and state your conclusion.

3.9 Rice (1987) cites an experiment that was performed to determine whether two forms of iron (Fe^{2+} and Fe^{3+}) are retained differently. The investigators divided 108 mice randomly into 6 groups of 18 each; 3 groups were given Fe^{2+} in 3 different concentrations, 10.2, 1.2, and 0.3 millimolar, and 3 groups were given Fe^{3+} at the same three concentrations. The mice were given the iron orally, and the percentage of iron retained was calculated by radioactively labeling the iron. The data for the six "treatments" of iron by each mouse are listed in the following table:

a. Construct side-by-side box plots using `proc sgplot` as in SAS Example C7 for the six treatments. Place the plots on the x-axis in the order high, medium, and low doses for each of the two forms of iron, respectively. Compare and comment on features such as shape, location, dispersion, and outliers of the six iron retention distributions.
b. Comment on any observed trend in the median % iron retention over the levels and the forms of iron. What is the observed trend in dispersion (as measured by, say, IQR)? That is, compare the distributions of % iron retention across the six treatments.
c. Statistical analysis of this data (e.g., using a one-way ANOVA model) may be complicated by failure of assumptions such as homogeneity of variance and/or non-normal distributions. Do the box plots show evidence of these problems? Explain. If there is reason to believe that the assumptions fail based on the plots, a possible explanation is that each distribution is related to the median level of % iron retention in some way. Discuss whether there appears to be such a relationship and describe the relationship algebraically.

Fe^{3+}			Fe^{2+}		
10.2	1.2	0.3	10.2	1.2	0.3
0.71	2.20	2.25	2.20	4.04	2.71
1.66	2.93	3.93	2.69	4.16	5.43
2.01	3.08	5.08	3.54	4.42	6.38
2.16	3.49	5.82	3.75	4.93	6.38
2.42	4.11	5.84	3.83	5.49	8.32
2.42	4.95	6.89	4.08	5.77	9.04
2.56	5.16	8.50	4.27	5.86	9.56
2.60	5.54	8.56	4.53	6.28	10.01
3.31	5.68	9.44	5.32	6.97	10.08
3.64	6.25	10.52	6.18	7.06	10.62
3.74	7.25	13.46	6.22	7.78	13.80
3.74	7.90	13.57	6.33	9.23	15.99
4.39	8.85	14.76	6.97	9.34	17.90
4.50	11.96	16.41	6.97	9.91	18.25
5.07	15.54	16.96	7.52	13.46	19.32
5.26	15.89	17.56	8.36	18.4	19.87
8.15	18.3	22.82	11.65	23.89	21.60
8.24	18.59	29.13	12.45	26.39	22.25

3.10 Insulin production from beta islets (insulin-producing cells) in the pancreas of obese rats, reported in Koopmans (1987), are reproduced below. In addition to measurements made at end of each of the first 3 weeks, data are also available for the first day of the experiment (labeled Week 0):

Week 0	Week 1	Week 2	Week 3
31.2	18.4	55.2	69.2
72.0	37.2	70.4	52.0
31.2	24.0	40.0	42.8
28.2	20.0	42.8	40.6
26.4	20.6	26.8	31.6
40.2	32.2	80.4	66.8
27.2	23.0	60.4	62.0
33.4	22.2	65.6	59.2
17.6	7.8	15.8	22.4

a. Construct side-by-side box plots using `proc sgplot`, as in SAS Example C13, for the four periods of study. Compare and comment on features such as shape, location, dispersion, and outliers of the distributions of insulin production for each week.
b. Comment on the observed trend in the median insulin production as time increases. What is the observed trend in dispersion (as measured by IQR)? That is, compare the distributions of insulin production across the period of study.
c. Statistical analysis of this data (e.g., using a one-way ANOVA model) may be complicated by failure of assumptions such as homogeneity of variance, non-normal distributions, or presence of outliers. Do the box plots show evidence of any of these problems? Explain. If any of the above-mentioned problems exist, can you relate these problems to the median level of insulin production? Explain.

3.11 Use the `fitness` data set (see Table B.4) and `proc sgpanel` to obtain vertical bar charts showing the means of the `BMI` variable for each of the three age groups defined by the `AgeGrp` variable. Obtain this plot in three panels, each panel corresponding to a level of the `WtGrp` variable. Make sure that the three panels all appear side by side in one row. Use appropriate user-defined formats for the grouping variables. (Note: In Exercise 2.11, the numeric category variable `WtGrp` that takes values 1, 2, or 3 accordingly as the subject's weight is $\leq$140 lbs, over 140 but $\leq$165 lbs, or above 165 lbs, respectively, and another numeric category variable `AgeGrp` that takes values 'A', 'B', or 'C' accordingly as the subject's age is $\leq$25, values between 25 and $\leq$45, or above 45, respectively, were created. You may use user-defined formats to enhance the appearance of these levels in the output graphics.)

3.12 Use the `fitness` data set (see Table B.4) and `proc sgplot` to obtain horizontal bar charts showing the means of the `Aero` variable for each of the three groups of people with oxygen uptake rates ($<$9.3, 9.3 to $\leq$10.5, and $>$10.5), defined as a user-defined format for `Oxygen` variable. Subdivide each bar into the three 1.5-mile runtime groups (45, 45 to $\leq$49, and $>$49) defined by a user-defined format for `RunTime` variable. Then use format statements in the `proc sgplot` step to associate these user-defined formats to the variables `Oxygen` and `Runtime`. Use the `keylegend` statement to position the legend in the top-right corner inside the outline. To make room for the legend includes the statement `yaxis offsetmin=.2;` in your proc step. (Note: As noted in Sect. 2.1.4, using user-defined formats directly in a proc step to assign continuous variable values into classes is much simpler method than creating a separate category variables for this purpose.)

3.13 Obtain a normal probability plot of the `Popurban` variable and a histogram of the the `Deathrat` variable in the demographic data set on countries (see Table B.5) used in SAS Example C12. Use `proc univariate`, as was done in the SAS Example C13 program. Use the `normal` option to add a reference line to the probabilty plot. Suppress all other output being produced from this step. (Note: You may use the SAS data set `world` created in Exercise 2.14 for this problem.)

3.14 Use the `world` SAS data set and `proc sgplot` to obtain horizontal bar charts showing the means of the per capita GNP variable for each of the two groups of countries with level of technology defined by the `Techgrp` variable. Subdivide each bar into the three groups, defined by the `Infgrp` variable. Use the `keylegend` statement to position the legend in the top-right corner inside the outline. To make room for the legend includes the statement `yaxis offsetmin=.2;` in your proc step. Use appropriate user-defined formats for all grouping variables. For this problem, assume that the `world` SAS data set created in Exercise 2.14 is available (otherwise, you may create this data set by executing SAS code as described in that exercise).

3.15 For this problem assume that the `world` SAS data set created in Exercise 2.14 is available (otherwise, you may create this data set by executing SAS code as described in that exercise). Write five proc steps using `statistical graphics (sg)` procedures in the same SAS program (or separate SAS programs) to access the *world* SAS data set and produce the following plots:

a. Scatterplot of crude birth rate against per capita GNP identifying each point by the country name. (Use `markerattrs=` option on the scatter statement to select the color, symbol, and size of the marker symbol and data labels)
b. Histogram of life expectancy with bins staring at 30 with binwidth of 4. The variable on the vertical axis must be the frequencies (counts).
c. Simple linear regression of life expectancy against crude birth rate showing confidence and prediction bands.
d. Scatterplot matrix of the four variables crude birth rate, crude death rate, infant mortality, and life expectancy identifying each point by technology group.

e. Obtain a dot plot of the `Lifeexp` variable to display life expectancy per country with countries for each technology group in separate panels (one panel per row). (Use `rowaxis` statement to control the color and size of the labels on the category variable axis.)

3.16 Use the the `world` SAS data set created in Exercise 2.14 for this problem as in the above problem. In order to study the effects of urbanization on crude death and birth rates, obtain side-by-side scatter plots of these variables against urban population variable. Identify the points by grouping them by the technology category variable. Change the attributes of the plot symbol so that they are filled circles of size 5 pixels. Also overlay each scatter plot with a loess smoother curve that will show the trend more clearly, making sure that the grouping variable is ignored for this fit and that a quadratic is used for each local regression. Also choose a dashed line type and magenta color for the smoother lines.

3.17 The data set available in the Excel file `heart.xlsx` is a selected subset of the SASHELP SAS data file *heart* containing data from the Framingham Heart Study. The variables of interest are `AgeAtDeath` (a numeric variable), `DeathCause`, and `Sex` which are character variables. Import the data using `proc import` and create a SAS data set named `heart.` Include statements to create a user-defined numeric format to convert values of age ≤ 29 30 to $\leq$50, 51 to $\leq$70, and >71 to character strings "Under 30," "30 to 50 years," "51 to 70 Years," and "Over 70," respectively. Use `proc sgpanel` to obtain a 2×2 panel of bar charts of numbers of deaths due each of four causes of death grouped by sex (i.e., bars for each value of sex to appear side by side (i.e., clustered)). Note carefully that the formatted variable `AgeAtDeath` serves as the category variable to be used in `proc sgpanel` statements.

3.18 Obtain a histogram of the `Nox` variable from the pollution data set (see Tables B.7). Examining the output from `proc univariate` for this variable, it is observed that `Nox` values are highly dispersed. So it is necessary to use unevenly spaced intervals to obtain a useful bar chart for this variable. To do this, center the bars at the values, first from 2.5 through 27.5, incremented by 5, and then at 35, 50, 80, and 200, respectively (i.e., 10 intervals in all). In this problem, use a user-defined numeric format and the `vbar` statement in `proc univariate.` Subdivide the bars by the category variable `Density` created in the data step so it has values "Low," "Medium," or "High," depending on whether the value for the variable `Popn` is less than 3000. A suggested range of values of the `Nox` variable to be converted to midpoints is as follows: low, <5="2.5"; 5, <10="7.5"; 10, <15="12.5"; 15, <20="17.5"; 20, <25="22.5"; 25, <30="27.5"; 30, <40="35"; 40, <60="50"; 60, <100="80"; and 100, high="200."

4

Statistical Analysis of Regression Models

4.1 An Introduction to Simple Linear Regression

In this section, a review of simple linear regression or straight-line regression is presented. Consider `bivariate data` consisting of ordered pairs of numerical values (x, y). Often such data arise by setting an X variable at certain fixed values and taking a random sample from the population of Y that, hypothetically, exists for each setting of X. The variable Y is called the dependent variable or the response variable and the variable X is called the independent variable or the predictor variable. The y-values observed at each x-value are assumed to be a random sample from a normal distribution with the mean $E(y) = \mu(x) = \beta_0 + \beta_1 x$; that is, the mean of the distribution is modeled as a linear function of x. The variance of the normal distributions at each x-value is assumed to have the same value σ^2. Thus, the y-values can be related to the x-values through the relationship

$$y = \beta_0 + \beta_1 x + \epsilon \tag{4.1}$$

where ϵ is a random variable (called `random error`) with mean zero (i.e., $E(\epsilon) = 0$) and variance σ^2. Equation 4.1 is called the *simple linear regression model* and β_0, β_1, and σ^2 are the *parameters* of the model. In the above representation, note that y is a random variable with mean $E(y) = \beta_0 + \beta_1 x$ and variance σ^2.

Estimation of Parameters

The first step in regression fitting is to obtain estimates of the model parameters using the observed data. The *method of least squares* selects a model that minimizes the total squared error of prediction. The error or the *residual* is the difference between an observed value y for a given value of x and $\hat{y}$ and

M. G. Marasinghe, K. J. Koehler, *Statistical Data Analysis Using SAS*, Springer Texts in Statistics, https://doi.org/10.1007/978-3-319-69239-5_4

the value of y predicted by the fitted model at that particular value of x. The method of least squares selects the line that produces the smallest value of the sum of squares of all residuals; that is, mathematically, it finds the estimates $\hat{\beta}_0$ and $\hat{\beta}_1$ that minimize the quantity

$$\sum_i (y_i - \hat{y}_i)^2 = \sum_i (y_i - \hat{\beta}_0 - \hat{\beta}_1 x_i)^2$$

where (x_i, y_i), $i = 1, 2, \ldots, n$, are all pairs of observations available. The values $\hat{\beta}_0$ and $\hat{\beta}_1$ are called the `least squares estimates` of β_0 and β_1, respectively. The fitted regression equation $\hat{y} = \hat{\beta}_0 + \hat{\beta}_1 x$ is usually called the prediction equation and is used to calculate the predicted value $\hat{y}$ for a specified value of x. Since the mean of y is $E(y) = \beta_0 + \beta_1 x$, the estimate of the slope $\hat{\beta}_1$ is the estimate of the change in the mean of y for a unit change in the x-value.

Statistical Inference

Following the fitting of a least squares line to the data, it is of interest to measure how well the line estimates the population means. This is achieved by the computation of quantities necessary to perform statistical inference about the parameters of the model, such as hypothesis tests and confidence intervals. The first step is usually to construct an analysis of variance that partitions the total sum of squares, `SSTot`$= \sum(y - \bar{y})^2$, into two parts: the sum of squares due to regression, `SSReg`$= \sum(\hat{y} - \bar{y})^2$, and the sum of squares of residuals, `SSE`$= \sum(y - \hat{y})^2$. This algebraic partition is represented in the following ANOVA table for regression:

Source	df	Sum of squares	Mean square	F
Regression	1	SSReg	MSReg=SSReg/1	MSReg/MSE
Error	$n-2$	SSE	MSE=SSE/$(n-2)$	
Total	$n-1$	SSTot		

This table provides several important statistics that are useful for assessing the fit of the model. First, the F-ratio, $F =$ MSReg/MSE, is used to test the hypothesis that $H_0 : \beta_1 = 0$ versus $H_a : \beta_1 \neq 0$. Second, the `MSE` from the above table provides the `least squares estimate` of σ^2. This is useful for calculating the standard errors of the estimates $\hat{\beta}_0$ and $\hat{\beta}_1$, commonly denoted by $s_{\hat{\beta}_0}$ and $s_{\hat{\beta}_1}$, respectively. Test statistics for performing t-tests and confidence intervals for the β_0 and β_1 coefficients are calculated using these standard errors. For example, a t-statistic for testing $H_0 : \beta_1 = 0$ versus $H_a : \beta_1 \neq 0$ is given by

$$t = \frac{(\hat{\beta}_1 - 0)}{s_{\hat{\beta}_1}}$$

and a $(1 - \alpha)100\%$ confidence interval for β_1 is given by

$$\hat{\beta}_1 \pm t_{\alpha/2,(n-2)} \times s_{\hat{\beta}_1}$$

Third, the ratio $R^2 = \text{SSReg}/\text{SSTot}$, called the *coefficient of determination*, measures the proportion of variation in y explained by using $\hat{y}$ to predict y. A simple linear regression model using the predictor variable x with a larger R^2 does better in predicting y than one with a smaller R^2. Note that R^2 does not say how *accurate* the prediction is nor does it say that a straight line is the best function of x that could be used to model the variation in y.

4.1.1 Simple Linear Regression Using PROC REG

Consider the following problem. An investigation of the relationship between traffic flow x (thousands of cars per 24 hours) and lead content y of bark on trees near the highway ($\mu g/g$ dry weight) yielded the data in Table 4.1 reported in Devore (1982).

Table 4.1. Lead content in trees near highways (Devore 1982)

Traffic flow, x	8.3	8.3	12.1	12.1	17.0	17.0	17.0	24.3	24.3	24.3	33.6
Lead content, y	227	312	362	521	640	539	728	945	738	759	1263

SAS Example D1

In the SAS Example D1 program (see Fig. 4.1), a simple linear regression model is fitted to these data and statistics for making some of the statistical inferences about the model discussed earlier are computed using `proc reg`. This program produces printed output that helps to perform an elementary regression analysis of the data. It contains SAS statements necessary to compute an analysis of variance table, estimated values of the parameters and their standard errors, associated t-statistics and confidence intervals, and the predicted values and residuals. It also produces, in ODS Graphics, a scatter plot of the data overlaid with the fitted regression line and several residual plots. These plots were selected by the use of the `plots=` option on the `proc reg` statement as discussed below. Various other diagnostic statistics and plots necessary for examining the adequacy of the model and accompanying assumptions will be discussed in later examples.

The data are entered in a straightforward manner, where a pair of values for the x variable named `Traffic` and the y variable named `Lead`, separated by a blank comprise each line of data, so that the statement `input Traffic Lead;` is all that is necessary to read the data and create the SAS data set (named `d1` in this example). The label statement

```
data d1;
input Traffic Lead;
label Traffic="Traffic Flow (Thousands/24 hours)" Lead="Lead Content (mcg/gm)";
datalines;
 8.3  227
 8.3  312
12.1  362
12.1  521
17.0  640
17.0  539
17.0  728
24.3  945
24.3  738
24.3  759
33.6 1263
;
proc reg data=d1 plots(only)=(fit(nolimits) residuals residualbypredicted qq);
  model Lead=Traffic/p r clb;
  title "Simple Linear Regression of Lead Content Data";
run;
```

Fig. 4.1. SAS Example D1: program

```
label Traffic="Traffic Flow (Thousands/24 hours)" Lead="Lead
                    Content (mcg/gm)";
```

adds descriptive labels to the variables `Traffic` and `Lead` as illustrated in SAS Example A10. In the `proc reg` step, the option `data=d1` specifies that the SAS data set to be processed is named `d1`. The use of such options is described in Sect. 1.8 of Chap. 1. The model statement `model y=model Lead=Traffic/r p clb;` specifies model (7.1) as `Lead=Traffic`. In this representation of the model, an intercept β_0 and the error term are assumed to be part of the model. The output resulting from the model statement is shown in the section of the SAS output (Fig. 4.2) titled *Analysis of Variance.*

Various other options may be included in the model statement, following a slash (solidus) symbol, as keywords separated by at least one blank. For example, the option `noint` may be used to specify that a model be fitted without an intercept β_0. In the present example, the options `p` (for `predicted`) and `r` (for `residual`) request that residuals and predicted values are to be computed and output along with the observed values for the response `Lead`, as shown in the *Output Statistics* section of the SAS output (Fig. 4.3). The `clb` option specifies that $(1-\alpha)100\%$ confidence intervals for the β coefficients be computed. By default, $\alpha = 0.05$ is used, but the `alpha=` option allows the user to specify an alternate confidence level. For example,

```
model Lead=Traffic/r p clb alpha=.1;
```

would produce 90% confidence intervals for the β's, instead of the 95% confidence intervals produced by default. The ANOVA table given below is constructed directly using the SAS output in Fig. 4.2.

Since the p-value is smaller than a level of significance selected for this study (say, $\alpha = 0.05$), the null hypothesis of $H_0 : \beta_1 = 0$ is rejected and it is concluded that the lead content on tree bark can be modeled by a simple linear regression using traffic flow as a predictor variable. The R^2 value for this fit, also extracted from this section of the output, is 0.9143, showing that 91.3% of the variation in the response is explained by the fitted model. This measure is used in practice as a measure of fit of the line and the close fit shown in Fig. 4.4a supports this interpretation.

Source	df	Sum of squares	Mean square	F	p-value
Regression	1	815,966	815,966	96.0	< 0.0001
Error	9	76,493	8499		
Total	10	892,459			

Additionally, the parameter estimates (estimates of the coefficients), their standard errors, and the t-statistics discussed in Sect. 4.1 are found in the section titled *Parameter Estimates* of the SAS output (see Fig. 4.2).

Simple Linear Regression of Lead Content Data

The REG Procedure
Model: MODEL1
Dependent Variable: Lead Lead Content (mcg/gm)

Number of Observations Read	11
Number of Observations Used	11

Analysis of Variance

Source	DF	Sum of Squares	Mean Square	F Value	Pr > F
Model	1	815966	815966	96.01	<.0001
Error	9	76493	8499.17298		
Corrected Total	10	892459			

Root MSE	92.19096	R-Square	0.9143
Dependent Mean	639.45455	Adj R-Sq	0.9048
Coeff Var	14.41712		

Parameter Estimates

Variable	Label	DF	Parameter Estimate	Standard Error	t Value	Pr > |t|	95% Confidence Limits	
Intercept	Intercept	1	-12.84155	72.14287	-0.18	0.8627	-176.04007	150.35696
Traffic	Traffic Flow (Thousands/24 hours)	1	36.18385	3.69290	9.80	<.0001	27.82994	44.53776

Fig. 4.2. SAS Example D1: analysis of variance and parameter estimates

The prediction equation is calculated to be $\hat{y} = -12.84 + 36.18\,x$; thus, the fitted line has a positive slope. The value of the t-statistic is 9.80 with a p-value < 0.0001, which again shows that the hypothesis $H_0 : \beta_1 = 0$ will be rejected. The 95% confidence interval for β_1 is computed to be (27.83, 44.54), as shown in the last two columns of the *Parameter Estimates* section. Thus with 95% confidence, the increase in mean lead content on tree bark that results from an increase in traffic flow by 1000 cars in 24 hours lies in the above interval.

The above prediction equation must be used with caution. In practice, it is recommended that predictions be made only within the range of x-values observed. That is because *extrapolation* outside this range may cause problems, as the model may no longer be valid. For example, the intercept of the fitted line is $\hat{y} = -12.84$ (i.e., the predicted value given by the prediction equation at $x = 0$). Thus, when there is no traffic flow, the lead content of bark on trees near the highway is predicted to be -12.84 when the model is extrapolated to $x = 0$! This argument shows that in such situations, the routine test of $\beta_0 = 0$ (i.e., the y-intercept at $x = 0$) also will not make sense. Although -12.84 is an unrealistic value, this value of the intercept is needed to pass the least squares regression line through the center of the data. Thus the results of the test of $\beta_0 = 0$ or the confidence interval calculated for β_0 should be disregarded here.

Simple Linear Regression of Lead Content Data

The REG Procedure
Model: MODEL1
Dependent Variable: Lead Lead Content (mcg/gm)

<table>
<tr><th colspan="10">Output Statistics</th></tr>
<tr><th>Obs</th><th>Dependent Variable</th><th>Predicted Value</th><th>Std Error Mean Predict</th><th>Residual</th><th>Std Error Residual</th><th>Student Residual</th><th>-2-1 0 1 2</th><th>Cook's D</th></tr>
<tr><td>1</td><td>227.0000</td><td>287.4844</td><td>45.4206</td><td>-60.4844</td><td>80.226</td><td>-0.754</td><td>| *| |</td><td>0.091</td></tr>
<tr><td>2</td><td>312.0000</td><td>287.4844</td><td>45.4206</td><td>24.5156</td><td>80.226</td><td>0.306</td><td>| | |</td><td>0.015</td></tr>
<tr><td>3</td><td>362.0000</td><td>424.9830</td><td>35.3804</td><td>-62.9830</td><td>85.132</td><td>-0.740</td><td>| *| |</td><td>0.047</td></tr>
<tr><td>4</td><td>521.0000</td><td>424.9830</td><td>35.3804</td><td>96.0170</td><td>85.132</td><td>1.128</td><td>| |** |</td><td>0.110</td></tr>
<tr><td>5</td><td>640.0000</td><td>602.2839</td><td>28.0543</td><td>37.7161</td><td>87.819</td><td>0.429</td><td>| | |</td><td>0.009</td></tr>
<tr><td>6</td><td>539.0000</td><td>602.2839</td><td>28.0543</td><td>-63.2839</td><td>87.819</td><td>-0.721</td><td>| *| |</td><td>0.026</td></tr>
<tr><td>7</td><td>728.0000</td><td>602.2839</td><td>28.0543</td><td>125.7161</td><td>87.819</td><td>1.432</td><td>| |** |</td><td>0.105</td></tr>
<tr><td>8</td><td>945.0000</td><td>866.4260</td><td>36.1835</td><td>78.5740</td><td>84.793</td><td>0.927</td><td>| |* |</td><td>0.078</td></tr>
<tr><td>9</td><td>738.0000</td><td>866.4260</td><td>36.1835</td><td>-128.4260</td><td>84.793</td><td>-1.515</td><td>| ***| |</td><td>0.209</td></tr>
<tr><td>10</td><td>759.0000</td><td>866.4260</td><td>36.1835</td><td>-107.4260</td><td>84.793</td><td>-1.267</td><td>| **| |</td><td>0.146</td></tr>
<tr><td>11</td><td>1263</td><td>1203</td><td>63.8739</td><td>60.0643</td><td>66.478</td><td>0.904</td><td>| |* |</td><td>0.377</td></tr>
</table>

Fig. 4.3. SAS Example D1: output statistics

The `proc reg` statement `plots=` option used in SAS Example D1

```
plots(only)=(fit(nolimits) residuals residualbypredicted qq)
```

is an example of using the `plots=` option for selecting ODS-based graphics created in `proc reg` for output. The option `fit(nolimits)` produces a scatter plot with the y-variable on the vertical axis and the x-variable on the horizontal axis overlaid with the fitted regression line. The `nolimits` suboption suppresses the plotting of confidence and prediction bands that will be discussed later. The resulting graph is shown in Fig. 4.4a. The statistics such as R^2 and `MSE` that appear by default outside the right margin of the plot may be suppressed using the suboption `stats=none` or output only selected statistics using a suboption such as `stats=(aic cp mse rsquare)`. The option `residuals` will produce a plot of the residuals against the x variable (`Traffic`) as shown in Fig. 4.4b. The option `residualbypredicted` will produce a plot of the residuals against the predicted values, a very useful diagnostic plot as will be discussed below. This graph is shown in Fig. 4.4c. The option `qq` produces a normal probability plot of the *residuals* shown in Fig. 4.4d.

Graphical tools may also be used to help identify cases for which assumptions about the distribution of ϵ is not valid. Most of the plots used for this purpose involve some form of the residuals obtained from fitting the model, $(y_i - \hat{y}_i)$, $i = 1, \ldots, n$. In the plot of residuals versus x shown in Fig. 4.4b, the residuals are expected to scatter evenly and randomly around the zero reference line as the value of x changes, because if the linear relationship specified by the model is correct, the residuals should not display any relationship to x. If some systematic nonlinear pattern is observed as x varies, it is usually an indication of a need for higher-order polynomial terms in x to be included in the model or for a model that is not linear in the parameters in some other way. That is, a systematic pattern of the residuals when plotted against the x variable is an indication of *model inadequacy*. Additionally, this plot may also show a departure from the *homogeneity of variance* assumption, as a marked decrease or increase of the spread of the residuals around zero may indicate a dependence of the variance of y (and therefore of ϵ), on the actual values of x. A few points that stand out in this plot due to a comparatively larger or smaller residual than the others may also highlight outliers.

The graph of residuals versus predicted values is shown in Fig. 4.4c. This scatter plot should also show no systematic pattern and should indicate random scatter of residuals around the zero reference line if the straight-line model is adequate. If the homogeneity of variance assumption is not plausible, a pattern indicating a steady increase or decrease in spread of the residuals around the zero reference line as $\hat{y}_i$ increases will also be evident. This pattern may show up along with a curved pattern of the residuals, in both this and the previous plot if nonlinearity is also present along with heterogeneous variance.

The graph in Fig. 4.4d displays a normal probability plot where *residuals* are plotted against the corresponding percentiles of the standard normal

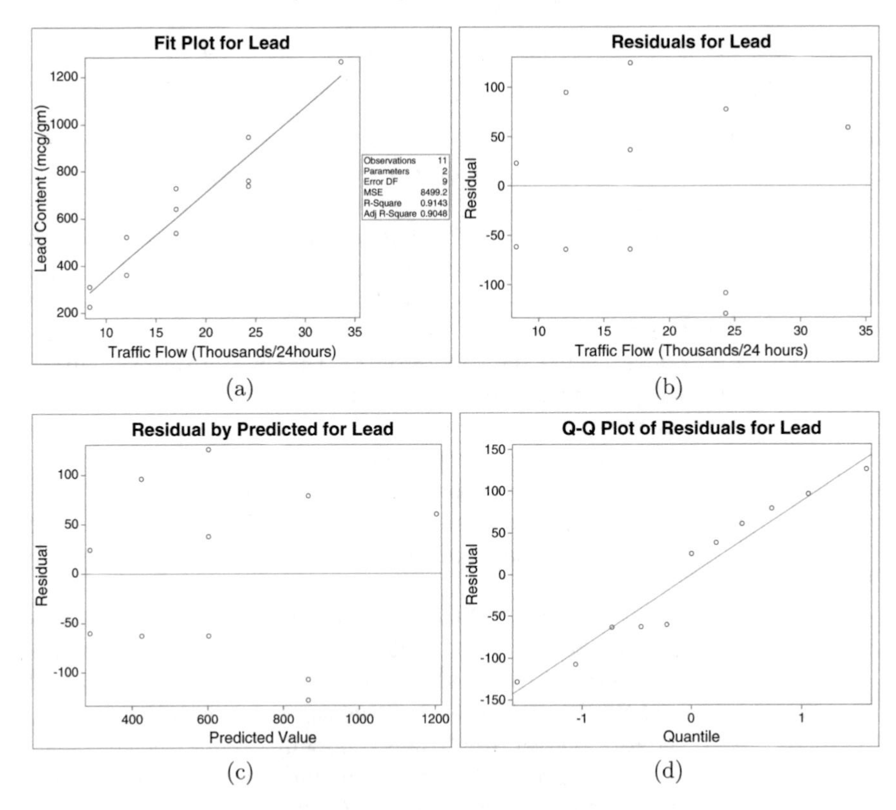

Fig. 4.4. (**a**) Graph of `Lead` versus `Traffic` overlaid by the fitted line, (**b**) graph of residuals versus `Traffic`, (**c**) graph of residuals versus predicted, (**d**) normal probability plot of the residuals

distribution. Although residuals are used here, *internally* studentized residuals, a version of standardized residuals and defined in Sect. 4.2.1 are also sometimes used in this plot. This plot will show an approximate straight-line pattern unless the normality assumption about the ϵ's is questionable. More discussion on how to interpret a normal probability plot appears in Chap. 2 in Sect. 2.2, as part of the discussion of graphics produced by the UNIVARIATE procedure.

The residual plots shown above do not appear to show any inadequacy of the model or any serious violation of the model assumptions made concerning the straight-line model fitted to the lead content data. Although a few points may appear to have a larger residual in magnitude than the others, there is clearly no overall pattern showing a systematic variation of the residuals as the values of the x variable vary.

4.1.2 Lack of Fit Test

Whenever regression data contain more than one response (y) value at one or more values of x, responses are said to be replicated. The data set shown in Table 4.1 contains replicated responses. In such cases, the sum of squares of residuals, SSE, can be partitioned into two parts: sum of squares representing pure experimental error, SSE_{Pure}, and sum of squares due to lack of fit, SS_{Lack}. Introducing a new notation, the responses within a subset i of observations with the same x-value are represented by y_{ij}, $j = 1, \ldots, n_i$, where n_i is the number of observations in the subset and $i = 1, \ldots, g$, where g is the number of such subsets. The partitioning of SSE is given by

$$\text{SSE} = \text{SSE}_{\text{Pure}} + \text{SS}_{\text{Lack}} \tag{4.2}$$

$$\sum_i \sum_j (y_{ij} - \hat{y}_{ij})^2 = \sum_i \sum_j (y_{ij} - \bar{y}_{i.})^2 + \sum_i \sum_j (\bar{y}_{i.} - \hat{y}_{ij})^2 \tag{4.3}$$

$$(n - 2) = (n - g) + (g - 2) \tag{4.4}$$

where $n = \sum n_i$ is the total number of observations, and Eq. 4.4 represents the *degrees of freedom* for each sum of squares. The hypotheses of interest are

$$\begin{aligned} H_0 &: \ E(y) = \beta_0 + \beta_1 x \\ H_a &: \ E(y) \neq \beta_0 + \beta_1 x \end{aligned}$$

The test for lack of fit is an F-test. The F-statistic is the ratio of `mean squares` for lack of fit and pure experimental error. The above partitioning can be used to motivate the lack of fit using the following argument. Consider the sum of squares due to lack of fit $\sum_i \sum_j (\bar{y}_i - \hat{y}_{ij})^2$. If the true relationship among Y population means is indeed $E(y) = \beta_0 + \beta_1 x$, then one would expect this sum of squares to be small because both $\bar{y}_{i.}$ and $\hat{y}_{ij}$ are estimates of the mean of the Y population at a given x_i. If $E(y)$ deviates from $\beta_0 + \beta_1 x$, then one would expect the sum of squares due to lack of fit to be larger. The F-test is usually performed using a supplementary ANOVA table as follows:

Source	df	Sum of squares	Mean square	F
Lack of fit	$g - 2$	SS_{Lack}	$\text{MS}_{\text{Lack}} = \text{SS}_{\text{Lack}}/(g - 2)$	$\text{MS}_{\text{Lack}}/\text{MSE}_{\text{Pure}}$
Pure error	$n - g$	SSE_{Pure}	$\text{MSE}_{\text{Pure}} = \text{SSE}_{\text{Pure}}/(n - g)$	
Total error	$n - 2$	SSE		

As an example, this table is first constructed by hand calculation for the data in Table 4.1 used previously in SAS Example D1. From the previous analysis,

SSE=76,493 with 9 degrees of freedom. The following table simplifies the computation of SSE_{Pure}, noting that $g = 5$.

i	x_i	y_i			$(y_i - \bar{y})^2$			$\sum(y_i - \bar{y})^2$	$n_i - 1$
1	8.3	227	312		1806.25	1806.25		3612.50	1
2	12.1	362	521		6320.25	6320.25		12,640.50	1
3	17.0	640	539	728	18.78	9344.44	8525.44	17,888.67	2
4	24.3	745	738	759	17,161.00	5776.00	3025.00	25,962.00	2
5	33.6	1263				0		0	0
Total								60,103.67	6

Thus, $\text{SSE}_{\text{Pure}} = 60{,}103.67$ with 6 degrees of freedom. The lack of fit F-statistic is computed in the following ANOVA table:

Source	df	Sum of squares	Mean square	F
Lack of fit	3	16,389.33	5463.11	0.55
Pure error	6	60,103.67	10,017.28	
Total error	9	76,493		

As expected, the test fails to reject $H_0: E(y) = \beta_0 + \beta_1 x$; thus, the means of the populations at each value of X are modeled adequately by a simple linear regression model.

SAS Example D2

The SAS Example D2 program (see Fig. 4.5) illustrates the use of the `lackfit` option for computing the lack of fit test in `proc reg`. Note that in this example `lackfit` is the only option used; thus, the *output statistics* table will not appear as a standard part of the output from `proc reg`. The ANOVA table shown in the section of the SAS output (Fig. 4.6) titled *Analysis of Variance* is a modified version of the original ANOVA table displayed in Fig. 4.2. The degrees of freedom for error is partitioned to Lack of Fit and Pure Error sums of squares and an F-statistic is calculated to test the Lack of Fit hypothesis, as discussed above. It is observed that the results are identical (up to the accuracy afforded by the hand calculations).

4.1.3 Diagnostic Use of Case Statistics

In addition to the *diagnostic statistics* such as residuals and studentized residuals, several other statistics that correspond to each observation (or *case statistics*, as they are commonly called in modern regression literature) may be computed and output on request. For example, `proc reg` in SAS will output

```
data d2;
input Traffic Lead;
label Traffic="Traffic Flow (Thousands/24 hours)" Lead="Lead Content (mcg/gm)";
datalines;
 8.3  227
 8.3  312
12.1  362
12.1  521
17.0  640
17.0  539
17.0  728
24.3  945
24.3  738
24.3  759
33.6 1263
;

proc reg data=d2 plots=none;
  model Lead=Traffic/lackfit;
  title "Simple Linear Regression of Lead Content Data";
run;
```

Fig. 4.5. SAS Example D2: program

case statistics labeled as `Cook's D`, `RStudent`, `Hat Diag H`, `DFFITS`, and `DFBETAS` when certain options are specified in the `model` statement. These case statistics, called *influence statistics*, measure how well a specific data point fits the regression line. If the point is a large distance away from the center of the fitted line in the x-direction, it is said to be a *high leverage point* and is called an *x-outlier.* A high leverage point will exhibit a comparatively large value for the `Hat Diag H` statistic.

If the point is a large distance away from the fitted line in the y-direction, it will have a large residual or studentized residual. A statistical test procedure is available to determine whether a studentized residual is sufficiently large for the case to be declared a *y-outlier.* The `Cook's D` case statistic measures the *influence* a data point has on the estimated parameters and/or overall fit statistics; that is, it measures whether the deletion of a data point will markedly change the estimates of the parameters β_0 and β_1, as well as the MSE and R^2 values. If a large Cook's D value is observed, then that particular data point is identified as an influential case.

A high leverage point that is also a y-outlier will most likely be a highly influential case and will have to be examined for validity by the experimenter. This is because including the specific case in the data set may have a substantial effect on the predictions made using the corresponding fitted model. A more detailed discussion follows in Sects. 4.2.1 and 4.2.2.

SAS Example D3

In this example, an artificial data set is used to illustrate the above concepts by examining, both numerically and graphically, the effects of changing a single case on a variety of statistical measures including case statistics. Four SAS

Simple Linear Regression of Lead Content Data

The REG Procedure
Model: MODEL1
Dependent Variable: Lead Lead Content (mcg/gm)

Number of Observations Read	11
Number of Observations Used	11

Analysis of Variance					
Source	DF	Sum of Squares	Mean Square	F Value	Pr > F
Model	1	815966	815966	96.01	<.0001
Error	9	76493	8499.17298		
Lack of Fit	3	16389	5462.96339	0.55	0.6691
Pure Error	6	60104	10017		
Corrected Total	10	892459			

Root MSE	92.19096	R-Square	0.9143
Dependent Mean	639.45455	Adj R-Sq	0.9048
Coeff Var	14.41712		

Parameter Estimates						
Variable	Label	DF	Parameter Estimate	Standard Error	t Value	Pr > |t|
Intercept	Intercept	1	-12.84155	72.14287	-0.18	0.8627
Traffic	Traffic Flow (Thousands/24 hours)	1	36.18385	3.69290	9.80	<.0001

Fig. 4.6. SAS Example D2: output

programs were used to obtain the SAS output of case statistics and regression plots used in this example.

Four different sets of values for case number 4 are used with 10 other cases with values that remain unchanged, to create Artificial Data Sets 1 to 4, analyzed in these SAS programs, respectively. For example, Artificial Data Set 1 used in the program for SAS Example D3 shown in Fig. 4.7 has case number 4 set to the pair of values (8.0 8.3). Note that this case has the value "D" for the variable `Name` and identifies case number 4. The variable `Name` is used to label points in the tables of case statistics as well as plots in the SAS output.

The other three SAS programs are all similar to SAS Example D3 displayed in Fig. 4.7, except that case number 4 is set respectively to the values (17 12.9), (8.0 5.8), and (17 8.5) in each. The residual plots for the fitted models are shown in Figs. 4.8a, b, c, and d, respectively, and the correspond-

```
data ds31;
input Name $ X Y;
datalines;
A 5.0  7.0
B 10.0  8.2
C 7.0  8.0
D 8.0  8.3
E 11.0 10.0
F 3.0  7.2
G 1.0  4.3
H 6.0  8.8
I 4.0  5.8
J 2.0  5.7
K 9.0 10.1
;

proc reg data=ds31 plots=none ;
  model Y=X/p r influence;
  id Name;
  title 'Artificial Data Set 1';
run;

proc sgplot data=ds31;
 reg x=X y=Y/datalabel=Name markerattrs= (Color=magenta Size=2 mm Symbol=CircleFilled)
            datalabelattrs=(Color=darkblue Family=arial Size=3 mm Weight=bold);
run;
```

Fig. 4.7. SAS Example D3: program 1

ing tables of case statistics are displayed in Figs. 4.9, 4.10, 4.11, and 4.12, respectively.

The SAS Example D3 program illustrates the use of the `influence` option with the `model` statement. This option must be used in conjunction with the `r` option and leads to the computation of additional diagnostic case statistics as seen in the output from this program (Fig. 4.9). The main case statistic of interest here is the column titled `Hat Diag H`, which lists the h_{ii}'s discussed later in Sect. 4.2.1. The case statistic tabulated under the column titled `RStudent` consists of the *externally* studentized residuals (t_i's) also discussed in Sect. 4.2.1. These can be used to test for outliers using the Bonferroni procedure described there.

In addition, in the SAS Example D3 program the ODS Statistical Graphics procedure `sgplot` is used to produce a plot of the data superimposed with the fitted regression line. See Sect. 3.1 for an example of `sgplot` procedure. In SAS Example D3, the plot is enhanced by changing the symbol attributes as well as by using the `datalabel=` option to use *labels* to identify each point by the corresponding value of the variable `Name`.

Figure 4.9 displays the case statistics resulting from fitting a simple linear regression model to Artificial Data Set 1. It is observed that none of the leverages (`Hat Diag H`) is numerically larger than 0.36 (i.e., twice the average of all leverage values $2/n$, here $4/11 \approx 0.36$ is used as a cutoff to identify high-leverage points). Thus, there are no cases indicated as possible x-outliers. This is also apparent from inspecting Fig. 4.8a, where the x-values are evenly spread out between 0 and 12. The largest externally studentized residual (`RStudent`) value of -1.96 is not significant at 0.05. Recall that these statistics have t-distributions and critical values (Bonferroni adjusted for multiple testing) are available in Tables B.11 and B.12 of Appendix B. The absolute value of `RStudent` is used to perform a two-sided test of a hypothesis whether a case is an outlier. An $\alpha = 0.05$ critical value for $k = 1$ and $n = 11$ from Table B.11 is 3.90. An ad hoc procedure is to flag cases with values greater than 2 as possible y-outliers. Here, the presence of y-outliers can thus be ruled out. It has been suggested that, as a rule of thumb, cases with influence values (displayed under the column labeled `Cook's D`) numerically larger than $4/n$ may be considered for further investigation. Here, only the case 2 (labeled B) exceeds that value. As discussed in Sect. 4.2.1, relatively large values for both the studentized residual and the leverage for this case result in an inflated value for Cook's D.

From inspecting the case statistics resulting from fitting a simple linear regression model to Artificial Data Set 2 displayed in Fig. 4.10, case number 4 is identified as an x-outlier since the leverage (`Hat Diag H`) is numerically

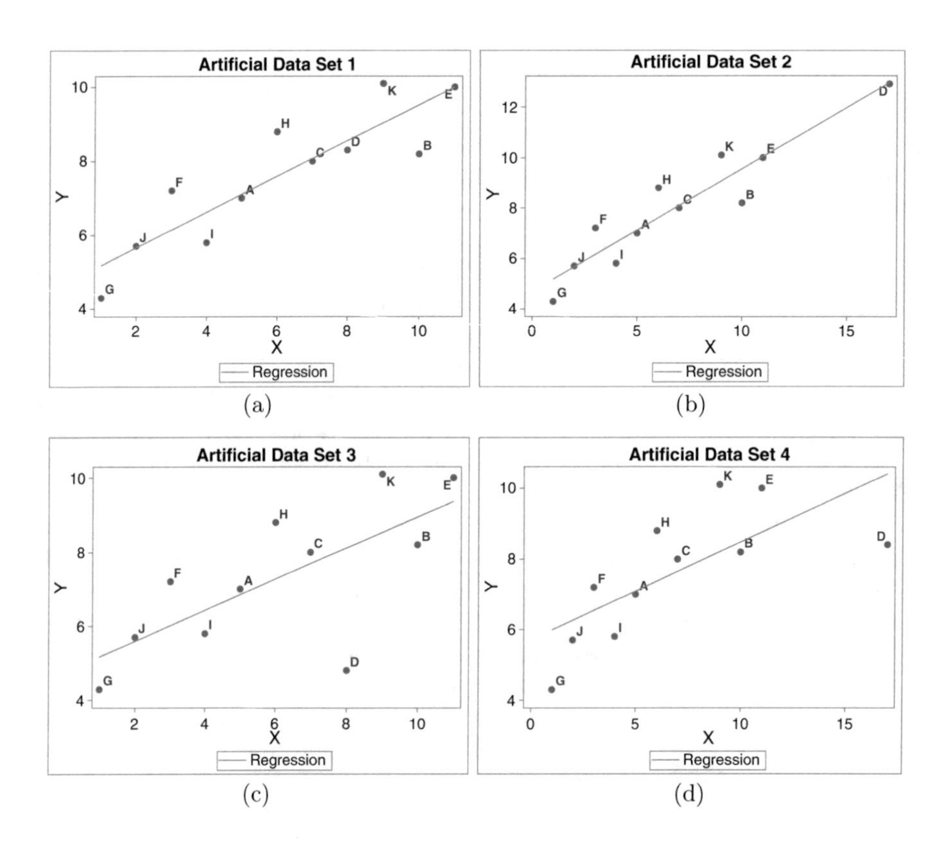

Fig. 4.8. (**a**) Artificial Data Set 1, (**b**) Artificial Data Set 2, (**c**) Artificial Data Set 3, (**d**) Artificial Data Set 4

larger than 0.36. As observed from Fig. 4.8b, the point labeled D is located away from the centroid of the other x-values, far to the right. An externally studentized residual (`RStudent`) value of -1.86 is again not significant clearly showing that there are no y-outliers. Influence (`Cook's D`) values are all small indicating that no observations is influential. Comparing the summary statistics of the fits of Artificial Data Sets 1 and 2 (see Fig. 4.13) shows that the fits are almost identical (only the MSE increased slightly), so the introduction of an x-outlier did not affect the fit appreciably, since it was not also a y-outlier.

The case statistics resulting from fitting a simple linear regression model to Artificial Data Set 3 (see Fig. 4.11) show that, again, the leverages (`Hat Diag H`) are all smaller than 0.36. Thus, there are no x-outliers, as confirmed by inspecting Fig. 4.8c, where the x-values are, again, evenly spread out between 0

Output Statistics

Obs	Name	Dependent Variable	Predicted Value	Std Error Mean Predict	Residual	Std Error Residual	Student Residual	-2-1 0 1 2
1	A	7.0000	7.0982	0.2786	-0.0982	0.836	-0.117	| | |
2	B	8.2000	9.5164	0.4284	-1.3164	0.770	-1.710	| ***| |
3	C	8.0000	8.0655	0.2786	-0.0655	0.836	-0.0783	| | |
4	D	8.3000	8.5491	0.3143	-0.2491	0.823	-0.303	| | |
5	E	10.0000	10.0000	0.4970	-1.78E-15	0.728	-24E-16	| | |
6	F	7.2000	6.1309	0.3662	1.0691	0.801	1.334	| |** |
7	G	4.3000	5.1636	0.4970	-0.8636	0.728	-1.187	| **| |
8	H	8.8000	7.5818	0.2657	1.2182	0.840	1.450	| |** |
9	I	5.8000	6.6145	0.3143	-0.8145	0.823	-0.990	| *| |
10	J	5.7000	5.6473	0.4284	0.0527	0.770	0.0685	| | |
11	K	10.1000	9.0327	0.3662	1.0673	0.801	1.332	| |** |

Output Statistics

Obs	Name	Cook's D	RStudent	Hat Diag H	Cov Ratio	DFFITS	DFBETAS	
1	A	0.001	-0.1108	0.1000	1.4019	-0.0369	-0.0263	0.0111
2	B	0.452	-1.9616	0.2364	0.7556	-1.0913	0.4418	-0.8561
3	C	0.000	-0.0739	0.1000	1.4043	-0.0246	-0.0044	-0.0074
4	D	0.007	-0.2868	0.1273	1.4208	-0.1095	0.0086	-0.0585
5	E	0.000	-2.3E-15	0.3182	1.8563	-0.0000	0.0000	-0.0000
6	F	0.186	1.4042	0.1727	0.9847	0.6416	0.6077	-0.4416
7	G	0.329	-1.2186	0.3182	1.3205	-0.8325	-0.8299	0.7036
8	H	0.105	1.5617	0.0909	0.8177	0.4938	0.2303	0.0000
9	I	0.071	-0.9883	0.1273	1.1518	-0.3774	-0.3272	0.2017
10	J	0.001	0.0646	0.2364	1.6556	0.0359	0.0353	-0.0282
11	K	0.185	1.4012	0.1727	0.9863	0.6403	-0.1733	0.4407

Fig. 4.9. Diagnostics for the fit of Artificial Data Set 1

Output Statistics									
Obs	Name	Dependent Variable	Predicted Value	Std Error Mean Predict	Residual	Std Error Residual	Student Residual	-2-1 0 1 2
1	A	7.0000	7.1184	0.2854	-0.1184	0.829	-0.143	| | |
2	B	8.2000	9.5428	0.3245	-1.3428	0.815	-1.649	| ***| |
3	C	8.0000	8.0882	0.2646	-0.0882	0.836	-0.105	| | |
4	D	12.9000	12.9371	0.6578	-0.0371	0.580	-0.0640	| | |
5	E	10.0000	10.0277	0.3621	-0.0277	0.799	-0.0347	| | |
6	F	7.2000	6.1486	0.3477	1.0514	0.805	1.306	| |** |
7	G	4.3000	5.1788	0.4340	-0.8788	0.762	-1.154	| **| |
8	H	8.8000	7.6033	0.2688	1.1967	0.835	1.434	| |** |
9	I	5.8000	6.6335	0.3126	-0.8335	0.819	-1.017	| **| |
10	J	5.7000	5.6637	0.3888	0.0363	0.786	0.0462	| | |
11	K	10.1000	9.0579	0.2942	1.0421	0.826	1.262	| |** |

Output Statistics									
Obs	Name	Cook's D	RStudent	Hat Diag H	Cov Ratio	DFFITS	DFBETAS	
1	A	0.001	-0.1348	0.1060	1.4092	-0.0464	-0.0382	0.0175
2	B	0.216	-1.8604	0.1370	0.7145	-0.7413	0.0286	-0.4300
3	C	0.001	-0.0995	0.0911	1.3890	-0.0315	-0.0162	-0.0013
4	D	0.003	-0.0603	0.5629	2.8930	-0.0685	0.0374	-0.0627
5	E	0.000	-0.0327	0.1705	1.5254	-0.0148	0.0025	-0.0101
6	F	0.159	1.3681	0.1573	0.9864	0.5910	0.5674	-0.3839
7	G	0.216	-1.1781	0.2450	1.2173	-0.6712	-0.6693	0.5323
8	H	0.107	1.5391	0.0940	0.8315	0.4956	0.3419	-0.0893
9	I	0.075	-1.0197	0.1271	1.1355	-0.3890	-0.3540	0.2075
10	J	0.000	0.0436	0.1966	1.5746	0.0215	0.0212	-0.0158
11	K	0.101	1.3110	0.1126	0.9663	0.4670	0.0586	0.2049

Fig. 4.10. Diagnostics for the fit of Artificial Data Set 2

and 12. An externally studentized residual (`RStudent`) value of -3.8 probably has a p-value close to 0.05. Thus case number 4 is close to being a y-outlier. The influence (`Cook's D`) value for case number 4 is quite large; thus, this case may also be considered influential. As discussed in Sect. 4.2.1, this is inflated due to the fact that it is a possible y-outlier (large **`Studentized Residual`**) although it is not an x-outlier. Comparing the summary statistics of the fits of Artificial Data Sets 1 and 3 (see Fig. 4.13) shows that both R^2 and MSE have changed substantially although the fitted line is almost identical. Thus, the presence of a y-outlier that is not an x-outlier may not substantially change the fitted line but may affect the estimate of the error variance, which, in turn, affects the hypothesis tests, confidence intervals, and prediction intervals.

In Fig. 4.12, the leverage (`Hat Diag H`) for case number 4 is again large indicating that case number 4 is an x-outlier. This is also verified from Fig. 4.8d, where this point lies far to the right of the rest of the data. An externally studentized residual (`RStudent`) value of -3.26 probably is not significant at 0.05 but is still quite large. The influence (`Cook's D`) value for case number 4 is extremely large and, thus, this case is highly influential. Again, this is inflated due to the fact that it is a y-outlier (large studentized residual) as well as an x-outlier (high leverage). Comparing the summary statistics of the fits of Artificial Data Sets 1 and 4 (see Fig. 4.13) shows that all summary statistics for the fit of data set 4 have changed substantially from those for the other fits. Thus, the presence of a y-outlier that is influential drastically affects the fit of a model.

Output Statistics

Obs	Name	Dependent Variable	Predicted Value	Std Error Mean Predict	Residual	Std Error Residual	Student Residual	-2-1 0 1 2
1	A	7.0000	6.8436	0.4645	0.1564	1.394	0.112	| | |
2	B	8.2000	8.9436	0.7142	-0.7436	1.284	-0.579	| *| |
3	C	8.0000	7.6836	0.4645	0.3164	1.394	0.227	| | |
4	D	4.8000	8.1036	0.5241	-3.3036	1.372	-2.407	| ****| |
5	E	10.0000	9.3636	0.8286	0.6364	1.213	0.525	| |* |
6	F	7.2000	6.0036	0.6105	1.1964	1.336	0.895	| |* |
7	G	4.3000	5.1636	0.8286	-0.8636	1.213	-0.712	| *| |
8	H	8.8000	7.2636	0.4429	1.5364	1.401	1.097	| |** |
9	I	5.8000	6.4236	0.5241	-0.6236	1.372	-0.454	| | |
10	J	5.7000	5.5836	0.7142	0.1164	1.284	0.0906	| | |
11	K	10.1000	8.5236	0.6105	1.5764	1.336	1.180	| |** |

Output Statistics

Obs	Name	Cook's D	RStudent	Hat Diag H	Cov Ratio	DFFITS	DFBETAS	
1	A	0.001	0.1059	0.1000	1.4023	0.0353	0.0251	-0.0106
2	B	0.052	-0.5566	0.2364	1.5361	-0.3097	0.1254	-0.2429
3	C	0.003	0.2146	0.1000	1.3902	0.0715	0.0127	0.0216
4	D	0.423	-3.8034	0.1273	0.1839	-1.4525	0.1145	-0.7764
5	E	0.064	0.5024	0.3182	1.7445	0.3432	-0.1711	0.2900
6	F	0.084	0.8845	0.1727	1.2694	0.4042	0.3828	-0.2782
7	G	0.118	-0.6910	0.3182	1.6530	-0.4721	-0.4706	0.3990
8	H	0.060	1.1111	0.0909	1.0448	0.3514	0.1638	0.0000
9	I	0.015	-0.4334	0.1273	1.3844	-0.1655	-0.1435	0.0885
10	J	0.001	0.0855	0.2364	1.6543	0.0476	0.0468	-0.0373
11	K	0.145	1.2098	0.1727	1.0932	0.5528	-0.1496	0.3805

Fig. 4.11. Diagnostics for the fit of Artificial Data Set 3

<table>
<tr><th colspan="9">Output Statistics</th></tr>
<tr><th>Obs</th><th>Name</th><th>Dependent Variable</th><th>Predicted Value</th><th>Std Error Mean Predict</th><th>Residual</th><th>Std Error Residual</th><th>Student Residual</th><th>-2-1 0 1 2</th></tr>
<tr><td>1</td><td>A</td><td>7.0000</td><td>7.0886</td><td>0.4354</td><td>-0.0886</td><td>1.265</td><td>-0.0700</td><td>| | |</td></tr>
<tr><td>2</td><td>B</td><td>8.2000</td><td>8.4700</td><td>0.4951</td><td>-0.2700</td><td>1.243</td><td>-0.217</td><td>| | |</td></tr>
<tr><td>3</td><td>C</td><td>8.0000</td><td>7.6411</td><td>0.4036</td><td>0.3589</td><td>1.275</td><td>0.281</td><td>| | |</td></tr>
<tr><td>4</td><td>D</td><td>8.4000</td><td>10.4040</td><td>1.0036</td><td>-2.0040</td><td>0.884</td><td>-2.266</td><td>| ****| |</td></tr>
<tr><td>5</td><td>E</td><td>10.0000</td><td>8.7463</td><td>0.5524</td><td>1.2537</td><td>1.218</td><td>1.029</td><td>| |** |</td></tr>
<tr><td>6</td><td>F</td><td>7.2000</td><td>6.5360</td><td>0.5305</td><td>0.6640</td><td>1.228</td><td>0.541</td><td>| |* |</td></tr>
<tr><td>7</td><td>G</td><td>4.3000</td><td>5.9834</td><td>0.6621</td><td>-1.6834</td><td>1.162</td><td>-1.448</td><td>| **| |</td></tr>
<tr><td>8</td><td>H</td><td>8.8000</td><td>7.3649</td><td>0.4100</td><td>1.4351</td><td>1.273</td><td>1.127</td><td>| |** |</td></tr>
<tr><td>9</td><td>I</td><td>5.8000</td><td>6.8123</td><td>0.4768</td><td>-1.0123</td><td>1.250</td><td>-0.810</td><td>| *| |</td></tr>
<tr><td>10</td><td>J</td><td>5.7000</td><td>6.2597</td><td>0.5931</td><td>-0.5597</td><td>1.199</td><td>-0.467</td><td>| | |</td></tr>
<tr><td>11</td><td>K</td><td>10.1000</td><td>8.1937</td><td>0.4488</td><td>1.9063</td><td>1.260</td><td>1.513</td><td>| |*** |</td></tr>
</table>

Output Statistics								
Obs	Name	Cook's D	RStudent	Hat Diag H	Cov Ratio	DFFITS	DFBETAS	
1	A	0.000	-0.0660	0.1060	1.4141	-0.0227	-0.0187	0.0086
2	B	0.004	-0.2054	0.1370	1.4512	-0.0818	0.0032	-0.0475
3	C	0.004	0.2665	0.0911	1.3680	0.0843	0.0433	0.0034
4	D	3.307	-3.2601	0.5629	0.5340	-3.6997	2.0186	-3.3878
5	E	0.109	1.0329	0.1705	1.1879	0.4683	-0.0802	0.3200
6	F	0.027	0.5183	0.1573	1.4058	0.2239	0.2150	-0.1455
7	G	0.340	-1.5594	0.2450	0.9859	-0.8884	-0.8859	0.7046
8	H	0.066	1.1467	0.0940	1.0303	0.3693	0.2547	-0.0665
9	I	0.048	-0.7931	0.1271	1.2462	-0.3026	-0.2753	0.1614
10	J	0.027	-0.4456	0.1966	1.5000	-0.2204	-0.2173	0.1616
11	K	0.145	1.6517	0.1126	0.7931	0.5883	0.0739	0.2581

Fig. 4.12. Diagnostics for the fit of Artificial Data Set 4

Model	$\hat{\beta}_0$	$\hat{\beta}_1$	MSE	R^2
1	4.69	0.48	0.78	0.78
2	4.69	0.48	0.77	0.88
3	4.74	0.42	2.16	0.50
4	5.70	0.28	1.79	0.50

Fig. 4.13. Summary of fit statistics for Artificial Data Sets 1–4

The `DFFITS` values are scaled measures of the change in each predicted value (or the *fit*) when a case is deleted and thus measure the influence of a deleted case on the *prediction* from the fitted model. It can be shown that the magnitude of a `DFFITS` tends to be large when the case is a y-outlier, x-outlier, or both (a la Cook's D). A suggested measure of high influence is a value larger than 1 for smaller data sets and larger than $2\sqrt{(k+1)/n}$ for large data sets. Whereas `Cook's D` measures the influence of a case on *all* fitted values jointly, `DFFITS` measures the influence of a case on an *individual* fitted value. It can be observed clearly that the interpretation of `DFFITS` for the above four model fits closely follow that of `Cook's D`.

The `DFBETAS` values are scaled measures of the change in each parameter estimate when a case is deleted and thus measure the influence of a deleted case on the *estimation*. The *sign* of a `DFBETAS` value indicates whether the inclusion of the case leads to an increase or a decrease of the estimated coefficient. The magnitude of a `DFBETAS` value indicates the impact or influence of the case on estimating a regression coefficient. A suggested measure of high influence is a value larger than 1 for smaller data sets and larger than $2/\sqrt{n}$ for large data sets. In Fig. 4.12 the case labeled D clearly has a large `DFBETAS` value; the estimation of the parameters would be clearly affected by deleting this observation from the data set. This is confirmed by observing the estimated parameter values for Model 4 in Fig. 4.13. Thus it is clear that both estimation of parameters and prediction are affected by the case labeled D in the fit of Artificial Data Set 4.

4.1.4 Prediction of New y Values Using Regression

There are two possible interpretations of a y prediction at a specified value of x. Recall that the prediction equation for the lead content data is $\hat{y} = -12.84 + 36.18\,x$, where x = traffic flow in thousands of cars per 24 hours and y = lead content of bark on trees near the highway in $\mu g/g$ dry weight. If $x = 10$ is substituted in this equation, the value $\hat{y} = 348.96$ is obtained. This `predicted value` of y can be interpreted as either

- The estimate of the average or mean lead content of bark $E(y)$ near all highways with traffic flow of 10,000 cars per 24 hours is 348.96 $\mu g/g$ dry weight,

or

- The lead content of bark y of a specific highway randomly selected from the set of all highways with a traffic flow of 10,000 cars per 24 hours is 348.96 $\mu g/g$ dry weight.

The difference in the two predictions is that the standard error of predictions will be different. Since it is possible to more accurately predict a mean than an individual value, the first type of prediction will have less error than the second type. Thus, the confidence interval calculated for the mean of y, $E(y)$, for a given x will be narrower than the prediction interval calculated for a new value y at a given x.

SAS Example D4

```
data d4;
input Sample $ X Y;
label X="Traffic Flow" Y="Lead Content";
datalines;
A 8.3  227
B 8.3  312
C 12.1  362
D 12.1  521
E 17.0  640
F 17.0  539
G 17.0  728
H 24.3  945
I 24.3  738
J 24.3  759
K 33.6 1263
L 10.0    .
M 15.0    .
;

proc reg data=d4 plots(only label)=(diagnostics fit);
  model Y=X/clm cli;
  id Sample;
  title 'Prediction Intervals: Lead Content Data';
run;
```

Fig. 4.14. SAS Example D4: program

In `proc reg`, these intervals are calculated for the observations in the input data set and printed as part of the output of case statistics. However, if these intervals are required for observations with new x-values, then these observations must be included as cases in the input data set with missing value indicators (periods) as the corresponding y-values. In the SAS Example D4 program (see Fig. 4.14), which is a modified version of the SAS Example D1 program, the original lead content data have been supplemented by adding two cases (Samples labeled L and M) each with values of 10.0 and 15.0 for `Traffic Flow` and missing values for `Lead Content`, respectively. `Proc reg` fits the regression model using only the first 11 cases and calculates the two types of prediction intervals for all cases, including the two cases with the new x-values. The output statistics from the SAS Example D4 program are displayed in Fig. 4.15.

Assuming normally distributed data and the first type of prediction, `proc reg` calculates $(1-\alpha)100\%$ *confidence interval* for $E(y) = \beta_0 + \beta_1 x$, for each x-value in the data, including those cases with missing values specified for y. The second type of prediction is used to calculate a $(1-\alpha)100\%$ *prediction interval* for a new observation y at each x-value in the data, including the new x-values specified. For a default value of $\alpha = 0.05$, these are obtained by specifying `clm` and/or `cli`, respectively, for the confidence intervals and the prediction intervals as model statement options, as shown in Fig. 4.14.

Output Statistics									
Obs	**Sample**	**Dependent Variable**	**Predicted Value**	**Std Error Mean Predict**	**95% CL Mean**		**95% CL Predict**		**Residual**
1	A	227.0000	287.4844	45.4206	184.7359	390.2328	54.9967	519.9721	-60.4844
2	B	312.0000	287.4844	45.4206	184.7359	390.2328	54.9967	519.9721	24.5156
3	C	362.0000	424.9830	35.3804	344.9470	505.0190	201.6021	648.3640	-62.9830
4	D	521.0000	424.9830	35.3804	344.9470	505.0190	201.6021	648.3640	96.0170
5	E	640.0000	602.2839	28.0543	538.8206	665.7471	384.2911	820.2767	37.7161
6	F	539.0000	602.2839	28.0543	538.8206	665.7471	384.2911	820.2767	-63.2839
7	G	728.0000	602.2839	28.0543	538.8206	665.7471	384.2911	820.2767	125.7161
8	H	945.0000	866.4260	36.1835	784.5731	948.2788	642.3876	1090	78.5740
9	I	738.0000	866.4260	36.1835	784.5731	948.2788	642.3876	1090	-128.4260
10	J	759.0000	866.4260	36.1835	784.5731	948.2788	642.3876	1090	-107.4260
11	K	1263	1203	63.8739	1058	1347	949.2204	1457	60.0643
12	L	.	348.9969	40.6376	257.0684	440.9255	121.0843	576.9095	.
13	M	.	529.9162	29.9605	462.1408	597.6915	310.6292	749.2032	.

Fig. 4.15. Prediction intervals: lead content data

The `alpha=` model statement option may be used to change the default value of the confidence coefficient, noting that this will affect the intervals calculated for the β coefficients using the `clb` option, as well, if it is used in the same model statement. These sets of intervals are tabulated under the headings `95% CL Mean` and `95% CL Predict`, respectively, in Fig. 4.15.

The `plots=` option in the `proc reg` statement is used SAS Example D1 (Fig. 4.1) to control the ODS Graphics output from the regression procedure. By default (that is , when no `plots=` options are used), `proc reg` outputs the *regression diagnostics panel*, shown in Fig. 4.16, the residual plot (that is a plot of the residuals against the explanatory variable, as shown in Fig. 4.4b, and the scatter plot of the data overlaid with the fitted regression line, called the *fit plot* and shown in Fig. 4.4a. Note that the fit plot is produced only in the case when there is only one explanatory variable. All graphical output will be suppressed if `plots=none` option is used.

In SAS Example D4, the `plots(only label)=(diagnostics fit)` option is used to select the diagnostics panel (Fig. 4.16) and the fit plot (Fig. 4.17) to be the only plots generated. The sub-option `only` suppresses all default plots; only plots specifically requested using sub-options, here `diagnostics` and `fit`, being output. The diagnostics panel includes the plots of residuals against predicted values, externally studentized residuals (`RStudent`) against the predicted values, `RStudent` against leverage, normal quantile plot of the residuals, dependent variable against the predicted values, and an index plot of `Cook's D`. Any subset of these plots may be obtained separately by specifying the appropriate sub-options, e.g., `plots(only)=(fit(nolimits)`

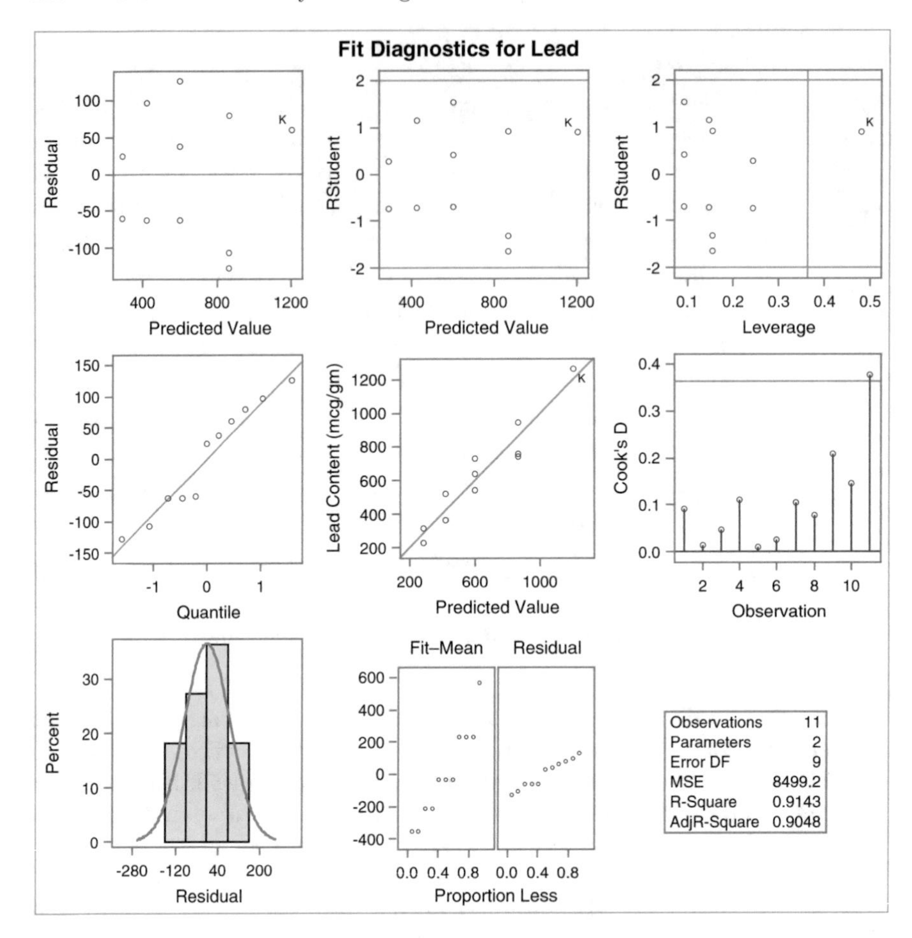

Fig. 4.16. Diagnostics panel: lead content data

`residuals residualbypredicted qq)` as illustrated in SAS Example D1 (see Fig. 4.1). The `label` sub-option identifies each point on various plots by placing a *label* near a plotted point on those plots when a point is deemed an outlier or influential on the appropriate plots. The label is the value of the `id` variable corresponding to the point, or if an `id` statement is not included in the proc step, the index value of the case corresponding to the point. Table B.17 lists a selected sample of `plots=` options available for generating specific ODS Graphics output in `proc reg`.

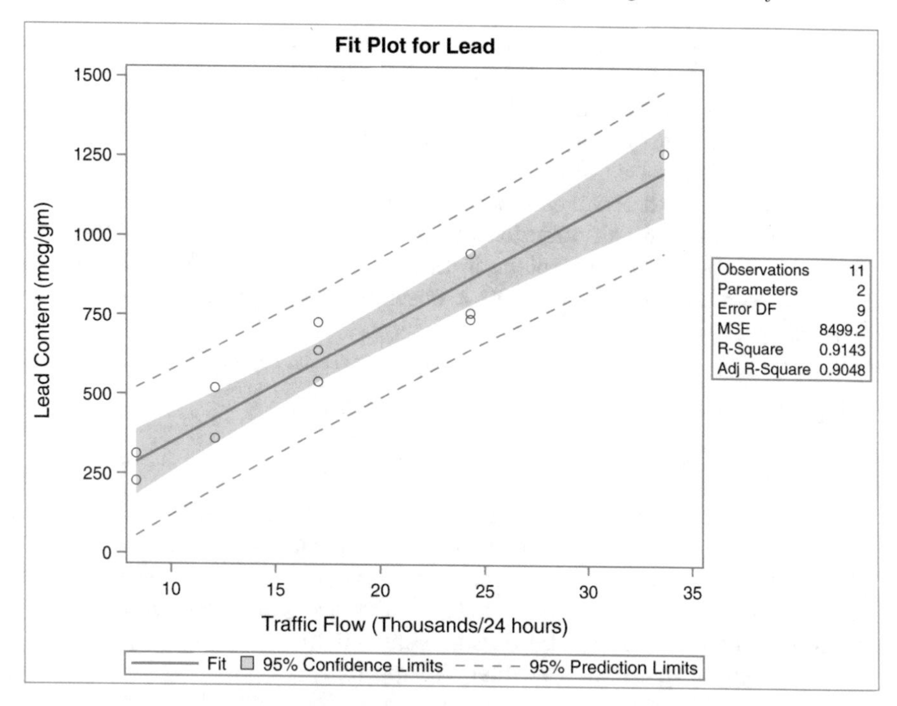

Fig. 4.17. Fit plot: lead content data

4.2 An Introduction to Multiple Regression Analysis

In this section the simple linear regression model introduced in Sect. 4.1 is extended to the multiple linear regression case to handle situations in which the dependent variable y is modeled by a relationship that involves more than a single independent or explanatory variable, say $x_1, x_2, \ldots, x_k$. The study of multiple regression analysis may thus be viewed as the approximation of the functional relationship that may exist between a variable y and another set of variables $x_1, x_2, \ldots, x_k$. This relationship may sometimes model a postulated theoretical relationship between y and the x variables, whereas at other times it may simply be a mathematical equation that approximates such a relationship. Such an equation is useful for prediction of a value for y when the values of the x variables are known.

Multiple Regression Model

When this approximation is linear in the unknown parameters, it is called a multiple linear regression model and is expressed in the form

$$y = \beta_0 + \beta_1 x_1 + \cdots + \beta_k x_k + \epsilon$$

The variable y is the response or the dependent variable, which is a realization of a random variable Y observed for a fixed set of values of the explanatory variables $x_1, x_2, \ldots, x_k$. The coefficients $\beta_0, \beta_1, \ldots, \beta_k$ are unknown constants and ϵ is an unobservable random variable representing the random error in observing y. Under this model, the y-values observed at each x-value are a random sample with the mean $E(y) = \mu(x) = \beta_0 + \beta_1 x_1 + \cdots + \beta_k x_k$, i.e., the mean is modeled as a function of the x values. This function is linear in $\beta_0, \beta_1, \ldots, \beta_k$, and is a hyperplane in the parameter space of the coefficients. Multiple regression data consist of n observations or cases of $k+1$ values denoted by

$$(y_i,\, x_{i1},\, x_{i2}, \ldots, x_{ik}), \quad i = 1, 2, \ldots, n$$

Using this full notation, the model may be rewritten as

$$y_i = \beta_0 + \beta_1\, x_{i1} + \beta_2\, x_{i2} + \cdots + \beta_k\, x_{ik} + \epsilon_i \quad i = 1, 2, \ldots, n.$$

In addition to the fact that the random errors for the individual observations have mean zero, they are also assumed to be uncorrelated random variables with the same variance σ^2. The parameters $\beta_0, \beta_1, \ldots, \beta_k$ and σ^2 can be estimated using the least squares method. For the purpose of making valid statistical inference, it may be further assumed that the errors $\epsilon_1, \epsilon_2, \ldots, \epsilon_n$ are a random sample from a normal distribution.

Note that the definition of the regression model above admits models that may include squared terms, product terms, etc. of *quantitative* explanatory variables (x variables). Thus, a model of the form

$$y = \beta_0 + \beta_1\, x_1 + \beta_2\, x_2 + \beta_3\, x_1^2 + \beta_4\, x_2^2 + \beta_5\, x_1^3 + \beta_6\, x_2^3 + \beta_7\, x_1\, x_2 + \epsilon$$

can also be expressed in the form of a multiple regression model

$$y = \beta_0 + \beta_1\, x_1 + \beta_2\, x_2 + \beta_3\, x_3 + \beta_4\, x_4 + \beta_5\, x_5 + \beta_6\, x_6 + \beta_7 x_7 + \epsilon$$

where the values for the variables x_3, x_4, x_5, x_6, and x_7 are obtained by substituting the values of x_1 and x_2 in the expressions $x_3 = x_1^2$, $x_4 = x_2^2$, $x_5 = x_1^3$, $x_6 = x_2^3$, and $x_7 = x_1\, x_2$, respectively. The inclusion of a product term such as $x_1\, x_2$ in a model allows the investigator to perform a statistical test of the absence (or presence) of *interaction* between two independent variables (say) x_1 and x_2. The concept of interaction will be discussed in a later section.

Estimation of Parameters

The *method of least squares* is used to obtain the estimates of the regression coefficients using the observed data. The least squares estimates of the β denoted by $\hat{\beta}_0, \hat{\beta}_1, \ldots, \hat{\beta}_k$ are obtained by minimizing the sum of squares of residuals:

$$Q = \sum_{i=1}^{n} \{y_i - (\beta_0 + \beta_1\, x_{1i} + \cdots + \beta_k\, x_{ki})\}^2$$

where $y_i, x_{1i}, x_{2i}, \ldots, x_{ki}$, $i = 1, \ldots, n$, denotes the n observations (or cases). The estimates are found by setting the partial derivatives of Q with respect to each of the β coefficients equal to zero. The resulting set of equations, called the *normal equations*, is linear in the β. These are solved to yield the estimates of the parameters denoted by $\hat{\beta}$. The prediction equation or the fitted regression model is then

$$\hat{y} = \hat{\beta}_0 + \hat{\beta}_1 x_1 + \cdots + \hat{\beta}_k x_k$$

The predicted or fitted values of y_i, $i = 1, \ldots, n$, that correspond to the n observations (or cases) are calculated by substituting the n sets of observed values of the explanatory variables $x_{1i}, x_{2i}, \ldots, x_{ki}$, $i = 1, \ldots, n$, in the prediction equation to obtain $\hat{y}_i$, $i = 1, \ldots, n$. The predicted or fitted value, say $\hat{y}_{new}$, that corresponds to a new case $x_{1,new}, x_{2,new}, \ldots, x_{k,new}$, is calculated by substituting these values in the prediction equation.

Matrix Notation

In matrix notation, the linear regression model may be expressed in the form

$$\mathbf{y} = X\boldsymbol{\beta} + \boldsymbol{\epsilon}$$

where

$$\mathbf{y} = \begin{bmatrix} y_1 \\ y_2 \\ \vdots \\ y_n \end{bmatrix}, \quad X = \begin{bmatrix} 1 & x_{11} & \cdots & x_{k1} \\ 1 & x_{12} & \cdots & x_{k2} \\ \vdots & \vdots & & \vdots \\ 1 & x_{1n} & & x_{kn} \end{bmatrix}, \boldsymbol{\beta} = \begin{bmatrix} \beta_0 \\ \beta_1 \\ \vdots \\ \beta_k \end{bmatrix}, \text{ and } \boldsymbol{\epsilon} = \begin{bmatrix} \epsilon_1 \\ \epsilon_2 \\ \vdots \\ \epsilon_n \end{bmatrix}$$

In this notation, the sum of squares to be minimized is

$$Q = (\mathbf{y} - X\boldsymbol{\beta})'(\mathbf{y} - X\boldsymbol{\beta})$$

and the resulting normal equations are

$$X'X\boldsymbol{\beta} = X'\mathbf{y}.$$

The solution to the normal equations gives the least squares estimate $\hat{\boldsymbol{\beta}}$ of $\boldsymbol{\beta}$:

$$\hat{\boldsymbol{\beta}} = (X'X)^{-1}X'\mathbf{y}$$

where $(X'X)^{-1}$ is the inverse of the $X'X$ matrix assumed to be nonsingular. Note that $X'X$, a $(k+1) \times (k+1)$ matrix, is an important quantity associated with multiple regression computations. An analysis of variance table for testing the hypothesis

$$H_0 : \beta_1 = \beta_2 = \cdots = \beta_k = 0 \qquad \text{versus} \qquad \mathrm{H_a} : \text{at least one } \beta \neq 0$$

where matrix expressions are used to define computational formulas for the various sums of square, is given as follows:

Source	df	SS	MS	F
Regression	k	$\hat{\boldsymbol{\beta}}' X'\mathbf{y} - n\bar{y}^2$	MSReg= SSReg/k	MSReg/MSE
Error	$n-k-1$	$\mathbf{y}'\mathbf{y} - \hat{\boldsymbol{\beta}}' X'\mathbf{y}$	MSE=SSE/$(n-k-1)$	
Total	$n-1$	$\mathbf{y}'\mathbf{y} - n\bar{y}^2$		

where $\bar{y} = \sum_{i=1}^{n} y_i/n$, $\mathbf{y}'\mathbf{y} = \sum_{i=1}^{n} y_i^2$, SSReg denotes the regression sum of squares given by $\hat{\boldsymbol{\beta}}' X'\mathbf{y} - n\bar{y}^2$, and SSE denotes the residual or error sum of squares given by $\mathbf{y}'\mathbf{y} - \hat{\boldsymbol{\beta}}' X'\mathbf{y}$. The null hypothesis is rejected if the above F-statistic exceeds the α level upper percentage point of the F-distribution with $(k, n-k-1)$ degrees of freedom. The mean square for error (MSE) denoted by $s^2 = \text{SSE}/(n-k-1)$ is an unbiased estimate of σ^2, the variance of the random errors, that is, $\hat{\sigma}^2 = s^2$. Another quantity of statistical interest computed from the above analysis is the coefficient of determination or the multiple correlation coefficient R^2 given by

$$R^2 = \text{ Regression SS/Total (Corrected) SS} = \text{SSReg/SSTot}$$

where SSTot $= \mathbf{y}'\mathbf{y} - n\bar{y}^2$. R^2 measures the proportion of the variability of y explained by the fitted regression model. Further, suppose that elements of the inverse of the $(X'X)^{-1}$ matrix are denoted by c_{ij}. Thus,

$$(X'X)^{-1} = \begin{bmatrix} c_{00} & c_{01} & c_{02} & \cdots & c_{0k} \\ c_{10} & c_{11} & c_{12} & \cdots & c_{1k} \\ \vdots & \vdots & & & \vdots \\ c_{k0} & c_{k1} & & \cdots & c_{kk} \end{bmatrix}$$

Then the standard error of the least squares estimate of the mth regression coefficients $\hat{\beta}_m$ is given by $s_{\hat{\beta}_m} = c_{mm}^{1/2} s$. Thus, a t-statistic for testing the hypothesis $H_0 : \beta_m = 0$ versus $H_a : \beta_m \neq 0$ is

$$t = \hat{\beta}_m / s_{\hat{\beta}_m}$$

and a $(1-\alpha)100\%$ confidence interval for β_m is

$$\hat{\beta}_m \pm t_{\alpha/2,(n-k-1)} \times s_{\hat{\beta}_m}$$

where $t_{\alpha/2,(n-k-1)}$ is the upper $\alpha/2$ critical value of the t-distribution with $(n-k-1)$ degrees of freedom.

4.2.1 Multiple Regression Analysis Using PROC REG

The data, popularly called Hald data and shown in Table B.8, are taken from Draper and Smith (1981). The explanatory variables X_1, X_2, X_3, and X_4 are percentages of the primary chemical components of clinkers from which cement is made and the response variable y is the heat evolved as the concrete hardens, measured in calories per gram of cement. The objective is to model the heat produced as a linear function of the composition of clinkers. These data will be used here to illustrate procedures in SAS that are useful in the process of regression model building. The purpose of SAS Example D5 is fitting a multiple regression model to these data and producing statistics discussed in Sect. 4.2. Diagnostics and residual plots necessary for examining the adequacy of the model and accompanying assumptions will be calculated for the same regression model in later sections.

```
data  cement;
input X1-X4 y;
datalines;
 7 26  6 60  78.5
 1 29 15 52  74.3
11 56  8 20 104.3
11 31  8 47  87.6
 7 52  6 33  95.9
11 55  9 22 109.2
 3 71 17  6 102.7
 1 31 22 44  72.5
 2 54 18 22  93.1
21 47  4 26 115.9
 1 40 23 34  83.8
11 66  9 12 113.3
10 68  8 12 109.4
;

proc sgscatter data=cement;
  title "Scatter plot Matrix for Hald Data";
  matrix X1-X4 y;
run;

proc reg data=cement simple corr plots=none;
  model y = X1 X2 X3 X4/clb xpx i;
  title 'Regression Analysis of Hald Data';
run;
```

Fig. 4.18. SAS Example D5: program

SAS Example D5

In the SAS Example D5 program (see Fig. 4.18), the multiple linear regression model

$$y = \beta_0 + \beta_1 X_1 + \beta_2 X_2 + \beta_3 X_3 + \beta_4 X_4 + \epsilon \tag{4.5}$$

is fitted to these data using `proc reg`. Preceding the `proc reg` step, `proc sgscatter` is used to create a scatter plot matrix of the four explanatory variables plus the response variable. The scatter plot matrix and `proc sgscatter` were discussed in Sect. 3.3.3 (refer to SAS Example C12). The scatter plot matrix is a two-dimensional array of scatter plots of all pairs of variables and enables the user to study the relationships between pairs of variables visually. The output from this step is shown in Fig. 4.19

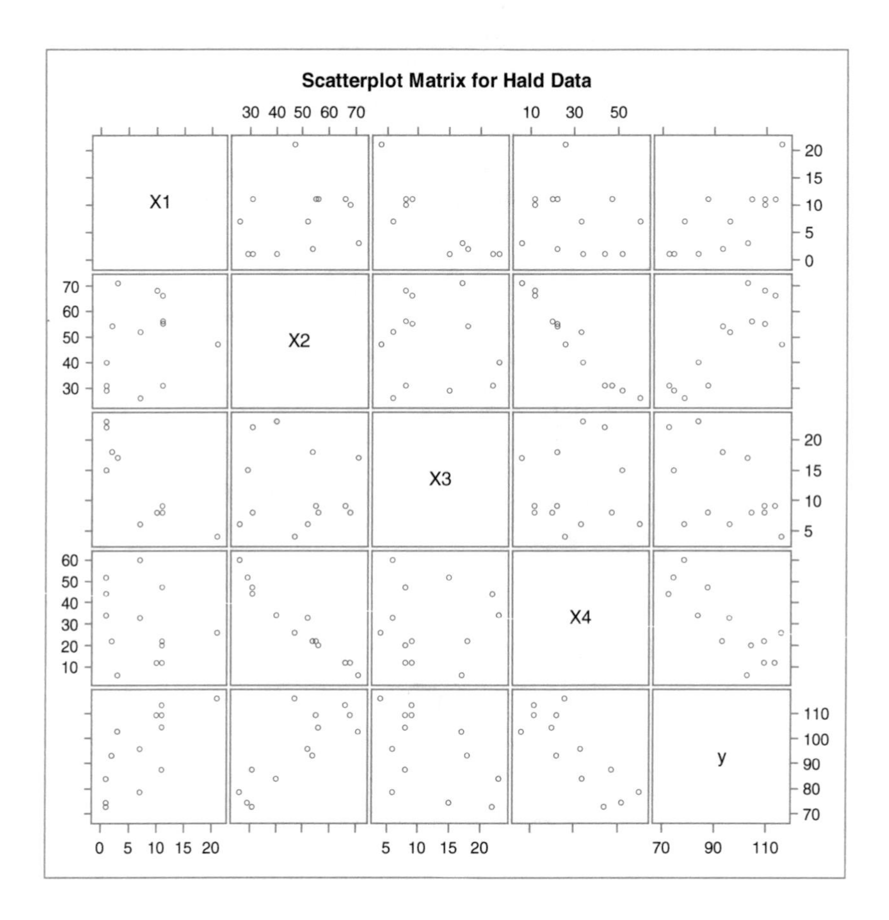

Fig. 4.19. Scatter plot matrix: Hald data

The `proc reg` step produces output containing statistics for making some of the statistical inferences about the model as discussed in the previous section. It contains SAS statements necessary to compute an analysis of variance, estimated values of the parameters, their standard errors, associated t-statistics, and confidence intervals. Additionally, options used with the `model` statement request that information about the normal equations be output.

The output from this step is displayed in Figs. 4.20, 4.21, 4.22, 4.23, and 4.24. The keyword options `simple` and `corr` used with `proc reg` result in the output of descriptive statistics for each of the variables and the correlation matrix displayed in Fig. 4.20.

Correlation					
Variable	**X1**	**X2**	**X3**	**X4**	**y**
X1	1.0000	0.2286	-0.8241	-0.2454	0.7307
X2	0.2286	1.0000	-0.1392	-0.9730	0.8163
X3	-0.8241	-0.1392	1.0000	0.0295	-0.5347
X4	-0.2454	-0.9730	0.0295	1.0000	-0.8213
y	0.7307	0.8163	-0.5347	-0.8213	1.0000

Fig. 4.20. Correlation matrix: Hald data

The model statement options `xpx` and `i` results in the output of the $X'X$ matrix shown in Fig. 4.21 and the inverse of the $X'X$ matrix shown in Fig. 4.22, respectively. The $X'X$ matrix and $X'\mathbf{y}$ vector displayed below are extracted from Fig. 4.21:

$$X'X = \begin{pmatrix} 13 & 97 & 626 & 153 & 390 \\ 97 & 1139 & 4922 & 769 & 2620 \\ 626 & 4922 & 33050 & 7201 & 15739 \\ 153 & 769 & 7201 & 2293 & 4628 \\ 390 & 2620 & 15739 & 4628 & 15062 \end{pmatrix}, \quad X'\mathbf{y} = \begin{pmatrix} 1240.5 \\ 10032 \\ 62027.8 \\ 13981.5 \\ 34733.3 \end{pmatrix}$$

The normal equations can thus be constructed as follows:

$$\begin{aligned} 13\beta_0 + 97\beta_1 + 626\beta_2 + 153\beta_3 + 390\beta_4 &= 1240.5 \\ 97\beta_0 + 1139\beta_1 + 4922\beta_2 + 769\beta_3 + 2620\beta_4 &= 10032 \\ 626\beta_0 + 4922\beta_1 + 33050\beta_2 + 7201\beta_3 + 15739\beta_4 &= 62027.8 \\ 153\beta_0 + 769\beta_1 + 7201\beta_2 + 2293\beta_3 + 4628\beta_4 &= 13981.5 \\ 390\beta_0 + 2620\beta_1 + 15739\beta_2 + 4628\beta_3 + 15062\beta_4 &= 34733.3 \end{aligned}$$

Note that the last element in the row (or the column) labeled y of the sums of squares and cross-products matrix (see Fig. 4.21) is the total sum of squares $\mathbf{y}'\mathbf{y} = \sum_{i=1}^{n} y_i^2 = 121{,}088.09$. The solutions to the normal equations provide the estimates of $\boldsymbol{\beta}$. They appear in the last column of the inverse of the $X'X$ matrix table from the `proc reg` output (see the last column under y in Fig. 4.22) or in the parameter estimates table (see Fig. 4.24). These values are reported below rounded to five significant digits:

$$\hat{\boldsymbol{\beta}} = (62.405,\ 1.5511,\ 0.51017,\ 0.10191,\ -0.14406)'.$$

Further, the last element in the row (or the column) labeled y in Fig. 4.22 is the SSE, computed using the formula $\mathbf{y}'\mathbf{y} - \hat{\boldsymbol{\beta}}' X'\mathbf{y}$. Rounded to seven digits, it is equal to 47.86364, giving $s^2 = 47.86364/(13-4-1) = 5.98295$ (same as the MSE in the analysis of variance table in Fig. 4.23).

Model Crossproducts X'X X'Y Y'Y						
Variable	**Intercept**	**X1**	**X2**	**X3**	**X4**	**y**
Intercept	13	97	626	153	390	1240.5
X1	97	1139	4922	769	2620	10032
X2	626	4922	33050	7201	15739	62027.8
X3	153	769	7201	2293	4628	13981.5
X4	390	2620	15739	4628	15062	34733.3
y	1240.5	10032	62027.8	13981.5	34733.3	121088.09

Fig. 4.21. Sums of squares and cross-products matrix: Hald data

The following analysis of variance table for the regression (where numbers are reported to 2 decimals) is constructed from the output shown in Fig. 4.23:

Source	df	Sum of squares	Mean square	F	p-Value
Regression	4	2667.90	666.97	111.48	< 0.0001
Error	8	47.86	5.98		
Total	12	2715.76			

Obviously, the null hypothesis $H_0 : \beta_1 = \beta_2 = \beta_3 = \beta_4 = 0$ is rejected at $\alpha = 0.01$, say, since the p-value is less than 0.01.

The estimate s^2 of σ^2 is given by the MSE$= 5.98$. The R^2 value is also reported in Fig. 4.23, here equal to 0.9824, meaning that about 98% of the variation in the heat evolved from various batches of concrete is explained by the fitted multiple regression model involving all four explanatory variables, each giving percentages of chemical components of clinkers.

The inverse of the $X'X$ matrix is the 5×5 matrix in upper left corner in Fig. 4.22 and is reproduced below with elements rounded to six significant digits:

X'X Inverse, Parameter Estimates, and SSE						
Variable	**Intercept**	**X1**	**X2**	**X3**	**X4**	**y**
Intercept	820.65457471	-8.441801862	-8.457779848	-8.634538775	-8.289743778	62.4053693
X1	-8.441801862	0.0927104019	0.0856862094	0.0926373566	0.0844549553	1.5511026475
X2	-8.457779848	0.0856862094	0.0875602572	0.0878666397	0.0855980995	0.5101675797
X3	-8.634538775	0.0926373566	0.0878666397	0.0952014097	0.0863919188	0.1019094036
X4	-8.289743778	0.0844549553	0.0855980995	0.0863919188	0.0840311912	-0.144061029
y	62.4053693	1.5511026475	0.5101675797	0.1019094036	-0.144061029	47.86363935

Fig. 4.22. The inverse of $X'X$ matrix from `proc reg`: Hald data

$$(X'X)^{-1} = \begin{pmatrix} 820.655 & -8.44180 & -8.45778 & -8.63454 & -8.28974 \\ -8.44180 & 0.0927104 & 0.0856862 & 0.0926374 & 0.0844550 \\ -8.45778 & 0.0856862 & 0.0875603 & 0.0878666 & 0.0855981 \\ -8.63454 & 0.0926374 & 0.0878666 & 0.0952014 & 0.0863919 \\ -8.28974 & 0.0844550 & 0.0855981 & 0.0863919 & 0.0840312 \end{pmatrix}$$

To illustrate the use of the elements of the inverse of the $X'X$ matrix, the standard error of, say, $\hat{\beta}_1$ is computed as follows:

$$s_{\hat{\beta}_1} = \sqrt{c_{11}}s = \sqrt{0.0927104} \cdot \sqrt{5.983} = 0.74477$$

and, thus, a 95% confidence interval for β_1 is given by

$$\hat{\beta}_1 \pm t_{.025,8} \cdot s_{\hat{\beta}_1} \equiv 1.5511 \pm (2.306)(0.74477) \quad \text{giving} \quad (-0.16634, 3.2685).$$

These values can be verified by comparing with the corresponding values in the *Parameter Estimates* table (see Fig. 4.24), where the parameter estimates and their associated statistics are tabulated (see under columns labeled `Parameter Estimate`, `Standard Error`, etc.).

By inspecting these, it is observed that none of the p-values for the t-test for testing $\beta_m = 0$ for $m = 1,\ 2,\ 3,$ and 4 in model (4.5) is less than 0.05; thus, none of these hypotheses can be rejected at $\alpha = 0.05$. This is also clearly reflected in the fact that the 95% confidence intervals for these coefficients all contain zero. Upon further examination, it is seen that the reason for both of these results is that the standard errors of the estimates of the coefficients are comparatively large (i.e., they are all near 0.7, larger than some of the estimates themselves!). Obviously, this indicates large sampling variability of the estimated regression coefficients.

It appears that the conclusions from these individual t-tests (or confidence intervals) contradict the result of the F-test of $H_0 : \beta_1 = \beta_2 = \beta_3 = \beta_4 = 0$. However, it is erroneous to infer from the results of the one-at-a-time t-tests that the coefficients in the model are all zero simultaneously. Rather, the contradictory nature of the results of the t-tests and the F-test must be taken as an indication that there are, possibly large, correlations among the explanatory variables. This condition, called *multicollinearity*, discussed in detail in Sect. 4.2.4, could lead to the situation that one or more explanatory variables may exhibit little or no effects on the response variable in the presence, in the same model, of other explanatory variables that are highly correlated with one or more of them. From the correlation matrix reported in Fig. 4.20, it is clear that pairs of variables ($X1$, $X3$) and ($X2$, $X4$) are highly (negatively) correlated with each other. This is also evident from the plots shown in Fig. 4.19. Therefore, a model containing only one of the variables in each of these pairs may turn out to be a model that exhibits less multicollinearity.

In general, this situation may indicate the need to select a subset of the explanatory variables to be included in a model for which multicollinearity does not have substantial effects on the sampling variability and therefore the accuracy of the estimated parameters. Procedures for *variable subset selection* are discussed in Sect. 4.4.

Analysis of Variance					
Source	DF	Sum of Squares	Mean Square	F Value	Pr > F
Model	4	2667.89944	666.97486	111.48	<.0001
Error	8	47.86364	5.98295		
Corrected Total	12	2715.76308			

Root MSE	2.44601	R-Square	0.9824
Dependent Mean	95.42308	Adj R-Sq	0.9736
Coeff Var	2.56333		

Fig. 4.23. The ANOVA table from `proc reg`: Hald data

Parameter Estimates							
Variable	DF	Parameter Estimate	Standard Error	t Value	Pr > |t|	95% Confidence Limits	
Intercept	1	62.40537	70.07096	0.89	0.3991	-99.17855	223.98929
X1	1	1.55110	0.74477	2.08	0.0708	-0.16634	3.26855
X2	1	0.51017	0.72379	0.70	0.5009	-1.15889	2.17923
X3	1	0.10191	0.75471	0.14	0.8959	-1.63845	1.84227
X4	1	-0.14406	0.70905	-0.20	0.8441	-1.77914	1.49102

Fig. 4.24. The parameter estimates from `proc reg`: Hald data

4.2.2 Case Statistics and Residual Analysis

Many other statistical quantities are computed for use in residual analysis or in diagnostic plots. Some of these were introduced and discussed earlier in the context of simple linear regression.

These are necessary for checking the adequacy of the model or for assessing the plausibility of assumptions made in formulating the model. Others allow testing for presence or absence of outliers and assessing their effects on the fitted model if they are present. Collectively, these statistics are called *diagnostic statistics* or *case statistics.* The programs for SAS Examples D1, D2, and D3 illustrated the use of SAS statements in `proc reg` to produce these case statistics and also obtain residual plots. In SAS Example D3, in particular, an artificial data set was used to illustrate the concepts associated with several of these case statistics. In this subsection, several of these statistics are formally defined and their use demonstrated in the case of the multiple regression model.

Predicted or Fitted Values

The predicted values that correspond to the observed explanatory variables are calculated using the prediction equation

$$\hat{\mathbf{y}} = X\hat{\boldsymbol{\beta}}$$

where

$$\hat{y}_i = \hat{\beta}_0 + \hat{\beta}_1 x_{1i} + \cdots + \hat{\beta}_k x_{ki}, \qquad i = 1, \ldots, n$$

Note that the regression sum of squares defined earlier as SSReg in the ANOVA table can also be expressed as

$$\sum_{i=1}^{n} (\hat{y}_i - \bar{y}_i)^2$$

Residuals

The residuals that correspond to the observed data are expressed in vector form as $\mathbf{e} = \mathbf{y} - \hat{\mathbf{y}}$, where the elements $\mathbf{e} = (e_1, e_2, \ldots, e_n)'$ are calculated as $e_i = y_i - \hat{y}_i$, $i = 1, \ldots, n$. Note that the residual sum of squares defined in the ANOVA table as SSE can also be expressed in the form

$$\text{SSE} = \sum_{i=1}^{n} (y_i - \hat{y}_i)^2 = \sum_{i=1}^{n} e_i^2 = \mathbf{e}'\mathbf{e}$$

Hat Matrix

The predicted values $\hat{\mathbf{y}} = X\hat{\boldsymbol{\beta}}$ may be expressed in the form

$$\begin{aligned}\hat{\mathbf{y}} &= X(X'X)^{-1}X'\mathbf{y}\\ &= H\mathbf{y}\end{aligned}$$

where $H = X(X'X)^{-1}X'$ is an $n \times n$ symmetric matrix called the *hat matrix.* Let h_{ij} denote the ijth element of H. It can be shown that the ith diagonal element of H satisfies

$$\frac{1}{n} \leq h_{ii} \leq \frac{1}{d}$$

where d is the number of times the ith observation is replicated. Thus, a specific value of h_{ii} is considered to be relatively small if it is near $\frac{1}{n}$ or relatively large if it is near $\frac{1}{d}$. Note that for a case that is not replicated, the upper bound is 1.

It is easier to visualize the relationship of the magnitude of h_{ii} to the position of a case in the space of explanatory variables in simple liner regression. In general, cases with relatively larger values of h_{ii} will correspond to those cases with x-values further away from the average of the x-values (i.e., center of the x-space).

The elements of the hat matrix are useful since the variance of the predicted values and the residuals among several other quantities can be expressed in terms of the elements of this matrix. For example, the variance of $\hat{y}_i$ is $\sigma^2\, h_{ii}$, the standard deviation of $\hat{y}_i$ is $\sigma\sqrt{h_{ii}}$, and, hence, the standard error of $\hat{y}_i$ is $s\sqrt{h_{ii}}$, for $i = 1, 2, \ldots, n$. Noting that the vector of residuals may be expressed in the form $\mathbf{e} = \mathbf{y} - \hat{\mathbf{y}}$,

$$\begin{aligned}\mathbf{e} &= \mathbf{y} - H\mathbf{y}\\ &= (I - H)\mathbf{y}.\end{aligned}$$

It is easily shown that the variance of e_i is $\sigma^2\,(1-h_{ii})$, the standard deviation of e_i is $\sigma\sqrt{(1-h_{ii})}$, and, hence, the standard error of e_i is $s\sqrt{(1-h_{ii})}$ for $i = 1, 2, \ldots, n$. It is clear that the magnitudes of the standard errors of both $\hat{y}_i$ and e_i for the ith case depend on the magnitude of h_{ii} as the value of s remains a fixed number for a fitted model. For example, standard errors for predicted values will be larger and the standard errors for the residuals will be smaller for cases for which the diagonal elements of the hat matrix are closer to 1 than it is for those cases for which they are small.

Confidence Interval for the Mean $E(y_i)$

A $(1-\alpha)100\%$ confidence interval for the mean of the ith observation $E(y_i) = \beta_0 + \beta_1\, x_{1i} + \cdots + \beta_k\, x_{ki}$ is

$$\hat{y}_i \pm t_{\alpha/2,(n-k-1)} \times s\sqrt{h_{ii}}$$

Prediction Interval for y_i

A $(1-\alpha)100\%$ prediction interval for ith observation y_i is

$$\hat{y}_i \pm t_{\alpha/2,(n-k-1)} \times s\sqrt{1+h_{ii}}$$

Studentized Residuals

Studentized residuals, denoted by r_i, $i = 1, \ldots, n$, are a standardized version of the ordinary residuals, which are useful for the detection of outliers. It is common practice in statistics to use standardization when comparing statistics that are heterogeneous in variance. An *internally studentized* version of the residuals is obtained by directly dividing the residuals by their respective standard errors, as

$$r_i = e_i/(s\sqrt{1-h_{ii}})$$

for $i = 1, \ldots, n$. The maximum in absolute value of studentized residuals can be used as a basis of a test for the presence of a single *y-outlier* (i.e., an outlier in the y-direction), using the tables of percentage points reproduced in Tables B.9 and B.10. The null hypothesis H_0 : No Outliers is rejected in favor of H_a : A Single Outlier Present, if the computed value of $\max_i |r_i|$ exceeds the appropriate percentage point obtained from this table. It is common practice to use studentized residuals for normal probability plots.

Externally Studentized Residuals

Related statistics, denoted usually by t_i, $i = 1, \ldots, n$, are another version of standardized residuals. Instead of s^2, the ordinary residuals are standardized using $s^2_{(i)}$, the error mean square obtained from a regression model fitted with the ith case deleted. That is, externally studentized residuals are defined as

$$t_i = e_i/(s_{(i)}\sqrt{1-h_{ii}})$$

for $i = 1, \ldots, n$. The advantage of this statistic is that since each t_i can be shown to have a t-distribution with $n-k-2$ degrees of freedom, it can be used to construct a test for *y-outliers* (i.e., outliers in the y-direction). Since it is not known in advance which of the observations may be outliers, n t-tests must be performed using each of the n externally studentized residuals; that is, each of the n t_i values is compared to an appropriate critical value obtained from the t-distribution to test if the case is a y-outlier.

One approach is to use the Bonferroni method to obtain a conservative critical value for this multiple testing procedure. The critical value chosen is the $(\alpha/n)\times 100\%$ percentage point of the t-distribution with $n-k-2$ degrees of freedom. This test guarantees only that the Type I error will not exceed α (as opposed to being exactly equal to α); thus, it will only provide a conservative test. To use this method, special t-tables are needed because percentile points for small probability values are not available in ordinary t-tables. For example, for $n = 25, p = 3$, and $\alpha = 0.05$, the critical value needed corresponds to the $(0.025/25) = 0.001$th upper percentile point of a t-distribution with 20 df. (Note that this will be a two-tailed test using $|t_i|$; hence, 0.025 must be used instead of 0.05.) Since the required percentage points are not available from ordinary t-table, a special table was constructed (Weisberg 1985). This table has been reconstructed by the authors and appear as Tables B.11 and B.12.

Leverage

Recall that

$$\hat{\mathbf{y}} = H\mathbf{y}$$

where H is the $n \times n$ hat matrix. Recall also that

$$\begin{aligned} \text{var}(\hat{y}_i) &= h_{ii}\,\sigma^2 \\ \text{var}(e_i) &= (1 - h_{ii})\,\sigma^2 \end{aligned}$$

These imply that "larger" h_{ii} causes $\text{var}(\hat{y}_i)$ to be larger and $\text{var}(e_i)$ to be smaller. Hence, by examining diagonal elements of the hat matrix, it can be determined that an observed value is going to be predicted well or not by the regression on the x-values. Rewriting $\hat{\mathbf{y}} = H\mathbf{y}$ as

$$\hat{y}_i = h_{ii}\,y_i + \sum_{j \neq i} h_{ij}\,y_j$$

it is observed that when h_{ii} is closer to 1, the predicted value $\hat{y}_i$ will be closer to y_i and, therefore, e_i will be closer to zero. (Remember that $\frac{1}{n} \leq h_{ii} \leq \frac{1}{d}$.) For this reason, h_{ii} is called the `leverage` of the ith observation or case: It measures the effect that y_i will have on determining $\hat{y}_i$. However, note that the actual magnitude of $\hat{y}_i$, and therefore that of e_i, depends on `both` h_{ii} `and` y_i. As a rule of thumb, cases with leverages numerically larger than $2(k+1)/n$ (i.e., twice the average of all h_{ii}) may be marked for further investigation. Note also that h_{ii} actually measures how far away the ith case, $\mathbf{x}_i = (x_{1i}, x_{2i}, \ldots, x_{pi})$ is from the other cases; that is, a large h_{ii} may indicate an *x-outlier*.

Influence Statistics: Cook's D

Cases whose deletion causes major changes in the fitted model are called `influential`. A diagnostic tool that measures the influence of the ith case on the fit of the model is known as the Cook's distance statistic and is defined by

$$\begin{aligned} D_i &= \frac{1}{k'} \left\{ \frac{e_i}{s\sqrt{1 - h_{ii}}} \right\}^2 \left(\frac{h_{ii}}{1 - h_{ii}} \right) \\ &= \frac{1}{k'} r_i^2 \left(\frac{h_{ii}}{1 - h_{ii}} \right) \end{aligned}$$

where $k' = k + 1$. D_i measures the importance of the ith case on the fitted model. The fact that D_i may be partitioned as

$$D_i = \text{ constant } \times \text{ studentized residual}^2 \times \text{ monotone increasing function of } h_{ii}$$

indicates that a large D_i may be due to a large r_i, or a large h_{ii}, or both. So some cases with large leverages may actually not be influential because r_i is small, indicating that these cases actually fit the model well. Cases with relatively large values for both h_{ii} and r_i should be of more concern. As demonstrated in SAS Example D3, some cases that are not x-outliers but are significant y-outliers may cause inflation of the predictive variance and thus may be influential. As a rule of thumb, cases with Cook's D numerically larger than $4/n$ are flagged for further investigation.

Influence Statistics: DFFITS

It is interesting to examine cases whose deletion causes major changes in the fit of that individual case, i.e., cases that are `influential` in their own prediction. This is a different concept from the idea of measuring how an individual case might affect the fit of the entire model. The statistic below is a scaled measure of the change in prediction of a response when that case is deleted:

$$\text{DFFITS} = \frac{\hat{y}_i - \hat{y}_{(i)}}{s_{(i)}\sqrt{h_{ii}}}$$

where the numerator is the change in the predicted value for the ith observation when the ith observation is deleted, $\hat{y}_{(i)}$ and $s_{(i)}$ were defined previously. The DFFITS statistic is very similar to Cook's D statistic defined above. Cook's D measures the influence of the ith case on all fitted values jointly while DFFITS measures the influence of the ith case on the ith fitted value, $\hat{y}_i$. Comparably large values of DFFITS indicate influential observations, in the sense defined here. A suggested measure of high influence is a value larger than 1 for smaller data sets and larger than $2\sqrt{(k+1)/n}$ for large data sets.

Influence Statistics: DFBETAS

Similarly, one might want to measure the effect of deletion of a case on the estimates of individual coefficients. A scaled measure of the change in the jth parameter estimate when the ith observation is deleted is given by:

$$\text{DFBETAS}_j = \frac{\hat{\beta}_j - \hat{\beta}_{(i)j}}{s_{(i)}\sqrt{c_{(jj)}}}$$

where c_{jj} denotes the jjth element (i.e., the jth diagonal element) of the $(X'X)^{-1}$ matrix, as defined earlier. In general, large values of DFBETAS indicate observations that are influential in estimating a specific parameter. A suggested measure of high influence is a value larger than 1 for smaller data sets and larger than $2/\sqrt{n}$ for large data sets.

SAS Example D5 (Continued)

Although a residual analysis is usually deferred until an appropriate model is selected by identifying a subset of explanatory variables, some options available in `proc reg` are used with the model fitted in SAS Example D5 to illustrate the options needed to produce various case statistics discussed in this section.

In addition to the options specified on the `model` statement in the SAS Example D5 program, the option `p` (for `predicted`) requests that predicted values and residuals be part of the case statistics output. If instead, or in addition, the option `r` (for `residual`) is specified, standard errors of the predicted values, standard errors of the residuals, studentized residuals, and Cook's D statistics (the column titled `Cook's D`) are calculated. If in addition, the option `influence` is added, i.e., if the following model statement is used

```
model y = X1 X2 X3 X4/clb xpx i r influence;
```

the entire set of case statistics shown in Fig. 4.25 is produced.

Using the criteria described in this subsection, it is observed that none of the studentized residuals (or externally studentized residuals) is large enough for a case to be judged as a y-outlier. The column labeled `Hat Diag H` also indicates that for none of the cases does h_{ii} exceed the value $(10/13 \approx 0.8)$. The only case with a large enough `Cook's D` value $(> 4/13 \approx 0.308)$ is case number 8, but it is so only because of the relatively large values for h_{ii} and r_{ii}, with neither of those indicating any abnormality. Although case number 8 has a somewhat of a large `Cook's D` value, the `DFFITS` values are small indicating that deleting this case will not cause a substantial change in the prediction of the response for this case. In Sect. 4.2.3, residuals from the above fit are analyzed using several residual plots.

4.2.3 Residual Plots

The use of residual plots as a part of the analysis of residuals from regression has been common practice during the past several decades. Many of these plots have been described in the regression texts by Draper and Smith (1981) and Weisberg (1985). The most useful of these plots are the scatter plots of residuals (or studentized residuals) versus each of the explanatory variables (i.e., e_i or r_i versus each of the x_i) and the scatter plot of residuals (or studentized residuals) versus the predicted values (i.e., e_i or r_i versus the $\hat{y}_i$).

<table>
<tr><th colspan="8">Output Statistics</th></tr>
<tr><th>Obs</th><th>Dependent Variable</th><th>Predicted Value</th><th>Std Error Mean Predict</th><th>Residual</th><th>Std Error Residual</th><th>Student Residual</th><th>-2-1 0 1 2</th></tr>
<tr><td>1</td><td>78.5000</td><td>78.4952</td><td>1.8145</td><td>0.004760</td><td>1.640</td><td>0.00290</td><td>| | |</td></tr>
<tr><td>2</td><td>74.3000</td><td>72.7888</td><td>1.4120</td><td>1.5112</td><td>1.997</td><td>0.757</td><td>| |* |</td></tr>
<tr><td>3</td><td>104.3000</td><td>105.9709</td><td>1.8579</td><td>-1.6709</td><td>1.591</td><td>-1.050</td><td>| **| |</td></tr>
<tr><td>4</td><td>87.6000</td><td>89.3271</td><td>1.3291</td><td>-1.7271</td><td>2.053</td><td>-0.841</td><td>| *| |</td></tr>
<tr><td>5</td><td>95.9000</td><td>95.6492</td><td>1.4627</td><td>0.2508</td><td>1.960</td><td>0.128</td><td>| | |</td></tr>
<tr><td>6</td><td>109.2000</td><td>105.2746</td><td>0.8619</td><td>3.9254</td><td>2.289</td><td>1.715</td><td>| |*** |</td></tr>
<tr><td>7</td><td>102.7000</td><td>104.1487</td><td>1.4820</td><td>-1.4487</td><td>1.946</td><td>-0.744</td><td>| *| |</td></tr>
<tr><td>8</td><td>72.5000</td><td>75.6750</td><td>1.5634</td><td>-3.1750</td><td>1.881</td><td>-1.688</td><td>| ***| |</td></tr>
<tr><td>9</td><td>93.1000</td><td>91.7217</td><td>1.3270</td><td>1.3783</td><td>2.055</td><td>0.671</td><td>| |* |</td></tr>
<tr><td>10</td><td>115.9000</td><td>115.6185</td><td>2.0471</td><td>0.2815</td><td>1.339</td><td>0.210</td><td>| | |</td></tr>
<tr><td>11</td><td>83.8000</td><td>81.8090</td><td>1.5956</td><td>1.9910</td><td>1.854</td><td>1.074</td><td>| |** |</td></tr>
<tr><td>12</td><td>113.3000</td><td>112.3270</td><td>1.2544</td><td>0.9730</td><td>2.100</td><td>0.463</td><td>| | |</td></tr>
<tr><td>13</td><td>109.4000</td><td>111.6943</td><td>1.3480</td><td>-2.2943</td><td>2.041</td><td>-1.124</td><td>| **| |</td></tr>
</table>

Output Statistics										
						DFBETAS				
Obs	Cook's D	RStudent	Hat Diag H	Cov Ratio	DFFITS	Intercept	X1	X2	X3	X4
1	0.000	0.002715	0.5503	4.3353	0.0030	-0.0011	0.0008	0.0011	0.0007	0.0012
2	0.057	0.7345	0.3332	2.0173	0.5193	0.1995	-0.2451	-0.1938	-0.2139	-0.1783
3	0.301	-1.0581	0.5769	2.1948	-1.2356	-1.0953	1.0331	1.0828	1.0837	1.0970
4	0.059	-0.8240	0.2952	1.7413	-0.5333	-0.2367	0.1781	0.2482	0.2215	0.2255
5	0.002	0.1198	0.3576	3.0041	0.0894	-0.0188	0.0028	0.0238	0.0025	0.0226
6	0.083	2.0170	0.1242	0.2252	0.7594	0.1905	-0.1239	-0.1854	-0.1776	-0.1980
7	0.064	-0.7218	0.3671	2.1514	-0.5497	0.0538	-0.0048	-0.0717	-0.0549	-0.0436
8	0.394	-1.9675	0.4085	0.3649	-1.6352	0.4186	-0.5144	-0.3601	-0.6437	-0.4081
9	0.038	0.6459	0.2943	2.0684	0.4171	0.2484	-0.2622	-0.2427	-0.2248	-0.2514
10	0.021	0.1973	0.7004	6.3297	0.3016	-0.0530	0.1169	0.0414	0.0789	0.0464
11	0.171	1.0859	0.4255	1.5583	0.9345	-0.3273	0.3787	0.2982	0.4630	0.3099
12	0.015	0.4394	0.2630	2.3089	0.2625	-0.1359	0.1353	0.1413	0.1280	0.1318
13	0.110	-1.1459	0.3037	1.1854	-0.7568	0.3569	-0.3000	-0.3865	-0.2897	-0.3552

Fig. 4.25. Residual and influence statistics: Hald data

SAS Example D6

As in the case of simple linear regression, `proc reg` produces a residual diagnostics panel by default (i.e., when `plots=` options are not used in the proc statement). In addition, it also produces the residual plots against each of the explanatory variables (regressors). The proc step in SAS Example D6 program (shown in Fig. 4.26) results in *the fit diagnostics panel* displayed in Fig. 4.27 and the four plots displayed in Fig. 4.28.

```
insert data step to create the SAS dataset 'cement' here

title 'Residual Plots: Regression Analysis of Hald Data';
proc reg data=cement;
  model y = X1-X4/r p;
run;
```

Fig. 4.26. SAS Example D6: program

The first plot in the diagnostics panel (Fig. 4.27) is the *residuals versus the predicted values* plot. This plot is a dual purpose diagnostic plot. The examination of the scatter of the residuals around zero is aided by a horizontal reference line drawn at $e = 0$. Since the e_i and the $\hat{y}_i$ are uncorrelated, this plot should show an even scatter of points around the zero reference line as the $\hat{y}_i$ values increase (or decrease), if the fitted model is the correct one. If the model assumptions about the errors (i.e., that ϵ_i are uncorrelated and randomly distributed) are valid, one would expect the residuals to be evenly distributed around zero irrespective of the values of x or $\hat{y}_i$. Sometimes a marginal box plot or a histogram of e_i is appended to this scatter plot as an aid to the examination of the distribution of the residuals.

Additionally, if the *spread* of the points around the line remains approximately the same throughout the range of values of $\hat{y}$, it would suggest that the variance of e_i (and, therefore, the variance of y_i) remains constant for all possible values of x_i, and thus is not dependent on x. If the *spread* of the residuals around the zero reference line has an increasing or a decreasing pattern as the value of $\hat{y}_i$ changes, then a dependence of the residual variance on the "mean" of the response variable may be suspected. For example, such would be the case if the response variable had a Poisson distribution when the variance of the response increases as the mean of the response increases.

A useful variation of the above plot is *RStudent versus the Predicted Values* plot shown as the second plot in the first row of the diagnostics panel. Because `RStudent` values are a standardized version of the residuals, they have t-distributions and their variances are approximately equal. Thus, their magnitudes may be compared with each other for determining if there are extreme values. The plot in Fig. 4.27 shows two horizontal reference lines drawn at the points ± 2 of the y-axis. These indicate the rough cutoff values used to determine cases that may be flagged as possible y-outliers. In this example, there are two points that fall on the reference lines, but they are not too far outside the cutoff value for those cases to be considered y-outliers.

The third plot on the first row of diagnostics panel (Fig. 4.27) allows the consideration of extreme values in both RStudent and Leverage (hat values)

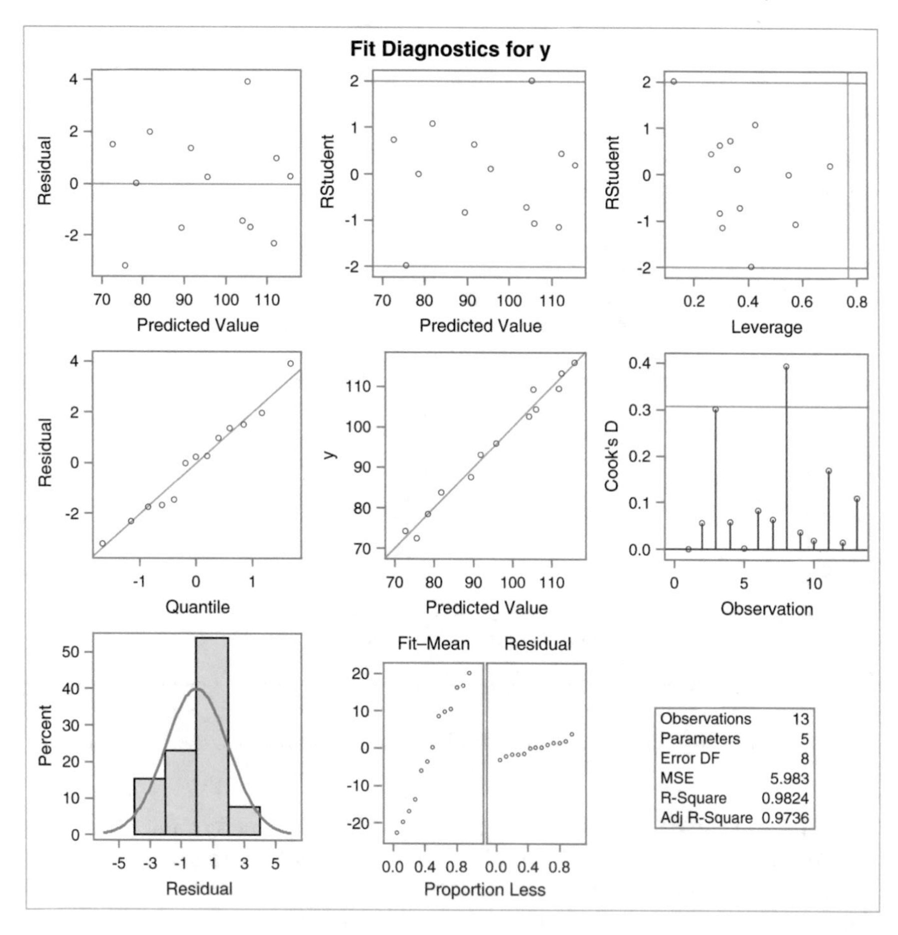

Fig. 4.27. Diagnostics panel: multiple regression of Hald data

by plotting these quantities against each other. In the y-direction, this plot can be interpreted in a manner similar to the previous plot. In the x-direction (where the values are all positive as they are hat values), the vertical reference line shows the approximate cutoff for x-outliers. In this example, this cutoff is equal to $10/13 \approx 0.77$. Since Cook'D statistic becomes inflated if either (or both) of Rstudent or Leverage becomes large, this is a useful plot for examining cases that are possibly influential. The third plot in the second row is an index plot of the Cook's D values and also contains a horizontal reference line, here drawn at $4/13 \approx 0.31$. In this example, the Cook's D for Case Number 8 exceeds this value (as can be visually ascertained from Fig. 4.29) and thus is shown to be an influential case. However, this observation is neither an x-outlier nor a y-outlier; thus, it may be of little concern to the investigator.

The points in the normal probability plot in Fig. 4.27 (first plot in the second row) do not deviate sufficiently from a straight line to question the

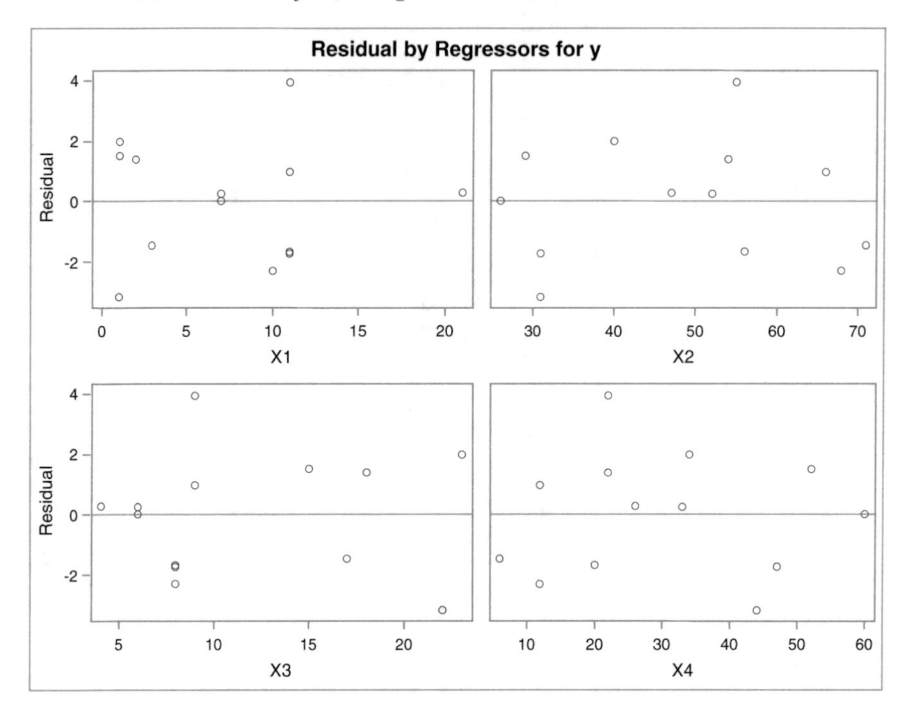

Fig. 4.28. Residuals vs. Regressors Plots: multiple regression of Hald data

assumption that the errors are normally distributed. The normal probability plot of the residuals (or, preferably, the studentized residuals) is useful for checking the model assumption of normality of the errors (ϵ_i) and also for determining the presence of any outliers. As described in Chap. 2, Sect. 2.2, a better approach for assessing a normal probability plot is to examine the plot to check whether a specific pattern of the points is identifiable that conforms to a long-tailed, short-tailed, or a skewed distribution relative to a normal distribution rather than judge whether the points deviate from a straight line.

Any curvature pattern in the scatter plot of e_i versus x_i may suggest a need for inclusion of higher-order terms in that particular x variable in the model or a transformation either of the x variable or the response y. This plot is also useful for determining if any independent variables that were observed but not included in the fitted model will make any additional contribution if added to the model. Simply plot the e_i's from the fitted model against these variables one at a time to check whether these variables exhibit any systematic relationship (such as linearity) with the residuals.

Furthermore, if any of the residuals versus x variable plots shows a systematically increasing or decreasing pattern in the spread around the reference line, it would be an indication of a dependence of the *residual variance* on

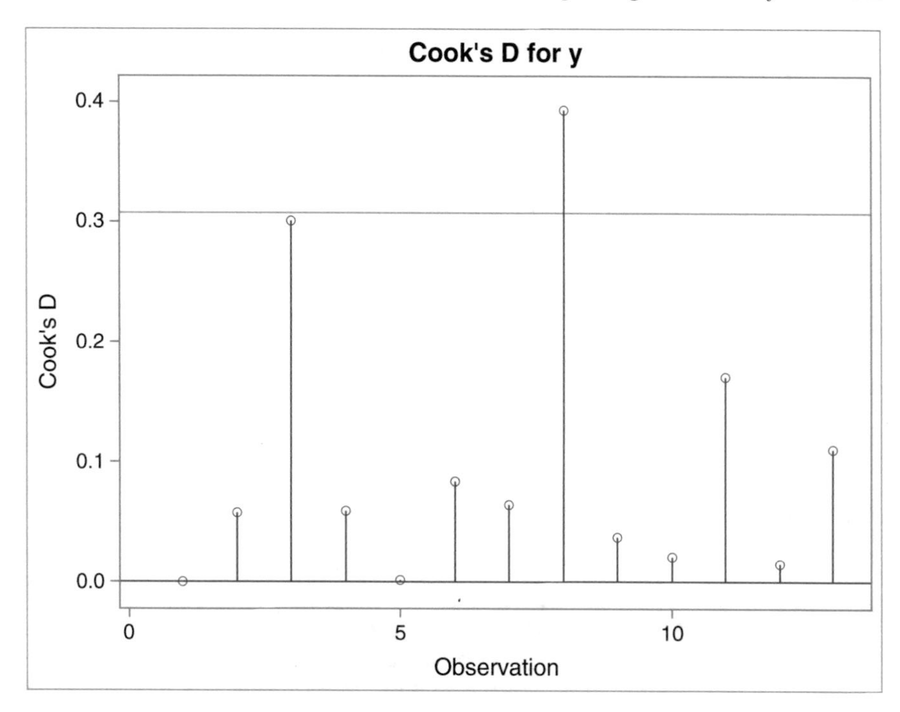

Fig. 4.29. Index plot of Cook's D: multiple regression of Hald data

the particular x variable (i.e., that the residual variance is function of that x variable). Examining the location of the points relative to the horizontal reference line drawn at a residual value of zero (i.e., $e = 0$) in the four plots shown in Fig. 4.28, it is seen that none of the residual plots indicate any discernible pattern, the residuals being distributed evenly as the x-values change and lie within a band equidistant from the reference line.

Sample residual plots resulting from artificially generated data serve as a rough guide to interpreting such plots. In practice, one rarely encounters plots that clearly indicate a pattern as recognizable as those illustrated in such plots. If a plot is difficult or ambiguous to interpret, it is perhaps best to consider it inconclusive rather than drawing a possibly erroneous conclusion. Moreover, the presence of one or two extreme values or outliers may affect the model fit and may lead to misleading interpretation of these plots.

In addition to the plots described so far, index plots of case statistics such as DFFITS and DFBETAS that include reference lines for indicating suggested cutoff values are useful devices for comparing the magnitudes of these statistics visually. It is easy to construct such plots using a `plots=` options such as `plots(only)=(CooksD DFFITS DFBETAS)`. This specification produced the graphs shown in Figs. 4.29, 4.30, and 4.31. Note that in the case of the DFFITS plot, the horizontal reference lines are drawn at $\pm 2\sqrt{(5/13)} = 1.24$, while in the DFBETAS plots, they are drawn at $\pm 2\sqrt{(1/n)} = 0.55$.

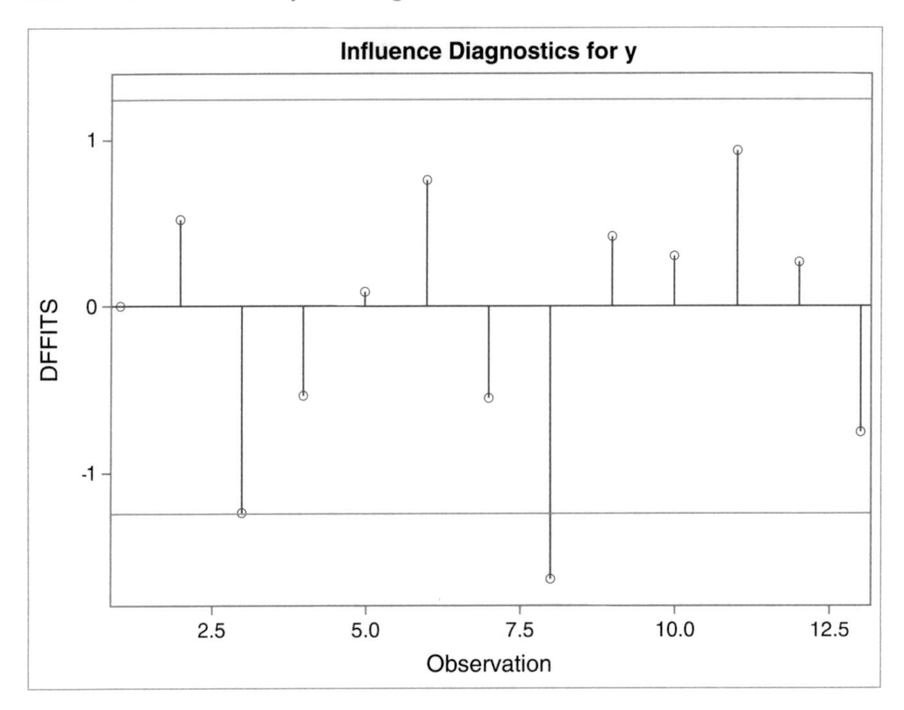

Fig. 4.30. Index plot of DFFITS: multiple regression of Hald data

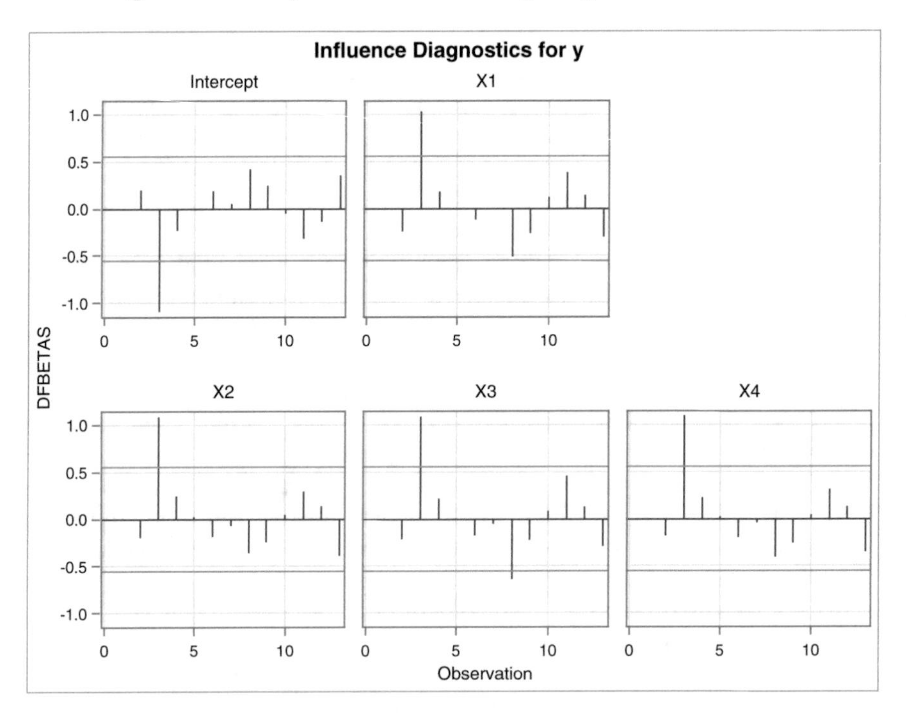

Fig. 4.31. Index plots of DFBETAS: multiple regression of Hald data

4.2.4 Examining Relationships Among Regression Variables

In simple linear regression, the relationship between the response y and the explanatory variable x can be visually examined easily using a scatter plot of y versus x. This plot gives a direct visual impression of the contribution of x to the regression as well as how well the data fit the regression. In the case of multiple regression, however, the scatter plot of y against any one of the x variables, called the *partial response plot*, may be less useful for this purpose. Although still a good indicator of the strength of the linear relationship between y and the particular x variable, this plot does not take into account the effects of the other explanatory variables and, therefore, may not help in determining the *contribution* of individual explanatory variables to the overall regression fit.

Recall that plotting residuals from a multiple regression against a variable that has been observed but not considered for the current model is a good way to determine whether there is evidence for the variable to be included in the model. Thus, to check whether an explanatory variable is useful, the residuals from fitting a multiple regression on the rest of the x variables may be plotted against the explanatory variable left out.

The *partial regression residual plot or the PRR plot* (sometimes also called the *added-variable plot*) is an improvement on this idea. It is a graphical tool that allows the display of the relationship between y and a single x variable when both variables are *adjusted for the effect* of the other explanatory variables in the model. Suppose that the interest is in the relationship between variables y and x_m, after both have been adjusted for the effect of the other xs in the full model. Define $x_{(m)} = (x_1, \ldots, x_{m-1}, x_{m+1}, \ldots, x_k)$. The *PRR plot* is obtained as follows:

- Compute the residuals from the regression fit of y on $x_{(m)}$. These, denoted here by $e_{y|x_{(m)}}$, represent the remaining variation in y after removing the effects of all x variables other than x_m.
- Compute the residuals from the regression fit of x_m on $x_{(m)}$. These, denoted here by $e_{x_m|x_{(m)}}$, represent the remaining variation in x_m after removing the effect of all x variables other than x_m.
- Plot $e_{y|x_{(m)}}$ against $e_{x_m|x_{(m)}}$ to obtain the *PRR plot* for the variable x_m. Thus, there is a PRR plot corresponding to each x_m.

The partial regression residual plot possesses several important properties that provide valuable insight about the multiple regression fit of y on the xs. The *intercept* of the regression of $e_{y|x_{(m)}}$ on $e_{x_m|x_{(m)}}$ is exactly zero and the *slope* of the fitted straight line is identical to the partial slope estimate $\hat{\beta}_m$ obtained from the full regression of y on all the x variables. Additionally, the residuals from the simple linear regression of $e_{y|x_{(m)}}$ on $e_{x_m|x_{(m)}}$ are identical to those from the full regression. Thus, the standard error of $\hat{\beta}_m$ and the MSE s^2 would also agree if the degrees of freedom for MSE of $n-k-1$ are used for its calculation (instead, degrees of freedom of $n-2$ was used for calculating the MSE because this is a straight line fit).

Thus, the PRR plot summarizes the regression of y on x_m adjusted for the other xs. A strong linear relationship indicated in this plot corresponds to a strong contribution by x_m to the overall model *even in the presence* of the other x variables. Similarly, the indication of a weak linear relationship in the PRR plot of x_m is evidence that x_m may not contribute significant additional predictive information to the regression when the other x variables are present in the model. This would indicate that x_m may be strongly correlated to one or more of the other x variables in the model, a condition identified as *multicollinearity*.

Multicollinearity

One would immediately suspect multicollinearity problems if a large sampling variance is observed for the estimated coefficient of an x variable in the statistics computed from a fitted multiple regression. The statistic, called the *variance inflation factor* (hereinafter called VIF), computed for the each coefficient β_m is a direct measure of the effect of multicollinearity in the estimation of the parameter. Computationally,

$$\text{VIF}_m = \frac{1}{(1 - R^2_{(m)})}$$

where $R^2_{(m)}$ is the coefficient of determination (or the multiple correlation coefficient) of the regression of x_m on $x_{(m)}$ (including an intercept). A high $R^2_{(m)}$ indicates that x_m is nearly a linear combination of the rest of the x variables (denoted by $x_{(m)}$) in the model.

The VIF_m value is the factor by which the variance is inflated over what it would be if the variable x_m were completely uncorrelated with all other x variables. A relatively large VIF for a coefficient (as a rule of thumb, a value in excess of 10) indicates that the multicollinearity that exists among the x variables is adversely affecting the estimation of that particular coefficient. If a variable is completely uncorrelated with all other x variables, the VIF value of its estimated coefficient will be exactly 1. On the other hand, if a predictor is highly collinear with one or more of the other predictors as indicated by a high R^2_m value, it will produce a large VIF_m. The options `collin, collioint`, and `tol` in `proc reg` provide statistics called *collinearity diagnostics* for detecting dependencies among the regressor variables and also determine when these may begin to affect the regression estimates.

```
insert data step to create the SAS dataset 'cement' here

ods select ANOVA ParameterEstimates PartialPlot;
title 'Regression Analysis of Hald Data';
proc reg data=cement;
  model y = X1 X2 X3 X4/clb vif partial;
  model y = X1 X2/clb vif partial;
run;
```

Fig. 4.32. SAS Example D7: program

SAS Example D7

In the SAS Example D7 program displayed in Fig. 4.32, the full model used in SAS Example D5 (Model 1) and a reduced model containing only the variables X1 and X2 (Model 2) are fitted to the Hald cement data using `proc reg`. The options `clb, vif,` and `partial` are specified requesting the 95% confidence intervals and the VIFs for the respective coefficients and partial regression residual plots for each variable in the respective model be output. The parts of the output giving the information on the parameter estimates and other fit statistics of interest are reproduced here in Figs. 4.33 and 4.34, respectively.

Analysis of Variance					
Source	DF	Sum of Squares	Mean Square	F Value	Pr > F
Model	4	2667.89944	666.97486	111.48	<.0001
Error	8	47.86364	5.98295		
Corrected Total	12	2715.76308			

Parameter Estimates								
Variable	DF	Parameter Estimate	Standard Error	t Value	Pr > |t|	Variance Inflation	95% Confidence Limits	
Intercept	1	62.40537	70.07096	0.89	0.3991	0	-99.17855	223.98929
X1	1	1.55110	0.74477	2.08	0.0708	38.49621	-0.16634	3.26855
X2	1	0.51017	0.72379	0.70	0.5009	254.42317	-1.15889	2.17923
X3	1	0.10191	0.75471	0.14	0.8959	46.86839	-1.63845	1.84227
X4	1	-0.14406	0.70905	-0.20	0.8441	282.51286	-1.77914	1.49102

Fig. 4.33. SAS Example D7: fit statistics for Model 1

From the fit statistics for Model 1 displayed in Fig. 4.33, it is seen that none of the coefficients is significantly different from zero at $\alpha = 0.05$. This is also evident from the fact that the 95% confidence intervals for these coefficients all contain zero. Of course, this is a consequence, as noted earlier, of the standard errors of the estimates of these coefficients being extremely large. The VIFs for the four estimates are all larger than 10, and extremely large for X2 and X4, indicating that although there is severe multicollinearity among all four variables, each of these two variables is highly collinear with the others.

The partial regression residual plots produced by the `partial` option may be *selected* for output using the ODS SELECT statement shown in Figs. 4.32 as an alternative to the `plots=` option used in previous SAS programs. Using ODS SELECT the user may request only those tables and graphical output required by the user to be included in the ODS output created in the proc step. The graphs as displayed in Figs. 4.35 and 4.36 were produced as part of the output produced by using the ODS graph name `PartialPlot` in the ODS SELECT statement. The graph names are listed in Table B.17. The options `ANOVA` and `ParameterEstimates` produced the tables in Figs. 4.33 and 4.34; some selected table names can be found in Table B.18.

Analysis of Variance					
Source	**DF**	**Sum of Squares**	**Mean Square**	**F Value**	**Pr > F**
Model	2	2657.85859	1328.92930	229.50	<.0001
Error	10	57.90448	5.79045		
Corrected Total	12	2715.76308			

Parameter Estimates								
Variable	**DF**	**Parameter Estimate**	**Standard Error**	**t Value**	**Pr > |t|**	**Variance Inflation**	**95% Confidence Limits**	
Intercept	1	52.57735	2.28617	23.00	<.0001	0	47.48344	57.67126
X1	1	1.46831	0.12130	12.10	<.0001	1.05513	1.19803	1.73858
X2	1	0.66225	0.04585	14.44	<.0001	1.05513	0.56008	0.76442

Fig. 4.34. SAS Example D7: fit statistics for Model 2

The partial regression residual plots for the full model are displayed in Fig. 4.35. It is clear that none of these plots shows strong linearity leading to the conclusion that, in the presence of the other variables in the model, none of these variables contribute significant additional information to the prediction of y, the heat evolved.

The second model analyzed in this example, Model 2, fits the model containing only the two variables X1 and X2, which were chosen because they showed only a small correlation (≈ 0.23) between them. As expected, this model displays no multicollinearity problems, as seen from examining the fit statistics for Model 2 appearing in Fig. 4.34. The VIFs for both variables are small and so are the standard errors of estimates of both coefficients. The p-values for the t-statistics are both extremely small and the confidence intervals also show that coefficients are both positive. The partial regression residual plots for the Model 2 displayed in Fig. 4.36 show clearly a significant linear relationship of each variable with y in the presence of the other, showing that any collinearity present will not affect the estimation of the respective coefficients in a multiple regression model.

The examples used above are intended to illustrate diagnostic tools available for identifying multicollinearity and how these can be obtained using `proc reg`. The possible remedies available for multicollinearity require further study of this problem. The remedy suggested here of excluding variables may not be advisable in situations when the omitted variable may provide predictive information for data not available for building the regression model. Although the presence of multicollinearity may affect the estimation of some coefficients as seen earlier, the fitted model is still useful for estimating the

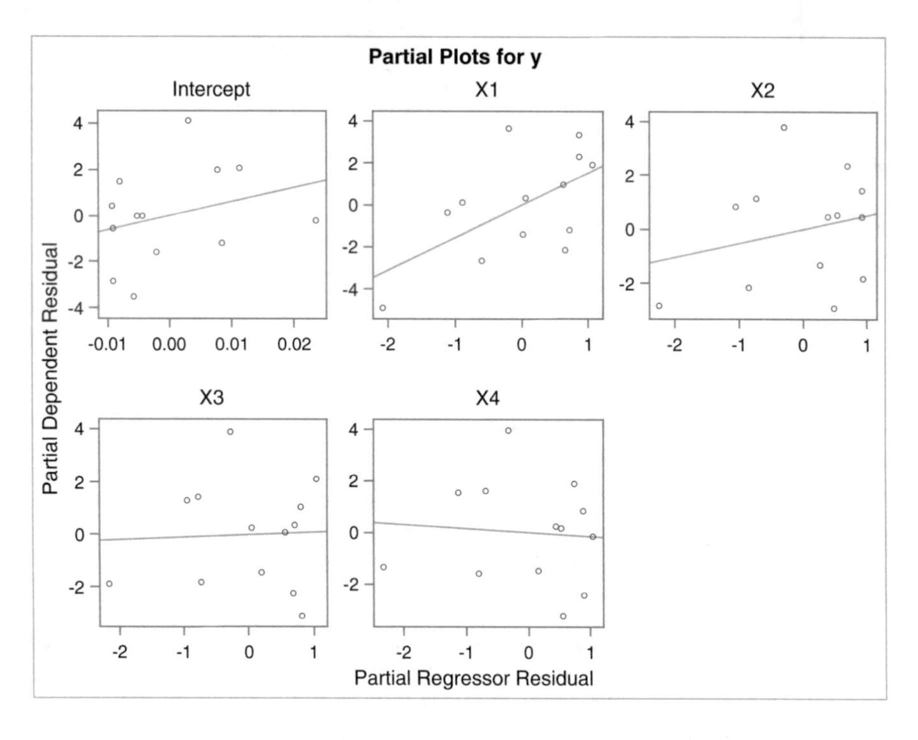

Fig. 4.35. Partial regression residual plots: Model 1

mean responses or making predictions. One consequence of high variability of a coefficient estimate is that it may be infeasible to use the actual estimate to measure the effect of the variable on the expected response.

4.3 Types of Sums of Squares Computed in PROC REG

4.3.1 Model Comparison Technique and Extra Sum of Squares

In Sect. 4.4, F-statistics based on Type II sum of squares will be used in `proc reg` to obtain F-tests for individual parameters. This approach for testing the significance of one or more parameters in a regression model is called the *model comparison technique* and the difference in the error sum of sum of squares of the two models thus compared is called the *extra sum of squares.*

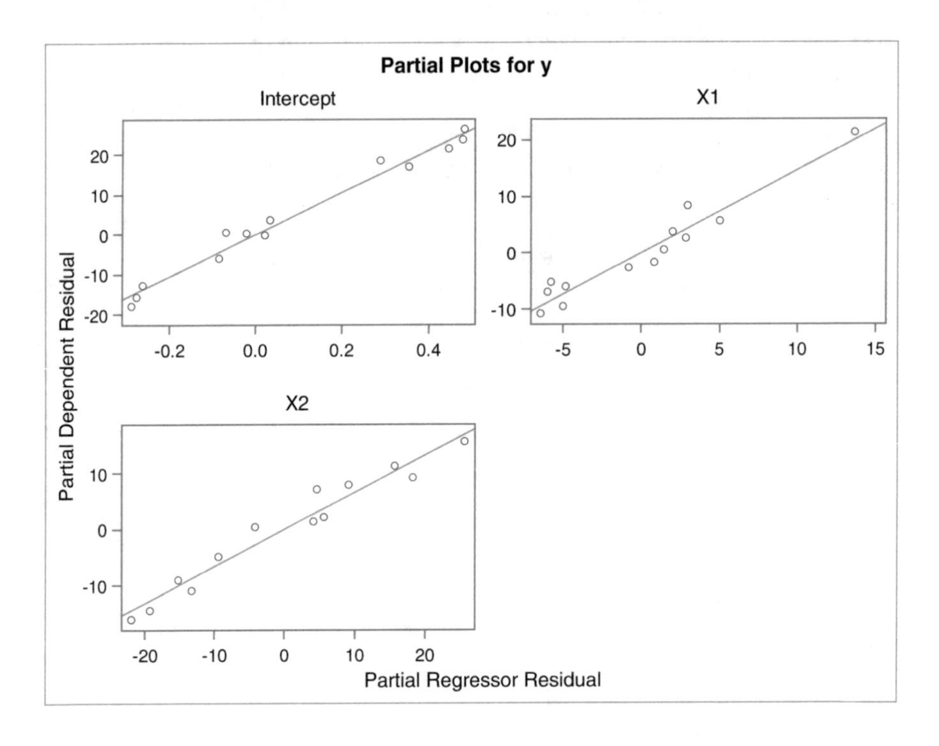

Fig. 4.36. Partial regression residual plots: Model 2

A hypothesis about any one of the β coefficients in the model, say $H_0 : \beta_2 = 0$, may be tested using a t-test. Sometimes a hypothesis may involve testing whether two or more of the β coefficients in a model are zero against the alternative hypothesis that at least one of these β coefficients is not zero. To introduce the model comparison technique, assume that k is the total number of explanatory variables (x's) in `Model1` and that ℓ is the number of explanatory variables considered to be in a smaller model, `Model2`, where, obviously, $\ell < k$. For convenience and ease of presentation, it is assumed here that the explanatory variables in the two models are ordered so that $x_{\ell+1}, x_{\ell+2}, \ldots, x_k$ is the subset of explanatory variables excluded from `Model1`. The two models are thus

$$\texttt{Model1:} \qquad y = \beta_0 + \beta_1 x_1 + \cdots + \beta_k x_k + \epsilon$$

$$\texttt{Model2:} \qquad y = \beta_0 + \beta_1 x_1 + \cdots + \beta_\ell x_\ell + \epsilon$$

Thus, $k-\ell$ represents the number of variables excluded from `Model1` to obtain `Model2`. It is of interest to test the hypothesis $H_0 : \beta_{\ell+1} = \beta_{\ell+2} = \cdots = \beta_k = 0$ against H_a : at least one of these is not zero, about the coefficients in `Model1`. To formulate a test of this hypothesis based on model comparison technique, first fit the two models and compute the respective residual sums of squares. The F-test statistic for testing H_0 is

$$F = \frac{[\text{SSE}(\texttt{Model2}) - \text{SSE}(\texttt{Model1}))]/(k-\ell)}{\text{SSE}(\texttt{Model1})/(n-k-1)}$$

The numerator and denominator degrees of freedom for the F-distribution associated with F-statistic are $k-\ell$ and $n-k-1$, respectively. When this computed F exceeds a critical value for a specified α level obtained from the F-tables, H_0 is rejected.

Reduction Notation

A system of notation for representing differences in residual sums of squares resulting from fitting two regression models was promoted by Searle (1971). Denoted by $R(\ \ )$ and called the reduction notation, this representation is useful for discussing types of sums of squares and for representing statistics useful in variable subset selection methods computed by computer programs. Consider, for example, the model

$$y = \beta_0 + \beta_1 x_1 + \beta_2 x_2 + \beta_3 x_3 + \beta_4 x_4 + \beta_5 x_5 + \epsilon\ .$$

To determine whether one of the explanatory variables $x_1, \ldots, x_5$ in this model does not contribute to the regression, a hypothesis of the form, say $H_0 : \beta_2 = 0$, may be tested. To put it in the framework of the model comparison technique, consider the above model to be `Model1`. The corresponding analysis of variance table (only the sources of variation, df, and the sums of squares are shown here) is

Source	df	SS
Regression	5	$\hat{\boldsymbol{\beta}}' X'\mathbf{y} - n\bar{y}^2 = \text{SSR}\ (\beta_1, \beta_2, \beta_3, \beta_4, \beta_5)$
Residual	$n-6$	$\sum y_i^2 - \hat{\boldsymbol{\beta}}' X'\mathbf{y} = \text{SSE}\ (\beta_0, \beta_1, \beta_2, \beta_3, \beta_4, \beta_5)$
Total	$n-1$	$\sum y_i^2 - n\bar{y}^2$

where n = number of observations in the data set. Suppose now that `Model2` consists of, say, the explanatory variables x_1, x_2, x_3, and x_4, so the model

$$y = \beta_0 + \beta_1 x_1 + \beta_2 x_2 + \beta_3 x_3 + \beta_4 x_4 + \epsilon$$

is fitted to the data. The corresponding analysis of variance table is

Source	df	SS
Regression	4	$\text{SSR}(\beta_1, \beta_2, \beta_3, \beta_4)$
Residual	$n - 5$	$\text{SSE}(\beta_0, \beta_1, \beta_2, \beta_3, \beta_4)$
Total	$n - 1$	$\sum y_i^2 - n\bar{y}^2$

Now, note that the residual sum of squares for the `Model1` is *always* smaller in magnitude than the corresponding sum of squares for `Model2`. The quantity representing the *reduction* in the residual sum of squares due to the addition of the variable x_5 to `Model2` is denoted by $R(\beta_5/\beta_0, \beta_1, \beta_2, \beta_3, \beta_4)$; that is,

$$\begin{aligned} R(\beta_5/\beta_0, \beta_1, \beta_2, \beta_3, \beta_4) &= \text{SSE}(\beta_0, \beta_1, \beta_2, \beta_3, \beta_4) - \text{SSE}(\beta_0, \beta_1, \beta_2, \beta_3, \beta_4, \beta_5) \\ &= \text{SSE}(\texttt{Model2}) - \text{SSE}(\texttt{Model1}) \end{aligned}$$

$R(\beta_5/\beta_0, \beta_1, \beta_2, \beta_3, \beta_4)$ can now be used to formulate the F-statistic for testing $H_0 : \beta_5 = 0$ versus $H_1 : \beta_5 \neq 0$ as

$$F = \frac{R(\beta_5/\beta_0, \beta_1, \beta_2, \beta_3, \beta_4)/(1)}{\text{SSE}(\beta_0, \beta_1, \beta_2, \beta_3, \beta_4, \beta_5)/(\text{n} - 6)}$$

The numerator and denominator degrees of freedom here for the F-statistic are 1 and $n - 6$, respectively. Note that when a single parameter is tested for zero, the numerator degrees of freedom of the F-test will always be equal to one and thus the F-test is equivalent to the t-test available for testing the same hypothesis. Note that this notation can be extended in an obvious way to the case when several variables are added or deleted.

4.3.2 Types of Sums of Squares in SAS

In the regression setting, SAS computes two types of sums of squares associated with testing hypotheses about coefficients in the model. These are referred to as `Type I` and `Type II` sums of squares and are output by specifying `ss1` or `ss2`, or both together, as options in the `model` statement.

Definition: Type I (or Sequential) Sums of Squares This is a partitioning of the regression sum of squares into component sums of squares each with one degree of freedom that represent the reduction in residual sum of squares (or, equivalently, the increase in regression sum of squares) as each variable is added to the model in the order they are specified in the `model` statement.

Definition: Type II (or Partial) Sum of Squares This is a partitioning of regression sum of squares into component sums of squares each with one degree of freedom that represent the reduction in residual sum of squares (or, equivalently, the increase in regression sum of squares) when each variable is added to a model containing rest of the variables specified in the `model` statement.

Type I and Type II Sums of Squares in Reduction Notation Consider the following `model` statement:

```
model y = X1 X2 X3 X4/ ss1 ss2;
```

The resulting sums of squares computed by `proc reg` can be identified using the reduction notation developed above, as indicated below:

Effect	Type I SS	Type II SS
X1	$R(\beta_1/\beta_0)$	$R(\beta_1/\beta_0, \beta_2, \beta_3, \beta_4)$
X2	$R(\beta_2/\beta_0, \beta_1)$	$R(\beta_2/\beta_0, \beta_1, \beta_3, \beta_4)$
X3	$R(\beta_3/\beta_0, \beta_1, \beta_2)$	$R(\beta_3/\beta_0, \beta_1, \beta_2, \beta_4)$
X4	$R(\beta_4/\beta_0, \beta_1, \beta_2, \beta_3)$	$R(\beta_4/\beta_0, \beta_1, \beta_2, \beta_3)$

These sums of squares are important and useful because they can be used to construct F-statistics (or equivalently, t-statistics) to test certain hypotheses. For example, the Type I SS $R(\beta_2/\beta_0, \beta_1)$ may be used to test whether X_2 should be added to the model $y = \beta_0 + \beta_1\text{X1} + \epsilon$, whereas the Type II SS $R(\beta_2/\beta_0, \beta_1, \beta_3, \beta_4)$ may be used to construct an F-statistic to test the hypothesis $H_0 : \beta_2 = 0$ versus $H_a : \beta_2 \neq 0$ in the model $y = \beta_0 + \beta_1\text{X1} + \beta_2\text{X2} + \beta_3\text{X3} + \beta_4\text{X4} + \epsilon$.

As will be discussed in Sect. 4.4, the Type II sums of squares computed in subset selection procedures in `proc reg` are used to construct both F-to-enter and F-to-delete statistics for adding variables to a model and removing variables from a model, respectively, as illustrated via an example in that section.

SAS Example D8

Figure 4.37 shows the model fit statistics and parameter estimates portions of the output if the model statement in the `proc reg` step in the SAS program in Fig. 4.18 is replaced by

```
model y = X1 X2 X3 X4/ss1 ss2;
```

It is informative to observe from this output, for example, that the reduction in the residual sum of squares for adding variable X2 to a model with only X1 currently in the model (Type I SS: $R(\beta_2/\beta_0, \beta_1) = 1207.78227$) is considerably larger than the reduction in the residual sum of squares for adding variable X2

Analysis of Variance					
Source	DF	Sum of Squares	Mean Square	F Value	Pr > F
Model	4	2667.89944	666.97486	111.48	<.0001
Error	8	47.86364	5.98295		
Corrected Total	12	2715.76308			

Root MSE	2.44601	R-Square	0.9824
Dependent Mean	95.42308	Adj R-Sq	0.9736
Coeff Var	2.56333		

Parameter Estimates							
Variable	DF	Parameter Estimate	Standard Error	t Value	Pr > |t|	Type I SS	Type II SS
Intercept	1	62.40537	70.07096	0.89	0.3991	118372	4.74552
X1	1	1.55110	0.74477	2.08	0.0708	1450.07633	25.95091
X2	1	0.51017	0.72379	0.70	0.5009	1207.78227	2.97248
X3	1	0.10191	0.75471	0.14	0.8959	9.79387	0.10909
X4	1	-0.14406	0.70905	-0.20	0.8441	0.24697	0.24697

Fig. 4.37. SAS Example D8: output

to a model that contains X1, X3, and X4 (Type II SS: $R(\beta_2/\beta_0, \beta_1, \beta_3, \beta_4) = 2.97248$).

F-statistics to test various hypotheses using the model comparison technique may be constructed using the Type II sums of squares in this output. For example, to test $H_0 : \beta_2 = 0$ versus $H_a : \beta_2 \neq 0$ in the model $y = \beta_0 + \beta_1\text{X1} + \beta_2\text{X2} + \beta_3\text{X3} + \beta_4\text{X4} + \epsilon$, the F-statistic can be calculated as:

$$F = \frac{R(\beta_2/\beta_0, \beta_1, \beta_3, \beta_4)/(1)}{\text{SSE}(\beta_0, \beta_1, \beta_2, \beta_3, \beta_4)/(13-5)} = \frac{2.97248}{47.86364/8} = 0.4967$$

Also, note that the p-values corresponding to the Type II F-tests are identical to those computed for the t-tests for the individual parameters given in the parameter estimates table in Fig. 4.37, because both sets of statistics test the same hypotheses about the coefficients, as discussed earlier. Thus the p-value associated with the above F-statistic is 0.5009.

Interactive Model Fitting Using PROC REG

Suppose that it is required to test the hypothesis $H_0 : \beta_3 = \beta_4 = 0$ versus $H_a : \beta_3$ and/or $\beta_4 \neq 0$ in the model $y = \beta_0 + \beta_1 \text{X1} + \beta_2 \text{X2} + \beta_3 \text{X3} + \beta_4 \text{X4} + \epsilon$ (Full Model). This would be the case if the interest is in excluding both variables $X3$ and $X4$ simultaneously, if they are not significant in the full model. To use the model comparison technique for doing this, it may require the model $y = \beta_0 + \beta_1 \text{X1} + \beta_2 \text{X2} + \epsilon$ (Reduced Model) to be fitted using a separate model statement in the `proc reg` step. However, in the interactive model-fitting facility in `proc reg`, once a model is fitted, the `add` and `delete` statements can be executed to perform computations pertaining to a modified model. For example, once the SAS Example D5 program (see Fig. 4.18) is executed to fit the full model, the following three SAS statements may be submitted to fit the reduced model given above:

```
delete X3 X4;
print;
run;
```

This produces the complete output from the fit of the reduced model. The analysis of variance table from this fit which is needed for the computation of the F-statistic is reproduced in Fig. 4.38. Thus the required F-statistic for testing $H_0 : \beta_3 = \beta_4 = 0$ versus $H_a : \beta_3$ and/or $\beta_4 \neq 0$ is computed as follows:

Analysis of Variance					
Source	DF	Sum of Squares	Mean Square	F Value	Pr > F
Model	2	2657.85859	1328.92930	229.50	<.0001
Error	10	57.90448	5.79045		
Corrected Total	12	2715.76308			

Fig. 4.38. SAS Example D8: fitting a reduced model

$$F = \frac{[\text{SSE}(\texttt{Reduced Model}) - \text{SSE}(\texttt{Full Model}))]/(4-2)}{\text{SSE}(\texttt{Full Model})/(13-4-1))}$$
$$= \frac{(57.90448 - 47.86364)/2}{47.86364/8} = 5.02042/5.982955 = 0.84$$

An alternative and easier computation of the above statistics can be accomplished by using the `test` statement in `proc reg`. The statements required to perform the F-tests for $H_0 : \beta_2 = 0$ versus $H_a : \beta_2 \neq 0$ and $H_0 : \beta_3 = \beta_4 = 0$ versus $H_a : \beta_3$ and/or $\beta_4 \neq 0$ are given in Fig. 4.39. The two tests are labeled B2 and B3B4 as shown in the SAS code and these labels are used in the output reproduced in Fig. 4.40.

```
insert data step to create the SAS dataset 'cement' here

title 'F-tests for Model Comparison: Hald Data';
proc reg data=cement plots=none;
  model y = X1 X2 X3 X4/ss1 ss2;
  B2:test X2=0;
  B3B4:test X3=0,X4=0;
run;
```

Fig. 4.39. SAS Example D8: program for hypothesis testing

Test B2 Results for Dependent Variable y				
Source	DF	Mean Square	F Value	Pr > F
Numerator	1	2.97248	0.50	0.5009
Denominator	8	5.98295		

Test B3B4 Results for Dependent Variable y				
Source	DF	Mean Square	F Value	Pr > F
Numerator	2	5.02042	0.84	0.4668
Denominator	8	5.98295		

Fig. 4.40. SAS Example D8: hypothesis tests for model comparison

4.4 Subset Selection Methods in Multiple Regression

An experimenter usually attempts to choose a subset of variables from a large number of explanatory variables (x variables) measured to construct a "good" regression model. Here, "good" may be taken to mean that the model is adequate for prediction and at the same time is economical to use. The model containing all explanatory variables measured is called the *full model*, whereas a model containing a subset of those is called a *subset model*.

For the purpose of selecting a good model, some criteria for selecting one model over another are needed. Usually, statistical test procedures available for this purpose are based on the residual sum of squares remaining after fitting each model. If the inclusion of additional independent variables does not "significantly" decrease the residual sum of squares, then the additional variables may be excluded to obtain a more parsimonious model. If one is dealing with a smaller number of explanatory variables, then it is perhaps best to look at all possible subset models. This can be done using an appropriate program that fits all combinations of the explanatory variables. Although such a procedure may be expensive if the number of explanatory variables that have to be considered is large, algorithms that reduce the amount of computations necessary to obtain the "best" model according to some criterion, containing different numbers of explanatory variables (*subset models of a specified size*), are available.

First, some classical methods for variable subset selection are summarized below.

Forward Selection Method

In the forward selection method, explanatory variables are entered into the model, one at a time, at each stage testing whether there is a significant decrease in the residual sum of squares. For example, suppose that k explanatory variables $x_1, x_2, \ldots, x_k$ are available. First, one independent variable is selected to obtain the best simple linear model. Suppose that each of the explanatory variables is used to fit simple linear regression models of the type

$$y_i = \beta_0 + \beta_j x_{ji} + e_i, \quad i = 1, 2, \ldots, n$$

for each $j = 1, 2, \ldots, k$ and the Type II F-statistic for each model determined. These F-statistics are a measure of each variable's contribution to the model if it is included in the model. The variable corresponding to the largest F-statistic is chosen to enter the model, if the p-value associated with the F-statistic exceeds some preassigned value called the *significance level for entry.* Note that since the largest F-statistic does not have an exact F-distribution, this is not equivalent to testing the hypothesis $H_0 : \beta_j = 0$ in the current model. Nevertheless, since the p-value is a monotone function of the denominator degrees of freedom (determined by the number of variables already in the model), it makes sense to compare it to some cutoff value to decide whether a variable enters the current model or not.

Suppose for simplicity that x_1 is the variable that is chosen to enter the model first. Thus, after the first step, the model is

$$y_i = \beta_0 + \beta_1 x_{1i} + e_i \tag{4.6}$$

The residual sum of squares for this model is denoted by $\text{SSE}(\beta_0, \beta_1)$. Now, the next variable to enter the above model is determined by considering each of the explanatory variables other than x_1, one at a time. Thus, a two-variable model is of the form

$$y_i = \beta_0 + \beta_1 x_{1i} + \beta_j x_{ji} + e_i$$

for $j \neq 1$. Suppose that the residual sum of squares for this model is denoted by $\text{SSE}(\beta_0, \beta_1, \beta_j)$. Then an F-statistic for testing $H_0 : \beta_j = 0$ is

$$F = \frac{\text{SSE}(\beta_0, \beta_1) - \text{SSE}(\beta_0, \beta_1, \beta_\text{j})}{\text{SSE}(\beta_0, \beta_1, \beta_\text{j})/(\text{n} - 3)} = \frac{R(\beta_j/\beta_0, \beta_1)}{\text{MSE}(\beta_0, \beta_1, \beta_j)}$$

which is distributed as $F(1, n-3)$. This statistic, called the *F-to-enter* statistic for variable x_j, effectively tests whether the reduction in the residual sum of squares by adding or "entering" the variable x_j to model (4.6) is significant. The variable that produces the largest of the F-to-enter statistics is determined by fitting each of these two-variable models using the variables $x_2, \ldots, x_p$. A p-value for this F-to-enter statistic is computed using the $F(1, n-3)$ distribution and if it is below the preselected significance level for entry, the variable is entered, and the two-variable model becomes the current model. This process is continued until, at any stage of the process, the p-value corresponding to the variable most recently considered as a candidate to enter the model exceeds the cutoff value.

Backward Elimination Method

A method that is the direct opposite of forward selection is the so-called backward elimination method. In this method, as a first step, a regression model with all variables in the model is computed:

$$y = \beta_0 + \beta_1 x_1 + \cdots + \beta_k x_k + \epsilon$$

Suppose that one variable, say x_j, is removed (or deleted) from the above model and the residual sum of squares determined. From the residual sums of squares of these two models, an F-statistic called *F-to-remove* (or *F-to-delete* can be computed. This F-statistic will have an $F(1, n-3)$ distribution. An F-to-remove statistic is computed for each variable x_j, $j = 1, \ldots, k$, contained in the current model. The p-value of the smallest F-to-remove value is determined and the corresponding variable deleted from the model if this p-value exceeds the preassigned cutoff called *significance level for deletion.* Again, note that the smallest F-statistic will not have an F-distribution; thus, this does not correspond to a test of the hypothesis $H_0 : \beta_j = 0$ in the above model.

Once a variable is removed, a regression model with the remaining variables is computed, and the entire process is repeated. The process is stopped when a variable with the smallest F-to-remove value fails to be removed from the current model.

Stepwise Method

Another more commonly used procedure in computerized regression model building is the stepwise method. This is a combination of both forward selection and backward elimination. Two preassigned cutoffs are selected: one for entry of variables and one for removal. The procedure is similar to forward selection except that after each new variable is entered in a forward selection step, a single backward elimination step is performed on the current model. Obviously, the cutoff for entry may not be greater than or equal to the cutoff for deletion, for, otherwise, the most recently entered variable will be a candidate for immediate deletion. Some computer programs allow this, but the procedure is terminated when a variable to be entered to the model is one just deleted from it. In any case, the choice of these cutoffs is quite arbitrary as they are not true significance levels but are used to determine the entry or removal of variables. In practice, the p-values printed for the F-statistics can be compared to the cutoff values selected to check whether model selected is sensitive to changes in the cutoffs chosen.

Other Stepwise Methods

Stepwise methods based on finding models that maximize the improvement in R^2 have been proposed.

Coefficient of Multiple Correlation R^2 This statistic, introduced first in Sect. 4.1, can be expressed in the form

$$R^2 = \text{SSReg}/\text{SSTot} = 1 - \text{SSE}/\text{SSTot}.$$

It measures the proportion of variance explained by fitted model and is obviously maximized for the full model. The objective is to find subset models with comparable R^2 values so that the inclusion of any of the explanatory variables left out will not increase it to any appreciable degree.

One such method begins by finding the one-variable model producing the highest R^2 and then adds another variable that yields the largest increase in R^2. Once the two-variable model is obtained, each of the variables in the model is swapped with the variables not yet in the model to find the swap that produces the largest increase in R^2. The process continues until the method finds that no switch could increase R^2 further at which time it is stopped and the "best" two-variable model is declared to be found. Another variable is then added to the model, and the swapping process is repeated to find the "best" three-variable model and so forth.

All-Subsets Methods

The variable subset selection methods discussed above are all based on finding a subset of variables such that the inclusion of further variables does not decrease the residual sum of squares significantly. However, since all these methods add and/or delete variables one at a time, it is clear that the order of entering or deleting variables may lead to different models being selected. Further, these methods may be affected to various degrees by multicollinearity, if present.

Each of the above methods has its own merits as well as deficiencies. Thus, although one can provide arguments in favor of one or more of these methods, in practice, it is recommended that criteria other than those based on the minimum sum of squares of residuals alone be used in selecting a "best" subset model. Some proposed criteria available in computer packages are briefly discussed below. Associated computational algorithms that determine the best subset of a given size, in the sense of minimizing or maximizing the specified criterion without computing all possible regressions, are available. These perform well at a reasonable cost.

A possible alternative to one-variable-at-a-time sequential methods of selecting a good subset of variables is considering all possible models of each "size"; that is, find the best two-variable model by fitting all two-variable models and so on. Algorithms that require performing only a fraction of the regressions required to find the best subsets of each size by doing a complete search have been developed. In practice, implementations of such methods in computer software usually incorporate the ability for the user to specify that the "best" models of each size to be selected according to some criterion, such as the R^2, R^2_{adj} or the C_p statistic (these statistics are described below). The user may also request that best models thus selected be limited to a specified number and that values of other statistics such as the MSE or the SSE be included among the information output for each of these models. Thus, the user has the option of selecting a model based by considering other criteria other than the criterion used to determine the optimal ones of each size chosen by such algorithms.

This brings up an important aspect of model selection. Since it is known that the full model will always have the smallest SSE and hence the largest R^2, it seems logical that more predictor variables are in a model, the better the model might be. However, including too many predictors in a regression model leads to the problem commonly called "over-fitting." This means that the model fits the data used for building the model (usually called *training data*) so closely that it will produce predictions for new data (usually called *test data*) that may be highly variable. Thus, including more predictors in a model will lead to a less-biased model but will produce predictions with high variance. Thus, for comparison of subset models, statistics other than SSE and R^2 have been proposed.

Adjusted R^2 As shown above $R^2 = 1 - \text{SSE}/\text{SSTot}$, and since SSTot remains a constant for a given data set, comparing models based on R^2 is equivalent to comparing models based on SSE. An adjusted R^2 statistic that adjusts for the degrees of freedom of the SSE has been proposed and is given by

$$R^2_{adj} = 1 - \frac{(n-1)}{(n-p-1)} \frac{\text{SSE}}{\text{SSTot}} = 1 - \frac{n-1}{n-p}(1 - R^2)$$

It can be shown that comparing models based on R^2_{adj} is equivalent to comparing models based on MSE. Since it is possible to have subset models with a smaller MSE than the full model, using R^2_{adj} or, equivalently, the MSE has been suggested as an alternative criterion, for selecting subset models that will have comparably lower prediction error than the full model.

Mallows' C_p *Statistic* For a subset model with p explanatory variables, this statistic is defined as

$$C_p = (\text{SSE}_p/s^2) - (n - 2p)$$

where $s^2 = \text{MSE}$ for the full model (i.e., the model containing all k explanatory variables of interest). SSE_p is the residual sum of squares for the subset model containing p explanatory variables *counting the intercept* (i.e., the number of parameters in the subset model). Usually C_p is plotted against p for the collection of subset models of various sizes under consideration. Acceptable models in the sense of minimizing the total bias of the predicted values are those models for which C_p approaches the value p (i.e., those subset models that fall near the line $C_p = p$ in the above plot).

To understand what is meant by *unbiased predicted values*, consider the full model to be

$$\mathbf{y} = X\boldsymbol{\beta} + \boldsymbol{\epsilon}$$

and the subset model to be of the form

$$\mathbf{y} = X_1\boldsymbol{\beta}_1 + \boldsymbol{\epsilon}$$

where $X = (X_1, \; X_2)$ and $\boldsymbol{\beta}' = (\boldsymbol{\beta}_1', \boldsymbol{\beta}_2')$ are conformable partitions of X and $\boldsymbol{\beta}$ from the full model. Let the ith predicted value from the full model be $\hat{y}_i$ and that from the subset model be denoted by $\hat{y}_i^*$. The mean squared error of a fitted value for the full model is given by the expression:

$$\text{mse}(\hat{y}_i) = \text{var}(y_i) + [E(\hat{y}_i) - E(y_i)]^2$$

where $[E(\hat{y}_i) - E(y_i)]$ is called the *bias* in predicting the observation y_i using $\hat{y}_i$. If it is assumed that the full model allows unbiased prediction, the bias term must be zero; that is,

$$E(\hat{y}_i) - E(y_i) \;=\; 0$$

The mean squared error of a fitted value for the subset model is given by the expression

$$\text{mse}(\hat{y}_i^*) = \text{var}(y_i) + [E(\hat{y}_i^*) - E(y_i)]^2$$

which gives the bias in predicting y_i using the subset model fitted value to be

$$E(\hat{y}_i^*) - E(y_i)$$

Under the assumption that the full model is "unbiased," this bias term thus reduces to

$$E(\hat{y}_i^*) - E(\hat{y}_i)$$

The statistic C_p, as defined above, is a measure of the total mean squared error of prediction (MSEP) of a subset model scaled by σ^2, given by

$$\frac{1}{\sigma^2}\sum_{i=1}^{n} \text{mse}(\hat{y}_i^*).$$

C_p has been constructed so that if the subset model is unbiased (i.e., if $E(\hat{y}_i^*) - E(\hat{y}_i) = 0$), then it follows that

$$\begin{aligned} C_p &= \frac{\text{SSE}_p}{s^2} - (n - 2p) \\ &\approx \frac{(n-p)\sigma^2}{\sigma^2} - (n - 2p) \\ &= p \end{aligned} \tag{4.7}$$

Recall that p here denotes the total number of parameters in the subset model (i.e., including the intercept). Thus only those subset models that have C_p values close to p must be considered if *unbiasedness* in the sense presented earlier is a desired criterion for selection of a subset model.

However, the construction of the C_p criterion is based on the assumption that s^2, the MSE from fitting the full model, is an unbiased estimate of σ^2. If the full model happens to contain a large number of insignificant parameters (β's that are possibly *not* significantly different from zero), s^2 will be inflated (i.e., larger than the estimate of σ^2 obtained from a model in which more variables are significant). This is because the variables that are not contributing to significantly decreasing the SSE are still counted toward the degrees of freedom when computing the MSE in the full model. If this is the case, C_p will not be a suitable criterion to use for determining a good model. Thus, models chosen by other methods that have lower MSE values than the models selected based on the C_p must be considered competitive.

The **AIC** *Criterion* Akaike's information criterion uses the fitted value $L \equiv L(\hat{\beta}, \hat{\sigma}^2)$ of the maximum likelihood function and the number of parameters p:

$$\text{AIC} = -2 \log L + 2p = n \log \text{SSE}/n + 2p$$

This is an unbiased estimate of the expectation of the log-likelihood function under a normally distributed errors assumption. The better models are those with smaller computed AIC values. Because of the form of the above formula, its values are negative, and as with statistics like R^2 and C_p, AIC tends toward a constant value as the number of variables p in the model increases. For normal errors AIC is proportional to C_p and thus only C_P needs to be considered.

The **BIC** *and the* **SBC** *Criteria* The BIC is derived from Bayesian theory and similar to the AIC criterion. The SBC (Schwarz's Bayesian information criterion) is easier to compute

$$\text{SBC} = n \log \text{SSE}/n + p \log n$$

The behavior of BIC (Bayesian information criterion) and AIC is similar, and thus BIC is not discussed further in this book. It is to be noted that SBC is the default criterion used in SAS procedures that perform model selection.

4.4.1 Subset Selection Using PROC REG

The Hald cement data (see Table B.8), used previously in SAS Example D5 to demonstrate multiple regression, is used again to illustrate the use of subset selection procedures using `proc reg`. The use of `proc reg` for subset selection is straightforward. The only action statement required is a `model` statement with accompanying options. The options enable a user to specify one of several variable subset selection procedures available in SAS, to specify other parameters required by these methods, to specify lists of statistics output by these methods, to coerce selected variables to be included in the final model, and to fit a model without an intercept. Any number of `model` statements may appear in the same procedure step.

SAS Example D9

In the program for SAS Example D9, displayed in Fig. 4.41, three subset selection methods, `forward`, `backward`, and `stepwise`, are specified in the first three model statements. With the `forward` selection (`selection=f`), the significance level for entry of variables, `sle` (or `slentry`), used is 0.1, and with `backward` (`selection=b`), an significance level for deletion, `sls` (or `slstay`), of 0.1 is used. The default settings of both `sle` and `sls` equal to 0.15 were used with the `stepwise` selection method.

```
insert data step to create the SAS dataset 'cement' here

title 'Regression : Variable Subset Selection Techniques';
proc reg data=cement plots=none;
   model y = X1-X4/selection=f sle=.1 ;
   model y = X1-X4/selection=b sls=.1;
   model y = X1-X4/selection=stepwise sle=.15 sls=.15;
run;
```

Fig. 4.41. SAS Example D9: program

The output from the program SAS Example D9 appears in Figs. 4.42, 4.43, 4.44, 4.45, 4.46, and 4.47, separated into six parts and edited (e.g., page titles and page numbers have been removed and some spacing reduced) for compactness and clarity, but retaining the original sequence of tables as in the original `proc reg` output. Since it is informative for the user to relate quantities in this output to the computational techniques discussed in Sect. 4.3, at various points in the discussion of this output, some of the results appearing in these tables are recomputed in the text using the notation developed in that section.

The relevant content extracted from the output for the `forward` selection method is shown in Figs. 4.42 and 4.43. Note that two statistics of interest computed for the fitted model in each step are R^2 and Mallows' C_p, which were discussed earlier. The first variable to enter the model in the `forward`

Forward Selection: Step 1

Variable X4 Entered: R-Square = 0.6745 and C(p) = 138.7308

Analysis of Variance					
Source	DF	Sum of Squares	Mean Square	F Value	Pr > F
Model	1	1831.89616	1831.89616	22.80	0.0006
Error	11	883.86692	80.35154		
Corrected Total	12	2715.76308			

Variable	Parameter Estimate	Standard Error	Type II SS	F Value	Pr > F
Intercept	117.56793	5.26221	40108	499.16	<.0001
X4	-0.73816	0.15460	1831.89616	22.80	0.0006

Forward Selection: Step 2

Variable X1 Entered: R-Square = 0.9725 and C(p) = 5.4959

Analysis of Variance					
Source	DF	Sum of Squares	Mean Square	F Value	Pr > F
Model	2	2641.00096	1320.50048	176.63	<.0001
Error	10	74.76211	7.47621		
Corrected Total	12	2715.76308			

Variable	Parameter Estimate	Standard Error	Type II SS	F Value	Pr > F
Intercept	103.09738	2.12398	17615	2356.10	<.0001
X1	1.43996	0.13842	809.10480	108.22	<.0001
X4	-0.61395	0.04864	1190.92464	159.30	<.0001

Fig. 4.42. SAS Example D9: forward selection method (Steps 1 and 2)

method is X4, whose F-to-enter value, 22.80, is the largest among the four independent variables. Its p-value is below 0.05, the `sle`, thus X4 is entered. Here, note that the Type II SS for X4 is equal to $R(\beta_4/\beta_0) = 1831.89616$.

With X4 in the model, X1 is found to have the largest F-to-enter value, 108.22, with a p-value that is also less than 0.05. Hence, X1 is the next variable to be entered. Thus, at the end of Step 2, the current model has variables X4 and X1. Computationally, note that Type II SS for X1 and X4 are, respectively, equal to $R(\beta_1/\beta_0, \beta_4) = 809.10480$ and $R(\beta_4/\beta_0, \beta_1) = 1190.92464$ and that F-to-enter X1 to the model in Step 1 is computed as

$$\text{F-to-enter(X1)} = \frac{R(\beta_1/\beta_0, \beta_4)/1}{\text{SSE}(\beta_0, \beta_1, \beta_4)/10} = 809.10480/7.4762 = 108.22$$

Note that the quantity F-to-enter X2 to the model in Step 2, for example, cannot be computed from the quantities that are available in Fig. 4.42. To obtain these values, the `details=all` option must be specified with the model statement, as will be illustrated in SAS Example D10.

With these two variables in the model, X2 is found to have the largest F-to-enter value, which is computed to be 5.03. The p-value for this variable does not exceed 0.1, and so X2 also enters the model at this stage. Hence, the final model selected by the `forward` method is the three-variable model containing X1, X2, and X4, as seen in Fig. 4.43.

In the backward elimination method in `proc reg`, the `significance level for deletion` is specified using the option `slstay=` (or `sls=`) as stated above. This means that the `F-to-delete` statistic of a selected variable must

Forward Selection: Step 3

Variable X2 Entered: R-Square = 0.9823 and C(p) = 3.0182

Analysis of Variance					
Source	DF	Sum of Squares	Mean Square	F Value	Pr > F
Model	3	2667.79035	889.26345	166.83	<.0001
Error	9	47.97273	5.33030		
Corrected Total	12	2715.76308			

Variable	Parameter Estimate	Standard Error	Type II SS	F Value	Pr > F
Intercept	71.64831	14.14239	136.81003	25.67	0.0007
X1	1.45194	0.11700	820.90740	154.01	<.0001
X2	0.41611	0.18561	26.78938	5.03	0.0517
X4	-0.23654	0.17329	9.93175	1.86	0.2054

Fig. 4.43. SAS Example D9: forward selection method (Step 3)

have a p-value below this value to be retained in the model; otherwise, it will be deleted from the current model.

In Step 0 of the output resulting from the specification `selection= backward` shown in Fig. 4.44, a model containing all explanatory variables X1, X2, X3, and X4 is fitted to the data. The Type II sums of squares and F-statistics (the partial sums of squares and accompanying F-statistics) correspond to the F-to-delete values for each of these variables.

The smallest of these is for variable X3, which is 0.02, and corresponding p-value of ≈ 0.9 is larger than 0.1, the slstay specification; thus, X3 is deleted from the model. Type II SS for X3 is $R(\beta3/\beta0, \beta1, \beta2, \beta4) = 0.10909$ and

$$\text{F-to-delete(X3)} = \frac{R(\beta_3/\beta_0, \beta_1, \beta_2, \beta_4)/1}{\text{SSE}(\beta_0, \beta_1, \beta_2, \beta_3, \beta_4)/8} = 0.10909/5.98295 = 0.02$$

verifying the value shown for F-to-delete X3 in Step 0. In Step 1, X4 is deleted, since the F-to-delete value for this variable is 1.86, and the associated p-value also exceeds 0.1. F-to-delete X2 from the model in Step 1 is computed as

$$\text{F-to-delete(X2)} = \frac{R(\beta_2/\beta_0, \beta_1, \beta_4)/1}{\text{SSE}(\beta_0, \beta_1, \beta_2, \beta_4)/9} = 26.78938/5.33030 = 5.03$$

and this is identical to F-to-enter X2 to the model containing only X1 and X4. In Step 2 (see Fig. 4.45), no variable qualifies for deletion since the F-to-delete values for both X1 and X2 are quite large, and the corresponding p-values are both much smaller than 0.1. Thus the final model selected by the `backward` method contains only the variables X1 and X2. Note that this model is different from the one selected by the `forward` selection method. It is informative to note that the value of F-to-enter X2 to the model in Step 2 in the forward selection method (see Fig. 4.43) is identical to the value for F-to-delete X2 (computed as 5.03) from the model $y = \beta_0 + \beta_1 \text{X1} + \beta_2 \text{X2} + \beta_4 \text{X4} + \epsilon$ in the backward elimination method (see Step 1 in the output displayed in Fig. 4.44).

Backward Elimination: Step 0

All Variables Entered: R-Square = 0.9824 and C(p) = 5.0000

Analysis of Variance					
Source	DF	Sum of Squares	Mean Square	F Value	Pr > F
Model	4	2667.89944	666.97486	111.48	<.0001
Error	8	47.86364	5.98295		
Corrected Total	12	2715.76308			

Variable	Parameter Estimate	Standard Error	Type II SS	F Value	Pr > F
Intercept	62.40537	70.07096	4.74552	0.79	0.3991
X1	1.55110	0.74477	25.95091	4.34	0.0708
X2	0.51017	0.72379	2.97248	0.50	0.5009
X3	0.10191	0.75471	0.10909	0.02	0.8959
X4	-0.14406	0.70905	0.24697	0.04	0.8441

Backward Elimination: Step 1

Variable X3 Removed: R-Square = 0.9823 and C(p) = 3.0182

Analysis of Variance					
Source	DF	Sum of Squares	Mean Square	F Value	Pr > F
Model	3	2667.79035	889.26345	166.83	<.0001
Error	9	47.97273	5.33030		
Corrected Total	12	2715.76308			

Variable	Parameter Estimate	Standard Error	Type II SS	F Value	Pr > F
Intercept	71.64831	14.14239	136.81003	25.67	0.0007
X1	1.45194	0.11700	820.90740	154.01	<.0001
X2	0.41611	0.18561	26.78938	5.03	0.0517
X4	-0.23654	0.17329	9.93175	1.86	0.2054

Fig. 4.44. SAS Example D9: backward elimination method (Steps 0 and 1)

In the output resulting from the `stepwise` option appearing in Figs. 4.46 and 4.47, variables are added based on F-to-enter statistics as in forward selection, but a backward elimination step is performed immediately after each forward step, based on F-to-delete statistics. Here, X4 is entered in Step 1 since the F-to-enter value of 22.80 is the largest and the p-value is less than 0.15. X4 is not considered for deletion since it is the variable just entered. In Step 2, variable X1 is entered with an F-to-enter value of 108.22; obviously, the p-value is again less than 0.15. The F-to-delete statistics for variables X1 and X4, as before, are given by corresponding Type II F-statistics in Step 2. Since both of the p-values exceed 0.15, no variable qualifies for deletion at this stage.

In Step 3, (see Fig. 4.47) variable X2 is added to the model with an F-to-enter statistic of 5.03 since the p-value is below 0.15. Here, $R(\beta_2/\beta_0, \beta_1, \beta_4) = 26.78938$, and thus F-to-enter X2 to the model in Step 2 is computed as

$$\text{F-to-enter(X2)} = \frac{R(\beta_2/\beta_0, \beta_1, \beta_4)/1}{\text{SSE}(\beta_0, \beta_1, \beta_2\beta_4)/9} = 26.78938/5.3303 = 5.03.$$

Backward Elimination: Step 2

Variable X4 Removed: R-Square = 0.9787 and C(p) = 2.6782

Analysis of Variance					
Source	DF	Sum of Squares	Mean Square	F Value	Pr > F
Model	2	2657.85859	1328.92930	229.50	<.0001
Error	10	57.90448	5.79045		
Corrected Total	12	2715.76308			

Variable	Parameter Estimate	Standard Error	Type II SS	F Value	Pr > F
Intercept	52.57735	2.28617	3062.60416	528.91	<.0001
X1	1.46831	0.12130	848.43186	146.52	<.0001
X2	0.66225	0.04585	1207.78227	208.58	<.0001

All variables left in the model are significant at the 0.1000 level.

Fig. 4.45. SAS Example D9: backward elimination method (Step 2)

Stepwise Selection: Step 1

Variable X4 Entered: R-Square = 0.6745 and C(p) = 138.7308

Analysis of Variance					
Source	DF	Sum of Squares	Mean Square	F Value	Pr > F
Model	1	1831.89616	1831.89616	22.80	0.0006
Error	11	883.86692	80.35154		
Corrected Total	12	2715.76308			

Variable	Parameter Estimate	Standard Error	Type II SS	F Value	Pr > F
Intercept	117.56793	5.26221	40108	499.16	<.0001
X4	-0.73816	0.15460	1831.89616	22.80	0.0006

Stepwise Selection: Step 2

Variable X1 Entered: R-Square = 0.9725 and C(p) = 5.4959

Analysis of Variance					
Source	DF	Sum of Squares	Mean Square	F Value	Pr > F
Model	2	2641.00096	1320.50048	176.63	<.0001
Error	10	74.76211	7.47621		
Corrected Total	12	2715.76308			

Variable	Parameter Estimate	Standard Error	Type II SS	F Value	Pr > F
Intercept	103.09738	2.12398	17615	2356.10	<.0001
X1	1.43996	0.13842	809.10480	108.22	<.0001
X4	-0.61395	0.04864	1190.92464	159.30	<.0001

Fig. 4.46. SAS Example D9: stepwise method (Steps 1 and 2)

The F-to-delete statistics for X1, X3, and X4, respectively, are 154.0, 5.03, and 1.86. The smallest value 1.86 has p-value of 0.2 which is larger than 0.15 and, thus, X4 is a candidate for deletion. Type II SS for X4 is $R(\beta_4/\beta_0, \beta_1, \beta_2) = 9.93175$, and the F-to-delete X4 from the model in Step 3 is computed as

$$\text{F-to-delete(X4)} = \frac{R(\beta_4/\beta_0, \beta_1, \beta_2)/1}{\text{SSE}(\beta_0, \beta_1, \beta_2, \beta_4)/9} = 9.93175/5.3303 = 1.86.$$

X4 is deleted in Step 4 giving the final model since no other variable qualifies for entry at 0.15 level when X1 and X2 are already in the model.

Stepwise Selection: Step 3

Variable X2 Entered: R-Square = 0.9823 and C(p) = 3.0182

Analysis of Variance					
Source	DF	Sum of Squares	Mean Square	F Value	Pr > F
Model	3	2667.79035	889.26345	166.83	<.0001
Error	9	47.97273	5.33030		
Corrected Total	12	2715.76308			

Variable	Parameter Estimate	Standard Error	Type II SS	F Value	Pr > F
Intercept	71.64831	14.14239	136.81003	25.67	0.0007
X1	1.45194	0.11700	820.90740	154.01	<.0001
X2	0.41611	0.18561	26.78938	5.03	0.0517
X4	-0.23654	0.17329	9.93175	1.86	0.2054

Stepwise Selection: Step 4

Variable X4 Removed: R-Square = 0.9787 and C(p) = 2.6782

Analysis of Variance					
Source	DF	Sum of Squares	Mean Square	F Value	Pr > F
Model	2	2657.85859	1328.92930	229.50	<.0001
Error	10	57.90448	5.79045		
Corrected Total	12	2715.76308			

Variable	Parameter Estimate	Standard Error	Type II SS	F Value	Pr > F
Intercept	52.57735	2.28617	3062.60416	528.91	<.0001
X1	1.46831	0.12130	848.43186	146.52	<.0001
X2	0.66225	0.04585	1207.78227	208.58	<.0001

All variables left in the model are significant at the 0.1500 level.

No other variable met the 0.1500 significance level for entry into the model.

Fig. 4.47. SAS Example D9: stepwise method (Steps 3 and 4)

SAS Example D10

If the option `details=all` is included in the model statement when using a model selection procedure such as `stepwise`, a more detailed output is obtained. The sample output page in Fig. 4.48 shows the portion of the output in Step 3 of the stepwise procedure discussed previously, if the model statement is of the form

```
model y = X1-X4/selection=stepwise sle=.15 sls=.15 details=all;
```

This illustrates the additional information available if the user wants to keep track of the decision-making process. For example, from this output, it can be seen immediately that the largest F-to-enter to the model in Step 2 is 5.03 with a p-value $= 0.0517$. This is smaller than 0.15, the `sle` value; thus, the variable X2 enters the model. The smallest F-to-delete from the model in Step 3 is 1.86 with a p-value $= 0.2054$. This is larger than the `sls` value 0.15 and thus the variable X4 is removed from the model in Step 3.

Stepwise Selection: Step 3

Statistics for Removal DF = 1,10				
Variable	Partial R-Square	Model R-Square	F Value	Pr > F
X1	0.2979	0.6745	108.22	<.0001
X4	0.4385	0.5339	159.30	<.0001

Statistics for Entry DF = 1,9				
Variable	Tolerance	Model R-Square	F Value	Pr > F
X2	0.053247	0.9823	5.03	0.0517
X3	0.289051	0.9813	4.24	0.0697

Variable X2 Entered: R-Square = 0.9823 and C(p) = 3.0182

Analysis of Variance					
Source	DF	Sum of Squares	Mean Square	F Value	Pr > F
Model	3	2667.79035	889.26345	166.83	<.0001
Error	9	47.97273	5.33030		
Corrected Total	12	2715.76308			

Variable	Parameter Estimate	Standard Error	Type II SS	F Value	Pr > F
Intercept	71.64831	14.14239	136.81003	25.67	0.0007
X1	1.45194	0.11700	820.90740	154.01	<.0001
X2	0.41611	0.18561	26.78938	5.03	0.0517
X4	-0.23654	0.17329	9.93175	1.86	0.2054

Fig. 4.48. SAS Example D10: Step 3 of stepwise output obtained with `details=all`

SAS Example D11

The `rsquare` option is used in a model statement in SAS Example D11 shown in Fig. 4.49 to fit all possible regression models using the R^2 criterion. Other options, `sse, mse, aic`, and `cp`, request that the specified statistics be output for the fitted models in addition to the R^2. In a second model statement, the number of models output is restricted to a specified range of subset *sizes* (i.e., number of independent variables to be included in the models) and using the R^2 criterion for selecting a specified number of *best* models of each subset size. In this example, the keyword options `start=, stop=`, and `best=` are used to specify that all possible subset models that include one variable, two variables, and three variables, be fitted but output information only on the best two models of each size, as determined by decreasing R^2 value.

```
insert data step to create the SAS dataset 'cement' here

proc reg corr plots(only label)=criteria;
  model y = X1-X4/selection=rsquare sse cp aic;
  title 'Regression : Variable Subset Selection Techniques';
run;
```

Fig. 4.49. SAS Example D11: rsquare method

The table displayed in Fig. 4.50 is the output resulting from the first model statement with the `rsquare` selection criterion. This table is the default output produced from the `rsquare` method and summarizes the fit statistics for all possible models of different subset sizes ordered by decreasing R^2 values. In this example, the statistics printed are those requested by using the options `sse` and `cp`.

By including the plot options `plots(only label)=criteria`, the ODS Graphics panel of fit criteria shown in Fig. 4.51 is also produced. Each plot in this panel graphically compares the values of a fit statistic (such as R^2 or

Model Index	Number in Model	R-Square	C(p)	AIC	MSE	Variables in Model
1	1	0.6745	138.7308	58.8516	80.35154	X4
2	1	0.6663	142.4864	59.1780	82.39421	X2
3	1	0.5339	202.5488	63.5195	115.06243	X1
4	1	0.2859	315.1543	69.0674	176.30913	X3
5	2	0.9787	2.6782	25.4200	5.79045	X1 X2
6	2	0.9725	5.4959	28.7417	7.47621	X1 X4
7	2	0.9353	22.3731	39.8526	17.57380	X3 X4
8	2	0.8470	62.4377	51.0371	41.54427	X2 X3
9	2	0.6801	138.2259	60.6293	86.88801	X2 X4
10	2	0.5482	198.0947	65.1167	122.70721	X1 X3
11	3	0.9823	3.0182	24.9739	5.33030	X1 X2 X4
12	3	0.9823	3.0413	25.0112	5.34562	X1 X2 X3
13	3	0.9813	3.4968	25.7276	5.64846	X1 X3 X4
14	3	0.9728	7.3375	30.5759	8.20162	X2 X3 X4
15	4	0.9824	5.0000	26.9443	5.98295	X1 X2 X3 X4

Fig. 4.50. SAS Example D11: output for all subset models

AIC) for all models of each subset size. The best models selected for each subset size (based on the R-square statistic) are indicated on the plots in this panel. The `label` sub-option specifies that these models are identified by the *model number* that appears in the summary table shown in Fig. 4.51.

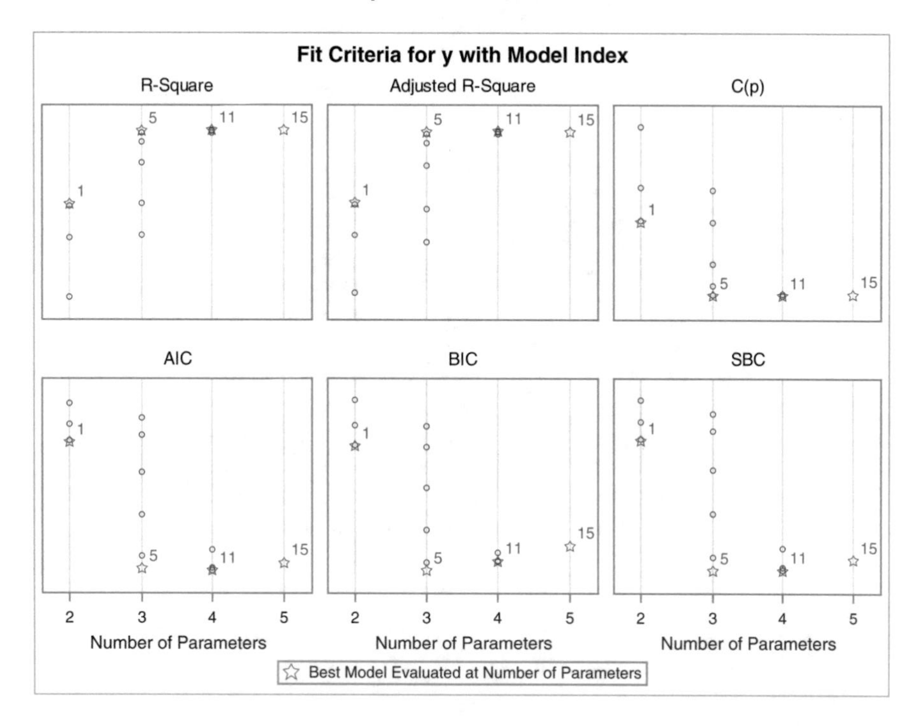

Fig. 4.51. SAS Example D11: comparison of fit criteria

SAS Example D12

SAS Example D12 further illustrates the use of all-subset selection options in `proc reg`. In this example, `adsrsq` is used instead of `rsquare` as the model selection criterion. The data used here came from a study that examined the correlation between the level of prostate-specific antigen and a number of clinical measures in men who were about to receive a radical prostatectomy. The goal was to predict log prostate-specific antigen (*lpsa*) from a number of measurements including log cancer volume (*lcavol*), log prostate weight (*lweight*), *age*, log benign prostatic hyperplasia amount (*lbph*), seminal vesicle invasion (*svi*), log capsular penetration (*lcp*), Gleason score (*gleason*), and percentage Gleason scores 4 or 5 (*pgg45*). The data (see Table B.14) appeared in Stamey et al. (1989). SAS Example D12 shown in Fig. 4.52 fits all possible regression models to develop a prediction equation relating lpsa to the other

```
data prostate;
infile "C:\users\user_name\Documents\...\prostate.txt";
input case lcavol lweight age lbph svi lcp gleason pgg45 lpsa;
;
title 'Variable Subset Selection: Prostate Data';
proc reg data=prostate plots(only)=(cp(label) aic(label));
  model lpsa=lcavol lweight age lbph svi lcp gleason pgg45/selection=adjrsq
                                        start=2 stop=5 best=12 sse mse aic cp b;
run;
```

Fig. 4.52. SAS Example D12: model selection with adjrsq option

Adjusted R-Square Selection Method

Model Index	Number in Model	Adjusted R-Square	R-Square	C(p)	AIC	MSE	SSE
1	5	0.6245	0.6441	5.7150	-61.3742	0.50028	45.52565
2	4	0.6208	0.6366	5.6264	-61.3516	0.50527	46.48490
3	5	0.6195	0.6394	6.9224	-60.0917	0.50694	46.13159
4	5	0.6184	0.6383	7.2029	-59.7961	0.50849	46.27237
5	5	0.6175	0.6374	7.4226	-59.5653	0.50970	46.38261
6	5	0.6164	0.6364	7.6828	-59.2925	0.51113	46.51321
7	4	0.6149	0.6309	7.0742	-59.8471	0.51317	47.21149
8	5	0.6144	0.6345	8.1574	-58.7971	0.51375	46.75137
9	3	0.6144	0.6264	6.2169	-60.6760	0.51382	47.78496
10	5	0.6141	0.6342	8.2352	-58.7161	0.51418	46.79042
11	4	0.6135	0.6296	7.4149	-59.4965	0.51503	47.38245
12	4	0.6134	0.6295	7.4418	-59.4688	0.51517	47.39598

Model Index	Number in Model	Adjusted R-Square	Parameter Estimates								
			Intercept	lcavol	lweight	age	lbph	svi	lcp	gleason	pgg45
1	5	0.6245	0.95102	0.56561	0.42369	-0.01489	0.11184	0.72096	.	.	.
2	4	0.6208	0.14556	0.54960	0.39087	.	0.09009	0.71174	.	.	.
3	5	0.6195	0.06648	0.53340	0.40526	.	0.08303	0.65376	.	.	0.00253
4	5	0.6184	-0.38641	0.53238	0.40752	.	0.08468	0.69305	.	0.07390	.
5	5	0.6175	0.12064	0.56898	0.38545	.	0.09156	0.76638	-0.03661	.	.
6	5	0.6164	0.25371	0.53946	0.56852	-0.01258	.	0.57784	.	.	0.00403
7	4	0.6149	-0.32687	0.53102	0.51505	.	.	0.59762	.	.	0.00319
8	5	0.6144	-0.58433	0.53462	0.57502	-0.01209	.	0.63676	.	0.12956	.
9	3	0.6144	-0.26807	0.55164	0.50854	.	.	0.66616	.	.	.
10	5	0.6141	-0.42258	0.56615	0.50988	.	.	0.69046	-0.08241	.	0.00451
11	4	0.6135	0.17197	0.56193	0.54671	-0.00929	.	0.66505	.	.	.
12	4	0.6134	-0.94665	0.52843	0.52135	.	.	0.64482	.	0.09889	.

Fig. 4.53. SAS Example D12: best subset models for prostate data

variables. As in the previous example, options `start=`, `stop=`, and `best=` are useful for controlling the number of models of interest when a large number of explanatory variables are involved. However, since `adjrsq` and `cp` are not monotonically increasing functions of the model size, if `best=` option is specified it just indicates the total number of models to be displayed, when these selection methods are in use. The MSE and AIC are additional statistics requested in this example using corresponding options, and the parameter estimates for each selected model are also requested to be output by specifying the `b` option.

In Fig. 4.53 information is displayed only for the *best* models of sizes 2, 3, and 4 explanatory variables, determined as those with the largest adjusted R^2 values, along with the variables selected for each model. Note that C_p must be compared to the number of independent variables plus one; thus, the value printed in this table as "Number in Model" must be incremented by one and taken as p for comparing it to C_p. By including the options `plots(only)=(cp(label) aic(label))` in the `proc reg` statement, the ODS Graphics plots of C_p versus p shown in Fig. 4.54 and AIC versus p shown in Fig. 4.55 are obtained.

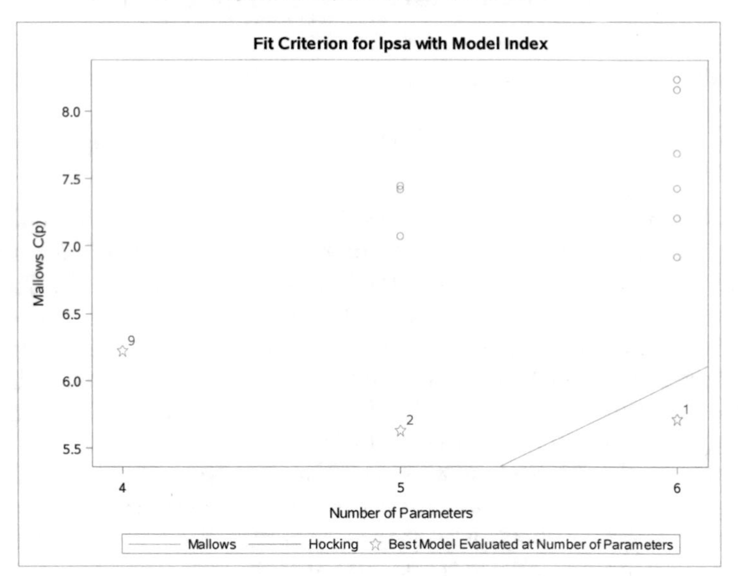

Fig. 4.54. SAS Example D12: plot of C_p versus p

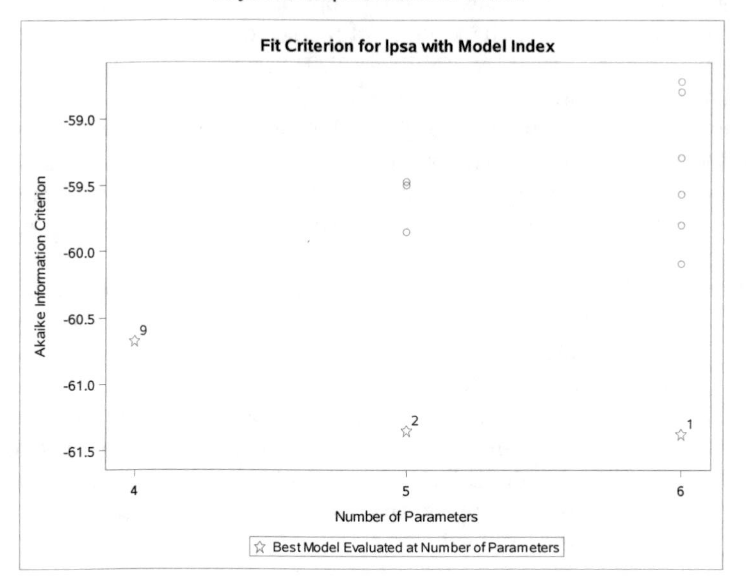

Fig. 4.55. SAS Example D12: plot of AIC versus p

4.4.2 Other Options Available in PROC REG for Model Selection

Several other model selection options are available in `proc reg`. Those selection methods are as follows:

`maxr` Maximum R^2 Improvement Selection Method: The `maxr` method begins by finding the one-variable model producing the highest R^2. Then it adds another variable that produces the largest improvement in R^2. Then each variable in the model is replaced with a variable not in the model to find if such a swap will produce an increase in R^2. The swap that produces the largest increase in R^2 is output as the "best" two-variable model by `maxr`. Another variable is then added to the model, and the comparing-and-swapping process is repeated to locate the "best" three-variable model, and so forth. The `maxr` method is more expensive computationally than the `stepwise` method because it evaluates all combinations of variables before making a swap; whereas the `stepwise` method may remove the "worst" variable in a sequential fashion without considering the effects of replacing any of the other variables in the model by the remaining variables. Consequently, `maxr` typically takes much longer to run than `stepwise`.

`minr` Minimum R^2 Improvement Selection Method: The `minr` method closely resembles the `maxr` method, but the swap chosen at each stage is the one that produces the smallest increase in R^2.

`adjrsq` Adjusted R^2 Selection Method: This method is similar to the `rsquare` method, except that the adjusted R^2 statistic is used as the criterion for selecting models, and the method finds the models with the highest adjusted R^2 within the range of model sizes. The option is `method=adjrsq`.

`cp` Mallows' C_p Selection Method: This method is similar to the `adjrsq` method, except that Mallows' C_p statistic is used as the criterion for model selection. Models are listed in ascending order of C_p.

`groupnames` Specifying Groups of Variables: Groups of variables can be specified to be treated as a single set during the selection process. For example,

```
model y=x1 x2 x3 /
              selection=stepwise groupnames='x1 x2' 'x3';
```

Another example is:

```
model mpg = cyl disp hp drat wt / selection=stepwise sle=.1
              sls=.2 groupnames='cyl' 'disp_hp_drat' 'wt';
```

4.5 Model Selection Using PROC GLMSELECT: Validation and Cross-Validation

In Sect. 4.4 methods for selecting subsets of variables from k potential explanatory variables were discussed. The selected models were compared using statistics such as R^2 and C_p that were calculated from the fitted models themselves. Although such methods are useful when the number of available observations is limited and the aim is identifying a useful and interpretable model, there is no guarantee that the selected models will perform well when accuracy of predicting new observations is of interest. A standard approach for assessing predictive ability of different regression models is to evaluate their performance on a new data set. When a sufficiently large data set is available, this is usually achieved by randomly splitting the data into a *training data set* and a *hold out data set* (often called the *validation data set*). The training data set is used to obtain a set of candidate models to be compared, presumably using methods discussed previously, while the validation data set is used to compare the performance of the selected models, using some criterion. In the case of quantitative response variables, a measure often used for estimating the error in prediction of a model fitted using the training data is the *average squared error* (ASE) of prediction, i.e.,

$$\frac{1}{N}\sum(y_i - \hat{y}_i)^2$$

where y_i is the ith case in the validation data set and $\hat{y}_i$ is its predicted value using the fitted model and N is the validation sample size. This estimate of prediction error obtained using a validation data set, called the validation ASE, is highly variable because it depends on the actual split (i.e., the cases that are actually included in the training and validation data, respectively).

An alternative approach is to use *K-fold cross-validation* (CV_K). Here the original data is first randomly divided into K equal-sized partitions. One of these partitions (say, the kth one, usually called the kth fold) is considered the hold out data and used to obtain the prediction error (ASE) of the model fitted to the remaining data (i.e., the other $K - 1$ partitions put together) considered as the training data. The whole procedure is repeated using the kth fold where $k = 1, \ldots, K$ as the hold out data set and remaining data as the training data and the ASEs resulting from each partition are then combined. Thus ASE of CV_K may be given by

$$\frac{1}{K} \sum_{k=1}^{K} \frac{(y_{ki} - \hat{y}_{ki})^2}{n_k}$$

where y_{ki} denotes ith case in the kth fold, n_k is the size of the kth fold, and $\hat{y}_{ki}$ is prediction using the model fitted to the corresponding remaining data without the kth fold. If $n_k = N/K$, then this reduces to

$$\frac{1}{N} \sum_{k=1}^{K} (y_{ki} - \hat{y}_{ki})^2$$

remembering that $\hat{y}_{ki}$ for each k are calculated using different prediction equations.

The K-fold cross-validation method is suitable when the size of the data set is not large enough to be split into training and validation data sets but large enough so that each fold is sufficiently large to obtain a good estimate of the prediction error and training data are quite adequate to obtain a good prediction equation. In practice, it is recommended that a value of 5 or 10 be used for K (i.e., fivefold or tenfold) for moderately large data sets. In many cases, a third data set called the *test set* is used to obtain an independent estimate of the prediction error, called the *test error* once the final model is selected using cross-validation. When K-fold cross-validation is used, the test set may be a hold out data set randomly selected before the model selection procedure begins.

In examples below, the SAS procedure `GLMSELECT` is used to illustrate how it can be used to perform validation and cross-validation for the purpose of model selection. Although a variety of model selection methods are available in GLMSELECT procedure, including more sophisticated methods such as the LASSO method of Tibshirani (1996) and the related LAR method of Efron et al. (2004), the emphasis in the examples presented will be on the methods discussed above. The user may elect to provide the training, validation, and test data as already prepared SAS data sets or may specify that a single input data set provided be randomly subdivided into a training and validation data sets (and optionally, a test data set) in given proportions. The user may opt to use traditional significance-level-based criteria such as stepwise or those based on other criteria such as R^2_{adj}, C_p, or AIC to arrive at candidate models and then choose the best model based on ASE computed using the validation data set. The procedure also provides graphical summaries that aids this process. Alternatively, the user may request that K-fold cross-validation be performed by selecting that the K folds be selected randomly or sequentially.

SAS Example D13

In the SAS Example D13 program (see Fig. 4.56), `proc glmselect` is used to perform model selection using the prostate data introduced in SAS Example D12 and used in the SAS program (see Fig 4.52) to illustrate the `adjrsq` model selection option in `proc reg`.

```
data prostate;
infile "C:\users\user_name\Documents\...\prostate.txt";
input case lcavol lweight age lbph svi lcp gleason pgg45 lpsa;
;
title 'Using validation ASE for model selection';
proc glmselect data=prostate
               seed=12345 plots(stepAxis=number)=(criteria ASE);
   partition fraction(validate=.35);
   model lpsa=lcavol lweight age lbph svi lcp gleason pgg45
           / selection=stepwise(choose = validate
                                select = sl)
             stb;
run;
```

Fig. 4.56. SAS Example D13: model selection using validation ASE

The full model used is the same model used in SAS Example D12 where `lpsa` is used as a response variable in a multiple regression model with the 8 variables `lcavol, lweight, age, lbph, svi, lcp, gleason,` and `pgg45` as the regressors. The `selection=` option specifies the model selection method optionally followed by parentheses enclosing suboptions available for the specified method. The model selection methods that may be specified are *none, forward, backward, stepwise, lar, lasso,* and *elasticnet.* The stepwise method specified in this example is similar to the same option in `proc reg`, except that it is used in `proc glmselect` as a general procedure rather than only for selecting models based on significance levels. It is similar in that individual effects are added sequentially and an effect may be removed in each cycle, but the selection of effects may be based on other criteria than significance levels. For specifying the selection criterion using the suboption `select=` in `proc glmselect`, the available options are *adjrsq, aic, aicc, bic, sbc, cp, cv, press, rsquare, sl,* and *validate.*

The `select=sl` used in this example specifies that stepwise method be performed using the significance level criterion for entering and removing variables, much the same way as in `proc reg` with the `selection=stepwise` in use. That is, `proc glmselect` will stop the process when a predictor cannot

The GLMSELECT Procedure

Data Set	WORK.PROSTATE
Dependent Variable	lpsa
Selection Method	Stepwise
Select Criterion	Significance Level
Stop Criterion	Significance Level
Choose Criterion	Validation ASE
Entry Significance Level (SLE)	0.15
Stay Significance Level (SLS)	0.15
Effect Hierarchy Enforced	None
Random Number Seed	12345

Number of Observations Read	97
Number of Observations Used	97
Number of Observations Used for Training	69
Number of Observations Used for Validation	28

Dimensions	
Number of Effects	9
Number of Parameters	9

Fig. 4.57. SAS Example D13: stepwise subset selection using SLE = 0.15 SLS = 0.15

be added or removed at the default significance levels. However, the difference is that by using the suboption `choose=`, the user may specify another criterion to be used for determining the "best" among models found in each step. For example, using `choose=cp` requests that the model with the smallest C_p value be chosen as the best regardless of where the procedure stopped. In this example, `choose=validate` is used to specify that the model with the smallest validation ASE be determined. Since `proc glmselect` produces tables and graphics that contains the validation ASE for models in each step when this option is used, the users have the opportunity to compare the relative magnitudes of the ASE among the models and make their own decisions. The option `stb` requests that standardized parameter estimates to be output for the selected model.

Figure 4.57 summarizes the model options used in the `proc glmselect` step. As indicated above, the standard stepwise method with significance levels of 0.15 for both entry and deletion of variable is used here and the "best" model is chosen using a validation data set obtained by randomly splitting the original prostate data set of 97 cases to obtain training data set with 69 cases and a validation set of 28 cases. The results of the stepwise method is summarized in Fig. 4.58, where it is clear that no variables were deleted at any

The GLMSELECT Procedure

Stepwise Selection Summary							
Step	Effect Entered	Effect Removed	Number Effects In	ASE	Validation ASE	F Value	Pr > F
0	Intercept		1	1.2421	1.5236	0.00	1.0000
1	lcavol		2	0.5592	0.7374	81.83	<.0001
2	lweight		3	0.4670	0.7839	13.02	0.0006
3	svi		4	0.4469	0.6793*	2.93	0.0918
* Optimal Value Of Criterion							

Selection stopped because the candidate for entry has SLE > 0.15 and the candidate for removal has SLS < 0.15.

Stop Details					
Candidate For	Effect	Candidate Significance		Compare Significance	
Entry	age	0.3321	>	0.1500	(SLE)
Removal	svi	0.0918	<	0.1500	(SLS)

Fig. 4.58. SAS Example D13: stepwise selection summary

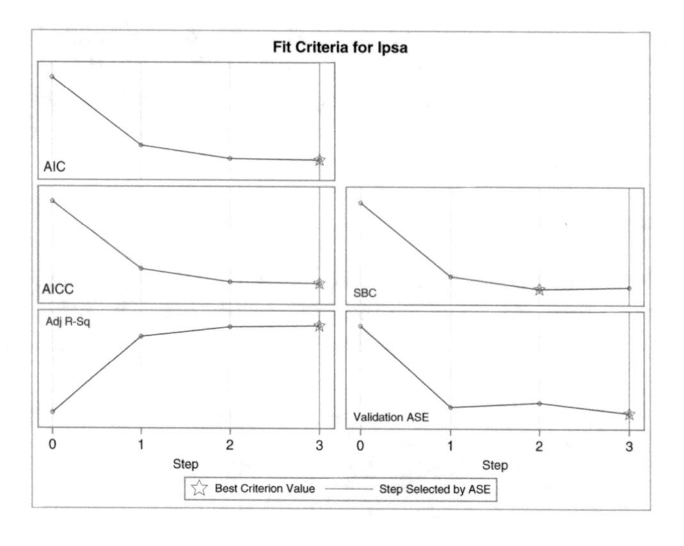

Fig. 4.59. SAS Example D13: panel of fit criteria plots

intermediate step and the procedure stopped in Step 3 when no other variable qualified for entry. A panel of six plots of several fit criteria values plotted against step numbers are shown in Fig. 4.59; one of these plots shows that the model in Step 2 has the smallest SBC value. Validation ASE was calculated in each step, and the model with an `intercept, lcavol, lweight,` and `svi` is shown to be the model with smallest ASE. This fact is also illustrated visually

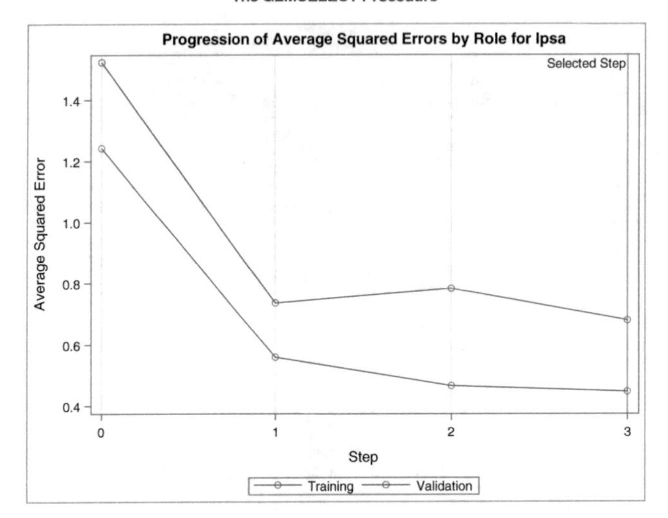

Fig. 4.60. SAS Example D13: ASE for models in each step for training and validation data

in Fig. 4.60 where the ASE calculated for the training and validation data sets are plotted in the same graph for each model fit for comparison. It is evident that the validation ASE is comparatively larger than the training ASE for all models and that 0.6793 is the lowest that can be obtained using the stepwise selection method with the specified significance levels.

SAS Example D14

In the SAS Example D14 program (see Fig. 4.61) `proc glmselect` is used to perform model selection using K-fold cross-validation. The prostate data set was used in SAS Example D13 to illustrate the use of a validation data set

```
data prostate;
infile "C:\users\user_name\Documents\...\prostate.txt";
input case lcavol lweight age lbph svi lcp gleason pgg45 lpsa;
;
proc glmselect data=prostate
               seed=12345 plots(stepAxis=number)=(criteria coefficients);
   model lpsa=lcavol lweight age lbph svi lcp gleason pgg45
            / cvmethod=random(5) selection=stepwise(select=adjrsq choose=cv)
              stats=(cp aic sbc) stb;
run;
```

Fig. 4.61. SAS Example D14: model selection using cross-validation

obtained via random data splitting for calculation of validation ASE. As an illustration, the model selection method used here is the adjusted R^2 criterion (that is, in each step the effect that yields the largest increase in adjusted R^2 is entered). The fivefold cross-validation is used to compute cross-validation estimate of prediction error for each fold and then averaged across the folds (see formula given above for CV_K). This statistic is called `CV PRESS` in `proc glmselect` and is computed for each model selected at each step.

The model statement option `cvmethod=random(5)` specifies that fivefold cross-validation with the folds selected randomly (each of size, approximately n/k rounded down to an integer) be used to perform cross-validation. The option `selection=stepwise` with suboptions `select=adjrsq` and `choose=cv` specifies that the adjusted R^2 criterion be used as described above to select a model at each step and compute *cv press* for each model by performing cross-validation using the method indicated in the `cvmethod=` option. Using the option `stats=(cp aic sbc)` requests that the statistics C_p, `AIC` and `SBC` to be output in the model summary in addition to R^2_{adj} and `CV PRESS`. The option `stb` requests that standardized parameter estimates to be output for the selected model and for the *coefficient progression plot* that results from the `plot=` option `coefficients`.

The model selection settings used in the `proc glmselect` step of Example D14 are detailed in Fig. 4.62. The model selection method specified was `stepwise` with adjusted R^2 as the selection criterion. The estimate of prediction error `cv press` is calculated for each model using fivefold cross-validation using random splitting.

The GLMSELECT Procedure

Data Set	WORK.PROSTATE
Dependent Variable	lpsa
Selection Method	Stepwise
Select Criterion	Adj R-Sq
Stop Criterion	Adj R-Sq
Choose Criterion	Cross Validation
Cross Validation Method	Random
Cross Validation Fold	5
Effect Hierarchy Enforced	None
Random Number Seed	12345

Number of Observations Read	97
Number of Observations Used	97

Dimensions	
Number of Effects	9
Number of Parameters	9

Fig. 4.62. SAS Example D14: GLMSELECT model selection settings

The GLMSELECT Procedure

Stepwise Selection Summary								
Step	Effect Entered	Effect Removed	Number Effects In	Adjusted R-Square	AIC	CP	SBC	CV PRESS
0	Intercept		1	0.0000	127.8376	159.8903	31.4123	131.5364
1	lcavol		2	0.5346	54.6340	24.3943	-39.2166	61.2748
2	lweight		3	0.5771	46.3098	14.5414	-44.9661	58.0141*
3	svi		4	0.6144	38.3240	6.2169	-50.3772*	60.3833
4	lbph		5	0.6208	37.6484	5.6264*	-48.4780	59.3252
5	age		6	0.6245	37.6258*	5.7150	-45.9259	58.1399
6	pgg45		7	0.6259	38.2115	6.4019	-42.7655	58.8810
7	lcp		8	0.6273*	38.7689	7.0822	-39.6334	58.6097
* Optimal Value Of Criterion								

Selection stopped at a local maximum of the AdjRSq criterion.

Stop Details				
Candidate For	Effect	Candidate Adj-RSq		Compare Adj-RSq
Entry	gleason	0.6234	<	0.6273
Removal	lcp	0.6259	<	0.6273

Fig. 4.63. SAS Example D14: stepwise selection using R^2_{adj} and cross-validation

The GLMSELECT Procedure

Fit Criteria for lpsa

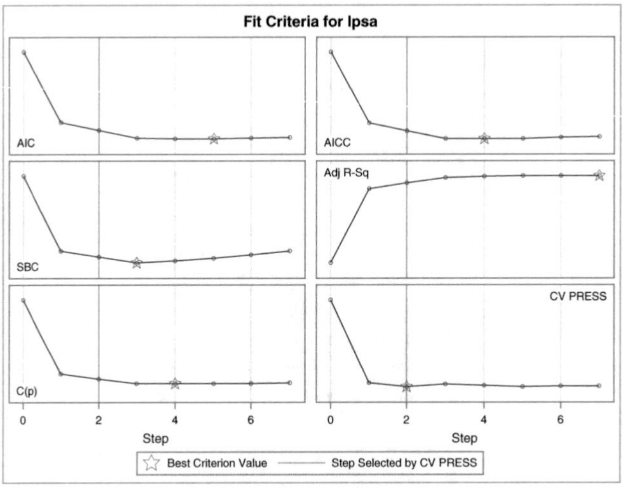

Fig. 4.64. SAS Example D14: panel of fit criteria plots

Figure 4.63 displays the results of the model selection procedure where seven different models are fitted and the statistics `AIC`, C_p, and `SBC` are calculated using the entire prostate data set. Fivefold cross-validation is performed using each model to obtain `CV PRESS`, the estimated prediction error from each model. The selected model with the largest value of R^2_{adj} is the model that does not include the regressor variable `gleason`.

Panels of fit criteria values plotted against step numbers are shown in Fig. 4.64, one of which shows that `CV PRESS` is minimum for Model 2 with an intercept and `lcavol` and `lweight` as the regressors. Final fit statistics for this model are summarized in Fig. 4.65.

However, the `CV PRESS` values for the models 2 through 7 are similar. Thus a possible good model may be one of the other models, say Model 4, that has the smallest `Cp` value of 5.6264 as well as several other good fit criteria. This model with an intercept and the regressors `lcavol, lweight,svi,` and `lbph` thus appears to have the smallest bias and comparably small error in prediction.

The selected model, based on Cross Validation, is the model at Step 2.

Effects:	Intercept lcavol lweight

Analysis of Variance				
Source	DF	Sum of Squares	Mean Square	F Value
Model	2	74.95130	37.47565	66.51
Error	94	52.96636	0.56347	
Corrected Total	96	127.91766		

Root MSE	0.75065
Dependent Mean	2.47839
R-Square	0.5859
Adj R-Sq	0.5771
AIC	46.30976
AICC	46.74454
BIC	-51.18891
C(p)	14.54143
SBC	-44.96610
CV PRESS	58.01412

Parameter Estimates					
Parameter	DF	Estimate	Standardized Estimate	Standard Error	t Value
Intercept	1	-0.302598	0	0.569040	-0.53
lcavol	1	0.677526	0.691786	0.066262	10.22
lweight	1	0.510944	0.219825	0.157256	3.25

Fig. 4.65. SAS Example D14: statistics for the selected model

An interesting plot that is produced by using the `plot=coefficients` option is shown in Fig. 4.66. This shows a panel of two plots showing how the standardized coefficients and the criterion used to choose the final model evolve as the selection progresses. The upper plot in the panel displays the standardized coefficients as a function of the step number. The lower plot shows how the value of the criterion value (here `CV PRESS`) used to choose the final model evolves as the selection progresses.

The standardized coefficients for each parameter are connected by lines in order for the user to track the changes in that parameter. Coefficients corresponding to effects that are not in the selected model at a given step are zero and thus do not yet appear in the plot. For example, at Step 2, the regressors `lcavol` and `lweight` are in the model, and their coefficient estimates are positive; all the other coefficients are zero. At Step 4 there are four coefficients that are nonzero, as shown in the plot and their standardized estimates stay about the same in subsequent steps.

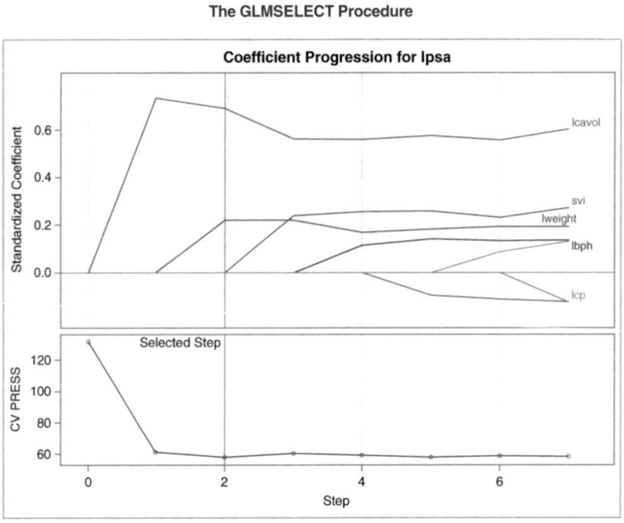

Fig. 4.66. SAS Example D14: coefficient estimates for subset models selected

4.6 Exercises

4.1 The following data are from an experiment that tested the performance of an industrial engine (Schlotzhauer and Littell 1997). The experiment used a mixture of diesel fuel and gas from distilling organic materials.

The horsepower (y) of the machine was measured at several speeds (x) measured in hundreds of revolutions per minute (rpm $\times$ 100).

x	22	20	18	16	14	12	15	17	19	21	22	20
y	64.03	62.47	54.94	48.84	43.73	37.48	46.85	51.17	58.00	63.21	64.03	59.63
x	18	16	14	12	10.5	13	15	17	19	21	23	24
y	52.90	48.84	42.74	36.63	32.05	39.68	45.79	51.17	56.65	62.61	65.31	63.89

Write and execute a SAS program to perform the computations to provide answers to the following questions. Extract material from the output and copy or attach them as the required answers.

a. Use the method of least squares to obtain estimates $\hat{\beta}_0$ and $\hat{\beta}_1$ of the parameters in the model $y = \beta_0 + \beta_1 x + \epsilon$.
b. Construct a plot that shows the scatter of (x, y) data points with y on the vertical axis.
c. Give the least squares prediction equation. Superimpose the least squares line on the scatter plot in part (b).
d. Compute a table of predicted values $\hat{y}$ and residuals $y - \hat{y}$ corresponding to the observed values y.
e. Identify the sums of squares $\sum(y-\bar{y})^2$, $\sum(\hat{y}-\bar{y})^2$, and $\sum(y-\hat{y})^2$ from your output. Verify that these give a decomposition of total variability in y into two parts and identify the parts by sources of variation.
f. Calculate the proportion of the total variability in the y-values that is accounted for by the linear regression model. Explain why this value is a measure of how well your model fits the data.
g. Give the point estimate s^2 of σ^2.
h. State the estimated standard errors of $\hat{\beta}_0$ and $\hat{\beta}_1$.
i. Construct 95% confidence intervals for β_0 and β_1.
j. Test the hypothesis $H_0 : \beta_1 = 0$ versus $H_a : \beta_1 \neq 0$ using a t-test. Give your conclusion using $\alpha = 0.05$ and the p-value.
k. Obtain plots of the residuals against x and the predicted values, respectively. Do these two plots suggest any inadequacies of this model? Explain why you reached your conclusion.
l. Obtain a normal probability plot of the *studentized residuals*. State the model assumption that you can verify using this plot. Is this a plausible assumption for this model?

4.2 The text McClave et al. (2000) discusses the following problem that relates the value of a home to its upkeep expenditure.

Home	Value of home, x ($1000s)	Upkeep expenditure, y (dollars)	Home	Value of home, x ($1000s)	Upkeep expenditure, y (dollars)
1	237.00	1412.08	21	153.04	849.14
2	153.08	797.20	22	232.18	1313.84
3	184.86	872.48	23	125.44	602.06
4	222.06	1003.42	24	169.82	642.14
5	160.68	852.90	25	177.28	1038.80
6	99.68	288.48	26	162.82	697.00
7	229.04	1288.46	27	120.44	324.34
8	101.78	423.08	28	191.10	965.10
9	257.86	1351.74	29	158.78	920.14
10	96.28	378.04	30	178.50	950.90
11	171.00	918.08	31	272.20	1670.32
12	231.02	1627.24	32	48.90	125.40
13	228.32	1204.78	33	104.56	479.78
14	205.90	857.04	34	286.18	2010.64
15	185.72	775.00	35	83.72	368.36
16	168.78	869.26	36	86.20	425.60
17	247.06	1396.00	37	133.58	626.90
18	155.54	711.50	38	212.86	1316.94
19	224.20	1475.18	39	122.02	390.16
20	202.04	1413.32	40	198.02	1090.84

Quality Home Improvement Center (QHIC) operates five stores in a large metropolitan area. The marketing department at QHIC wishes to study the relationship between x, home value (in thousands of dollars), and y, yearly expenditure on home upkeep (in dollars). A random sample of 40 homeowners is taken and asked to estimate their expenditures during the previous year on the types of home upkeep products and services offered by QHIC. Public records of the county auditor are used to obtain the previous year's assessed values of the homeowner's homes. The resulting x- and y-values are given in the following table. Use a SAS program to fit a simple linear regression model to this data. You must use SAS statements to produce output necessary to answer questions given below.

a. Obtain a scatter plot of the dependent variable y against the values of the independent variable x. Does this plot suggest that simple linear regression model might relate y to x?
b. Use SAS to fit the model

$$y = \beta_0 + \beta_1 x + \epsilon$$

to these data. From the output, report the following information on a separate sheet. You must extract this information from the output and write them in the form discussed in the text.

c. Give the least squares estimates of β_0 and β_1 and their standard errors, respectively. What is your prediction equation?
d. Use your prediction equation to estimate the expected increase in yearly upkeep expenditure for an additional $10,000 increase in home value. Show your work clearly.
e. What is the coefficient of determination for your regression equation? In you own words, explain what this means to you in terms of variability in yearly upkeep expenditure.
f. Construct a 95% confidence interval for β_1. State in words what this interval says about the expected increase in yearly upkeep expenditure.
g. Test the hypothesis $H_0 : \beta_1 = 0$ against $H_a : \beta_1 \neq 0$ at $\alpha = 0.05$ using the p-value printed. State your decision.
h. Find the point estimate of the mean yearly upkeep expenditure of all homes worth $220,000 and a 95% confidence interval for this mean.
i. Obtain a graph with plots of the 95% confidence interval and 95% prediction interval curves for the fitted regression line overlaid on a scatter plot of the original data.
j. Obtain a scatter plot of residuals against the x variable. Does the assumption of a constant error variance appears to be satisfied? Explain.
k. Obtain a normal probability plot of residuals and a plot of residuals against the predicted values variable. Do these plots indicate that any of the model assumptions are not plausible? In particular, is the assumption of normal errors reasonable? Explain.

4.3 Athletes are constantly seeking measures of the degree of their cardiovascular fitness prior to a major race, because they want to ensure that they are training at a level which will produce peak performance. One such measure of fitness is the time to exhaustion from running on a treadmill at a specified angle and speed. Twenty experienced distance runners who professed to be at top condition were evaluated on the treadmill and then had their times recorded in a 10-km race. The resulting x- and y-values are given in the following table. You must use SAS statements in a SAS program to produce output necessary to answer questions given below. Attach your SAS output (or cut-and-paste tables or graphics as appropriate) but provide answers in the form discussed in the text to each of the questions using information from the output.

Treadmill time (minutes), x	7.5	7.8	7.9	8.1	8.3	8.7	8.9	9.2	9.4	9.8
10-km time (minutes), y	43.5	45.2	44.9	41.1	43.8	44.4	42.7	43.1	41.8	43.7
x (*continued...*)	10.1	10.3	10.5	10.7	10.8	10.9	11.2	11.5	11.7	11.8
y	39.5	38.2	43.9	37.1	37.7	39.2	35.7	37.2	34.8	38.5

Use the regression model $y = \beta_0 + \beta_1 x + \epsilon$ in `proc reg` step(s) to produce a Normal quantile plot of residuals, a residuals versus predicted values plot, a plot containing a regression line, confidence limits, and prediction limits overlaid on a scatter plot of data, and a plot of residuals versus the explanatory variable. Use the `plots=` option to select plots to be output; do not use the *diagnostics panel* of plots. Also use the ANOVA table, the estimates table, and the output statistics table as necessary to answer *all* questions given below.

a. Using numbers output from the SAS output, construct an analysis of variance table including a column for the F-statistic to test $H_0 : \beta_1 = 0$ against $H_a : \beta_1 \neq 0$ at $\alpha = 0.05$. State your decision using the p-value.
b. Give the least squares estimates of β_0 and β_1 and their standard errors, respectively. What is your prediction equation?
c. Use your *prediction equation* to estimate the expected decrease in the 10-km time for an increase of 2 minutes in treadmill time. Show your work clearly.
d. What is the coefficient of determination for your regression equation? In you own words, explain what this means to you in terms of the variability in 10-km time.
e. Give a 95% confidence interval for β_1. State in words what this interval says about the expected increase in 10-km time.
f. Test the hypothesis $H_0 : \beta_1 = 0$ against $H_a : \beta_1 \neq 0$ at $\alpha = 0.05$ using the p-value from the estimates table. State your decision.
g. Find the point estimate of the mean 10-km time for the population of athletes with a treadmill time of 10 minutes. Calculate a 95% confidence interval for the mean 10-km time for the population of athletes with a treadmill time of 10 minutes.
h. The following plots must be produced as part of the graphical output from your SAS program as described at the beginning of this question. Attach these plots to your solution and answer the questions relating to them, if any.
   a. Obtain a graph with plots of the 95% confidence interval and 95% prediction interval curves for the fitted regression line overlaid on a scatter plot of the original data.
   b. Obtain a scatter plot of residuals against the x variable. Using this plot say whether the assumption of a constant error variance for the response y appear to be plausible. Explain.
   c. Obtain a normal probability plot residuals and a plot of residuals against the predicted values variable. Do these plots indicate that any of the model assumptions are not plausible? In particular, is the assumption of normal errors reasonable? Explain.

4.4 It is reasonable to expect that more miles an automobile has been driven, the higher will be the levels of pollutants it emits. The following data give hydrocarbon (HC) emissions (y) at idling speed, in parts per million (ppm), and the distances traveled by each automobile in thousands of miles (x). Obtain statistics and the plots necessary to answer the following:

Automobile	Distance x (1000 miles)	HC emissions y (ppm)
A	34	270
B	312	2352
C	84	1058
D	50	1035
E	89	2305
F	50	658
G	33	588
H	109	811
I	38	1200
J	2	600
K	71	540
L	52	247
M	31	129

Use a SAS program to fit a single variable regression model and obtain all *residual case statistics* and *diagnostic plots* discussed in class. In this problem you may use the *diagnostic panel* of plots and the *fit plot* (obtained by default) with appropriate *labelling* of points. Use both of the statistics and the plots in your answers to each of the following parts.

a. Are there any cases that are x-outliers? Explain.
b. Are there any cases that are y-outliers? Explain.
c. The Cook's D statistic for some of these cases are "large." Explain reasons for this by using the fact Cook's D is a product of functions of studentized residuals and hat diagonals.
d. Use the appropriate case statistics for cars labeled B and E from the above fit to say what would happen to the model fit if these cases are deleted (one at a time), explaining what each of these statistics indicates.
e. Refit the model after the cars labeled B and E are deleted one at a time (separately). Discuss the model fit for each of these compared with the fit of the original model, using statistics from the output whenever possible.

4.5 A realtor studied the relationship between x = annual income (in 1000's of dollars) of home purchasers and y = sale price of the house (in 1000's of dollars). The realtor gathered the data from mortgage applications for 24 recent sales in the realtor's sales area in one season (text file provided).

Home	Annual income x (in \$1000's)	Sale price y (in \$1000's)	Home	Annual income x (in \$1000's)	Sale price y (in \$1000's)
A	25.0	84.9	M	36.0	110.0
B	28.5	94.0	N	39.0	125.0
C	29.2	96.5	O	39.0	119.9
D	30.0	93.5	P	40.5	130.6
E	31.0	102.9	Q	40.9	120.8
F	31.5	99.5	R	42.5	129.9
G	31.9	101.0	S	44.0	135.5
H	32.0	105.0	T	45.0	140.0
I	33.0	99.9	U	50.0	150.7
J	33.5	110.0	V	54.6	170.0
K	34.0	100.0	W	65.0	110.0
L	35.9	116.0	X	70.0	185.0

Use a SAS program to fit a single variable regression model and obtain all *residual case statistics* and *diagnostic plots* discussed in class. In this problem you may use the *diagnostic panel* of plots and the *fit plot* (obtained by default) with appropriate *labelling* of points. Use both of the statistics and the plots in your answers to each of the following parts.

a. Are there any cases that are x-outliers? Explain.
b. Are there any cases that are y-outliers? Explain.
c. The Cook's D statistic for some of these cases are "large." Explain reasons for this by using the fact Cook's D is a product of functions of studentized residuals and hat diagonals.
d. Suppose that the model is refitted after the homes labeled X and W are deleted one at a time. Discuss the fit of each of these models compared with the fit of the original model.
e. Explain the effect of removing these cases by using appropriate case statistics for these cases obtained from the model fit to the original data.

4.6 To model the relationship between the dose level of a drug product and its potency, a pharmaceutical firm inoculated each of 15 test tubes were with a virus culture and incubated for 5 days at 30 °C. Three test tubes were randomly assigned to each of the five dose levels to be investigated (2, 4, 8, 16, 32 mg). Each tube was injected with one dose level and a measure of the antiviral strength of the product was obtained. The experiment was described in Ott and Longnecker (2001) and the data are reproduced here:

x, dose level (mg)	y, response
2	5, 7, 3
4	10, 12, 14
8	18, 15, 18
16	20, 21, 19
32	23, 24, 28

Write and execute a SAS program to perform the computations to provide answers to the following questions. You may use as many steps in your program as necessary.

a. Plot the data in a scatter plot. Does it appear that a simple linear regression model would be a good fit?
b. Compute an analysis of variance table for regression.
c. Compute a lack of fit test for this model and report the results in an ANOVA table where SS_{Lack} and SSE_{Pure} are shown as a partition of SSE. What is your conclusion from this test? Use $\alpha = 0.05$.
d. Compute the predicted values and residuals. Plot the residuals against the dose level and the predicted values, respectively. Do these two plots suggest any inadequacies of this model? Explain why you reached your conclusion.
e. Many times a logarithmic transformation can be used on the dose levels to *linearize the response* with respect to the dose level. Plot the response against the logarithms (to the base 10) of the five dose levels and comment on this plot.

4.7 Experience with a certain type of plastic indicates that a relation exists between the hardness (measured in Brinell units) of items molded from the plastic (y) and the elapsed time since termination of the molding process (x). In a study to examine this relationship, 16 batches of the plastic were made, and from each batch, 1 test item was molded (Kutner et al. 2004; data slightly modified for ease of hand calculation without affecting results of the analysis). Each test item was randomly assigned to one of four predetermined time levels, and the hardness was measured after the assigned time had elapsed. The results are shown as follows:

x, elapsed time (hours)	y, hardness
16	199, 205, 196, 200
24	218, 220, 215, 223
32	237, 234, 235, 230
40	250, 248, 253, 245

Answer the following questions. You must execute an appropriate SAS program and extract portions of the output to provide the answers.

a. Plot the data in a scatter plot. Does it appear that a simple linear regression model would be a good fit?
b. Use the least squares method to fit a simple linear regression model. What is your prediction equation? Add the straight line of your prediction equation to the plot in part (a).
c. Construct an analysis of variance table for the regression. Use the F-ratio to perform a test of $H_0 : \beta_1 = 0$ versus $H_a : \beta_1 \neq 0$ using $\alpha = 0.05$

d. Perform a lack of fit test for this model. Report the results in an ANOVA table where SS_{Lack} and SSE_{Pure} are shown as a partition of SSE. What is your conclusion from this test? Use $\alpha = 0.05$.
e. Compute the predicted values and residuals. Plot the residuals against `elapsed time` and the predicted values, respectively. Do these two plots suggest any inadequacies of this model? Explain how you reached your conclusion.

4.8 Thirteen specimens of Cu–Ni alloys with varying degrees of iron content in percent were submerged in sea water for 60 days and the weight loss due to corrosion recorded in units of milligrams per square decimeter per day. In a study to examine the dependency of corrosion (y) on iron content (x), a simple linear regression model was fitted to the data. The data as reported in Draper and Smith (1998) are given below:

Specimen	x, Fe %	y, Weight loss (mg/dm)
1	0.01	127.6
2	0.48	124.0
3	0.71	110.8
4	0.95	103.9
5	1.19	101.5
6	0.01	130.2
7	0.48	122.0
8	1.44	92.3
9	0.71	113.2
10	1.96	83.7
11	0.01	128.0
12	1.44	91.7
13	1.96	86.3

Answer the following questions using output from an appropriate SAS program.

a. Plot the data in a y vs. x scatter plot. Does it appear that a simple linear regression model would be a good fit?
b. Use the LS method to fit a simple linear regression model. What is your prediction equation? Add the plot of your prediction equation to the plot in part (a).
c. Construct an analysis of variance table for the regression. Use the F-ratio to perform a test of $H_0 : \beta_1 = 0$ vs. $H_a : \beta_1 \neq 0$ using $\alpha = 0.05$
d. Perform a lack of fit test for this model. Report the results in an ANOVA table where SS_{Lack} and SSE_{Pure} are shown as a partition of SSE. What is your conclusion from this test? Use $\alpha = 0.05$.
e. Compute the predicted values and residuals. Obtain plots of the residuals against `Fe %` and the predicted values ($\hat{y}$), respectively. Do these two plots suggest any inadequacies of this model? Explain why you reached your conclusion.

4.9 In an experiment to study the problem of predicting the tensile strength (y) of concrete beams from measurements of their specific gravity (x_1) and moisture content (x_2), a multiple regression was fitted to data for 10 specimens (Devore 1982). The data are given as follows:

Obs.	Tensile strength y	Specific gravity x_1	Moisture content x_2
1	11.24	0.499	11.1
2	12.64	0.558	8.9
3	12.93	0.604	8.8
4	11.32	0.441	8.9
5	11.68	0.510	8.8
6	11.90	0.528	9.9
7	10.73	0.418	10.1
8	11.70	0.480	10.5
9	11.12	0.406	10.5
10	11.41	0.467	10.7

Use a SAS program to fit the model $y = \beta_0 + \beta_1 x_1 + \beta_2 x_2 + \epsilon$. Extract and/or calculate answers for the following from the SAS output:

a. Give the analysis of variance including the F-ratio and the associated p-value. State the null and alternative hypotheses about the coefficients you will test using the F-ratio. Use the p-value to perform this test using $\alpha = 0.05$

b. Report the prediction equation and standard errors of $\hat{\beta}_1$ and $\hat{\beta}_2$. If the specific gravity is kept constant at 0.5, estimate the change in mean tensile strength of a concrete beam if its moisture content increases from 8.0 to 10.0 units. Does the mean tensile strength increase or decrease?

c. Using the estimate of β_1 and its standard error given in the SAS output, compute a 95% confidence interval for β_1. Test the hypothesis $H_0 : \beta_1 = 0$ versus $H_a : \beta_1 \neq 0$ using this interval, specifying the α level of the test.

d. The specifications stipulate that the mean tensile strength must increase by at least 1 unit for an increase in specific gravity of 0.1 if the moisture content is unchanged. Test this hypothesis using the above confidence interval.

e. Compute the predicted tensile strength $\hat{y}_{11}$ of a beam with $x_1 = 0.5$ and $x_2 = 9.0$. Extract the 95% confidence interval for $E(y_{11})$ from the SAS output. What does this interval say about the tensile strength of beams with the given specific gravity and moisture content values?

4.10 The data in the following table (Kutner et al. 2005) are from a study relating the amount of body fat (y) to several possible predictor variables:

triceps skinfold thickness (x_1), thigh circumference (x_2), and midarm circumference (x_3), measured on a random sample of healthy females 25–34 years old.

Triceps skinfold thickness	Thigh circumference	Midarm circumference	Body fat
19.5	43.1	29.1	11.9
24.7	49.8	28.2	22.8
30.7	51.9	37.0	18.7
29.8	54.3	31.1	20.1
19.1	42.2	30.9	12.9
25.6	53.9	23.7	21.7
31.4	58.5	27.6	27.1
27.9	52.1	30.6	25.4
22.1	49.9	23.2	21.3
25.5	53.5	24.8	19.3
31.1	56.6	30.0	25.4
30.4	56.7	28.3	27.2
18.7	46.5	23.0	11.7
19.7	44.2	28.6	17.8
14.6	42.7	21.3	12.8
29.5	54.4	30.1	23.9
27.7	55.3	25.7	22.6
30.2	58.6	24.6	25.4
22.7	48.2	27.1	14.8
25.2	51.0	27.5	21.1

The amount of body fat is obtained using a cumbersome and expensive procedure involving the immersion of a person in water, so it would be helpful if a reliable prediction equation based on easy-to-measure explanatory variables is available to estimate body fat. Use a SAS program to fit the full model

$$y = \beta_0 + \beta_1 x_1 + \beta_2 x_2 + \beta_3 x_3 + \epsilon$$

Write answers to the following questions extracting numbers from a SAS output. Highlite or circle values you use from the output and label them carefully.

a. Report $\hat{\beta}_0$, $\hat{\beta}_1$, $\hat{\beta}_2$, and $\hat{\beta}_3$. Explain what the estimate $\hat{\beta}_3$ tells you about the mean body fat.
b. Report s_ϵ, $s_{\hat{\beta}_0}$, $s_{\hat{\beta}_1}$, $s_{\hat{\beta}_2}$, and $s_{\hat{\beta}_3}$. Calculate by hand the t-statistic for testing $H_0 : \beta_3 = 0$ versus $H_a : \beta_3 \neq 0$, using $\hat{\beta}_3$ and its standard error. Compare it with the value in the output.
c. Construct an analysis of variance for the above regression. Report the coefficient of determination and interpret its value in the context of this problem.

d. Use the F-test statistic for testing $H_0 : \beta_1 = \beta_2 = \beta_3 = 0$ versus H_a : at least one β is not zero, and report the p-value for the test. State your decision using the p-value.

e. Use the t-test statistic for testing $H_0 : \beta_2 = 0$ versus $H_a : \beta_2 \neq 0$ and report the p-value for the test. State your decision based on the p-value. What does this say about the role of the variable `thigh circumference` in your model?

f. Construct a 95% confidence interval for β_2. Use this interval to test $H_0 : \beta_2 = 0$ versus $H_a : \beta_2 \neq 0$. What is the α level of this test.

g. Construct a 95% confidence interval for $E(y)$ at $x_1 = 20$, $x_2 = 50$, and $x_3 = 30$. Describe in words what this interval tells you about the population of individuals with these values for triceps skinfold thickness, thigh circumference, and midarm circumference.

h. Construct a 95% prediction interval for body fat of a new individual on whom the values $x_1 = 20$, $x_2 = 50$, and $x_3 = 30$ have been measured. Describe in words what this interval tells you.

i. Obtain plots of the residuals versus predicted values, x_1, x_2, and x_3, respectively. Does any pattern of the types discussed in class observed in the plots? Give your interpretation of each plot.

j. Obtain a normal probability plot of the *studentized residuals.* State the model assumption that you can verify using this plot. Is this a plausible assumption for this model?

k. Add a model statement to your SAS program to fit the model

$$y = \beta_0 + \beta_1 x_1 + \beta_3 x_3 + \epsilon$$

to the above data. Use the results of the t-tests and R^2 value to state why this model may be preferred compared to the first model.

4.11 A laundry detergent manufacturer wished to test a new product prior to market release. One area of concern was the relationship between the height of the detergent suds in a washing machine to the amount of detergent added and the degree of agitation in the wash cycle. For a standard-size washing machine tub filled to the full level, different agitation levels (measured in minutes) and amounts of detergent were assigned in random order, and sud heights measured. This problem appears in Ott and Longnecker (2001) and the data are reproduced in the table presented. Write and execute a SAS program to perform the computations needed to provide answers to the following questions. Add SAS statements and/or steps necessary to obtain all answers required. Highlite or circle values you use from the output and label them carefully.

a. By examining only the plot of `height`, y, against `agitation`, x_1, at each value of `amount`, x_2, suggest a multiple regression model to explain the variation in sud height. Be specific about whether you think a higher-order term x_1^2 or an interaction term $x_1 x_2$ is needed or not and give reasons for your choices.

b. The first order model

$$y = \beta_0 + \beta_1 x_1 + \beta_2 x_2 + \epsilon$$

was fitted to the data. Using the SAS output, construct an analysis of variance table and test the hypothesis of $H_0 : \beta_1 = \beta_2 = 0$ versus H_a : at least one of β_1 or $\beta_2 \neq 0$. Give the R^2 value. Is there a reason to look beyond the first-order model because R^2 is extremely high? Explain.

Height (y)	Agitation (x_1)	Amount (x_2)
28.1	1	6
32.3	1	7
34.8	1	8
38.2	1	9
43.5	1	10
60.3	2	6
63.7	2	7
65.4	2	8
69.2	2	9
72.9	2	10
88.2	3	6
89.3	3	7
94.1	3	8
95.7	3	9
100.6	3	10

c. Use the residual plots to examine the plausibility of the model assumptions. By examining the residual plots, can you find reasons to suspect that the first-order model is not adequate? If so, what other terms would you consider adding to the model? Explain how you reached your conclusion.

d. Based on the above residual analysis, the model

$$y = \beta_0 + \beta_1 x_1 + \beta_2 x_2 + \beta_3 x_1^2 + \beta_4 x_2 x_1^2 + \epsilon$$

was fitted to the data. Using the sums of squares from the SAS output, construct an F-statistic to test the hypothesis $H_0 : \beta_3 = \beta_4 = 0$. Use a percentile from the F-table to test this hypothesis at $\alpha = 0.05$. What does the result of this test tell you?

e. Use residual plots from fitting the model given in part (d) to verify the model assumptions. What improvements do these plots show compared to the first-order model, if any?

f. Use a t-statistic and the corresponding p-value from the SAS output of the fit of the model in part (d) to test $H_0 : \beta_4 = 0$. What does the result of this test suggest?

Problems 4.12–4.17 concern the following example presented in Bowerman and O'Connell (2004):

A multiple regression analysis conducted to determine labor needs of U.S. Navy hospitals is reproduced from Bowerman and O'Connell (2004). The data set consists of five independent variables: average daily patient load (LOAD), monthly X-ray exposures (XRAY), monthly occupied bed days (a hospital has one occupied bed day if one bed is occupied for an entire day) (BEDDAYS), eligible population in the area (in 1000s) (POP), average length of patients' stay (in days) (LENGTH), and a dependent variable, monthly labor hours required (HOURS).

H	LOAD	XRAY	BEDDAYS	POP	LENGTH	HOURS
1	15.57	2463	472.92	18.0	4.45	566.52
2	44.02	2048	1339.75	9.5	6.92	696.82
3	20.42	3940	620.25	12.8	4.28	1033.15
4	18.74	6505	568.33	36.7	3.90	1603.62
5	49.20	5723	1497.60	35.7	5.50	1611.37
6	44.92	11,520	1365.83	24.0	4.60	1613.27
7	55.48	5779	1687.00	43.3	5.62	1854.17
8	59.28	5969	1639.92	46.7	5.15	2160.55
9	94.39	8461	2872.33	78.7	6.18	2305.58
10	128.02	20,106	3655.08	180.5	6.15	3503.93
11	96.00	13,313	2912.00	60.9	5.88	3571.89
12	131.42	10,771	3921.00	103.7	4.88	3741.40
13	127.21	15,543	3865.67	126.8	5.50	4026.52
14	252.90	36,194	7684.10	157.7	7.00	10,343.81
15	409.20	34,703	12,446.33	169.4	10.78	11,732.17
16	463.70	39,204	14,098.40	331.4	7.05	15,414.94
17	510.22	86,533	15,524.00	371.6	6.35	18,854.45

Write and execute SAS programs to perform the computations necessary to provide answers to the following questions. Add SAS statements and/or steps needed to obtain all answers required.

4.12 Consider relating y (HOURS) to x_1 (XRAY), x_2 (BEDDAYS), and x_3 (LENGTH) by using the model

$$y = \beta_0 + \beta_1 x_1 + \beta_2 x_2 + \beta_3 x_3 + \epsilon$$

Plot y versus each of x_1, x_2, and x_3. Do the plots indicate that the above model might appropriately relate y to x_1, x_2, and x_3? Explain your answer. You may compute other statistics to support your answer.

4.13 The main objective of this regression analysis is to help the Navy evaluate the performance of its hospitals in terms of how many labor hours are used relative to how many labor hours are needed. The Navy selected hospitals 1 through 17 from hospitals that it thought were efficiently run and wishes to use a regression model based on efficiently run hospitals to evaluate the efficiency of questionable hospitals. For hospital 14, note that x_1 = 36,194, x_2 = 7684.10, and x_2= 7.00. Using the SAS output, discuss the case diagnostics for hospital 14. Discuss statistical evidence to show that this case might not fit the above model very well.

4.14 Since the Navy wishes to use a regression model based on efficiently run hospitals, it follows that hospital 14 be removed from the data set if it is concluded that hospital 14 was inefficiently run. Using the results from fitting model defined in Exercise 4.9 to the data set modified by removing hospital 14, answer the following:
   a. Do all of the residuals on the output appear to have come from the same distribution? Provide evidence for supporting or rejecting your claim.
   b. Use the leverage values for cases 14, 15, and 16 (which are the original hospitals 15, 16, and 17) to determine if these hospitals are outliers with respect to their x-values? How does being an x-outlier affect other diagnostics of each of these cases?
   c. Use the Cook's distance measure for case 14 (the original hospital 15) to explain why removing hospital 14 from the data set has made the original hospital 15 noninfluential?
   d. Which hospital had the largest Cook's D when all 17 hospitals were used to perform the regression analysis? Does this hospital appear to be less influential after removing hospital 14 from the data set? If so, explain why.

4.15 For two hospitals not used in the above analysis whose efficiency the Navy questions, the values of XRAY, BEDDAYS, and LENGTH are 56,194, 14,077.88, and 6.89 and 6021, 1651.42, and 5.41, respectively. Use SAS to obtain predictions of the number of monthly labor hours used for these hospitals. Use the model in Exercise 4.9 fitted to data excluding hospital 14. Use the observed number of labor hours for these hospitals, y = 17,207.3 and y = 1823.4, respectively, to comment on the efficiency of these two hospitals.

4.16 Consider relating y (HOURS) to x_1 (LOAD), x_2 (XRAY), x_3 (BEDDAYS), x_4(POP), and x_5(LENGTH) by fitting the model

$$y = \beta_0 + \beta_1 x_1 + \beta_2 x_2 + \beta_3 x_3 + \beta_4 x_4 + \beta_5 x_5 + \epsilon$$

using all 17 hospitals.
   a. Use the scatter plot matrix and correlation matrices to carry out a preliminary assessment of the relationship between y and each of x_1, x_2, x_3, x_4, and x_5. Based on your assessment, which independent variables do you judge might be most strongly involved in multicollinearity?
   b. Do any least squares estimates of the regression coefficients have a sign (positive or negative) that is different from what you would intu-

itively expect (another consequence of multicollinearity)? Which two variables have the largest variance inflation factors? What is conspicuous about these variables?

c. Obtain the partial regression residual plots for each of the variables in the model. Use these to comment on the effects of multicollinearity on the estimation of each coefficient in the fitted model.

d. If the independent variables x_1 and x_4 are removed from the five-variable model above and use regression to relate y to x_2, x_3, and x_5 alone, the model fitted is identical to that in Exercise 4.9. Is the fit of that model less affected by multicollinearity than the five-variable model? Explain. Does x_3 seem to have additional importance in the smaller model than the larger one? Justify your answers.

4.17 Since the previous analyses indicated that hospital 14 might have been inefficiently run, use SAS procedures to obtain the "best" model resulting from using possible combinations of the five explanatory variables after hospital 14 is removed from the data set.

a. Use a SAS procedure to do all possible regressions containing no less than two and no more than four independent variables. Print statistics for only the four best models in each case. Construct a plot of the C_p statistic for all models with "reasonable" C_p values. Select "good" models each with two, three, and four independent variables, respectively, for the purpose of predicting monthly labor hours, *indicating your reasons for selection of each model.* There may be several possible choices; that is, there may be many "good" models but give arguments for each of your choices. Primarily, use s^2, R^2, and C_p in your arguments. Select one of these models as your final model and provide arguments supporting your choice.

b. Use the following subset selection procedures:

   a. backward elimination, with significance level of 0.05 for deleting variables

   b. stepwise, with significance levels of 0.20 for entry and 0.10 for deletion of variables

   State the model(s) selected in each case and report estimates of parameters and the analysis of variance table for these models. Compare models selected from each procedure with final model selected in part (a). Comment on whether changing the cutoff levels up or down by small amount would change the models selected by these procedures. (You do not need to re-run programs; examine the p-values from outputs from the SAS programs used for procedures A and B.)

4.18 This problem may be considered a complete SAS project. Import an Excel data set named `air_pollution.xls` using `proc import` to create a SAS data set. The data shown in Table B.7 consists of air pollution measured as SO2 content in the air and related other variables for 41 U.S. cities. SO2 is the response variable y, and the explanatory variables

are AvTemp (x_1), NumFirms (x_2), Population (x_3), WindSpeed (x_4), AvPrecip (x_5), and PrecipDays(x_6), respectively.
In a second SAS program, access this SAS dataset and perform a variable subset selection procedure using `proc reg` as described below.

a. Use `proc sgscatter` to obtain a scatter plot matrix and `proc reg` to perform a preliminary assessment of the pairwise relationships between y and x_1, x_2, x_3, x_4, x_5, and x_6. On this basis alone, select a few explanatory variables that may good predictors in a multiple regression model. Using the plot and the correlation matrix, find the four variables that are most strongly correlated among the explanatory variables. Based on above analysis alone suggest the explanatory variables that are most strongly involved in multicollinearity when fitting the full model.
b. Use a SAS procedure to fit a first-order multiple regression model to all 41 cities. `Discuss` the fit of this model using the ANOVA table, R^2, and the estimates table. Use other diagnostic tools including output statistics and plots of residuals to examine the adequacy of the model (use the diagnostic panel of plots). Examining the plots, Obs #31 to be an influential y-outlier. Use the diagnostic statistics to confirm this conjecture.
c. Identify any least squares estimates of the regression coefficients ($\hat{\beta}$'s) from the fit of the model in part b), that have a sign (positive or negative) that is different from what you would expect for the parameter—an indication of multicollinearity? Use the standard errors of the parameter estimates to show that these are poorly estimated. Do the variance inflation factors (VIF's) identify these parameters?
d. Remove the explanatory variables x_3 and x_6 from the five-variable model and use a multiple regression model to relate y to x_1, x_2, x_4 and x_5 only. What can you observe about the multicollinearity in the new model? Is there improvement in the accuracy of estimation of parameters of this model (e.g., decreases standard errors, more t-statistics are significant etc.)? Justify your answers.
e. Remove the case you determined above in part b) to be a possible outlier from the data and use all 6 variables for the analyses described in the following three parts:
    i. Use a SAS procedure to do all possible regressions containing no less than 2 and no more than 4 explanatory variables. Print statistics for only the 4 best models in each case. Construct a plot of the C_p statistic for all models with "reasonable" C_p values. Select a single model, each with 2, 3, and 4 explanatory variables, respectively, for the purpose of predicting annual mean concentration of sulfur dioxide in a city, `indicating your reasons for selection of each model.` There may be several possible choices, i.e., there may be many "good" models but give arguments for each of your choices. Primarily, use s^2, R^2, and C_p in

your arguments. Select one of these models as your final model and provide arguments supporting your choice.

ii. Use the backward elimination subset selection procedure with significance level of 0.05 for deleting variables to select a possible model. State the model selected and report estimates of parameters and the analysis of variance table for this model.

iii. Use the stepwise subset selection procedure, with significance levels of 0.10 for entry and 0.05 for deletion of variables, respectively, to select a possible model. State the model selected and report estimates of parameters and the analysis of variance table for this model.

A selected subset of the SAS data set *baseball* available from the SASHELP library containing data on baseball player salaries in 1986/87 is used in the following two problems. There are 21 variables and 71 observations in this data set. To obtain information about the variables, run the SAS code *ods select position;proc contents data=baseball varnum; run;*. Note that the variable attributes table is named *position* when the option *varnum* is used with `proc contents`. You may print the first 5 observations by running the SAS code *proc print data=baseball(obs=5);run;* to observe a few values for the above variables. Fit `logsalary` as a first order multiple regression model of the variables `nAtBat, nHits, nHome, nRuns, nRBI, nBB, yrMajor, crAtBat, crHits, crHome, crRuns, crRbi, crBB, nOuts, nAssts` and `nError`. Use the `glmselect` procedure to perform model selection using two different approaches as described below:

4.19 In this problem use the significance level criterion with the *stepwise* method model selection and the *validate* method for selecting the best model (in terms of smallest validation ASE) in each step. Partition the data randomly so that 35% of the data are used for validation. Request plots of the fit criteria by iteration step and a plot for comparison of validation and training ASE by iteration step. Discuss the results of this analysis.

4.20 In this problem use the adjusted R^2 criterion with the *stepwise* method model selection and the *cv* method for selecting the best model (in terms of the smallest cross-validation prediction error, called CV PRESS) in each step. Specify that fivefold cross-validation method to be used. Also specify that, in addition to the R^2_{adj} and CV PRESS, the statistics C_p, AIC and SBC to be included in the model fit summaries. Discuss the results of this analysis and compare it to the results of the stepwise selection using the significance level criterion.

5

Analysis of Variance Models

5.1 Introduction

In Chap. 4, multiple regression models discussed involved quantitative variables as explanatory variables. As discussed in Sect. 4.2, the least squares method was used to obtain the estimates of the parameters of the model. Using the matrix form of the multiple regression model

$$\mathbf{y} = X\boldsymbol{\beta} + \boldsymbol{\epsilon}$$

this was done by solving the normal equations

$$X'X\boldsymbol{\beta} = X'\mathbf{y},$$

the solution to which gave the least squares estimate $\hat{\boldsymbol{\beta}}$ of $\boldsymbol{\beta}$:

$$\hat{\boldsymbol{\beta}} = (X'X)^{-1}X'\mathbf{y}.$$

The *analysis of variance models* introduced in this chapter, although conceptually different, can also be represented in this framework, where the X matrix now represents the *design matrix*. The design matrix is constructed from the linear model describing the responses observed from a specific experiment. Linear models describing various experiments discussed in this chapter will be called analysis of variance models. Analysis of variance models used for describing responses from several experimental situations is discussed in detail in separate sections in this chapter. Here, a linear model is used as an example demonstrating how a design matrix is constructed from it and for discussing properties of the resulting normal equations. Consider the linear model

$$y_{ij} = \mu + \alpha_i + \tau_j + \epsilon_{ij}, \qquad i = 1, 2, 3;\ j = 1, 2, 3, 4$$

M. G. Marasinghe, K. J. Koehler, *Statistical Data Analysis Using SAS*,
Springer Texts in Statistics, https://doi.org/10.1007/978-3-319-69239-5_5

This model, for example, may be used to describe the yield of corn, y_{ij}, observed from 12 1-acre plots where combinations of 3 different levels of nitrogen ($i = 1, 2, 3$) and 4 levels of irrigation ($j = 1, 2, 3, 4$) were applied. In this model, α_i and τ_j represent the *effects* of nitrogen level i and irrigation level j, respectively, expressed as deviations from an overall mean μ. When formulating this model for this situation, it is assumed that the above effects and the random component ϵ_{ij} representing *experimental error* are *additive.* In this chapter, the $\epsilon_{11}, \epsilon_{12}, \ldots, \epsilon_{34}$ are assumed to be a random sample from a normal distribution with mean 0 and variance σ^2. The matrix form of the model is

$$\mathbf{y} = X\boldsymbol{\beta} + \boldsymbol{\epsilon}$$

where

$$\mathbf{y} = \begin{bmatrix} y_{11} \\ y_{12} \\ y_{13} \\ y_{14} \\ y_{21} \\ y_{22} \\ y_{23} \\ y_{24} \\ y_{31} \\ y_{32} \\ y_{33} \\ y_{34} \end{bmatrix}, \quad X = \begin{bmatrix} 1&1&0&0&1&0&0&0 \\ 1&1&0&0&0&1&0&0 \\ 1&1&0&0&0&0&1&0 \\ 1&1&0&0&0&0&0&1 \\ 1&0&1&0&1&0&0&0 \\ 1&0&1&0&0&1&0&0 \\ 1&0&1&0&0&0&1&0 \\ 1&0&1&0&0&0&0&1 \\ 1&0&0&1&1&0&0&0 \\ 1&0&0&1&0&1&0&0 \\ 1&0&0&1&0&0&1&0 \\ 1&0&0&1&0&0&0&1 \end{bmatrix}, \boldsymbol{\beta} = \begin{bmatrix} \mu \\ \alpha_1 \\ \alpha_2 \\ \alpha_3 \\ \tau_1 \\ \tau_2 \\ \tau_3 \\ \tau_4 \end{bmatrix}, \text{ and } \boldsymbol{\epsilon} = \begin{bmatrix} \epsilon_{11} \\ \epsilon_{12} \\ \epsilon_{13} \\ \epsilon_{14} \\ \epsilon_{21} \\ \epsilon_{22} \\ \epsilon_{23} \\ \epsilon_{24} \\ \epsilon_{31} \\ \epsilon_{32} \\ \epsilon_{33} \\ \epsilon_{34} \end{bmatrix}$$

For example, the first line of this model is $y_{11} = \mu + \alpha_1 + \tau_1 + \epsilon_{11}$, the second line is $y_{12} = \mu + \alpha_1 + \tau_2 + \epsilon_{12}$, and so on. There are certain obvious differences from the matrix form of the multiple regression model of Sect. 4.2. The design matrix X, for example, consists entirely of 0's and 1's, their positions determined by the subscripts i and j. The 12 rows of X correspond to the 12 observations $y_{11}, y_{12}, \ldots, y_{34}$ in that order. Note that the ordering is determined by letting the first subscript i remain fixed and the second subscript j take values 1 through 4 representing the four different irrigation methods. The first subscript i takes the values 1 through 3 in that order and represents the three nitrogen levels. The eight columns of X represent the eight parameters μ, α_1, α_2, α_3, τ_1, τ_2, τ_3, τ_4, respectively. Thus, in any row of X, the presence of a 1 or 0 in any column indicates that the corresponding parameter, as determined by the column position, appears or not in the model for the observation represented by that row. For example, in row 7 of the above design matrix, there is a 1 in columns 1, 3, and 7, and the rest of the columns are 0, indicating the presence of the parameters μ, α_2, and τ_3 in the model; thus, the model for y_{23}, the observation represented by row 7 of X, is $y_{23} = \mu + \alpha_2 + \tau_3 + \epsilon_{23}$.

The special structure of X results in a situation not usually encountered when attempting to solve the normal equations $X'X\boldsymbol{\beta} = X'\mathbf{y}$ to obtain the least squares estimates of the parameters in the multiple regression model. In that case, it is assumed that the matrix $X'X$ is nonsingular; that is, the inverse of the matrix can be calculated, and therefore, a unique solution to the normal equations exists that is given by $(X'X)^{-1}X'\mathbf{y}$. In analysis of variance models, the rank of the $X'X$ matrix is less than p, the number of parameters, and therefore, there is an infinite number of solutions to the normal equations. Hence, these models are also called *less than full rank* models. The outcome of this is that the parameters μ, α_1, α_2, α_3, τ_1, τ_2, τ_3, and τ_4 cannot be uniquely estimated. In general, a nonunique solution to such a system of linear equations may be found by setting some of the unknown parameters to a constant (such as zero) and obtaining a solution for the rest of the parameters. This solution will be unique up to the constants chosen. Thus, when computer software produce *estimates* of parameters in an analysis of variance model, the numbers output are not unique and depend on the procedure adopted by the software to obtain a solution to the normal equations.

As an example, it can be easily shown that the $X'X$ matrix for the linear model given earlier is

$$X'X = \begin{bmatrix} 12 & 4 & 4 & 4 & 3 & 3 & 3 & 3 \\ 4 & 4 & 0 & 0 & 1 & 1 & 1 & 1 \\ 4 & 0 & 4 & 0 & 1 & 1 & 1 & 1 \\ 4 & 0 & 0 & 4 & 1 & 1 & 1 & 1 \\ 3 & 1 & 1 & 1 & 3 & 0 & 0 & 0 \\ 3 & 1 & 1 & 1 & 0 & 3 & 0 & 0 \\ 3 & 1 & 1 & 1 & 0 & 0 & 3 & 0 \\ 3 & 1 & 1 & 1 & 0 & 0 & 0 & 3 \end{bmatrix}$$

and that the normal equations required for obtaining the least squares estimates are given by

$$\begin{aligned}
12\mu + 4\alpha_1 + 4\alpha_2 + 4\alpha_3 + 3\tau_1 + 3\tau_2 + 3\tau_3 + 3\tau_4 &= y_{..} \\
4\mu + 4\alpha_1 \qquad\qquad\qquad + \tau_1 + \tau_2 + \tau_3 + \tau_4 &= y_{1.} \\
4\mu + \qquad 4\alpha_2 + \qquad\quad \tau_1 + \tau_2 + \tau_3 + \tau_4 &= y_{2.} \\
4\mu + \qquad\qquad 4\alpha_3 + \tau_1 + \tau_2 + \tau_3 + \tau_4 &= y_{3.} \\
3\mu + \alpha_1 + \alpha_2 + \alpha_3 + 3\tau_1 \qquad\qquad\qquad &= y_{.1} \\
3\mu + \alpha_1 + \alpha_2 + \alpha_3 + \qquad 3\tau_2 \qquad\qquad &= y_{.2} \\
3\mu + \alpha_1 + \alpha_2 + \alpha_3 + \qquad\qquad + 3\tau_3 \qquad &= y_{.3} \\
3\mu + \alpha_1 + \alpha_2 + \alpha_3 + \qquad\qquad\qquad 3\tau_4 &= y_{.4}
\end{aligned}$$

where $y_{..} = \sum_{i=1}^{3}\sum_{j=1}^{4} y_{ij}$, $y_{i.} = \sum_{j=1}^{4} y_{ij}$ for $i = 1, 2, 3$, and $y_{.j} = \sum_{i=1}^{3} y_{ij}$ for $j = 1, 2, 3, 4$. It can be shown that the rank of $X'X$ is 6, so only two of the parameters need to be set to a constant to obtain a solution to the normal equations. Another way to obtain a solution to the normal equations in a *balanced design model* such as in the above example is to impose restrictions on parameters. For example, in the above example, two such restrictions are needed. The so-called *sum-to-zero* restrictions are $\sum_{i=1}^{3}\alpha_i = 0$ and $\sum_{j=1}^{4}\tau_j = 0$ of this type. These can also be written as $\alpha_3 = -\alpha_1 - \alpha_2$ and $\tau_4 = -\tau_1 - \tau_2 - \tau_3$; thus, a solution for the other parameters can be obtained by eliminating α_3 and τ_4 from the equations. The solution to the normal equations under these restrictions is $\tilde{\mu} = \bar{y}_{..}$, $\tilde{\alpha_i} = \bar{y}_{i.} - \bar{y}_{..}$, for $i = 1, 2, 3$, and $\tilde{\tau_j} = \bar{y}_{.j} - \bar{y}_{..}$ for $j = 1, 2, 3, 4$. For the experiment described, most experimenters consider these quantities as the "estimates" of the respective parameters. Essentially, setting some parameters equal to a constant is also a restriction on the parameters. The method of computation used in `proc glm` in SAS produces estimates equivalent to those obtained by setting the last parameter for each effect equal to zero. Thus, in the above model, setting $\alpha_3 = 0$ and $\tau_4 = 0$ and solving the normal equations will result in the same estimates as those produced by SAS. These solutions are $\tilde{\mu} = -\bar{y}_{..} + \bar{y}_{3.} + \bar{y}_{.4}$, $\tilde{\alpha_i} = \bar{y}_{i.} - \bar{y}_{3.}$ for $i = 1, 2$, and $\tilde{\tau_j} = \bar{y}_{.j} - \bar{y}_{.4}$ for $j = 1, 2, 3$.

Thus, it is evident that the solutions to the normal equations are not unique, and therefore, it is more useful to obtain estimates of "interesting" functions of the parameters that are unique. Linear functions of the parameters for which unique estimates exist are called *estimable functions*, and fortunately for the experimenter, some of the estimable functions of parameters of a given model can be usefully interpreted. Estimates of these functions will be the same no matter which solution to the normal equations is used to compute them.

Analysis of variance models may be used to analyze data from

- Designed experiments
- Observational studies

The statistical analyses of these data based on analysis of variance models require some familiarity with the basic concepts associated with such studies. In the subsections that follow, a few of these ideas are briefly reviewed.

5.1.1 Treatment Structure

Treatments (or more generally, *factor levels*) are various settings of the conditions that are being compared in an experiment. Generally, treatments are applied to *experimental units*, and a *response* (a value of the dependent variable) is measured from each experimental unit.

Example 1: Study effects of baking temperature on a commercial cake mixture

- Levels of temperature: 150 °C, 170 °C, 190 °C, 210 °C
- Response variable: Area of a cross section of cake
- Replications: The number of cakes baked at each temperature

Example 2: Same experiment as in Example 1 with an additional factor: a chemical additive

- Treatments: All combinations of 4 levels of temperature and 3 chemical additives forming 12 treatment combinations in a 4×3 *factorial* treatment structure

- Response Variable: Area of a cross section of cake
- Replications: It is needed to bake at least 24 cakes (i.e., 2 `Replications` per treatment combination in order to be able to estimate *experimental error variance.*

Example 3: Study the variation of number of traffic tickets issued in different precincts in a large U.S. city

- Levels of factor: Select 10 precincts at random.
- Response variable: Number of traffic tickets issued in a 6-month period
- Replications: From each precinct, select several police officers at *random* for each of whom a response is measured.
- Note: In this experiment, the interest is in measuring the variability of the number of traffic tickets issued from precinct to precinct within the city, rather than estimating the mean number of traffic tickets issued in a particular precinct.

Factors can be categorized into two basic types:

Fixed Factors

- Experimenter selects the levels of each of the factors to be included in the experiment.
- Interest is in the estimation and comparison of differences among these selected levels.

Example: Cake-Baking Experiment
Both `Temperature` and `Additive` are fixed factors because the experimenter selected these levels to be studied in the experiment. The effects of the levels of Temperature and Additive will be compared in the analysis of the data resulting from this experiment.

Random Factors

- Experimenter randomly samples the population of levels of each of the random factors.
- Interest here is in measuring the variability of the response over the population.

Example: Study of Traffic Tickets
Both officers and precincts are random factors because the levels of these factors were selected from the available population of levels. The experimenter is not interested in the effects of a particular officer or a particular precinct but the variation of the number of traffic tickets issued.

5.1.2 Experimental Designs

Experimental designs describe how treatments (treatment combinations) are assigned to the experimental units for application to the experimental units and in the order in which observations are taken.

Example 1: Completely Randomized Design

Compare four fertilizers (A, B, C, D) on corn yield in a field experiment. Suppose 20 plots are available in a field as experimental units. If a *completely randomized design* (CRD) is used, the following design could be used to allocate the treatments to the plots:

A	B	C	D	D
B	D	A	A	C
C	A	B	D	C
B	A	B	C	D

Here, the four fertilizers are assigned at random to the 20 plots so that each fertilizer is applied to five plots, giving five replications of each treatment.

Example 2: Randomized Blocks Design

If, on the other hand, a *randomized complete block design* (RCBD) is used, the following allocation of the fertilizers to the plots may result:

Blocks				
1	2	3	4	5
A	B	D	A	C
C	C	C	B	D
B	D	A	D	A
D	A	B	C	B

Here, the 20 plots are first grouped into 5 blocks (numbered from 1 to 5 in the diagram) of 4 plots each. The four fertilizers are assigned at random to the four plots in each block. Note that in the actual field layout, the plots within blocks may not be aligned across the blocks as shown above.

5.1.3 Linear Models

Data arising from designed experiments is represented by a linear model for the purpose of statistical analysis of such data. One advantage accrued from using such a model is that all effects to be estimated and hypotheses that need to be tested to answer the research questions related to the experiment may be formulated in terms of the *estimable functions* of the parameters of the model. In various linear models introduced in the chapter, parameters are estimated by the *least squares method*; that is, the solutions to the normal equations minimize the sum of squared deviations of the observations from their expected values. These estimates are equivalent to *maximum likelihood estimators* obtained by maximizing the likelihood function under the assumption that the observations are normally distributed. In either case, the estimators of the parameters (i.e., estimable functions of the parameters) are called "best" since they have the properties of being unbiased and having minimum variance in the class of unbiased estimators.

Example: Cake-Baking Experiment

In the cake-baking experiment, for example, a model that may be used to describe the observations y_{ijk} is of the form

$$y_{ijk} = \underbrace{\mu + \alpha_i + \beta_j + \gamma_{ij}}_{\mu_{ij}} + \epsilon_{ijk},\ i = 1,2,3,4;\quad j = 1,2,3;\quad k = 1,2.$$

where y_{ijk} is the area of cross section of the kth replication of the ijth temperature-additive combination and ϵ_{ijk} is a random error assumed to be normally distributed with zero mean and constant variance σ^2. $\mu_{ij} =$ expected mean response of treatment combination ij; that is, this model says that $E(y_{ijk}) = \mu_{ij}$ for each k.

From this formulation, μ_{ij}, $\alpha_i - \alpha_{i'}$, $\beta_j - \beta_{j'}$, σ^2, etc. may be estimated. For example, $\bar{y}_{ij.}$ is the "best" estimate of μ_{ij}. Further, hypotheses about model parameters such as $H_0 : \gamma_{ij} = 0$ for all i, j versus $H_a :$ not H_0, or $H_0 : \alpha_i = \alpha_{i'}$ versus $H_a : \alpha_i \neq \alpha_{i'}$, etc. may be tested.

Example: Study of Traffic Tickets

In the second example concerning the number of traffic tickets, the model may be represented by

$$y_{ij} = \mu + A_i + \epsilon_{ij},\ i = 1,2,\ldots,I;\ j = 1,2,\ldots,n_i$$

where A_i is the effect of the ith precinct with $A_i \sim$ iid $N(0, \sigma_A^2)$, ϵ_{ij} is the effect of the jth officer in the ith precinct with $\epsilon_{ij} \sim$ iid $N(0, \sigma^2)$, and y_{ij} is the number of traffic tickets issued by the jth officer in the ith precinct.

Since the "I" precincts were chosen at random, "precinct" is considered to be a random factor and it is assumed that the effects of the precincts, A_i, are independently distributed as $N(0, \sigma_A^2)$ random variables. The officers were selected randomly within each precinct; hence, "officer within precinct" denoted by officer/precinct or officer(precinct) is independently distributed as a random factor, and it is assumed that ϵ_{ij} are

$N(0, \sigma^2)$ random variables. The above is a one-way random model where the "officer" effect is nested within factor A (precinct); that is, levels of factor B (say) are nested within levels of factor A.

Using this model the *variance components* σ_A^2 and σ^2 may be estimated, and hypotheses about them such as $H_0 : \sigma_A^2 = 0$ versus $H_a : \sigma_A^2 > 0$ may be tested.

5.2 One-Way Classification

Data generated from a study of several levels of a single factor in a completely randomized design are said to be in a one-way classification. In this situation, the levels of the factor are also sometimes called "treatments."

Model

A linear model appropriate for the response y_{ij} observed from the jth replication of the ith treatment is given by

$$y_{ij} = \mu + \alpha_i + \epsilon_{ij} \quad i = 1, 2, \ldots, t, \quad j = 1, \ldots, n_i$$

where α_i is the effect of the ith treatment expressed as the deviation of the *treatment mean* μ_i from an overall mean μ (i.e. $\alpha_i = \mu_i - \mu$) and ϵ_{ij} is the random experimental error associated with the ijth observation assumed to have normal distribution with zero mean and variance σ^2, usually expressed as $\epsilon_{ij} \sim \text{iid}\, N(0, \sigma^2)$. This model is equivalent to assuming that the observations for each treatment i, $y_{i1}, y_{i2}, \ldots, y_{in_i}$, is a random sample from the $N(\mu_i, \sigma^2)$ distribution where μ_i is the ith treatment mean. Thus, it is implied that μ_i is the mean of the population from which the sample corresponding to the ith treatment was drawn. Note that this also incorporates the assumption of *homogeneity of variance*; that is, populations corresponding to each treatment have a common variance σ^2. The above model may be reexpressed in terms of the treatment means as follows:

$$y_{ij} = \mu_i \ + \epsilon_{ij} \quad i = 1, 2, \ldots, t, \quad j = 1, \ldots, n_i$$

where $\mu_i = \mu + \alpha_i$. This model is called the "means model" and the previous model the "effects model."

Estimation

The best estimates of μ_i and σ^2 are, respectively,

$$\hat{\mu}_i = \bar{y}_{i\cdot} = (\textstyle\sum_j y_{ij})/n_i, \quad i = 1, \ldots, t$$

$$\hat{\sigma}^2 = s^2 = \frac{\sum_i \sum_j (y_{ij} - \bar{y}_{i\cdot})^2}{N - t}, \quad N = \textstyle\sum_i n_i$$

and the best estimate of the difference between the effects of two treatments labeled p and q is

$$\widehat{\alpha_p - \alpha_q} = \bar{y}_{p.} - \bar{y}_{q.}, \; p \neq q$$

with standard error given by

$$s_d = s\sqrt{\frac{1}{n_p} + \frac{1}{n_q}}$$

A $(1 - \alpha)100\%$ confidence interval (C.I.) for $\alpha_p - \alpha_q$ (or $\mu_p - \mu_q$) is

$$(\bar{y}_{p.} - \bar{y}_{q.}) \pm t_{\alpha/2,\nu} \cdot s_d$$

where $t_{\alpha/2,\nu}$ = upper $\alpha/2$ percentage point of the t-distribution with ν df where $\nu = N - t$.

Testing Hypotheses

An analysis of variance (ANOVA) table corresponding to the above model is

SV	df	SS	MS	F	p-Value
Trt	$t-1$	$\texttt{SS}_{\texttt{Trt}}$	$\texttt{MS}_{\texttt{Trt}}$	$F_c = \texttt{MS}_{\texttt{Trt}}/\texttt{MSE}$	$\Pr(F > F_c)$
Error	$N-t$	SSE	MSE$(= s^2)$		
Total	$N-1$	$\texttt{SS}_{\texttt{Tot}}$			

The above F-statistic tests the hypothesis of equality of treatment means

$$H_0: \; \mu_1 = \mu_2 = \cdots = \mu_t \text{ versus } \; H_a: \; \text{at least one inequality}$$

or, equivalently, the hypothesis of equality of treatment effects

$$H_0: \; \alpha_1 = \alpha_2 = \cdots = \alpha_t \text{ versus } H_a: \; \text{at least one inequality}$$

To test the equality of means of two treatments labeled p and q (i.e., $H_0: \mu_p = \mu_q$ versus $H_a: \; \mu_p \neq \mu_q$) or, equivalently, to test the equality of effects of two treatments labeled p and q (i.e., $H_0: \; \alpha_p = \alpha_q$ versus $H_a: \; \alpha_p \neq \alpha_q$), use the following t-statistic:

$$t_c = \frac{|\bar{y}_{p\cdot} - \bar{y}_{q\cdot}|}{s_d}$$

The null hypothesis H_0 is rejected if $t_c > t_{\alpha/2,\nu}$, where $t_{\alpha/2,\nu}$ is the upper $\alpha/2$ percentage point of the t-distribution with $\nu = N - t$ degrees of freedom.

Preplanned or A Priori Comparisons of Means

When the hypothesis $H_0 : \mu_1 = \mu_2 = \cdots = \mu_t$ is rejected using the ANOVA F-test, the inference is that at least one of the t population means differs from the rest. The next step is to identify the means that are different from each other. If the researcher had planned to compare treatment effects or means by suggesting specific questions about them based on the treatment structure, this would be a much simpler task. For example, "Is the average $(\mu_1 + \mu_2)/2$ different from $(\mu_3 + \mu_4)/2$?" would be a meaningful question if treatments 3 and 4 contained a component or an ingredient, say, that treatments 1 and 2 did not have.

Many times these questions may not result in a simple comparison of whether a difference like $\mu_2 - \mu_3$ is significant or not. It may be a question that requires a more complex comparison such as $\mu_1 - (\mu_2 + \mu_3)/2$ to be made. Not all questions can be formulated in the form of comparisons. To understand what kinds of questions can be formulated as comparisons, define a `linear comparison` as a linear combination of the means $\mu_1, \mu_2, \ldots, \mu_t$ to be of the form

$$\ell = a_1\mu_1 + a_2\mu_2 + \cdots + a_t\mu_t = \sum_{i=1}^{t} a_i\mu_i$$

for given numbers $a_1, a_2, \ldots, a_t$ with the restriction that the sum of these numbers is zero (i.e., $\sum_{i=1}^{t} a_i = 0$). Several examples of linear comparisons of means are provided to illustrate this definition.

Examples: Suppose the number of treatments is $t = 5$; that is, consider the population means μ_1, μ_2, μ_3, μ_4, and μ_5.

- The linear combination $\ell = \mu_2 - \mu_3$ has a_i values

$$a_1 = 0,\ a_2 = 1,\ a_3 = -1,\ a_4 = 0,\ a_5 = 0$$

  Note that $\sum a_i = 0$ as required; thus, ℓ is a linear comparison.

- The linear combination $\ell = (\mu_1 + \mu_2)/2 - (\mu_3 + \mu_4)/2$ has a_i values

$$a_1 = 1/2,\ a_2 = 1/2,\ a_3 = -1/2,\ a_4 = -1/2, a_5 = 0$$

  Again, $\sum a_i = 0$; thus, ℓ is a linear comparison.

An estimate of a linear comparison is called a `linear contrast` of the means and is given by

$$\hat{\ell} = a_1\bar{y}_{1.} + a_2\bar{y}_{2.} + a_3\bar{y}_{3.} + \cdots + a_t\bar{y}_{t.}$$

where $\sum a_i = 0$. Recall that the sample treatment means $\bar{y}_{1.}, \bar{y}_{2.}, \ldots, \bar{y}_{t.}$ are the best estimates of the population means $\mu_1, \mu_2, \ldots, \mu_t$.

A linear comparison is a way to express a hypothesis among population means that are naturally suggested by the choice of levels of the factor under study. It is not surprising that an experimenter would like to make more than one such comparison for a given experiment. A comparison is *made* by the testing of the hypotheses,

$$H_0 : \ell = 0 \quad \text{versus} \quad \mathrm{H_a} : \ell \neq 0$$

where ℓ is a linear combination of $\mu_1, \mu_2, \ldots, \mu_t$ at a specified level α.

To compute an F-statistic to test the above hypotheses, a "SS due to the contrast" needs to be computed. In the most general case, when the sample sizes for the treatments are different, say, $n_1, n_2, \ldots, n_t$, this SS is calculated using the formula

$$\mathrm{SSC} = \frac{\hat{\ell}^2}{\sum_{i=1}^{t}(a_i^2/n_i)}$$

When the sample sizes are all equal to n, the above reduces to the form

$$\mathrm{SSC} = \frac{n\hat{\ell}^2}{\sum_{i=1}^{t} a_i^2}$$

Example 5.2.1

The data shown in Table 5.1 are from an experiment in plant physiology described in Sokal and Rohlf (1995) and give the length (in coded units) of pea sections grown in tissue culture. The purpose of the experiment was to compare the effects of the addition of various sugars on growth as measured by this length. Four treatments representing three different sugars and one mixture of sugars plus one control treatment with no sugar were used. Ten independent samples were obtained for each treatment in a completely randomized design. The model for the observations in terms of the means $\mu_1, \ldots, \mu_5$ of

Table 5.1. Effect of sugars on growth of peas (Sokal and Rohlf 1995)

		Sugars		
Control	2% glucose	2% fructose	1% glucose + 1% fructose	2% sucrose
75	57	58	58	62
67	58	61	59	66
70	60	56	58	65
75	59	58	61	63
65	62	57	57	64
71	60	56	56	62
67	60	61	58	65
67	57	60	57	65
76	59	57	57	62
68	61	58	59	67

populations that represent the five samples is

$$y_{ij} = \mu_i + \epsilon_{ij}, \quad i = 1, 2, \ldots, 5, \quad j = 1, \ldots, 10$$

and the analysis of variance (ANOVA) table that is required for the analysis of this data is

SV	df	SS	MS	F	p-Value
Sugars	4	1077.32	269.33	49.37	<0.0001
Error	45	245.50	5.46		
Total	49	1322.82			

Labeling the five treatments as `Trt1,...,Trt5`, respectively, in the order they appear in the data table, the five treatment means are

Trt1	Trt2	Trt3	Trt4	Trt5
70.1	59.3	58.2	58.0	64.1

It would be more useful to test the preplanned (or a priori) comparisons given below rather than, say, test all pairwise differences among the five treatment means (or effects) to determine the means that are different. In this study, the experimenter could have planned to (i) compare the effect of the control with the average effect of the sugars, (ii) compare the effect of the mixed sugars with average effects of pure sugars, and (iii) compare the differences among the effects of the three pure sugars. The linear combinations

(i) $\mu_1 - \frac{1}{4}(\mu_2 + \mu_3 + \mu_4 + \mu_5)$
(ii) $\mu_4 - \frac{1}{3}(\mu_2 + \mu_3 + \mu_5)$
(iii) $\mu_2 - \mu_3$, $\mu_2 - \mu_5$, or $\mu_3 - \mu_5$

represent these comparisons, in terms of the five respective population means, $\mu_1, \ldots, \mu_5$ Consider testing

$$H_0 : \mu_1 - \frac{1}{4}(\mu_2 + \mu_3 + \mu_4 + \mu_5) = 0 \quad \text{versus} \quad H_a : \mu_1 - \frac{1}{4}(\mu_2 + \mu_3 + \mu_4 + \mu_5) \neq 0$$

or, equivalently,

$$H_0 : 4\mu_1 - \mu_2 - \mu_3 - \mu_4 - \mu_5 = 0 \quad \text{versus} \quad H_a : 4\mu_1 - \mu_2 - \mu_3 - \mu_4 - \mu_5 \neq 0$$

Here $\ell_1 = 4\mu_1 - \mu_2 - \mu_3 - \mu_4 - \mu_5$. The coefficients of the corresponding contrast are therefore

a_1	a_2	a_3	a_4	a_5
4	−1	−1	−1	−1

Thus,

$$\begin{aligned} \hat{\ell}_1 &= 4\bar{y}_1 - \bar{y}_2 - \bar{y}_3 - \bar{y}_4 - \bar{y}_5 \\ &= 4 \times 70.1 - 59.3 - 58.2 - 58.0 - 64.1 = 40.8 \end{aligned}$$

where

$$\sum_{i=1}^{5} \frac{a_i^2}{n_i} = \frac{4^2}{10} + \frac{1^2}{10} + \frac{1^2}{10} + \frac{1^2}{10} + \frac{1^2}{10} = \frac{20}{10}$$

Thus,

$$\text{SSC}_1 = \frac{\hat{\ell}_1^2}{\sum \frac{a_i^2}{n_i}} = \frac{(40.8)^2}{2} = 832.32$$

Therefore,

$$F_c = \frac{\text{SSC}_1/1}{\text{MSE}} = \frac{832.32}{5.456} = 152.56$$

Since $F_{.05,1,45} \approx 4.0$, we reject H_0 at $\alpha = 0.05$. Now, consider testing

$$H_0 : \mu_4 - \frac{1}{3}(\mu_2 + \mu_3 + \mu_5) = 0 \quad \text{versus} \quad H_a : \mu_4 - \frac{1}{3}(\mu_2 + \mu_3 + \mu_5) \neq 0$$

Since H_0 is equivalent to $\mu_2 + \mu_3 + \mu_5 - 3\mu_4 = 0$, the problem is equivalent to testing

$$H_0 : \ell_2 = 0 \quad \text{versus} \quad H_a : \ell_2 \neq 0$$

where $\ell_2 = \mu_2 + \mu_3 + \mu_5 - 3\mu_4$. Here, the contrast coefficients are

a_1	a_2	a_3	a_4	a_5
0	1	1	−3	1

giving $\hat{\ell}_2 = 59.3 + 58.2 + 64.1 - 3 \times 58.0 = 7.6$ and the divisor is

$$\frac{a_i^2}{n_i} = \frac{0^2}{10} + \frac{1^2}{10} + \frac{1^2}{10} + \frac{3^2}{10} + \frac{1^2}{10} = \frac{12}{10} = 1.2$$

Thus, similar to the above,

$$\text{SSC}_2 = \frac{\hat{\ell}_2^2}{\sum \frac{a_i^2}{n_i}} = \frac{(7.6)^2}{1.2} = 48.1333$$

Therefore,

$$F_c = \frac{\text{SSC}_2/1}{\text{MSE}} = \frac{48.1333}{5.456} = 8.82$$

which leads us to reject H_0 at $\alpha = 0.05$, the same result as $F_{.05,1,45} \approx 4.0$. Computations of the F-tests for the other two comparisons follow in a similar fashion. The above F-tests performed for making comparisons of interest also may be carried out by equivalent t-tests. In the equal sample size case (i.e., $n_1 = n_2 = \cdots = n$), a test for a preplanned (or an *a priori*) comparison

$$H_0 : \textstyle\sum_i c_i \mu_i = 0 \quad \text{versus} \quad H_a : \textstyle\sum_i c_i \mu_i \neq 0$$

is given by the t-statistic

$$t_c = \frac{|\sum_i c_i \bar{y}_{i\cdot}|}{s\sqrt{\sum c_i^2/n}}$$

and we reject $H_0:$ if $t_c > t_{\alpha/2,(N-t)}$ for a two-tailed test. For testing

$$H_0: 4\mu_1 - \mu_2 - \mu_3 - \mu_4 - \mu_5 = 0 \quad \text{versus} \quad H_a: 4\mu_1 - \mu_2 - \mu_3 - \mu_4 - \mu_5 \neq 0$$

the t-statistic is

$$t_c = \frac{40.78}{\sqrt{5.456} \times \sqrt{2}} = 12.35$$

and H_0 is rejected as $t_{.025,45} \approx 2.0$

Two contrasts $\hat{\ell}_1 = \sum_i a_i \bar{y}_{i.}$ and $\hat{\ell}_2 = \sum_i b_i \bar{y}_{i.}$ are said to be *orthogonal* whenever $\sum_i a_i b_i = 0$. This is defined only when $n_1 = n_2 = \cdots = n_t = n$. If all linear contrasts in a set

$$\hat{\ell}_1, \hat{\ell}_2, \ldots, \hat{\ell}_{t-1}$$

are pairwise orthogonal (i.e., every possible pair is orthogonal), then the set is said to be a *mutually orthogonal* set of linear contrasts. Given t means $\mu_1, \mu_2, \ldots, \mu_t$ and sample means $\bar{y}_{1.}, \bar{y}_{2.}, \ldots, \bar{y}_{t.}$ (all based on the same number n of observations), it is the case that the *maximum number of mutually orthogonal contrasts that exist is* $(t-1)$. There are many $(t-1)$ sets of contrasts that are mutually orthogonal. Thus, in the previous example, a set of four comparisons that are mutually orthogonal could be found since $t = 5$.

Pairwise Comparisons of Means

One-at-a-time comparisons between pairs of mean μ_p and μ_q control *the per-comparison error rate.* These comparisons are carried out simply by:

- doing a t-test of $H_0: \mu_p = \mu_q$, or
- constructing a confidence interval for $\mu_p - \mu_q$, or
- equivalently, *when sample sizes are equal,* using the least significant difference (LSD) procedure. Note that the rejection region for the t-test for $H_0: \mu_p - \mu_q = 0$ is equivalent to Reject H_0 if

$$|\bar{y}_{p\cdot} - \bar{y}_{q\cdot}| > \underbrace{t_{\alpha/2,(N-t)} \cdot s \cdot \sqrt{2/n}}_{\text{LSD}_\alpha}, \quad \text{where } n = \text{sample size}$$

The right member of this inequality is not a function of i or j. It is constant and is called the *least significant difference* or LSD and is denoted here as LSD_α. In the LSD procedure, differences of pairs of the sample means are compared to the single computed value of LSD_α, to determine the pairs that are significantly different. To minimize the number of comparisons needed to be made, the $\bar{y}_{i.}$s are first arranged smallest to largest in value. Using the notation $\bar{y}_{(i)}$ for the ith smallest $\bar{y}$, the ordered means may be represented as

$$\bar{y}_{(1)} \leq \bar{y}_{(2)} \leq \bar{y}_{(3)} \leq \cdots \leq \bar{y}_{(t)}$$

Now, note that if the difference $\bar{y}_{(t)} - \bar{y}_{(1)}$, for example, does not exceed the LSD value, then all the differences $\bar{y}_{(t)} - \bar{y}_{(2)}$, $\bar{y}_{(t)} - \bar{y}_{(3)} \ldots$, $\bar{y}_{(t)} - \bar{y}_{(t-1)}$ will not exceed the LSD. It follows that computing all the above differences and comparing them to the LSD are avoided. The LSD procedure is based on this idea.

Often the findings are reported using the following scheme called the *underscoring procedure*:

- The ordered list of the computed values of the means is identified (in a separate line above them) by the treatment numbers (or the names identifying the corresponding treated populations) corresponding to the ordered means. For example, suppose the means are

Trt5	Trt3	Trt1	Trt4	Trt2
9.5	10.5	11.6	12.2	13.5

- Connect the means by underscoring those pairs of means whose differences are less than the LSD_α in the following manner.
- Consider each mean in turn beginning from the smallest and moving right to the next largest mean and so on.
- On a separate line below the list starting from column 1, begin the underscore connecting means until a mean is found that is significantly different (i.e., difference is larger than the LSD_α) from the mean in column 1. Extend the line all the way and stop to the left of this mean.
- This line implies that those means that are connected with this line are not significantly different from the mean in column 1.
- Now, the procedure is restarted at column 2 and is repeated the same way as described above.
- For example, for an LSD_α value of 2.72, the underscoring procedure produces the following:

Trt5	Trt3	Trt1	Trt4	Trt2
9.5	10.5	11.6	12.2	13.5

- Now, the populations whose identifying numbers or labels are joined by an underscore have means that are found to be not significantly different. To simplify the display, any line that is completely covered by (i.e., overlaps) another line and therefore is not needed can be deleted. Thus, the above reduces to

Trt5	Trt3	Trt1	Trt4	Trt2
9.5	10.5	11.6	12.2	13.5

When an experimenter wants to make all possible pairwise comparisons, one of the multiple comparison procedures such as the Tukey procedure is recommended, because such procedures control the experimentwise error rate.

Multiple Comparisons of Pairs of Means

Following an ANOVA F-test that rejects the hypothesis that the population means are equal to each other, one may conduct tests of equality of all pairs of the population means in order to ascertain which of these are actually different from each other. One-at-a-time comparisons (i.e., individual t-tests or confidence intervals for differences in pairs of means) may be used for this purpose. However, these tests are not adjusted for multiple inference; that is, the error rate controlled is the Type I error rate for each individual test. When pairwise comparisons of population means are made, *multiple comparison procedures* such as the Bonferroni method attempt to control the probability of making at least one Type I error. These procedures protect the experimenter against declaring too many pairs of means significantly different when they are actually not, when making all possible pairwise comparisons; that is, they ensure that the probability of making at least one Type I error is controlled and thus is more conservative than one-at-a-time comparisons. They are said to control the *experimentwise error rate.* Although SAS makes available several procedures through a variety of options, just three such procedures are discussed here.

- The Bonferroni method involves the use of a t-percentile corrected for the total number of pairwise comparisons. This procedure controls the probability of making at least one Type I error to be less than or equal to α.
- The Tukey procedure (also called Tukey–Kramer method when an adjustment for unequal sample sizes is incorporated) for all possible pairwise comparisons simultaneously, also called the HSD (honestly significant difference) procedure. This procedure controls the maximum experimentwise error rate.
- The Scheffé procedure is used for testing a set of comparisons (contrasts of the type $\sum c_i\mu_i$) simultaneously and controls the maximum experimentwise error rate for the set. The set of comparisons may include pairwise comparisons and any other contrast of interest, as well.

Instead of performing tests for all comparisons among t means, these procedures may also be carried out in a manner similar to the LSD procedure to minimize the number of comparisons. For example, LSD_α is replaced by

$$\text{HSD}_\alpha = q_{\alpha,t,\nu}\sqrt{\frac{s^2}{n}}$$

where $q_{\alpha,t,\nu}$ is the upper α percentage point of the Studentized range distribution with ν degrees of freedom for performing the Tukey procedure (for equal sample sizes). For the Bonferroni method, the t-percentile used in the LSD procedure $t_{\alpha/2,\nu}$ is replaced by $t_{\alpha/2m,\nu}$ where m is the total number of comparisons made. For all pairwise comparisons among t means, $m = t(t-1)/2$. For the Scheffé procedure, LSD_α is replaced by

$$S_\alpha = \sqrt{\widehat{\text{Var}}(\hat{\ell})}\ \sqrt{(t-1)F_{\alpha,df_1,df_2}}$$

where $df_1 = t - 1$, $df_2 = \nu$, and $\widehat{\text{Var}}(\hat{\ell}) = s^2 \sum_i \frac{a_i^2}{n_i}$, and s^2 is the error mean square with ν degrees of freedom. The Tukey–Kramer method is more powerful than either the Bonferroni or the Scheffé methods when only pairwise comparisons are made.

Confidence intervals for all pairwise differences can also be adjusted for each of these methods by changing the percentile value used in their construction. For example, 95% Bonferroni-corrected confidence intervals for m pairwise differences are obtained by substituting $t_{0.025/m,\nu}$ in place of $t_{0.025,\nu}$. In conclusion, a word of caution about drawing inferences from pairwise comparisons is warranted. The transitivity that one expects from logical relationships may not exist among the results of these hypothesis tests; for example, μ_A may be found to be not significantly different from μ_B and μ_B from μ_C, but it is possible that μ_A and μ_C are declared significantly different.

5.2.1 Using PROC ANOVA to Analyze One-Way Classifications

Consider the data appearing in Table 5.2 (Box et al. 1978). These are the observed coagulation times (in seconds) of blood drawn from 24 animals randomly allocated to 4 different diets, labeled A, B, C, and D.

Table 5.2. Blood Coagulation Data (in seconds): Example E1

Diet			
A	B	C	D
62	63	68	56
60	67	66	62
63	71	71	60
59	64	67	61
	65	68	63
	66	68	64
			63
			59

The one factor in this experiment is `Diet` with four levels, the levels being the four types of diet (may also be called four treatments). The experiment was conducted in a completely randomized design. SAS Example E1 illustrates the analysis of these data using `proc anova`.

SAS Example E1

The SAS Example E1 program shown in Fig. 5.1 is used to obtain the necessary analysis of the above data. The `input` statement with the trailing @@

is useful for inputting this type of data. This allows several observations to be continued on the same data line rather than using a new line for each observation. Notice that it is necessary to separate data values by at least one blank as in the case of list input. Although the sample sizes are unequal, `proc anova` may be used for the analysis of this data since it is a one-way classification. The `class` statement identifies the variables that appear in the `model` statement, which are classification variables. In this case, the variable `Diet` is the classification variable with four classes (the four diets), and `Time` is the dependent variable (y). Note that in specifying the model this way, a mean μ as well as an error term are implicitly assumed to be part of the model and thus are omitted from the statement.

The first part of the SAS output (see Fig. 5.2) is a result of the `proc print` statement and illustrates the appearance of the SAS data set produced by the data step. The class level information page resulting from the `class` statement and the analysis of variance (ANOVA) table produced by `proc anova` as a result of the `model` statement follows (see Fig. 5.3). The experimenter may construct an analysis of variance table in the standard form given in statistics textbooks by extracting information from the above output. The analysis of variance (ANOVA) table for the coagulation time data is

SV	df	SS	MS	F	p-Value
Diet	3	228.0	76.0	13.57	0.0001
Error	20	112.0	5.6		
Total	23	340.0			

The `means` statement produces side-by-side boxplots of the observations at each level of the classification variable (here diets) reproduced below in

```
data blood;
input Diet Time @@;
datalines;
1 62 1 60 1 63 1 59
2 63 2 67 2 71 2 64 2 65 2 66
3 68 3 66 3 71 3 67 3 68 3 68
4 56 4 62 4 60 4 61 4 63 4 64 4 63 4 59
;

proc print data=blood;
  title "Analysis of Blood Coagulation Data";
run;

proc anova data=blood;
  class Diet;
  model Time=Diet;
  means Diet/t cldiff;
  means Diet/tukey alpha=.05;
  means Diet/hovtest;
run;
```

Fig. 5.1. SAS Example E1: program

Analysis of Blood Coagulation Data

Obs	Diet	Time
1	1	62
2	1	60
3	1	63
4	1	59
5	2	63
6	2	67
7	2	71
8	2	64
9	2	65
10	2	66
11	3	68
12	3	66
13	3	71
14	3	67
15	3	68
16	3	68
17	4	56
18	4	62
19	4	60
20	4	61
21	4	63
22	4	64
23	4	63
24	4	59

Fig. 5.2. SAS Example E1: output from `proc print`

Fig. 5.4. Note that this will be reproduced for each means statement present in the step and may be suppressed entirely by including the option `plots=none` in the `proc anova` statement.

The output from the `proc anova` step resulting from the first `means` statement and containing the pairwise comparisons of means is shown in Fig. 5.5. The `cldiff` option on the `means` statement requests that the comparisons be given in the form of 95% confidence intervals on pairwise differences of means constructed using the t-percentage points (`t` or `lsd` option). For example, the 95% confidence interval on $\mu_1 - \mu_2$ is $-5.0 \pm (2.086)(1.5275) = (-8.186, -1.814)$, where $t_{.025}(20) = 2.086$, $s\sqrt{1/n_1 + 1/n_2} = 1.5275$, and $s^2 = 5.6$.

Since the sample sizes n_i are not the same, `proc anova` would have produced confidence intervals instead of LSD pairwise comparisons, in any case, by default. Note that the `lines` option in the means statement allows the user to request that an approximate procedure be carried out when sample sizes are unequal, by calculating an LSD value using a "sample size" equal to the harmonic mean of actual sample sizes. Since the harmonic mean is always smaller than the average, this procedure will lead to more liberal tests of the differences than if exact confidence intervals are used to make the pairwise comparisons.

Analysis of Blood Coagulation Data

The ANOVA Procedure

Class Level Information		
Class	Levels	Values
Diet	4	1 2 3 4

Number of Observations Read	24
Number of Observations Used	24

Dependent Variable: Time

Source	DF	Sum of Squares	Mean Square	F Value	Pr > F
Model	3	228.0000000	76.0000000	13.57	<.0001
Error	20	112.0000000	5.6000000		
Corrected Total	23	340.0000000			

R-Square	Coeff Var	Root MSE	Time Mean
0.670588	3.697550	2.366432	64.00000

Source	DF	Anova SS	Mean Square	F Value	Pr > F
Diet	3	228.0000000	76.0000000	13.57	<.0001

Fig. 5.3. SAS Example E1: output tables from `proc anova`

By specifying `alpha=p` as a `means` statement option, $(1-p)100\%$ confidence intervals may be calculated by request ($p = 0.05$ is the default when this option is omitted). The following is the set of 95% confidence intervals on the six pairwise differences of means extracted from the SAS output (Fig. 5.5):

$$\mu_1 - \mu_2 : (-8.186,\ -1.814)$$
$$\mu_1 - \mu_3 : (-10.186,\ -3.814)$$
$$\mu_1 - \mu_4 : (-3.023,\ 3.023)$$
$$\mu_2 - \mu_3 : (-4.850,\ 0.850)$$
$$\mu_2 - \mu_4 : (2.334,\ 7.666)$$
$$\mu_3 - \mu_4 : (4.334,\ 9.666)$$

The intervals for $\mu_1 - \mu_4$ and $\mu_2 - \mu_3$ include zero, thus indicating that those pairs of means are not significantly different at an α level of 0.05. The main conclusion to be drawn is that mean coagulation times due to Diets B and C are similar but significantly larger than those due to Diets A and D, which are also similar. The LSD procedure may be replaced by one of several other more conservative procedures. `bon`, `tukey`, and `scheffe` are examples of options that may replace the `t` or `lsd` option for this purpose.

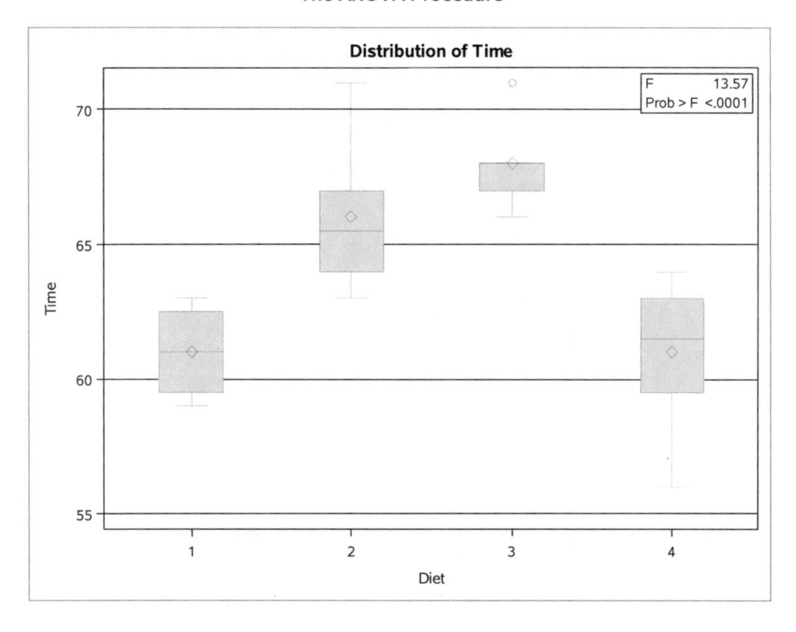

Fig. 5.4. SAS Example E1: side-by-side box plots produced in `proc anova`

The SAS output resulting from the statement `means Diet/t lines` is shown in Fig. 5.6. Observe that the harmonic mean of the sample sizes is ≈5.65 and that based on that an LSD value of 2.9377 was calculated. Note that, in this example, using the approximation did not affect the results as the same conclusion that there is no significant difference between each of the pairs of means 1 and 4 and the pair of means 2 and 3, respectively. Thus, if the sample sizes do not deviate substantially, the approximate procedure may be used.

The SAS output resulting from the statement `means Diet/tukey` is shown in Fig. 5.7. The following is the set of 95% confidence intervals on the six pairwise differences of means extracted from this output:

$$\mu_1 - \mu_2 : (-9.275,\ -0.725)$$
$$\mu_1 - \mu_3 : (-11.275,\ -2.725)$$
$$\mu_1 - \mu_4 : (-4.056,\ 4.056)$$
$$\mu_2 - \mu_3 : (-5.824,\ 1.824)$$
$$\mu_2 - \mu_4 : (1.423,\ 8.577)$$
$$\mu_3 - \mu_4 : (3.423,\ 10.577)$$

Analysis of Blood Coagulation Data

The ANOVA Procedure

t Tests (LSD) for Time

Note: This test controls the Type I comparisonwise error rate, not the experimentwise error rate.

Alpha	0.05
Error Degrees of Freedom	20
Error Mean Square	5.6
Critical Value of t	2.08596

Comparisons significant at the 0.05 level are indicated by *.**				
Diet Comparison	**Difference Between Means**	**95% Confidence Limits**		
3 - 2	2.000	-0.850	4.850	
3 - 1	7.000	3.814	10.186	***
3 - 4	7.000	4.334	9.666	***
2 - 3	-2.000	-4.850	0.850	
2 - 1	5.000	1.814	8.186	***
2 - 4	5.000	2.334	7.666	***
1 - 3	-7.000	-10.186	-3.814	***
1 - 2	-5.000	-8.186	-1.814	***
1 - 4	0.000	-3.023	3.023	
4 - 3	-7.000	-9.666	-4.334	***
4 - 2	-5.000	-7.666	-2.334	***
4 - 1	0.000	-3.023	3.023	

Fig. 5.5. SAS Example E1: output from the `t` (or `lsd`) option

Notice that this option results in wider confidence intervals than the ones produced using the `lsd` option. This is because the Tukey procedure controls the *experimentwise error rate* resulting in a more conservative procedure; that is, there is less of a chance of finding significant differences using this procedure. The confidence intervals based on the t-statistics (`t` or `lsd` option) control the *per-comparison error rate* that guarantees only that the Type I error of each comparison will be controlled at the specified `alpha` value.

Procedures based on controlling experimentwise error rate are recommended for use when the fact that the experimenter will be making inferences using $t-1$ comparisons among the means has to be taken into account. Otherwise, the Type I error rate for all comparisons made will be more than

Analysis of Blood Coagulation Data

The ANOVA Procedure

t Tests (LSD) for Time

Note: This test controls the Type I comparisonwise error rate, not the experimentwise error rate.

Alpha	0.05
Error Degrees of Freedom	20
Error Mean Square	5.6
Critical Value of t	2.08596
Least Significant Difference	2.9377
Harmonic Mean of Cell Sizes	5.647059

Note: Cell sizes are not equal.

Means with the same letter are not significantly different.			
t Grouping	**Mean**	**N**	**Diet**
A	68.000	6	3
A			
A	66.000	6	2
B	61.000	4	1
B			
B	61.000	8	4

Fig. 5.6. SAS Example E1: output from the `t` (or `lsd`) and `lines` options

the nominal significance level specified for an individual comparison. Which procedure is to be used depends on many factors. As a rule of thumb, it is recommended that one use the Bonferroni or Tukey procedure when all pairwise comparisons are being made and use the Scheffé procedure when, in addition to all pairwise comparisons, other contrasts or comparisons among the means are also considered when inferences are being made. Note that each of these three procedures is progressively more conservative than the LSD procedure, and therefore, the corresponding confidence intervals will be progressively wider. Note, however, that in this example, the conclusions drawn from using the Tukey procedure are identical to those drawn using the LSD procedure.

The option `hovtest=` used in a `means` statements allows the user to specify that one of several tests for homogeneity of variance be calculated. The available selections are `bartlett, bf, levene`, and `obrien`. Although, traditionally, experimenters have used Bartlett's test for this purpose in practice, currently Levene's test is widely recognized to be the standard procedure for testing homogeneity of variance.

Analysis of Blood Coagulation Data

The ANOVA Procedure

Tukey's Studentized Range (HSD) Test for Time

Note: This test controls the Type I experimentwise error rate.

Alpha	0.05
Error Degrees of Freedom	20
Error Mean Square	5.6
Critical Value of Studentized Range	3.95825

Comparisons significant at the 0.05 level are indicated by ***.				
Diet Comparison	Difference Between Means	Simultaneous 95% Confidence Limits		
3 - 2	2.000	-1.824	5.824	
3 - 1	7.000	2.725	11.275	***
3 - 4	7.000	3.423	10.577	***
2 - 3	-2.000	-5.824	1.824	
2 - 1	5.000	0.725	9.275	***
2 - 4	5.000	1.423	8.577	***
1 - 3	-7.000	-11.275	-2.725	***
1 - 2	-5.000	-9.275	-0.725	***
1 - 4	0.000	-4.056	4.056	
4 - 3	-7.000	-10.577	-3.423	***
4 - 2	-5.000	-8.577	-1.423	***
4 - 1	0.000	-4.056	4.056	

Fig. 5.7. SAS Example E1: output from the `tukey` option

Analysis of Blood Coagulation Data

The ANOVA Procedure

Levene's Test for Homogeneity of Time Variance ANOVA of Squared Deviations from Group Means					
Source	DF	Sum of Squares	Mean Square	F Value	Pr > F
Diet	3	89.6667	29.8889	0.60	0.6237
Error	20	999.7	49.9833		

Fig. 5.8. SAS Example E1: output from the `hovtest` option

Figure 5.8 shows the results of Levene's test (produced by default) as an F-test organized in an ANOVA table format. In this example, the p-values for the test are large indicating that the null hypothesis of equal variances will not be rejected. One may examine this assumption visually by constructing a side-by-side box plot as illustrated earlier (see Fig. 5.4).

5.2.2 Making Preplanned (or A Priori) Comparisons Using PROC GLM

Consider the data (see Table. 5.1) introduced in Sect. 5.2. These were used earlier in Sect. 5.2 to illustrate how to calculate F-statistics or t-statistics for testing hypotheses of the type

$$H_0 : \ell = 0 \quad \text{versus} \quad \mathrm{H_a} : \ell \neq 0,$$

where $\ell = \sum_{i=1}^{t} a_i \mu_i$ for given numbers $a_1, a_2, \ldots, a_t$, which satisfy $\sum_{i=1}^{t} a_i = 0$ at a specified significance level of α.

The one factor in this experiment is labeled `Sugar` with five levels (treatments), four treatments representing three different sugars, one consisting of a mixture of sugars, and one a control treatment not containing any sugars. Random samples of size 10 were obtained for each treatment in a completely randomized design. SAS Example E2 illustrates the SAS program used for the analysis of these data.

SAS Example E2

The SAS Example E2 program (see Fig. 5.9) is used primarily to illustrate the use of `contrast` and `estimate` statements in `proc glm` to obtain F-test and t-tests, respectively, for making the comparisons suggested in Sect. 5.2. These comparisons were represented by the following linear combinations of the five respective population means, $\mu_1, \mu_2, \ldots, \mu_5$, as follows:

(i) $\mu_1 - \frac{1}{4}(\mu_2 + \mu_3 + \mu_4 + \mu_5)$
(ii) $\mu_4 - \frac{1}{3}(\mu_2 + \mu_3 + \mu_5)$
(iii) $\mu_2 - \mu_3$
(iv) $\mu_2 - \mu_5$

The `input` statements with trailing @ and a do loop were used for inputting the data in a straightforward way, with the *levels* for the classification variable `Sugar` being identified by the numbers $1, 2, \ldots, 5$ in the data. The use of the trailing @ was described in Sect. 1.7.2 in Chap. 1. The data step first reads the level of sugar from a data line, holds the line, and then reads ten numbers successively as values of the variable `Length`, writing a pair of values for `Sugar` and `Length` each time through the loop as observations into the SAS data set named `peas`. The SAS output from `proc print` (not shown) may be examined to make sure that the data set has been created in the required format.

The analysis of variance (ANOVA) tables produced by `proc glm` as a result of the `model` statement, appear in Fig. 5.10. By default, `proc glm` computes two types of sums of squares (called Types I and III) for each of the independent variables (i.e., variables appearing to the right of the equal sign) included in the model statement. In the case of one-way classification, these two sets of sums of squares are identical in magnitude, as is the case here. The experimenter may construct an analysis of variance table in the standard form by extracting information from the above output:

SV	df	SS	MS	F	p-Value
Sugars	4	1077.32	269.33	49.37	<0.0001
Error	45	245.50	5.46		
Total	49	1322.82			

The next part of the SAS output from `proc glm` shown in Fig. 5.11 results from the first `means` statement and contains the pairwise comparisons of means using the LSD procedure. The means are arranged in decreasing order

```
data peas;
input Sugar @;
do i=1 to 10;
 input Length @;
 output;
end;
drop i;
datalines;
1 75 67 70 75 65 71 67 67 76 68
2 57 58 60 59 62 60 60 57 59 61
3 58 61 56 58 57 56 61 60 57 58
4 58 59 58 61 57 56 58 57 57 59
5 62 66 65 63 64 62 65 65 62 67
;
proc print data=peas;
  title ' Effect of Sugars on the Growth of Peas';
run;

proc glm data=peas;
  class Sugar ;
  model Length =  Sugar;
  means Sugar/lsd alpha = .05;
  means Sugar/tukey alpha = .05;

  contrast 'CONTROL VS. Sugars' Sugar 4 -1 -1 -1 -1;
  contrast 'Sugars VS. MIXED  ' Sugar 0  1  1 -3  1;
  contrast 'GLUCOSE=FRUCTOSE'   Sugar 0  1 -1  0  0;
  contrast 'FRUCTOSE = SUCROSE' Sugar 0  1  0  0 -1;

  estimate 'CONTROL VS. SugarS' Sugar 4 -1 -1 -1 -1;
  estimate 'Sugars VS. MIXED  ' Sugar 0  1  1 -3  1;
  estimate 'GLUCOSE=FRUCTOSE'   Sugar 0  1 -1  0  0;
  estimate 'FRUCTOSE = SUCROSE' Sugar 0  1  0  0 -1;
run;
```

Fig. 5.9. SAS Example E2: program

Effect of Sugars on the Growth of Peas

The GLM Procedure

Dependent Variable: Length

Source	DF	Sum of Squares	Mean Square	F Value	Pr > F
Model	4	1077.320000	269.330000	49.37	<.0001
Error	45	245.500000	5.455556		
Corrected Total	49	1322.820000			

R-Square	Coeff Var	Root MSE	Length Mean
0.814412	3.770928	2.335713	61.94000

Source	DF	Type I SS	Mean Square	F Value	Pr > F
Sugar	4	1077.320000	269.330000	49.37	<.0001

Source	DF	Type III SS	Mean Square	F Value	Pr > F
Sugar	4	1077.320000	269.330000	49.37	<.0001

Fig. 5.10. SAS Example E2: output tables from `proc glm`

of magnitude down the page, and the level of the corresponding treatment (sugar) is shown in the last column. The $\text{LSD}_{.05}$ is computed as 2.1039, and means that are not significantly different are grouped by the same letter in the first column of the output. This is comparable to the underscoring procedure described at the beginning of Sect. 5.2, which results in

$$\begin{array}{ccccc} \text{Trt4} & \text{Trt3} & \text{Trt2} & \text{Trt5} & \text{Trt1} \\ 58.0 & 58.2 & 59.3 & 64.1 & 70.1 \\ \cline{1-3} \end{array}$$

This can be interpreted to indicate that the mean lengths for treatments 1 (Control) and 5 (Sucrose) are significantly different from each other and from the other three treatments (Glucose, Fructose, and Mixed Sugars) but that there is no significant difference among those three. The output resulting from the second means statement `means Diet/tukey` is shown in Fig. 5.12. In this case the differences are compared to the $\text{HSD}_{.05}$ value calculated as 2.9681 and labeled Minimum Significant Difference. This turns out to be larger than the LSD$_{.05}$ value as expected, but the outcome of the underscoring procedure remains unchanged from that of the LSD procedure.

The four `contrast` statements result in the computation of F-statistics for testing the four `single degree of freedom` comparisons of interest. The syntax of these statements is of the form

```
contrast 'label' effect_name contrast_coefficients
                                        < / options > ;
```

These results usually appear below the ANOVA table in the SAS output but on a separate table (see Fig. 5.13). The divisor mean square used for

Effect of Sugars on the Growth of Peas

The GLM Procedure

t Tests (LSD) for Length

Note: This test controls the Type I comparisonwise error rate, not the experimentwise error rate.

Alpha	0.05
Error Degrees of Freedom	45
Error Mean Square	5.455556
Critical Value of t	2.01410
Least Significant Difference	2.1039

Means with the same letter are not significantly different.			
t Grouping	**Mean**	**N**	**Sugar**
A	70.100	10	1
B	64.100	10	5
C	59.300	10	2
C			
C	58.200	10	3
C			
C	58.000	10	4

Fig. 5.11. SAS Example E2: output from LSD procedure

Effect of Sugars on the Growth of Peas

The GLM Procedure

Tukey's Studentized Range (HSD) Test for Length

Note: This test controls the Type I experimentwise error rate, but it generally has a higher Type II error rate than REGWQ.

Alpha	0.05
Error Degrees of Freedom	45
Error Mean Square	5.455556
Critical Value of Studentized Range	4.01842
Minimum Significant Difference	2.9681

Means with the same letter are not significantly different.			
Tukey Grouping	**Mean**	**N**	**Sugar**
A	70.100	10	1
B	64.100	10	5
C	59.300	10	2
C			
C	58.200	10	3
C			
C	58.000	10	4

Fig. 5.12. SAS Example E2: Tukey procedure

constructing the F-statistics is the same as Error MS from the ANOVA table. The four `estimate` statements result in the computation of t-statistics for testing the same four comparisons. The F-tests above are equivalent to these t-tests as the numerator degrees of freedom are equal to 1 for each F-statistic. This is reflected by the observation that the p-values for corresponding tests are identical.

The p-values of the four F-tests and four t-tests, respectively, are identical, as they are testing the same hypotheses. These p-values indicate that the hypothesis equating the effect of the control with the average effect of all sugars and the hypothesis equating the effect of the mixed sugars with average effects of pure sugars are rejected at an α level of 0.05. This results in the

Effect of Sugars on the Growth of Peas

The GLM Procedure

Dependent Variable: Length

Contrast	DF	Contrast SS	Mean Square	F Value	Pr > F
CONTROL VS. Sugars	1	832.3200000	832.3200000	152.56	<.0001
Sugars VS. MIXED	1	48.1333333	48.1333333	8.82	0.0048
GLUCOSE=FRUCTOSE	1	6.0500000	6.0500000	1.11	0.2979
FRUCTOSE = SUCROSE	1	115.2000000	115.2000000	21.12	<.0001

Parameter	Estimate	Standard Error	t Value	Pr > |t|
CONTROL VS. SugarS	40.8000000	3.30319710	12.35	<.0001
Sugars VS. MIXED	7.6000000	2.55864547	2.97	0.0048
GLUCOSE=FRUCTOSE	1.1000000	1.04456264	1.05	0.2979
FRUCTOSE = SUCROSE	-4.8000000	1.04456264	-4.60	<.0001

Fig. 5.13. SAS Example E2: making preplanned comparisons

finding that all sugars depress the mean pea lengths and the effect of the mixed sugars is less than the average effect of the pure sugars. The two hypotheses comparing the differences among the effects of the three pure sugars result in the findings that there is a significant difference between the Fructose and Sucrose means but no significant difference between the Fructose and Glucose means.

Finally, it is important to recognize, for example, that when an estimate statement such as

```
estimate 'CONTROL VS. SUGARS' Sugar 4 -1 -1 -1 -1;
```

is used, the numerical value output by `proc glm` is the estimate of $4\mu_1 - \mu_2 - \mu_3 - \mu_4 - \mu_5)$ and not of $\mu_1 - \frac{1}{4}(\mu_2 + \mu_3 + \mu_4 + \mu_5)$, as one might mistakenly consider the estimate (and its standard error) output to be. Although this will not affect the computed value of the t-statistic and the associated p-value, there may be instances when the actual estimate may be needed. One could use the `divisor=` option of the `estimate` statement to obtain the correct estimate without affecting the t-test by writing the two statements as follows:

```
estimate 'CONTROL VS. SUGARS' Sugar 4 -1 -1 -1 -1/divisor=4;

estimate 'SUGARS VS. MIXED   ' Sugar 0  1  1 -3  1/divisor=3;
```

These changes will result in estimates of 10.20 and 2.53 for the two comparisons with standard errors 0.8258 and 0.8529, respectively.

5.2.3 Testing Orthogonal Polynomials Using Contrasts

An experimenter may be interested in determining whether the observed mean response of a factor under study is related to the levels of the factor in some way. This relationship can be linear or curved. Orthogonal polynomials may be used to partition the sums of squares due to the factor into components that allow the experimenter to construct F-statistics with 1 degree of freedom each for the numerators. Each of these can then be used to test whether the relationship can be represented by a linear, quadratic, cubic, etc. function of the factor levels in a sequential fashion. To make the computations easier, this partitioning of the treatment sum of squares can be performed using appropriate orthogonal contrasts.

Table 5.3. Effect of engine size on gasoline consumption

Car	Engine size			
	300	350	400	450
1	16.6	14.4	12.4	11.5
2	16.9	14.9	12.7	12.8
3	15.8	14.2	13.3	12.1
4	15.5	14.1	13.6	12.0
Mean	16.1	14.4	13.0	12.1

Example 5.2.2

It must be understood that the reparameterized linear model obtained using orthogonal polynomials in this fashion is not exactly equivalent to a regression model with the (centered) factor levels as regressor variables. Thus, care is needed about how the results of tests involving orthogonal polynomials are interpreted. A suggested procedure is to start with a test of a linear trend of the mean response on the factor levels and use a lack of fit test to check if the remaining treatment sum of squares is significant. If there is lack of fit, proceed by successively increasing the order of the polynomial and performing lack of fit tests.

To illustrate the technique, consider the following example (Morrison 1983). Suppose that a consumer research group wishes to study the gasoline consumption of large eight-cylinder passenger cars of a given model year. The cars of interest have been classified by their engine sizes of approximately 300, 350, 400, and 450 cubic inches. Four cars were drawn at random from each engine size, and each car is driven over a standard urban route three times. These miles per gallon of fuel, recorded for the 16 cars, are shown in Table 5.3.

The analysis of variance computed for this data gave the following statistics:

SV	df	SS	MS	F
Engine size	3	38.25	12.7833	44.59
Error	12	3.44	0.2867	

According to this analysis, significant differences among the engine size means exist. It is now of interest to determine whether the decreasing gas mileage is a linear function of the engine volume or related in a more complex way to the engine volume.

The contrasts corresponding to the linear, quadratic, and cubic orthogonal polynomials have the coefficients given by

$$\begin{aligned} c_1' &= [-3, -1, 1, 3] \\ c_2' &= [1, -1, -1, 1] \\ c_3' &= [-1, 3, -3, 1] \end{aligned}$$

as can be obtained from a standard table of orthogonal polynomials (see Table B.13 of Appendix B). The sums of squares and F-statistics corresponding to the three single-degree of freedom contrasts are calculated to be

Contrast	df	SS	F
Linear	1	37.538	131.00
Quadratic	1	0.810	2.83
Cubic	1	0.002	0.01
Total	3	38.350	–

Since the three contrasts are mutually orthogonal, the sum of squares corresponds to a partitioning of the sum of squares for treatment (here Engine Size) with three degrees of freedom as evident from the analysis of variance table given above.

The linear trend contrast is significant with 1 and 12 degrees of freedom. The lack of fit F-statistic is not significant. The differences among the means seem to be explainable by a linear trend alone. These contrast sum of squares and corresponding F-tests are usually included in a complete ANOVA table as follows:

```
data mileage;
input Size $ @;
    do   i=1 to 4;
    input  MPG @;
    output;
    end;
drop i;
datalines;
300 16.6 16.9 15.8 15.5
350 14.4 14.9 14.2 14.1
400 12.4 12.7 13.3 13.6
450 11.5 12.8 12.1 12.0
;

proc glm data=mileage plots=Diagnostics;
  class Size;
  model MPG = Size;
  contrast 'Linear Trend' Size -3 -1 1 3;
  estimate 'Linear Trend' Size -3 -1 1 3;
  title 'Analysis of Gas Mileage Data';
  run;
```

Fig. 5.14. SAS Example E3: program (Part 1)

SV	df	SS	MS	F
Engine size	3	38.350	12.7833	44.59
Linear	1	37.538	37.5380	131.00
Lack of fit	2	0.812	0.4060	1.42
Error	12	3.440	0.2867	

Since the lack of fit is not significant, a higher-order polynomial is not needed. The experimenter can conclude that there is a significant decreasing linear trend in the mean gas mileage as the engine size increases.

SAS Example E3

Part 1 of the SAS Example E3 program (see Fig. 5.14) illustrates how the sum of squares needed for testing linear trend can be obtained using a `contrast` statement in `proc glm`. In this example, the gas mileage data are read using a similar approach to that used in the program for SAS Example E2. The trailing @ symbol is used to hold the data line after engine size is input. Then successive `input` and `output` statements are executed in a `do-end` loop to read each of the gas mileage values and write new observations into the SAS data set. Each observation output will have the current value of engine size and the gas mileage as they appear in the program data vector (PDV).

The `model` statement, similar to Example E1, codes the appropriate model for one-way classification. The `contrast` statement contains the coefficients taken from Table B.13 and corresponds to values tabulated for *Number of levels*=4 and *Degree of polynomial*=1. These are $-3, -1, 1$, and 3, respectively, and each coefficient corresponds to a level of `Size`. The `contrast` statement is

Analysis of Gas Mileage Data

The GLM Procedure

Dependent Variable: MPG

Source	DF	Sum of Squares	Mean Square	F Value	Pr > F
Model	3	38.35000000	12.78333333	44.59	<.0001
Error	12	3.44000000	0.28666667		
Corrected Total	15	41.79000000			

R-Square	Coeff Var	Root MSE	MPG Mean
0.917684	3.844974	0.535413	13.92500

Source	DF	Type I SS	Mean Square	F Value	Pr > F
Size	3	38.35000000	12.78333333	44.59	<.0001

Source	DF	Type III SS	Mean Square	F Value	Pr > F
Size	3	38.35000000	12.78333333	44.59	<.0001

Fig. 5.15. SAS Example E3: analysis of variance

```
contrast 'Linear Trend' Size -3 -1 1 3;
```

The resulting output is shown in Fig. 5.15. Part of the output showing class level information, which is omitted here, shows the levels of the treatment factor (here `Size`) that must always be checked to verify that they are in the correct sequence. The tables in Fig. 5.15, as in SAS Example E2, provide the information necessary to construct the analysis of variance table. Results of both the `contrast` and `estimate` statements are shown in Fig. 5.16. The F-test has a p-value of < 0.0001; thus, the linear trend is significant. The output from the `estimate` statement can be used to calculate the slope of the straight line fitted to the levels of engine size, using the formula

$$b = \frac{\sum c_i \bar{y}_{i.}}{\sum c_i^2}$$

where $\bar{y}_{i.}$ are the treatment means and $c_1, c_2, \ldots, c_t$ are the contrast coefficients. The estimate output is $\sum c_i \bar{y}_{i.} = -13.7$. Thus, the slope is calculated as $-13.7/(3^2 + 1^2 + 1^2 + 3^2) = -13.7/20 = -0.685$ per unit in the coded scale of the x variable (i.e., a decrease of 0.685 miles per gallon for every increase of 25 cubic inches in engine size). This estimate may be directly calculated in

SAS by modifying the estimate statement to

```
estimate 'Linear Trend' Size -3 -1 1 3/divisor=20;
```

Thus, the slope is estimated as -0.685 with a standard error of 0.05986.

Contrast	DF	Contrast SS	Mean Square	F Value	Pr > F
Linear Trend	1	37.53800000	37.53800000	130.95	<.0001

Parameter	Estimate	Standard Error	t Value	Pr > |t|
Linear Trend	-13.7000000	1.19721900	-11.44	<.0001

Fig. 5.16. SAS Example E3: contrast and estimate

Using the `plots=` option in the `proc glm` statement as shown in Fig. 5.17, a panel of useful diagnostic plots can be produced. Instead, several of the plots appearing in the above panel may be produced as separate graphs using an ODS SELECT statement such as

```
ods select ResidualPlots ResidualByPredicted FitPlot QQPlot;
```

as illustrated in SAS Example C3 (see Fig. 3.4). While this is an easy way to obtain these plots, it is more useful to collect selected plots in a panel setting as displayed in Fig. 5.18.

A method to extract plots appearing in the Fit Diagnostics panel, modify them as required, to produce the panel shown in Fig. 5.18 will be discussed in Appendix A. For the moment note that the option `p` used in the `model` statement results in producing new variables containing predicted values and residuals. By including an `output` statement in the SAS program (as shown in Fig. 5.17), the new SAS data set `stats1` that contains these new variables named `Predicted` and `Residual`, respectively, along with the variables in the original SAS data set (named `mileage`) is created. This data set is saved as a permanent file in a library (in this case, a folder under the Windows system) for later use. This data set will be used to obtain the panel displayed here by modifying the original graphical template that produces the complete `Fit Diagnostics` panel.

In Fig. 5.18, the standard residual plots of residuals against the levels of engine size and the residuals against the predicted values appear on the top two panels. Both these plots do not exhibit any outliers or trends in the dispersion of the points around zero (the reference line drawn at residual value equal to 0 is useful for visually ascertaining this) as values plotted on the x-axis change. Thus, there is evidence supporting the model assumption of homogeneity of variance and the adequacy of a first-order model in engine size to describe the variation in gas mileage.

```
proc glm data=mileage plots=Diagnostics;
  class Size;
  model MPG = Size/p;
  output out=stats1 p=Fitted r=Residual;
  title 'Analysis of Gas Mileage Data';
  run;
```

Fig. 5.17. SAS Example E3: program (Part 2)

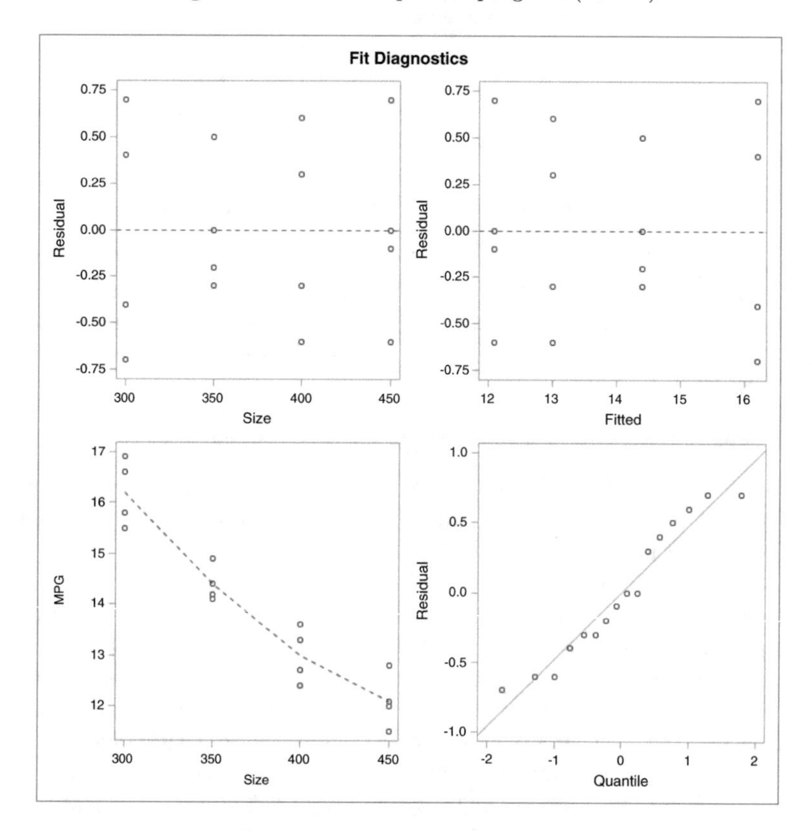

Fig. 5.18. SAS Example E3: plots. (**a**) Residuals versus engine size; (**b**) residuals versus predicted values; (**c**) predicted values overlaid on observed values; (**d**) normal probability plot of the residuals

The plot showing the observed values plotted against the levels of engine size also displays the predicted values (as connected by dashed line segments, so that they are easier to pick out). Note that the predicted value for all observations at each factor level is their sample mean. The data in `stats1` are also used for producing the `normal probability plot` of the residuals. The method used for obtaining this plot will be described in Appendix A (in Sect. A.4) and modifies the original template used to produce the regression diagnostic panel and uses `proc sgrender` to produce the plot.

5.3 One-Way Analysis of Covariance

The technique of measuring an additional variable, say x, called a *covariate*, in addition to the response variable y on each experimental unit in designed experiments can be used to increase precision of an experiment. The analysis of data from such experiments involves adjusting the analysis of variance and estimation procedures to account for the regression variable x. The resulting model for y thus contains the measured variable x in addition to the usual effects for the treatment factor.

Model

A single-factor experiment in a completely randomized design where a single covariate is measured is considered below. Equal replication of sample size n is assumed for the purpose of discussion.

$$y_{ij} = \mu + \tau_i + \beta(x_{ij} - \bar{x}_{..}) + \epsilon_{ij} \qquad i = 1, \ldots, t; \quad j = 1, \ldots, n$$

where τ_i is the ith treatment effect as in Sect. 5.2. It is also assumed that the random error ϵ_{ij} is distributed as iid $N(0, \sigma^2)$. The above model stipulates straight-line regression models for each treatment with the same slope β and different intercepts α_i for $i = 1, \ldots, t$, where $\alpha_i = \mu + \tau_i - \beta\,\bar{x}_{..}$, because the model expresses the relationship between y_{ij} and x_{ij} for each ith treatment as

$$\begin{array}{ll}
\text{Treatment 1:} & y_{1j} = \alpha_1 + \beta x_{1j} + \epsilon_{1j}\ ,\ j = 1, \ldots, n \\
\text{Treatment 2:} & y_{2j} = \alpha_2 + \beta x_{2j} + \epsilon_{2j}\ ,\ j = 1, \ldots, n \\
\vdots & \qquad\qquad\qquad\qquad\qquad \vdots \\
\text{Treatment t:} & y_{tj} = \alpha_t + \beta x_{tj} + \epsilon_{tj}\ ,\ \ j = 1, \ldots, n
\end{array}$$

Because of the stipulation that the straight lines have the same slope β, the above model is often called the *equal slopes* model.

Estimation

Note that since $E(y_{ij}) = \mu + \tau_i + \beta(x_{ij} - \bar{x}_{..})$, unlike the model in Sect. 5.2, the ith treatment mean now depends on different values of x_{ij}. For the equal slopes model, one treatment mean evaluated at $x_{ij} = \bar{x}_{..}$ and denoted by μ_i is of interest. This is usually called the "adjusted mean." Note that $\mu_i = \mu + \tau_i$. The best linear unbiased estimate of μ_i is

$$\hat{\mu}_i = \bar{y}_{i.}(\text{Adj.}) = \bar{y}_i - b(\bar{x}_{i.} - \bar{x}_{..}) \text{ for } i = 1, \ldots, t$$

These estimates are called *adjusted treatment means* because the usual estimate $\bar{y}_i$ is adjusted for regression on x_i using the estimated common slope b, where b is the usual least squares estimate of β given by

$$b = \frac{\sum_i \sum_j (x_{ij} - \bar{x}_{i.})(y_{ij} - \bar{y}_{i.})}{\sum_i \sum_j (x_{ij} - \bar{x}_{i.})^2} = \frac{S_{xy}}{S_{xx}}$$

where $\bar{y}_{i.} = (\sum_j y_{ij})/n$, $\bar{x}_{i.} = (\sum_j x_{ij})/n$, and $\bar{x}_{..} = (\sum_i \sum_j x_{ij})/tn$. Also, an unbiased estimate of the error variance σ^2 is given by $\hat{\sigma}^2 = s^2$, where s^2 is the Error MS from the analysis of covariance table below. It can be shown that the adjusted treatment means $\hat{\mu}_i$ are the predicted responses $\hat{y}_{ij}$ computed at the value of $x_{ij} = \bar{x}_{..}$ using the fitted regression models, for each $i = 1, \ldots, t$.

A $(1-\alpha)100\%$ confidence interval for $\mu_p - \mu_q$, the difference between the effects of two treatments labeled p and q, is

$$(\bar{y}_{p.}(Adj.) - \bar{y}_{q.}(Adj.)) \pm t_{\alpha/2,\nu} s_d$$

where s_d, the standard error of the difference between two adjusted means $\bar{y}_{p.}(Adj.) - \bar{y}_{q.}(Adj.)$, is given by

$$s_d = s\left\{\frac{2}{n} + \frac{(\bar{x}_{p.} - \bar{x}_{q.})^2}{S_{xx}}\right\}^{1/2}$$

and $t_{\alpha/2,\nu}$ is the upper $\alpha/2$ percentile of the t-distribution with $\nu = t(n-1)-1$ degrees of freedom.

Testing Hypotheses

The presentation of the results of the analysis is somewhat complicated because the total sum of squares is partitioned in two ways: one partition shows the test of the main effects without covariate adjustment and one shows it with covariate adjustment. It is convenient to present the results of both in a single compact analysis of variance table rather than two separate tables. In the following table, the analysis above the `Total SS` line shows the treatment sum of squares *unadjusted* for the covariate and the partition of the regression sum of squares from the `Error SS` to form the test of the regression parameter. The second analysis shown below the `Total SS` line contains the treatment sum of squares *adjusted* for the covariate. This table is called the *analysis of covariance* table:

SV	df	SS	MS	F
Trt	$t-1$	SS_{Trt}	MS_{Trt}	$\text{MS}_{\text{Trt}}/\text{MSE}_{\text{Unadj.}}$
Error(Unadj.)	$t(n-1)$	$\text{SSE}_{\text{Unadj.}}$	$\text{MSE}_{\text{Unadj.}}$	
Regression	1	SS_{Reg}	MS_{Reg}	$\text{MS}_{\text{Reg}}/\text{MSE}$
Error(Adj.)	$t(n-1)-1$	SSE	$\text{MSE}(=s^2)$	
Total	$tn-1$	SS_{Tot}		
Trt(Adj.)	$t-1$	SS_{Trt}	MS_{Trt}	$\text{MS}_{\text{Trt}}/\text{MSE}$
Error(Adj.)	$t(n-1)-1$	SSE	$\text{MSE}(=s^2)$	

The F-statistic for `Trt` tests the hypothesis

$$H_0 : \mu_1 = \mu_2 = \cdots = \mu_t \quad \text{versus} \quad H_a : \text{ at least one inequality}$$

when the covariate is not present in the model (i.e., without taking into account any adjustment due to the covariate). The divisor for computing the F-statistic is the MS for `Error(Unadj.)` with $t(n-1)$ df. The F-statistic for `Regression` tests the hypothesis

$$H_0 : \beta = 0 \quad \text{versus} \quad H_a : \beta \neq 0$$

and thus is a test of whether the covariate has an effect on the response as an explanatory variable in a linear regression.

The F-statistic for Trt(Adj.) tests the hypothesis that the adjusted treatment means are the same or, equivalently, the treatment effects

$$H_0 : \tau_1 = \tau_2 = \cdots = \tau_t \quad \text{versus} \quad H_a : \text{ at least one inequality}$$

when β is not zero (i.e., when the analysis of variance is adjusted for the covariate). This test is also equivalent to comparing the intercepts of the regression lines, i.e.,

$$H_0 : \alpha_1 = \alpha_2 = \cdots = \alpha_t \quad \text{versus} \quad H_a : \text{ at least one inequality}$$

If this hypothesis is rejected, then at least one pair of treatment effects (equivalently, adjusted treatment means) is different. One can proceed to make preplanned or pairwise comparisons of these means as in SAS Example E1 but using the adjusted treatment means. Note carefully that SS_{Trt} (and thus MS_{Trt} and the F-statistics) in the bottom table will be different in magnitude from those in the top table since those take into account that a covariate is present in the model.

5.3.1 Using PROC GLM to Perform One-Way Covariance Analysis

The data displayed in Table 5.4 are results from an experiment on the use of two treatments, slow-release fertilizer (S) and a fast-release fertilizer (F), on the yield (grams) of peanut plants compared to a control (C), a standard fertilizer, described in Ott and Longnecker (2001). Ten replications of each treatment were grown in a greenhouse study.

Since the researcher recognized that the 30 peanut plants used were different in their development and health, the height (in centimeters) of each plant was recorded at the start of the experiment to be used as a covariate to adjust for this variation. This experiment is an example of a single-factor experiment in a completely randomized design in which a covariate is also measured on each experimental unit.

The graph shown in Fig. 5.19 is a simple scatter plot of the yield of peanuts against the heights, identified by the fertilizer (slow release, fast release, or

Table 5.4. Yield of peanut plants from three fertilizer treatments and their initial heights

Fertilizer Treatments					
Control (C)		Slow Release (S)		Fast Release (F)	
Yield	Height	Yield	Height	Yield	Height
12.2	45	16.6	63	9.5	52
12.4	52	15.8	50	9.5	54
11.9	42	16.5	63	9.6	58
11.3	35	15.0	33	8.8	45
11.8	40	15.4	38	9.5	57
12.1	48	15.6	45	9.8	62
13.1	60	15.8	50	9.1	52
12.7	61	15.8	48	10.3	67
12.4	50	16.0	50	9.5	55
11.4	33	15.8	49	8.5	40

the control) received by individual plants. The `proc sgplot` step in the SAS program (see Fig. 5.20) produced this graph. It is obvious that straight lines could be fitted to data for each type of fertilizer.

Let y_{ij} and x_{ij} be the yield and the pretreatment measure of height of the jth plant treated with the ith fertilizer, respectively. The model is then

$$y_{ij} = \mu_i + \beta(x_{ij} - \bar{x}_{..}) + \epsilon_{ij}, \quad i = 1, 2, 3; \quad j = 1, \ldots, 10,$$

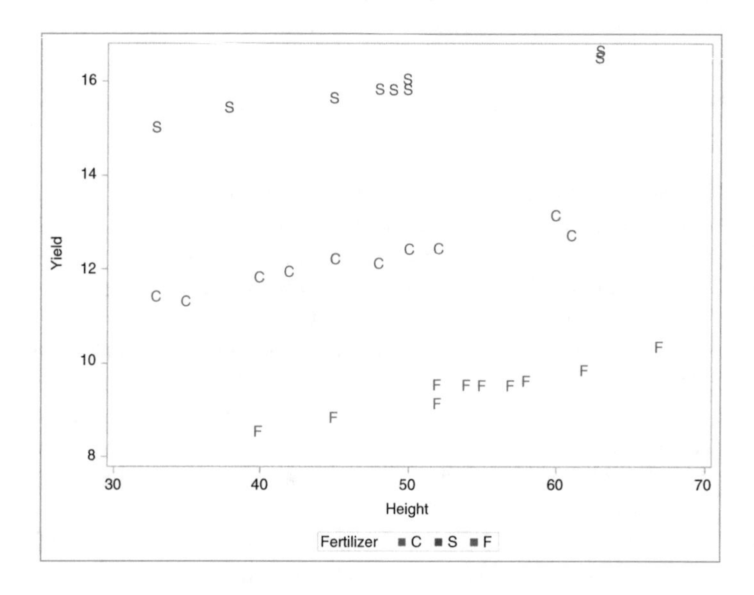

Fig. 5.19. SAS Example E4: plot of yield versus height by fertilizer

where μ_i are the mean yields from the three fertilizers. Note that with the notation used earlier, $\mu_i = \mu + \tau_i$ where τ_i represents the effects of the three fertilizers. Thus, three regression lines (corresponding to the three fertilizers) with the same slope parameter β are stipulated by this model. The hypothesis of equality of the mean yields due to the three fertilizers, $H_0 : \mu_1 = \mu_2 = \mu_3$ versus H_a : at least one inequality, is tested using the analysis of covariance discussed earlier.

SAS Example E4

The SAS Example E4 program (see Fig. 5.20) is used to obtain the necessary analysis of the above data. In the SAS program, once again the data are input in a straightforward format. The "effects" form of the model $E(y_{ij}) = \mu + \alpha_i + \beta x_{ij}$ is used to specify the model for analysis by `proc glm`. The `model` statement thus includes the term `Fertilizer` to represent the treatment effect α_i and the term `Height` to represent the covariate x_{ij}. Note that this variable does not appear in the `class` statement; thus, when it occurs on the right side of the model statement, it is recognized to be a regression-type variable and not a classificatory variable, by default. It is also important to note that the covariate appears after the treatment variable in the model statement. The design matrix

$$X = \begin{bmatrix} 1 & 1 & 0 & 0 & 45 \\ 1 & 1 & 0 & 0 & 52 \\ 1 & 1 & 0 & 0 & 42 \\ \vdots & \vdots & \vdots & \vdots & \vdots \\ 1 & 1 & 0 & 0 & 33 \\ 1 & 0 & 1 & 0 & 63 \\ 1 & 0 & 1 & 0 & 50 \\ 1 & 0 & 1 & 0 & 63 \\ \vdots & \vdots & \vdots & \vdots & \vdots \\ 1 & 0 & 1 & 0 & 49 \\ 1 & 0 & 0 & 1 & 52 \\ 1 & 0 & 0 & 1 & 54 \\ 1 & 0 & 0 & 1 & 58 \\ \vdots & \vdots & \vdots & \vdots & \vdots \\ 1 & 0 & 0 & 1 & 40 \end{bmatrix}$$

that results from the model statement is a 30×5 matrix as shown. The columns of X correspond to the parameters μ, α_1, α_2, α_3, and β, respectively.

The `lsmeans` statement is required for `proc glm` to generate the *adjusted treatment means* and their standard errors, instead of the ordinary sample means. Just as in the `means` statement, `lsmeans` statement lists effects that involve only classification variables. Note that the sample means produced by the `means` statement are not adjusted for the covariate. The class level information in Fig. 5.21 shows that the ordering of the levels of factor `Fertilizer`

```
data peanuts;
input Fertilizer $ Yield Height;
datalines;
C  12.2     45
C  12.4     52
.    .       .
.    .       .
.    .       .
C  12.4     50
C  11.4     33
S  16.6     63
S  15.8     50
.    .       .
.    .       .
.    .       .
S  16.0     50
S  15.8     49
F   9.5     52
F   9.5     54
.    .       .
.    .       .
.    .       .
F   9.5     55
F   8.5     40
;
proc sgplot data=peanuts;
 scatter x=Height y=Yield/markerchar=Fertilizer
                    markercharattrs=(family="centb" size=10 pt weight=bold)
                    group=Fertilizer;
run;

proc glm data=peanuts order=data;
  class Fertilizer;
  model Yield = Fertilizer Height;
  lsmeans Fertilizer/stderr cl tdiff pdiff;
  contrast 'Modified vs. Standard' Fertilizer 1 -.5 -.5;
  contrast 'Slow-release vs. Fast-release' Fertilizer 0  1 -1;
  title 'Covariance Analysis of Peanut Fertilizer Data';
  run;
```

Fig. 5.20. SAS Example E4: program

Covariance Analysis of Peanut Fertilizer Data

The GLM Procedure

Class Level Information		
Class	**Levels**	**Values**
Fertilizer	3	C S F

Number of Observations Read	30
Number of Observations Used	30

Fig. 5.21. SAS Example E4: class levels

seen in the input data is retained as requested by including the `order=data` as a `proc glm` option.

The information needed to construct the analysis of covariance table described earlier is found in the output resulting from the model statement. Specifically, the `Total` and `Error(Adj.)` degrees of freedom, sums of squares, and mean squares needed are extracted from those found in the corresponding columns for `Corrected Total` and `Error`, respectively, given in the table shown on top part of the SAS output (see Fig. 5.22). Values for `Fertilizer` and `Height`, respectively, from the Type I SS part on the same SAS output, provide the degrees of freedom, sums of squares, mean squares, and the corresponding F-statistics for both the `Fertilizer(Unadj.)` and `Regression` lines in the top portion of the analysis of covariance table.

Covariance Analysis of Peanut Fertilizer Data

The GLM Procedure

Dependent Variable: Yield

Source	DF	Sum of Squares	Mean Square	F Value	Pr > F
Model	3	214.3759539	71.4586513	4447.85	<.0001
Error	26	0.4177128	0.0160659		
Corrected Total	29	214.7936667			

R-Square	Coeff Var	Root MSE	Yield Mean
0.998055	1.017537	0.126751	12.45667

Source	DF	Type I SS	Mean Square	F Value	Pr > F
Fertilizer	2	207.6826667	103.8413333	6463.47	<.0001
Height	1	6.6932872	6.6932872	416.62	<.0001

Source	DF	Type III SS	Mean Square	F Value	Pr > F
Fertilizer	2	213.9038045	106.9519022	6657.08	<.0001
Height	1	6.6932872	6.6932872	416.62	<.0001

Fig. 5.22. SAS Example E4: output from the `model` statement

The "unadjusted" Error SS is then obtained by summing the regression and adjusted error sums of squares. Thus, the top portion of the table is complete. To complete the `Fertilizer(Adj.)` line in the bottom portion of the table, the degrees of freedom, sums of squares, mean squares, and the corresponding F-statistics for `Fertilizer` are obtained from the Type III SS part on the same SAS output. The completed table is

SV	DF	SS	MS	F	p-Value
Fertilizer(Unadj.)	2	207.6827	103.8414	394.23	< 0.0001
Error	27	7.1110	0.2634		
Regression	1	6.6933	6.6933	416.62	<0.0001
Error(Adj.)	26	0.4177	0.0161		
Total	29	214.7937			
Fertilizer(Adj.)	2	213.9038	106.9519	6657.08	<0.0001
Error(Adj.)	26	0.4177	0.0161		

First, the p-value for the F-test for `Regression` clearly shows that the hypothesis of $H_0 : \beta = 0$ is rejected, thus confirming that plant height is linearly related to seed yield. Second, from the p-value for the F-statistic for `Fertilizer(Adj.)`, the hypothesis of no difference in fertilizer effects is also rejected.

In the analysis of variance table shown earlier, the `Total` line is the result of ignoring the covariate. In this table, the F-statistic for testing no difference in fertilizer effects hypothesis (shown in the `Fertilizer(Unadj.)` line) actually is much smaller than the F-statistic for `Fertilizer(Adj.)`. It can be easily seen that this is due to the inflated error variance estimate given by the MSE (0.2634), because the mean squares for `Fertilizer(Adj.)` and `Fertilizer(Unadj.)` are similar in magnitude. This shows that, in other situations, it is possible for differences that may exist among the treatments to go undetected if covariance adjustment is not taken into account if the effect of the adjustment is substantial.

The SAS output shown in Fig. 5.23 is produced as a result of the `lsmeans` statement used in the current proc step. By default, the statistics computed are identical to the adjusted treatment means $\bar{y}_{i.}$(Adj.) discussed previously in this section. These are displayed under `LSMEAN` in Fig. 5.23 along with their standard errors. It is important to note that `proc glm` computes the `lsmeans` by setting the covariate values equal to their mean (i.e., $x_{ij} = \bar{x}_{..}$) as discussed previously. This implies that, implicitly, the option `at means` is in effect, as the default. This is appropriate, as the regression lines are parallel when the equal slopes model holds and, thus, the differences in `lsmeans` are the same at any value of x, but those computed at the means have the smallest standard errors. If the slopes were different, however, the `at` option would enable the user to request these to be computed at different covariate values considered interesting for comparison of the predicted responses at those values (e.g., by using an option like `at x=10`).

The `stderr` option on the `lsmeans` statement resulted in the standard errors of the adjusted treatment means to be also output. Using the option `tdiff` on the `lsmeans` statement produces t-statistics for comparing the pairwise differences in the means (i.e., hypotheses of the form $H_0 : \mu_i = \mu_j$ versus

Covariance Analysis of Peanut Fertilizer Data

The GLM Procedure
Least Squares Means

Fertilizer	Yield LSMEAN	Standard Error	Pr > |t|	LSMEAN Number
C	12.3141728	0.0410853	<.0001	1
S	15.8858099	0.0401754	<.0001	2
F	9.1700172	0.0417711	<.0001	3

Least Squares Means for Effect Fertilizer
t for H0: LSMean(i)=LSMean(j) / Pr > |t|

Dependent Variable: Yield

i/j	1	2	3
1		-62.6244 <.0001	52.07806 <.0001
2	62.62441 <.0001		114.7842 <.0001
3	-52.0781 <.0001	-114.784 <.0001	

Fertilizer	Yield LSMEAN	95% Confidence Limits	
C	12.314173	12.229721	12.398625
S	15.885810	15.803228	15.968392
F	9.170017	9.084155	9.255879

Least Squares Means for Effect Fertilizer

i	j	Difference Between Means	95% Confidence Limits for LSMean(i)-LSMean(j)	
1	2	-3.571637	-3.688869	-3.454405
1	3	3.144156	3.020055	3.268256
2	3	6.715793	6.595528	6.836058

Fig. 5.23. SAS Example E4: output from the `lsmeans` statement

Covariance Analysis of Peanut Fertilizer Data

The GLM Procedure

Dependent Variable: Yield

Contrast	DF	Contrast SS	Mean Square	F Value	Pr > F
Modified vs. Standard	1	0.2830511	0.2830511	17.62	0.0003
Slow-release vs. Fast-release	1	211.6745206	211.6745206	13175.4	<.0001

Fig. 5.24. SAS Example E4: output from the `contrast` statements

$H: \mu_i \neq \mu_j$ for all pairs (i, j)). The `pdiff` option produced the p-values associated with these tests.

The last portion of this output consists of the 95% confidence intervals for individual adjusted means and their pairwise differences produced as a result of the `cl` option. The `alpha=` keyword option may be added to specify a confidence coefficient different from 95%.

It is possible to use the `contrast` statement to test single-degree of freedom comparisons of interest about treatment means. Here, the average effect of the fertilizers with the control is compared using the comparison $\tau_1 - (\tau_2 + \tau_3)/2$. Note that the sums of squares and the F-tests are also adjusted for the covariate. From the SAS output shown in Fig. 5.24, this hypothesis is clearly rejected, implying that the two, on the average, lower the mean yield compared to the control. In addition, by examining the tests and confidence intervals shown in Fig. 5.23, it is found that there is a significant difference between the two fertilizers.

The `lsmeans` statement may be modified as follows:

```
lsmeans Fertilizer/stderr cl tdiff pdiff adjust=bon;
```

Part of the output is shown in Fig. 5.25 as only the pairwise tests and confidence limits are affected by this modification. These are different from the previous output because of the `adjust=bon` option included in the modified `lsmeans` statement. This option causes a multiple comparison adjustment to be made to the the p-values and confidence limits for the pairwise differences. Here, the adjustment requested is the Bonferroni adjustment. `tukey` and `scheffe` are two other adjustments available. The default is `adjust=t`, which really signifies no adjustment made for doing multiple comparisons.

The graph shown in Fig. 5.26 is produced automatically by `proc glm` as a result of the above analysis. This plot confirms that it is feasible to model the yield as a linear function of height for each fertilizer and that the assumption of equal slopes is reasonable. The intercepts are clearly different, thus supporting the result of the test of equal treatment means.

Covariance Analysis of Peanut Fertilizer Data

The GLM Procedure
Least Squares Means
Adjustment for Multiple Comparisons: Bonferroni

Least Squares Means for Effect Fertilizer t for H0: LSMean(i)=LSMean(j) / Pr > |t| Dependent Variable: Yield			
i/j	1	2	3
1		-62.6244 <.0001	52.07806 <.0001
2	62.62441 <.0001		114.7842 <.0001
3	-52.0781 <.0001	-114.784 <.0001	

Least Squares Means for Effect Fertilizer				
i	j	Difference Between Means	Simultaneous 95% Confidence Limits for LSMean(i)-LSMean(j)	
1	2	-3.571637	-3.717580	-3.425694
1	3	3.144156	2.989662	3.298649
2	3	6.715793	6.566074	6.865511

Fig. 5.25. SAS Example E4: multiple testing using `adjust=bon` option

The regression lines fitted under the equal slope assumption are superimposed on this plot. The parameter estimates of these regression lines can be output from the SAS program. To do this, modify the `model` statement in the `proc glm` step as follows:

```
model Yield = Fertilizer Height/noint solution;
```

This results in the output shown in Fig. 5.27.

5.3.2 One-Way Covariance Analysis: Testing for Equal Slopes

In the beginning of Sect. 5.3, the equal slopes model for a single-factor experiment in a completely randomized design in which a single covariate is measured was introduced as

$$y_{ij} = \alpha_i + \beta x_{ij} + \epsilon_{ij}, \quad i = 1, \ldots, t;\ j = 1, \ldots, n$$

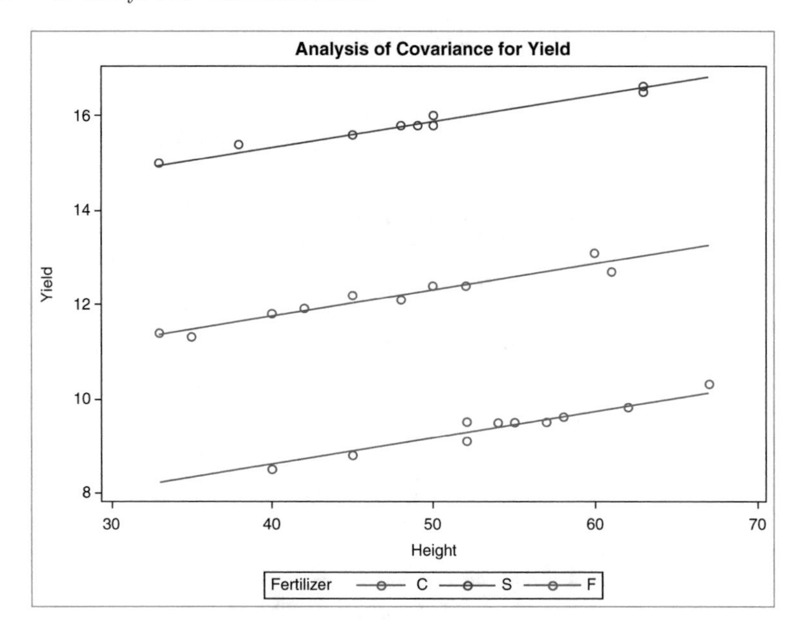

Fig. 5.26. SAS Example E4: analysis of covariance plot

The above model is not the most general model for the analysis of data from this experiment because of the often unrealistic assumption of equal slopes, although it is useful in certain situations. A more general model is the unequal slopes model

$$y_{ij} = \alpha_i + \beta_i x_{ij} + \epsilon_{ij}, \quad i = 1, \ldots, t; \; j = 1, \ldots, n$$

where different slopes β_i, $i = 1, \ldots, t$, are assumed for the t regression lines relating the responses y_{ij} to the covariates x_{ij}. The advantage of this model over the equal slopes model is that a hypothesis of whether the slopes are indeed the same (i.e., $H_0 : \beta_1 = \beta_2 = \cdots = \beta_t = \beta$ versus H_a : at least one inequality) may be tested as part of the inference from this model.

Parameter	Estimate	Standard Error	t Value	Pr > |t|
Fertilizer C	9.52925636	0.13357349	71.34	<.0001
Fertilizer S	13.10089348	0.13958529	93.86	<.0001
Fertilizer F	6.38510075	0.15352310	41.59	<.0001
Height	0.05580995	0.00273429	20.41	<.0001

Fig. 5.27. SAS Example E4: regression parameter estimates

If this hypothesis is rejected, then a hypothesis of equal treatment means $\mu_{ij} = E(y_{ij}) = \alpha_i + \beta_i x_{ij}$ is of interest. In this case, however, such a test is not equivalent to the test of equality of the intercepts $H_0 : \alpha_1 = \alpha_2 = \cdots = \alpha_t$ as was the case in the equal slopes model. This is because the differences in means now depend on the value of β_i. Thus, the difference in intercepts will actually depend on the value x_{ij} at which the intercepts are compared. These comparisons are therefore dependent on the value of the covariate at which the comparisons are made. In practice, the treatment means are compared at several choices of the covariate values such as the mean, the median, the minimum, or the maximum or at values of the covariate that are of special interest to the experimenter.

Since the above model may be expressed as

$$y_{ij} = \alpha_i + \bar{\beta} x_{ij} + (\beta_i - \bar{\beta}) x_{ij} + \epsilon_{ij}, \quad i = 1, \ldots, t; j = 1, \ldots, n$$

this model is used in `proc glm` to obtain the sum of squares for testing the equal slopes hypothesis. This model is fitted to the cholesterol data taken from Milliken and Johnson (2001) in SAS Example E5. Thirty-two female subjects were assigned completely at random to one of four different diets. The response variable is the cholesterol level determined after being on the diet for 8 weeks. The cholesterol levels of the subjects measured before the experiment began were used as a covariate. The data appear in Table 5.5.

Table 5.5. Prediet and postdiet cholesterol levels by diet

Cholesterol Measurements							
Diet 1		Diet 2		Diet 3		Diet 4	
Post-	Pre-	Post-	Pre-	Post-	Pre-	Post-	Pre-
174	221	211	203	199	249	224	297
208	298	211	223	229	178	209	279
210	232	201	164	198	166	214	212
192	182	199	194	233	223	218	192
200	258	209	248	233	274	253	151
164	153	172	268	221	234	246	191
208	293	224	249	199	271	201	284
193	283	222	297	236	207	234	168

SAS Example E5

The variables named `Diet`, `PostChol`, and `PreChol`, in the SAS Example E5 program (see Fig. 5.28) identify the four diets, the response variable, and the covariate, respectively. In order to fit the unequal slopes model using `proc glm`, the `model` statement from the `proc glm` step in the SAS program

```
data women;
input Diet $ PostChol PreChol;
datalines;
1 174 221
1 208 298
.  .  .
.  .  .
.  .  .
1 208 293
1 193 283
2 211 203
2 211 223
.  .  .
.  .  .
.  .  .
2 224 249
2 222 297
3 199 249
3 229 178
.  .  .
.  .  .
.  .  .
3 199 271
3 236 207
4 224 297
4 209 279
.  .  .
.  .  .
.  .  .
4 201 284
4 234 168
;

proc glm data=women;
class Diet;
model PostChol = Diet PreChol PreChol*Diet;
run;
```

Fig. 5.28. SAS Example E5: SAS program

shown in Fig. 5.20 is modified in this program as shown. Recall that `Diet` is a classificatory variable and `PreChol` is a regression variable. The `PreChol*Diet` term in the above model is called a *discrete by continuous* interaction because of this reason. The resulting output is shown in Fig. 5.29.

The `proc glm` output in Fig. 5.29 can be used to obtain tests for three hypotheses of interest. First, the Type III SS for `PreChol*Diet` and the corresponding F-statistic provide a test of the hypothesis $H_0 : \beta_1 = \beta_2 = \cdots = \beta_t = \beta$ versus H_a : at least one inequality (i.e., that the slopes are all the same). Second, the Type I SS for `PreChol` and the corresponding F-statistic provide a test of the hypothesis $H_0 : \beta = 0$ if the equal slopes model is used following the result of the previous test. This can be used to determine if the `PostChol` can be modeled as a linear function of the `PreChol` at all.

The Type III SS for `Diet` and the corresponding F-statistic provide a test of the hypothesis $H_0 : \mu_{1j} = \mu_{2j} = \cdots = \mu_{tj}$ versus H_a : at least one inequality at the value $x_{ij} = 0$ for all i. Thus, this is

The GLM Procedure

Dependent Variable: PostChol

Source	DF	Sum of Squares	Mean Square	F Value	Pr > F
Model	7	6930.55635	990.07948	4.06	0.0045
Error	24	5852.91240	243.87135		
Corrected Total	31	12783.46875			

R-Square	Coeff Var	Root MSE	PostChol Mean
0.542150	7.408809	15.61638	210.7813

Source	DF	Type I SS	Mean Square	F Value	Pr > F
Diet	3	4593.843750	1531.281250	6.28	0.0027
PreChol	1	1.903672	1.903672	0.01	0.9303
PreChol*Diet	3	2334.808924	778.269641	3.19	0.0417

Source	DF	Type III SS	Mean Square	F Value	Pr > F
Diet	3	3718.772130	1239.590710	5.08	0.0073
PreChol	1	1.824611	1.824611	0.01	0.9318
PreChol*Diet	3	2334.808924	778.269641	3.19	0.0417

Fig. 5.29. SAS Example E5: unequal slopes model ANOVA

equivalent to the test that the intercepts of the regression lines are the same (i.e., $H_0 : \alpha_1 = \alpha_2 = \cdots = \alpha_t$ versus H_a : at least one inequality at the value $x_{ij} = 0$ for all i), that is, a test whether the regression lines for the diets intersect at the `PreChol` value of zero. Since the value of `PreChol` can never be zero, this test is not particularly useful. Thus, the above test is omitted from the following adjusted analysis of covariance table.

SV	df	SS	MS	F	p-Value
Diet(Unadj.)	3	4593.84			
Regression	1	1.90	1.90	0.01	0.9374
Error	27	8187.72	303.25		
Regression	3	2334.81	778.27	3.19	0.0417
Error(Adj.)	24	5852.91	243.87		
Total	31	12,783.47			

However, recall that $\mu_{ij} = E(y_{ij}) = \alpha_i + \beta_i x_{ij}$. Thus, comparisons of the means μ_{ij} (and therefore the intercepts α_i) may be made at any other selected value(s) of x_{ij} using adjusted least squares means. To compare adjusted least squares means at values of `PreChol`, values equal to 190 and 250 (say) modify the `model` statement in the `proc glm` step as follows:

```
model PostChol = Diet  PreChol*Diet/noint solution;
```

and include the two `lsmeans` statements

```
lsmeans Diet/stderr cl pdiff adjust=bon at PreChol=190;

lsmeans Diet/stderr cl pdiff adjust=bon at PreChol=250;
```

The results from the `solution` option in the modified model statement are shown in Fig. 5.30. The estimates of the coefficients available in this output are used for drawing the fitted lines shown in Fig. 5.33.

Parameter	Estimate	Standard Error	t Value	Pr > |t|
Diet 1	137.6323082	27.26507534	5.05	<.0001
Diet 2	195.7360754	32.03129844	6.11	<.0001
Diet 3	223.7329893	33.70599704	6.64	<.0001
Diet 4	276.6032780	23.66995633	11.69	<.0001
PreChol*Diet 1	0.2333029	0.11125081	2.10	0.0467
PreChol*Diet 2	0.0450224	0.13673614	0.33	0.7448
PreChol*Diet 3	-0.0232319	0.14761695	-0.16	0.8763
PreChol*Diet 4	-0.2332730	0.10379713	-2.25	0.0341

Fig. 5.30. SAS Example E5: regression parameter estimates in the unequal slopes model

Options available for the `lsmeans` statement are used to perform multiple comparisons of the adjusted PostChol means by constructing Bonferroni-adjusted confidence intervals at two different values of the prediet cholesterol levels using the `at` option. The values of 190 and 250 were selected because they are values just below the "desired" level of 200 and just above the "high-risk" level of 240, respectively. Extracts from the SAS output from the `lsmeans` statements are shown in Figs. 5.31 and 5.32.

All pairs of Diet means are found to be not significantly different at the prediet cholesterol level of 250 (since the intervals for all differences included zero); however, at the prediet cholesterol level of 190, Diet 1 PostChol mean is found to be significantly lower than both Diets 3 and 4 means. Note that the default confidence coefficient was 95% for these intervals.

The GLM Procedure
Least Squares Means at PreChol=190
Adjustment for Multiple Comparisons: Bonferroni

Diet	PostChol LSMEAN	Standard Error	Pr > |t|	LSMEAN Number
1	181.959856	7.837460	<.0001	1
2	204.290336	7.844175	<.0001	2
3	219.318925	7.586851	<.0001	3
4	232.281416	6.429979	<.0001	4

Least Squares Means for Effect Diet

i	j	Difference Between Means	Simultaneous 95% Confidence Limits for LSMean(i)-LSMean(j)	
1	2	-22.330480	-54.211226	9.550266
1	3	-37.359069	-68.720812	-5.997326
1	4	-50.321560	-79.468042	-21.175079
2	3	-15.028589	-46.404206	16.347028
2	4	-27.991080	-57.152490	1.170330
3	4	-12.962491	-41.555581	15.630598

Fig. 5.31. SAS Example E5: comparison of means in the unequal slopes model

A graph containing plots of the fitted models of the postdiet cholesterol values as straight lines of prediet cholesterol predictors superimposed on the scatter plots of the data values was also produced as part of the output from `proc glm` and is shown in Fig. 5.33. As suggested from the results of the multiple comparisons procedure of the adjusted diet means, it is observed from this graph that the differences of average postdiet cholesterol among the diets appear to be lower for those individuals with higher prediet cholesterol than those with lower prediet levels.

As observed from the graph, the fitted lines converge to a point as prediet cholesterol levels increase, indicating that standard errors of the adjusted means decrease, a fact shown by narrower intervals for the differences in post-diet means at `PreChol=250` compared to those at `PreChol=190`.

The GLM Procedure
Least Squares Means at PreChol=250
Adjustment for Multiple Comparisons: Bonferroni

Diet	PostChol LSMEAN	Standard Error	Pr > |t|	LSMEAN Number
1	195.958029	5.632193	<.0001	1
2	206.991682	6.116555	<.0001	2
3	217.925010	6.620583	<.0001	3
4	218.285039	6.251569	<.0001	4

Least Squares Means for Effect Diet				
i	j	Difference Between Means	Simultaneous 95% Confidence Limits for LSMean(i)-LSMean(j)	
1	2	-11.033653	-34.939131	12.871825
1	3	-21.966981	-46.957773	3.023811
1	4	-22.327010	-46.519475	1.865455
2	3	-10.933328	-36.848180	14.981524
2	4	-11.293357	-36.439236	13.852521
3	4	-0.360029	-26.539850	25.819792

Fig. 5.32. SAS Example E5: comparison of means in the unequal slopes model

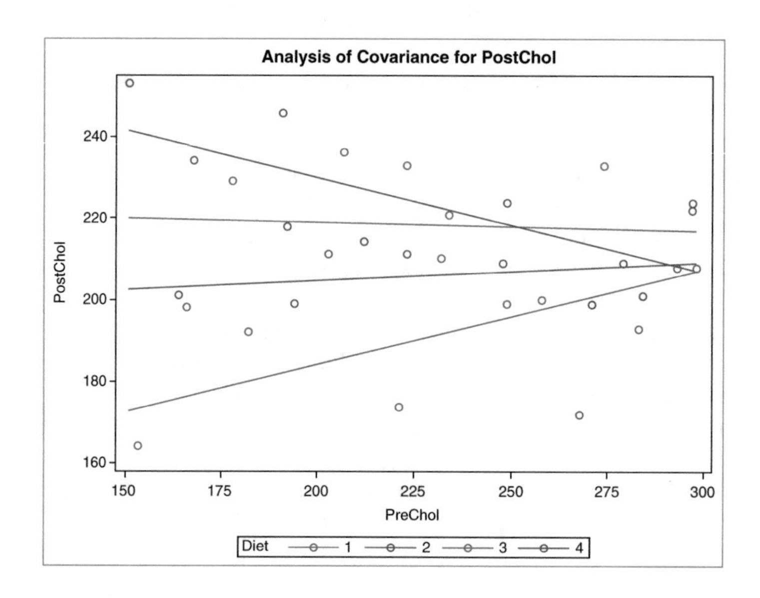

Fig. 5.33. SAS Example E5: plot of postdiet cholesterol versus prediet cholesterol by diet

5.4 A Two-Way Factorial in a Completely Randomized Design

A two-way factorial treatment structure consists of all combinations of levels of two factors under study in the experiment. The design employed is a completely randomized design (CRD) if these treatment combinations have been applied completely randomly to the experimental units. In the following discussion, it is assumed that all combinations of a two-way factorial with "a" levels of factor A and "b" levels of factor B are used, and an equal number of replications per each treatment combination are obtained. The cake-baking experiment discussed earlier is an example of this setup.

Model

The model is

$$y_{ijk} = \underbrace{\mu + \alpha_i + \beta_j + \gamma_{ij}}_{\mu_{ij}} + \epsilon_{ijk}, \quad i = 1, \ldots, a; \quad j = 1, \ldots, b; \quad k = 1, \ldots, n$$

where $\mu_{ij} = E(y_{ijk})$ is the mean of an observation in the ijth cell of the two-way classification and is called the *cell means model.* It is also assumed that the random error ϵ_{ijk} is distributed as iid $N(0, \sigma^2)$. A model expressed in terms of the cell means is called the "means model." If the cell means are partitioned into the sum of effects α_i of level i of A and β_j of level j of B and an interaction effect γ_{ij} of the ith level of A and the jth level of B, an "effects model" is said to be in use.

The n observations corresponding to the ijth treatment combination y_{ijk}, $k = 1, \ldots, n$, are assumed to be a random sample from the $N(\mu_{ij}, \sigma^2)$ distribution under this model. It is convenient and useful to express the hypotheses of interest to the experimenter in terms of "averaged" means or marginal means of the cell means that are defined as follows:

$$\text{Factor A Means: } \bar{\mu}_{i.} = \Big(\sum_j \mu_{ij}\Big)/b, \quad i = 1, \ldots, a$$

$$\text{Factor B Means: } \bar{\mu}_{.j} = \Big(\sum_i \mu_{ij}\Big)/a, \quad j = 1, \ldots, b$$

A table of cell means as shown in Fig. 5.34 is a visual illustration of the two-way classification model with equal sample sizes in each cell. The "averaged" means defined above for the two factors appear in the margins of this table.

Levels of Factor B

		1	2	$\cdots$	j	$\cdots$	b	
	1	μ_{11}	μ_{12}	$\cdots$		$\cdots$	μ_{1b}	$\bar{\mu}_{1.}$
	2	μ_{21}	μ_{22}				μ_{2b}	$\bar{\mu}_{2.}$
Levels of	$\vdots$	$\vdots$					$\vdots$	$\vdots$
Factor A	i				μ_{ij}			$\bar{\mu}_{i.}$
	$\vdots$	$\vdots$					$\vdots$	$\vdots$
	a	μ_{a1}	μ_{a2}	$\cdots$		$\cdots$	μ_{ab}	$\bar{\mu}_{a.}$
		$\bar{\mu}_{.1}$	$\bar{\mu}_{.1}$	$\cdots$	$\bar{\mu}_{.j}$	$\cdots$	$\bar{\mu}_{.b}$	

Fig. 5.34. Two-way factorial: cell means and marginal means

Hypotheses Testing

The usual format of the analysis of variance (ANOVA) table for computing the required F-statistics for testing hypotheses of interest in a two-way classification is

SV	df	SS	MS	F	
Treatment	$ab-1$	$\mathtt{SS_{Trt}}$			
A	$a-1$	$\mathtt{SS_A}$	$\mathtt{MS_A}$	$\mathtt{MS_A/MSE}$	(1)
B	$b-1$	$\mathtt{SS_B}$	$\mathtt{MS_B}$	$\mathtt{MS_B/MSE}$	(2)
A*B	$(a-1)(b-1)$	$\mathtt{SS_{AB}}$	$\mathtt{MS_{AB}}$	$\mathtt{MS_{AB}/MSE}$	(3)
Error	$ab(n-1)$	$\mathtt{SSE}$	$\mathtt{MSE}$		
Total	$abn-1$	$\mathtt{SS_{Tot}}$			

The F-statistics from the ANOVA table are used to perform tests of the following hypotheses of interest:

(1) Use this F-statistic to test main effects of Factor A. Using the marginal means for Factor A, these are expressed as

$$H_0 : \bar{\mu}_{1.} = \bar{\mu}_{2.} = \cdots = \bar{\mu}_{a.} \quad \text{versus} \quad H_a\text{: at least one inequality}$$

(2) Use this F-statistic to test main effects of Factor B. Using the marginal means for Factor B, these are expressed as

$$H_0 : \bar{\mu}_{.1} = \bar{\mu}_{.2} = \cdots = \bar{\mu}_{.b} \quad \text{versus} \quad H_a\text{: at least one inequality}$$

(3) Use this F-statistic to test interaction of Factors A and B. Using the marginal means for both Factors A and B and the cell means, these are expressed as

$$H_0 : (\mu_{ij} - \bar{\mu}_{i.} - \bar{\mu}_{.j} + \bar{\mu}_{..}) = 0 \text{ for all combinations of } (i,j) \quad \text{versus}$$
$$H_a\text{: at least one } (\mu_{ij} - \bar{\mu}_{i.} - \bar{\mu}_{.j} + \bar{\mu}_{..}) \neq 0$$

(equivalent to H_0 : no interaction present versus H_a : interaction present)

An approach for using these in practical situations and how to proceed based on the result of each test is discussed below and in the several examples to follow. Although the interpretation of the results of the experiment appears to be simpler using the "means model" and associated means, in practice many experimenters resort to the "effects model" for such purposes. SAS procedures available for the analysis of data from designed experiments usually require that the effects models be used to describe the model equation to the program. An understanding of the theory of the linear model is necessary to clarify complications that result from the usage of the "effects model." An attempt will be made to illustrate some of these differences in the course of the discussions of the examples below.

Estimation

The best estimates of the cell means μ_{ij} and the marginal means $\bar{\mu}_{i.}$ and $\bar{\mu}_{.j}$, respectively, are given by

$$\hat{\mu}_{ij} = \bar{y}_{ij.} = \Big(\sum_k y_{ijk}\Big)/n$$

$$\hat{\bar{\mu}}_{i.} = \bar{y}_{i..} = \Big(\sum_j \sum_k y_{ijk}\Big)/bn$$

$$\hat{\bar{\mu}}_{.j} = \bar{y}_{.j.} = \Big(\sum_i \sum_k y_{ijk}\Big)/an$$

An estimate of the error variance σ^2 is $\hat{\sigma}^2 = s^2$, where s^2 is the MSE value obtained from the ANOVA table. The standard error of the difference in the pair of Factor A means at levels i and i' is

$$\text{s.e.}(\bar{y}_{i..} - \bar{y}_{i'..}) = s\sqrt{2/bn}$$

and the standard error of the difference in the pair of Factor B means at levels j and j' is

$$\text{s.e.}(\bar{y}_{.j.} - \bar{y}_{.j'.}) = s\sqrt{2/an}$$

Thus, $(1-\alpha)100\%$ confidence intervals for the differences in a pair of Factor A and B means are, respectively, given by

$$\bar{\mu}_{i.} - \bar{\mu}_{i'.} : \qquad (\bar{y}_{i..} - \bar{y}_{i'..}) \pm t_{\alpha/2,\nu} \cdot s \cdot \sqrt{2/bn}$$

$$\bar{\mu}_{.j} - \bar{\mu}_{.j'} : \qquad (\bar{y}_{.j.} - \bar{y}_{.j'.}) \pm t_{\alpha/2,\nu} \cdot s \cdot \sqrt{2/an}$$

where $t_{\alpha/2,\nu}$ is the upper $\alpha/2$ percentile of the t-distribution with ν df and ν is the degrees of freedom for MSE equal to $ab(n-1)$.

Differences in factor means ($\bar{\mu}_{i.}$ or $\bar{\mu}_{.j}$) may not measure actual differences in the *cell means* for Factor A or Factor B, respectively, at levels of the other factor when interaction is present (i.e., when the model is nonadditive). Thus, the interpretation of main effects depends on whether interaction effects are found to be significant or not.

- The F-test for interaction in the ANOVA table is a test whether the model is additive. It is recommended that this test be performed prior to making inferences from the main effects tests.
- If the interaction effects turn out to be not significant, then essentially the effects of the two factors A and B may be interpreted independently of each other. The F-tests for Factors A and B in the ANOVA table are then used to test for main effects of A and B. If either or both of these main effect F-tests are significant, then the averaged marginal means may be compared (say, using preplanned comparisons or multiple comparison procedures) and significant comparisons interpreted as usual.
- If interaction F-test is significant, then the model is nonadditive. This implies that care must be taken in interpreting main effect hypotheses of Factors A and B because there is significant interaction. The F-tests for main effects may still be performed, but the results may not be meaningful because differences in averaged means may not reflect the differences of the effects of one factor at each level of the other factor.
- An interaction plot may be useful for identifying whether differences in main effect means (marginal means) are affected significantly by interaction. If it is found that this is the case, comparisons of cell means of one factor over the levels of the other factor (e.g., $\mu_{12}-\mu_{13}$), may be necessary. If preplanned comparisons of the factor means are available, interesting interaction comparisons may be constructed that will aid in interpreting the significant interaction.

5.4.1 Analysis of a Two-Way Factorial Using PROC GLM

The data shown in Table 5.6 are survival times of groups of four animals randomly allocated to each of all combinations of three poisons and four drugs. The experiment was an investigation to combat the effects of certain toxic agents. This example is from Box et al. (1978) where a standard analysis as well as an analysis based on transforming the data using a variance-stabilizing transformation is performed. It is assumed that the observations in each cell of the above classification are random samples of size 4 from normal distributions with means μ_{ij} and the same variance σ^2. The model is thus

$$y_{ijk} = \mu_{ij} + \epsilon_{ijk}, \quad i = 1,2,3; \quad j = 1,2,3,4; \quad k = 1,2,3,4$$

where $\epsilon_{ij} \sim$ iid $N(0,\sigma^2)$. A nonadditive model $\mu_{ij} = \mu + \alpha_i + \beta_j + \gamma_{ij}$ for the cell means is considered for partitioning the treatment sum of squares given the two-way factorial treatment structure.

Table 5.6. Survival times data

Poison	Drug			
	A	B	C	D
I	0.31	0.82	0.43	0.45
	0.45	1.10	0.45	0.71
	0.46	0.88	0.63	0.66
	0.43	0.72	0.76	0.62
II	0.36	0.92	0.44	0.56
	0.29	0.61	0.35	1.02
	0.40	0.49	0.31	0.71
	0.23	1.24	0.40	0.38
III	0.22	0.30	0.23	0.30
	0.21	0.37	0.25	0.36
	0.18	0.38	0.24	0.31
	0.23	0.29	0.22	0.33

SAS Example E6

In SAS Example E6, the `glm` procedure in SAS is used to obtain the appropriate analysis of variance table. The SAS program is shown in Fig. 5.35. The `class` statement must precede the `model` statement and declare the classification variables, here the two factors `Poison` and `Drug`. Note that the "effects model" is used for formulating the `model` statement, by including a term for each effect (except a term for μ that is assumed to be in the model by default). The `Poison`× `Drug` interaction term is specified as `Poison*Drug` in the `model` statement. The `means` statement with the `lsd` option requests that all pairwise comparisons be made for both `Poison` and `Drug` means. In this case, by default, the `lsd` procedure is performed because the sample sizes are equal. On the other hand, if the sample sizes were unequal, confidence intervals would be constructed for all pairwise differences of the main effects. These could be specifically requested by including the option `cldiff` in the `means` statement. The confidence coefficient used by default is 95%; this could be changed by using the option `alpha=`.

In the course of analyzing two-way factorial data using SAS, the `means` and `sgplot` procedure may be used prior to the `glm` procedure to obtain a scatter plot of the cell means. This plot, used as an example of an `interaction plot` in Chap. 3 (see SAS Example C8 in Sect. 3.1) and displayed in Fig. 3.19, shows a *profile* of the means across the levels of one factor at the same level of the other factor. However, this is no longer necessary as this plot is automatically produced as graphical output from `proc glm`. This plot is shown at the end of this discussion in Fig. 5.36.

```
data mice;
input  Poison 1. @;
   do Drug=1 to 4;
      input Time 3.2 @;
      output;
   end;
datalines;
1 31 82 43 45
1 45110 45 71
1 46 88 63 66
1 43 72 76 62
2 36 92 44 56
2 29 61 35102
2 40 49 31 71
2 23124 40 38
3 22 30 23 30
3 21 37 25 36
3 18 38 24 31
3 23 29 22 33
;

proc print ;
  title 'Analysis of Survival Times of Mice: Original Data';
run;

proc glm data=mice;
  class Poison Drug;
  model Time = Poison Drug Poison*Drug;
  means Poison Drug/lsd;
run;
```

Fig. 5.35. SAS Example E6: program

In Fig. 5.36, levels of `Poison` are plotted on the x-axis, and the points corresponding to the same levels of `Drug` are connected with line segments. Not only do the line segments allow the pattern of the mean response of each drug to the three poisons to be observed visually, but they also allow the mean responses to be compared across the four drugs. Thus, it is useful for interpreting and locating any significant interaction that may exist between the two factors. The class level information provided in Fig. 5.37 is useful for checking whether the factor levels are as expected and are in proper order.

Since there are equal sample sizes for each treatment combination (number of observations in each cell), the Type I and III sums of squares are the same as expected (Fig. 5.38). However, it is recommended that Type III sums of squares be always used in situations where the model does not contain any terms other than those representing fixed classificatory factors and their interactions. From the output from SAS Example E6 shown in Fig. 5.38, the following analysis of variance table is constructed.

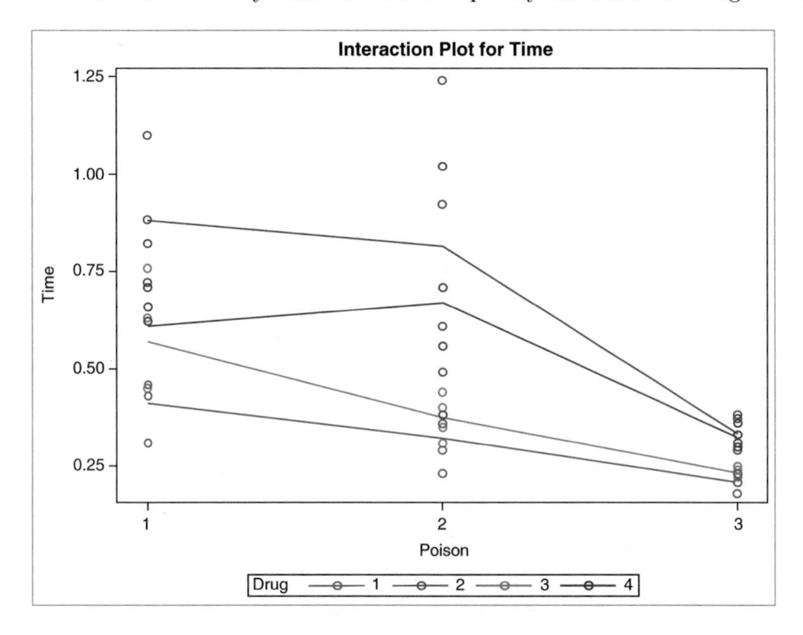

Fig. 5.36. SAS Example E6: interaction plot from `proc glm`

Analysis of Survival Times of Mice: Original Data

The GLM Procedure

Class Level Information		
Class	**Levels**	**Values**
Poison	3	1 2 3
Drug	4	1 2 3 4

Number of Observations Read	48
Number of Observations Used	48

Fig. 5.37. SAS Example E6: Class Levels

Analysis of Survival Times of Mice: Original Data

The GLM Procedure

Dependent Variable: Time

Source	DF	Sum of Squares	Mean Square	F Value	Pr > F
Model	11	2.20435625	0.20039602	9.01	<.0001
Error	36	0.80072500	0.02224236		
Corrected Total	47	3.00508125			

R-Square	Coeff Var	Root MSE	Time Mean
0.733543	31.11108	0.149139	0.479375

Source	DF	Type I SS	Mean Square	F Value	Pr > F
Poison	2	1.03301250	0.51650625	23.22	<.0001
Drug	3	0.92120625	0.30706875	13.81	<.0001
Poison*Drug	6	0.25013750	0.04168958	1.87	0.1123

Source	DF	Type III SS	Mean Square	F Value	Pr > F
Poison	2	1.03301250	0.51650625	23.22	<.0001
Drug	3	0.92120625	0.30706875	13.81	<.0001
Poison*Drug	6	0.25013750	0.04168958	1.87	0.1123

Fig. 5.38. SAS Example E6: analysis of variance

SV	df	SS	MS	F	p-Value
Treatment	11	2.2043			
Poison	2	1.0330	0.51651	23.22	<0.0001
Drug	3	0.9212	0.30707	13.81	<0.0001
Poison × Drug	6	0.2501	0.04169	1.87	0.1123
Error	36	0.8007	0.02224		
Total	47	3.0051			

Since the interaction between `Poison` and `Drug` is not significant at 5%, one may conclude that these two factors act additively. Thus, it may be reasonable to examine the main effects and test the hypotheses $H_{01} : \bar{\mu}_{1.} = \bar{\mu}_{2.} = \bar{\mu}_{3.}$ and $H_{02} : \bar{\mu}_{.1} = \bar{\mu}_{.2} = \bar{\mu}_{.3} = \bar{\mu}_{.4}$ independently, in order to determine the effects of poisons and drugs, respectively. As seen from the extremely small p-values, both `Poison` and `Drug` effects are highly significant. The LSD procedure output produced from the `means` statement, shown in Figs. 5.39 and 5.40

for Poison and Drug means, respectively, finds that at $\alpha = 0.05$, the mean survival times for:

(i) Poison 3 is significantly lower than those of Poisons 1 and 2,
(ii) Poisons 1 and 2 are not significantly different,
(iii) Drugs 1 and 3 are not significantly different but are significantly lower than those of Drugs 2 and 4, respectively,
(iv) Drug 4 is significantly lower than that of Drug 2.

The GLM Procedure

t Tests (LSD) for Time

Note: This test controls the Type I comparisonwise error rate, not the experimentwise error rate.

Alpha	0.05
Error Degrees of Freedom	36
Error Mean Square	0.022242
Critical Value of t	2.02809
Least Significant Difference	0.1069

Means with the same letter are not significantly different.			
t Grouping	**Mean**	**N**	**Poison**
A	0.61750	16	1
A			
A	0.54438	16	2
B	0.27625	16	3

Fig. 5.39. SAS Example E6: LSD procedure for Poison means

The analysis would have greatly improved if preplanned comparisons considered important were suggested by the experimenter. Due to the lack of such comparisons, the interpretation of the results relied on multiple comparison procedures. Adjustments for making multiple comparisons can be made by using methods such as those based on Bonferroni or Tukey procedures. These methods will be used in other examples to follow.

5.4.2 Residual Analysis and Transformations

A residual analysis shows that the variance of the data increases with the expected mean of the observed data. This is clearly evident in the plot of residuals against the predicted values shown in Fig. 5.41. This plot is obtained by first modifying the `proc glm` step in the SAS Example E6 program

The GLM Procedure

t Tests (LSD) for Time

Note: This test controls the Type I comparisonwise error rate, not the experimentwise error rate.

Alpha	0.05
Error Degrees of Freedom	36
Error Mean Square	0.022242
Critical Value of t	2.02809
Least Significant Difference	0.1235

Means with the same letter are not significantly different.			
t Grouping	**Mean**	**N**	**Drug**
A	0.67667	12	2
B	0.53417	12	4
C	0.39250	12	3
C			
C	0.31417	12	1

Fig. 5.40. SAS Example E6: LSD procedure for Drug means

by including the following statement:

```
output out=new r=Residuals p=Predicted ;
```

This results in the residuals and the predicted values from the fitted model being added to the original data in the SAS data set named `mice` and saved in a new data set named `new`. This data set is then accessed using the data= option in the subsequent `proc sgplot` step to obtain the plot shown in Fig. 5.41. The new proc step is shown in Fig. 5.42. Although a similar plot can be obtained using the option `plots=diagnostics(unpack)` and the `ods select residualbypredicted;` statement, it is a better alternative to use `proc sgplot` to construct this plot as illustrated here.

This plot suggests that a *variance-stabilizing transformation* may be attempted to increase the sensitivity of the experiment as well as for easier interpretation of the results. If the error variance σ is proportional to a power α of the mean μ, a power transformation of the response of the form y^λ may be attempted to stabilize the variance.

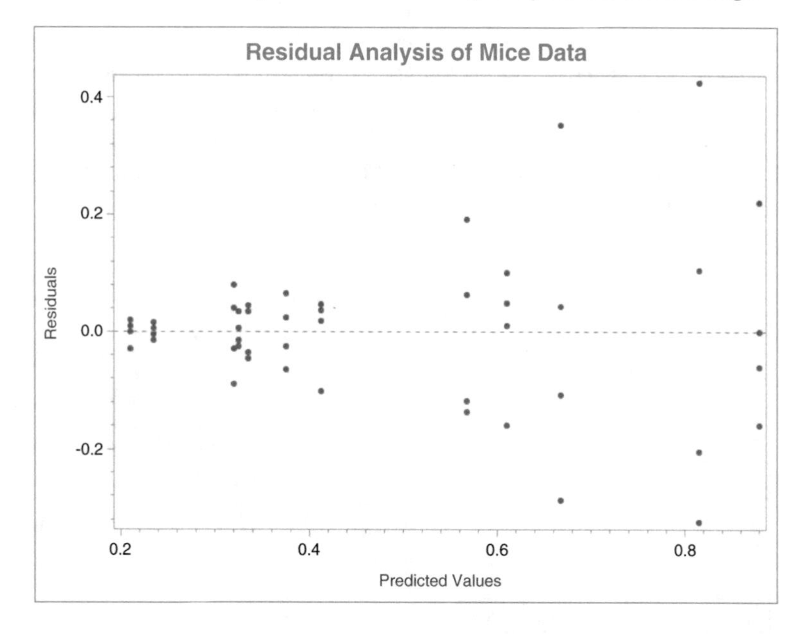

Fig. 5.41. SAS Example E6: plot of residuals versus predicted values

```
proc sgplot data=new;
 title c=steelblue bold h=2 "Residual Analysis of Mice Data";
 scatter x=Predicted y=Residuals/markerattrs=(color=magenta size= 5 pt symbol=CircleFilled);
 refline 0/lineattrs=(color=cornflowerblue pattern=2);
 xaxis minor minorcount=9 label="Predicted Values"
                          labelattrs=(color=blueviolet family=arial size=10);
 yaxis minor label="Residuals"
             labelattrs=(color=blueviolet family=arial size=10);
run;
```

Fig. 5.42. SAS Example E6: proc step to create a residual plot

An empirical method for obtaining an estimate of λ for replicated data is to plot the logarithm of the sample standard deviation for each treatment combination against the logarithm of the sample mean. By fitting a straight line, an interval estimate of the *slope* of the line α can be obtained. Estimates of α thus obtained can be used to determine an appropriate transformation λ since the required power transformation can be shown to be given by $\lambda = 1 - \alpha$ where α is the slope of regression of the logarithm of the cell standard deviation on the logarithm of the cell mean. Most common of such transformations are *reciprocal, inverse square root, logarithmic, and square root*, respectively, for estimates of λ close to $-1, -\frac{1}{2}, 0$, and $\frac{1}{2}$. The plot of the $\log s_{ij}$ versus $\log \bar{y}_{ij}$ values for the 12 cells of the mice data is shown in Fig. 5.43. The mean and standard deviation of data in each cell were first calculated, log transformations were performed, and then a `proc sgplot` step was used to obtain this plot as shown in Fig. 5.44.

The estimated slope of the regression is 1.977, and a 95% interval is (1.39, 2.56). Taking $\alpha \approx 2$ suggests that a reciprocal transformation of survival times (also called an inverse transformation) may be appropriate. Generally, for data measured in time units, an inverse transformation is often found to be appropriate for stabilizing the variance; the data values are transformed into *survival rates*, a natural unit of measure for this study.

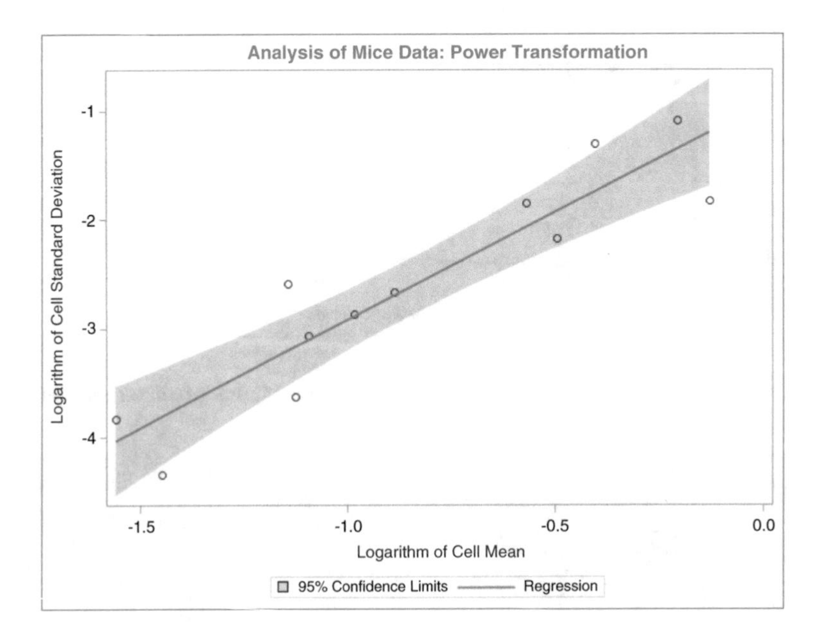

Fig. 5.43. SAS Example E6: empirical estimation of variance-stabilizing transformation

SV	df	SS	MS	F	p-Value
Treatment	11	56.8622			
Poison	2	34.8771	17.4386	72.63	<0.0001
Drug	3	20.4143	6.8048	28.34	<0.0001
Poison × Drug	6	1.5708	0.2618	1.09	0.3867
Error	36	8.6431	0.2401		
Total	47	65.5053			

The SAS Example E6 program is modified to include the statement `time = 1/time` in the data step to effect this transformation of the response variable. The analysis of variance obtained from this analysis is shown above.

It is observed that the `Poison` × `Drug` interaction has become even less significant, thus allowing the experimenter to be more confident of the suitability of an additive model. Further, mean squares for both `Poison` and `Drug` effects are now much larger relative to the error mean square, implying increased sensitivity compared to the previous analysis. Note that for presentation of the results of the analysis, statistics calculated using the transformed data such as confidence intervals are preferably transformed back to the original units for easier interpretation by the experimenter.

```
proc sort data=mice;
by Poison Drug;
run;

proc means mean std noprint;
by Poison Drug;
var Time;
output out=new mean=Mean_Time std=SD_Time;
run;

data trans;
set new;
logSD=log(SD_Time);
logMean=log(Mean_Time);
run;

proc sgplot data=trans;
title color=steelblue "Analysis of Mice Data: Power Transformation";
reg x=logMean y=logSD/clm;
xaxis label="Logarithm of Cell Mean" labelattrs=(color=indigo);
yaxis label="Logarithm of Cell Standard Deviation" labelattrs=(color=indigo);
run;
```

Fig. 5.44. SAS Example E6: program for power transformation plot

5.5 Two-Way Factorial: Analysis of Interaction

In Sect. 5.4 it was shown how the treatment sum of squares (SS) with $(ab-1)$ degrees of freedom (df) was subdivided into sums of squares corresponding to main effects A and B and their interaction effect with $(a-1)$, $(b-1)$, and $(a-1)(b-1)$ df, respectively. This subdivision was suggested by the effects model and enabled the testing of hypotheses appropriate for determining the presence or absence of interaction and main effects. In Sect. 5.2 it was shown how the treatment SS in a single-factor experiment may be partitioned into one df SS that are appropriate for making inferences about preplanned or a priori comparisons among the factor means.

In general, for a factor with a levels, the associated df of $(a-1)$ implies that the SS may be partitioned into a set of $(a-1)$ orthogonal comparisons and thus into $(a-1)$ SS each with a single degree of freedom. In the case of a two-factor experiment, each of the two main effects and interaction SS may be partitioned into several one df SS corresponding to comparisons of interest. In particular, the $(a-1)(b-1)$ df for interaction may also be partitioned into $(a-1)(b-1)$ one df SS. The partitioning of the interaction SS may be derived on the basis of the main effect comparisons. For example, if a comparison of interest among Factor A means was "Control versus Others," it might be of interest to the experimenter to examine whether this comparison differs among the levels of Factor B. The resulting comparisons constitute a subset of A × B interaction comparisons.

Generally, one df interaction comparisons that make sense may be formulated by considering one df main effect comparisons of the two factors. For example, in SAS Example E6 (see Sect. 5.4), the contrast coefficients corresponding to a possible comparison of interest $3\bar{\mu}_{.1} - \bar{\mu}_{.2} - \bar{\mu}_{.3} - \bar{\mu}_{.4}$ among the Poison means are (3 –1 –1 –1). Possible interaction comparisons may be those obtained by making the same comparison of the cell means at each level of the drug and then comparing them among the levels of the drug.

For example, to test that the above comparison is the same between levels 2 and 3 of the drug factor, the two comparisons of the cell means $3\mu_{21} - \mu_{22} - \mu_{23} - \mu_{24}$ and $3\mu_{31} - \mu_{32} - \mu_{33} - \mu_{34}$ must be compared. This comparison can be written using the contrast coefficients (0 0 0 0 +3 –1 –1 –1 –3 1 1 1), giving an interaction contrast of possible interest. Note that the coefficients may be obtained via the "elementwise product" of the main effect contrasts (0 1 –1) × (3 –1 –1 –1).

SAS Example E7

An example taken from Snedecor and Cochran (1989) illustrates the use of preplanned comparisons in two-way factorial experiments for analyzing interactions. The data shown below are gains in weight of male rats under six feeding treatments in a completely randomized design. The two factors were

A (2 levels) : Level of protein (high, low)

B (3 levels) : Source of protein (beef, cereal, pork)

The data are shown in Table 5.7.

Table 5.7. Weight gain in rats under six diets

High protein			Low protein		
Beef	Cereal	Pork	Beef	Cereal	Pork
73	98	94	90	107	49
102	74	79	76	95	82
118	56	96	90	97	73
104	111	98	64	80	86
81	95	102	86	98	81
107	88	102	51	74	97
100	82	108	72	74	106
87	77	91	90	67	70
117	86	120	95	89	61
111	92	105	78	58	82

A standard analysis for a two-way factorial in a completely randomized design would start with computing an analysis variance table as described in Sect. 5.4. Using the output from a `proc glm` step in SAS, the following analysis of variance table was constructed:

SV	df	SS	MS	F	p-Value
Treatments	5	4612.93	922.59	4.30	0.0023
Level	1	3168.27	3168.27	14.77	0.0003
Source	2	266.53	133.27	0.62	0.5411
Level × Source	2	1178.13	589.07	2.75	0.0732
Error	54	11,586.00	214.56		
Total	59	16,198.93			

Since this analysis of variance table shows that the p-value value for the Level × Source interaction is between 10% and 5%, it would not be possible for the experimenter to be entirely comfortable in assuming an additive model. The null hypothesis of no interaction will be rejected at $\alpha = 0.1$. This demonstrates the dilemma an experimenter might encounter if one attempts to interpret results from a factorial experiment using the usual partitioning of treatment sum of squares found in an analysis of variance table for two-way classifications. Should one proceed with an analysis of main effects assuming an additive model or attempt to make sense of the interaction that may be present?

On the other hand, in many experiments of this nature, the structure of the treatment factors may suggest a priori comparisons among levels of factors, which might be more helpful for making useful interpretations. These may lead to a more natural explanation of any interaction that may be present among these factor levels. In the above experiment, comparisons of interest among the sources of protein means may be:

- Average of beef and pork with cereal, and
- Beef versus pork.

These comparisons are suggested since beef and pork are animal sources of protein, whereas cereal is a vegetable source. When the main effect sums of squares are subdivided into single degree of freedom sums of squares, a comparable subdivision of the interaction sum of squares can also be made. To illustrate the procedure, as usual, assume the model for observations be

$$y_{ijk} = \mu_{ij} + \epsilon_{ijk}, \quad i = 1, 2; \quad j = 1, 2, 3; \quad k = 1, \ldots, 10$$

The means in the above model are the population means of the observations obtained from each combination of the two factors `Level of Protein` and `Source of Protein` as illustrated in the following table:

		Source			
		Beef	Cereal	Pork	
Level	High	μ_{11}	μ_{12}	μ_{13}	$\bar{\mu}_{1.}$
	Low	μ_{21}	μ_{22}	μ_{23}	$\bar{\mu}_{2.}$
		$\bar{\mu}_{.1}$	$\bar{\mu}_{.2}$	$\bar{\mu}_{.3}$	

The sample cell means and marginal means that are the best estimates of the above parameters are found in the following table:

		Source			
		Beef	Cereal	Pork	
Level	High	100.0	85.9	99.5	95.13
	Low	79.2	83.9	78.7	80.6
		89.6	84.9	89.1	

The usual main effect hypotheses based on the marginal means are H_{01} : $\bar{\mu}_{1.} = \bar{\mu}_{2.}$ and $H_{02} : \bar{\mu}_{.1} = \bar{\mu}_{.2} = \bar{\mu}_{.3}$. However, instead of testing the main effect hypotheses, the corresponding df may be partitioned using orthogonal one df comparisons among the main effect means (marginal means). For example, the 2 df of the Protein sum of squares may be split into two one df sums of squares that represent two orthogonal comparisons. For example, the two comparisons of the `protein means` stated above may be formulated as the contrast of the marginal means: "Average of beef and pork with cereal" comparison is expressed in the form $(\bar{\mu}_{.1} + \bar{\mu}_{.3})/2 - \bar{\mu}_{.2}$ and "Beef with pork" comparison as $\bar{\mu}_{.1} - \bar{\mu}_{.2}$. Notice that these two comparisons are orthogonal to each other. They are summarized in the following table of contrast coefficients:

Comparison	Coefficients		
Animal versus Vegetable	+1	–2	+1
Beef versus Pork	+1	0	–1

The above comparisons are called "main effect" comparisons, as the contrasts involve only the marginal means. The corresponding subdivision of the interaction sum of squares is obtained by the following comparisons of the cell means. These correspond to the comparisons of the cell means $\mu_{11} + \mu_{13} - 2\mu_{12} = \mu_{21} + \mu_{23} - 2\mu_{22}$ and $\mu_{11} - \mu_{13} = \mu_{21} - \mu_{23}$. The comparisons such as $\mu_{11} - \mu_{13}$ and $\mu_{11} + \mu_{13} - 2\mu_{12}$ are called "simple effects." Thus, interaction comparisons are comparisons of simple effects.

Comparison	Coefficients					
Animal versus Vegetable × Level of Protein	+1	–2	+1	–1	+2	–1
Beef versus Pork × Level of Protein	+1	0	–1	–1	0	+1

The first interaction comparison tests if the "Animal versus Vegetable" effect, if any, is the same at both levels of protein, whereas the second comparison tests if the "Beef versus Pork" effect, if any, is the same at both levels of protein. The coefficients for the contrast labeled **Animal vs. Vegetable x Level of Protein** are obtained by the elementwise product $(1 \; -1) \times (+1 \; -2 \; 1)$ and those of **Beef vs. Pork x Level of Protein** by the product $(1 \; -1) \times (1 \; 0 \; -1)$.

```
data rats;
input Level Source @;
   do i=1 to 10;
      input Weight @;
      output;
   end;
drop i;
datalines;
1 1  73 102 118 104  81 107 100  87 117 111
1 2  98  74  56 111  95  88  82  77  86  92
1 3  94  79  96  98 102 102 108  91 120 105
2 1  90  76  90  64  86  51  72  90  95  78
2 2 107  95  97  80  98  74  74  67  89  58
2 3  49  82  73  86  81  97 106  70  61  82
;

proc glm data=rats plots=none;;
  class Level Source;
  model Weight = Level Source Level*Source;
  contrast 'animal vs. vegetable'
             Source 1 -2 1;
  contrast 'beef vs. pork'
             Source 1 0 -1;
  contrast 'an. vs. veg. by Level'
             Level*Source 1 -2 1 -1 2 -1;
  contrast 'an. vs. veg. by Level'
             Level*Source 1 0 -1 -1 0 1;
run;
```

Fig. 5.45. SAS Example E7: program

Although the "means model" was used in the discussion making statistical inferences from the two-way classification, the use of the "effects model" is required for specification of the model as well as contrast coefficients in SAS programs. The "effects model" for the two-way classification is the partitioning of the mean μ_{ij} into main effects and interaction parameters $\mu + \alpha_i + \beta_j + \gamma_{ij}$, as introduced in Sect. 5.4.

In SAS Example E7 (see program in Fig. 5.45), `proc glm` is used for the purpose of subdivision of the `Source of Protein` and the `Level x Source` sums of squares into single degree of freedom sums of square corresponding to the preplanned comparisons of the protein means discussed earlier. These single degree of freedom sums of squares are obtained using the `contrast` statement in the `proc glm` step.

Note that data arranged in this format may be handled conveniently by the use of `input`, `do`, and `output` statements as illustrated in the program.

In fact, this structure of the input data is preferable to the more traditional method where a line of input data corresponds to each cell value.

The right-hand side of the model statement

```
model Weight = Level Source Level*Source;
```

codes the effects model formulation for μ_{ij} given earlier with the terms `Level` and `Source` representing the two classificatory factors "Level of Protein" and "Source of Protein," respectively, and the term `Level*Source` representing the interaction effect.

It is important to note that because of the use of the "effects model" parameters in the specification of the single degree of freedom hypotheses in the above program, it is necessary to ensure that the same ordering of the subscripts for the levels of the factors present in the data be maintained in the program, so that the contrast coefficients (1, –2, 1, etc.) used in defining the comparisons of interest may be used to specify those hypotheses.

This is accomplished by continuing to use the "Level of Protein" as the first factor and the "Source of Protein" as the second factor within the program. This is specified by the order of appearance of these effects in the `class` statement. In the SAS program (see Fig. 5.45), `Level` appears before `Source` in the `class` statement. With this `class` statement in effect, the `Level*Source` interaction contrast coefficients are ordered so that the second subscript corresponds to the levels of `Source` and changes faster than the first subscript, which corresponds to the levels of the factor `Level`. Otherwise, the coefficients as given in the previous section may define different comparisons among the means. Using the "effects model," it is seen that

$$\begin{aligned}
&\bar{\mu}_{.1} - 2\bar{\mu}_{.2} + \bar{\mu}_{.3} \\
&= \frac{1}{2}\left\{\sum_i (\mu + \alpha_i + \beta_1 + \gamma_{i1})\right\} - \left\{\sum_i (\mu + \alpha_i + \beta_2 + \gamma_{i2})\right\} \\
&\quad + \frac{1}{2}\left\{\sum_i (\mu + \alpha_i + \beta_3 + \gamma_{i3})\right\} \\
&= \beta_1 - 2\beta_2 + \beta_3 + \bar{\gamma}_{.1} - 2\bar{\gamma}_{.2} + \bar{\gamma}_{.3}
\end{aligned}$$

Thus, it is clear that a comparison that can simply be specified as $\bar{\mu}_{.1} - 2\bar{\mu}_{.2} + \bar{\mu}_{.3}$ using the cell means involves both the main effect and interaction parameters, in terms of the effects parameters. Fortunately, for making a main effects comparisons using `proc glm`, when the sample sizes are equal, as in this example, one needs to specify only the main effect portion of the contrast. This is because `proc glm` completes the rest of the specification as a convenience for the user. In practice, most users are not aware of this occurrence. Because of this behavior, the statements

```
contrast 'animal vs. vegetable' Source 1 -2 1
                            Level*Source .5 -1 .5 .5  -1 .5 ;
```

```
contrast 'animal vs. vegetable' Source 1 -2 1 ;
```

will produce identical results. However, when a comparison involves the cell means (as opposed to marginal means), such as the comparison of `Beef vs. Pork x Level of Protein`, the `Level*Source` portion of the coefficients is important. Using the "effects model," it may be verified that

$$\begin{aligned}
&\mu_{11} - 2\mu_{12} + \mu_{13} - \mu_{21} + 2\mu_{22} - \mu_{23} \\
&\quad = (\mu + \alpha_1 + \beta_1 + \gamma_{11}) - 2(\mu + \alpha_1 + \beta_2 + \gamma_{12}) + (\mu + \alpha_1 + \beta_3 + \gamma_{13}) \\
&\qquad -(\mu + \alpha_2 + \beta_1 + \gamma_{21}) - 2(\mu + \alpha_2 + \beta_2 + \gamma_{22}) + (\mu + \alpha_2 + \beta_3 + \gamma_{23}) \\
&\quad = \gamma_{11} - 2\gamma_{12} + \gamma_{13} + \gamma_{21} - 2\gamma_{22} + \gamma_{23}
\end{aligned}$$

is a contrast among the interaction parameters and does not involve any main effect parameters.

The following analysis of variance table is constructed from the SAS output shown in Fig. 5.46. A recommended format for the ANOVA table is

SV	DF	SS	MS	F	p-Value
Treatments	5	4612.93	922.59	4.30	0.0023
Level	1	3168.27	3168.27	14.77	0.0003
Animal versus Vegetable	1	264.03	264.03	1.23	0.2722
Beef versus Pork	1	2.50	2.50	0.01	0.9144
(Animal versus Vegetable)× Level	1	1178.13	1178.13	5.49	0.0228
(Beef versus Pork) × Level	1	0.00	0.00	0.00	1.0000
Error	54	11,586.00			
Total	59	16,198.93			

Of the five comparisons tested, two are significant at $\alpha = 0.05$. There appears to be no difference in the mean gains between beef and pork at either levels of protein as well as no difference between the animal and vegetable sources on the average. However, the interaction of animal versus vegetable comparison with level of protein is significant. This indicates that the animal versus vegetable simple effect is different at one level of protein compared to the other. The addition of the estimate statements

```
estimate 'an. vs. veg. at low'
                    Source 1 -2 1 Level*Source 1 -2 1 0 0 0;
estimate 'an. vs. veg. at high'
                    Source 1 -2 1 Level*Source 0 0 0 1 -2 1;
```

results in the t-tests shown in Fig. 5.47 for testing the animal versus vegetable simple effect at each level of protein: This clearly shows that the animal versus vegetable simple effect is only significant at the high level of protein.

If preplanned comparisons were not available and interaction turns out to be significant in a two-way classification, one option available is to compare

The SAS System

The GLM Procedure

Dependent Variable: Weight

Source	DF	Sum of Squares	Mean Square	F Value	Pr > F
Model	5	4612.93333	922.58667	4.30	0.0023
Error	54	11586.00000	214.55556		
Corrected Total	59	16198.93333			

R-Square	Coeff Var	Root MSE	Weight Mean
0.284768	16.67039	14.64772	87.86667

Source	DF	Type I SS	Mean Square	F Value	Pr > F
Level	1	3168.266667	3168.266667	14.77	0.0003
Source	2	266.533333	133.266667	0.62	0.5411
Level*Source	2	1178.133333	589.066667	2.75	0.0732

Source	DF	Type III SS	Mean Square	F Value	Pr > F
Level	1	3168.266667	3168.266667	14.77	0.0003
Source	2	266.533333	133.266667	0.62	0.5411
Level*Source	2	1178.133333	589.066667	2.75	0.0732

Contrast	DF	Contrast SS	Mean Square	F Value	Pr > F
animal vs. vegetable	1	264.033333	264.033333	1.23	0.2722
beef vs. pork	1	2.500000	2.500000	0.01	0.9144
an. vs. veg. by Level	1	1178.133333	1178.133333	5.49	0.0228
an. vs. veg. by Level	1	0.000000	0.000000	0.00	1.0000

Fig. 5.46. SAS Example E7: output

means at each level of one factor. For example, the three protein source means may be compared at each level of protein. The `slice=` option available with the `lsmeans` statement produces F-tests for these hypotheses. In the above example, F-tests for testing the hypotheses $H_0 : \mu_{11} = \mu_{12} = \mu_{13}$ and $H_0 : \mu_{21} = \mu_{22} = \mu_{23}$ are produced by the statement

```
lsmeans Level*Source/slice=Level;
```

Parameter	Estimate	Standard Error	t Value	Pr > |t|
an. vs. veg. at low	27.7000000	11.3460713	2.44	0.0179
an. vs. veg. at high	-9.9000000	11.3460713	-0.87	0.3868

Fig. 5.47. SAS Example E7: animal versus vegetable at levels of protein

and are shown in Fig. 5.48.

One difference in the `lsmeans` statement from the `means` statement is that interaction effects can also be specified as arguments. For example, the statement

```
lsmeans Level*Source/cl pdiff adjust=tukey;
```

produces tests and confidence intervals for all pairwise differences of the form $\mu_{ij} - \mu_{i'j'}$. The output from this statement is not reproduced here. One caution is that output from such statements usually is quite extensive because of the large number of pairs of differences among the means, even between factors with smaller number of levels such as in the above example.

Level*Source Effect Sliced by Level for Weight					
Level	DF	Sum of Squares	Mean Square	F Value	Pr > F
1	2	1280.066667	640.033333	2.98	0.0590
2	2	164.600000	82.300000	0.38	0.6833

Fig. 5.48. SAS Example E7: `slice=` option in the `lsmeans` statement

5.6 Two-Way Factorial: Unequal Sample Sizes

In this section a detailed analysis of a two-way classification with unequal sample sizes is presented. The intent is to show the complexities that arise in the analysis because of the unbalanced data structure compared to the analysis of the complete data set used in the previous analysis of the two-way classification data as discussed in Sects. 5.4 and 5.5. The model used for the purpose of the analysis in those sections is

$$y_{ijk} = \mu_{ij} + \epsilon_{ijk}$$

called the "means model." This was extended to an "effects model"

$$\mu_{ij} = \mu + \alpha_i + \beta_j + \gamma_{ij}$$

by partitioning each of the cell means μ_{ij}, as discussed in those sections. In this section, `proc glm` is used for the analysis of the same model and includes several SAS procedure information statements available with `proc glm` for illustrating their use. Some of these statements were used in Sect. 5.5 for the analysis of interaction comparisons. Although it is instructive to learn the syntax of these statements, an attempt is also made to illustrate typical examples in which these statements may provide useful information to the experimenter.

SAS Example E8

The data set used in SAS Example E6 is modified by deleting some data values to introduce unequal sample sizes while ensuring that there are no missing cells. The data used are as shown in Table 5.8.

Table 5.8. Survival times data: unequal sample sizes

Poison	Drug			
	A	B	C	D
I	0.31	0.82	0.43	0.71
	0.45	1.10	0.45	0.66
	0.46		0.63	0.62
	0.43		0.76	
II	0.36	0.61	0.44	0.56
	0.40	0.49	0.35	0.71
			0.31	0.38
			0.40	
III	0.22	0.30	0.23	0.30
	0.21	0.37	0.25	0.36
	0.18	0.38	0.22	
	0.23			

The following "effects model" is used to specify the model for analysis by `proc glm`,

$$y_{ijk} = \mu + \alpha_i + \beta_j + \gamma_{ij} + \epsilon_{ijk}$$

with $i = 1, 2, 3$ (Poison levels), $j = 1, 2, 3, 4$ (Drug levels), $k = 1,\ldots,n_{ij}$ (cell sample sizes), and $N = \sum_i \sum_j n_{ij}$ is the total number of observations. Thus, 20 $(= 1+3+4+12)$ parameters are utilized to model the mean μ_{ij} as a linear function of parameters representing main effects and interaction. However, only 12 $(= 1 + 2 + 3 + 6)$ degrees of freedom are available to be partitioned from the treatment sum of squares for this purpose. Hence, the model is said to be overspecified (or overparameterized); that is, since more parameters are being used than the available degrees of freedom, 8 out of the 20 parameters cannot be estimated. Thus, the normal equations to be solved to obtain the least squares estimates of the 20 parameters will not have a unique solution. To obtain a solution to the normal equations, restrictions (or constraints) on the parameters must be imposed. These may take the form of setting some of the parameters equal to functions of others or simply equating some of them to a baseline value such as zero. In this example, `proc glm` uses the following eight restrictions:

$$\alpha_3 = \beta_4 = \gamma_{14} = \gamma_{24} = \gamma_{31} = \gamma_{32} = \gamma_{33} = \gamma_{34} = 0$$

This implies that the normal equations may be solved to obtain "estimates" of the rest of the parameters, and "estimates" of the above parameters will be set to the baseline value of zero. Obviously, this is not a unique solution to the normal equations, as other solutions may be obtained using different sets of restrictions.

In the SAS Example E8 (program shown in Fig. 5.49), the `model` statement in `proc glm` includes options `ss1, ss2, ss3,` and `ss4`, requesting that all four types of sums of squares computed by `proc glm` be output. In two-way classification models, the Types II, III, and IV sums of squares are all exactly the same in magnitude when the sample sizes are equal, but Type I sums of squares differ from those. When the sample sizes are unequal (with no completely empty cells), Type II sums of squares differ from Types III and IV sums of squares (which are still of the same magnitude). This is observed from part of the output from this program displayed in Fig. 5.50.

Type I sums of squares that correspond to each effect in the model are computed by fitting each term in the order listed in the `model` statement sequentially and calculating the increase in the treatment sum of squares. Thus, Type I sums of squares are appropriate for testing the significance of adding an effect sequentially into the current model. Type III sum of squares, on the other hand, measures the increase in the treatment sum of squares when the effect is added to a model with all other effects already in the model. Thus, Type I sums of squares are appropriate for testing the significance of an effect in the full model. It is recommended that Type III sums of squares be used to construct an analysis of variance table because these correspond to those calculated in Yates' weighted squares of means analysis. That method is recommended for use for unbalanced data when main effects need to be tested in the presence of interaction. In addition, the `solution` option requests that a solution to the normal equations be produced. As expected, parameter estimates corresponding to those eight parameters discussed earlier are equal to zero as printed in the output resulting from this option shown in Fig. 5.51.

The estimates of other parameters shown in this output are not unique (these are flagged with the letter "B" to indicate this); this means that there is no direct interpretation of the magnitudes of these estimates. However, these values may be used to construct estimates of specific *estimable functions of the parameters* that are useful for interpreting the results of the experiment. Some of these estimates could be obtained directly by using `lsmeans`, `contrast`, and `estimate` statements. Results obtained from some of the these statements included in the SAS program are discussed next.

The `means` statement provides the "unadjusted" Poison and Drug means, whereas the `lsmeans` statement produces the "least squares means" or the adjusted means together with their standard errors. The unadjusted means are shown in Fig. 5.52. These quantities estimate functions of μ_{ij}. For example, consider the estimates printed for `Poison 1`. The unadjusted mean 0.6023 with sample size 13 is an estimate of the *weighted* marginal mean $(4\,\mu_{11} + 2\,\mu_{12} + 4\,\mu_{13} + 3\,\mu_{14})/(4+2+4+3)$, a "weighted average of cell means." As

```
data mice2;
input  Poison 1. @;
  do Drug=1 to 4;
      input Time 3.2 @;
      output;
  end;
datalines;
1 31 82 43  .
1 45110 45 71
1 46  . 63 66
1 43  . 76 62
2 36  . 44 56
2  . 61 35  .
2 40 49 31 71
2  .  . 40 38
3 22 30 23 30
3 21 37 25 36
3 18 38  .  .
3 23  . 22  .
;

proc glm data=mice2;
  class Poison Drug;
  model Time = Poison Drug Poison*Drug/ss1 ss2 ss3 ss4 solution;

  means Poison Drug;
  lsmeans Poison/stderr cl pdiff tdiff adjust=tukey;
  lsmeans Drug /stderr cl pdiff tdiff adjust=tukey;

  contrast 'Poison 1 vs 2' Poison -1 1;
  contrast 'Poison 1 vs 2 *' Poison -1 1
                Poison*Drug  -.25 -.25 -.25 -.25 .25 .25 .25 .25;
  contrast 'Drug A&B vs C&D' Drug 1 1 -1 -1;

  estimate 'Poison 1 vs 2' Poison  -1 1;
  estimate 'Poison 1 mean' intercept 1
                    Poison 1 0 0
                    Drug .25 .25 .25 .25
                    Poison*Drug .25 .25 .25 .25;
  estimate 'Drug A @ Poison 1-2' Poison -1 1 Poison*Drug -1 0 0 0 1;

  title 'Analysis of  Twoway Data : Unequal Sample Sizes';
run;
```

Fig. 5.49. SAS Example E8: program

Source	DF	Type I SS	Mean Square	F Value	Pr > F
Poison	2	0.68676873	0.34338436	39.02	<.0001
Drug	3	0.39430953	0.13143651	14.94	<.0001
Poison*Drug	6	0.14850230	0.02475038	2.81	0.0323

Source	DF	Type II SS	Mean Square	F Value	Pr > F
Poison	2	0.72205463	0.36102731	41.03	<.0001
Drug	3	0.39430953	0.13143651	14.94	<.0001
Poison*Drug	6	0.14850230	0.02475038	2.81	0.0323

Source	DF	Type III SS	Mean Square	F Value	Pr > F
Poison	2	0.79692581	0.39846290	45.28	<.0001
Drug	3	0.38425215	0.12808405	14.56	<.0001
Poison*Drug	6	0.14850230	0.02475038	2.81	0.0323

Source	DF	Type IV SS	Mean Square	F Value	Pr > F
Poison	2	0.79692581	0.39846290	45.28	<.0001
Drug	3	0.38425215	0.12808405	14.56	<.0001
Poison*Drug	6	0.14850230	0.02475038	2.81	0.0323

Fig. 5.50. SAS Example E8: Types I, II, III, and IV sums of squares

can be observed, this definition makes the marginal mean depend on the cell sample sizes. These are computed by the `means` statement and given on the SAS output in Fig. 5.52. The adjusted mean 0.6508 (extracted from part of Fig. 5.53 not shown), however, is an estimate of $\bar{\mu}_{1.}(=(\mu_{11}+\mu_{12}+\mu_{13}+\mu_{14})/4)$, an "unweighted average of cell means," defined assuming cell sample sizes are all equal as specified in the model definition. These are called *least squares means* and are estimates of cell means and marginal means defined using the population means irrespective of the sample sizes. Part of the results from the `lsmeans` statement are reproduced in Fig. 5.53.

Perhaps estimates provided by the `lsmeans` statement are more appropriate if the interest is in the differences in Poison means of the treatment populations as defined by the model. This is justified from the point of view that for the purpose of comparing factor means, all cell means (treatment means) should be regarded as equally important and, therefore, equally weighted, regardless of the sample sizes.

The best estimates of the cell means μ_{ij} and the marginal means $\bar{\mu}_{i.}$ and $\bar{\mu}_{.j}$, respectively, are given by

$$\hat{\mu}_{ij} = \bar{y}_{ij.} = \left(\sum_{k} y_{ijk}\right)/n_{ij}$$

Analysis of Twoway Data : Unequal Sample Sizes

The GLM Procedure

Dependent Variable: Time

Parameter		Estimate		Standard Error	t Value	Pr > |t|
Intercept		0.3300000000	B	0.06632988	4.98	<.0001
Poison	1	0.3333333333	B	0.08563150	3.89	0.0007
Poison	2	0.2200000000	B	0.08563150	2.57	0.0168
Poison	3	0.0000000000	B	.	.	.
Drug	1	-.1200000000	B	0.08123718	-1.48	0.1526
Drug	2	0.0200000000	B	0.08563150	0.23	0.8173
Drug	3	-.0966666667	B	0.08563150	-1.13	0.2701
Drug	4	0.0000000000	B	.	.	.
Poison*Drug	1 1	-.1308333333	B	0.10831624	-1.21	0.2389
Poison*Drug	1 2	0.2766666667	B	0.12110124	2.28	0.0315
Poison*Drug	1 3	0.0008333333	B	0.11164982	0.01	0.9941
Poison*Drug	1 4	0.0000000000	B	.	.	.
Poison*Drug	2 1	-.0500000000	B	0.11803488	-0.42	0.6756
Poison*Drug	2 2	-.0200000000	B	0.12110124	-0.17	0.8702
Poison*Drug	2 3	-.0783333333	B	0.11164982	-0.70	0.4897
Poison*Drug	2 4	0.0000000000	B	.	.	.
Poison*Drug	3 1	0.0000000000	B	.	.	.
Poison*Drug	3 2	0.0000000000	B	.	.	.
Poison*Drug	3 3	0.0000000000	B	.	.	.
Poison*Drug	3 4	0.0000000000	B	.	.	.

Note: The X'X matrix has been found to be singular, and a generalized inverse was used to solve the normal equations. Terms whose estimates are followed by the letter 'B' are not uniquely estimable.

Fig. 5.51. SAS Example E8: parameter estimates

$$\hat{\hat{\mu}}_{i.} = \sum_j \left(\sum_k y_{ijk}/n_{ij} \right)/b = \left(\sum_j \bar{y}_{ij.} \right)/b$$

$$\hat{\hat{\mu}}_{.j} = \sum_i \left(\sum_k y_{ijk}/n_{ij} \right)/a = \left(\sum_i \bar{y}_{ij.} \right)/a$$

where the sample sizes n_{ij} are given in the following table:

		Poison			
		1	2	3	4
	1	4	2	4	3
Drug	2	2	2	4	3
	3	4	3	3	2

The standard error of the difference in the estimates of Factor A means at levels i and i' is

$$\text{s.e.}(\hat{\hat{\mu}}_{i.} - \hat{\hat{\mu}}_{i'.}) = \frac{s}{b}\sqrt{\sum_j \left(\frac{1}{n_{ij}} + \frac{1}{n_{i'j}}\right)}$$

and the standard error of the difference in the estimates of pair of Factor B means at levels j and j' is

$$\text{s.e.}(\hat{\hat{\mu}}_{.j} - \hat{\hat{\mu}}_{.j'}) = \frac{s}{a}\sqrt{\sum_i \left(\frac{1}{n_{ij}} + \frac{1}{n_{ij'}}\right)}$$

where an estimate of the error variance $\hat{\sigma}^2 = s^2$, where s^2 is the MSE obtained from the ANOVA table, as earlier. Thus, a $(1 - \alpha)100\%$ confidence interval for the differences in a pair estimates Factor A and B means are, respectively,

$$(\hat{\hat{\mu}}_{i.} - \hat{\hat{\mu}}_{i'.}) \pm t_{\alpha/2,\nu} \cdot \frac{s}{b}\sqrt{\sum_j \left(\frac{1}{n_{ij}} + \frac{1}{n_{i'j}}\right)}$$

and

$$(\hat{\hat{\mu}}_{.j} - \hat{\hat{\mu}}_{.j'}) \pm t_{\alpha/2,\nu} \cdot \frac{s}{a}\sqrt{\sum_i \left(\frac{1}{n_{ij}} + \frac{1}{n_{ij'}}\right)}$$

where $t_{\alpha/2,\nu}$ is the upper $\alpha/2$ percentage point of the t-distribution with ν degrees of freedom, where $\nu = (N - ab)$ is the degrees of freedom for MSE.

Table 5.9. Survival times cell means

Poison	Drug			
	A	B	C	D
I	0.4125	0.9600	0.5675	0.6633
II	0.3800	0.5500	0.3750	0.5500
III	0.2100	0.3500	0.2333	0.3300

From Table 5.6 the least squares means for Poisons 1 and 2 are, respectively, $\hat{\hat{\mu}}_{1.} = (0.4125 + 0.9600 + 0.5675 + 0.6633)/4 = 0.6508$ and $\hat{\hat{\mu}}_{2.} =$

$(0.3800 + 0.5500 + 0.3750 + 0.5500)/4 = 0.4638$. The standard error of the difference $\hat{\bar{\mu}}_{2.} - \hat{\bar{\mu}}_{1.}$ is

$$\frac{s}{b}\sqrt{\sum_j \left(\frac{1}{n_{1j}} + \frac{1}{n_{2j}}\right)} = \sqrt{0.0088 \times \left(\frac{1}{4} + \frac{1}{2} + \frac{1}{2} + \frac{1}{2} + \frac{1}{4} + \frac{1}{4} + \frac{1}{3} + \frac{1}{3}\right)/4}$$
$$= 0.04$$

The SAS output resulting from the `lsmeans` statements contains estimates, tests, and confidence intervals for individual least squares means of `Poison` and `Drug` levels, respectively, and is shown in Figs. 5.53 and 5.54. These also contain estimates, tests, and confidence intervals for *all pairwise differences* of least squares means for the two factors. Moreover, those comparisons incorporate the Tukey adjustment of confidence levels for making multiple comparisons, as requested by using the `lsmeans` statement option `adjust=tukey`. Results of the calculations associated with the difference $\hat{\bar{\mu}}_{2.} - \hat{\bar{\mu}}_{1.}$ illustrated earlier can be checked from this output.

Analysis of Twoway Data : Unequal Sample Sizes

The GLM Procedure

Level of Poison	N	Time Mean	Time Std Dev
1	13	0.60230769	0.21358659
2	11	0.45545455	0.12420657
3	12	0.27083333	0.06894772

Level of Drug	N	Time Mean	Time Std Dev
1	10	0.32500000	0.10865337
2	7	0.58142857	0.28852663
3	11	0.40636364	0.16776607
4	8	0.53750000	0.16671190

Fig. 5.52. SAS Example E8: unadjusted means

The `contrast` statements are similar to those one might use in the balanced case. For example, the first `contrast` statement computes an F-statistic to test the hypothesis

$$H_0 : \bar{\mu}_{2.} - \bar{\mu}_{1.} = 0$$

and the second to test

$$H_0 : (\bar{\mu}_{.1} + \bar{\mu}_{.2})/2 = (\bar{\mu}_{.3} + \bar{\mu}_{.4})/2$$

These functions are *estimable* because they are linear functions of the means μ_{ij}; however, they must first be expressed in terms of effects model parameters α_i's (Poison effects), β_j's (Drug effects), and γ_{ij}'s (interaction effects), so that they could be specified in `proc glm` statements. As discussed in Sect. 5.5, both main effect and interaction parameters are needed to express the contrast for the mean comparison $\bar{\mu}_{2.} - \bar{\mu}_{1.}$. The required expression is obtained by substituting

$$\mu_{ij} = \mu + \alpha_i + \beta_j + \gamma_{ij}$$

in $\bar{\mu}_{2.} - \bar{\mu}_{1.}$ as follows:

$$\begin{aligned}
&(\mu_{21} + \mu_{22} + \mu_{23} + \mu_{24})/4 - (\mu_{11} + \mu_{12} + \mu_{13} + \mu_{14})/4 \\
&\quad = \{(\mu + \alpha_2 + \beta_1 + \gamma_{21}) + (\mu + \alpha_2 + \beta_2 + \gamma_{22}) + (\mu + \alpha_2 + \beta_3 + \gamma_{23}) \\
&\qquad + (\mu + \alpha_2 + \beta_4 + \gamma_{24})\}/4 - \{(\mu + \alpha_1 + \beta_1 + \gamma_{11}) + (\mu + \alpha_1 + \beta_2 + \gamma_{12}) \\
&\qquad + (\mu + \alpha_1 + \beta_3 + \gamma_{13}) + (\mu + \alpha_1 + \beta_4 + \gamma_{14})\}/4 \\
&\quad = (\alpha_2 - \alpha_1) + (\gamma_{21} + \gamma_{22} + \gamma_{23} + \gamma_{24})/4 - (\gamma_{11} + \gamma_{12} + \gamma_{13} + \gamma_{14})/4
\end{aligned}$$

Thus, the contrast statement needed is

```
contrast 'Poison 1 vs 2 *' Poison -1 1

                    Poison*Drug  -.25 -.25 -.25 -.25 .25 .25 .25 .25;
```

However, SAS allows the user to specify the comparison using only the main effect portion (and so completes the interaction portion of the coefficients needed). Thus, the statement

```
contrast 'Poison 1 vs 2' Poison -1 1;
```

produces identical results. The contrast statement for the second comparison, $(\mu_{.1} + \mu_{.2})/2 - (\mu_{.3} + \mu_{.4})/2$, that is equivalent to the comparison $\mu_{.1} + \mu_{.2} - \mu_{.3} - \mu_{.4})$ is

```
contrast 'Drug A&B vs C&D' Drug 1 1 -1 -1;
```

The two `estimate` statements request estimates of the functions of the cell means μ_{ij} : $(\mu_{11} + \mu_{12} + \mu_{13} + \mu_{14})/4$ and $\mu_{21} - \mu_{11}$. The first `estimate` statement is used to directly compute the Poison 1 least squares mean, in order to illustrate the differences between those resulting from the `means` and `lsmeans` statements. The second is simply the difference in cell means between Poison levels II and I at Drug level A. Such differences may be of interest if the interaction is found to be significant.

As before, these are expressed as functions of effects model parameters:

Analysis of Twoway Data : Unequal Sample Sizes

The GLM Procedure
Least Squares Means
Adjustment for Multiple Comparisons: Tukey-Kramer

Least Squares Means for Effect Poison
t for H0: LSMean(i)=LSMean(j) / Pr > |t|

Dependent Variable: Time

i/j	1	2	3
1		4.67119 0.0003	9.514176 <.0001
2	-4.67119 0.0003		4.503275 0.0004
3	-9.51418 <.0001	-4.50327 0.0004	

Least Squares Means for Effect Poison

i	j	Difference Between Means	Simultaneous 95% Confidence Limits for LSMean(i)-LSMean(j)	
1	2	0.187083	0.087066	0.287101
1	3	0.370000	0.272882	0.467118
2	3	0.182917	0.081480	0.284353

Fig. 5.53. SAS Example E8: Poison comparisons

(i) First,

$$\begin{aligned}
&(\mu_{11} + \mu_{12} + \mu_{13} + \mu_{14})/4 \\
&= \mu + \alpha_1 + (\beta_1 + \beta_2 + \beta_3 + \beta_4)/4 + (\gamma_{11} + \gamma_{12} + \gamma_{13} + \gamma_{14})/4 \\
&= \mu + \alpha_1 + \bar{\beta}_. + \bar{\gamma}_{1.}
\end{aligned}$$

The term μ in the above expression requires "intercept 1" to be included in the `estimate` statement, and the term α_1 is coded as "Poison 1 0 0" for representing level 1 of the Poison effect. The term $(\beta_1+\beta_2+\beta_3+\beta_4)/4$ is coded as "Drug .25 .25 .25 .25," and $(\gamma_{11} + \gamma_{12} + \gamma_{13} + \gamma_{14})/4$ is coded as "Poison*Drug .25 .25 .25 .25 0 0 0 0 0 0 0 0." Thus, the complete statement needed to perform this computation is

```
estimate 'Poison 1 mean'
                intercept 1
                Poison 1 0 0
                Drug .25 .25 .25 .25
                Poison*Drug .25 .25 .25 .25 0 0 0 0 0 0 0 0;
```

Analysis of Twoway Data : Unequal Sample Sizes

The GLM Procedure
Least Squares Means
Adjustment for Multiple Comparisons: Tukey-Kramer

Least Squares Means for Effect Drug
t for H0: LSMean(i)=LSMean(j) / Pr > |t|

Dependent Variable: Time

i/j	1	2	3	4
1		-5.98441 <.0001	-1.3647 0.5326	-3.91691 0.0034
2	5.984413 <.0001		4.954977 0.0003	2.13505 0.1708
3	1.3647 0.5326	-4.95498 0.0003		-2.77024 0.0488
4	3.916906 0.0034	-2.13505 0.1708	2.770245 0.0488	

Least Squares Means for Effect Drug

i	j	Difference Between Means	Simultaneous 95% Confidence Limits for LSMean(i)-LSMean(j)	
1	2	-0.285833	-0.417593	-0.154074
1	3	-0.057778	-0.174570	0.059014
1	4	-0.180278	-0.307244	-0.053311
2	3	0.228056	0.101089	0.355022
2	4	0.105556	-0.030828	0.241939
3	4	-0.122500	-0.244485	-0.000515

Fig. 5.54. SAS Example E8: drug comparisons

The zeros at the end of the estimate statement may be omitted. Of course, the above statement may also be entered in the form

```
estimate 'Poison 1 mean' intercept 4
                         Poison 4 0 0
                         Drug 1 1 1 1
                         Poison*Drug 1 1 1 1/divisor=4;
```

using the `divisor=` option.

(ii) Similarly,

$$\mu_{21} - \mu_{11} = -\alpha_1 + \alpha_2 + \gamma_{21} - \gamma_{11}$$

The term $-\alpha_1 + \alpha_2$ requires that "Poison –1 1 0" be included in the `estimate` statement, and $\gamma_{21} - \gamma_{11}$ requires that "Poison*Drug –1 0 0 0 1 0 0 0 0 0 0 0" be included. Thus, the complete statement is

Analysis of Twoway Data : Unequal Sample Sizes

The GLM Procedure

Dependent Variable: Time

Contrast	DF	Contrast SS	Mean Square	F Value	Pr > F
Poison 1 vs 2	1	0.19200095	0.19200095	21.82	<.0001
Poison 1 vs 2 *	1	0.19200095	0.19200095	21.82	<.0001
Drug A&B vs C&D	1	0.00474103	0.00474103	0.54	0.4700

Parameter	Estimate	Standard Error	t Value	Pr > |t|
Poison 1 vs 2	-0.18708333	0.04005047	-4.67	<.0001
Poison 1 mean	0.65083333	0.02707906	24.03	<.0001
Drug A @ Poison 1-2	-0.03250000	0.08123718	-0.40	0.6926

Fig. 5.55. SAS Example E8: output from contrast and estimate statements

```
'Drug A @ Poison 2 -1' Poison -1 1 Poison*Drug  -1 0 0 0 1;
```

again omitting the trailing zeros.

The results of the `contrast` and `estimate` statements used in SAS Example E8 are found in Fig. 5.55.

5.7 Two-Way Classification: Randomized Complete Block Design

When experimental units tend not to be homogeneous (as usually the case in many studies), the experimenter can often employ a different design than a completely randomized design (CRD) that may be more efficient by helping to control the error variance. One such design is called the *randomized complete block design* (RCBD). Recall that the analysis of data from an experiment carried out in a completely randomized design involves the estimation of the error variance by combining (or pooling) the sample variances calculated from responses from experimental units assigned the same treatment. This is called the *within sample variance* or *within treatment variance.*

How different experimental units respond to the application of the same treatment may depend on the nature of the units. For example, the yields of a variety of cereal crop from plots of land may vary widely if the plots are very different from each other in their soil constitution, moisture content, degree of drainage, exposure to sunlight or shade, or in some other way. The yields obtained from such plots planted with the same variety may lead to an inflated estimate of error variance than if the plots were more homogeneous.

By grouping experimental units that are considered similar, the contribution to experimental error due to this variation can be reduced or eliminated. Groups of units that are similar in this way comprise what are called *blocks*. In an agricultural experiment, these might be plots that are contiguously located in the field. Once blocks are formed, a *complete* set of treatments are applied to the experimental units in a block. Thus, the *block size* (the number of experimental units in a block) must be exactly the same as the number of treatments (i.e., the number of levels of the factor under study). This process is repeated for every block available. The treatments are assigned randomly to the experimental units within each block. The number of blocks is determined by the number of experimental units available. For example, if 4 varieties of corn were under study, the availability of 20 plots would ensure that 5 complete blocks could be formed.

The "blocks' are also called "reps" (for replications) because every treatment is repeated once in each block. A variation is that the complete set of treatments is applied more than once in a block. This would require at least twice the number of experimental units than would be used in a regular RCBD. Another variation is that less than the full set of treatments is used in each block. Such designs are called incomplete block designs and are not discussed in this book. Data from a one-factor experiment performed using a RCBD also form a two-way classification. In the following discussion, the more common practice of each treatment (or combination of treatments, if a factorial arrangement of treatments is used) appearing once in each block is considered.

Model

The model for observations from an RCBD is

$$y_{ij} = \underbrace{\mu + \tau_i}_{\mu_i} + \beta_j + \epsilon_{ij}, \quad i = 1, \ldots, t; \quad j = 1, \ldots, r$$

It is assumed that the random error ϵ_{ij} is distributed as iid $N(0, \sigma^2)$. In the "means model," μ_i is the mean of ith treatment and τ_i is effect of ith treatment. In this presentation as well as in many textbooks, for the purpose of analyzing results from an RCBD, the effect of the jth block, β_j, is considered a fixed effect. This may not be reasonable, as the blocks are selected randomly and thus are better represented in the model by random effects. In Chap. 6, an analysis of an RCBD as a *mixed model* will be presented. Although blocks are considered fixed effects, an additive model is used to represent the observations; that is, an interaction term is not included in the model. Thus, the assumption that treatment differences remain the same from block to block is built into the model. This allows the error variance to be estimated even though treatments are not replicated within each block.

Estimation

The best estimates of μ_i and σ^2 are, respectively,

$$\hat{\mu}_i = \bar{y}_{i.} = \left(\sum_j y_{ij}\right)/r$$

$$\hat{\sigma}^2 = s^2$$

where s^2 is the error mean square from the analysis of variance table presented below. Since the difference in treatment means of two treatments labeled p and q is $\mu_p - \mu_q = \mu + \tau_p - (\mu - \tau_q) = \tau_p - \tau_q$, it is the same as the difference in the corresponding treatment effects. Similarly, a comparison in treatment means is identical to the same comparison in treatment effects.

The best estimate of the difference between the effects of two treatments labeled p and q is

$$\widehat{\mu_p - \mu_q} = \widehat{\tau_p - \tau_q} = \bar{y}_{p.} - \bar{y}_{q.}$$

with standard error given by

$$s_d = \text{s.e.}(\bar{y}_{p.} - \bar{y}_{q.}) = s\sqrt{\frac{2}{r}}$$

A $(1-\alpha)100\%$ confidence interval for $\mu_p - \mu_q$ (or, equivalently, $\tau_p - \tau_q$) is

$$(\bar{y}_{p.} - \bar{y}_{q.}) \pm t_{\alpha/2,\nu} \cdot s\sqrt{\frac{2}{r}}$$

where $t_{\alpha/2,\nu}$ is the upper $\alpha/2$ percentile point of a t-distribution with $\nu = (t-1)(r-1)$ degrees of freedom.

Testing Hypotheses

An analysis of variance (ANOVA) table corresponding to the above model is

SV	df	SS	MS	F	p-Value
Blocks	$r-1$	SS_{Blk}	MS_{Blk}		
Trts	$t-1$	SS_{Trt}	MS_{Trt}	MS_{Trt}/MSE	
Error	$(r-1)(t-1)$	SSE	$MSE(=s^2)$		
Total	$rt-1$	SS_{Tot}			

The above F-statistic tests the hypothesis of equality of treatment means

$$H_0:\ \mu_1 = \mu_2 = \cdots = \mu_t \text{ versus } H_a:\ \text{at least one inequality}$$

or, equivalently, the hypothesis of equality of treatment effects

$$H_0:\ \tau_1 = \tau_2 = \cdots = \tau_t \text{ versus } H_a:\ \text{at least one inequality.}$$

H_0 is rejected if the observed F-value exceeds the α upper percentile of an F-distribution with $\text{df}_1 = t - 1$ and $\text{df}_2 = (r - 1)(t - 1)$ or, equivalently, if the computed p-value is less than α, where level α controls the Type I error of the test and is selected by the experimenter prior to the experiment. For a treatment mean $\bar{y}_{i.}$, the model can be used to show that

$$\begin{aligned}\bar{y}_{i.} &= \sum_{j=1}^{b}(\mu + \tau_i + \beta_j + \epsilon_{ij})/b \\ &= \mu + \tau_i + \bar{\beta} + \bar{\epsilon}_{i.}\end{aligned}$$

Thus, the difference $\bar{y}_{p.} - \bar{y}_{q.}$ has the form

$$\bar{y}_{p.} - \bar{y}_{q.} = (\tau_p - \tau_q) + (\bar{\epsilon}_{p.} - \bar{\epsilon}_{q.})$$

Thus, the expected value of $\bar{y}_{p.} - \bar{y}_{q.}$ is

$$E(\bar{y}_{p.} - \bar{y}_{q.}) = \tau_p - \tau_q.$$

It is important to note that the mean difference estimates the difference in treatment effects only since the effect of blocks cancels out. These differences are called "within block" comparisons because they are averages of treatment differences *within each block* (i.e., $\bar{y}_{p.} - \bar{y}_{q.} = \sum_j (y_{pj} - y_{qj})/r$). Thus, tests and confidence intervals for differences in treatment means $\mu_p - \mu_q$ may be constructed using these sample mean differences. For example, the null hypothesis H_0 is rejected if $t_c > t_{\alpha/2,\nu}$, where

$$t_c = \frac{|\bar{y}_{p\cdot} - \bar{y}_{q\cdot}|}{s_d}$$

$t_{\alpha/2,\nu}$ is the upper $\alpha/2$ percentile of the t-distribution with $\nu = N - t$ degrees of freedom. Equivalently, a difference τ_p and τ_q is declared significantly different at the α level if

$$|\bar{y}_{p.} - \bar{y}_{q.}| > \text{LSD}_\alpha$$

where $\text{LSD}_\alpha = t_{\alpha/2,\nu} \cdot s_d$. This is used to perform tests of all pairwise differences of treatment means (or treatment effects). Hypotheses of the type $H_0 : \tau_1 - (\tau_2 + \tau_3 + \tau_4 + \tau_5)/4 = 0$ or, equivalently, $H_0 : 4\tau_1 - \tau_2 - \tau_3 - \tau_4 - \tau_5 = 0$ may be tested using contrasts of the τ's. These are single df preplanned comparisons discussed in detail in Sect. 5.2.

5.7.1 Using PROC GLM to Analyze a RCBD

SAS Example E9

It is standard agronomic practice to treat seeds chemically to increase germination rate prior to planting. In SAS Example E10, an experiment described in Snedecor and Cochran (1989) in which four seed treatments are compared with a control of no treatment (labeled as "Check" in the data table below) on soybeans is considered. The responses are number of plants that failed to emerge out of 100 seeds planted in each plot. The set of five treatments is replicated five times, each replication representing a block. The data are shown in Table 5.10. The model for the response from the ith treatment in the jth block, y_{ij}, is

$$y_{ij} = \mu + \tau_i + \beta_j + \epsilon_{ij}, \quad i = 1, \ldots, 5; \quad j = 1, \ldots, 5$$

where τ_i is the effect of ith treatment, β_j is the effect of jth block, and the random error ϵ_{ij} is distributed as iid $N(0, \sigma^2)$.

Table 5.10. Effect of seed treatments on germination of soybeans (Snedecor and Cochran 1989)

Treatment	Replication				
	1	2	3	4	5
Check	8	10	12	13	11
Arasan	2	6	7	11	5
Spergon	4	10	9	8	10
Samesan	3	5	9	10	6
Fermate	9	7	5	5	3

```
data soybean;
input Trt $ @;
do Rep=1 to 5;
   input Yield @;
   output;
end;
datalines;
check 8 10 12 13 11
arasan 2 6 7 11 5
spergon 4 10 9 8 10
samesan 3 5 9 10 6
fermate 9 7 5 5 3
;

proc print data=soybean;
run;

proc glm order=data plots=none;
   class Trt Rep;
   model Yield = Trt Rep;
   means Trt/lsd cldiff alpha=.05;
   contrast 'Check vs Chemicals' Trt 4 -1 -1 -1 -1;
run;
```

Fig. 5.56. SAS Example E9: analysis of seed treatments

In the SAS program displayed in Fig. 5.56, an `input` statement with a `trailing @` combined with an `iterative do` loop is used to input the data. The trailing @ "holds" the data line after accessing a character value for the variable `Trt`. SAS then reads the next numeric value for the variable `Yield` from the data line. Each time through the loop, the `output` statement causes an observation to be written containing the current values for `Trt`, `Rep`, and `Yield` to the data set `soybean` and then returns to read another value for `Yield`. This process is repeated five times for each data line. The first five observations in the SAS data set are thus

The SAS System

The GLM Procedure

Class Level Information		
Class	**Levels**	**Values**
Trt	5	check arasan spergon samesan fermate
Rep	5	1 2 3 4 5

Number of Observations Read	25
Number of Observations Used	25

Dependent Variable: Yield

Source	DF	Sum of Squares	Mean Square	F Value	Pr > F
Model	8	133.6800000	16.7100000	3.09	0.0262
Error	16	86.5600000	5.4100000		
Corrected Total	24	220.2400000			

R-Square	Coeff Var	Root MSE	Yield Mean
0.606974	30.93006	2.325941	7.520000

Source	DF	Type I SS	Mean Square	F Value	Pr > F
Trt	4	83.84000000	20.96000000	3.87	0.0219
Rep	4	49.84000000	12.46000000	2.30	0.1032

Source	DF	Type III SS	Mean Square	F Value	Pr > F
Trt	4	83.84000000	20.96000000	3.87	0.0219
Rep	4	49.84000000	12.46000000	2.30	0.1032

Fig. 5.57. SAS Example E9: output, analysis of variance

Obs	Trt	Rep	Yield
1	check	1	8
2	check	2	10
3	check	3	12
4	check	4	13
5	check	5	11

The `class` statement specifies the classification variables that will appear in the `model` statement. Ordinarily, when `proc ANOVA` or `proc glm` processes data sets with classification variables, levels of variables listed in a class statement are lexically ordered prior to the processing of the data by the procedure. For levels that are numeric, the values are usually in the increasing order and thus will not affect the analysis. However, when the values of levels are character type, such as the names of chemicals or "check" as values of the variable `Trt` in this example, the ordering might be affected. To determine the ordering of levels used in the procedure, the user must check part of the SAS output page titled `class level information`. In Fig. 5.57, the levels shown are `check`, `arasan`, `spergon`, `samesan`, and `fermate`, in that order. Thus, it is seen that `proc glm` retained the same ordering found in the data set for these levels. This is a result of inserting the option `order=data` in the `proc glm` statement. If, however, this option is omitted, the `proc glm` would have reordered the levels to be `arasan`, `check`, `fermate`, `samesan`, and `spergon`; as can be observed, the levels are in increasing alphabetical order.

Knowledge of the actual ordering of the levels used in the procedure prior to executing the program is important because the user needs to specify, for example, coefficients of contrasts in the correct order. For example, the coefficients of the contrast for the comparison of "Check versus Chemicals" specified as 4 –1 –1 –1 –1 with option `order=data` present (see Fig. 5.56) would have to be changed to –1 4 –1 –1 –1 if this option is omitted.

The analysis of variance table for the seed treatment data constructed using information on the SAS output shown in Fig. 5.57 under Type III SS is

SV	df	SS	MS	F	p-Value
Treatment	4	83.84	20.96	3.87	0.0219
Rep	4	49.84	12.46	2.30	0.1032
Error	16	86.56	5.41		
Total	24	220.24			

The hypothesis of no difference among the five seed treatments (the check and four chemicals) $H_0 : \tau_1 = \tau_2 = \cdots = \tau_5$ is rejected in favor of $H_a :$ at least one inequality at $\alpha = 0.05$ since the p-value is smaller. Since the blocks are considered fixed effects, there is a comparable test for block effects. Rather than performing a standard test for differences in block effects (which is not a meaningful hypothesis considering that the labeling of the blocks is done randomly), some practitioners use the F-statistic as a nominal measure of whether the mean square for blocks is inflated compared to the error mean square. This may indicate whether using a blocked experiment was more efficient compared to a completely randomized design. There are other efficiency measures that may be calculated based on the mean squares from the ANOVA table. It now remains to determine the effects of the seed treatments that are actually different.

The `lsd` option (or, equivalently, the `t` option) with the `means` statement is used to perform all comparisons of pairwise means using the percentage points of the t-distribution. The concurrent use of the `cldiff` option specifies that these comparisons be given in the form of 95% confidence intervals on pairwise differences of the means. The output is in Fig. 5.58 and shows confidence intervals for 20 pair differences. The relevant set of 95% confidence intervals on the relevant pairwise differences of means extracted from this SAS output table is

$$
\begin{aligned}
\texttt{check - spergon} &: (-0.518,\ 5.718)\\
\texttt{check - samesan} &: (1.082,\ 7.318)\\
\texttt{check - arasan} &: (1.482,\ 7.718)\\
\texttt{check - fermate} &: (1.882,\ 8.118)\\
\texttt{spergon - samesan} &: (-1.518,\ 4.718)\\
\texttt{spergon - arasan} &: (-1.118,\ 5.118)\\
\texttt{spergon - fermate} &: (-0.718,\ 5.518)\\
\texttt{samesan - arasan} &: (-2.718,\ 3.518)\\
\texttt{samesan - fermate} &: (-2.318,\ 3.918)\\
\texttt{arasan - fermate} &: (-2.718,\ 3.518)\\
\texttt{fermate - arasan} &: (-3.518,\ 2.718)
\end{aligned}
$$

Since zero is included in every interval except (check–samesan), (check–arasan), and (check–fermate), the conclusion that might be drawn is that whereas spergon is not found to be different from the control of no seed treatment on germination rate, samesan, arasan, and fermate are found to produce significantly higher germination rates than the control. Also, there appears to be no significant differences among the chemicals on their effect on germination. It must be noted that these are individual comparisons, and no adjustment is made for making multiple comparisons.

The result of the contrast statement also confirmed that the hypothesis $H_0 : 4\tau_1-\tau_2-\tau_3-\tau_4-\tau_5 = 0$ is rejected with a p-value of 0.0028 (see Fig. 5.59), and thus, the average effect of the chemicals on emergence is different from the effect of the control of no chemical treatment. Also, note that if the hypothesis tested is one tailed (e.g., $H_0 : \tau_1 < (\tau_2 - \tau_3 - \tau_4 - \tau_5)/4$ versus $H_a : \tau_1 \geq (\tau_2 - \tau_3 - \tau_4 - \tau_5)/4$), the same set of contrast coefficients may be used. In this instance, the output will be the same, but the p-value for the one-tailed test is one-half of the p-value printed, which is 0.0014. The remainder of the treatment sum of squares after subtracting the sum of squares due to

The GLM Procedure

t Tests (LSD) for Yield

Note: This test controls the Type I comparisonwise error rate, not the experimentwise error rate.

Alpha	0.05
Error Degrees of Freedom	16
Error Mean Square	5.41
Critical Value of t	2.11991
Least Significant Difference	3.1185

Comparisons significant at the 0.05 level are indicated by *.**				
Trt Comparison	**Difference Between Means**	**95% Confidence Limits**		
check - spergon	2.600	-0.518	5.718	
check - samesan	4.200	1.082	7.318	***
check - arasan	4.600	1.482	7.718	***
check - fermate	5.000	1.882	8.118	***
spergon - check	-2.600	-5.718	0.518	
spergon - samesan	1.600	-1.518	4.718	
spergon - arasan	2.000	-1.118	5.118	
spergon - fermate	2.400	-0.718	5.518	
samesan - check	-4.200	-7.318	-1.082	***
samesan - spergon	-1.600	-4.718	1.518	
samesan - arasan	0.400	-2.718	3.518	
samesan - fermate	0.800	-2.318	3.918	
arasan - check	-4.600	-7.718	-1.482	***
arasan - spergon	-2.000	-5.118	1.118	
arasan - samesan	-0.400	-3.518	2.718	
arasan - fermate	0.400	-2.718	3.518	
fermate - check	-5.000	-8.118	-1.882	***
fermate - spergon	-2.400	-5.518	0.718	
fermate - samesan	-0.800	-3.918	2.318	
fermate - arasan	-0.400	-3.518	2.718	

Fig. 5.58. SAS Example E9: pairwise comparisons

this contrast will be the total sum of squares from any set of three orthogonal comparisons one can make among the four chemicals and, thus, will have three degrees of freedom. The compound hypothesis tested by the corresponding

The GLM Procedure

Dependent Variable: Yield

Contrast	DF	Contrast SS	Mean Square	F Value	Pr > F
Check vs Chemicals	1	67.24000000	67.24000000	12.43	0.0028

Fig. 5.59. SAS Example E9: F-test for preplanned comparison

F-statistic is $H_0 : \tau_2 = \tau_3 = \tau_4 = \tau_5$ versus $H_a :$ at least one inequality (i.e., whether there are any differences among the effects of the chemicals). This sum of squares can be obtained by subtraction or directly by including the following contrast statement in the above SAS program (i.e., the SAS program shown in Fig. 5.56):

```
contrast 'Among Chemicals' Trt 0 3 -1 -1 -1,
                           Trt 0 0  2 -1 -1,
                           Trt 0 0  0  1 -1;
```

The results of the partitioning of the treatment sum of squares resulting from the preplanned comparison may be usefully included in the ANOVA table as follows:

SV	df	SS	MS	F	p-Value
Treatment	4	83.84	20.96	3.87	0.0219
Check versus Chemical	1	67.24	67.24	12.43	0.0028
Among Chemicals	3	16.60	5.53	1.02	0.4087
Rep	4	49.84	12.46	2.30	0.1032
Error	16	86.56	5.41		
Total	24	220.24			

It is perhaps instructive to note that the sum of squares for the "Among Chemicals" hypothesis may also be obtained by specifying contrasts that are nonorthogonal but distinct as follows:

```
contrast 'Among Chemicals2' Trt 0 1 -1  0  0,
                            Trt 0 0  1 -1  0,
                            Trt 0 0  0  1 -1;
```

The earlier conclusion of no significant differences among the chemicals on their effect on germination is confirmed from the p-value of 0.4087 above.

5.7.2 Using PROC GLM to Test for Nonadditivity

In Sect. 5.7.1, an additive model is used to analyze data generated from experiments using RCBDs. As discussed there, this is based on the assumption that differences in responses to the treatments are unaffected by the block effects. In many experiments, this assumption may not be plausible because an interaction may occur between the treatment factor and the blocking factor. This is possible in cases in which the responses are not necessarily obtained from experimental units that are grouped together prior to the experiment to form blocks. For example, a block may consist of responses from a complete replicate of experimental trials performed by a person (or a group of persons such as a team), a time unit (such as a day, a week, or a month), a space unit (such as a lab, a location, a growth chamber, a bench, or an oven), and so forth.

In some instances, it may be that the second factor may not be considered a blocking factor (such as when experimental trials are run within an enclosure to control an environmental factor such as temperature or humidity). If independent replications of the levels of the first factor are not obtained, this arrangement would result in a two-way factorial experiment with only a single response observed per cell. In such cases, a nonadditive model may not be appropriate.

Tukey (1949) proposed a test for nonadditivity in nonreplicated two-way classifications. It was shown later that this Tukey's F-statistic is a test of $H_0: \lambda = 0$ in the model

$$y_{ij} = \mu + \tau_i + \beta_j + \lambda\tau_i\beta_j + \epsilon_{ij}, \quad i = 1, \ldots, t; \quad j = 1, \ldots, r$$

where the interaction is modeled as $\gamma_{ij} = \lambda\tau_i\beta_j$, which is a scalar multiple of the product of the main effects τ_i and β_j. This test is called the *single degree of freedom test of nonadditivity* because the sum of squares due to the null hypothesis has one degree of freedom. This test is popular among practitioners

```
insert data step to create the SAS dataset 'soybean' here

proc glm order=data;
   class Trt Rep;
   model Yield = Trt Rep;
   output out=new p=Yhat;
run;

proc glm order=data;
class Trt Rep;
model Yield = Trt Rep Yhat*Yhat/ss1;
title "Tukey's Test for Nonadditivity";
run;
```

Fig. 5.60. SAS Example E10: program for Tukey's test of nonadditivity

for the important reason that if the data display this type of nonadditivity, it is possible to find a power transformation of the response variable that may restore additivity so that one can proceed with further analysis of the treatment effects. Since this test and associated computations are detailed in many textbooks, they will not be repeated here. However, a method for obtaining Tukey's statistic for testing nonadditivity and an associated p-value using SAS is presented.

SAS Example E10

In the SAS program displayed in Fig. 5.60, the first `proc glm` step is a modified version of the program used in Sect. 5.7.1. Here, the predicted values from fitting the additive model to the soybean data are saved as a variable name `yhat` using the `predicted=` (or, equivalently, `p=` as used here) option in an `output` statement.

This results in the creation of new SAS data set (named `new` here), which is the same as the original data set `soybeans`, augmented by the addition of the new variable named `yhat`. This data set is used as the input data set to a second `proc glm` step where the model statement

```
model Yield = Trt Rep Yhat*Yhat/ss1;
```

is used to fit a new model. In this model, the term `Yhat*Yhat` is equivalent to a regression variable (i.e., a covariate). The `ss1` option requests that only

Tukey's Test for Nonadditivity

The GLM Procedure

Dependent Variable: Yield

Source	DF	Sum of Squares	Mean Square	F Value	Pr > F
Model	9	135.1756858	15.0195206	2.65	0.0461
Error	15	85.0643142	5.6709543		
Corrected Total	24	220.2400000			

R-Square	Coeff Var	Root MSE	Yield Mean
0.613765	31.66724	2.381377	7.520000

Source	DF	Type I SS	Mean Square	F Value	Pr > F
Trt	4	83.84000000	20.96000000	3.70	0.0275
Rep	4	49.84000000	12.46000000	2.20	0.1187
Yhat*Yhat	1	1.49568580	1.49568580	0.26	0.6150

Fig. 5.61. SAS Example E10: output from PROC GLM

Type I SS be computed and output. From the output in Fig. 5.61, the Tukey's single degree of freedom sum of squares, the F-statistic for performing Tukey's test of nonadditivity, and the corresponding p-value are the Type I SS values pertaining to the regression term `yhat*yhat` and are observed to be 1.4956, 0.26, and 0.6150, respectively. Thus, the hypothesis of no interaction (i.e.,

$H_0 : \lambda = 0$) is not rejected based on this p-value; hence, an additive model is appropriate for this data.

5.8 Exercises

5.1 An experiment was carried out in a *completely randomized design* to compare the differences in the levels of physiologically active polyunsaturated fatty acids (PAPUFA, in percentages) of six different brands of diet margarine, resulting in the following data (Devore 1982):

Imperial	14.1	13.6	14.4	14.3	
Parkay	12.8	12.5	13.4	13.0	12.3
Blue Bonnet	13.5	13.4	14.1	14.3	
Chiffon	13.2	12.7	12.6	13.9	
Mazola	16.8	17.2	16.4	17.3	18.0
Fleischmann's	18.1	17.2	18.7	18.4	

Prepare and run a SAS program to obtain the output necessary to provide all of the following information. *You must extract numbers from the output and write answers on a separate sheet of paper.*

a. Assuming the *fixed effects one-way classification model* for these data, give estimates of true mean PAPUFA percentages μ_1, μ_2, μ_3, μ_4, μ_5, and μ_6 and the error variance σ^2. Write down the corresponding analysis of variance table including the p-value. State the hypothesis tested by the F-statistic and your decision based on the p-value.
b. Mazola and Fleischmann's are corn based, and the others are soybean based. Use an appropriate `contrast` statement to compute an F-statistic for testing the hypothesis that the average of true mean PAPUFA percentages for corn-based brands is the same as the average for soybean-based brands. Include the corresponding sum of squares and the F-statistic in the above ANOVA table. Based on the p-value, state your conclusions from the experiment in words.
c. Compute Tukey 95% confidence intervals for all pairwise differences in true mean PAPUFA percentages, i.e., $(\mu_r - \mu_s)$s. Explain why these intervals are wider than the t-distribution based 95% confidence intervals (other than the fact that the Tukey percentiles are larger than the corresponding t values).

5.2 Six samples of each of four types of cereal grain grown in a certain region were randomly selected and analyzed to determine the thiamin content (mcg/gm) in an experiment reported in Devore (1982). The data were input to the following SAS program:

```
data ex2;
input cereal $ @;
```

```
do i=1 to 6;
   input thiamin @;
   output;
end;
datalines;
Wheat 5.2 4.5 6.0 6.1 6.7 5.8
Barley 6.5 7.0 6.1 7.5 5.9 5.7
Maize 5.8 4.7 6.4 4.9 6.0 5.2
Oats 8.3 6.7 7.8 7.0 5.9 7.2
;
run ;
proc print;
   title ' Thiamin Content in Cereal Grains';
run;
proc glm;
   class cereal ;
   model thiamin = cereal ;
run;
```

Assuming the one-way fixed effects model

$$y_{ij} = \mu + \tau_i + \epsilon_{ij}, \qquad i = 1, \ldots, 4; \quad j = 1, \ldots, 6$$

where $\mu_i = \mu + \tau_i$ are the population mean thiamin content for each cereal and ϵ_{ij} are iid $N(0, \sigma^2)$ errors, complete the SAS program as needed to answer the following:

a. It is thought that oats are higher in thiamin content than other cereals. Add a `contrast` statement for testing the appropriate comparison to test this hypothesis.
b. Add an option to the `proc` statement for the levels of the `cereal` variable to retain the ordering present in the data.
c. Add an option to the `model` statement to obtain the estimates of the parameters μ, τ_1, τ_2, τ_3, and τ_4.
d. Compute 95% confidence intervals on all pairwise differences (i.e., $(\mu_p - \mu_q)$'s) adjusted for multiple testing using the Bonferroni method.

5.3 The following are clotting times of plasma (in minutes) for four different levels of a drug compared in a *completely randomized design* (modified from a different experiment reported in Armitage and Berry 1994). Blood samples were taken from each of eight subjects randomly assigned to each of the four levels of the drug.

Drug			
0.1%	0.2%	0.3%	0.4%
8.4	9.4	9.8	12.2
12.8	15.2	12.9	14.4
9.6	9.1	11.2	9.8
9.8	8.8	9.9	12.0
8.4	8.2	8.5	8.5
8.6	9.9	9.8	10.9
8.9	9.0	9.2	10.4
7.9	8.1	8.2	10.0

a. Write a SAS data step to read the data (that you enter instream), and create a SAS data set in the form necessary to be used by `proc glm`.
b. Write a complete SAS proc step to obtain the ANOVA table necessary, and include a statement to compute confidence intervals for differences in all pairwise drug means. Report the ANOVA table and the confidence intervals and use them to summarize the results of the experiment.
c. Add a statement to obtain the necessary F-statistic to test whether the average clotting times have an increasing linear trend with the level of the drug. What is your conclusion?

5.4 A marketing consultant conducted an experiment to compare four different package designs for a new breakfast cereal (this problem is described in Kutner et al. (2005); however, the data given are artificial). Twenty-four stores with approximately similar sales volumes were selected, and each store was required to carry only one of the package designs. Thus, each package design was randomly assigned to six stores. Other relevant conditions such as price, amount and location of shelf space, and advertising were kept roughly similar for all stores participating in the experiment. Sales, in number of cases, were observed for the study period. The data are

Package design			
1	2	3	4
12	14	19	24
18	12	17	30
14	13	21	27
15	10	23	28
17	15	16	32
15	12	20	30

Write and execute a SAS program to obtain the output necessary to provide answers to all of the following questions:

a. Construct an analysis of variance using numbers from the SAS output. Test $H_0 : \mu_1 = \mu_2 = \mu_3 = \mu_4$ versus H_a : at least one μ_i is different from the others, using $\alpha = 0.05$. Use the p-value from the SAS output to make a decision.
b. Use the `lsd` procedure to test all possible differences of $H_0 : \mu_i - \mu_j = 0$ versus $H_a : \mu_i - \mu_j \neq 0$ using $\alpha = 0.05$. Report the results using the underscoring display. Summarize your conclusions from this procedure in a sentence or two.
c. Use the `Tukey` procedure to test all possible differences of $H_0 : \mu_i - \mu_j = 0$ versus $H_a : \mu_i - \mu_j \neq 0$ using $\alpha = 0.05$. Report the results using the underscoring display. Point out any conclusions that are different from those made using the `lsd` procedure. Explain why they are different.
d. It would be more useful to test preplanned comparisons than testing pairwise differences among the four packaging designs given the following additional information on the package designs used.

Package design	Design style
1	3 colors, with cartoons
2	3 colors, without cartoons
3	5 colors, with cartoons
4	5 colors, without cartoons

Thus, the experimenter could have planned to

i. compare the the average effect of the three-color designs with the average effect of the five-color designs,
ii. compare the average effect of the designs with cartoons with the average effect of the designs without cartoons,
iii. compare the effect of the three-color design with cartoons with the effect of the three-color design without cartoons, and
iv. compare the the effect of the five-color design with cartoons with the effect of the five-color design without cartoons.

Compute appropriate t-statistics to test the comparisons given above controlling the error rate for each comparison at $\alpha = 0.05$. State your conclusion in each case.

5.5 In an experiment conducted to compare the effects of sleep deprivation on reaction time to onset of light (Kirk 1982), 32 subjects were randomly divided into 4 groups of 8 subjects each. Four levels of sleep deprivation (12, 24, 36, and 48 hours) were randomly assigned to the four groups. The reaction times of the 32 subjects (in hundredths of a second) are tabulated below. Prepare and run a SAS program to obtain the output necessary to provide all of the following information. You must extract numbers from the output and write answers on a separate sheet of paper. Model this program after SAS Example E3.

Duration of sleep deprivation			
12 hours	24 hours	36 hours	48 hours
20	21	25	26
20	20	23	27
17	21	22	24
19	22	23	27
20	20	21	25
19	20	22	28
21	23	22	26
19	19	23	27

a. Obtain a scatter plot of the data with the levels of duration on the horizontal axis. Overlay line segments connecting the averages. What kind of trend does the averages indicate?
b. Construct an ANOVA table and use the F-statistics to test the hypothesis that the population means of reaction times at each sleep deprivation are all equal against the alternative that there is at least one difference, using $\alpha = 0.05$
c. Test the hypothesis that there is no linear trend in the population means using $\alpha = 0.05$
d. Test the hypothesis that the nonlinear components of the trend (i.e., the deviation from linear components) equal zero using $\alpha = 0.05$. Note that this is the same as a test of lack of fit of a linear trend.

5.6 Researchers conducted an experiment to compare the effectiveness of four new weight-reducing agents to that of an existing agent (Ott and Longnecker 2001). The researchers randomly divided a random sample of 50 males into 5 equal groups, with preparation A assigned to the first group, B to the second group, and so on. They then gave a prestudy physical to each person in the experiment and told him how many pounds overweight he was. A comparison of the mean number of pounds overweight for the groups showed no significant differences. The researchers then began the study program, and each group took the prescribed preparation for a fixed period of time. The weight losses recorded at the end of the study period are as follows:

Agent	Weight loss									
$A1$	12.4	10.7	11.9	11.0	12.4	12.3	13.0	12.5	11.2	13.1
$A2$	9.1	11.5	11.3	9.7	13.2	10.7	10.6	11.3	11.1	11.7
$A3$	8.5	11.6	10.2	10.9	9.0	9.6	9.9	11.3	10.5	11.2
$A4$	12.7	13.2	11.8	11.9	12.2	11.2	13.7	11.8	11.5	11.7
S	8.7	9.3	8.2	8.3	9.0	9.4	9.2	12.2	8.5	9.9

The standard agent is labeled agent S, and the four new agents are labeled $A1$, $A2$, $A3$, and $A4$:

$A1$: Drug therapy with exercise and counseling

$A2$: Drug therapy with exercise but no counseling
$A3$: Drug therapy with counseling but no exercise
$A4$: Drug therapy with no exercise and no counseling
Denoting the means for the five treated populations by μ_1, μ_2, μ_3, μ_4, and μ_5, respectively, linear combinations for making comparisons among the agent means that will address the following questions can be constructed:

a. Compare the mean for the standard to the average of the four agent means:
$$(\mu_1 + \mu_2 + \mu_3 + \mu_4)/4 - \mu_5$$
b. Compare the average mean for the agents with counseling to average mean for those without counseling (ignoring the standard):
$$(\mu_1 + \mu_3)/2 - (\mu_2 + \mu_4)/2$$
c. Compare the average mean for the agents with exercise to average mean for those without exercise (ignoring the standard):
$$(\mu_1 + \mu_2)/2 - (\mu_3 + \mu_4)/2$$
d. Compare the mean for the agents with counseling to the standard:
$$(\mu_1 + \mu_3)/2 - \mu_5$$

5.7 An experiment, carried out in a *completely randomized design* (Devore 1982), to compare the effects of five different plate lengths of 4, 6, 8, 10, and 12 inches on axial stiffness of metal plate-connected trusses used for roof support, yielded the following observations on axial stiffness index (kips/in.).

Plate length				
4	6	8	10	12
309.2	402.1	392.4	346.7	407.4
409.5	347.2	366.2	452.9	441.8
311.0	361.0	351.0	461.4	419.9
326.5	404.5	357.1	433.1	410.7
349.8	331.0	409.9	410.6	473.4
309.7	348.9	367.3	384.2	441.2
316.8	381.7	382.0	362.6	465.8

Apart from the overall differences of the effects of the five plate lengths on axial stiffness, the experimenter was interested in determining whether the mean stiffness index depends `linearly` on the actual plate length. This may be done using an appropriate orthogonal polynomial of the means. Prepare and run a SAS program to obtain the output necessary to provide all of the following information. You must extract numbers from the output and write answers on a separate sheet of paper.

a. Assuming the fixed effects one-way classification model for the data, give estimates of μ_1, μ_2, μ_3, μ_4, μ_5, and σ_2.

b. Write down the corresponding analysis of variance table including the p-value. State the hypothesis tested by the F-statistic and your decision based on the p-value.
c. Use a `contrast` statement to compute an F-statistic for testing whether the mean stiffness index is linearly related to the plate length. Include the corresponding sum of squares and the F-statistic in an expanded ANOVA table. Based on the p-values, summarize your conclusions.
d. Include an `estimate` statement for the same comparison as in part (c). If the mean stiffness index is shown to have a straight-line relationship with the plate length, then estimate the slope of this line by the estimated value of the comparison divided by the sum of squares of its coefficients. Interpret this slope as an increase or decrease of axial index per inch of plate length.

5.8 A researcher wants to evaluate the difference in mean film thickness of a coating placed on silicon wafers using three different processes (Ott and Longnecker 2001). Six wafers are randomly assigned to each of the processes. The film thickness and the temperature in the lab during the coating process are recorded for each wafer. The researcher is concerned that fluctuations in temperature may affect the film thickness. The results were analyzed using a one-way covariance model $y_{ij} = \mu_i + \beta(x_{ij} - \bar{x}) + \epsilon_{ij}$ where temperature was used as the covariate. Complete and execute the following SAS program and use the output to answer the questions that follow.

```
data ex8;
input process @;
do i=1 to 6;
  input temperature thickness @@;
  output;
end;
datalines;
1 26 100 35 150 28 106 31  95 29 113 34 144
2 24 118 28 134 29 138 32 147 36 165 35 159
3 37 124 31  95 34 120 27  86 28  98 25  81
;
run;

proc glm data=ex7;
 class ;
 model thickness = ;
 title 'Film thickness adjusted by Temperature';
run;
```

a. Construct an adjusted ANOVA table.

b. Using the above ANOVA table, test the hypothesis of $H_0 : \mu_1 = \mu_2 = \mu_3$ (use the p-value and state decision).
c. Compute 95% confidence intervals for all differences in pairs of means (e.g., $\mu_1 - \mu_2$) adjusted for multiple testing using the Bonferroni method.
d. What does the test of $H_0 : \beta = 0$ tell you? Test this hypothesis using the above adjusted ANOVA table and state your conclusion.
e. Give the sum of squares for the variety effect that is not adjusted for the moisture effect.

5.9 A process engineer is interested in determining if there is a difference in the breaking strength of a monofilament fiber produced using three different machines for a textile company (Montgomery 1991). However, the strength of the fiber is also related to its diameter, with thicker fibers being stronger than thinner ones. Random samples of five fiber specimens each were selected from each machine, and the breaking strength y (in pounds) and the diameter x (in 10^{-3} inches) (to be used as a covariate) are measured.

Machine 1		Machine 2		Machine 3	
y	x	y	x	y	x
36	20	45	22	35	21
41	25	52	28	37	23
39	24	44	22	42	26
42	25	49	30	34	21
48	32	51	28	32	15

Use `proc glm` in SAS and *the one-way covariance model* to analyze this data. *Extract numbers from the SAS output and write your own answers to the following questions*:

a. Write an appropriate model for analyzing these data so that it is possible to perform a test of the equality of the slopes from your analysis. Construct an "adjusted" analysis of variance table to test the hypothesis of equal means (or effects) corresponding to the machines. State your conclusion based on the p-value.
b. Provide estimates of the regression coefficients of the straight lines if it is determined that their slopes are different.
c. Calculate 95% confidence intervals on all differences in pair of means (e.g., $\mu_i - \mu_j$) adjusted at the mean value of x.
d. Include a `proc` step in your SAS program to obtain a scatter plot of y versus x, superimposed by the fitted regression lines.

5.10 The data displayed below are results from an experiment described in Snedecor and Cochran (1989) on the use of drugs in the treatment of leprosy. The drugs were A and D, which were antibiotics, and F is an inert drug used as a control. The dependent variable Y was a score of leprosy bacilli measured on each patient after several months of treatment. The covariate X was a pretreatment score of leprosy bacilli.

Drugs					
A		D		F	
X	Y	X	Y	X	Y
11	6	6	0	16	13
8	0	6	2	13	10
5	2	7	3	11	18
14	8	8	1	9	5
19	11	18	18	21	23
6	4	8	4	16	12
10	13	19	14	12	5
6	1	8	9	12	16
11	8	5	1	7	1
3	0	15	9	12	20

a. Use `proc glm` and the one-way covariance (equal slopes) model to analyze this data. Construct an adjusted ANOVA table.
b. Using the above ANOVA table, test the hypothesis $H_0 : \mu_1 = \mu_2 = \mu_3$ (use the p-value and state decision).
c. Construct 95% confidence intervals for all differences in pairs of means (e.g., $\mu_1 - \mu_2$) adjusted for multiple testing using the Bonferroni method.
d. What does the test of $H_0 : \beta = 0$ tell you? Test this hypothesis using the above adjusted ANOVA table and state your conclusion.
e. Construct an analysis of variance that is not adjusted for the pre-score. What conclusion can you draw from this ANOVA table.

5.11 An experiment conducted to study the friction properties of lubricants is described in Mason et al. (1989). A key constituent of lubricants that is of interest to the researchers is the additive that is mixed with the base lubricant. In order to ascertain whether two competing additives produce a different effect on the friction properties of lubricants, ten mixtures of a base lubricant and each of the additives were made. The mixtures of base lubricant cannot be made sufficiently uniform to ensure that all batches have identical physical properties. Consequently, the plastic viscosity, an important characteristic of the base lubricant that is related to its friction-reducing capability, was measured for each mixture prior to the addition of the additives. This measures the variation among batches due to the base lubricant alone. Analyze the following data to determine whether the additives differ in the mean friction measurements associated with each.

Additive A		Additive B	
Plastic viscosity(x)	Friction measurement(y)	Plastic viscosity(x)	Friction measurement(y)
12	27.1	15	28.6
13	26.6	13	37.1
15	28.9	14	37.9
14	27.1	14	30.6
10	23.6	13	33.6
10	26.4	13	34.9
13	28.1	13	33.1
14	26.1	14	34.4
12	24.4	14	32.6
14	29.1	12	35.6

Use `proc glm` and `the one-way covariance model` to analyze this data. A plot of the data suggests that the slopes of the regression lines for the two additives may be different. The data were analyzed using a one-way covariance model

$$y_{ij} = \alpha_i + \beta_i x_{ij} + \epsilon_{ij}$$

Thus, it is possible to perform a test of the equality of the slopes as a part of your analysis. Extract numbers from the SAS output to write your own answers to the following questions.

a. Construct an "adjusted" analysis of variance table to test the hypothesis $H_0 : \beta_1 = \beta_2$ (or, equivalently, $H_0 : \tau_1 = \tau_2$). Use the p-value to draw a conclusion.
b. If the slopes are found to be unequal, obtain estimates of parameters for the two regressions.
c. Use the `lsmeans` statement in `proc glm` to obtain a confidence interval on the difference in the slopes $\alpha_1 - \alpha_2$ adjusted at the mean plastic viscosity.
d. Use the `lsmeans` statements in `proc glm` to obtain a confidence interval on the difference in the slopes $\alpha_1 - \alpha_2$ adjusted at plastic viscosity values of 12 and 14.

5.12 In an experiment described in Kirk (1982), four methods for teaching arithmetic were being evaluated. Thirty-two students were randomly assigned to four classrooms, each with eight students. An intelligence test was administered to each student at the beginning of the experiment. The resulting scores (x) are used to adjust the arithmetic achievement scores (y) obtained at the conclusion of the experiment for differences in intelligence among the students. The results are recorded as follows:

Method 1		Method 2		Method 3		Method 4	
y	x	y	x	y	x	y	x
3	42	4	47	7	61	7	65
6	57	5	49	8	65	8	74
3	33	4	42	7	64	9	80
3	47	3	41	6	56	8	73
1	32	2	38	5	52	10	85
2	35	3	43	6	58	10	82
2	33	4	48	5	53	9	78
2	39	3	45	6	54	11	89

a. Use `proc glm` and `the one-way covariance (equal slopes) model` to analyze this data. Construct an adjusted ANOVA table.
b. Using the above ANOVA table, test the hypothesis of $H_0 : \mu_1 = \mu_2 = \mu_3 = \mu_4$ (use the p-value and state decision).
c. Construct 95% confidence intervals for all differences in pairs of means (e.g., $\mu_a - \mu_b$) adjusted for multiple testing using the Tukey method.
d. What does the test of $H_0 : \beta = 0$ tell you? Test this hypothesis using the above adjusted ANOVA table and state your conclusion.
e. Construct an analysis of variance that is not adjusted for the intelligence score. What conclusion can you draw from this ANOVA table.

5.13 A medical experiment is run to determine the side effects on children when they take various dosages of a drug administered by different methods. A two-way factorial with four dosages (0.5, 1.0, 1.5, and 2.0 milligrams) and three methods of administering (oral, extended release, intravenous) is used in a completely randomized design, with each treatment combination replicated twice. The response variable is the amount of a certain chemical present in the lever after 24 hours. The data are as follows:

Method	Dosage			
	0.5	1.0	1.5	2.0
1	0.414	0.541	0.592	0.672
	0.312	0.423	0.575	0.610
2	0.537	0.513	0.595	0.709
	0.451	0.580	0.573	0.623
3	0.572	0.622	0.613	0.695
	0.554	0.597	0.650	0.751

Use appropriate SAS procedures to analyze these data. Use the output from program to answer all of the following on a separate sheet:

a. Estimate the cell means μ_{ij} and report these in a two-way table. Obtain a graph using `proc sgplot` of the cell means, with dosage on the x-axis, and using different symbols for each method. Join the points for each method by line segments of different colors and line types.
b. Construct the ANOVA table to test the hypotheses of no interaction and main effects. What are the conclusions from each test? Does the graph in part (a) support your conclusion from the test for interaction. Discuss.
c. Obtain 95% confidence intervals for the pairwise differences in methods means and make an overall conclusion.
d. The hypothesis that the average effects of dosage are linearly related to the level of dosage can be examined by constructing a contrast of the dosage means with appropriate coefficients. Include a `contrast` statement in your program. Partition the dosage sum of squares from the results and determine if there is evidence for such a linear trend.

5.14 A mechanical engineer is comparing the thrust force produced by a drill press under various combinations of drill speed and feed rate. A two-way factorial with four feed rates (0.015, 0.030, 0.045, and 0.060 in./min) and two drill speeds (125 and 200 rpm) is used in a completely randomized design with each treatment combination replicated twice (Montgomery 1991). The results are as follows:

Drill speed	Feed rate			
	0.015	0.030	0.045	0.060
125	2.70	2.45	2.60	2.75
	2.78	2.49	2.72	2.86
200	2.83	2.85	2.86	2.94
	2.86	2.80	2.87	2.88

Use appropriate SAS procedures to analyze these data. Use the output from program to answer all of the following:

a. Use `proc means` to obtain estimates of the cell means μ_{ij}. Construct a two-way table of means showing the Feed Rate and Drill Speed cell means.
b. Obtain a graph using `proc sgplot` of the cell means with Feed Rate on the x-axis. Join the points for each Drill Speed by line segments. (This is the interaction plot of means discussed in the text.)
c. Construct the ANOVA table to test the hypotheses of no interaction and zero main effects. What are the conclusions from each test using $\alpha = 0.05$?
d. Use the graph in part (b) to explain your conclusion from the test for interaction. Comment on the variation in mean response across the levels of Feed Rate at each level of Drill Speed, and use it to explain any significant interaction.

e. Compute a 95% confidence interval for the difference in Drill Speed means. Use this interval to say if these means are significantly different.
f. The hypothesis that the average effects of Feed Rate are linearly related to the levels of Feed Rate can be tested by constructing a contrast of the Feed Rate means with coefficients to be obtained from the table of orthogonal polynomial coefficients. Include a `contrast` statement in your program to do this computation. Insert a line into the ANOVA table to report the results. What do you conclude from this test?

5.15 Ostle (1963) presented an example of an agronomic experiment to assess the effects of date of planting and type of fertilizer on the yield of soybeans. Thirty-two experimental plots were randomly assigned to the treatment combinations so that each combination was replicated four times. The yield data are reported in the following table. Write a SAS program with appropriate procedure steps to obtain the output necessary to answer all of the following questions. Extract or write answers on separate pages and attach any graphs requested.

Date of planting	Fertilizer type	Yields from plots (bushels/acre)			
Early	Check	28.6	36.8	32.7	32.6
	Aero	29.1	29.2	30.6	29.1
	Na	28.4	27.4	26.0	29.3
	K	29.2	28.2	27.7	32.0
Late	Check	30.3	32.3	31.6	30.9
	Aero	32.7	30.8	31.0	33.8
	Na	30.3	32.7	33.0	33.9
	K	32.7	31.7	31.8	29.4

a. Use `proc means` to obtain estimates of the cell means μ_{ij}. Construct a two-way table showing the date of planting by fertilizer type mean yields.
b. Obtain a scatter plot of the cell means using `proc sgplot` with type of fertilizer on the horizontal axis. Join the points for each date of planting by line segments. (This will produce an interaction plot of treatment means discussed in the text.)
c. Construct the ANOVA table to test the hypotheses of no interaction and zero main effects. What are the conclusions from each test using $\alpha = 0.05$?
d. Use the graph in part (b) to explain your conclusion from the test for interaction. Comment on the trends in mean response across the levels of type of fertilizer for each date of planting, and use it to explain significant interaction, if any.

e. The hypothesis comparing the check with the average of the types of fertilizers (a main effect comparison) can be tested by constructing a contrast of the fertilizer means. Use a `contrast` statement in your program to do this computation. Add a line to the ANOVA table to report the results. What do you conclude from this test?

f. Use a contrast statement to test the interaction comparison that the comparison in part (e) is the same at both dates of planting. What is your conclusion and does it support your conclusion for parts (c) and (d)? Add a line to the ANOVA table to report the results.

5.16 The yield (grams per plant) of beetroots grown in pots in response to two crossed factors, wood chips from three different sources and nitrogen at three levels, in a completely randomized design with three replications, is given below (Bliss 1970). The rate of application of the wood chips was 10 tons/acre, and nitrogen levels used were 0, $\frac{1}{2}$, and 1 gram/100 grams of organic matter added as chips. There were nine replications of pots with a control treatment of no chips and, therefore, no additional nitrogen. Use appropriate SAS procedures to analyze the data. Provide answers to the following questions on separate sheets, extracting material from the SAS output as needed.

No chips			Pine chips			Oak-hickory			Aspen-birch		
0	0	0	0	$\frac{1}{2}$	1	0	$\frac{1}{2}$	1	0	$\frac{1}{2}$	1
16.9	15.5	16.8	13.3	20.8	21.2	3.6	10.0	18.3	1.2	6.0	7.8
17.5	20.0	19.7	16.1	16.5	18.4	3.8	10.5	18.2	1.2	3.3	9.7
20.0	19.0	20.5	14.3	14.0	16.3	1.8	9.9	14.2	4.1	5.1	10.5

a. Estimate the cell means μ_{ij} and report these in a two-way table. Obtain a graph using `proc sgplot` of the cell means with nitrogen levels on the x-axis and using different symbols for each type of wood chip. Join the points for each chip type by line segments of different colors and line types.

b. Construct the ANOVA table to test the hypotheses of no interaction and main effects. What are the conclusions from each test? Does the graph in part (a) support your conclusion from the test for interaction. Discuss.

c. The hypothesis that the average beet yields are linearly related to the levels of nitrogen can be examined by constructing a contrast of the dosage means with appropriate coefficients. Include a `contrast` statement in your program. Partition the nitrogen main effect sum of squares into linear and lack-of-fit components, and determine if there is evidence for such a linear trend.

d. Partition the wood chip main effect sum of squares into sums of squares corresponding to the following three orthogonal comparisons: chips versus control, hard versus soft woods, and between hard woods. (Note: Pine is a soft wood, whereas the others are all hard woods.) Include `contrast` statements in your program to obtain F-statistics to test the corresponding hypotheses. Report the results of these tests and use these to make conclusions about the wood chip main effects.

e. Use a contrast statement to extract the interaction sum of squares that corresponds to the interaction comparison of linear $\times$ between hard woods. What hypothesis does this comparison test? What is your conclusion?

5.17 Use the following data set, which is similar to that used in SAS Example E8 (Fig. 5.49) except that different data values are missing.

Poison	Drug			
	A	B	C	D
I	0.31	0.82	.	0.45
	0.45	.	0.45	0.71
	.	0.88	0.63	0.66
	0.43	0.72	.	0.62
II	0.36	0.92	0.44	0.56
	.	.	0.35	.
	0.40	0.49	0.31	0.71
	.	.	0.40	0.38
III	0.22	0.30	0.23	0.30
	.	0.37	0.25	.
	0.18	0.38	0.24	.
	0.23	0.29	0.22	0.33

You are required to add a `proc glm` step to this program to do the computations described below. Use the usual two-way classification model for this analysis.

a. Obtain the least squares estimates of μ, α_i, β_j, and γ_{ij} for $i = 1, 2, 3$ and $j = 1, 2, 3, 4$ computed by `proc glm`.
b. Use `lsmeans` statements to obtain 95% confidence intervals on pairwise differences in Poison and Drug means, adjusted for multiple testing.
c. It can be shown that $\bar{\mu}_{.1} - \bar{\mu}_{.2} = \beta_1 - \beta_2 + \bar{\gamma}_{.1} - \bar{\gamma}_{.2}$. Estimate $\bar{\mu}_{.1} - \bar{\mu}_{.2}$ by substituting estimates of the parameters from part (a) in this expression, using hand computation. Include an `estimate` statement in the `proc glm` step to verify this estimate and to compute its standard error.
d. Use the results of the `estimate` statement to compute a 95% confidence interval for $\bar{\mu}_{.1} - \bar{\mu}_{.2}$.
e. Test $H_0 : \bar{\mu}_{.1} = \bar{\mu}_{.2}$ using an appropriate `contrast` statement.

f. Include a statement to obtain estimates of the cell means μ_{ij} and their standard errors for all i and j.
g. Construct an analysis of variance table to test main effects and interaction for these data, and use the p-values to state conclusions. Use the values output from `proc glm` and write your own complete ANOVA table.

5.18 Kutner et al. (2005) discussed an example in which human growth hormone was administered at a clinical research center to short prepubescent children. The investigator was studying the effects of gender and bone development levels on the rate of growth induced by hormone administration. Three children were randomly selected from each gender-bone development group, but 4 of the 18 children dropped out of the yearlong study. The data are as follows:

Gender	Bone development		
	Severely depressed	Moderately depressed	Mildly depressed
Male	1.4	2.1	0.7
	2.4	1.7	1.1
	2.2		
Female	2.4	2.5	0.5
		1.8	0.9
		2.0	1.3

Use a SAS program with a `proc glm` step to this program to perform the computations described below. Use the usual two-way classification model for this analysis.

a. Use `proc means` to obtain estimates of the cell means μ_{ij}. Construct a two-way table of gender by bone development levels showing the growth rate cell means.
b. Obtain an interaction plot using `proc gplot` of the growth rate cell means with bone development levels on the x-axis. Join the points for each gender by line segments. Use colors and line types to identify gender.
c. Obtain the least squares estimates of μ, α_i, β_j, and γ_{ij} for $i = 1, 2$ and $j = 1, 2, 3$ computed by `proc glm`.
d. Construct an analysis of variance table to test main effects and interaction for these data. Use the values output from `proc glm` and write your own complete ANOVA table. Using the p-values with $\alpha = 0.05$, make conclusions from each test.
e. Use `lsmeans` statements to obtain 90% confidence intervals on pairwise differences in `bone development` means adjusted for multiple testing using the Tukey method. What do you conclude from these intervals?

f. Kutner et al. (2005) tested the hypothesis whether the average growth rate of children with only mildly depressed bone development is significantly larger than zero. Add an `estimate` statement to obtain a t-statistic to test this one-sided hypothesis.

g. Test the hypothesis that the growth rates of children with severely depressed bone development are different in males and females using an appropriate `contrast` statement.

5.19 Four brands of airplane tires are compared to assess the differences in the rate of tread wear. The data were collected on eight planes, with two tires used under each wing. The researcher uses each airplane as a block, mounting four test tires, one of each brand, in random order on each airplane. Thus, a randomized complete block design with "airplane" as a blocking factor is the design used. The amount of tread is measured initially, and after 6 months, the following wear rates were obtained. A larger value indicates greater wear.

Airplane	Brand			
	A	B	C	D
1	4.02	2.46	2.06	3.49
2	4.50	3.39	2.91	3.18
3	2.73	1.69	2.37	1.48
4	3.74	1.95	3.39	3.09
5	3.21	1.20	1.72	2.65
6	2.53	1.04	2.52	1.23
7	3.07	2.55	2.42	2.07
8	3.10	1.09	2.22	2.57

Brand A is currently used by the airline, and Brands B, C, and D from three different competitors are being evaluated to replace A. Thus, the management is interested in the following:

1. Comparing Brand A with the average wear of Brands B, C, and D,
2. Comparing Brands B, C, and D

Prepare and run a SAS/GLM program necessary and provide answers to the following questions (on a separate sheet) assuming the model SAS Example E9 (see Fig. 5.56):

a. Construct an analysis of variance table and test the hypothesis $H_0 : \tau_A = \tau_B = \tau_C = \tau_D$. State your conclusion based on the p-value.

b. Use a `contrast` statement for making comparison 1 by testing $H_0 : \tau_A = (\tau_B + \tau_C + \tau_D)/3$.

c. Use a `contrast` statement for making the comparison 2 by testing $H_0 : \tau_B = \tau_C = \tau_D$. One way to test this hypothesis is to make the comparisons $\tau_B - \tau_C$ and $\tau_B - \tau_D$ simultaneously in a single `contrast` statement. This results in the computation of a SS with 2 df and, therefore, an F-test with 2 df for the numerator. Add the results of (b) and (c) to the ANOVA table as additional lines and summarize conclusions from this analysis.

d. Construct 95% confidence intervals for $\mu_B - \mu_C$ and $\mu_B - \mu_D$, using the output from appropriate `means` statements used with `proc glm`.
e. Include the statement `output out=new p=fitted r=residual;` in the `proc glm` step. Then use `proc sgplot` and the SAS data set `new` to obtain scatter plots of Residuals versus Machine and Residuals versus Fitted. Add a reference line at zero value on the residuals axis to each of these plots. Use these plots to comment briefly on whether your model assumptions were reasonable.
f. Perform Tukey's test of nonadditivity using a `proc glm` step and the SAS data set `new` created in part (e). What is your conclusion?

5.20 Four machines are compared to assess the differences in the rate of production of a certain part (Part No. Z-15) (Ostle 1963). The data were collected over 5 days. All four machines were run each day (in random order), thus using a *randomized complete block design* with "Day" as a blocking factor. The following data are the number of units produced per day.

Day	Machine			
	A	B	C	D
1	293	308	323	333
2	298	353	343	363
3	280	323	350	368
4	288	358	365	345
5	260	343	340	330

Machine A is currently in use in a factory, and Machines B, C, and D from three different competitors are being evaluated to replace A. Thus, the management is interested in the following:

1. comparing Machine A with the average production of Machines B, C, and D,
2. comparing B, C, and D

Prepare and run a `proc glm` step necessary and provide complete answers, including hypotheses tested and statistics used, to the following questions (on a separate sheet as before). Use the model shown for the RCBD for analyzing these data.

a. Construct an analysis of variance table and test the hypothesis $H_0: \tau_A = \tau_B = \tau_C = \tau_D$. State your conclusion based on the p-value.
b. Use a `contrast` statement for making comparison 1 by testing $H_0: \tau_A = (\tau_B + \tau_C + \tau_D)/3$. What is your conclusion?
c. Use a `contrast` statement for making comparison 2 by testing $H_O: \tau_B = \tau_C = \tau_D$. One way to test this hypothesis is to make the comparisons $\tau_B - \tau_C$ and $\tau_B + \tau_C - 2\tau_D$ simultaneously, in a single `contrast` statement. This results in the computation of a SS with 2 df and an F-test with 2 df for the numerator. Add these tests as lines in an expanded ANOVA table and summarize the results from your analysis.

d. Construct 95% confidence intervals for $\tau_B - \tau_C$, $\tau_B - \tau_D$, and $\tau_C - \tau_D$, using an appropriate statement in the `proc glm` step. Use the results of parts (c) and (d) to make a statement about the new machines being tried out assuming higher production rate is of interest.
e. Include the statement `output out=stats p=fitted r=residual;` in `proc glm` step. Then use `proc sgplot` and the SAS data set `stats` to obtain scatter plots of Residuals versus Machine and Residuals versus Fitted. State the purpose for which these plots may be used. Do these plots identify any problems with your model assumptions?
f. Perform Tukey's test of nonadditivity using a `proc glm` step and the SAS data set `stats` created in part (e). What is your conclusion?

5.21 A consumer product-testing organization wished to compare the annual power consumption of five different brands of dehumidifier (Devore 1982). Because power consumption depends on the prevailing humidity level, it was decided to monitor each brand at four different areas of humidity, ranging from moderate to heavy. Within each area, the five brands were randomly assigned to five different locations for testing, resulting in a randomized complete block experiment with the areas as blocks. The resulting power consumption (annual kWh) values are as follows:

Brand	Blocks			
	1	2	3	4
1	685	792	838	875
2	722	806	893	953
3	733	802	880	941
4	811	888	952	1005
5	828	920	978	1023

Use a SAS procedure and the model given for the RCBD for analyzing these data.

a. Construct an analysis of variance table and test the hypothesis $H_0 : \tau_1 = \tau_2 = \tau_3 = \tau_4 = \tau_5$. State your conclusion based on the p-value.
b. Use Tukey's underscoring procedure to compare all pairwise treatment effects. Make a concluding statement about the annual power consumption of the five different brands of dehumidifiers.
c. Although comparing the block effects is not of interest, use the F-test to comment about the variability among the blocks in this experiment.
d. Add a proc step to your SAS program to perform Tukey's test for nonadditivity.

5.22 The following experiment is described in (Montgomery 2013). A resin (PFTE) is used to produce artificial vascular grafts by extruding into tubes. A study is performed to determine the cause of hard protrusions

called *flicks*. The extrusion pressure is suspected to be a factor in the occurrence of flicks. Four levels of extrusion pressure (8500, 8700, 8900, 9100 *psi*) are the treatments, and the response is the percentage of tubing produced in the run that did not contain any flicks.
The experimenter also knows that there is batch-to-batch variation among the batches of resin delivered by the supplier and thus plans on using a RCBD with batches of resin as blocks. Six batches of resin are used in the experiment with the four levels of extrusion pressure as described above used with resin samples taken from each batch, the order of runs within each block being randomly determined separately for each batch. Thus, there are a total of 24 runs in the experiment. The responses, expressed as percentages of usable tubing, were as follows:

Extrusion	Batch of resin (block)					
pressure (PSI)	1	2	3	4	5	6
8500	90.3	89.2	98.2	93.9	87.4	97.9
8700	92.5	89.5	90.6	94.7	87.0	95.8
8900	85.5	90.8	89.6	86.2	88.0	93.4
9100	82.5	89.5	85.6	87.4	78.9	90.7

Use a SAS procedure and the model given for the RCBD for analyzing these data.

a. Construct an analysis of variance table and test the hypothesis H_0 : $\mu_1 = \mu_2 = \mu_3 = \mu_4$ where $\mu_i = \mu + \tau_i$ represent the mean percentage for the ith extrusion pressure. State your conclusion based on the p-value.
b. Use Fisher's LSD procedure to determine which of the above treatment (extrusion pressure) means differ. Can you conclude from this analysis that lower extrusion pressures lead to fewer defects in the tubes on the average?
c. Note that the levels of extrusion pressures selected by the experimenter are equally spaced. Use the orthogonal polynomial method to partition the treatment sum of squares to a linear effect sum of squares and a lack of fit sum of squares. Use these to perform a test of whether there is a linear trend in the mean percentages as extrusion pressure increases. State your conclusions.

6

Analysis of Variance: Random and Mixed Effects Models

6.1 Introduction

In Chap. 5, the data sets considered were produced from experiments involving treatment factors that were regarded as *fixed*. The levels of factors studied in such experiments were those that the experimenter was interested in comparing and were not a random sample from a population of all possible levels. As discussed in Sect. 5.1, *random* factors were defined as those for which the levels of factors in the experiment consisted of a random sample from a population of levels. When random factors are present, the interest of the experimenter is to study the variance of the hypothetical population of factors rather than the differences among the effects of different factor levels. Thus, the two types of factors are different not only in the way the treatment levels are selected but also in the way they affect the objectives of the study and, therefore, in the type of inferences made.

Whereas *random models* involve only random effects, *mixed models* are models that incorporate both fixed and random effects. Different variations of these are useful for modeling data generated from many experimental and observational studies. In this chapter, several applications of these models will be discussed and analyzed using SAS software. In the first few sections, one-way and two-way random models are considered, followed by several sections presenting different applications of the mixed model. In the latter sections, data from an RCBD are reanalyzed and a split-plot experiment presented regarding the blocks as a random factor, instead of a fixed factor. Several SAS procedures will be used in the analyses, primarily `proc glm` and `proc mixed`, and the differences identified and compared.

M. G. Marasinghe, K. J. Koehler, *Statistical Data Analysis Using SAS*, Springer Texts in Statistics, https://doi.org/10.1007/978-3-319-69239-5_6

It is necessary to note some of the differences in the analyses presented here of models that include random effects from those that involve only fixed effects. A primary difference will be the inclusion of a column for *expected mean squares* in the analysis of variance table. An expected mean square is the linear function of the parameters that the particular mean square is expected to estimate unbiasedly and is usually derived mathematically using the computational formula for the mean square (called a *quadratic form*) and the model used for the observations. They are typically used for construction of F-ratios that are used to test whether a particular function of the parameters of interest is zero or not. For example, in the one-way classification model with a fixed effect used in Sect. 5.2, the expected mean square for the source of variation (SV), labeled `Trt`, is determined to be $E(\text{MS}_{\text{Trt}}) = \sigma^2 + \sum_i(\alpha_i - \bar{\alpha}_.)^2/(t-1)$ and the expected mean squares for `Error` is $E(\text{MSE}) = \sigma^2$. Now the fact that $\sum_i(\alpha_i - \bar{\alpha}_.)^2/(t-1)$ is zero if the null hypothesis of $H_0 : \alpha_1 = \alpha_2 = \cdots = \alpha_t$ is true and larger than zero if it is false implies that the ratio of mean squares $\texttt{MS}_{\texttt{Trt}}/\text{MSE}$ is an appropriate measure for constructing a statistical test of whether H_0 is true or not.

It was considered unnecessary to include this column as part of the analyses in Chap. 5, partly because the hypothesis being tested using the F-statistic for a particular effect was unambiguous. This was so because the models in that chapter did not contain random effects or nested effects. In experiments using complex designs (e.g., split-plots) that involve only fixed treatment effects, it may not be obvious how F-ratios that test particular hypotheses may be constructed. It may be helpful for the analysis of such experiments to include the expected mean square column in the ANOVA table.

In experiments that involve random factors, the variance of the response is usually partitioned into parts called *variance components*, explained as variation due to each of the random effects appearing in the model. Apart from determining appropriate test statistics, expected mean squares are also used to estimate variance components using the *method of moments*. This involves equating the expected mean squares of each source of variation in the ANOVA table, to the respective observed mean square, and solving the resulting set of linear equations for the variance components. The resulting estimates are called method of moments estimates or ANOVA estimates. These estimates have the useful properties of being unbiased and having minimum variance.

Another difference in the analyses of models that include random effects is that in many situations, closed-form solutions to the normal equations for obtaining maximum likelihood estimates do not exist and, thus, iterative optimization techniques need to be employed to obtain the estimates. Section 6.5 contains an introduction to the mixed model that includes a brief discussion of estimation of fixed effects and variance components that is illustrated with a simplified example. Here, a brief introduction to the iterative methods available is given. The SAS procedure recommended for analyzing mixed models is `proc mixed`. It incorporates two popular likelihood-based methods: maximum likelihood (ML) and restricted maximum likelihood (REML). A detailed theoretical discussion of these methods is beyond the scope of this book. However, at least a brief explanation will help users of SAS programs like `proc mixed` understand the basic principles involved. The presentation below (supplemented as needed later) is intended to provide users with a minimal explanation necessary to understand some of the options available in the usage of these procedures as well as help make choices among the possible values that may be specified for them. Readers are urged to obtain additional information from more advanced textbooks on the topic as well as from the detailed descriptions provided in the manuals.

The end result of the models that will be described in this chapter is the specification of the joint distribution of the observations, say y_{ijk}. Generally, it is easier to describe this as the multivariate distribution of an n-dimensional data vector $\mathbf{y}$. In the notation used in Chaps. 4 and 5, a regression or a fixed effects ANOVA model was represented by $\mathbf{y} = X\boldsymbol{\beta} + \boldsymbol{\epsilon}$ where the errors (ϵ_is) were assumed to be iid $N(0, \sigma^2)$ random variables. Another way to express this model is to say that $\mathbf{y}$ is distributed as an n-dimensional multivariate normal with mean vector $\boldsymbol{\mu} = X\boldsymbol{\beta}$ and variance–covariance matrix $\sigma^2 I$ where I is an $n \times n$ identity matrix. In order to allow other variance–covariance structures, this matrix may be represented by the symbol Σ (an element of this matrix is represented by σ_{ij}). For the two-way model used in Sect. 5.1 in Chap. 5, the vector $\boldsymbol{\beta} = (\mu, \alpha_1, \alpha_2, \alpha_3, \tau_1, \tau_2, \tau_3, \tau_4)'$ and $\Sigma = \sigma^2 I$. Thus, the parameters of that model are $\mu, \alpha_1, \alpha_2, \alpha_3, \tau_1, \tau_2, \tau_3, \tau_4$, and σ^2. The joint density function of the elements of $\mathbf{y}$, using matrix notation, is represented by

$$f(\mathbf{y}) = \frac{1}{\sqrt{(2\pi)^n|\Sigma|}} \exp-\frac{1}{2}\left[(\mathbf{y} - X\boldsymbol{\beta})'\Sigma^{-1}(\mathbf{y} - X\boldsymbol{\beta})\right]$$

The likelihood function of the parameters of this model is the same function given on the right-hand side of the above equation but is considered as a function of the parameters (as opposed to a function of the elements in $\mathbf{y}$). It is denoted by $L(\boldsymbol{\theta})$, where $\boldsymbol{\theta}$ is a vector containing all unknown parameters in the density function. Since it is easier to manipulate mathematically, the logarithm of L, called the *log-likelihood* and denoted by $\ell(\boldsymbol{\theta})$, is often used. The log-likelihood for the above model is

$$\ell(\boldsymbol{\theta}) = -\frac{n}{2}\log(2\pi) - \frac{1}{2}\log|\Sigma| - \frac{1}{2}\left[(\mathbf{y} - X\boldsymbol{\beta})'\Sigma^{-1}(\mathbf{y} - X\boldsymbol{\beta})\right]$$

where $\boldsymbol{\theta} = (\boldsymbol{\beta}, \sigma_{11}, \ldots, \sigma_{nn})'$. The *maximum likelihood estimates* (MLEs) are those values of the parameters that maximize the log-likelihood function $\ell(\boldsymbol{\theta})$ over the *parameter space*. For unbalanced data sets, calculating the MLEs usually requires numerical optimization methods that involve iterative procedures. Inference procedures for the parameters based on MLEs usually involve large sample properties of these estimates. Usually, for construction of test statistics and interval estimates, an approximate estimate of the variance–covariance matrix of the estimated parameter vector that results from the optimization procedure is used.

So far this description has only included models that involved regression-type or ANOVA fixed-effects-type parameters. The general model for random and mixed models will be described in other sections of this chapter. In the most general form, the mixed model is given by

$$\mathbf{y} = X\boldsymbol{\beta} + Z\,\mathbf{u} + \boldsymbol{\epsilon}$$

where the random vectors $\mathbf{u}$ and $\boldsymbol{\epsilon}$ have independent multivariate normal distributions $N(\mathbf{0},\, G)$and $N(\mathbf{0},\, R)$, respectively, and the variance–covariance matrices G and R are fixed unknown constants. Using this specification, the variance–covariance matrix of $\mathbf{y}$ is of the form $V = ZGZ^{'} + R$, and the marginal distribution of $\mathbf{y}$ is multivariate normal with mean vector $\boldsymbol{\mu} = X\boldsymbol{\beta}$ and variance–covariance matrix V (i.e., $N(X\boldsymbol{\beta},\, V)$).

In the classical variance component model, the random subvectors $\mathbf{u}_1$, $\mathbf{u}_2$, $\ldots, \mathbf{u}_k$ of $\mathbf{u}$ (say, corresponding to k random effects) and $\boldsymbol{\epsilon}$ have the multivariate normal distributions $N(\mathbf{0},\, \sigma_1^2\, I)$, $N(\mathbf{0},\, \sigma_2^2\, I), \ldots,$ $N(\mathbf{0},\, \sigma_k^2\, I)$, and $N(\mathbf{0},\, \sigma^2\, I)$, respectively, where the matrices $\sigma_1^2 I$ and so forth are diagonal matrices with the diagonal elements all equal to the respective variance components. The variance–covariance matrix of $\mathbf{y}$ is thus given by the matrix V:

$$V \;=\; \sum_i Z_i Z_i^{'} \sigma_i^2 + \sigma^2\, I.$$

where Z_i is the design matrix for the ith random effect; that is, in this case, both G and R turn out to be diagonal matrices, whose diagonal elements are the variance components $(\sigma_1^2,\, \sigma_2^2, \ldots, \sigma_k^2,\, \sigma^2)$.

The log-likelihood function of the parameters in the mixed model is obtained using the marginal distribution of $\mathbf{y}$ given above and is

$$\ell(\boldsymbol{\beta}, V) = -\frac{n}{2}\log{(2\pi)} - \frac{1}{2}\log{|V|} - \frac{1}{2}\left[(\mathbf{y} - X\boldsymbol{\beta})'V^{-1}(\mathbf{y} - X\boldsymbol{\beta})\right]$$

In the classical variance component model described above, V is of the form given above and, thus, is a function of the variance components $\boldsymbol{\sigma}^2 = (\sigma_1^2,\, \sigma_2^2, \ldots, \sigma_k^2,\, \sigma^2)'$. The MLEs of $\boldsymbol{\beta}$ and the variance components are obtained by equating the first derivatives of $\ell(\boldsymbol{\beta}, V)$ with respect to $\boldsymbol{\beta}$ and V to zero and solving the resulting equations for $\boldsymbol{\beta}$ and $\boldsymbol{\sigma}^2$. The usual strategy is to first solve the set of equations

$$X'V^{-1}X\tilde{\boldsymbol{\beta}} = X'V^{-1}\mathbf{y}$$

for $\tilde{\boldsymbol{\beta}}$, assuming that the variance components are known. The solution can be obtained in closed form as

$$\tilde{\boldsymbol{\beta}} = (X'V^{-1}X)^{-1}X'V^{-1}\mathbf{y}$$

where $\tilde{\boldsymbol{\beta}}$ is a function of the unknown variance components. Substituting $\tilde{\boldsymbol{\beta}}$ in $\ell(\boldsymbol{\beta}, V)$ and using an iterative procedure to maximize the resulting *profile log-likelihood*, $\hat{\boldsymbol{\sigma}}^2$, the maximum likelihood estimate of $\boldsymbol{\sigma}^2$ is obtained. A procedure to ensure that the variance components are in the parameter space (i.e., they are nonnegative) is incorporated. The MLE of the variance components $\hat{\boldsymbol{\sigma}}^2$ is then used to compute the MLE of $\boldsymbol{\beta}$ using

$$\hat{\boldsymbol{\beta}} = (X'V(\hat{\boldsymbol{\sigma}}^2)^{-1}X)^{-1}X'V(\hat{\boldsymbol{\sigma}}^2)^{-1}\mathbf{y}$$

where $V(\hat{\boldsymbol{\sigma}}^2)$ is the estimated variance–covariance matrix of $\mathbf{y}$ using the MLE $\hat{\boldsymbol{\sigma}}^2$ of $\boldsymbol{\sigma}^2$.

Even balanced data ML estimates are not identical to the method of moments estimates. This is mainly because the ML estimation method does not "adjust for" using the estimate of the fixed part of the model to estimate the variance components. The so-called restricted maximum likelihood (REML) estimation method overcomes this problem. It is easiest to understand the REML estimation as based on maximizing the log-likelihood of the transformed data vector $K\mathbf{y}$ instead of the log-likelihood of $\mathbf{y}$. The rows of the matrix K, $\mathbf{k}'$, are such that $E(\mathbf{k}'\mathbf{y}) = \mathbf{k}'X\boldsymbol{\beta} = 0$. These linear combinations of the observations are called *error contrasts* in the literature and can be obtained by selecting $n - r$ linearly independent rows of the matrix $I - X(X'X)^{-1}X'$, where r is the rank of X. Once the matrix K is constructed, it can be employed to transform the setup used for the previous maximization problem by transforming $\mathbf{y}$ to $K\mathbf{y}$, $X\boldsymbol{\beta}$ to zero, Zs to KZ, and V to KVK'. Note that the new objective function is the log-likelihood function only of the variance components, but the new observed values are the transformed data values in $K\mathbf{y}$. The results of maximizing the transformed likelihood give the REML estimates of the variance components, $\hat{\boldsymbol{\sigma}}^2_R$. The estimates of the fixed effect parameters are then obtained from

$$\hat{\boldsymbol{\beta}}_R = (X'V(\hat{\boldsymbol{\sigma}}^2_R)^{-1}X)^{-1}X'V(\hat{\boldsymbol{\sigma}}^2_R)^{-1}\mathbf{y}$$

where $V(\hat{\boldsymbol{\sigma}}^2_R)$ is the estimated variance–covariance matrix using the REML estimate of $\boldsymbol{\sigma}^2$.

6.2 One-Way Random Effects Model

Experiments involving one random factor are considered in this section. This type of experiment is similar to the "Traffic Tickets" example discussed in Chap. 5. The random factor of interest was "precinct," that is, precincts in a large city were selected randomly for comparing the number of tickets issued for traffic-related violations. The main interest in this experiment is concerned with making inferences (i.e., estimation and hypothesis tests), about the variance among the precincts in the number of tickets issued. Suppose that a precincts are under study and n officers are randomly sampled in each precinct. Let y_{ij} be the number of tickets issued by the jth officer in the ith precinct.

Model

The one-way random effects model is given by

$$y_{ij} = \mu + A_i + \epsilon_{ij}, \quad i = 1, \ldots, a; \quad j = 1, \ldots, n$$

where the random effects A_i, $i = 1, \ldots, a$, are distributed as iid $N(0, \sigma_A^2)$ random variables independent of the random errors ϵ_{ij}. As usual, the random errors ϵ_{ij}, $i = 1, \ldots, a$; $j = 1, \ldots, n$, are distributed independently as $N(0, \sigma^2)$ random variables. If this is formulated as a "means model" where $\mu_i = \mu + A_i$, then μ_i, $i = 1, \ldots, a$, are assumed to be distributed as iid $N(\mu, \sigma_A^2)$ random variables, where μ represents the mean of the population of the factor levels. In the traffic ticket example, this would correspond to the mean number of tickets issued by all police officers regardless of the precinct. The random effect A_i models the random increment the ith precinct would add to (or subtract from) μ to give the mean number of tickets μ_i issued in that precinct.

It is important to note that the mean of A_i is zero; that is, on average, the incremental mean number of tickets issued in different precincts cancels out. However, σ_A^2, the variance of A_i, and hence of μ_i, measures the variance among the mean numbers of tickets issued in different precincts. If precinct means are all the same, then it will be zero; otherwise, it will be positive and will be larger the more variable the mean numbers of tickets issued among different precincts. On the other hand, the variance among the numbers of tickets issued by officers within each precinct is assumed to be the same for all precincts. This measures the "error" variance σ^2 among the experimental units (police officers) in this study.

Estimation and Hypothesis Testing

An analysis of variance that corresponds to the above model is constructed using the same computational formulas used for the computation of the ANOVA for the one-way fixed effects model. However, as discussed in Sect. 6.1, an additional column displaying the expected mean squares is included in the ANOVA table for the one-way random effects model:

SV	df	SS	MS	F	E(MS)
A	$a-1$	$\mathtt{SS_A}$	$\mathtt{MS_A}$	$\mathtt{MS_A/MSE}$	$\sigma^2 + n\,\sigma_A^2$
Error	$a(n-1)$	SSE	MSE($= s^2$)		σ^2
Total	$an-1$				

The computation of the expected mean squares does not require the distributional assumption of normality of the random effects. However, it is required for performing hypothesis tests and constructing confidence intervals using the F-, t-, and the chi-square distributions.

The hypothesis of main interest in the above model is

$$H_0 : \sigma_A^2 = 0 \quad \texttt{versus} \quad H_a : \sigma_A^2 > 0$$

The two mean squares needed to construct an appropriate F-ratio are determined so that both the numerator and the denominator mean squares will have the same expectation if the null hypothesis of $\sigma_A^2 = 0$ holds and the numerator will have a larger expectation if $\sigma_A^2 > 0$. By examining the $E(\mathrm{MS})$ column, it can be observed that the F-statistic shown satisfies this requirement. The null hypothesis is rejected if the computed F-statistic exceeds the upper $(1-\alpha)$ percentile of the F-distribution with $a-1$ and $a(n-1)$ degrees of freedom or, equivalently, if the p-value is less than α for an α level selected by the experimenter to control the Type I error rate.

As usual, the estimate of the error variance σ^2 is the MSE from the ANOVA table $\hat{\sigma}^2 = s^2$. If the hypothesis $H_0 : \sigma_A^2 = 0$ is rejected in favor of $H_a : \sigma_A^2 > 0$, the mean squares may also be used to estimate σ_A^2. To do this, the expected mean square (which is a linear combination of the variance components) is set equal to its observed value $\mathrm{MS_A}$ from the ANOVA table; that is, set

$$\sigma^2 + n\,\sigma_A^2 = \mathrm{MS_A}$$

and the resulting equation is solved for σ_A^2 after substituting the estimate s^2 for σ^2. This gives the result

$$\hat{\sigma}_A^2 = \frac{\mathrm{MS_A} - s^2}{n}$$

where the right-hand side consists only of quantities computed from the data and obtained from the ANOVA table. This method of estimation is called the *method of moments*. These estimates are identical to maximum likelihood estimates when the sample sizes are the same, as is the case in this example.

One approach for obtaining approximate confidence intervals requires the normality assumptions stated in the model definition. A $(1-\alpha)100\%$ confidence interval for σ_A^2 is provided by

$$\frac{\nu\hat{\sigma}_A^2}{\chi^2_{1-\alpha/2,\nu}} < \sigma_A^2 < \frac{\nu\hat{\sigma}_A^2}{\chi^2_{\alpha/2,\nu}} \tag{6.1}$$

where $\chi^2_{1-\alpha/2,\nu}$ and $\chi^2_{\alpha/2,\nu}$ are the $1-\alpha/2$ and $\alpha/2$ percentile points of the chi-squared distribution with ν degrees of freedom, respectively, and the value of ν is obtained using the Satterthwaite approximation. This approximation is required because, as seen earlier, $\hat{\sigma}_A^2 = \frac{1}{n}\mathrm{MS_A} - \frac{1}{n}\mathrm{MSE}$, a linear combination of two mean squares, and thus it does not have an exact chi-square distribution. The approximation defines ν as

$$\nu = \frac{(n\hat{\sigma}_A^2)^2}{(\mathrm{MS_A})^2/(a-1) + (s^2)^2/a(n-1)} \tag{6.2}$$

Note that no approximation is required to obtain a $(1-\alpha)100\%$ confidence interval for σ^2 since the interval is based on a single mean square (i.e., MSE). It is given by

$$\frac{\nu s^2}{\chi^2_{1-\alpha/2,\nu}} < \sigma^2 < \frac{\nu s^2}{\chi^2_{\alpha/2,\nu}} \tag{6.3}$$

where $\nu = a(n-1)$, the degrees of freedom for error. Also, note that the above confidence intervals are not symmetrical around the estimate of the corresponding variance component. The confidence intervals for the standard deviations σ_A or σ are found by taking square roots of both end points of each interval.

Although factor levels are independent, the observations from the same factor are correlated. This correlation is another important quantity that may be estimated from this type of an experiment and is called the *intraclass correlation*:

$$\text{Corr}(y_{ij}, y_{ij'}) = \frac{\sigma_A^2}{\sigma_A^2 + \sigma^2} \qquad \text{for } j \neq j'$$

This ratio also measures the proportion of the total variation in y_{ij} only due to the random effect, since obviously, $\text{Var}(y_{ij}) = \sigma_A^2 + \sigma^2$. In plant breeding experiments, for example, investigators might be interested in selecting inbred lines that have large intraclass correlations, as that would indicate variation due to genetic influences of those breeds that have a larger effect than, say, environmental effects on the trait being measured.

6.2.1 Using PROC GLM to Analyze One-Way Random Effects Models

The following example is taken from Snedecor and Cochran (1989). An experiment was conducted at the Iowa Agricultural Experiment Station to determine if there is significant variation of average daily gain of pigs from litter to litter. Average daily gain in weight is an indicator of growth rate in animals. For the study, four litters were chosen at random from a single inbred line of swine. The average daily gains of two animals selected at random from each litter were measured. The data are shown in Table 6.1.

Table 6.1. Average daily gain of swine

Litter	1	2	3	4
Gain	1.18	1.36	1.37	1.07
	1.11	1.65	1.40	0.90

The model for average daily gain is

$$y_{ij} = \mu + A_i + \epsilon_{ij}, \quad i = 1, \ldots, 4; \quad j = 1, 2$$

where A_i is the effect of the ith litter and is assumed to be an iid $N(0, \sigma_A^2)$ random variable and ϵ_{ij}, the sampling error associated with pigs within each litter, an iid $N(0, \sigma^2)$ random variable.

```
data hogs;
input Litter Gain;
datalines;
1  1.18
1  1.11
2  1.36
2  1.65
3  1.37
3  1.40
4  1.07
4  0.90
;

proc glm data=hogs;
  class Litter;
  model Gain = Litter;
  random Litter/test;
  title 'Average Daily Gain in Swine';
run;
```

Fig. 6.1. SAS Example F1: program

SAS Example F1

The SAS Example F1 program (see Fig. 6.1) illustrates how `proc glm` can be used to perform the necessary computations. The data may be input using methods used for one-way fixed effects experiments (see, e.g., Fig. 5.1 in Chap. 5). However, in this example, since the sample sizes are small and equal, a straightforward approach can be used. The data are entered exactly in the same format as required by `proc glm`. That is, values for a classification variable `Litter` and the response variable `Gain` are entered in the lines of data separated by blanks so that they are accessed easily using the *list input* method. The `class` and the `model` statements are exactly as for a fixed effects model; however, an additional statement `random Litter/test` is included here. This statement indicates that the effect `Litter` in the model is a random effect and also requests that a test be performed to test the hypothesis that the corresponding variance component is significantly different from zero.

For this example, the SAS output produced in the default style *HTMLBlue* is reproduced here for the purpose of illustration. In the rest of the chapter, the SAS output displayed will be those produced using the ODS destination *rtf*. From the `proc glm` output reproduced in Fig. 6.2, the construction of the following ANOVA table is straightforward:

SV	df	SS	MS	F	p-Value	E(MS)
Litter	3	0.3288	0.1096	7.38	0.0416	$\sigma^2 + 2\,\sigma_A^2$
Boar (Litter)	4	0.0594	0.01485			σ^2
Total	7	0.3882				

Note carefully that the information necessary for completing the additional column titled E(MS) containing the expected mean squares is available from

Average Daily Gain in Swine

The GLM Procedure

Class Level Information		
Class	Levels	Values
Litter	4	1 2 3 4

Number of Observations Read	8
Number of Observations Used	8

Dependent Variable: Gain

Source	DF	Sum of Squares	Mean Square	F Value	Pr > F
Model	3	0.32880000	0.10960000	7.38	0.0416
Error	4	0.05940000	0.01485000		
Corrected Total	7	0.38820000			

R-Square	Coeff Var	Root MSE	Gain Mean
0.846986	9.710006	0.121861	1.255000

Source	DF	Type I SS	Mean Square	F Value	Pr > F
Litter	3	0.32880000	0.10960000	7.38	0.0416

Source	DF	Type III SS	Mean Square	F Value	Pr > F
Litter	3	0.32880000	0.10960000	7.38	0.0416

Fig. 6.2. SAS Example F1: anova output

part of this output subtitled `Type III Expected Mean Square` (see Fig. 6.3). The expected MS for the Litter effect is a simple linear combination of the variance components σ_A^2 and σ^2. Also, recall that the expected mean square for error, MSE, is always σ^2 under the model assumptions. These two expectations form the equations that are used to obtain the method of moments estimates of the variance components, as shown below.

The results of the F-test of $H_0 : \sigma_A^2 = 0$ is displayed on a table titled `Tests of Hypotheses for Random Model Analysis of Variance` (see Fig. 6.4) of the output. Note that the F-statistic is the ratio $\text{MS}_{\text{Litter}}/\text{MSE}$. The denominator of the F-ratio is the MSE (identified as MS(Error) in the output). Also note that, in the ANOVA table above, the corresponding source of variation is labeled Boar(Litter).

If $\sigma_A^2 = 0$, the expectations of both the numerator and the denominator of the ratio would have the same value of σ^2. If the F-statistic is found to be significantly large, it should lead to the conclusion that σ_A^2 is greater than zero. Thus, in more complex experiments, information from the $E(\text{MS})$ column

Average Daily Gain in Swine

The GLM Procedure

Source	Type III Expected Mean Square
Litter	Var(Error) + 2 Var(Litter)

Fig. 6.3. SAS Example F1: expected mean squares

could be used to identify ratios of mean squares needed to test hypotheses about different variance components.

The p-value from the ANOVA table is smaller than 0.05, and, hence, the null hypothesis $H_0 : \sigma_A^2 = 0$ is rejected at $\alpha = 0.05$; that is, evidence exists in the data from this experiment that there is a significant variation of average daily gain among the litters. Since the litters were a random sample, this result would apply to all litters from the inbred line of swine from which these litters were sampled.

Since the hypothesis $H_0 : \sigma_A^2 = 0$ is rejected, it is useful to quantify this variation by estimating the variance component σ_A^2. The method of moments requires setting the computed values of the mean squares equal to the corresponding expressions found in the $E(\text{MS})$ column and solving the resulting linear equations for the variance components. In this case, the equations are

$$\sigma^2 + 2\sigma_A^2 = 0.1096$$

$$\sigma^2 = 0.01485$$

and solving these equations gives the required estimates

$$\hat{\sigma}^2 = 0.01485$$

$$\hat{\sigma}_A^2 = \frac{0.1096 - 0.01485}{2} = 0.0474$$

The GLM Procedure
Tests of Hypotheses for Random Model Analysis of Variance

Dependent Variable: Gain

Source	DF	Type III SS	Mean Square	F Value	Pr > F
Litter	3	0.328800	0.109600	7.38	0.0416
Error: MS(Error)	4	0.059400	0.014850		

Fig. 6.4. SAS Example F1: hypothesis test about variance component

Finally, to compute a $(1 - \alpha)100\%$ confidence interval for σ_A^2, instead of using formula (6.1) for hand computation, it may be coded in SAS as shown in the following simple data step:

```
data cint;
   alpha=.05; n=2; a=4;
   msa= 0.1096; s2=0.01485; sa2=.0474;
   nu=(n*sa2)**2/(msa**2/(a-1)+ s2**2/a*(n-1));
   L= (nu*sa2)/cinv(1-alpha/2,nu);
   U= (nu*sa2)/cinv(alpha/2,nu);
   put nu=   L=    U=;
run;
```

Note that the first two lines have been completed using information obtained from Fig. 6.2 and that the last three lines use these values for the computation of the confidence interval with the required confidence coefficient. The results of executing this code are output on the log page/window instead of the output page/window because of the use of the `put` statement. The 95% confidence interval for σ_A^2 is (0.01342, 1.3809). This calculation is shown here for illustrative purposes only as the same confidence interval is computed as part of the output from an analysis of this data using `proc mixed` in the next section.

6.2.2 Using PROC MIXED to Analyze One-Way Random Effects Models

Although the use of the `random` statement in `proc glm` gives the user the capability to compute expected mean squares and perform F-tests about variance components, other statements in `proc glm` do not make use of this information. For example, `lsmeans` and `estimate` statements assume that all effects are fixed irrespective of whether the `random` statement is present or not. Thus, it is recommended that one use `proc mixed` to analyze both mixed effects models and random effects models. Among other advantages, `proc mixed` gives the user the option of choosing among several estimation methods in addition to the method of moments as well as the ability to use `estimate` statements to estimate *best linear unbiased predictors* (BLUPs) (i.e., predictable linear combinations of fixed and random effects).

SAS Example F2

The SAS Example F2 program (see Fig. 6.5) illustrates how `proc mixed` may be used to perform an analysis of a one-way random model. The essential difference from a `proc glm` step is that in the model statement in `proc mixed`, only the fixed part of the model needs to be specified; thus, `model Gain = ;` implies that only an overall mean μ is present in the model in addition to any random effects. The `random` statement specifies the random effect terms:

```
insert data step to create the SAS dataset 'hogs' here

proc mixed data=hogs cl;
  class Litter ;
  model Gain = ;
  random Litter/solution;
  estimate 'Litter 1 Effect' | Litter 1 0 0 0;
  estimate 'Litter 2 Effect' | Litter 0 1 0 0;
  estimate 'Litter 3 Effect' | Litter 0 0 1 0;
  estimate 'Litter 4 Effect' | Litter 0 0 0 1;
  title 'Average Daily Gain in Swine';
run;
```

Fig. 6.5. SAS Example F2: program using PROC MIXED

in this case, the term `Litter` and a random error term. The variances of the random effects constitute the *variance components* including the `error` variance component that is assumed by default.

The default method of estimation is *restricted maximum likelihood* (commonly known as REML), which assumes that random effects are normally distributed. Other estimation methods available include *maximum likelihood* and the method known as MIVQUE(0). These can be requested using the proc statement options `method=ml` or `method=mivque0`, respectively. Since the REML estimation is popular among practitioners, that method is set as the default. Finally, the `cl` option specified on the `proc mixed` statement requests the calculation of confidence limits for the variance components.

What follows is a brief explanation of the contents of output (see Fig. 6.6) from `proc mixed`. In the `Model Information` table, the phrase *variance components* describing the covariance structure implies that the model specified by the user has been identified as a traditional mixed model in which variances of the random effects parameters (or the variance components) are the only covariance parameters present. The `Dimensions` section gives the sizes of the design matrices (described in Sect. 6.1 where mixed model theory is introduced). The `noinfo` option on the proc statement may be used to suppress the above two tables. `Iteration History` table provides details of the convergence of the iterative procedure used to optimize the objective function used in the case of REML or maximum likelihood methods, respectively. The column labeled "−2 Log Like" for maximum likelihood or "−2 Res Log Like" for REML lists the value of −2 times the log-likelihood or −2 times the log residual likelihood function. This statistic, called the deviance, is used for testing hypotheses about parameters by model comparison. The column labeled "Evaluations" lists the number of times the objective function was evaluated during each iteration. Using the `noitprint` option in the proc statement suppresses the "Iteration History" table.

The next part of the output (see Fig. 6.7) contains the estimates of the variance components in a table subtitled `Covariance Parameter Estimates`. Here they are, respectively, $\hat{\sigma}_A^2 = 0.04738$ and $\hat{\sigma}^2 = 0.01485$. The Satterth-

Average Daily Gain in Swine

The Mixed Procedure

Model Information	
Data Set	WORK.HOGS
Dependent Variable	Gain
Covariance Structure	Variance Components
Estimation Method	REML
Residual Variance Method	Profile
Fixed Effects SE Method	Model-Based
Degrees of Freedom Method	Containment

Class Level Information		
Class	Levels	Values
Litter	4	1 2 3 4

Dimensions	
Covariance Parameters	2
Columns in X	1
Columns in Z	4
Subjects	1
Max Obs per Subject	8

Number of Observations	
Number of Observations Read	8
Number of Observations Used	8
Number of Observations Not Used	0

Iteration History			
Iteration	Evaluations	-2 Res Log Like	Criterion
0	1	1.69956771	
1	1	-1.52719436	0.00000000

Fig. 6.6. SAS Example F2: output (page 1)

waite approximation introduced previously is used to construct confidence intervals for the variance components appearing here. This is true for all iterative methods, as variance components are constrained to be nonnegative. As will be seen in SAS Example F3, when the method of moments estimation method is used, this constraint is not used; instead large sample methods will be used to calculate these confidence limits (except for the error variance). `Solution for Random Effects` table contains predicted values of the `Litter` random effects, which results from using the `solution` option in the `random`

Average Daily Gain in Swine

The Mixed Procedure

Convergence criteria met.

Covariance Parameter Estimates				
Cov Parm	**Estimate**	**Alpha**	**Lower**	**Upper**
Litter	0.04738	0.05	0.01341	1.3843
Residual	0.01485	0.05	0.005331	0.1226

Fit Statistics	
-2 Res Log Likelihood	-1.5
AIC (Smaller is Better)	2.5
AICC (Smaller is Better)	5.5
BIC (Smaller is Better)	1.2

Solution for Random Effects						
Effect	**Litter**	**Estimate**	**Std Err Pred**	**DF**	**t Value**	**Pr > |t|**
Litter	1	-0.09510	0.1291	4	-0.74	0.5021
Litter	2	0.2161	0.1291	4	1.67	0.1693
Litter	3	0.1124	0.1291	4	0.87	0.4330
Litter	4	-0.2334	0.1291	4	-1.81	0.1448

Estimates					
Label	**Estimate**	**Standard Error**	**DF**	**t Value**	**Pr > |t|**
Litter 1 Effect	-0.09510	0.1291	4	-0.74	0.5021
Litter 2 Effect	0.2161	0.1291	4	1.67	0.1693
Litter 3 Effect	0.1124	0.1291	4	0.87	0.4330
Litter 4 Effect	-0.2334	0.1291	4	-1.81	0.1448

Fig. 6.7. SAS Example F2: output (page 2)

statement. They are estimates of the BLUPs of the random effect for each litter. These predictions may also be obtained using the `estimate` statements:

```
estimate 'Litter 1 Effect' | litter 1 0 0 0;
estimate 'Litter 2 Effect' | litter 0 1 0 0;
estimate 'Litter 3 Effect' | litter 0 0 1 0;
estimate 'Litter 4 Effect' | litter 0 0 0 1;
```

Note that the syntax of the estimate statement is similar to that used in the analysis of fixed effects models with `proc glm`, except that the specification of the random effects must appear after the vertical bar ("|"). For mixed models, both fixed effects and random effects may appear in the same estimate statement: fixed effects before | and random effects after. The output from the above set of estimate statements (not shown) is identical to that of the output from the `solution` option seen in Fig. 6.7.

SAS Example F3

It is important to note that estimates produced by maximum likelihood estimation methods will be different from the method of moments estimates calculated using `proc glm` *if the sample sizes are not equal.* However, `proc mixed` may be used to also obtain the method of moments estimates as illustrated in SAS Example F3 (see Fig. 6.8). The proc statement options `noclprint` suppress the class-level information table and `noinfo` suppress several other tables that are not relevant here. The option `method=type3` specifies the type of the mean squares (and the corresponding expected values) that are to be used to estimate the variance components. Usually, Types 1 and 3 are used; however, they will be identical in equal sample size case. The option `asycov` requests the asymptotic covariance matrix of the estimated variance components, and the option `cl` requests the confidence intervals on the variance components (based on asymptotic standard errors of the estimates of the variance components).

```
insert data step to create the SAS dataset 'hogs' here

proc mixed data=hogs noclprint noinfo method=type3 asycov cl;
  class Litter ;
  model Gain = ;
  random Litter/solution;
  title 'Average Daily Gain in Swine';
run;
```

Fig. 6.8. SAS Example F3: method of moments estimates using PROC MIXED

The output is shown in Fig. 6.9. The ANOVA table showing the expected mean squares is given in the table titled `Type 3 Analysis of Variance`. It is exactly the same as that produced by `proc glm` where the F-ratio for testing the litter effect is constructed using the MS(Residual) as the divisor. The confidence intervals for the variance components are calculated using the estimated variance given in the table titled `Asymptotic Covariance Matrix of Estimates`. For example, an asymptotic 95% confidence interval for σ_A^2 is calculated as

$$\hat{\sigma}_A^2 \pm z_{0.025} \times \texttt{s.e.}(\hat{\sigma}_A^2)$$
$$0.04738 \pm 1.96 \times \sqrt{0.00203}$$

Average Daily Gain in Swine

The Mixed Procedure

Type 3 Analysis of Variance								
Source	DF	Sum of Squares	Mean Square	Expected Mean Square	Error Term	Error DF	F Value	Pr > F
Litter	3	0.328800	0.109600	Var(Residual) + 2 Var(Litter)	MS(Residual)	4	7.38	0.0416
Residual	4	0.059400	0.014850	Var(Residual)	.	.	.	.

Covariance Parameter Estimates				
Cov Parm	Estimate	Alpha	Lower	Upper
Litter	0.04738	0.05	-0.04092	0.1357
Residual	0.01485	0.05	0.005331	0.1226

Asymptotic Covariance Matrix of Estimates			
Row	Cov Parm	CovP1	CovP2
1	Litter	0.002030	-0.00006
2	Residual	-0.00006	0.000110

Fit Statistics	
-2 Res Log Likelihood	-1.5
AIC (Smaller is Better)	2.5
AICC (Smaller is Better)	5.5
BIC (Smaller is Better)	1.2

Solution for Random Effects						
Effect	Litter	Estimate	Std Err Pred	DF	t Value	Pr > |t|
Litter	1	-0.09510	0.1291	4	-0.74	0.5021
Litter	2	0.2161	0.1291	4	1.67	0.1693
Litter	3	0.1124	0.1291	4	0.87	0.4330
Litter	4	-0.2334	0.1291	4	-1.81	0.1448

Fig. 6.9. SAS Example F3: output

giving $(-0.04093, 0.1357)$. Since the number of litters is small, approximating the sampling distribution of $\hat{\sigma}_A^2$ by the normal distribution is questionable. So an interval based on the Satterthwaite approximation may be more appropriate here. However, the interval for σ^2 given here is not based on the asymptotic standard error; it is calculated using the formula (6.3). The estimated BLUPs of the litter random effects are the same as those obtained previously.

The model used in the beginning of this section assumed equal sample sizes, for each level of the random factor. The expressions given in the ANOVA table for the expected value of mean square (i.e., E(MS)) for effect A was based on this assumption. In the case of unequal sample sizes , this expression would be different. However, the investigator needs not know this formula, since, as observed previously, when the `random` statement is present, `proc glm` provides Type III expected mean squares as part of the output, and when the `method=type3` option is present, `proc mixed` computes and outputs this expectation as a part of the Type 3 analysis of variance. Thus, the user may proceed with an analysis based on the method of moments as usual.

SAS Example F4

In SAS Example F4, the program shown in Fig. 6.10 is used to illustrate the analysis of a data set with unequal sample sizes. An experiment on artificial insemination in which semen samples from six different bulls were used to inseminate different numbers of cows is described in Snedecor and Cochran (1989). The data are percentages of conceptions and are recorded in Table 6.2.

Table 6.2. Artificial insemination of cows (Snedecor and Cochran 1989)

Percentages of conceptions produced from a series of semen samples from six different bulls					
Bull 1	Bull 2	Bull 3	Bull 4	Bull 5	Bull 6
46	70	52	47	42	35
31	59	44	21	64	68
37		57	70	50	59
62		40	46	69	38
30		67	14	77	57
		64		81	76
		70		87	57
					29
					60

The one-way random effects model for the `percent` variable is

$$y_{ij} = \mu + A_i + \epsilon_{ij}, \quad i = 1, \ldots, 6; \quad j = 1, \ldots, n_i$$

where the random effects $A_i, \quad i = 1, \ldots, 6$ are distributed as iid $N(0, \sigma_A^2)$ random variables independent of the random errors ϵ_{ij}, which are distributed independently as $N(0, \sigma^2)$. The n_is represent the different sample sizes used in the experiment.

In the SAS program, the data are inputted using `trailing @@` to input pairs of data values for the variables `Bull` and `Percent`. A `proc mixed` step with a `method=type3` option used in this SAS program is similar to the one used for the analysis of the previous example. The SAS output pages are reproduced in Figs. 6.11 and 6.12.

```
data cows;
input Bull Percent @@;
datalines;
1 46 1 31 1 37  1 62 1 30
2 70 2 59
3 52 3 44 3 57 3 40 3 67 3 64 3 70
4 47 4 21 4 70 4 46 4 14
5 42 5 64 5 50 5 69 5 77 5 81 5 87
6 35 6 68 6 59 6 38 6 57 6 76 6 57 6 29 6 60
;

proc mixed data=cows noclprint noinfo method=type3 asycov cl;
  class Bull ;
  model Percent = ;
  random Bull/solution;
  title "Artificial Insemination of Cows";
run;
```

Fig. 6.10. SAS Example F4: program using PROC MIXED

From the table titled `Type 3 Analysis of Variance`, $E(\text{MS})$ for `Bull` is seen to be $\sigma^2 + 5.6686\sigma_A^2$. Thus, the equations to be solved for obtaining method of moments are

Artificial Insemination of Cows

The Mixed Procedure

Type 3 Analysis of Variance								
Source	DF	Sum of Squares	Mean Square	Expected Mean Square	Error Term	Error DF	F Value	Pr > F
Bull	5	3322.058730	664.411746	Var(Residual) + 5.6686 Var(Bull)	MS(Residual)	29	2.68	0.0416
Residual	29	7200.341270	248.287630	Var(Residual)	.	.	.	.

Covariance Parameter Estimates				
Cov Parm	Estimate	Alpha	Lower	Upper
Bull	73.4090	0.05	-75.7325	222.55
Residual	248.29	0.05	157.48	448.70

Asymptotic Covariance Matrix of Estimates			
Row	Cov Parm	CovP1	CovP2
1	Bull	5790.30	-783.21
2	Residual	-783.21	4255.22

Fit Statistics	
-2 Res Log Likelihood	292.5
AIC (Smaller is Better)	296.5
AICC (Smaller is Better)	296.9
BIC (Smaller is Better)	296.1

Fig. 6.11. SAS Example F4: output (variance components)

$$\sigma^2 + 5.6686\sigma_A^2 = 664.411746$$

$$\sigma^2 = 248.287630$$

and solving these equations give, the estimates $\hat{\sigma}^2 = 248.287630$ and $\hat{\sigma}_A^2 = (664.411746 - 248.287630)/5.6686 = 73.40862$. These values are confirmed from the table of `Covariance Parameter Estimates` that also include asymptotic standard errors and 95% confidence intervals. The interval for σ_A^2 is based on the normal distribution and the asymptotic standard error of its estimate, and the interval for σ^2 is calculated using the formula (6.3). See comment made earlier (see SAS Example F3) concerning intervals based on the large sample approximation. More importantly, the estimated BLUPs of the bull random effects are now displayed in the table `Solution for Random Effects` for which the standard errors are now different for each estimate.

Solution for Random Effects							
Effect	Bull	Estimate	Std Err Pred	DF	t Value	Pr > |t|
Bull	1	-7.2278	6.0660	29	-1.19	0.2431
Bull	2	4.1555	6.9940	29	0.59	0.5570
Bull	3	2.0016	5.7517	29	0.35	0.7304
Bull	4	-8.1822	6.0660	29	-1.35	0.1878
Bull	5	9.3218	5.7517	29	1.62	0.1159
Bull	6	-0.06890	5.5414	29	-0.01	0.9902

Fig. 6.12. SAS Example F4: output (BLUPs)

6.3 Two-Way Crossed Random Effects Model

An experiment with two random factors that are crossed is considered in this section. This situation is similar to a two-way factorial experiment in a completely randomized design discussed in Sect. 5.4, except that the levels of the two factors are selected randomly from populations of all possible levels.

Consider an experiment in which two factors that influence the breaking strength of plastic sheeting is under study. Four production machines (Factor A) and five operators (Factor B) are selected for the study. These factor levels are to be considered as random samples from populations of machines and operators used in a typical factory that produces plastic sheeting, and every machine will be used by all operators. A machine–operator combination performs a production run that will produce a measurement of breaking strength. As in the fixed effects case, a number of runs are performed by each machine–operator combination in random order, so that replications are available for estimating the random error variance. To have equal sample sizes, the number of replications carried out per factor combination is kept the same. Assuming that the number of replications is 2, the 40 experimental runs required are performed in a completely randomized design. Again, the main interest in this experiment will be estimation and hypothesis tests about the variance components (i.e., the variances of the random effects).

Model

The two-way crossed random effects model is given by

$$y_{ijk} = \mu + A_i + B_j + AB_{ij} + \epsilon_{ijk}, \quad i = 1, \ldots, a; \quad j = 1, \ldots, b; k = 1, \ldots, n$$

where the effects of Factor A, A_i, are assumed to be iid $N(0, \sigma_A^2)$ random variables; the effects of Factor B, B_j, are assumed to be iid $N(0, \sigma_B^2)$ random variables; the effects of the interaction between the two factors, denoted by AB_{ij}, are assumed to be iid $N(0, \sigma_{AB}^2)$; and the random errors ϵ_{ijk} are assumed to be iid $N(0, \sigma^2)$ random variables. In addition, A_i, B_j, AB_{ij}, and ϵ_{ijk} are pairwise independent. If this is formulated as a "means model," where $\mu_{ij} = \mu + A_i + B_j + AB_{ij}$, then $\mu_{ij}, \quad i = 1, \ldots, a; j = 1, \ldots, b$ are iid $N(\mu, \sigma_A^2 + \sigma_B^2 + \sigma_{AB}^2 + \sigma^2)$ random variables, where μ represents the mean of the population of the observations (i.e., $E(y_{ijk}) = \mu$).

Thus, the objective of the experiment is to identify the components of variance that contribute significantly to the total variance of the observations. A consequence of the above model is that observations realized from the same level of Factor A (or Factor B or both) are correlated. For example, it can be shown that the covariance between y_{111} and y_{122} is σ_A^2 and that between y_{111} and y_{112} is $\sigma_A^2 + \sigma_B^2 + \sigma_{AB}^2$. Thus, the model defines the "covariance structure" of the observed data vector.

Estimation and Hypothesis Testing

An analysis of variance that corresponds to the above model is constructed using the same computational formulas used for the computation of the anova that correspond to the two-way classification model with fixed effects discussed in Sect. 5.4. As discussed in Sect. 6.1, an additional column displaying the expected mean squares is included in the ANOVA table for the two-way random effects model:

SV	df	SS	MS	F	E(MS)
A	$a-1$	$\mathtt{SS_A}$	$\mathtt{MS_A}$	$\mathtt{MS_A}/\mathtt{MS_{AB}}$	$\sigma^2+n\,\sigma^2_{AB}+bn\,\sigma^2_A$
B	$b-1$	$\mathtt{SS_B}$	$\mathtt{MS_B}$	$\mathtt{MS_B}/\mathtt{MS_{AB}}$	$\sigma^2+n\,\sigma^2_{AB}+an\,\sigma^2_B$
AB	$(a-1)(b-1)$	$\mathtt{SS_{AB}}$	$\mathtt{MS_{AB}}$	$\mathtt{MS_{AB}}/\mathtt{MSE}$	$\sigma^2+n\,\sigma^2_{AB}$
Error	$ab(n-1)$	SSE	MSE($=s^2$)		σ^2
Total	$abn-1$				

Again, the computation of the expected mean squares does not require the distributional assumption of normality of the random effects, but normality is required for performing hypothesis tests and constructing confidence intervals. F-statistics are constructed for sources of variation A, B, and AB as shown in the analysis of variance table and are used to test the following hypotheses:

(i) $H_0:\sigma^2_A=0$ versus $H_a:\sigma^2_A>0$

(ii) $H_0:\sigma^2_B=0$ versus $H_a:\sigma^2_B>0$

(iii) $H_0:\sigma^2_{AB}=0$ versus $H_a:\sigma^2_{AB}>0$

respectively. Note, carefully, that these ratios are not the same as the F-ratios shown in the two-way fixed effects ANOVA table (see Sect. 5.4). As in Sect. 6.2, suitable F-ratios are determined so that both the numerator and the denominator mean squares will have the same expectations if the null hypothesis holds, but the numerator will have a larger expectation under the alternative. For example, by examining the E(MS) column, it can be observed that both MS_A and MS_AB will have expectation equal to $\sigma^2+n\,\sigma^2_{AB}$ if $\sigma^2_A=0$; however, the numerator will have expectation equal to $\sigma^2+n\,\sigma^2_{AB}+bn\,\sigma^2_A$ if $\sigma^2_A>0$. Thus, the F-statistic for effect A satisfies this requirement for testing the hypotheses stated in (i). The F-statistics for testing hypotheses (ii) and (iii) are also constructed in a similar fashion.

The estimate of the error variance σ^2 is the MSE from the ANOVA table (i.e., $\hat{\sigma}^2=s^2$). To estimate the other variance components, the expected mean squares are set equal to the corresponding observed values, and the resulting set of equations is solved. However, if any of the above null hypotheses fails to be rejected, these parameters may be set equal to zero in the above expressions for E(MS) before they are used to estimate the rest of the variance components.

If the hypothesis $H_0:\sigma^2_A=0$ is rejected in favor of $H_a:\sigma^2_A>0$, then σ^2_A may be estimated by solving

$$\sigma^2+bn\,\sigma^2_A=\text{MS}_\text{A},$$

Substituting the estimate s^2 for σ^2 gives the estimate

$$\hat{\sigma}^2_A=\frac{\text{MS}_\text{A}-s^2}{bn}$$

where the right-hand side is computed using values obtained from the ANOVA table. As earlier, these are called the *method of moments* estimates.

A $(1-\alpha)100\%$ confidence interval for σ_A^2 is provided by

$$\frac{\nu\hat{\sigma}_A^2}{\chi^2_{1-\alpha/2,\nu}} < \sigma_A^2 < \frac{\nu\hat{\sigma}_A^2}{\chi^2_{\alpha/2,\nu}} \tag{6.4}$$

where $\chi^2_{1-\alpha/2,\nu}$ and $\chi^2_{\alpha/2,\nu}$ are the $1-\alpha/2$ and $\alpha/2$ percentile points of the chi-squared distribution with ν degrees of freedom, respectively. The degrees of freedom for $\hat{\sigma}_A^2 = \frac{1}{bn}\text{MS}_\text{A} - \frac{1}{bn}\text{MSE}$ are obtained using the Satterthwaite approximation and are given by

$$\nu = \frac{(bn\hat{\sigma}_A^2)^2}{(\text{MS}_\text{A})^2/(a-1) + (s^2)^2/ab(n-1)}$$

Formulas for constructing confidence intervals for the other variance components can be similarly obtained.

6.3.1 Using PROC GLM and PROC MIXED to Analyze Two-Way Crossed Random Effects Models

The data shown in Table 6.3 appear in Kutner et al. (2005). An automobile manufacturer studied the effects of differences between drivers (factor A) and differences between cars (factor B) on gasoline consumption. Four drivers were selected at random, and five cars of the same model with manual transmission were also randomly selected from the assembly line. Each driver drove each car twice over a 40-mile test course, and the miles per gallon were calculated. The actual trials were run in completely random order.

Table 6.3. Automobile mileage data

Factor A	Factor B (car)				
(driver)	1	2	3	4	5
1	25.3	28.9	24.8	28.4	27.1
	25.2	30.0	25.1	27.9	26.6
2	33.6	36.7	31.7	35.6	33.7
	32.9	36.5	31.9	35.0	33.9
3	27.7	30.7	26.9	29.7	29.2
	28.5	30.4	26.3	30.2	28.9
4	29.2	32.4	27.7	31.8	30.3
	29.3	32.4	28.9	30.7	29.9

The interest here is in explaining the variation in gasoline consumption in terms of the variance components and determining whether their contributions to the total variation in the response are significant. The model is

$$y_{ijk} = \mu + A_i + B_j + AB_{ij} + \epsilon_{ijk}, \quad i = 1, \ldots, 4; \quad j = 1, \ldots, 5; \quad k = 1, 2$$

where A_i, B_j, and AB_{ij} are random effects of driver, car, and their interaction, distributed as independent normal random variables with mean zero and variances σ_A^2, σ_B^2, and σ_{AB}^2, respectively, and the random errors ϵ_{ijk} are distributed independently as $N(0, \sigma^2)$ random variables.

SAS Example F5

The SAS Example F5 program (see Fig. 6.13) illustrates how `proc glm` is used to fit the above model to the gasoline mileage data. The `proc glm` step carries

```
data auto;
input  Driver 1. @;
   do Car=1 to 5;
      input MPG @;
      output;
   end;
datalines;
1  25.3  28.9  24.8  28.4  27.1
1  25.2  30.0  25.1  27.9  26.6
2  33.6  36.7  31.7  35.6  33.7
2  32.9  36.5  31.9  35.0  33.9
3  27.7  30.7  26.9  29.7  29.2
3  28.5  30.4  26.3  30.2  28.9
4  29.2  32.4  27.7  31.8  30.3
4  29.3  32.4  28.9  30.7  29.9
;

proc glm data=auto;
class Driver Car;
model MPG =Driver Car Driver*Car;
random Driver Car Driver*Car/test;
title "Study of Variation in Gasoline Consumption";
run;
```

Fig. 6.13. SAS Example F5: program

out a standard analysis based on the method of moments for estimation of variance components and F-tests that are valid under the condition that the random effects have independent normal distributions as described earlier.

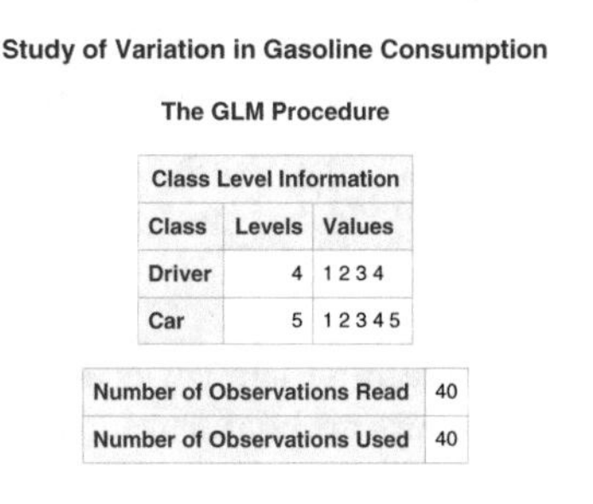

Study of Variation in Gasoline Consumption

The GLM Procedure

Class Level Information		
Class	Levels	Values
Driver	4	1 2 3 4
Car	5	1 2 3 4 5

Number of Observations Read	40
Number of Observations Used	40

Fig. 6.14. SAS Example F5: output from PROC GLM

Dependent Variable: MPG

Source	DF	Sum of Squares	Mean Square	F Value	Pr > F
Model	19	377.4447500	19.8655132	113.03	<.0001
Error	20	3.5150000	0.1757500		
Corrected Total	39	380.9597500			

R-Square	Coeff Var	Root MSE	MPG Mean
0.990773	1.395209	0.419225	30.04750

Source	DF	Type I SS	Mean Square	F Value	Pr > F
Driver	3	280.2847500	93.4282500	531.60	<.0001
Car	4	94.7135000	23.6783750	134.73	<.0001
Driver*Car	12	2.4465000	0.2038750	1.16	0.3715

Source	DF	Type III SS	Mean Square	F Value	Pr > F
Driver	3	280.2847500	93.4282500	531.60	<.0001
Car	4	94.7135000	23.6783750	134.73	<.0001
Driver*Car	12	2.4465000	0.2038750	1.16	0.3715

Fig. 6.15. SAS Example F5: output from PROC GLM continued

Part of the SAS output (see Figs. 6.14 and 6.15) displays the Types I and III analyses of variance table (which should be identical for balanced data) and is a part of the standard `proc glm` output, independent of whether the factors were fixed or random.

The ANOVA table that follows is constructed using the information in the SAS output. The total and error sums of squares and their respective degrees of freedom and the table of `Type III Expected Mean Squares` are available from the output (see Fig. 6.16). The F-tests for the variance components `Driver` (σ_A^2), `Car` (σ_B^2), and `Driver` $\times$ `Car` interaction (σ_{AB}^2), respectively, are available from the tables in the output titled `Tests of Hypotheses for Random Model Analysis of Variance`.

SV	df	SS	MS	F	p-Value	E(MS)
Driver	3	280.285	93.428	458.26	<0.0001	$\sigma^2 + 2\,\sigma_{AB}^2 + 10\,\sigma_A^2$
Car	4	94.714	23.678	116.14	<0.0001	$\sigma^2 + 2\,\sigma_{AB}^2 + 8\,\sigma_B^2$
Driver× Car	12	2.446	0.204	1.16	0.3715	$\sigma^2 + 2\,\sigma_{AB}^2$
Error	20	3.515	3.7917			σ^2
Total	39	380.960				

A p-value of 0.3715 for the interaction effects leads to the conclusion that σ^2_{AB} is not different from zero, whereas the p-values of less than 0.0001 for the `Driver` and `Car` random effects respectively show that the corresponding two variance components are significantly larger than zero. Their estimates can be obtained by the method of moments by setting the expressions for expected mean squares found on page 3 (see Fig. 6.16) for each effect equal to its computed mean square (Type I or Type III) from page 2. This gives the set of equations

$$\sigma^2 + 10\sigma^2_A + 2\sigma^2_{AB} = 93.428250$$

$$\sigma^2 + 8\sigma^2_B + 2\sigma^2_{AB} = 23.678375$$

$$\sigma^2 + 2\sigma^2_{AB} = 0.203875$$

$$\sigma^2 = 0.175750$$

Solving these gives the estimates

$$\hat{\sigma}^2 = 0.175750$$

$$\hat{\sigma}^2_{AB} = (0.203875 - 0.175750)/2 = 0.0140625$$

Study of Variation in Gasoline Consumption

The GLM Procedure

Source	Type III Expected Mean Square
Driver	Var(Error) + 2 Var(Driver*Car) + 10 Var(Driver)
Car	Var(Error) + 2 Var(Driver*Car) + 8 Var(Car)
Driver*Car	Var(Error) + 2 Var(Driver*Car)

Study of Variation in Gasoline Consumption

The GLM Procedure
Tests of Hypotheses for Random Model Analysis of Variance

Dependent Variable: MPG

Source	DF	Type III SS	Mean Square	F Value	Pr > F
Driver	3	280.284750	93.428250	458.26	<.0001
Car	4	94.713500	23.678375	116.14	<.0001
Error	12	2.446500	0.203875		
Error: MS(Driver*Car)					

Source	DF	Type III SS	Mean Square	F Value	Pr > F
Driver*Car	12	2.446500	0.203875	1.16	0.3715
Error: MS(Error)	20	3.515000	0.175750		

Fig. 6.16. SAS Example F5: output from PROC GLM (pages 3 and 4)

$$\hat{\sigma}_B^2 = (23.678375 - 2 \times 0.0140625 - 0.175750)/8 = 2.9343125$$

$$\hat{\sigma}_A^2 = (93.428250 - 2 \times 0.0140625 - 0.175750)/10 = 9.3224375$$

Confidence intervals for each nonzero variance component based on the Satterthwaite approximation can be computed by hand or by modifying the SAS code given in Sect. 6.2. For example, a 95% confidence interval for σ_A^2 is given by executing the SAS code

```
data cint;
  alpha=.05; n=2; a=4; b=5;
  msa= 93.428250; msab=0.203875; sa2=9.3224375;
  nu=(n*b*sa2)**2/(msa**2/(a-1)+ msab**2/(a-1)*(b-1));
  L= (nu*sa2)/cinv(1-alpha/2,nu);
  U= (nu*sa2)/cinv(alpha/2,nu);
  put nu=   L=    U=;
run;
```

which results in the interval (2.9864, 130.7911). A confidence interval for σ_B^2 can be similarly calculated. `Proc mixed` produces these intervals by default when an iterative method such as `ML` or `REML` is used along with the proc statement option `cl`.

SAS Example F6

In the SAS Example F6 program (see Fig. 6.17), `proc mixed` is used to fit the above model to the gasoline mileage data. In the proc statement, `method=type3` is specified, so instead of using an iterative algorithm for calculating the likelihood estimates, the method of moments estimators using the Type III expected mean squares are computed.

```
insert data step to create the SAS dataset 'auto' here

proc mixed  data=auto noclprint noinfo method=type3 cl;
class Driver Car;
model MPG = /;
random Driver Car Driver*Car;
run;
```

Fig. 6.17. SAS Example F6: method of moments estimation using PROC MIXED

The results, both estimates, F-statistics, and their p-values shown in Fig. 6.18 are identical to those obtained using `proc glm`. The variance component estimates are also the same as those calculated by hand using results from `proc glm`. However, the confidence intervals calculated by `proc mixed` are those based on large sample standard errors and the standard normal percentiles except those for σ^2, which are based on chi-square percentiles. Confidence intervals on the other variance components based on Satterthwaite approximation can be calculated using formulas similar to (6.1) in Sect. 6.2, as illustrated for SAS Example F5.

Effects of Drivers and Cars on Mileage

The Mixed Procedure

Type 3 Analysis of Variance								
Source	DF	Sum of Squares	Mean Square	Expected Mean Square	*Error Term*	Error DF	F Value	Pr > F
Driver	3	280.28475	93.428250	Var(Residual) + 2 Var(Driver*Car) + 10 Var(Driver)	MS(Driver*Car)	12	458.26	<.0001
Car	4	94.71350	23.678375	Var(Residual) + 2 Var(Driver*Car) + 8 Var(Car)	MS(Driver*Car)	12	116.14	<.0001
Driver*Car	12	2.44650	0.203875	Var(Residual) + 2 Var(Driver*Car)	MS(Residual)	20	1.16	0.3715
Residual	20	3.51500	0.175750	Var(Residual)	.	.	.	.

Covariance Parameter Estimates				
Cov Parm	Estimate	Alpha	Lower	Upper
Driver	9.3224	0.05	-5.6289	24.2738
Car	2.9343	0.05	-1.1677	7.0364
Driver*Car	0.01406	0.05	-0.08402	0.1121
Residual	0.1757	0.05	0.1029	0.3665

Fit Statistics	
-2 Res Log Likelihood	86.8
AIC (Smaller is Better)	94.8
AICC (Smaller is Better)	96.0
BIC (Smaller is Better)	92.3

Fig. 6.18. SAS Example F6: output

SAS Example F7

Finally, the same data are reanalyzed in the SAS Example F7 program (see Fig. 6.19) using REML, the default method in `proc mixed`. In the proc statement, `method` is unspecified, so an iterative algorithm is used to compute the restricted maximum likelihood estimates because the default method is REML.

The resulting SAS output (see Fig. 6.20) contains the estimates of the variance components, estimates of their asymptotic standard errors, z-tests, and associated p-values. These are output as a result of the `covtest` option. Confidence intervals computed using the Satterthwaite approximation are produced as a result of the `cl` option, because in this case, the variance components are constrained to be nonnegative. The `solution` option provides the estimates of the fixed effects (see table titled `Solution for Fixed Effects` in Fig. 6.20). Recall that the only fixed effect in the model is $E(y_{ijk}) = \mu$, the population mean of the observations. This is identified as the `Intercept` effect in the output.

```
insert data step to create the SAS dataset 'auto' here

proc mixed  data=auto noinfo noitprint covtest cl ;
class Driver Car;
model MPG = /solution;
random Driver Car Driver*Car;
run;
```

Fig. 6.19. SAS Example F7: REML estimation using PROC MIXED

The estimates of the variance components are identical to the method of moments estimates, as expected for balanced data. However, the standard errors have been calculated using large-sample results for maximum likelihood estimators, which assume that the numbers of levels for the two factors (i.e., sample sizes) are infinitely large. Thus, the results of the z-tests do not coincide with those of the F-tests based on assuming normal distributions for the random effects.

Covariance Parameter Estimates

Cov Parm	Estimate	Standard Error	Z Value	Pr > Z	Alpha	Lower	Upper
Driver	9.3224	7.6284	1.22	0.1108	0.05	2.9864	130.79
Car	2.9343	2.0929	1.40	0.0805	0.05	1.0464	24.9038
Driver*Car	0.01406	0.05004	0.28	0.3893	0.05	0.001345	3.592E17
Residual	0.1757	0.05558	3.16	0.0008	0.05	0.1029	0.3665

Fit Statistics

-2 Res Log Likelihood	86.8
AIC (Smaller is Better)	94.8
AICC (Smaller is Better)	96.0
BIC (Smaller is Better)	92.3

Solution for Fixed Effects

Effect	Estimate	Standard Error	DF	t Value	Pr > |t|
Intercept	30.0475	1.7096	3	17.58	0.0004

Fig. 6.20. SAS Example F7: output

The confidence interval for σ_A^2, on the other hand, agrees with that computed earlier using the Satterthwaite approximation. A $(1-\alpha)100\%$ confidence interval for σ_A^2 is given by formula (6.1) in Sect. 6.2. The computation is simplified because degrees of freedom are $\nu = 2z^2$, where z is the Wald statistic, given by $z = \sigma_A^2/\text{s.e.}(\sigma_A^2)$. Substituting the values needed to compute z for each variance component, the confidence intervals given in Fig. 6.20 can be verified.

For example, using the estimate and its standard error (9.3224 and 7.6284, respectively) for the `driver` variance component in the following SAS data step

```
data;
  alpha=.05; s2= 9.3224; ses2= 7.6284;
  z=s2/ses2;
  nu=2*z**2;
  L= (nu*s2)/cinv(1-alpha/2,nu);
  U= (nu*s2)/cinv(alpha/2,nu);
  put z=  nu=   L=    U=;
run;
```

results in the 95% confidence interval (2.9864, 130.7887).

6.3.2 Randomized Complete Block Design: Blocking When Treatment Factors Are Random

In the discussion of the RCBD presented in Sect. 5.7 of Chap. 5, both the treatment and block effects were considered to be fixed effects. It may be more reasonable to consider the block effects to be random effects. In Sect. 6.5, RCBDs with blocks as random effects will be discussed as a special case of the mixed effects model.

In this subsection, a model with both block and treatment effects random is presented. One of the consequences of the way blocks are formed is that, conceptually, it is not feasible for differences in treatment effects to be different from block to block because blocks are formed by grouping experimental units. Therefore, although blocks were considered to be fixed effects, an additive model was used in Sect. 5.7 to represent the observations; that is, an interaction term was not included in the model for observations from an experiment carried out as an RCBD. The same argument holds for the case when the treatments are random effects; thus, an interaction term is omitted from the model.

Montgomery (1991) discussed an experiment that uses `subjects` as the *blocking factor* and `analysts` that perform DNA analyses on three samples taken from each subject as *the treatment factor*. As the analysts are a random sample from a population and the samples from each subject are randomly assigned to the analysts, the experimental design is an RCBD, and the observed data y_{ij} may be modeled as the additive model

$$y_{ij} = \mu + A_i + B_j + \epsilon_{ij}, \quad i = 1, \ldots, a; \quad j = 1, \ldots, b$$

where A_i, the effects of analysts (Factor A), are iid $N(0, \sigma_A^2)$ random variables; B_j, effects of subjects (Factor B), are iid $N(0, \sigma_B^2)$ random variables; and random errors ϵ_{ij} are iid $N(0, \sigma^2)$ random variables, and these three set of random variables are pairwise independent. The analysis of the data is similar to that of the two-way crossed random effects model except that there is no interaction term in the model. The ANOVA table for this model is

SV	df	SS	MS	F	E(MS)
A	$a-1$	SS_A	MS_A	MS_A/MSE	$\sigma^2 + b\,\sigma_A^2$
B	$b-1$	SS_B	MS_B	MS_B/MSE	$\sigma^2 + a\,\sigma_B^2$
Error	$(a-1)(b-1))$	SSE	MSE$(=s^2)$		σ^2
Total	$ab-1$				

A SAS example showing the analysis of data from this model is omitted, as it is straightforward and follows in the same lines as the analysis of data for SAS Example F5, except that determining whether σ_A^2 is nonzero is of interest here.

6.4 Two-Way Nested Random Effects Model

In this section, a two-way random model for responses from an experiment with two random factors when one of the factors is *nested* in the other factor is considered. Consider two factors A and B. Factor B is said to be nested in Factor A if levels of B are different at each level of A. For example, in an extended version of the "traffic ticket" example discussed in Chap. 5, suppose that the number of tickets issued by officers in randomly selected precincts in several cities is under study. In this case, suppose also that both cities and precincts are randomly sampled. The factor `precinct` is nested within the factor `city` because the levels of `precinct` are different from city to city. This factor is called the `precinct within city`, with its levels defined using combinations of the levels of both city and precinct factors. In general, in a two-way nested classification, the sampling of the levels takes place in a *hierarchical* manner: First, the levels of one factor (Factor A) are randomly sampled, and then the levels of the *nested* factor (Factor B) are randomly sampled within each level of A. Although in this section both factors are considered random, it is also possible that at least one of them is a fixed factor.

Model

An appropriate model for the situation described above is

$$y_{ijk} = \mu + A_i + B_{ij} + \epsilon_{ijk}, \quad i = 1, \ldots, a;\ j = 1, \ldots, b;\ k = 1, \ldots, n$$

where it is assumed that A_i, the Factor A effect, is iid $N(0, \sigma_A^2)$; B_{ij}, the Factor B within A effect, is iid $N(0, \sigma_B^2)$; and the random error ϵ_{ijk} is iid

$N(0, \sigma^2)$. Further, it is assumed that A_i, B_{ij}, and ϵ_{ijk} are pairwise independently distributed. The parameters σ_A^2, σ_B^2, and σ^2 will constitute the "variance components" in this problem. It is important to note that it is assumed that all B_{ij}, irrespective of the level i, have the same variance σ_B^2. For example, in the motivating problem introduced earlier, this is equivalent to assuming that the variances in the mean number of tickets issued among the precincts are the same for all cities. It is possible to examine whether this is a plausible assumption using the observed data.

Estimation and Hypothesis Testing

An analysis of variance that corresponds to the model is constructed using the same computational formulas used for the computation of sums of squares of an ANOVA table if the factors `A` and `B within A` are considered to be fixed. These sums of squares would be the same for effect A as in the ANOVA table for the two-way crossed random effects model (given in Sect. 6.3). For the effect `B within A` (denoted in the following ANOVA table as B(A)), the sum of squares is obtained by combining (or pooling) the sum of squares for effects B and AB from the ANOVA table for the two-way crossed random effects model. That is $\mathrm{SS}_{\mathrm{B(A)}} = \mathrm{SS}_{\mathrm{B}} + \mathrm{SS}_{\mathrm{AB}}$ with df($\mathrm{SS}_{\mathrm{B(A)}}$)= df($\mathrm{SS}_{\mathrm{B}}$) + df($\mathrm{SS}_{\mathrm{AB}}$)= $(b-1) + (a-1)(b-1) = a(b-1)$. As with the random models considered so far, an additional column displaying the expected mean squares is included in the ANOVA table:

SV	df	SS	MS	F	E(MS)
A	$a-1$	$\mathtt{SS_A}$	$\mathtt{MS_A}$	$\mathtt{MS_A/MS_{B(A)}}$	$\sigma^2 + n\,\sigma_B^2 + bn\,\sigma_A^2$
B(A)	$a(b-1)$	$\mathtt{SS_{B(A)}}$	$\mathtt{MS_{B(A)}}$	$\mathtt{MS_{B(A)}/MSE}$	$\sigma^2 + n\,\sigma_B^2$
Error	$ab(n-1)$	SSE	MSE($= s^2$)		σ^2
Total	$abn-1$				

F-statistics are constructed for sources of variation A and B(A) as shown in the analysis of variance table and are used to test the hypotheses:

(i) $H_0 : \sigma_A^2 = 0$ versus $\mathrm{H_a} : \sigma_{\mathrm{A}}^2 > 0$

(ii) $H_0 : \sigma_B^2 = 0$ versus $\mathrm{H_a} : \sigma_{\mathrm{B}}^2 > 0$

respectively. Again, note, carefully, that these ratios are not the same as the F-ratios shown in the two-way crossed random effects ANOVA table (see Sect. 6.3), although the appropriate F-ratios are determined in the same principle described there. For example, by examining the E(MS) column, it can be observed that both $\mathrm{MS_A}$ and $\mathrm{MS_{B(A)}}$ will have expectation equal to $\sigma^2 + n\,\sigma_B^2$ if $\sigma_A^2 = 0$, irrespective of the value of σ_B^2. However, the numerator $\mathrm{MS_A}$ will have expectation equal to $\sigma^2 + n\,\sigma_B^2 + bn\,\sigma_A^2$ if $\sigma_A^2 > 0$. Thus, the F-statistic for effect A meets the requirement for testing the hypotheses stated in (i). The F-statistic for testing hypotheses (ii) is also constructed in a similar manner.

The *method of moments* estimates of variance components are obtained by setting the computed mean squares equal to their corresponding expected values and solving the resulting equations, as usual:

$$\begin{aligned} \text{MS}_{\text{A}} &= \sigma^2 + n\,\sigma_B^2 + bn\,\sigma_A^2 \\ \text{MS}_{\text{B(A)}} &= \sigma^2 + n\,\sigma_B^2 \\ \text{MSE} &= \sigma^2 \end{aligned}$$

The method of moments estimates of σ_A^2, σ_B^2, and σ^2 are thus given by

$$\begin{aligned} \hat{\sigma}_A^2 &= (\text{MS}_{\text{A}} - \text{MS}_{\text{B(A)}})/bn \\ \hat{\sigma}_B^2 &= (\text{MS}_{\text{B(A)}} - \text{MSE})/n \end{aligned}$$

and $\hat{\sigma}^2 = \text{MSE}$, respectively. When the sample sizes are equal (i.e., for balanced data), these estimators are unbiased and have minimum variance. As earlier, a $(1-\alpha)100\%$ confidence interval for σ_A^2 is provided by

$$\frac{\nu\hat{\sigma}_A^2}{\chi^2_{1-\alpha/2,\nu}} < \sigma_A^2 < \frac{\nu\hat{\sigma}_A^2}{\chi^2_{\alpha/2,\nu}}$$

where $\chi^2_{1-\alpha/2,\nu}$ and $\chi^2_{\alpha/2,\nu}$ are the $1-\alpha/2$ and $\alpha/2$ percentile points of the chi-square distribution with ν degrees of freedom, respectively. The degrees of freedom for $\hat{\sigma}_A^2 = \frac{1}{bn}\text{MS}_{\text{A}} - \frac{1}{bn}\text{MS}_{\text{B(A)}}$ are obtained using the Satterthwaite approximation and are given by

$$\nu = \frac{(bn\hat{\sigma}_A^2)^2}{(\text{MS}_{\text{A}})^2/(a-1) + (\text{MS}_{\text{B(A)}})^2/a(b-1)}$$

Formulas for constructing confidence intervals for the other variance components can be similarly obtained.

6.4.1 Using PROC GLM to Analyze Two-Way Nested Random Effects Models

In order to study the variation of the calcium content in turnip greens, four plants were selected at random. From each plant, three leaves were randomly selected, and from each leaf, two samples of 100 mg each were taken and the calcium content determined. This experiment is described in Snedecor and Cochran (1989). The data appear in Table 6.4.

The experimenter is interested in verifying whether there is a significant variation in calcium content from plant to plant compared to the variation within a plant. If so, it is also of interest to obtain an estimate of this variation. The model is

$$y_{ijk} = \mu + A_i + B_{ij} + \epsilon_{ijk} \; i = 1,4; \; j = 1,3; \; k = 1,2$$

where A_i, the effect of plant i, is assumed to be iid $N(0, \sigma_A^2)$; B_{ij}, the leaf i within plant j effect, is assumed to be iid $N(0, \sigma_B^2)$; and ϵ_{ijk}, the samples within leaf within plant effect, is iid $N(0, \sigma^2)$. It is also assumed that A_i, B_{ij}, and ϵ_{ijk} are pairwise independent.

Table 6.4. Calcium content in turnip greens (Snedecor and Cochran 1989)

Plant	Leaf	Determinations of Ca	
1	1	3.28	3.09
	2	3.52	3.48
	3	2.88	2.80
2	1	3.46	2.44
	2	1.87	1.92
	3	2.19	2.19
3	1	2.77	2.66
	2	3.74	3.44
	3	2.55	2.55
4	1	3.78	3.87
	2	4.07	4.12
	3	3.31	3.31

SAS Example F8

The SAS Example F8 program (see Fig. 6.21) illustrates how `proc glm` is used to fit the above model to the calcium in turnip data. The data are entered into SAS in a straightforward way. However, note how two separate observations are created in the SAS data set from the two sample values from a leaf entered in the same line of data, using the `output` statements. It is important to recognize that the levels of leaf are specified in the data as if the two factors were crossed; that is, the levels of leaf are labeled 1, 2, and 3 for every level of plant.

Both `plant` and `leaf` are declared as classification variables in the `class` statement, but in the `model` statement, a `leaf(plant)` term is used to define the `leaf within plant` effect; that is, the `leaf(plant)` notation represents the "B_{ij}" term in the model. When this model specification is used to code the necessary design matrices that correspond to the random effects, the levels of the factor `leaf within plant` are identified as the levels of leaves within each plant (i.e., 11, 12, 13, 21, 22, ..., etc.).

The `random` statement declares that `plant` and `leaf(plant)` are random effects, whereas the `test` option requests `proc glm` to construct suitable F-statistics for testing hypotheses about the variance components specified in the `random` statement. The expected values of the mean squares in the ANOVA table determine the ratios of sums of squares to be used to test the two

```
data turnip;
input Plant Leaf X1-X2;
drop X1-X2;
Calcium=X1; output;
Calcium=X2; output;
cards;
1 1 3.28 3.09
1 2 3.52 3.48
1 3 2.88 2.80
2 1 2.46 2.44
2 2 1.87 1.92
2 3 2.19 2.19
3 1 2.77 2.66
3 2 3.74 3.44
3 3 2.55 2.55
4 1 3.78 3.87
4 2 4.07 4.12
4 3 3.31 3.31
;

proc glm data=turnip;
  class Plant Leaf;
  model Calcium= Plant Leaf(Plant);
  random Plant Leaf(Plant)/test;
  title 'Analysis of a Two way Nested Random Model: Turnip Data';
run;
```

Fig. 6.21. SAS Example F8: program

hypotheses of interest: $H_0 : \sigma_A^2 = 0$ versus $H_a : \sigma_A^2 > 0$ and $H_0 : \sigma_B^2 = 0$ versus $H_a : \sigma_B^2 > 0$, as discussed previously.

The Types I and III sums of squares are used to compute the standard output from `proc glm` (see bottom portion of Fig. 6.22), and, by default, the error mean square is used as the denominator of the F-ratios constructed to test the above hypotheses. Although this will produce the correct F-statistic to test the `leaf(plant)` effect, the F-statistic calculated for testing the `plant` effect is incorrect. The inclusion of the `test` option in the `random` statement will result in the use of `leaf within plant` mean square as the denominator to test the `plant` effect, and this produces the correct F-statistic, as observed from the output shown in Fig. 6.23. The F-statistic for testing the `leaf(plant)` effect uses the MSE as the denominator and is identical to the statistic for testing this effect in Fig. 6.22.

The following ANOVA table for the turnip green data is constructed from the Type III sums of squares output from `proc glm`.

SV	df	SS	MS	F	p-Value	E(MS)
Plant	3	7.56035	2.52012	7.67	0.0097	$\sigma^2 + 2\sigma_B^2 + 6\sigma_A^2$
Leaf (Plant)	8	2.63020	0.32878	49.41	<0.0001	$\sigma^2 + 2\sigma_B^2$
Error	12	0.07985	0.00665			
Total	23	10.27040				

The expected mean squares are those derived by `proc glm` and displayed on page 3 in Fig. 6.23 in the table titled "Type III Expected Mean Square." To test $H_0 : \sigma_A^2 = 0$ versus $H_a : \sigma_A^2 > 0$, the statistic $F_1 = 2.52012/0.32878 = 7.67$ is used. Since the p-value is 0.0097, the null hypothesis is rejected at

Analysis of a Two way Nested Random Model: Turnip Data

The GLM Procedure

Class Level Information		
Class	Levels	Values
Plant	4	1 2 3 4
Leaf	3	1 2 3

Number of Observations Read	24
Number of Observations Used	24

Dependent Variable: Calcium

Source	DF	Sum of Squares	Mean Square	F Value	Pr > F
Model	11	10.19054583	0.92641326	139.22	<.0001
Error	12	0.07985000	0.00665417		
Corrected Total	23	10.27039583			

R-Square	Coeff Var	Root MSE	Calcium Mean
0.992225	2.708195	0.081573	3.012083

Source	DF	Type I SS	Mean Square	F Value	Pr > F
Plant	3	7.56034583	2.52011528	378.73	<.0001
Leaf(Plant)	8	2.63020000	0.32877500	49.41	<.0001

Source	DF	Type III SS	Mean Square	F Value	Pr > F
Plant	3	7.56034583	2.52011528	378.73	<.0001
Leaf(Plant)	8	2.63020000	0.32877500	49.41	<.0001

Fig. 6.22. SAS Example F8: output from PROC GLM (pages 1 and 2)

$\alpha = 0.05$. The F-statistic $F_2 = 0.32878/0.00665 = 49.41$ is associated with a p-value of < 0.0001. Thus, the null hypothesis of $H_0 : \sigma_B^2 = 0$ versus $H_a : \sigma_B^2 > 0$ is also rejected at $\alpha = 0.05$. The variance components are estimated by setting the computed mean squares equal to their expected values and solving the resulting set of equations:

$$\begin{aligned} \sigma^2 + 2\,\sigma_B^2 + 6\,\sigma_A^2 &= 2.52012 \\ \sigma^2 + 2\,\sigma_B^2 &= 0.32878 \\ \sigma^2 &= 0.00665 \end{aligned}$$

The solutions are $\hat{\sigma}^2 = 0.00665$, $\hat{\sigma}_B^2 = (0.32875 - 0.00665)/2 = 0.16105$, and $\hat{\sigma}_A^2 = (2.52012 - 0.32878)/6 = 0.36522$. The conclusion is that the variation

Analysis of a Two way Nested Random Model: Turnip Data

The GLM Procedure

Source	Type III Expected Mean Square
Plant	Var(Error) + 2 Var(Leaf(Plant)) + 6 Var(Plant)
Leaf(Plant)	Var(Error) + 2 Var(Leaf(Plant))

Tests of Hypotheses for Random Model Analysis of Variance

Dependent Variable: Calcium

Source	DF	Type III SS	Mean Square	F Value	Pr > F
Plant	3	7.560346	2.520115	7.67	0.0097
Error	8	2.630200	0.328775		
Error: MS(Leaf(Plant))					

Source	DF	Type III SS	Mean Square	F Value	Pr > F
Leaf(Plant)	8	2.630200	0.328775	49.41	<.0001
Error: MS(Error)	12	0.079850	0.006654		

Fig. 6.23. SAS Example F8: output from PROC GLM (pages 3 and 4)

in calcium content among the leaves within a plant is about 24 times as large, and among the plants, it is about 55 times as large as the variation among samples within leaves.

6.4.2 Using PROC MIXED to Analyze Two-Way Nested Random Effects Models

In this subsection, `proc mixed` is used to fit the model discussed in Sect. 6.4.1 to the turnip green data. The method of moments is used, mainly so that the results can be compared with the analysis obtained previously using `proc glm`. The differences between the `proc mixed` analysis obtained by this method and those obtained using MLE and REML methods are indicated at the end of this subsection.

SAS Example F9

In the SAS Example F9 program (see Fig. 6.24), the `method=type3` used in the proc statement requests that the method of moments estimators using the Type III expected mean squares are to be calculated. The resulting variance component estimates, F-statistics, and their p-values are shown in Fig. 6.25, and they are identical to those from `proc glm`. The variance component estimates are the same as those calculated by hand using results from `proc glm`. The confidence intervals calculated by `proc mixed` are again those based on large sample standard errors and the normal percentiles except those for σ^2, which are based on chi-square percentiles.

```
insert data step to create the SAS dataset 'turnip' here

proc mixed data=turnip noclprint noinfo method=type3 cl; ;
  class Plant Leaf;
  model Calcium= /solution;
  random Plant Leaf(Plant);
  title "Analysis of a Two way Nested Random Model: Turnip Data";
run;
```

Fig. 6.24. SAS Example F9: method of moments estimation using PROC MIXED

Analysis of a Two way Nested Random Model: Turnip Data

The Mixed Procedure

Type 3 Analysis of Variance								
Source	DF	Sum of Squares	Mean Square	Expected Mean Square	Error Term	Error DF	F Value	Pr > F
Plant	3	7.560346	2.520115	Var(Residual) + 2 Var(Leaf(Plant)) + 6 Var(Plant)	MS(Leaf(Plant))	8	7.67	0.0097
Leaf(Plant)	8	2.630200	0.328775	Var(Residual) + 2 Var(Leaf(Plant))	MS(Residual)	12	49.41	<.0001
Residual	12	0.079850	0.006654	Var(Residual)	.	.	.	.

Covariance Parameter Estimates				
Cov Parm	Estimate	Alpha	Lower	Upper
Plant	0.3652	0.05	-0.3091	1.0395
Leaf(Plant)	0.1611	0.05	-0.00006	0.3222
Residual	0.006654	0.05	0.003422	0.01813

Fit Statistics	
-2 Res Log Likelihood	2.2
AIC (Smaller is Better)	8.2
AICC (Smaller is Better)	9.4
BIC (Smaller is Better)	6.3

Solution for Fixed Effects					
Effect	Estimate	Standard Error	DF	t Value	Pr > |t|
Intercept	3.0121	0.3240	3	9.30	0.0026

Fig. 6.25. SAS Example F9: output

If the option `method=reml` (the default value) or `method=ml` is specified as the method of estimation, the same values as those obtained from the method of moments will be obtained for balanced data. However, the standard errors will be calculated using large-sample results for maximum likelihood estimators that assume the numbers of levels for the two factors (sample sizes) are large. The inferences made from the resulting z-tests will not be the same as those made from the F-tests based on assuming normal distributions for the random effects. The confidence intervals for the variance components will be those based on the chi-square distribution and the Satterthwaite approximation.

6.5 Two-Way Mixed Effects Model

The mixed model is a linear model that involves both fixed and random effects. In the following subsections, several applications of this model will be discussed. In Chaps. 4 and 5, the least squares method was used to obtain the estimates of the parameters of the regression and ANOVA models, respectively, using the matrix form of the respective models:

$$\mathbf{y} = X\boldsymbol{\beta} + \boldsymbol{\epsilon}.$$

In the case of full-rank regression models, the solution to the normal equations

$$X'X\boldsymbol{\beta} = X'\mathbf{y}$$

gave the least squares estimate $\hat{\boldsymbol{\beta}}$ of $\boldsymbol{\beta}$ as

$$\hat{\boldsymbol{\beta}} = (X'X)^{-1}X'\mathbf{y}$$

In Chap. 5, analysis of variance models was also represented in the same matrix model setup where the X matrix, now called the *design matrix*, was constructed from the linear model describing the responses observed from an experiment and the parameter vector consisted of the model effects. A specific example in Sect. 5.1 illustrated how the X matrix is constructed for a typical experimental situation.

The matrix representation of a mixed model will be described in this section and the methods of estimation briefly summarized. As an example of a two-factor mixed model with interaction, consider the machine–operator example discussed in Sect. 6.3. To keep the dimensions of the matrices involved in the example within manageable limits, instead of four production machines (Factor A), consider that the breaking strengths of plastic sheeting from two specific brands of machines were of interest, that three operators were randomly selected, and that two trials were performed by each machine–operator combination in a completely randomized design. Thus, the applicable model may be expressed as the two-way crossed mixed effects model given by

$$y_{ijk} = \mu + \alpha_i + B_j + \alpha B_{ij} + \epsilon_{ijk}, \quad i = 1, \ldots, 2; \quad j = 1, \ldots, 3; \quad k = 1, \ldots, 2$$

where α_i are fixed effects due to the two levels of Factor A (machines); B_j, the random effects of Factor B (operators), are iid $N(0, \sigma_B^2)$ random variables; the interaction effects between the two factors denoted by αB_{ij} are iid $N(0, \sigma_{\alpha B}^2)$; and the random errors ϵ_{ijk} are iid $N(0, \sigma^2)$ random variables. The interaction effects are random because the levels depend on the levels of Factor B, which are randomly sampled. The matrix form of the model is

$$\mathbf{y} = X\boldsymbol{\beta} + Z_1\,\mathbf{u}_1 + Z_2\,\mathbf{u}_2 + \boldsymbol{\epsilon}$$

where

$$\mathbf{y} = \begin{bmatrix} y_{111} \\ y_{112} \\ y_{121} \\ y_{122} \\ y_{131} \\ y_{132} \\ y_{211} \\ y_{212} \\ y_{221} \\ y_{222} \\ y_{231} \\ y_{232} \end{bmatrix}, \quad X = \begin{bmatrix} 1 & 1 & 0 \\ 1 & 1 & 0 \\ 1 & 1 & 0 \\ 1 & 1 & 0 \\ 1 & 1 & 0 \\ 1 & 1 & 0 \\ 1 & 0 & 1 \\ 1 & 0 & 1 \\ 1 & 0 & 1 \\ 1 & 0 & 1 \\ 1 & 0 & 1 \\ 1 & 0 & 1 \end{bmatrix}, \quad \boldsymbol{\beta} = \begin{bmatrix} \mu \\ \alpha_1 \\ \alpha_2 \\ \alpha_3 \end{bmatrix}, \quad Z_1 = \begin{bmatrix} 1 & 0 & 0 \\ 1 & 0 & 0 \\ 0 & 1 & 0 \\ 0 & 1 & 0 \\ 0 & 0 & 1 \\ 0 & 0 & 1 \\ 1 & 0 & 0 \\ 1 & 0 & 0 \\ 0 & 1 & 0 \\ 0 & 1 & 0 \\ 0 & 0 & 1 \\ 0 & 0 & 1 \end{bmatrix}, \quad \mathbf{u}_1 = \begin{bmatrix} B_1 \\ B_2 \\ B_3 \end{bmatrix},$$

$$Z_2 = \begin{bmatrix} 1 & 0 & 0 & 0 & 0 & 0 \\ 1 & 0 & 0 & 0 & 0 & 0 \\ 0 & 1 & 0 & 0 & 0 & 0 \\ 0 & 1 & 0 & 0 & 0 & 0 \\ 0 & 0 & 1 & 0 & 0 & 0 \\ 0 & 0 & 1 & 0 & 0 & 0 \\ 0 & 0 & 0 & 1 & 0 & 0 \\ 0 & 0 & 0 & 1 & 0 & 0 \\ 0 & 0 & 0 & 0 & 1 & 0 \\ 0 & 0 & 0 & 0 & 1 & 0 \\ 0 & 0 & 0 & 0 & 0 & 1 \\ 0 & 0 & 0 & 0 & 0 & 1 \end{bmatrix}, \quad \mathbf{u}_2 = \begin{bmatrix} \alpha B_{11} \\ \alpha B_{12} \\ \alpha B_{13} \\ \alpha B_{21} \\ \alpha B_{22} \\ \alpha B_{23} \end{bmatrix}, \quad \boldsymbol{\epsilon} = \begin{bmatrix} \epsilon_{111} \\ \epsilon_{112} \\ \epsilon_{121} \\ \epsilon_{122} \\ \epsilon_{131} \\ \epsilon_{132} \\ \epsilon_{211} \\ \epsilon_{212} \\ \epsilon_{221} \\ \epsilon_{222} \\ \epsilon_{231} \\ \epsilon_{232} \end{bmatrix}.$$

The observed data vector is arranged so that, as in Sect. 5.1, the subscripts of the observations are lexically ordered; that is, digits to the right change faster than the ones on the left. For the observation y_{ijk}, the model terms are obtained by scalar products of the appropriate vectors in the corresponding rows of the matrices X, Z_1, and Z_2 and the parameter vectors $\boldsymbol{\beta}$, $\mathbf{u}_1$, and $\mathbf{u}_1$, respectively. By locating the 1s in the appropriate rows of X, Z_1, and Z_2 and selecting the terms in the same positions in the parameter vectors $\boldsymbol{\beta}$, $\mathbf{u}_1$, and $\mathbf{u}_1$, the relevant model terms can be easily extracted. For example, for observation y_{132}, the relevant elements are found in row 6 of these matrices (these are highlighted), giving the model terms to be μ, α_1, B_3, αB_{13}, and ϵ_{132}. (For compactness, the above model can be reduced to the form

$$\mathbf{y} = X\boldsymbol{\beta} + Z\,\mathbf{u} + \boldsymbol{\epsilon}$$

where Z_1 and Z_2 are combined to form Z and $\mathbf{u}_1$ and $\mathbf{u}_2$ are stacked together to form $\mathbf{u}$, but the expanded form is helpful in expressing the relevant variance–covariance matrices in easily expressible forms.)

In the classical variance component model, the random vectors $\mathbf{u}_1$, $\mathbf{u}_2$, and $\boldsymbol{\epsilon}$ have the multivariate normal distributions $N(\mathbf{0},\ \sigma_B^2\, I)$, $N(\mathbf{0},\ \sigma_{\alpha B}^2\, I)$, and $N(\mathbf{0},\ \sigma^2\, I)$, respectively, where the matrices $\sigma_B^2\, I$ and so forth are diagonal matrices with the diagonal elements all equal to the respective variance components. The variance–covariance matrix of $\mathbf{y}$ is thus given by the 12×12 matrix V:

$$V \;=\; Z_1 Z_1^{'} \sigma_B^2 + Z_2 Z_2^{'} \sigma_B^2 + \sigma^2\, I$$

By performing the necessary matrix multiplications $Z_1 Z_1^{'}$ and $Z_2 Z_2^{'}$, the form of V can be determined, from which important features of the covariance structure of the observations can be obtained. Thus, the variance of an observation y_{ijk} is $\sigma^2 + \sigma_B^2 + \sigma_{\alpha B}^2$ and covariance between pairs of observations resulting from a replication using

- the same machine but different operators is $\text{Cov}(y_{111}, y_{121}) = 0$
- different machines but the same operator is $\text{Cov}(y_{111}, y_{211}) = \sigma_1^2$
- the same machine and the same operator is $\text{Cov}(y_{111}, y_{112}) = \sigma_B^2 + \sigma_{\alpha B}^2$
- different machines and different operators is $\text{Cov}(y_{111}, y_{221}) = 0$

A clear distinction exists between estimating a fixed parametric function, such as $\mu + \alpha_i$, and *predicting* a random variable such as B_j. To gain a little insight into the theoretical implications, it is necessary to have a minimal understanding of how statistical inferences are made from a mixed model. In general, the mixed model is expressed in the form

$$\mathbf{y} = X\boldsymbol{\beta} + Z\,\mathbf{u} + \boldsymbol{\epsilon}$$

where the random vectors $\mathbf{u}$ and $\boldsymbol{\epsilon}$ have the multivariate normal distributions $N(\mathbf{0}, G)$and $N(\mathbf{0}, R)$, respectively, where the variance–covariance matrices G and R are fixed unknown constants. Using this form, the variance–covariance matrix of $\mathbf{y}$ is given by $V = ZGZ' + R$. Thus, the covariance structure of $\mathbf{y}$ is determined by the random effects design matrix Z, and the covariance structures are defined by the matrices G and R. For the model discussed above, these matrices take simple forms: R is a 12×12 identity matrix, and G is a 9×9 diagonal matrix with the diagonal elements $\sigma_B^2, \sigma_B^2, \sigma_B^2, \sigma_{\alpha B}^2, \ldots, \sigma_{\alpha B}^2$. The matrices X and Z consist of constants (usually 0s and 1s) because they are the usual design matrices for the two types of effects. By writing the likelihood function for the parameters, $\boldsymbol{\beta}$, G, and R (i.e., the joint density function of $\mathbf{y}$ and $\mathbf{u}$) and taking derivatives with respect to $\boldsymbol{\beta}$ and $\mathbf{u}$, the following set of equations known as the *mixed model equations* are obtained:

$$\begin{bmatrix} X'R^{-1}X & X'R^{-1}Z \\ Z'R^{-1}X & Z'R^{-1}Z + G^{-1} \end{bmatrix} \begin{bmatrix} \tilde{\boldsymbol{\beta}} \\ \tilde{\mathbf{u}} \end{bmatrix} = \begin{bmatrix} X'R^{-1}\mathbf{y} \\ Z'R^{-1}\mathbf{y} \end{bmatrix}$$

The solutions to mixed model equations are given by $\tilde{\boldsymbol{\beta}} = (X'V^{-1}X)^{-1}X'V^{-1}\mathbf{y}$ and $\tilde{\mathbf{u}} = GZ'V^{-1}(\mathbf{y} - X\tilde{\boldsymbol{\beta}})$. Using these, best linear unbiased estimates of estimable linear functions of $\boldsymbol{\beta}$ as well as best linear unbiased predictors (BLUPs) of *predictable functions* of $\boldsymbol{\beta}$ and $\mathbf{u}$ may be constructed.

A predictable function of $\boldsymbol{\beta}$ and $\mathbf{u}$ is a linear combination of the form $\ell'\boldsymbol{\beta} + m'\mathbf{u}$, where $\ell'\boldsymbol{\beta}$ is an estimable function of $\boldsymbol{\beta}$. Recall that estimable functions were defined in Sect. 5.1 of Chap. 5. It is clear that if fixed parameters are not involved, it is possible to predict virtually any linear function of $\mathbf{u}$; however, in practice, predictable functions considered are only those that are interpretable as part of the inference made from a particular experiment. For example, the mixed model equations are due to Henderson (in Henderson et al. (1959); also see Searle et al. (1992)), who developed a procedure for predicting breeding values (defined as a predictable function, say $\mu + B_i$), for randomly selected sires (i.e., sire is the random Factor B) in an animal genetic experiment. See Searle et al. (1992) for a detailed presentation on BLUPs.

In the above machine–operator experiment, $E(y_{ijk}) = \mu + \alpha_i$ is an estimable function of the fixed parameters and estimates the mean strength of plastic sheeting from machine i, averaged over all operators in the population of operators. On the other hand, the function $\mu + \alpha_i + B_. + \alpha B_{i.}$ (where $B_. = \frac{1}{3}\sum_j B_j$ and $\alpha B_{i.} = \frac{1}{3}\sum_j \alpha B_{ij}$) is the expectation of y_{ijk} averaged over the three operators in the experiment. This is a predictable function different from the estimable function above and estimates the mean strength for Machine i given that the effects of the operators are fixed. Another example of a predictable function is $\alpha_i - \alpha_j + (\alpha B_{i.} - \alpha B_{j.})$, which measures the difference between the two machines i and j.

Although, BLUPs were not discussed for random models considered in Sects. 6.2, 6.3, and 6.4, they can also be defined for those models by treating them as mixed models by taking μ as the only fixed effect in each of the models. Thus, for example, in the one-way random model $y_{ij} = \mu + A_i + \epsilon_{ij}$ considered in Sect. 6.2, the BLUP of $\mu + A_i$ is of the form $\delta\bar{y}_{..} + (1-\delta)\bar{y}_{i.}$, where $\delta = \sigma^2/(\sigma^2 + n\sigma_A^2)$. From this example, it is clear that the BLUP is a function of the unknown variance components. Thus, for BLUPs to be practically useful, they need to be estimated because the values of the variance components involved in the expressions for the predictors are unknown. In practice, variance components are first estimated and then plugged into the BLUPs to obtain *estimated* BLUPs (eBLUPs).

6.5.1 Two-Way Mixed Effects Model: Randomized Complete Block Design

In Sect. 5.7, the analysis of a randomized block design with block effects considered as fixed effects was discussed. However, in practice, block effects need to be considered as random effects because statistical inferences that will be made about the differences in treatment effects from such a design must be valid regardless of the choice of blocks. By considering the blocks used in the experiment as a random sample from a hypothetical population of blocks, the effects of the blocks can be specified in the model as random effects. From such a model, inferences regarding differences in the treatment effects can be made using the (unconditional) means of the observations, with the variance of the block effects, and then accounting for the variability among blocks present.

With random block effects, using the same arguments presented when blocks were considered fixed, the model may still be specified as an additive model since the block effects do not interact with the treatment effects; thus, no interaction term is necessary in this case, as well.

Model

An appropriate model for an RCBD with random block effects is

$$y_{ij} = \mu + \tau_i + B_j + \epsilon_{ij}, \quad i = 1, \ldots, t;\ j = 1, \ldots, r$$

where τ_i is the effect of the ith treatment; the effect of the jth block, B_j, is assumed to be iid $N(0, \sigma_B^2)$; and the random error ϵ_{ij} iid $N(0, \sigma^2)$ is distributed independently of B_j. As a consequence of this model, the mean of a response to treatment i is $E(y_{ij}) = \mu + \tau_i$, and the variance is $\text{Var}(y_{ij}) = \sigma^2 + \sigma_B^2$. Further, the covariance between two observations in the same block is $\text{Cov}(y_{ij}, y_{i'j}) = \sigma_B^2$, but observations from different blocks are uncorrelated.

Estimation and Hypothesis Testing

An analysis of variance that corresponds to the model is constructed using the same computational formulas as when blocks were considered fixed but the expected mean squares must be calculated using the assumptions described above. As with the random models, an additional column displaying the expected mean squares is included in the ANOVA table for a mixed model:

SV	df	SS	MS	F	E(MS)
Blocks	$r-1$	$\mathtt{SS_A}$	$\mathtt{MS_A}$	$\mathtt{MS_A/MSE}$	$\sigma^2 + r\,\sigma_B^2$
Trts	$t-1$	$\mathtt{SS_{Trt}}$	$\mathtt{MS_{Trt}}$	$\mathtt{MS_{Trt}/MSE}$	$\sigma^2 + r\frac{\sum_i(\tau_i - \bar{\tau})^2}{(t-1)}$
Error	$(r-1)(t-1)$	SSE	MSE$(= s^2)$		σ^2
Total	$rt-1$				

The F-statistic for Trts tests the hypothesis of equality of treatment effects:

$$H_0: \tau_1 = \tau_2 = \cdots = \tau_t \text{ versus } H_a: \text{ at least one inequality}$$

or, equivalently,

$$H_0: \mu_1 = \mu_2 = \cdots = \mu_t \text{ versus } H_a: \text{ at least one inequality}$$

where $\mu_i = \mu + \tau_i$, the ith treatment mean. H_0 is rejected if the observed F-value exceeds the α upper percentile of an F-distribution with $\text{df}_1 = t-1$ and $\text{df}_2 = (r-1)(t-1)$. The best estimate of the difference between the effects of two treatments labeled p and q is

$$\widehat{\mu_p - \mu_q} = \widehat{\tau_p - \tau_q} = \bar{y}_{p.} - \bar{y}_{q.}$$

with standard error given by

$$s_d = \text{s.e.}(\bar{y}_{p.} - \bar{y}_{q.}) = s\sqrt{\frac{2}{r}}$$

where $s = \sqrt{\text{MSE}}$. A $(1-\alpha)100\%$ confidence interval for $\mu_p - \mu_q$ (or, equivalently, $\tau_p - \tau_q$) is

$$(\bar{y}_{p.} - \bar{y}_{q.}) \pm t_{\alpha/2,\nu} \cdot s\sqrt{\frac{2}{r}}$$

where $t_{\alpha/2,\nu}$ is the upper $\alpha/2$ percentile point of a t-distribution with $\nu = (r-1)(t-1)$ degrees of freedom. Thus, none of these results regarding the treatment effects is different from the fixed block effect case. Similarly, standard errors for linear comparisons of treatment means and the corresponding t-tests may be calculated.

SAS Example F10

The data from the experiment comparing five seed treatments in five replications described in Snedecor and Cochran (1989) and used in SAS Example E11 (see Sect. 5.7) is reanalyzed in this example, but considering the blocks as random effects.

```
data soybean;
input Trt $ @;
do Rep=1 to 5;
   input Yield @;
   output;
end;
datalines;
check 8 10 12 13 11
arasan 2 6 7 11 5
spergon 4 10 9 8 10
samesan 3 5 9 10 6
fermate 9 7 5 5 3
;

proc glm order=data;
   class Trt Rep;
   model Yield = Trt Rep;
   contrast 'Check vs Chemicals' Trt 4 -1 -1 -1 -1;
   random Rep/Q;
   lsmeans Trt/pdiff cl adjust=tukey;
run;
```

Fig. 6.26. SAS Example F10: analysis of seed treatments

The data were shown in Table 5.10 in Sect. 5.7. The model is thus

$$y_{ij} = \mu + \tau_i + B_j + \epsilon_{ij}, \quad i = 1, \ldots, 5;\ j = 1, \ldots, 5$$

where τ_i is the effect of the ith seed treatment; the effect of the jth block, B_j, is assumed to be iid $N(0, \sigma_B^2)$, and the random error ϵ_{ij} iid $N(0, \sigma^2)$ is distributed independently of B_j.

The `proc glm` step (see Fig. 6.26) is similar to that used in SAS Example E11 except for the inclusion of the `random` statement and use of the `lsmeans` statement instead of the `means` statement. This `random` statement leads to the computation of the expected mean squares for terms in the model statement.

The `Q` option causes the matrix for the quadratic forms (described below) that appear in the expected mean squares for fixed effects to be explicitly displayed. In this example, there is a single such quadratic form for the treatment effect.

The `lsmeans` statement requests 95% confidence intervals for pairwise differences in seed treatment means (effects) that are adjusted for simultaneous inference using the Tukey method. The results would be exactly the same as those resulting from the use of the `means trt/tukey cldiff;` statement. Note also that the `contrast` statement is placed ahead of the `random` statement so that the expected value of the mean square for testing the contrast hypothesis is also calculated.

Analysis of Seed Treatments in a RCBD

Dependent Variable: Yield

Source	DF	Sum of Squares	Mean Square	F Value	Pr > F
Model	8	133.6800000	16.7100000	3.09	0.0262
Error	16	86.5600000	5.4100000		
Corrected Total	24	220.2400000			

R-Square	Coeff Var	Root MSE	Yield Mean
0.606974	30.93006	2.325941	7.520000

Source	DF	Type I SS	Mean Square	F Value	Pr > F
Trt	4	83.84000000	20.96000000	3.87	0.0219
Rep	4	49.84000000	12.46000000	2.30	0.1032

Source	DF	Type III SS	Mean Square	F Value	Pr > F
Trt	4	83.84000000	20.96000000	3.87	0.0219
Rep	4	49.84000000	12.46000000	2.30	0.1032

Contrast	DF	Contrast SS	Mean Square	F Value	Pr > F
Check vs Chemicals	1	67.24000000	67.24000000	12.43	0.0028

Fig. 6.27. SAS Example F10: output (page 2)

Edited forms of the output from the SAS Example F10 program appears in Figs. 6.27, 6.28, and 6.29. Figure 6.27 contains the default Type III F-statistics for the `trt` effects and the results of the test of the contrast hypothesis. For the RCBD (considered as a mixed model), these F-statistics are the correct statistics for testing for fixed effects. The contrast hypothesis is rejected (p-value= 0.0028); thus, the average effect of the seed treatments on germination is found to be different from the effect of the control of no treatment used.

The GLM Procedure
Quadratic Forms of Fixed Effects in the Expected Mean Squares

Source: Type III Mean Square for Trt					
	Trt check	Trt arasan	Trt spergon	Trt samesan	Trt fermate
Trt check	4.00000000	-1.00000000	-1.00000000	-1.00000000	-1.00000000
Trt arasan	-1.00000000	4.00000000	-1.00000000	-1.00000000	-1.00000000
Trt spergon	-1.00000000	-1.00000000	4.00000000	-1.00000000	-1.00000000
Trt samesan	-1.00000000	-1.00000000	-1.00000000	4.00000000	-1.00000000
Trt fermate	-1.00000000	-1.00000000	-1.00000000	-1.00000000	4.00000000

Source: Contrast Mean Square for Check vs Chemicals					
	Trt check	Trt arasan	Trt spergon	Trt samesan	Trt fermate
Trt check	4.00000000	-1.00000000	-1.00000000	-1.00000000	-1.00000000
Trt arasan	-1.00000000	0.25000000	0.25000000	0.25000000	0.25000000
Trt spergon	-1.00000000	0.25000000	0.25000000	0.25000000	0.25000000
Trt samesan	-1.00000000	0.25000000	0.25000000	0.25000000	0.25000000
Trt fermate	-1.00000000	0.25000000	0.25000000	0.25000000	0.25000000

Source	Type III Expected Mean Square
Trt	Var(Error) + Q(Trt)
Rep	Var(Error) + 5 Var(Rep)

Contrast	Contrast Expected Mean Square
Check vs Chemicals	Var(Error) + Q(Trt)

Fig. 6.28. SAS Example F10: quadratic form for E(MS) for treatment effects

On part of the output displayed in Fig. 6.28, the Type III expected mean squares for the model effects and the contrast are displayed. Both the expressions for the expected mean square of the `trt` and the contrast (see tables toward the bottom) contain a term with a quadratic form (labeled `Q(trt)`). For the RCBD, `Q(trt)` is of the form $r\frac{\sum_i(\tau_i - \bar{\tau})^2}{(t-1)}$ (as shown in the ANOVA table given earlier in this subsection).

This can be verified using the matrix of the quadratic form displayed at the top table of Fig. 6.28 (under the title `Type III Mean Square for trt`). To obtain the form of Q for an effect, one needs to calculate the quadratic form $\boldsymbol{\tau}'A\boldsymbol{\tau}$ and divide by the degrees of freedom for the effect. Here, A is the matrix the columns of which are printed, and $\boldsymbol{\tau}$ is the vector of fixed effects parameters $\boldsymbol{\tau} = (\tau_1, \tau_2, \tau_3, \tau_4, \tau_5)'$. Thus, the computation requires the matrix multiplication

$$\begin{bmatrix}\tau_1 \ \tau_2 \ \tau_3 \ \tau_4 \ \tau_5\end{bmatrix}\begin{bmatrix} 4 & -1 & -1 & -1 & -1 \\ -1 & 4 & -1 & -1 & -1 \\ -1 & -1 & 4 & -1 & -1 \\ -1 & -1 & -1 & 4 & -1 \\ -1 & -1 & -1 & -1 & 4 \end{bmatrix}\begin{bmatrix}\tau_1 \\ \tau_2 \\ \tau_3 \\ \tau_4 \\ \tau_5\end{bmatrix} = 5(\textstyle\sum_i(\tau_i - \bar{\tau})^2)$$

giving `Q(trt)`$= 5(\sum_i(\tau_i - \bar{\tau})^2)/4$. This expected mean square is not different from the case where blocks were considered fixed effects.

In a similar fashion, the Q for the contrast expected mean square may be calculated using the corresponding matrix displayed in the table in the middle of Fig. 6.28 (under the title `Contrast Mean Square for Check vs. Chemicals`). In balanced data situations for common experimental designs, such as the RCBD, it is not necessary to use the `Q` option since the form of the expected mean squares for the fixed effects is available from many textbooks. It was used in this example for illustrating how the output matrix from the `Q` option is used to construct the quadratic form. This option is mainly useful for determining the expected mean squares in complex situations where standard forms are not available.

The GLM Procedure
Least Squares Means
Adjustment for Multiple Comparisons: Tukey

Trt	Yield LSMEAN	LSMEAN Number
check	10.8000000	1
arasan	6.2000000	2
spergon	8.2000000	3
samesan	6.6000000	4
fermate	5.8000000	5

Least Squares Means for effect Trt
Pr > |t| for H0: LSMean(i)=LSMean(j)

Dependent Variable: Yield

i/j	1	2	3	4	5
1		0.0443	0.4242	0.0740	0.0261
2	0.0443		0.6602	0.9987	0.9987
3	0.4242	0.6602		0.8102	0.4999
4	0.0740	0.9987	0.8102		0.9812
5	0.0261	0.9987	0.4999	0.9812	

Trt	Yield LSMEAN	95% Confidence Limits	
check	10.800000	8.594891	13.005109
arasan	6.200000	3.994891	8.405109
spergon	8.200000	5.994891	10.405109
samesan	6.600000	4.394891	8.805109
fermate	5.800000	3.594891	8.005109

Least Squares Means for Effect Trt

i	j	Difference Between Means	Simultaneous 95% Confidence Limits for LSMean(i)-LSMean(j)	
1	2	4.600000	0.093171	9.106829
1	3	2.600000	-1.906829	7.106829
1	4	4.200000	-0.306829	8.706829
1	5	5.000000	0.493171	9.506829
2	3	-2.000000	-6.506829	2.506829
2	4	-0.400000	-4.906829	4.106829
2	5	0.400000	-4.106829	4.906829
3	4	1.600000	-2.906829	6.106829
3	5	2.400000	-2.106829	6.906829
4	5	0.800000	-3.706829	5.306829

Fig. 6.29. SAS Example F10: output (edited results of lsmeans statement)

Figure 6.29 contains the results of the `lsmeans` statement. The `pdiff` option produced the second table on this SAS output, which gives p-values associated with testing pairwise differences in means (i.e., hypotheses of the form $H_0 : \mu_i = \mu_j$ versus $H : \mu_i \neq \mu_j$ for all pairs (i, j)). The 95% confidence intervals for pairwise differences displayed in the table at the bottom of

Fig. 6.29 were produced as a result of the cl option. The confidence intervals are *adjusted* for multiple comparisons using Tukey's procedure.

Zero is included in every interval except 1–2 (i.e., check–arasan) and 1–5 (i.e., check–fermate). This represents one less pair of means not found to be different than when t-based confidence intervals (i.e., those unadjusted for multiple comparisons) were used in Sect. 5.7. Thus, this procedure is slightly more conservative than using t-based intervals. The conclusions are similar to those drawn in Sect. 5.7 except that both *spergon* and *samesan* are found to be different from the control.

SAS Example F11

In this program (displayed in Fig. 6.30), proc mixed is used to analyze the data from the experiment comparing five seed treatments using Type 3 sums of squares and the method of moments. As discussed in Sect. 6.2, there are several advantages to using proc mixed instead of proc glm even when iterative estimation methods are not used. Although using the random statement in proc glm expected mean squares and F-tests about variance components can be computed, other statements in proc glm do not make use of this information. For example, the standard ANOVA tables proc glm produces still regard all effects as fixed effects, whereas in proc mixed, separate tests are performed for the variance components. Moreover, proc mixed allows the user to choose among several estimation methods as well as the ability to use estimate statements to estimate BLUPs involving fixed and random effects.

```
insert data step to create the SAS data set 'soybean' here

proc mixed order=data method=type3 covtest cl;
   class Trt Rep;
   model Yield = Trt;
   random Rep;
   lsmeans Trt/diff cl adjust=tukey;
   contrast 'Check vs Chemicals' Trt 4 -1 -1 -1 -1;
run
```

Fig. 6.30. SAS Example F11: analysis of seed treatments using PROC MIXED

The method=type3 option in the proc statement specifies that the variance components are to be estimated by the method of moments using the Type 3 sums of squares. The standard errors of the estimated variance components and associated confidence intervals are computed as a result of the covtest and the cl options. As discussed in Sect. 6.2, the model statement in proc mixed requires only the fixed part of the model to be specified; thus, model yield = trt; implies an overall mean μ and the fixed effect trt are in the model. The random statement specifies the random portion of the model: here, the random effect rep and a random error term. The variance of this effect

Randomized Blocks with PROC MIXED

The Mixed Procedure

Model Information	
Data Set	WORK.SOYBEAN
Dependent Variable	Yield
Covariance Structure	Variance Components
Estimation Method	Type 3
Residual Variance Method	Factor
Fixed Effects SE Method	Model-Based
Degrees of Freedom Method	Containment

Class Level Information		
Class	Levels	Values
Trt	5	check arasan spergon samesan fermate
Rep	5	1 2 3 4 5

Dimensions	
Covariance Parameters	2
Columns in X	6
Columns in Z	5
Subjects	1
Max Obs per Subject	25

Number of Observations	
Number of Observations Read	25
Number of Observations Used	25
Number of Observations Not Used	0

Type 3 Analysis of Variance								
Source	DF	Sum of Squares	Mean Square	Expected Mean Square	Error Term	Error DF	F Value	Pr > F
Trt	4	83.840000	20.960000	Var(Residual) + Q(Trt)	MS(Residual)	16	3.87	0.0219
Rep	4	49.840000	12.460000	Var(Residual) + 5 Var(Rep)	MS(Residual)	16	2.30	0.1032
Residual	16	86.560000	5.410000	Var(Residual)	.	.	.	.

Fig. 6.31. SAS Example F11: model fit output

and the error variance constitute the two `variance components` specified by this model. Thus, the `model` and `random` statements (in addition to the `class` statement) are needed to specify a mixed model in `proc mixed`. The `lsmeans` statement requests 95% confidence intervals for pairwise differences in seed treatment effects adjusted for simultaneous inference using the Tukey method. The `diff` option is redundant here as `adjust=` option implies the `diff` option.

The information on page 1 (see Fig. 6.31) is the same as that described earlier for SAS Examples F2 and F3 in Sect. 6.2. The table titled `Type 3 Analysis of Variance` provides the expected mean squares for all effects and is exactly the same as those produced by `proc glm`. The actual F-tests for both fixed and random effects are shown on page 2 (see Fig. 6.32). These are the same as those in the standard `Type III SS` table from `proc glm`. However, a separate table for the F-tests of the *fixed effects* is also provided lower on page 2. The table titled `Covariance Parameter Estimates` gives estimates of variance components obtained via the method of moments. As in Sect. 6.2, these are obtained by solving

$$\sigma^2 + 5\sigma_B^2 = 12.46$$

$$\sigma^2 = 5.41$$

which give $\hat{\sigma}^2 = 5.41$ and $\hat{\sigma}_B^2 = (12.46 - 5.41)/5 = 1.41$. The confidence intervals calculated by `proc mixed` for the variance components are those based on large-sample standard errors and the normal percentiles except those for σ^2. For example, the Wald statistic z is given by $z = \hat{\sigma}_B^2/\text{s.e.}(\hat{\sigma}_B^2)$, and a 95% confidence interval for σ_B^2 is thus $11.4478 \pm (1.96)(8.7204) = (-5.6442, 28.5398)$.

The Mixed Procedure

Covariance Parameter Estimates

Cov Parm	Estimate	Standard Error	Z Value	Pr Z	Alpha	Lower	Upper
Rep	1.4100	1.8032	0.78	0.4342	0.05	-2.1241	4.9441
Residual	5.4100	1.9127	2.83	0.0023	0.05	3.0008	12.5310

Fit Statistics

-2 Res Log Likelihood	101.9
AIC (Smaller is Better)	105.9
AICC (Smaller is Better)	106.6
BIC (Smaller is Better)	105.1

Type 3 Tests of Fixed Effects

Effect	Num DF	Den DF	F Value	Pr > F
Trt	4	16	3.87	0.0219

Contrasts

Label	Num DF	Den DF	F Value	Pr > F
Check vs Chemicals	1	16	12.43	0.0028

Fig. 6.32. SAS Example F11: estimates and tests

If a $(1-\alpha)100\%$ confidence interval based on the Satterthwaite approximation is desired, a formula similar to (6.1) in Sect. 6.2 could be used. The relevant SAS data step is

```
data cint;
   alpha=.05; a=5; r=5;
   msb= 12.46; s2=5.41; sb2=1.41;
   nu=(a*sb2)**2/(msb**2/(r-1)+ s2**2/((a-1)*(r-1)));
   L= (nu*sb2)/cinv(1-alpha/2,nu);
   U= (nu*sb2)/cinv(alpha/2,nu);
   put nu=   L=    U=;
run;
```

Executing the above gives the interval (0.3075, 430.516). However, since the Wald statistic z and the degrees of freedom $\nu = 2z^2$ are already available, the code given at the end of SAS Example F7 is simpler to use.

Note, however, that the variance component σ_B^2 is not of major interest in this experiment, but an estimate and a valid hypothesis test of the variation among the blocks are available to the experimenter. The output from `lsmeans` is of main interest and appears in pages 3 and 4 (see Figs. 6.33 and 6.34 for extracted parts from these pages). In Fig. 6.33, estimates, standard errors, and confidence intervals for the treatment means $\mu_i = \mu + \tau_i$ are given. It is important to note that the standard errors are computed (correctly) using the formula $\sqrt{(\hat{\sigma}^2 + \hat{\sigma}_B^2)/5} = \sqrt{(5.41 + 1.41)/5} = 1.1679$. Note that in SAS Example F10, `proc glm` would have used the formula $\sqrt{(\hat{\sigma}^2/5}$ (if the `stderr` option was specified requesting it) because `rep` is considered fixed in `proc glm` for the purpose of this computation.

The results of the t-tests are shown in Fig. 6.34. Besides the p-values computed for the standard t-tests, an additional column (titled `Adj P`) provides the p-values adjusted for multiple testing. Here, the adjustment is based on Tukey's studentized range distribution, the p-values being calculated are probabilities that studentized range random variable $q(t,\nu)$ exceeds $|\bar{y}_i - \bar{y}_j|/\sqrt{s^2/r}$, where $\nu = (a-1)(r-1)$ and $\bar{y}_i$ and $\bar{y}_j$are a pair of `trt`

Least Squares Means									
Effect	Trt	Estimate	Standard Error	DF	t Value	Pr > |t|	Alpha	Lower	Upper
Trt	check	10.8000	1.1679	16	9.25	<.0001	0.05	8.3242	13.2758
Trt	arasan	6.2000	1.1679	16	5.31	<.0001	0.05	3.7242	8.6758
Trt	spergon	8.2000	1.1679	16	7.02	<.0001	0.05	5.7242	10.6758
Trt	samesan	6.6000	1.1679	16	5.65	<.0001	0.05	4.1242	9.0758
Trt	fermate	5.8000	1.1679	16	4.97	0.0001	0.05	3.3242	8.2758

Fig. 6.33. SAS Example F11: estimates and standard errors of means

means. The SAS function probmc can be used to verify this by executing a data step such as

```
data pval;
   a=5; r=5; diff=4.6; nu=(a-1)*(r-1); s2=5.41;
   q=diff/sqrt(s2/r);
   p=1-probmc("Range", q, ., nu, 5);
   put  df= q=  p= ;
run;
```

which gives $q = 4.42226$ and $p = 0.04429$. The adjusted p-values do not change the results obtained previously for these comparisons.

Two sets of confidence intervals are shown in Fig. 6.34, one set based on the t-distribution, and the second set adjusted for multiple comparisons using Tukey's studentized range statistic. The results are exactly the same as those obtained from the previous analysis using proc glm.

Differences of Least Squares Means

Effect	Trt	_Trt	Estimate	Standard Error	DF	t Value	Pr > |t|	Adjustment	Adj P	Alpha
Trt	check	arasan	4.6000	1.4711	16	3.13	0.0065	Tukey-Kramer	0.0443	0.05
Trt	check	spergon	2.6000	1.4711	16	1.77	0.0962	Tukey-Kramer	0.4242	0.05
Trt	check	samesan	4.2000	1.4711	16	2.86	0.0115	Tukey-Kramer	0.0740	0.05
Trt	check	fermate	5.0000	1.4711	16	3.40	0.0037	Tukey-Kramer	0.0261	0.05
Trt	arasan	spergon	-2.0000	1.4711	16	-1.36	0.1928	Tukey-Kramer	0.6602	0.05
Trt	arasan	samesan	-0.4000	1.4711	16	-0.27	0.7892	Tukey-Kramer	0.9987	0.05
Trt	arasan	fermate	0.4000	1.4711	16	0.27	0.7892	Tukey-Kramer	0.9987	0.05
Trt	spergon	samesan	1.6000	1.4711	16	1.09	0.2929	Tukey-Kramer	0.8102	0.05
Trt	spergon	fermate	2.4000	1.4711	16	1.63	0.1223	Tukey-Kramer	0.4999	0.05
Trt	samesan	fermate	0.8000	1.4711	16	0.54	0.5941	Tukey-Kramer	0.9812	0.05

Differences of Least Squares Means

Effect	Trt	_Trt	Lower	Upper	Adj Lower	Adj Upper
Trt	check	arasan	1.4815	7.7185	0.09317	9.1068
Trt	check	spergon	-0.5185	5.7185	-1.9068	7.1068
Trt	check	samesan	1.0815	7.3185	-0.3068	8.7068
Trt	check	fermate	1.8815	8.1185	0.4932	9.5068
Trt	arasan	spergon	-5.1185	1.1185	-6.5068	2.5068
Trt	arasan	samesan	-3.5185	2.7185	-4.9068	4.1068
Trt	arasan	fermate	-2.7185	3.5185	-4.1068	4.9068
Trt	spergon	samesan	-1.5185	4.7185	-2.9068	6.1068
Trt	spergon	fermate	-0.7185	5.5185	-2.1068	6.9068
Trt	samesan	fermate	-2.3185	3.9185	-3.7068	5.3068

Fig. 6.34. SAS Example F11: adjusted t-tests and confidence intervals for pairwise differences of means

6.5.2 Two-Way Mixed Effects Model: Crossed Classification

The machine–operator example at the beginning of Sect. 6.5, introducing two-factor mixed models, discussed an experiment in which a fixed factor (two brands of machines) and a random factor (three randomly selected operators) were used. The general setup of such experiments will have "a" levels of a fixed Factor A and "b" levels of a random Factor B, where each factorial combination of levels of A and B is replicated n times in a completely randomized design. The interaction effect between A and B is random because the levels of the interaction effect involve the levels of Factor B which are randomly selected.

Model

The two-way crossed mixed effects model is given by

$$y_{ijk} = \mu + \alpha_i + B_j + \alpha B_{ij} + \epsilon_{ijk}, \quad i = 1, \ldots, a; \quad j = 1, \ldots, b; \quad k = 1, \ldots, n$$

where α_i are fixed effects (levels of Factor A), B_j are random effects (levels of Factor B) that are iid $N(0, \sigma_B^2)$ random variables, the interaction effects between the two factors denoted by αB_{ij} are iid $N(0, \sigma_{\alpha B}^2)$, and the random errors ϵ_{ijk} are iid $N(0, \sigma^2)$ random variables. The random variables B_j, αB_{ij}, and ϵ_{ijk} are pairwise independent. The mean of the responses is $E(y_{ijk}) = \mu + \alpha_i$ and the variance $\text{Var}(y_{ijk}) = \sigma^2 + \sigma_B^2 + \sigma_{\alpha B}^2$. The responses from the same level of random Factor B have covariance σ_B^2 if they are from different levels of Factor A and $\sigma_B^2 + \sigma_{\alpha B}^2$ if they are from the same level of Factor A. If they are from different levels of Factor B, they are uncorrelated.

A Special Comment The above model, sometimes called the *unconstrained parameters* (UP) model in the literature, is adopted by several authors as well as SAS software for the analysis of data from two-way crossed mixed effects experiments. However, other authors favor an alternative form of the model called the *constrained parameters* (CP) model, where the so-called summation restrictions are imposed on the fixed effects as well as the fixed-by-random interaction parameters. Using these constraints, in effect, imposes a covariance structure among the observations that is different from the one prescribed by the UP model. Although a relationship exists between the two sets of variance components, the meaning assigned to the parameters by experimenters, hence the interpretation of statistical inferences made, may be different. The two major differences between the two models are (i) the CP model allows for negative covariance among observations from different levels of Factor A, whereas the UP model does not, and (ii) expected mean squares for effect B are different for the two models. The second fact results in two different denominators for the F-statistic for testing the variance of effect B. The UP model is adopted for the rest of the discussion in this book.

Estimation and Hypothesis Testing

The usual format of the ANOVA table for computing the required F-statistics for testing hypotheses of interest in a two-way classification is

SV	df	SS	MS	F	E(MS)
A	$a-1$	$\mathtt{SS_A}$	$\mathtt{MS_A}$	$\mathtt{MS_A/MS_{AB}}$	$\sigma^2 + n\sigma^2_{\alpha B} + bn\frac{\sum_i(\alpha_i - \bar{\alpha})^2}{(a-1)}$
B	$b-1$	$\mathtt{SS_B}$	$\mathtt{MS_B}$	$\mathtt{MS_B/MS_{AB}}$	$\sigma^2 + n\sigma^2_{\alpha B} + an\sigma^2_B$
AB	$(a-1)(b-1)$	$\mathtt{SS_{AB}}$	$\mathtt{MS_{AB}}$	$\mathtt{MS_{AB}/MSE}$	$\sigma^2 + n\sigma^2_{\alpha B}$
Error	$ab(n-1)$	$\mathtt{SSE}$	$\mathtt{MSE}$		σ^2
Total	$abn-1$	$\mathtt{SS_{Tot}}$			

The F-statistic for A tests the hypothesis of equality of treatment effects:

$$H_0: \alpha_1 = \alpha_2 = \cdots = \alpha_t \text{ versus } H_a: \text{ at least one inequality}$$

or, equivalently,

$$H_0: \mu_1 = \mu_2 = \cdots = \mu_t \text{ versus } H_a: \text{ at least one inequality}$$

where $\mu_i = \mu + \alpha_i$, the ith treatment mean. H_0 is rejected if the observed F-value exceeds the α upper percentile of an F-distribution with $\text{df}_1 = a-1$ and $\text{df}_2 = (a-1)(b-1)$. It is informative to compare the expected mean squares of the denominator and numerator of this F-ratio, and note that they will have the same expectation if the null hypothesis above holds and the numerator will have a larger expectation if the null hypothesis is not true. It can be verified that $bn\frac{\sum_i(\alpha_i - \bar{\alpha})^2}{(t-1)}$ is zero if $\alpha_1 = \alpha_2 = \cdots = \alpha_t$ and is positive otherwise.

Inferences about estimable functions of the fixed effects can be made using the appropriate mean squares to calculate their standard errors. The best estimate of μ_i, the ith treatment mean, and its standard error are

$$\hat{\mu}_i = \bar{y}_{i..} = (\textstyle\sum_j \sum_k y_{ijk})/bn,$$
$$\text{s.e.}(\bar{y}_{i..}) = s_{\text{AB}}/\sqrt{bn}$$

where $s^2_{\text{AB}} = \text{MS}_{\text{AB}}$ and $i = 1, \ldots, a$. This result follows since it can be shown that the variance of $\bar{y}_{i..}$ is $(\sigma^2 + n\sigma^2_{\alpha B})/bn$ and $E(\text{MS}_{\text{AB}}) = \sigma^2 + n\sigma^2_{\alpha B}$; that is, MS_{AB} estimates $\sigma^2 + n\sigma^2_{\alpha B}$. Note that the above means would be identical to the "Factor A means" calculated in the two-way classification with fixed effects case discussed in Sect. 5.4, but the standard errors are obviously not the same.

The best estimate of the difference between the effects of two treatments labeled p and q is

$$\widehat{\mu_p - \mu_q} = \widehat{\alpha_p - \alpha_q} = \bar{y}_{p..} - \bar{y}_{q..}$$

with standard error given by

$$s_d = \text{s.e.}(\bar{y}_{p..} - \bar{y}_{q..}) = s_{\text{AB}}\sqrt{\frac{2}{bn}}$$

where $s^2_{\text{AB}} = \text{MS}_{\text{AB}}$. The standard error of linear contrasts of the effects (or the means) is similarly obtained by replacing s^2 with s^2_{AB} in the fixed effects model formulas.

A $(1-\alpha)100\%$ confidence interval for $\mu_p - \mu_q$ (or equivalently, $\alpha_p - \alpha_q$) is

$$(\bar{y}_{p..} - \bar{y}_{q..}) \pm t_{\alpha/2,\nu} \cdot s_{\text{AB}}\sqrt{\frac{2}{bn}}$$

where $t_{\alpha/2,\nu}$ is the upper $\alpha/2$ percentile point of a t-distribution with $\nu = (a-1)(b-1)$ degrees of freedom. Thus, these results regarding the treatment effects are similar to those for the fixed effects case except that here s is replaced by $s_{\mathbf{AB}}$. Similarly, standard errors for other linear comparisons of treatment means and the corresponding t-tests may be calculated.

F-statistics shown in the ANOVA table for sources of variation B and AB are used to test the hypotheses

(i) $H_0 : \sigma^2_B = 0$ versus $\text{H}_\text{a} : \sigma^2_\text{B} > 0$

(ii) $H_0 : \sigma^2_{\alpha B} = 0$ versus $\text{H}_\text{a} : \sigma^2_{\alpha\text{B}} > 0$

respectively. The null hypotheses in (i) or (ii) are rejected if the corresponding F-values exceed the α upper percentiles of F-distributions with $\text{df}_1 = b - 1$ and $\text{df}_2 = (a-1)(b-1)$ or $\text{df}_1 = (a-1)(b-1)$ and $\text{df}_2 = ab(n-1)$, respectively. Note that these F-statistics have different denominators: for the test of (i), the denominator is MS_{AB}, whereas for (ii) it is MSE. Comparing the expected mean squares of the denominator and numerator of these F-ratios allows one to verify whether these are the appropriate F-ratios. (With respect to the above special comment regarding the alternative model, note that the expected mean square for effect B under the CP model is $\sigma^2 + an\,\sigma^2_B$, suggesting that the appropriate F-statistic for testing hypothesis (i) is $\text{MS}_\text{B}/\text{MSE}$ under that model. Thus, the denominator for testing (i) under the CP model will be different.)

Method of moments estimators are obtained as usual by setting the expected mean squares to their respective computed values. Thus, estimates of σ^2_B, $\sigma^2_{\alpha B}$, and σ^2 are given by

$$\hat{\sigma}^2_B = (\text{MS}_\text{B} - \text{MS}_{\text{AB}})/an$$
$$\hat{\sigma}^2_{\alpha B} = (\text{MS}_{\text{AB}} - \text{MSE})/n$$

and $\hat{\sigma}^2 = \text{MSE}$, respectively. When the data are balanced, these estimators are unbiased and have minimum variance. Confidence intervals for the variance components are obtained as described in Sect. 6.4. For example, a $(1-\alpha)100\%$ confidence interval for σ_B^2 is provided by

$$\frac{\nu\hat{\sigma}_B^2}{\chi^2_{1-\alpha/2,\nu}} < \sigma_B^2 < \frac{\nu\hat{\sigma}_B^2}{\chi^2_{\alpha/2,\nu}}$$

where $\chi^2_{1-\alpha/2,\nu}$ and $\chi^2_{\alpha/2,\nu}$ are the $1-\alpha/2$ and $\alpha/2$ percentile points of the chi-square distribution with ν degrees of freedom, respectively. The degrees of freedom for $\hat{\sigma}_B^2 = \frac{1}{an}\mathtt{MS}_\mathtt{B} - \frac{1}{an}\mathtt{MS}_\mathtt{AB}$ are obtained using the Satterthwaite approximation and are given by

$$\nu = \frac{(an\hat{\sigma}_B^2)^2}{(\mathtt{MS}_\mathtt{B})^2/(b-1) + (\mathtt{MS}_\mathtt{AB})^2/(a-1)(b-1)}$$

SAS Example F12

The two most crucial factors that influence the strength of solders used in cementing computer chips into the motherboard of guidance systems of airplanes are identified to be the machine used to insert the solder and the operator of the machine. Four types of solder machine used in the plant were selected for a study planned to examine this dependence. Each of the three operators selected at random from the operators available at the company's plants made two solders on each of the four machines in random order. The data, taken from Ott and Longnecker (2001), appear in Table 6.5.

Table 6.5. Strength of solder in computer chips

Operator	Machine			
	1	2	3	4
1	204	205	203	205
	205	210	204	203
2	205	205	206	209
	207	206	204	207
3	211	207	209	215
	209	210	214	212

From the description, it is clear that `machine` is a fixed factor with four levels, that the factor `operator` is random with three levels, and that the two factors are crossed. The two-way crossed mixed effects model used for the analysis of these data is given by

$$y_{ijk} = \mu + \alpha_i + B_j + \alpha B_{ij} + \epsilon_{ijk}, \quad i = 1, \ldots, 4;\ j = 1, \ldots, 3;\ k = 1, \ldots, 2$$

where α_i is the effect of machine i, B_j is the effect of operator j distributed as iid $N(0, \sigma_B^2)$ random variables, and the interaction effects between the two factors are denoted by αB_{ij} distributed as iid $N(0, \sigma_{\alpha B}^2)$. The random errors ϵ_{ijk} are distributed as iid $N(0, \sigma^2)$ random variables, and the random variables B_j, αB_{ij}, and ϵ_{ijk} are pairwise independent.

```
data solder;
infile "C:\users\user_name\Documents\...\ex17-22.txt";
input Operator Machine Strength;
run;

proc glm data=solder;
   class Machine Operator;
   model Strength = Machine Operator Machine*Operator;
   random  Operator Machine*Operator/test;
   lsmeans Machine/stderr;
   title 'Analysis of Strength of Solder in Computer Chips using PROC GLM';
run;
```

Fig. 6.35. SAS Example F12: analysis of solder strength using PROC GLM

In the SAS Example F12 program, displayed in Fig. 6.35, `proc glm` is used to perform a conventional analysis of a mixed model. The data are read from a text file using the `list` input style. The `model` statement contains both the fixed and random effects as usual. The `random` statement identifies the `operator` and the `machine*operator` interaction as random effects. The `test` option requests `proc glm` to construct appropriate F-statistics for testing hypotheses about the variance components corresponding to these effects. The expected values of the mean squares in the ANOVA table (see below) determine the ratios of sums of squares that need to be used for testing the two hypotheses of interest regarding the variances of `operator` and `machine*operator` effects: $H_0 : \sigma_B^2 = 0$ versus $H_a : \sigma_B^2 > 0$ and $H_0 : \sigma_{\alpha B}^2 = 0$ versus $H_a : \sigma_{\alpha B}^2 > 0$ as discussed previously. The ANOVA table is

SV	df	SS	MS	F	p-Value	E(MS)
Machine	3	12.458	4.1528	0.56	0.6619	$\sigma^2 + 2\,\sigma_{\alpha B}^2 + 2\sum_i(\alpha_i - \bar{\alpha})^2$
Operator	2	160.333	80.1667	10.77	0.0103	$\sigma^2 + 2\,\sigma_{\alpha B}^2 + 8\,\sigma_B^2$
Machine× Operator	6	44.667	7.4444	1.96	0.1507	$\sigma^2 + 2\,\sigma_{\alpha B}^2$
Error	12	45.500	3.7917			σ^2
Total	23	262.958				

The `lsmeans` statement illustrates the estimation of an interesting BLUP. BLUPs for a mixed model were discussed in the introduction to this section.

Analysis of Strength of Solder in Computer Chips using PROC GLM

The GLM Procedure

Source	Type III Expected Mean Square
Machine	Var(Error) + 2 Var(Machine*Operator) + Q(Machine)
Operator	Var(Error) + 2 Var(Machine*Operator) + 8 Var(Operator)
Machine*Operator	Var(Error) + 2 Var(Machine*Operator)

Tests of Hypotheses for Mixed Model Analysis of Variance

Dependent Variable: Strength

Source	DF	Type III SS	Mean Square	F Value	Pr > F
Machine	3	12.458333	4.152778	0.56	0.6619
Operator	2	160.333333	80.166667	10.77	0.0103
Error	6	44.666667	7.444444		
Error: MS(Machine*Operator)					

Source	DF	Type III SS	Mean Square	F Value	Pr > F
Machine*Operator	6	44.666667	7.444444	1.96	0.1507
Error: MS(Error)	12	45.500000	3.791667		

Least Squares Means

Machine	Strength LSMEAN	Standard Error	Pr > |t|
1	206.833333	0.794949	<.0001
2	207.166667	0.794949	<.0001
3	206.666667	0.794949	<.0001
4	208.500000	0.794949	<.0001

Fig. 6.36. SAS Example F12: selected output

A BLUP consists of the sum of an estimable linear function for the fixed parameters (e.g., here $\mu + \alpha_i$) and a different function of the random effect parameters. The `lsmeans` statement in `proc glm` estimates the expectation of y_{ijk} averaged over the levels of the random factor (operator): $\mu + \alpha_i + \bar{B}_. + \overline{\alpha B}_{i.}$ (i.e., the conditional expectation conditioned on the operator and the interaction effects), where $\bar{B}_. = \frac{1}{3}\sum_j B_j$ and $\overline{\alpha B}_{i.} = \frac{1}{3}\sum_j \alpha B_{ij}$. Here, the interest is only in the effects of operators used in the study; conditioning on them is equivalent to considering their effects to be "fixed." (The `lsmeans` statement options `pdiff`, `cl`, and `adj=` may be used to obtain confidence intervals for pairs of differences of the above BLUPs, adjusted for multiple testing using a method of choice.)

The standard output from `proc glm` (factor level information) that precedes the tables containing Type I and III SS and associated F-tests (considering all factors to be fixed) are omitted. Figure 6.36 contains a table of the Type III Expected Mean Squares and a table of F-tests for all three effects in the model. The F-tests are constructed using the appropriate denominator mean squares. The denominator for the F-statistics for testing both the `machine` effect hypothesis (i.e., $H_0: \alpha_1 = \alpha_2 = \alpha_3$) and the hypothesis about the variance of the `operator` random effect (i.e., $H_0: \sigma_B^2 = 0$), respectively, is the mean square for the interaction effect `machine*operator`. The denominator for the F-statistic, for testing the hypothesis about the variance of the `machine*operator` random effect (i.e., $H_0: \sigma_{\alpha B}^2 = 0$), is the mean square for error. These tests fail to reject either the machine main effect hypothesis or the interaction hypothesis at level $\alpha = 0.05$; however, the operator variance component, σ_B^2, is found to be significantly different from zero. As demonstrated previously, the method of moments estimates can be obtained as usual. They are $\hat{\sigma}^2 = 3.792$, $\hat{\sigma}_B^2 = (7.444 - 3.792)/2 = 1.826$, and $\hat{\sigma}_{\alpha B}^2 = (80.16667 - 7.444)/8 = 9.0903$. The interaction plot of the mean solder strengths, shown in Fig. 6.37, is discussed later.

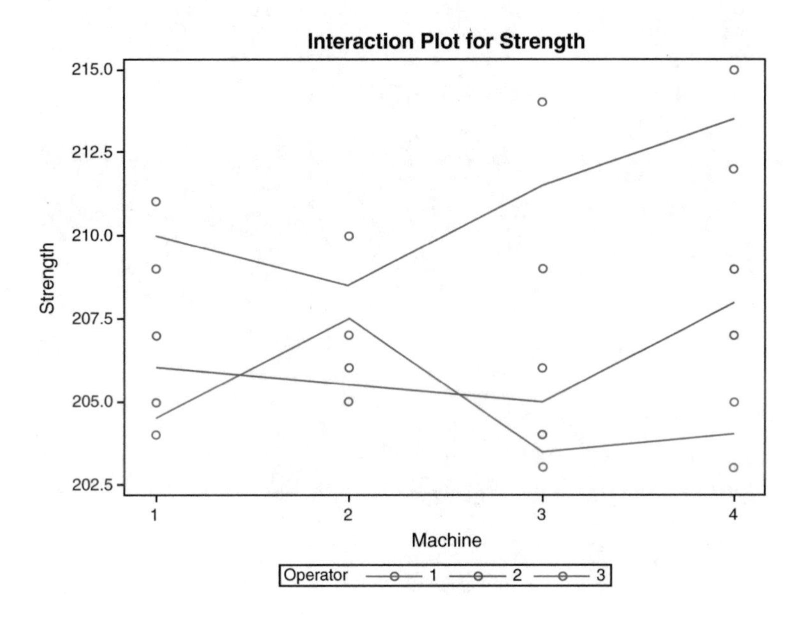

Fig. 6.37. SAS Example F12: interaction plot of solder strength data

SAS Example F13

In this program (displayed in Fig. 6.38), `proc mixed` is used to analyze the solder strength using Type 3 sums of squares and the method of moments so that the results may be compared with the previous analysis using `proc glm`. In addition to the advantages discussed in Sect. 6.2, `proc mixed` allows the use of `estimate` statements to obtain estimates of BLUPs with the correct standard errors. The only term required on the right-hand side of the `model` statement is the fixed effect term `machine`. The `random` statement declares the `operator` and `machine*operator` effects to be random effects.

The option `ddfm=` is required to specify the method that must be used by `proc mixed` for the computation of denominator degrees of freedom F-tests, t-tests, confidence intervals, etc. for fixed effects or any function of fixed effects such as contrasts or BLUPs. Here, the value specified is `satterth`. To understand what this means, recall that the Satterthwaite approximation was used in Sects. 6.3 and 6.4 for the construction of confidence intervals of variance components in random models. In those sections, this approximation was required when the denominator happened to be a linear combination of mean squares (rather than a single mean square).

```
data solder;
infile "C:\users\user_name\Documents\...\ex17-22.txt";
input Operator Machine Strength;
run;

proc mixed data=solder noclprint noinfo method=type3 cl;
   class Machine Operator;
   model Strength = Machine /ddfm=satterth;
   random  Operator Machine*Operator;
   lsmeans Machine;
   estimate 'BLUP_1: Oper 3'
            intercept 4
                        Machine 1 1 1 1|
                        Operator 0 0 4
            Machine*Operator 0 0 1 0 0 1 0 0 1 0 0 1/divisor=4;

   estimate 'BLUP_2: Oper 3'
            intercept 4
                        Machine 1 1 1 1|
                        Operator 0 0 4/divisor=4;

   estimate 'LSMEAN for Mach 1'
            intercept 3
                        Machine 3 0 0 0|
                        Operator 1 1 1
            Machine*Operator 1 1 1 0 0 0 0 0 0 0 0 0/divisor=3;
title "Analysis of Strength of Solder in Computer Chips using PROC MIXED";
run;
```

Fig. 6.38. SAS Example F13: analysis of solder strength using PROC MIXED

In the case of the two-way mixed model, if the sample sizes are unequal, the denominator of tests associated with fixed effects will be a linear combination of mean squares. Thus, the Satterthwaite approximation is needed for inference associated with the fixed effects. It could be, in the balanced data case, that this approximation may never be needed.

It is recommended that `ddfm=kr` be used for models with multiple random effects when the sample sizes are unequal. The reason is that the `kr` option, which stands for Kenward–Roger, employs a method developed by Kenward and Roger (1997), which adjusts the standard errors as well as the degrees of freedom when an approximation is needed. In many balanced models, no standard error adjustment is required. When it is needed, not only the degrees of freedom associated with the respective statistics but also the value of the t-statistics (or F-statistics) themselves are affected.

The profiles of mean solder strengths in Fig. 6.37 indicate several attributes of the random effects in the model. First, the profiles are roughly parallel, indicating that there is no appreciable machine–operator interaction. Second, there is a variation of the profiles from an average, an indication of the variability among operators. The management of the company, always interested in efficiency, might be intrigued by Operator 3, who appears to have a higher performance level than the other operators in the study over all four machines, as observed from this graph. By conditioning on both the `operator` and `machine*operator` random effects, a predictable function measuring Operator 3's expected mean strength is obtained as $\mu+\bar{\alpha}_{.}+B_3+\overline{\alpha B}_{.3}$, where $\bar{\alpha}_{.}=\frac{1}{4}\sum_i \alpha_i$ and $\overline{\alpha B}_{.3}=\frac{1}{4}\sum_i \alpha B_{i3}$. The `estimate` statement labeled "BLUP_1: Oper 3" requests that an estimate and a standard error of this BLUP be computed. Note that the option `divisor=4` used with the estimate statement enables the user to specify the linear combination of the parameters needed by entering the coefficients as whole numbers (instead of fractions). The coefficients needed to specify the estimation of $\mu+\bar{\alpha}_{.}+B_3+\overline{\alpha B}_{.3}$ are

μ
1

α_1	α_2	α_3	α_4
1/4	1/4	1/4	1/4

B_1	B_2	B_3
0	0	1

αB_{11}	αB_{12}	αB_{13}	αB_{21}	αB_{22}	αB_{23}	αB_{31}	αB_{32}	αB_{33}	αB_{41}	αB_{42}	αB_{43}
0	0	1/4	0	0	1/4	0	0	1/4	0	0	1/4

Since the two random effects are independent, it is possible to condition only on one of the random effects. By conditioning on the `operator` effect alone (i.e., averaging the interaction over the population of all operators), another measure of mean strength which can be constructed for Operator 3 is the expected mean $\mu+\bar{\alpha}_{.}+B_3$. The `estimate` statement labeled "BLUP_2: Oper 3" results in the computation of its estimate and standard error. One would expect estimates of these two BLUPs to be similar, given that the interaction is not significant. As noted earlier, these BLUPs cannot be estimated using the `estimate` statement in `proc glm`. Finally, the `estimate` statement labeled "LSMEAN for Mach 1" requests that an estimate and a standard error of the predictable function $\mu+\alpha_1+\bar{B}_{.}+\overline{\alpha B}_{1.}$ be computed. See the discussion on the output from this example below for an explanation for the inclusion of this statement.

Analysis of Strength of Solder in Computer Chips using PROC MIXED

The Mixed Procedure

Type 3 Analysis of Variance					
Source	**DF**	**Sum of Squares**	**Mean Square**	**Expected Mean Square**	**Error Term**
Machine	3	12.458333	4.152778	Var(Residual) + 2 Var(Machine*Operator) + Q(Machine)	MS(Machine*Operator)
Operator	2	160.333333	80.166667	Var(Residual) + 2 Var(Machine*Operator) + 8 Var(Operator)	MS(Machine*Operator)
Machine*Operator	6	44.666667	7.444444	Var(Residual) + 2 Var(Machine*Operator)	MS(Residual)
Residual	12	45.500000	3.791667	Var(Residual)	.

Covariance Parameter Estimates				
Cov Parm	**Estimate**	**Alpha**	**Lower**	**Upper**
Operator	9.0903	0.05	-10.5784	28.7590
Machine*Operator	1.8264	0.05	-2.6505	6.3032
Residual	3.7917	0.05	1.9497	10.3320

Fig. 6.39. SAS Example F13: model fit output from PROC MIXED

Except for fit statistics and the confidence intervals for the variance components, all results that appear on the output from SAS Example F13 (shown in Figs. 6.39 and 6.40) agree with those from `proc glm`. The estimates of the variance components (see table titled `Covariance Parameter Estimates`) are also the same as those obtained from `proc glm`. The confidence intervals, except those for σ^2, are based on large sample results and may not be appropriate for small sample sizes used in this example. Intervals based on Satterthwaite approximation can be constructed as illustrated earlier in this section. SAS code given previously may be used for this purpose; however, it is recommended that the `covtest` option be added to the `proc` statement to obtain the required standard errors of the variance component estimates.

The estimates of the two BLUPs for Operator 3 are given in the table titled `Estimates` (shown in Fig. 6.40), and as conjectured, they are similar in value. However, the standard errors differ, and the reason for this can be surmised from observing that their degrees of freedom are fractions. The fractions result from the fact that the standard errors are calculated using the Satterthwaite approximation. The two standard errors are obtained using different linear combinations of mean squares, and their degrees of freedom are computed using approximations as illustrated in previous examples.

The "LSMEAN for Mach 1" is the BLUP $\mu + \alpha_1 + \bar{B}_. + \overline{\alpha B}_{1.}$. This is a BLUP for Machine 1 and is the expectation of the observations for Machine 1 conditioned on the observed operators. The `lsmeans machine;` statement, on the other hand, estimates the mean for Machine 1 as $\mu+\alpha_1$, the unconditional expectation of the observations for Machine 1. Thus, both estimated values are the same, 206.83, but the standard errors of the two estimators are different. The `lsmeans` statement gives the standard error as 2.0666 (see `Least Squares Means` table), and the `estimate` statement calculates it as 0.7949 (see `Estimates` table). Both are correct since they are estimating standard errors of different BLUPs. In comparison, the `lsmeans` statement in `proc glm` (see Fig. 6.36) computes the standard error as 0.7949 by (incorrectly) considering both `operator` and `machine*operator` as fixed effects.

The above analysis was repeated using the default estimation method of REML as well. The covariance estimates and their standard errors results of the F-tests for the variance components, the t-tests for the estimates of BLUPs, and the lsmeans estimates all remain unchanged, as the data set is

The Mixed Procedure

Fit Statistics	
-2 Res Log Likelihood	100.7
AIC (Smaller is Better)	106.7
AICC (Smaller is Better)	108.2
BIC (Smaller is Better)	104.0

Type 3 Tests of Fixed Effects

Effect	Num DF	Den DF	F Value	Pr > F
Machine	3	6	0.56	0.6619

Estimates

Label	Estimate	Standard Error	DF	t Value	Pr > |t|
BLUP_1: Oper 3	210.71	0.6775	12.7	311.00	<.0001
BLUP_2: Oper 3	210.54	0.9343	6.78	225.34	<.0001
LSMEAN for Mach 1	206.83	0.7949	12	260.18	<.0001

Least Squares Means

Effect	Machine	Estimate	Standard Error	DF	t Value	Pr > |t|
Machine	1	206.83	2.0666	3.19	100.08	<.0001
Machine	2	207.17	2.0666	3.19	100.25	<.0001
Machine	3	206.67	2.0666	3.19	100.00	<.0001
Machine	4	208.50	2.0666	3.19	100.89	<.0001

Fig. 6.40. SAS Example F13: estimates and BLUPs from PROC MIXED

balanced. The output includes an F-test for the fixed effect (machine) using Type 3 ANOVA which is also exactly the same as obtained earlier. The only difference is that confidence intervals for the variance components calculated by `proc mixed` with `method=type3` are those based on large-sample standard errors and normal percentiles, whereas with REML, these are based on the chi-square distribution and the Satterthwaite approximation as displayed in Fig. 6.41.

Covariance Parameter Estimates				
Cov Parm	Estimate	Alpha	Lower	Upper
Operator	9.0903	0.05	2.2609	719.61
Machine*Operator	1.8264	0.05	0.4064	441.64
Residual	3.7917	0.05	1.9497	10.3320

Fig. 6.41. SAS Example F13: confidence intervals for the variance components with `method=reml` in PROC MIXED

6.5.3 Two-Way Mixed Effects Model: Nested Classification

A two-way random model for an experiment with two random factors where one of the factors is *nested* in the other factor was considered in Sect. 6.4. In this subsection, a mixed model for an experiment with two factors A and B, where Factor B is nested in Factor A, is discussed. Recall that the sampling of the levels of a nested factor takes place in a *hierarchical* manner: Thus, levels of Factor B are randomly sampled within each level of A, which are fixed. Suppose that in the machine–operator example, the primary interest is in the performance characteristics of, say, four types of machine and that it is possible to randomly sample 12 operators from all available operators. If three operators are assigned randomly to work on each machine and each of them performs the production operation on the assigned machine only, then the operators are nested within each machine.

This type of mixed effects model occurs naturally in animal and plant breeding experiments. As an example, consider the weight gain data from a pig-raising experiment described in Snedecor and Cochran (1989). The breeding values of five sires (bulls) are being evaluated. Each sire is mated to a random group of dams. Each mating produced a litter of pigs whose average daily gain was measured. Thus, the dams are nested within each sire (i.e., levels of "dam" are different for each level of "sire"). Here, the `Sire` genetic effect is a fixed effect, and the additive `Dam within Sire` genetic effect is considered a random effect. Perhaps the sires are selected from several breeding lines being evaluated. A trait such as weight gain of offspring produced from each mating is usually used to compare genetic merits of breeding lines. The information gained from the variation of the Dam within Sire effect will lead to more precise estimation of the sire effects.

Model

An appropriate model for the situation described above is

$$y_{ijk} = \mu + \alpha_i + B_{ij} + \epsilon_{ijk}, \quad i = 1, \ldots, a;\ j = 1, \ldots, b;\ k = 1, \ldots, n$$

where α_i is the effect of level i Factor A, and it is assumed that the Factor B within A effect, B_{ij}, is an iid $N(0, \sigma_B^2)$ random variable, the random error ϵ_{ijk} is an iid $N(0, \sigma^2)$ random variable, and B_{ij} and ϵ_{ijk} are independently distributed. The parameters σ_B^2 and σ^2 are the "variance components" to be estimated in this model.

Estimation and Hypothesis Testing

An ANOVA table that corresponds to the model is constructed using the same computational formulas used for the computation of the ANOVA if the factors `A` and `B within A` were considered fixed. The sums of squares would be the same for effect A as in the ANOVA table for the two-way crossed random effects model (given in Sect. 6.3). As noted in Sect. 6.4, the sum of squares for the effect B(A) is obtained by pooling the sum of squares for the effects B and AB and thus has $a(b-1)$ degrees of freedom. As with other models containing random effects considered so far, an additional column displaying the expected mean squares is included in the ANOVA table:

SV	df	SS	MS	F	E(MS)
A	$a-1$	$\texttt{SS}_\texttt{A}$	$\texttt{MS}_\texttt{A}$	$\texttt{MS}_\texttt{A}/\texttt{MS}_{\texttt{B(A)}}$	$\sigma^2 + n\sigma_B^2 + bn\frac{\sum_i(\alpha_i-\bar{\alpha})^2}{(a-1)}$
B(A)	$a(b-1)$	$\texttt{SS}_{\texttt{B(A)}}$	$\texttt{MS}_{\texttt{B(A)}}$	$\texttt{MS}_{\texttt{B(A)}}/\texttt{MSE}$	$\sigma^2 + n\sigma_B^2$
Error	$ab(n-1)$	SSE	MSE$(= s^2)$		σ^2
Total	$abn-1$				

The F-statistic for the source A in the ANOVA table tests the hypothesis of equality of Factor A effects:

$$H_0: \alpha_1 = \alpha_2 = \cdots = \alpha_a \text{ versus } H_a: \text{ at least one inequality.}$$

or, equivalently,

$$H_0: \mu_1 = \mu_2 = \cdots = \mu_a \text{ versus } H_a: \text{ at least one inequality}$$

where $\mu_i = \mu + \alpha_i$, the mean of an observation at the ith level of A. The F-ratios shown in the analysis of variance table for sources of variation A and B(A) are constructed using the same principle described in Sect. 6.3 or 6.5.2. The expected mean squares of the denominator and numerator of the F-ratio for A will have the same expectation if the null hypothesis above holds, and the numerator will have a larger expectation if it is not true since $\sum_i(\alpha_i - \bar{\alpha})^2$ is zero if $\alpha_1 = \alpha_2 = \cdots = \alpha_a$ and is positive otherwise. H_0 is rejected if the observed F-value exceeds the α upper percentile of an F-distribution with $\text{df}_1 = a-1$ and $\text{df}_2 = a(b-1)$. The standard errors for means, pairwise mean comparisons, and linear contrasts of the fixed effects (or the means) are obtained by replacing s^2 with $s^2_{\text{B(A)}}$ (and the corresponding degrees of freedom with $a(b-1)$ where needed) in the fixed effects model formulas, where

$s^2_{\text{B(A)}} = \text{MS}_{\text{B(A)}}$. For example, a $(1-\alpha)100\%$ confidence interval for $\mu_p - \mu_q$ (or, equivalently, $\alpha_p - \alpha_q$) is

$$(\bar{y}_{p..} - \bar{y}_{q..}) \pm t_{\alpha/2,\nu} \cdot s_{\text{B(A)}} \sqrt{\frac{2}{bn}}$$

where $t_{\alpha/2,\nu}$ is the upper $\alpha/2$ percentile point of a t-distribution with $\nu = a(b-1)$.

To test the hypothesis $H_0 : \sigma^2_B = 0$ versus $H_a : \sigma^2_B > 0$, observe that the mean square for the B(A) effect, $\text{MS}_{\text{B(A)}}$, has expectation equal to σ^2 if $\sigma^2_B = 0$ and, thus, MSE is the appropriate denominator for testing the nested effect. The null hypothesis is rejected for F-values that exceed the α upper percentile of an F-distribution with $\text{df}_1 = a(b-1)$ and $\text{df}_2 = ab(n-1)$. The *method of moments* estimates of variance components are obtained the usual way by setting the computed mean squares equal to their corresponding expected values. Solving the resulting equations

$$\begin{aligned} \text{MS}_{\text{B(A)}} &= \sigma^2 + n\,\sigma^2_B \\ \text{MSE} &= \sigma^2 \end{aligned}$$

give, the estimates $\hat{\sigma}^2 = \text{MSE}$ and $\hat{\sigma}^2_B = (\text{MS}_{\text{B(A)}} - \text{MSE})/n$.

SAS Example F14

The data for the animal breeding experiment discussed in the introduction are given in Table 6.6. Recall that each of the five sires from different breeding lines is mated to two randomly chosen dams, each mating producing a litter of pigs whose average daily gain was measured. Thus, the dams are nested within each sire. The Sire effect is a fixed effect, and the Dam within Sire effect is a random effect.

An appropriate model for analyzing the average daily gain is given by

$$y_{ijk} = \mu + \alpha_i + B_{ij} + \epsilon_{ijk} \quad i = 1, \ldots, 5;\ j = 1, \ldots, 2;\ k = 1, \ldots, 2.$$

where the fixed effect of Sire i is α_i; the effect of Dam j within Sire i, B_{ij}, is assumed to be iid $N(0, \sigma^2_B)$; the random error ϵ_{ijk} is assumed to be iid $N(0, \sigma^2)$; and B_{ij} and ϵ_{ijk} are independent. `Proc glm` is used in the SAS Example F14 program, displayed in Fig. 6.42, to fit the above model to the average daily gain data, using the method of moments. The data are entered with the values for each sire–dam combination in separate data lines. Thus, each line can be held with a trailing @ modifier for the two replications to

Table 6.6. Average daily gain of two pigs in each litter (Snedecor and Cochran 1989)

Sire	Dam	Average daily gain	
1	1	2.77	2.38
	2	2.58	2.94
2	1	2.28	2.22
	2	3.01	2.61
3	1	2.36	2.71
	2	2.72	2.74
4	1	2.87	2.46
	2	2.31	2.24
5	1	2.74	2.56
	2	2.50	2.48

```
data pigs;
input Sire Dam @;
do Rep=1 to 2;
 input Gain @;
 output;
end;
datalines;
1  1  2.77 2.38
1  2  2.58 2.94
2  1  2.28 2.22
2  2  3.01 2.61
3  1  2.36 2.71
3  2  2.72 2.74
4  1  2.87 2.46
4  2  2.31 2.24
5  1  2.74 2.56
5  2  2.50 2.48
;

proc glm data=pigs;
   class Sire Dam;
   model Gain = Sire Dam(Sire);
   random  Dam(Sire)/test;
   lsmeans Sire/stderr;
   lsmeans Sire/stderr e=Dam(Sire);
   title "Analysis of Average Daily Gain using PROC GLM";
run;
```

Fig. 6.42. SAS Example F14: analysis of average daily gain using PROC GLM

be read. They are written, along with the current sire–dam values, as two separate records in the SAS data set using a `do-end` structure and an `output` statement. In the proc step, both `sire` and `dam` are declared in the `class` statement. The nested effect B_{ij} is represented in the `model` statement as `dam(sire)`, the notation used in the SAS language to represent a nested effect. It is also identified as a random effect by including it in the `random` statement as any other random effect (see SAS Example F8 in Sect. 6.4).

Analysis of Average Daily Gain using PROC GLM

The GLM Procedure

Source	Type III Expected Mean Square
Sire	Var(Error) + 2 Var(Dam(Sire)) + Q(Sire)
Dam(Sire)	Var(Error) + 2 Var(Dam(Sire))

Tests of Hypotheses for Mixed Model Analysis of Variance

Dependent Variable: Gain

Source	DF	Type III SS	Mean Square	F Value	Pr > F
Sire	4	0.099730	0.024932	0.22	0.9155
Error: MS(Dam(Sire))	5	0.563550	0.112710		

Source	DF	Type III SS	Mean Square	F Value	Pr > F
Dam(Sire)	5	0.563550	0.112710	2.91	0.0707
Error: MS(Error)	10	0.387000	0.038700		

Fig. 6.43. SAS Example F14: analysis of average daily gain using PROC GLM

Part of the output from the SAS Example F14 program contains factor level information, total SS, and the standard output that include Types I and III SS and corresponding F-statistics calculated by `proc glm` and are not shown. The Type III Expected Mean Square of random effects is shown in Fig. 6.43. The two ANOVA tables in Fig. 6.43 show the results of tests for the `Sire` and the `Dam(Sire)` random effects. The `test` option in the random statement will result in the use of `Dam(Sire)` mean square as the denominator to obtain the appropriate statistic to test the `Sire` effect. This will produce the correct F-statistic to test the `Sire` effects hypothesis $H_0: \alpha_1 = \alpha_2 = \cdots = \alpha_5$ versus $H_a:$ at least one inequality. The ANOVA table is

SV	df	SS	MS	F	p-Value	E(MS)
Sire	4	0.09973	0.02493	0.22	0.9155	$\sigma^2 + 2\sigma_B^2 + \frac{10}{3}\sum_i(\alpha_i - \bar{\alpha})^2$
Dam(Sire)	5	0.56355	0.11271	2.91	0.0707	$\sigma^2 + 2\sigma_B^2$
Error	10	0.38700	0.03870			
Total	19	1.05028				

The p-value of 0.92 indicates that there are no significant differences among the sire effects. However, the p-value for the test of `Dam(Sire)` variance component $H_0: \sigma_B^2 = 0$ versus $H_a: \sigma_B^2 > 0$ is less than $\alpha = 0.10$ so it is rejected at 1%. The estimates of the variance components are obtained using the method of moments as usual. These are obtained by solving

$$\sigma^2 + 2\sigma_B^2 = 0.11271$$

$$\sigma^2 = 0.0387$$

which give $\hat{\sigma}^2 = 0.0387$ and $\hat{\sigma}_B^2 = (0.112710 - 0.0387)/2 = 0.037$. Finally, the first `lsmeans` statement produces estimates of the BLUPs $\mu + \alpha_i + \bar{B}_{i.}$ (same as the *conditional means* of the five sires) for $i = 1, \ldots, 5$ and their standard errors, as displayed in Fig. 6.44. By overriding the default error term using the `e=Dam(Sire)` option, the second `lsmeans` statement produces the estimates of $\mu + \alpha_i$ and the correct standard errors; that is, these will be identical to the estimates of unconditional means and their standard errors. Note that for the general case, the unconditional variance of $\bar{y}_{i..}$ is equal to $\sigma_B^2/b + \sigma^2/bn$. Thus, the standard error of $\bar{y}_{i..}$ is $\sqrt{0.037/2 + 0.0387/4} = 0.1679$ as given in this output (see Fig. 6.44).

A $(1-\alpha)100\%$ confidence interval based on the Satterthwaite approximation can be constructed for σ_B^2 using a formula similar to (6.1) in Sect. 6.2. This is implemented in the SAS data step

```
data cint;
   alpha=.05; a=5; b=2; n=2;
   msb=0.112710; s2=0.0387; sb2=0.037;
   nu=(n*sb2)**2/(msb**2/(a*(b-1))+ s2**2/(a*b*(n-1)));
   L= (nu*sb2)/cinv(1-alpha/2,nu);
   U= (nu*sb2)/cinv(alpha/2,nu);
   put nu=   L=    U=;
run;
```

and gives the 95% confidence interval (0.010106, 1.38352) for this example.

Analysis of Average Daily Gain using PROC GLM

Least Squares Means

Sire	Gain LSMEAN	Standard Error	Pr > |t|
1	2.66750000	0.09836158	<.0001
2	2.53000000	0.09836158	<.0001
3	2.63250000	0.09836158	<.0001
4	2.47000000	0.09836158	<.0001
5	2.57000000	0.09836158	<.0001

Standard Errors and Probabilities Calculated Using the Type III MS for Dam(Sire) as an Error Term

Sire	Gain LSMEAN	Standard Error	Pr > |t|
1	2.66750000	0.16786155	<.0001
2	2.53000000	0.16786155	<.0001
3	2.63250000	0.16786155	<.0001
4	2.47000000	0.16786155	<.0001
5	2.57000000	0.16786155	<.0001

Fig. 6.44. SAS Example F14: analysis of average daily gain using PROC GLM

```
proc mixed data=pigs noclprint noinfo cl;;
   class Sire Dam;
   model Gain = Sire/ddfm=satterth;
   random  Dam(Sire);
   lsmeans Sire;
   estimate 'Sire 1 BLUE'
            intercept 1 Sire 1 0 0 0 0;
   estimate 'Sire 1 BLUP'
            intercept 2
                        Sire  2 0 0 0 0|
                        Dam(Sire) 1 1 0 0 0 0 0 0 0 0/divisor=2 cl;
   title "Analysis of Average Daily Gain using PROC MIXED";
run;
```

Fig. 6.45. SAS Example F14: analysis of average daily gain using PROC MIXED

A SAS `proc mixed` step (see Fig. 6.45) is added to the SAS Example F14 program to illustrate the use of the `estimate` statements in `proc mixed` to obtain the best estimates of the linear function $\mu + \alpha_i$ and the predictable function $\mu+\alpha_i+\bar{B}_{i.}$. Only those parts of the output that are relevant are reproduced here (see Fig. 6.46). First, note that, by default, `proc mixed` uses the REML method and calculates the confidence intervals on variance components based in the Satterthwaite approximation (see the table titled `Covariance Parameter Estimates`). The 95% interval for σ_B^2 is close to the one calculated previously using the results from `proc glm`. Second, note that `lsmeans` produces the same estimates as with `proc glm`; that is, the standard error is that of the unconditional estimate of $\mu + \alpha_1$, 0.1679 (see the table titled `Least Squares Means`). The two estimate statements further clarify how the two BLUPs differ: The standard error of the BLUP for Sire 1 $\mu + \alpha_1 + \bar{B}_{1.}$ is 0.09836 (see the first `Estimates` table) and is different from that of the above estimate. This estimate gives the performance of a particular sire averaged only over the set of dams he was mated with. If the `estimate` statement option `cl` is included as in the following

```
estimate 'Sire 1 BLUP'
          intercept 2
          sire  2 0 0 0 0|
          dam(sire) 1 1 0 0 0 0 0 0 0 0/divisor=2 cl;
```

a t-based confidence interval (95%, by default) is calculated instead of the t-test. Recall that the `divisor=` option allows the user to avoid entering fractional coefficients when formulating estimate statements. In this example,

Analysis of Average Daily Gain using PROC MIXED

The Mixed Procedure

Covariance Parameter Estimates				
Cov Parm	Estimate	Alpha	Lower	Upper
Dam(Sire)	0.03701	0.05	0.01011	1.3825
Residual	0.03870	0.05	0.01889	0.1192

Type 3 Tests of Fixed Effects				
Effect	Num DF	Den DF	F Value	Pr > F
Sire	4	5	0.22	0.9155

Estimates								
Label	Estimate	Standard Error	DF	t Value	Pr > |t|	Alpha	Lower	Upper
Sire 1 BLUE	2.6675	0.1679	5	15.89	<.0001	0.05	2.2360	3.0990
Sire 1 BLUP	2.6675	0.09836	10	27.12	<.0001	0.05	2.4483	2.8867

Least Squares Means						
Effect	Sire	Estimate	Standard Error	DF	t Value	Pr > |t|
Sire	1	2.6675	0.1679	5	15.89	<.0001
Sire	2	2.5300	0.1679	5	15.07	<.0001
Sire	3	2.6325	0.1679	5	15.68	<.0001
Sire	4	2.4700	0.1679	5	14.71	<.0001
Sire	5	2.5700	0.1679	5	15.31	<.0001

Fig. 6.46. SAS Example F14: analysis of average daily gain data with PROC MIXED

every coefficient specified must be divided by two to obtain the actual linear function of the parameters estimated. This output (obtained in separate run) is reproduced in the second `Estimates` table of Fig. 6.46. Obviously, the interval for the BLUP is narrower because the estimate is an average only over the three dams nested in sire 1 and not over the entire population of dams.

SAS Example F15

Four manufacturing processes of comparable costs are being tried out for obtaining increased surface quality of a precision machine part. The aim is to obtain the best roughness average values (known as R_a in the industry) at the lowest possible cost. The costs depend on the milling, grinding, and polishing activities involved in each of the four processes. Each of the processes is being run at each of four different plants by three different teams selected at each plant at random from the plant's workforce. Three specimens of the part are produced by each team using a process in random order and the surface finish is measured in microinches. The data are displayed in Table 6.7.

Table 6.7. Surface finish (in μ-inches)

Process	Team		
	1	2	3
1	29	34	16
	12	24	17
	21	29	24
2	42	35	36
	37	39	18
	32	30	23
3	38	23	16
	35	36	27
	28	30	19
4	16	10	12
	13	16	17
	25	20	18

The model is

$$y_{ijk} = \mu + \alpha_i + B_{ij} + \epsilon_{ijk}, \quad i = 1, \ldots, 4;\ j = 1, \ldots, 3;\ k = 1, \ldots, 3$$

where α_i represents the fixed effect of Process i, B_{ij} is the effect of Team j within each Process i assumed to be iid $N(0, \sigma_B^2)$, and the random error ϵ_{ijk} is assumed to be iid $N(0, \sigma^2)$. Also B_{ij} and ϵ_{ijk} are assumed to be independent. The SAS Example F15 program, displayed in Fig. 6.47, is used to fit the above model to the surface finish data, using the default method REML. Since the data set is balanced, the estimates will be the same as those obtained from the method of moments.

The first part of the output is shown in Fig. 6.48. The estimates of the variance components, t-tests, and confidence intervals based on the Satterthwaite approximation are provided in the table titled `Covariance Parameter Estimates`. According to the p-value (0.1317), the variance component `Team(Process)` is not significantly different from zero. However, the z-tests for variance components based on large-sample results, may not be valid for

```
data parts;
input Process $ @;
do Team =1 to 3;
input Finish @;
output;
end;
datalines;
1 29 34 16
1 12 24 17
1 21 29 24
2 42 35 36
2 37 39 18
2 32 30 23
3 38 23 16
3 35 36 27
3 28 30 19
4 16 10 12
4 13 16 17
4 25 20 18
;

proc mixed data=parts noclprint noinfo cl covtest;
   class Process Team;
   model Finish = Process /ddfm=satterth;
   random Team(Process);
   lsmeans Process/diff cl adj=bon;
   estimate  'Process 4 mean:'
            intercept 3
                        Process  0 0 0 3|
                        Team(Process) 0 0 0 0 0 0 0 0 0 1 1 1/divisor=3;
   title "Analysis of Surface Finish by Process using PROC MIXED";
run;
```

Fig. 6.47. SAS Example F15: analysis of surface finish data with PROC MIXED

the numbers of levels of random factors used in this experiment. From an analysis (not shown) using `proc mixed` using the `method=type3` option, the following ANOVA table was constructed:

SV	df	SS	MS	F	p-Value	E(MS)
Process	3	1295.639	431.880	5.19	0.0279	$\sigma^2 + 3\,\sigma_B^2 + 3\sum_i(\alpha_i - \bar{\alpha})^2$
Team(Process)	8	665.778	83.222	2.36	0.0498	$\sigma^2 + 3\,\sigma_B^2$
Error	24	847.333	35.306			σ^2
Total	35	2808.750				

The resulting F-test for the variance component `team(process)` rejects $\sigma_B^2 = 0$ at $\alpha = 0.05$.

The four process means are found to be significantly different at $\alpha = 0.05$ using the F-test based on Type 3 SS (p-value=0.0279) (see the table titles `Type 3 Tests of Fixed Effects`). The coefficients specified in the `estimate` statement correspond to the linear combination of parameters needed to estimate the conditional mean $\mu + \alpha_4 + \bar{B}_{4.}$ for Process 4 conditioned on the nested effect. The standard error of this estimate is

Analysis of Surface Finish by Process using PROC MIXED

The Mixed Procedure

Iteration History			
Iteration	Evaluations	-2 Res Log Like	Criterion
0	1	222.99895630	
1	1	220.51004767	0.00000000

Convergence criteria met.

Covariance Parameter Estimates							
Cov Parm	Estimate	Standard Error	Z Value	Pr > Z	Alpha	Lower	Upper
Team(Process)	15.9722	14.2804	1.12	0.1317	0.05	4.7596	335.95
Residual	35.3056	10.1918	3.46	0.0003	0.05	21.5255	68.3270

Fit Statistics	
-2 Res Log Likelihood	220.5
AIC (Smaller is Better)	224.5
AICC (Smaller is Better)	224.9
BIC (Smaller is Better)	225.5

Type 3 Tests of Fixed Effects				
Effect	Num DF	Den DF	F Value	Pr > F
Process	3	8	5.19	0.0279

Estimates					
Label	Estimate	Standard Error	DF	t Value	Pr > |t|
Process 4 mean:	16.3333	1.9806	24	8.25	<.0001

Fig. 6.48. SAS Example F15: analysis of surface finish data with PROC MIXED

$\sqrt{35.3056/9} = 1.9806$ (matches the value in `Estimates` table). This is different from the standard error estimate of the Process 4 unconditional mean as illustrated below.

The second part of the output, shown in Fig. 6.49, displays the results of the `lsmeans` statement. The `Least Squares Means` table shows the standard errors, t-tests, and t-based confidence intervals of the estimates of the unconditional means $\mu+\alpha_i$. The `diff cl adj=bon` options request t-tests and 95% confidence intervals for pairwise differences in process means (or effects). The p-values reported in the table of `Differences of Least Squares Means` for *pairwise t-tests* and *confidence intervals for the differences* are both adjusted for simultaneous inference using the Bonferroni method.

Analysis of Surface Finish by Process using PROC MIXED

The Mixed Procedure

Least Squares Means									
Effect	**Process**	**Estimate**	**Standard Error**	**DF**	**t Value**	**Pr > |t|**	**Alpha**	**Lower**	**Upper**
Process	1	22.8889	3.0409	8	7.53	<.0001	0.05	15.8766	29.9012
Process	2	32.4444	3.0409	8	10.67	<.0001	0.05	25.4322	39.4567
Process	3	28.0000	3.0409	8	9.21	<.0001	0.05	20.9877	35.0123
Process	4	16.3333	3.0409	8	5.37	0.0007	0.05	9.3211	23.3456

Differences of Least Squares Means											
Effect	**Process**	**_Process**	**Estimate**	**Standard Error**	**DF**	**t Value**	**Pr > |t|**	**Adjustment**	**Adj P**	**Alpha**
Process	1	2	-9.5556	4.3004	8	-2.22	0.0570	Bonferroni	0.3420	0.05
Process	1	3	-5.1111	4.3004	8	-1.19	0.2687	Bonferroni	1.0000	0.05
Process	1	4	6.5556	4.3004	8	1.52	0.1659	Bonferroni	0.9955	0.05
Process	2	3	4.4444	4.3004	8	1.03	0.3316	Bonferroni	1.0000	0.05
Process	2	4	16.1111	4.3004	8	3.75	0.0057	Bonferroni	0.0339	0.05
Process	3	4	11.6667	4.3004	8	2.71	0.0265	Bonferroni	0.1592	0.05

Differences of Least Squares Means						
Effect	**Process**	**_Process**	**Lower**	**Upper**	**Adj Lower**	**Adj Upper**
Process	1	2	-19.4724	0.3613	-24.5163	5.4052
Process	1	3	-15.0280	4.8057	-20.0718	9.8496
Process	1	4	-3.3613	16.4724	-8.4052	21.516
Process	2	3	-5.4724	14.3613	-10.5163	19.405
Process	2	4	6.1943	26.0280	1.1504	31.071
Process	3	4	1.7498	21.5835	-3.2941	26.627

Fig. 6.49. SAS Example F15: analysis of surface finish data with PROC MIXED

Recall that the standard errors are computed using the formula

$$\sqrt{(\hat{\sigma}^2 + 3\hat{\sigma}_B^2)/9} = \sqrt{83.222/9} = 3.0409$$

since it is easily shown that $\mathrm{Var}(\bar{y}_{4..}) = (\sigma^2 + 3\sigma_B^2)/9$ and $\mathrm{MS}_{\mathrm{B(A)}}$ estimates $\sigma^2 + 3\sigma_B^2$. In the part of the output in Fig. 6.49 containing the t-tests, the `Adj P` column contains the p-values Bonferroni-adjusted for multiple testing. This is done by simply multiplying the standard p-values by the number of comparisons made (i.e., $a(a-1)/2 = 6$, in this example). Similarly, for calculating the Bonferroni adjusted confidence intervals, the upper tail $(1 - \alpha/12)100$ percentile of the t-distribution with eight degrees of freedom replaces the $(1 - \alpha/2)100$ percentile used for calculating the usual one-at-time t-based intervals. The SAS function call `quantile('T',1-alpha,df)` can be used to compute these percentiles in a data step.

Using the adjusted p-values, it is found that only processes 2 and 4 means are significantly different. The Bonferroni procedure controls the *maximum experimentwise error rate* (i.e., the probability of making at least one type 1 error at the specified α value). This is somewhat conservative, but the same result is obtained with Tukey's method. If no adjustment is made for multiple testing, Processes 3 and 4 are significantly different in addition to Processes 2 and 4. The conclusion that may be made from this experiment is that Process 4 produced a significantly better surface finish than Process 2, which produced the worst result. There is not enough evidence in the data to differentiate the other two process means from those of either Processes 2 or 4.

6.6 Models with Random and Nested Effects for More Complex Experiments

Several random and mixed models commonly used for the analysis of experiments were discussed in previous sections. The levels of factors studied in such experiments were combinations of fixed, random, or nested effects. In order to keep the introduction to these models to a reasonable level of complexity but be still informative, the examples of experiments considered were somewhat straightforward. In this section, several experiments that require more complicated models than those discussed so far are considered in order to build on the knowledge acquired from that introduction. In particular, several experiments that involve different combinations of factors or different randomization procedures from those discussed previously are introduced in several subsections. The presentation is slightly different from the previous sections in that instead of introducing the general model and inference procedures for a class of models and then illustrating with an example, the discussion in the following subsections is motivated by a specific example provided to illustrate the class of models.

6.6.1 Models for Nested Factorials

The so-called nested factorial experiments involve various combinations of crossed and nested factors that are either fixed or random. In the simplest set up, two factors are crossed in a factorial arrangement and a third factor is nested within combinations of those two factors. Instead, the third factor may be nested within only one of the two factors, or the third factor may be nested in a completely different factor. In any case, using the arrangement of the treatment factors and the experimental design structure, one should be able to formulate an appropriate model using the principles illustrated in previous sections of this chapter. Once the appropriate model is determined and the status of each effect in the model, whether fixed or random, is declared, `proc mixed` may be used to perform the computations necessary to analyze the data. The following experiment discussed in Montgomery (1991), provides a typical example of this type of model.

SAS Example F16

A study is designed to find ways to improve the speed of the assembly operation involved in the manual insertion of electronic components on printed circuit boards. An industrial engineer is comparing three assembly fixtures and two workplace layouts. It was decided to select four operators randomly to perform the assembly operation for each fixture-layout combination. Since the workplaces are in different locations within the plant, it is not feasible to use the *same* four operators for each layout. Thus, four different operators are chosen for each of the layouts. Two replications are obtained for each

Table 6.8. Circuit assembly time data

Fixture	Layout							
	1				2			
	Operator				Operator			
	1	2	3	4	1	2	3	4
1	22	23	28	25	26	27	28	24
	24	24	29	23	28	25	25	23
2	30	29	30	27	29	30	24	28
	27	28	32	25	28	27	23	30
3	25	24	27	26	27	26	24	28
	21	22	25	23	25	24	27	27

treatment combination. The assembly times are measured in seconds and are shown in Table 6.8.

In this experiment, operators are nested within levels of `Layouts` (i.e., different set of operators for each layout). The effect of `Operator within Layout` is random. The two factors `Fixture` and `Layout` are crossed because levels of layout occur with each level of fixture. The effects of fixture, layout, and their interaction `Fixture`×`Layout` are fixed effects. Thus, it is clear that operators are nested in one (layout) of those two crossed factors.

Since operators are nested in layouts, `Layout` × `Operator(Layout)` interaction is not tenable; however, it is possible for `Fixture` to interact with `Operator(Layout)`; hence, to be able to test whether it is significant, a term corresponding to this interaction is included in the model. Thus, the appropriate model is

$$y_{ijk\ell} = \mu + \alpha_i + \beta_j + G_{k(j)} + \gamma_{ij} + \alpha G_{ik(j)} + \epsilon_{ijk\ell}$$

with $i = 1, 2, 3$ (fixtures), $j = 1, 2$ (layouts), $k = 1, 2, 3, 4$ (operators), $\ell = 1, 2$ (reps), where α_i is the effect of the ith operator, β_j is the effect of the jth layout, $G_{k(j)}$ is the effect of the operator k within layout j, assumed to be iid $N(0, \sigma_G^2)$ random variables, γ_{ij} is the interaction effect of the ith level of fixture and the jth level of layout, $\alpha G_{ik(j)}$ is the interaction effect of the ith level of fixture and the operator k within layout j, assumed to be iid $N(0, \sigma^2_{\alpha G})$ random variables, and $\epsilon_{ijk\ell}$ are the usual random errors assumed to be iid $N(0, \sigma^2)$ random variables. In addition, the three random effects are mutually independent. Thus, the unconditional mean $\mu_{ij} = E(y_{ijk\ell}) = \alpha_i + \beta_j + \gamma_{ij}$ contains the fixed effects only, and the covariance matrix of $\mathbf{y} = (y_{ijk\ell})$ is defined using the variance components.

```
proc mixed data=circuit noclprint noinfo method=type3 cl;
   class Fixture Layout Operator;
   model Time = Fixture Layout Fixture*Layout/ddfm=satterth;
   random Operator(Layout) Fixture*Operator(Layout);
   lsmeans Fixture/pdiff cl adj=tukey;
   estimate 'F1-F2 for Op 4(1)'
            Fixture 1 -1 0|
            Fixture*Operator(Layout) 0 0 0 1  0 0 0 0
                                                      0 0 0 -1  0 0 0 0
                                                            0 0 0 0  0 0 0 0;
   estimate 'F1-F2 for Op 1(2)'
            Fixture 1 -1 0|
            Fixture*Operator(Layout) 0 0 0 0  1 0 0 0
                                     0 0 0 0 -1 0 0 0
                                     0 0 0 0  0 0 0 0;
   title 'Analysis of Nested Factorials using PROC MIXED';
run;
```

Fig. 6.50. SAS Example F16: analysis of circuit assembly time data with PROC MIXED

The SAS Example F16 program, displayed in Fig. 6.50, is used to fit the above model to the circuit assembly data, using the method of moments. The data are inputted using the method illustrated in several examples previously. The set of responses for the four operators for each fixture-layout combination, are read from a data line using a do-loop and each combination of fixture, layout, operator, and the response values written to the SAS data set using an output statement. The trailing @ symbol is used to hold the line after accessing the fixture and layout values as well as for repeatedly reading the responses for the four operators. The `method=type3` option in the proc statement requests the method of moments to be used. The `cl` option calculates confidence intervals for variance components based on the Satterthwaite approximation. The class statement in `proc mixed` step declares `Fixture`, `Layout`, and `Operator` as classification variables, in that order. The `model` statement includes the fixed effects `Fixture`, `Layout` and `Fixture*Layout` and the `random` statement, the random effects `Operator(Layout)` and `Fixture*Operator(Layout)`. Explanations of the `lsmeans` and `estimate` statements included in the proc step are provided in the discussion of the output produced from this program.

The following ANOVA table is constructed using the output from SAS Example F16 displayed in Fig. 6.51.

SV	df	SS	MS	F	p-Value	E(MS)
F	2	82.7917	41.3958	7.55	0.0076	$\sigma^2 + 2\,\sigma^2_{\alpha G} + Q(\alpha, \gamma)$
L	1	4.0833	4.0833	0.34	0.5807	$\sigma^2 + 2\,\sigma^2_{\alpha G} + 6\,\sigma^2_G + Q(\beta, \gamma)$
F×L	2	19.0417	9.5208	1.74	0.2178	$\sigma^2 + 2\,\sigma^2_{\alpha G} + Q(\gamma)$
O(L)	6	71.9167	11.9861	2.18	0.1174	$\sigma^2 + 2\,\sigma^2_{\alpha G} + 6\,\sigma^2_G$
F×O(L)	12	65.8333	5.4861	2.35	0.0360	$\sigma^2 + 2\,\sigma^2_{\alpha G}$
Error	24	56.0000	2.3333			σ^2
Total	47	299.6667				

Analysis of Nested Factorials using PROC MIXED

The Mixed Procedure

Type 3 Analysis of Variance				
Source	**DF**	**Sum of Squares**	**Mean Square**	**Expected Mean Square**
Fixture	2	82.791667	41.395833	Var(Residual) + 2 Var(Fixtu*Operat(Layout)) + Q(Fixture,Fixture*Layout)
Layout	1	4.083333	4.083333	Var(Residual) + 2 Var(Fixtu*Operat(Layout)) + 6 Var(Operator(Layout)) + Q(Layout,Fixture*Layout)
Fixture*Layout	2	19.041667	9.520833	Var(Residual) + 2 Var(Fixtu*Operat(Layout)) + Q(Fixture*Layout)
Operator(Layout)	6	71.916667	11.986111	Var(Residual) + 2 Var(Fixtu*Operat(Layout)) + 6 Var(Operator(Layout))
Fixtu*Operat(Layout)	12	65.833333	5.486111	Var(Residual) + 2 Var(Fixtu*Operat(Layout))
Residual	24	56.000000	2.333333	Var(Residual)

Type 3 Analysis of Variance				
Source	**Error Term**	**Error DF**	**F Value**	**Pr > F**
Fixture	MS(Fixtu*Operat(Layout))	12	7.55	0.0076
Layout	MS(Operator(Layout))	6	0.34	0.5807
Fixture*Layout	MS(Fixtu*Operat(Layout))	12	1.74	0.2178
Operator(Layout)	MS(Fixtu*Operat(Layout))	12	2.18	0.1174
Fixtu*Operat(Layout)	MS(Residual)	24	2.35	0.0360
Residual	.	.	.	.

Covariance Parameter Estimates				
Cov Parm	**Estimate**	**Alpha**	**Lower**	**Upper**
Operator(Layout)	1.0833	0.05	-1.2927	3.4593
Fixtu*Operat(Layout)	1.5764	0.05	-0.7156	3.8684
Residual	2.3333	0.05	1.4226	4.5157

Fig. 6.51. SAS Example F16: analysis of circuit assembly time data (variance components)

The Type 3 sums of squares, Expected Mean Squares, F-statistics and p-values available in relevant tables of the output in Fig. 6.51 are used for this purpose. The method of moments estimates for the variance components can be obtained using the equations

$$\sigma^2 + 2\,\sigma^2_{\alpha G} + 6\,\sigma^2_{G} = 11.9861$$

$$\sigma^2 + 2\,\sigma^2_{\alpha G} = 5.4861$$

$$\sigma^2 = 2.3333$$

giving the estimates $\hat{\sigma}^2 = 2.3333$, $\hat{\sigma}_{\alpha G} = 1.5764$, and $\hat{\sigma}^2_{G} = 1.0833$, agreeing with the values in Fig. 6.51. Large-sample confidence intervals are also available from the `Covariance Parameter Estimates` table; intervals based on Satterthwaite approximations can be constructed as usual.

From the ANOVA table, the effects of assembly fixtures are significantly different at $\alpha = 0.05$. However, the workplace layouts have no significant effects on mean assembly times and there is no significant interaction between layouts and fixtures. The variance component $\sigma^2_{\alpha G}$ measuring the interaction between fixtures and operators(layout) is found to be significantly different from zero. This indicates that the differences in effects of fixtures varies among operators within layouts although when averaged over the fixtures it is not significantly different from zero.

Although it is possible to compare the differences among fixtures unconditionally (i.e., averaging over the population of operators), the significant interaction suggests that some operators may be more effective than others. Thus, to compare mean performance using the assembly fixtures p and q (say) for each operator $k(j)$ in the experiment, predictable functions of the following type are needed:

$$\alpha_p - \alpha_q + \bar{\gamma}_{p.} - \bar{\gamma}_{q.} + \alpha G_{pk(j)} - \alpha G_{qk(j)}$$

Two examples of this type of BLUP are included in the SAS program for illustration. Note that the estimable function of the fixed effects parameters includes the interaction terms $\bar{\gamma}_{p.} - \bar{\gamma}_{q.}$; however, specifying this part can be omitted from the `estimate` statements, as SAS automatically includes the coefficients for the interaction term. Note that the combination $k(j)$ signifies a different operator for all combinations of values of k (levels of operator) and j (levels of layout). Again, the subscripts change in accordance with lexical ordering. Here, note that the effect `Layout` occurs before `Operator` in the `class` statement; thus, operator subscripts change faster. Thus, the values of $k(j)$ for the random effect parameter $\alpha G_{ik(j)}$ occur in the order 1(1), 2(1), 3(1), 4(1), 1(2), 2(2), 3(2), 4(2) for each $i = 1, 2, 3$; that is, the `layout` $\times$ `operator(layout)` interaction parameter vector will have 24 elements. These 24 positions are coded as either $0, 1$, or -1 when coding the $\alpha G_{14(1)} - \alpha G_{24(1)}$ portion in the comparison:

Analysis of Nested Factorials using PROC MIXED

The Mixed Procedure

Fit Statistics	
-2 Res Log Likelihood	187.3
AIC (Smaller is Better)	193.3
AICC (Smaller is Better)	194.0
BIC (Smaller is Better)	193.6

Type 3 Tests of Fixed Effects				
Effect	Num DF	Den DF	F Value	Pr > F
Fixture	2	12	7.55	0.0076
Layout	1	6	0.34	0.5807
Fixture*Layout	2	12	1.74	0.2178

Estimates					
Label	Estimate	Standard Error	DF	t Value	Pr > |t|
F1-F2 for Op 4(1)	-3.3340	1.4441	11.7	-2.31	0.0402
F1-F2 for Op 1(2)	-2.8605	1.8744	6.79	-1.53	0.1721

Least Squares Means									
Effect	Fixture	Estimate	Standard Error	DF	t Value	Pr > |t|	Alpha	Lower	Upper
Fixture	1	25.2500	0.6916	15.5	36.51	<.0001	0.05	23.7801	26.7199
Fixture	2	27.9375	0.6916	15.5	40.40	<.0001	0.05	26.4676	29.4074
Fixture	3	25.0625	0.6916	15.5	36.24	<.0001	0.05	23.5926	26.5324

Fig. 6.52. SAS Example F16: analysis of circuit assembly time data (fixed effects)

```
estimate 'F1-F2 for Op 4(1)'
         fixture 1 -1 0|
         fixture*operator(layout) 0 0 0 1  0 0 0 0
                                  0 0 0 -1 0 0 0 0
                                  0 0 0 0  0 0 0 0;
```

The t-statistics and p-values from the estimate statements used in SAS Example F16 program are shown in Fig. 6.52. These tests show that there is a significant difference (at $\alpha = 0.05$) between the effects of fixtures 1 and 2 on the average speed of assembly for Operator 4 in Layout 1 but not for Operator 1 in Layout 2. Other simple effects of this type may be similarly tested.

Analysis of Nested Factorials using PROC MIXED

The Mixed Procedure

Differences of Least Squares Means										
Effect	Fixture	_Fixture	Estimate	Standard Error	DF	t Value	Pr > |t|	Adjustment	Adj P	Alpha
Fixture	1	2	-2.6875	0.8281	12	-3.25	0.0070	Tukey-Kramer	0.0179	0.05
Fixture	1	3	0.1875	0.8281	12	0.23	0.8247	Tukey-Kramer	0.9722	0.05
Fixture	2	3	2.8750	0.8281	12	3.47	0.0046	Tukey-Kramer	0.0119	0.05

Differences of Least Squares Means						
Effect	Fixture	_Fixture	Lower	Upper	Adj Lower	Adj Upper
Fixture	1	2	-4.4918	-0.8832	-4.8968	-0.4782
Fixture	1	3	-1.6168	1.9918	-2.0218	2.3968
Fixture	2	3	1.0707	4.6793	0.6657	5.0843

Fig. 6.53. SAS Example F16: analysis of circuit assembly time data (differences in means)

Figure 6.53 contains the results of the `lsmeans` statement that produce t-tests and confidence intervals for pairwise differences of fixture means (or effects) adjusted for multiple testing using Tukey's method. As usual these are estimates of differences of the unconditional means for pairs of fixtures p and q averaged over the layouts:

$$\alpha_p - \alpha_q + \bar{\gamma}_{p.} - \bar{\gamma}_{q.}.$$

`Proc mixed` calculates the standard errors of these differences correctly. Since the $F \times L$ interaction is not significant, the above results can be usefully interpreted. They show that Fixture 2 mean is significantly larger than the means of both Fixtures 1 and 3 but that those of Fixtures 1 and 3 are similar.

6.6.2 Models for Split-Plot Experiments

The split-plot design is often used when one factor is more readily applied to large experimental units, called whole-plots (or main plots), and another factor can be applied to smaller experimental units within the whole-plot called subplots. Another frequent use of a split-plot design is when more precision is needed for comparisons among the levels of one factor than for the other factor. To ensure that a factor is more accurately estimated, its levels are applied to the subplots so that it will naturally have more replications. Note that in this introduction, only two-way factorials are considered, though treatments at each level (whole-plots or subplots) may be in any arrangement. For example, the whole-plot treatments (or sub-plot treatments or both) themselves may be factorial combinations of other factors.

A typical example is a field experiment in which irrigation levels are applied to larger plots and a factor like crop varieties or levels of fertilizer are randomized among smaller plots within each larger plot assigned a type of irrigation. The proper analysis of a split-plot design recognizes that treatments applied to whole-plots are subject to a different experimental errors than treatments applied to subplots. This results in the use of different mean squares as denominators for the F-ratios used to test the respective treatment effects.

Although the split-plot design originated in field experiments, it has found useful applications in many other areas. Consider the following experiment described in Montgomery (1991). A paper manufacturer is interested in studying the effects of three different pulp preparation methods and four different cooking temperatures on the tensile strength of paper. If a randomized complete block design (RCBD) is to be used for this experiment, the 12 method by treatment combinations (a 3×4 factorial), would need to be applied within each block (or replication). This requires that 12 pulp batches be prepared for each block, using each of the 3 methods of pulp preparation and each batch assigned to one of the 4 cooking temperatures.

However, the experiment was actually conducted in the following manner. Since the pilot plant is only capable of making 12 runs per day, the experimenter ran 1 replicate on each of 3 days and considered the runs performed per day as a block. On each day, he prepared three large batches of pulp using each of the three preparation methods (in random order). Then he divided each batch into four smaller samples and randomly assigned each sample to be cooked using one of the four temperatures. This is repeated for each of the three large batches. This procedure is repeated in each of the 3 days thus producing 36 tensile strength measurements.

The treatment arrangement continues to be a 3×4 factorial; however, the design is no longer a randomized block design. Rather, it is a split-plot design because the 12 treatment combinations are not assigned completely at random to 12 pulp batches; the 3 methods are assigned to the 3 large batches of pulp (whole-plots), which are then subdivided into 4 smaller samples (subplots) and assigned the temperatures. Thus, a split-plot design also introduces a different randomization scheme for factorial treatment combinations.

Rep I			Rep II			Rep III		
A3	A1	A2	A1	A3	A2	A3	A2	A1
B1	B3	B2	B3	B3	B4	B2	B1	B4
B4	B1	B4	B2	B1	B1	B3	B2	B3
B3	B2	B1	B4	B2	B2	B4	B3	B1
B2	B4	B3	B1	B4	B3	B1	B4	B2

Fig. 6.54. SAS Example F17: plan for the strength of paper experiment

The preparation method is the whole-plot factor, whereas the cooking temperature is the subplot factor. The experimental procedure is simplified because the experimenter makes 3 large batches of pulp using the 3 methods of preparation instead of making 12 small batches. The experimental plan is sketched out in Fig. 6.54, in which the levels of method are denoted by A1, A2, and A3 and the levels of temperature are denoted by B1, B2, B3, and B4.

The model for observations from this experiment reflects the fact that there are two types of experimental unit in the experiment and therefore two types of experimental error. The model and the analysis of variance for a general split-plot experiment similar to the above example is described below. It is easier to consider the model as representing two distinct parts of the design. The whole-plot design is an RCBD because the whole-plot treatment levels (Factor A) are replicated r times; thus, it is easier to visualize the model for the whole-plot part. The subplot part is also analyzed as an RCBD, where the whole-plots serve as blocks for the levels of the subplot treatment (Factor B). Assume that Factor A and Factor B have a and b levels, respectively. Let y_{ijk} represent an observation from the ith replication, the jth level of Factor A, and the kth level of Factor B. Then the model that describes this observation is given by

$$y_{ijk} = \mu + \beta_i + \tau_j + \epsilon_{ij} + \alpha_k + \delta_{jk} + \epsilon^*_{ijk}$$

where μ is the overall mean, β_i is the effect of the ith block (or replication), $i = 1, \ldots, r$, τ_j is the effect of the jth whole-plot factor (i.e., Factor A), $j = 1, \ldots, a$, ϵ_{ij} is the experimental error associated with the ijth whole-plot, α_k is the effect of the kth subplot factor (i.e., Factor B), $k = 1, \ldots, b$, δ_{jk} is the interaction between method j and temperature k, and ϵ^*_{iij} is the experimental error associated with ijkth subplot. As usual the random errors ϵ_{ij} are assumed to be iid $N(0, \sigma^2_W)$ and ϵ^*_{ijk} are assumed to be iid $N(0, \sigma^2_S)$ random variables. The ANOVA table associated with this model is

SV	df	SS	MS	F
Rep	$r-1$	SS_{REP}	MS_{REP}	
A	$a-1$	SS_A	MS_A	MS_A/MSE_W
Error A	$(r-1)(a-1)$	SSE_W	MSE_W	
B	$(b-1)$	SS_B	MS_B	MS_B/MSE_S
AB	$(a-1)(b-1)$	SS_{AB}	MS_{AB}	MS_{AB}/MSE_S
Error B	$a(r-1)(b-1)$	SSE_S	MSE_S	

Note carefully that whereas the subplot part of the experiment is, by necessity, an RCBD, the whole-plot part need not be so. In the above experiment, it was decided to use "days" as a blocking factor and, thus, an RCBD was used for the whole-plot part of the experiment. However, it is common to use a completely randomized design for the whole-plot part of the experiment. For example, if the a levels of the whole-plot factor were applied in a completely random manner to the ar whole-plots, then the whole-plot design would be a

CRD. In this case, the source Rep will not appear in the above ANOVA table, and the degrees of freedom for Error A will change to $a(r-1)$.

6.6.3 Analysis of Split-Plot Experiments Using PROC GLM

SAS Example F17

The observed data from the tensile strength of paper experiment described in Sect. 6.6.2 are found in Table 6.9. In the SAS Example F17 program, shown in Fig. 6.55 the data are inputted by the nesting of three do loops to create the values of the variables, Temp, Rep and Method. Then it uses the list input style to read each of nine values for each temperature contained in a line of data. Again a trailing @ symbol is used to read the data values repeatedly from the same data line. The output statement writes a record into the SAS data set after reading each data value along with the current values of each of Temp, Rep, and Method variables. Thus, only the data values need to be entered in the appropriate sequence in the four lines of data.

Table 6.9. Tensile strength of paper data

	Replication								
	I			III			III		
	Method			Method			Method		
Temperature	1	2	3	1	2	3	1	2	3
200	30	34	29	28	31	31	31	35	32
225	35	41	26	32	36	30	37	40	34
250	37	38	33	40	42	32	41	39	39
275	36	42	36	41	40	40	40	44	45

In SAS Example F17, proc glm is used to perform a traditional analysis in which the block (or replication) effect is considered a fixed effect in the model. Since the random statement cannot be used because there are no random effects in the model, the test option used for testing fixed effects is not available The user must provide an appropriate effect (or combination of effects) to be used as the error term (or denominator) of the F-ratio appropriate for testing hypotheses (and for constructing confidence intervals) about the effects of the whole-plot factor. This is the most important difference from an analysis of data from a similar experiment performed as a randomized blocks design from that of a split-plot experiment. Recall that under standard models with only fixed effects, SAS does not require the user to specify a term to represent the random error term in the model statement. This implies that the degrees of freedom remaining after sum of squares for the terms specified in the model are taken into account are automatically used to compute the residual or error sum of squares.

However, in the model shown for the split-plot experiment, there are two error terms. Thus, the user is required to determine how the sum of squares for the whole-plot error (labeled Error A) is to be calculated and specify it

```
data paper;
do Temp=200,225,250,275;
  do Rep=1 to 3;
    do Method=1 to 3;
       input Strength @;
       output;
    end;
  end;
end;
datalines;
30 34 29 28 31 31 31 35 32
35 41 26 32 36 30 37 40 34
37 38 33 40 42 32 41 39 39
36 42 36 41 40 40 40 44 45
;

proc glm data=paper;
  class Rep Method Temp;
  model Strength = Rep Method Rep*Method Temp Method*Temp;
  test h=Method e=Rep*Method;
  lsmeans Method*Temp/slice=Method;
  lsmeans Method/pdiff cl adj=tukey e=Rep*Method;
  contrast 'Temp:Linear Trend' Temp -3 -1 1 3;
  title 'Analysis of a Split-Plot Experiment using PROC GLM';
run;
```

Fig. 6.55. SAS Example F17: analysis of strength of paper data with PROC GLM

in a `test` statement available in `proc glm`. Usually the error term must be specified as a function of effects already present in the model statement. Recall that in this example, the whole-plot part of the design is an RCBD with replications as blocks and the levels of the whole-plot factor `Method` as treatment. Thus, the appropriate error sum of squares for the whole-plot design is equivalent to the sum of squares that correspond to the replication by method interaction. Thus, as illustrated in Fig. 6.55, by including a term that corresponds to the replication by method interaction in the model statement one is able to specify it as the error term to be used for testing the `Method` effects.

The data are classified according to `Rep, Method,` and `Temp`, so these are included in the `class` statement as usual. The `model` statement is specified as

```
strength = Rep Method Rep*Method Temp Method*Temp;
```

The effects in the `model` statement correspond to the terms in the model definition above except for the term that corresponds to Error A. As noted above, the term `Rep*Method` is included because it is intended that the sum of squares corresponding to this effect will be used as Error A. It is seen that Error B is the error (or residual) sum of squares after the other terms in the model are accounted for. The `test` statement

```
test h=Method  e=Rep*Method;
```

Analysis of a Split-Plot Experiment using PROC GLM

The GLM Procedure

Dependent Variable: Strength

Source	DF	Sum of Squares	Mean Square	F Value	Pr > F
Model	17	751.4722222	44.2042484	11.13	<.0001
Error	18	71.5000000	3.9722222		
Corrected Total	35	822.9722222			

R-Square	Coeff Var	Root MSE	Strength Mean
0.913120	5.531963	1.993043	36.02778

Source	DF	Type I SS	Mean Square	F Value	Pr > F
Rep	2	77.5555556	38.7777778	9.76	0.0013
Method	2	128.3888889	64.1944444	16.16	<.0001
Rep*Method	4	36.2777778	9.0694444	2.28	0.1003
Temp	3	434.0833333	144.6944444	36.43	<.0001
Method*Temp	6	75.1666667	12.5277778	3.15	0.0271

Source	DF	Type III SS	Mean Square	F Value	Pr > F
Rep	2	77.5555556	38.7777778	9.76	0.0013
Method	2	128.3888889	64.1944444	16.16	<.0001
Rep*Method	4	36.2777778	9.0694444	2.28	0.1003
Temp	3	434.0833333	144.6944444	36.43	<.0001
Method*Temp	6	75.1666667	12.5277778	3.15	0.0271

Tests of Hypotheses Using the Type III MS for Rep*Method as an Error Term					
Source	DF	Type III SS	Mean Square	F Value	Pr > F
Method	2	128.3888889	64.1944444	7.08	0.0485

Fig. 6.56. SAS Example F17: analysis of strength of paper data

specifies that the F-ratio for testing the `Method` effect to be constructed using the mean square corresponding to the term `Rep*Method` as the denominator (or error term). Recall that, ordinarily, the mean square corresponding to the residual (i.e., Error B here), would be used to construct this F-ratio in the Type I or III analysis of variance table in `proc glm`.

The investigator therefore must recognize that there will be two F-ratios for `Method` (and associated p-values) that will appear in the output and that the proper F ratio to be used for `Method` is the one that results from the `test` statement, appearing separately in the bottom table of the output shown in Fig. 6.56. The F ratio for `Method` appearing in the usual Type I or III ANOVA tables above it will be incorrect for testing the `Method` effects in the split-plot experiment. The ANOVA table for the split-plot design is thus completed using the values from this output:

SV	df	SS	MS	F	p-Value
Rep	2	77.55	38.78		
Method	2	128.39	64.20	7.08	0.0485
Error A	4	36.28	9.07		
Temp	3	434.08	144.69	36.45	0.0001
Method × Temp	6	75.17	12.53	3.15	0.0271
Error B	18	71.50	3.97		
Total	35	822.97			

From the ANOVA table `Method` × `Temp` interaction as well as the two main effects `Method` and `Temp` are significant at $\alpha = 0.05$. Further analysis of these results depend on the intentions of the experimenter. Making conclusions about method and temperature effects is complicated by the fact that interaction is significant. An interaction plot was obtained by default as part of the output from the SAS program in Fig. 6.55.

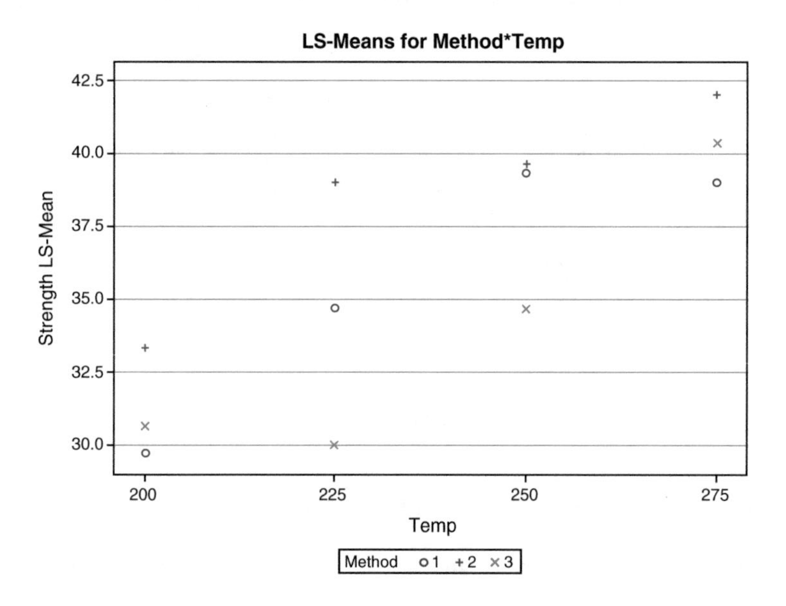

Fig. 6.57. SAS Example F17: plot of mean strength versus temperature by method

The plot, displayed in Fig. 6.57, suggests that, on the average, the mean strength appears to increase with temperature. However, the pattern of increase is different for each method. For example, as the temperature increases from 200 to 225, the mean strength increases for methods 1 and 2, whereas for Method 3, the mean strength stays about the same. However, for the temperature range over 225, a similar rate of increase in mean strength is seen for Method 3 also. However, for Method 2, mean strength does not show an appreciable change as the temperature increases from 225 to 250, as also observed for Method 1 as the temperature increases from 250 to 275. These differences in the effect of temperature for each method manifests itself as a significant interaction effect.

One may use the `proc glm` statement

```
lsmeans Method*Temp/slice=Method;
```

to examine temperature effect at each level of method. The output in Fig. 6.58 confirms the interaction plot: that the mean strengths are significantly different among the temperature levels for each method. Note that each test is associated with three degrees of freedom that correspond to comparing the four temperatures for each method; that is, the F-statistic for each slice tests a hypothesis of the form $H_0 : \mu_{j1} = \mu_{j2} = \mu_{j3} = \mu_{j4}$ versus H_a : at least one inequality, for each j, where $\mu_{jk} = \mu + \tau_j + \alpha_k + \delta_{jk}$.

Analysis of a Split-Plot Experiment using PROC GLM

The GLM Procedure
Least Squares Means

Method*Temp Effect Sliced by Method for Strength					
Method	DF	Sum of Squares	Mean Square	F Value	Pr > F
1	3	184.666667	61.555556	15.50	<.0001
2	3	121.666667	40.555556	10.21	0.0004
3	3	202.916667	67.638889	17.03	<.0001

Fig. 6.58. SAS Example F17: analysis of strength of paper data

However, this finding by itself is not sufficient to make a useful conclusion. One may need to use multiple comparisons to determine, say, the best method to use at each temperature. Thus, the type of inferences one may make depend on the purpose of the experiment. If the intention was to find the best method over all temperatures, one could compare the effects of method averaged over temperatures (thus ignoring the interaction).

The statement

```
lsmeans Method/pdiff cl adj=tukey e=Rep*Method;
```

compares the method marginal means using 95% confidence intervals adjusted using Tukey's method. It is important to note that in `proc glm`, the error term used for calculating the standard errors needs to be specified for the whole-plot effect: in this case, `Method`. Otherwise, the error mean square will be used by default. Thus, the option `e=Rep*Method` is included in the `lsmeans` statement.

The confidence intervals in Fig. 6.59 (displaying part of the output page 5) show that only methods 2 and 3 are significantly different and that Method 2 produces the strongest paper averaged over the temperature range, a finding that appears to be confirmed by the graph in Fig. 6.57. Finally, the experimenter may have selected equispaced levels for `Temperature` so that the trend in the increase or decrease in mean strength could be examined using orthogonal polynomials. Again, the interaction may be ignored, and the effects averaged over the three methods for this purpose. (This trend in temperature could also be examined for each method using appropriate contrast statements; however, that is a useful option only if the experimenter is interested in a particular method or if the trend appears to be different for each method.)

The contrast statement in Fig. 6.55 tests the hypothesis of linear trend in the marginal means averaged over the methods. The results of this statement are shown in Fig. 6.60. The appropriate partitioning of the `Temp` SS is thus given by

Analysis of a Split-Plot Experiment using PROC GLM

The GLM Procedure
Least Squares Means
Adjustment for Multiple Comparisons: Tukey

Standard Errors and Probabilities Calculated Using the Type III MS for Rep*Method as an Error Term

Least Squares Means for effect Method
Pr > |t| for H0: LSMean(i)=LSMean(j)

Dependent Variable: Strength

i/j	1	2	3
1		0.1660	0.4129
2	0.1660		0.0434
3	0.4129	0.0434	

Least Squares Means for Effect Method

i	j	Difference Between Means	Simultaneous 95% Confidence Limits for LSMean(i)-LSMean(j)	
1	2	-2.833333	-7.215044	1.548377
1	3	1.750000	-2.631711	6.131711
2	3	4.583333	0.201623	8.965044

Fig. 6.59. SAS Example F17: analysis of strength of paper data

Analysis of a Split-Plot Experiment using PROC GLM

The GLM Procedure

Dependent Variable: Strength

Contrast	DF	Contrast SS	Mean Square	F Value	Pr > F
Temp:Linear Trend	1	432.4500000	432.4500000	108.87	<.0001

Fig. 6.60. SAS Example F17: analysis of strength of paper data

SV	df	SS	MS	F	p-Value
Temp	3	434.08	144.69	36.45	0.0001
Linear	1	432.45	432.45	108.87	<0.0001
Lof	2	1.63	0.82	0.21	$\approx$1
Error B	18	71.50	3.97		

The results of the F-tests confirm a linear trend in the mean strength (averaged over the methods) as the cooking temperature increases.

6.6.4 Analysis of Split-Plot Experiments Using PROC MIXED

SAS Example F18

In previous discussions of the RCBD (in Sect. 6.5.1 and others), reasons were provided for regarding block effects as random effects in experiments that make use of blocks. In the analysis of the paper data in Sect. 6.6.3, neither the `rep` effect nor the `rep*method` effects were considered random effects. Recall that for computing the standard errors of `method` mean comparisons in SAS Example F17, the whole-plot error term was required to be explicitly identified in the `lsmeans` statement using the `e=rep*method` option. As mentioned previously, the `random` statement in `proc glm` does not result in the correct standard errors, as it does not set up the variance structure required for analyzing a general mixed model. Thus, `proc mixed` is the appropriate SAS procedure available for analyzing data from split-plot experiments. In SAS Example F18, the analysis of the strength of paper data is repeated using `proc mixed`. A discussion of the changes needed if the whole-plot part of the experiment is conducted as a completely randomized design appears following this example.

The specification of the model for the observations from a split-plot experiment with blocks as a random factor is slightly different. It is restated again with only the differences from the previous model highlighted:

$$y_{ijk} = \mu + B_i + \tau_j + \epsilon_{ij} + \alpha_k + \delta_{jk} + \epsilon^*_{ijk}$$

where B_i is the effect of the ith block (or replication) assumed to be iid $N(0, \sigma_B^2)$, ϵ_{ij} is the experimental error associated with the ijth whole-plot

```
insert data step to create the SAS dataset 'paper' here

proc mixed data=paper noclprint noinfo Method=type3 cl;;
  class Rep Method Temp;
  model Strength =  Method  Temp Method*Temp/ddfm=satterth;
  random Rep Rep*Method;
  lsmeans Method*Temp/slice=Method;
  lsmeans Method/diff cl adj=tukey;
  contrast 'Temp:Linear Trend' Temp -3 -1 1 3;
  title "Analysis of a Split-Plot Experiment using PROC MIXED";
run;
```

Fig. 6.61. SAS Example F18: analysis of strength of paper data with PROC MIXED

assumed to be iid $N(0, \sigma_W^2)$, and ϵ^*_{iij} is the experimental error associated with ijkth subplot assumed to be iid $N(0, \sigma_S^2)$. The SAS Example F18 program is displayed in Fig. 6.61. As earlier, the fixed effects are specified in the `model` statement and the random effects in the `random` statement. The user is again required to determine how the whole-plot error is calculated and then declare it as a random effect. Note also that the method of computation of variance components is specified as `type3`. The specification of the option `ddfm=satterth` requests that the Satterthwaite approximation be calculated to obtain degrees of freedom for any statistic for which it is required. Computation of the denominator sum of squares for certain F-statistics requires this approximation, as will be illustrated later.

As can be observed, every statement used with `proc glm` is available for use with `proc mixed`; however, the user is not required to specify the error terms to be used for tests of fixed effects or for computation of standard errors for comparisons of fixed effects. The output from `proc mixed` (note that the options `noclprint` and `noinfo` are in effect, so those parts of the output are suppressed) is shown in Fig. 6.62. It contains tables containing the Type 3 sums of squares and F-tests for all effects, expected mean squares, estimates, and asymptotic confidence intervals for variance components. The following ANOVA table that includes an expected mean squares column is assembled using this information:

SV	df	SS	MS	F	p-Value	$E(\text{MS})$
Rep	2	77.55	38.78	4.28	0.1016	$\sigma_S^2 + 4\sigma_W^2 + 12\sigma_b^2$
Method	2	128.39	64.20	7.08	0.0485	$\sigma_S^2 + 4\sigma_W^2 + Q(\beta, \tau)$
Error A	4	36.28	9.07	2.28	0.1003	$\sigma_S^2 + 4\sigma_W^2$
Temp	3	434.08	144.69	36.45	0.0001	$\sigma_S^2 + Q(\alpha, \delta)$
Method × Temp	6	75.17	12.53	3.15	0.0271	$\sigma_S^2 + Q(\delta)$
Error B	18	71.50	3.97			σ_S^2
Total	35	822.97				

Analysis of a Split-Plot Experiment using PROC MIXED

The Mixed Procedure

Type 3 Analysis of Variance					
Source	**DF**	**Sum of Squares**	**Mean Square**	**Expected Mean Square**	**Error Term**
Method	2	128.388889	64.194444	Var(Residual) + 4 Var(Rep*Method) + Q(Method,Method*Temp)	MS(Rep*Method)
Temp	3	434.083333	144.694444	Var(Residual) + Q(Temp,Method*Temp)	MS(Residual)
Method*Temp	6	75.166667	12.527778	Var(Residual) + Q(Method*Temp)	MS(Residual)
Rep	2	77.555556	38.777778	Var(Residual) + 4 Var(Rep*Method) + 12 Var(Rep)	MS(Rep*Method)
Rep*Method	4	36.277778	9.069444	Var(Residual) + 4 Var(Rep*Method)	MS(Residual)
Residual	18	71.500000	3.972222	Var(Residual)	.

Type 3 Analysis of Variance			
Source	**Error DF**	**F Value**	**Pr > F**
Method	4	7.08	0.0485
Temp	18	36.43	<.0001
Method*Temp	18	3.15	0.0271
Rep	4	4.28	0.1016
Rep*Method	18	2.28	0.1003
Residual	.	.	.

Covariance Parameter Estimates				
Cov Parm	**Estimate**	**Alpha**	**Lower**	**Upper**
Rep	2.4757	0.05	-3.9439	8.8953
Rep*Method	1.2743	0.05	-1.9343	4.4829
Residual	3.9722	0.05	2.2679	8.6869

Fig. 6.62. SAS Example F18: analysis of strength of paper data using PROC MIXED(page 1)

As seen here, the F-statistics for the whole-plot effects, subplot effects, and their interactions agree with those produced by `proc glm`. However, note that for unbalanced data, this will be not the case. The focus here is in comparison of the fixed effects and not the estimation or testing of the variance components. However, the magnitudes of the variance components should give the experimenter an idea about the precision of the experiment. For example, the subplot error variance with 18 degrees of freedom is 3.9722, quite large compared to the whole-plot error variance of 1.2743 with four degrees of freedom. Thus, there is a substantial variation among the smaller batches that is not accounted for by the different cooking temperatures. Confidence intervals based on the chi-square percentiles can be constructed for these as usual using the Satterthwaite approximation.

Type 3 Tests of Fixed Effects				
Effect	Num DF	Den DF	F Value	Pr > F
Method	2	4	7.08	0.0485
Temp	3	18	36.43	<.0001
Method*Temp	6	18	3.15	0.0271

Contrasts				
Label	Num DF	Den DF	F Value	Pr > F
Temp:Linear Trend	1	18	108.87	<.0001

Fig. 6.63. SAS Example F18: analysis of strength of paper data using PROC MIXED, fixed effects

A separate table showing the F-tests for the fixed effects and a table containing the results of the contrast statement extracted from the output are shown in Fig. 6.63. These results are identical to those from SAS Example F17 using `proc glm`. Relevant portions of results from the `lsmeans` statement for making pairwise comparisons of `method` means, extracted from output and edited, are included in Fig. 6.64. The results from the `lsmeans` statement with the `slice=` are also shown here. Again, these results are identical to those obtained from `proc glm`.

Other contrast statements can be used to analyze the interactions by making comparisons among the interaction means. The use of the `slice=` option produced an F-test to compare the four `Temp` means for a given method simultaneously (i.e., to test $H_0 : \mu_{j1} = \mu_{j2} = \mu_{j3} = \mu_{j4}$), as discussed previously. A more detailed analysis involves pairwise comparison of interaction means. For example, the contrast statements

```
contrast 'T1 vs. T2 @ M1' Temp 1 -1  0  0
                    Method*Temp  1 -1  0  0  0 0 0 0  0 0 0 0;
contrast 'T1 vs. T3 @ M1' Temp 1  0 -1  0
                    Method*Temp  1  0 -1  0  0 0 0 0  0 0 0 0;
contrast 'T1 vs. T4 @ M1' Temp 1  0  0 -1
                    Method*Temp  1  0  0 -1  0 0 0 0  0 0 0 0;
```

are each single-degree of freedom comparisons that compare cell means μ_{ij} for pairs of `Temp` levels at `Method` level 1. Similarly, comparison made earlier of the linear trend of `Temp` means can also be performed for each `Method` using similar contrast statements. The F-statistics produced for these contrasts using `proc glm` are identical to those produced by `proc mixed` as the denominator uses the subplot error mean square in both procedures.

However, for comparisons of cell means μ_{ij} for pairs of `Method` levels at fixed `Temp` levels, the two SAS procedures produce different F-statistics. This is because the F-statistics in these cases are obtained via the Satterthwaite

Differences of Least Squares Means										
Effect	Method	Temp	_Method	_Temp	Estimate	Standard Error	DF	t Value	Pr > |t|	Adjustment
Method	1		2		-2.8333	1.2295	4	-2.30	0.0825	Tukey-Kramer
Method	1		3		1.7500	1.2295	4	1.42	0.2277	Tukey-Kramer
Method	2		3		4.5833	1.2295	4	3.73	0.0203	Tukey-Kramer

Differences of Least Squares Means										
Effect	Method	Temp	_Method	_Temp	Adj P	Alpha	Lower	Upper	Adj Lower	Adj Upper
Method	1		2		0.1660	0.05	-6.2469	0.5802	-7.2150	1.5484
Method	1		3		0.4129	0.05	-1.6635	5.1635	-2.6317	6.1317
Method	2		3		0.0434	0.05	1.1698	7.9969	0.2016	8.9650

Tests of Effect Slices					
Effect	Method	Num DF	Den DF	F Value	Pr > F
Method*Temp	1	3	18	15.50	<.0001
Method*Temp	2	3	18	10.21	0.0004
Method*Temp	3	3	18	17.03	<.0001

Fig. 6.64. SAS Example F18: analysis of means of strength of paper data using PROC MIXED

approximation. More precisely, the denominator of the required F-statistic, in general, is an estimate of a linear combination of σ_S^2 and σ_W^2. This is a weighted average of the two mean squares MSE_W and MSE_W, the distribution of which is approximated by a chi-square distribution with degrees of freedom approximately obtained using the method of Satterthwaite, as illustrated previously. Thus, `proc mixed`, along with the `ddfm=satterth` option in the `model` statement is required for these F-statistics to be correctly computed. As mentioned elsewhere, it is recommended that the `ddfm=kr` be used for models with multiple random effects when the sample sizes are unequal. In SAS Example F17 and SAS Example F18 programs, the following contrast statements are included to illustrate this computation:

```
contrast 'M1 vs. M2 @ T1' Method 1 -1  0
                    Method*Temp  1 0 0 0  -1 0 0 0   0 0 0 0;
contrast 'M1 vs. M3 @ T1' Method 1  0 -1
                    Method*Temp  1 0 0 0   0 0 0 0  -1 0 0 0;
```

These compare the cell means for `Method` 1 with `Method` 2 at `Temp` 1 and cell means for `Method` 1 with `Method` 3 at `Temp` 1. The two sets of F-statistics produced by `proc glm` and `proc mixed`, respectively, for these contrasts are produced in Figs. 6.65 and 6.66. Note carefully that the degrees of freedom calculated for the denominator by `proc mixed` are obtained using the Satterthwaite approximation.

Contrast	DF	Contrast SS	Mean Square	F Value	Pr > F
M1 vs M2 @ T1	1	20.16666667	20.16666667	5.08	0.0370
M1 vs M3 @ T1	1	1.50000000	1.50000000	0.38	0.5466

Fig. 6.65. Simple effect contrasts of whole-plot factor with PROC GLM

Using the model for the split-plot experiments introduced in Sect. 6.6.2, it can be shown, in general, that the variance of the sample cell mean of Method 1 at Temperature 1, $\bar{y}_{\cdot 11}$, is $(\sigma_W^2 + \sigma_S^2)/r$. Thus, the mean square appropriate for the denominator of F-statistics for making comparisons of this type of means is an estimate of $\sigma_W^2 + \sigma_S^2$. By examining the ANOVA table for this experiment, it is easy to observe that a mean square with an expected value of $\sigma_W^2 + \sigma_S^2$ is not directly available. However, it is possible to *synthesize* a mean square (say, MS^*) by using a linear combination of MSE_W and MSE_S. To derive this linear combination by examining the expected values of these mean squares, note that

$$\frac{E(\text{MSE}_\text{W}) + 3E(\text{MSE}_\text{S})}{4} = \sigma_W^2 + \sigma_S^2 .$$

Thus, the linear combination needed is

$$\text{MS}^* = \frac{1}{4}\text{MSE}_\text{W} + \frac{3}{4}\text{MSE}_\text{S}$$

because the $E(\text{MS}^*)$ will then be equal to $\sigma_W^2 + \sigma_S^2$. The Satterthwaite approximation gives the degrees of freedom for a synthesized mean square of this type. Using this approximation, the degrees of freedom for $\text{MS}^* = \frac{1}{4}\text{MSE}_\text{W} + \frac{3}{4}\text{MSE}_\text{S} = 0.25\text{MSE}_\text{W} + 0.75\text{MSE}_\text{S}$ is given by

$$\nu = \frac{(\text{MS}^*)^2}{\frac{(0.25\text{MSE}_\text{W})^2}{\text{df}_1} + \frac{(0.75\text{MSE}_\text{S})^2}{\text{df}_2}},$$

where df_1 and df_2 are the degrees of freedom for MSE_W and MSE_S, respectively. The calculation can be done in a SAS data step similar to those used for computations of confidence intervals for components of variance in the previous sections. The execution of the data step results in the values $\text{MS}^* = 5.245$ and $\nu = 15.47$. This value for the denominator degrees of freedom, rounded to 15.5, is identical to the value reported in Fig. 6.66.

```
data df;
    mse1=9.07; mse2=3.97; df1=4; df2=18;  p=.25; q = .75;
    ms_star= p*mse1 +q*mse2;
    nu=ms_star**2/((p*mse1)**2/df1+ (q*mse2)**2/df2);
    put ms_star=  nu=  ;
run;
```

Contrasts				
Label	Num DF	Den DF	F Value	Pr > F
M1 vs M2 @ T1	1	15.5	3.84	0.0682
M1 vs M3 @ T1	1	15.5	0.29	0.6005

Fig. 6.66. Simple effect contrasts of whole-plot factor with PROC MIXED

Finally, if the whole-plot design is a CRD instead of an RCBD as used in the strength of paper experiment described in SAS Example F17, the model is modified as follows:

$$y_{ijk} = \mu + \tau_j + \epsilon_{ij} + \alpha_k + \delta_{jk} + \epsilon^*_{ijk}.$$

The whole-plot error is now estimated by the *replication within method* mean square usually denoted by the term `rep(method)` in the model statement. The partitioning of the degrees of freedom for the whole-plot analysis is adjusted as follows:

Whole plot design: RCBD

SV	df
Rep	$r-1$
A	$a-1$
Error A	$(r-1)(a-1)$

Whole plot design: CRD

SV	df
A	$a-1$
Error A	$a(r-1)$

In this case, for the use in SAS procedures such as `proc glm` or `proc mixed`, a variable denoting the *replication number* is also included as part of the data. Levels of this variable usually identify experimental units used for the whole-plot treatments. Suppose that the variable name `Rep` is used for this purpose in the paper example. And then the terms `Rep` and `Rep*Method` are replaced by the single term `Rep(Method)` in the model statement, and the error term for testing the `Method` main effect using `proc glm` (as in SAS Example F17) becomes `Rep(Method)`. Thus, the test statement changes to `test h=Method e=Rep(Method);`. The `random` statement is modified to `random Rep(Method);` in the `proc mixed` step.

6.7 Exercises

6.1 A textile mill weaves a fabric on a large number of looms (Montgomery 1991). To investigate whether there is an appreciable variation among the output of cloth per minute by the looms, the process engineer selects five looms at random and measured their output on five randomly chosen days. The following data are obtained:

Loom	Output (lbs/min)				
1	14.0	14.1	14.2	14.0	14.1
2	13.9	13.8	13.9	14.0	14.0
3	14.1	14.2	14.1	14.0	13.9
4	13.6	13.8	14.0	13.9	13.7
5	13.8	13.6	13.9	13.8	14.0

a. Write the *one-way random model* you will use to analyze this data stating assumptions about each parameter in the model and tell what each parameter represents. Construct the corresponding analysis of variance using `proc glm`. Write the ANOVA table including a column for expected mean square ($E(\text{MS})$).
b. Express the hypothesis that there is no variability in output among the looms, in terms of the model parameters. Perform a test of this hypothesis using the analysis in part (a) using $\alpha = 0.05$.
c. If the hypothesis in part (b) is rejected, estimates of the variance components associated with the model in part (a) may be desired. Use the method of moments to obtain these estimates from the results of parts (a) and (b).
d. Calculate 95% confidence intervals for the variance components that are found to be nonzero.

6.2 A sugar manufacturer wants to determine whether there is significant variability in purity of batches of raw cane among batches obtained from different suppliers as well as among different batches obtained from the same supplier (Montgomery 1991). Four batches of raw cane are obtained at random from each of three randomly selected suppliers. Three determinations of purity are made using random samples from each batch. The data are given as follows (note that the original data were given in coded form):

a. Use a two-way random model to analyze these data. Write an appropriate model explaining what each term in the model represents and stating any assumptions made. Prepare and run a `proc mixed` program necessary to test hypotheses and estimate variance components using this model.
b. Construct an analysis of variance table on a separate sheet (including expected mean squares) using the output from the program. Test all hypotheses concerning variance components of interest and interpret the results of these tests.

Batch	Supplier		
	1	2	3
	94	94	95
1	92	91	97
	93	90	93
	91	93	91
2	90	97	93
	89	95	95
	91	92	94
3	93	93	92
	94	91	95
	94	93	96
4	97	96	95
	93	95	94

c. Provide estimates of parameters of interest (variance components) depending on the outcome of each of the hypotheses tested in part (b); that is, only nonzero variance components need to be estimated. Show work.

d. Calculate 95% confidence intervals for the variance components that are found to be nonzero.

6.3 A manufacturer of diet foods suspects that the batches of raw material furnished by her supplier differ significantly in sodium content. There is a large number of batches currently in the warehouse, and the variability of sodium content among these batches is of interest. Five of these are randomly selected for study. Determinations of sodium in five samples taken from each batch were made, and the data obtained are reported in the following table.

Batch 1	Batch 2	Batch 3	Batch 4	Batch 5
23.36	23.59	23.51	23.28	23.29
23.48	23.46	23.64	23.40	23.46
23.56	23.42	23.46	23.37	23.37
23.39	23.49	23.52	23.46	23.32
23.40	23.50		23.39	23.38
	23.48			23.41
				23.35

a. Write the one-way random model you will use to analyze this data stating assumptions made about each parameter in the model and what each parameter represents. Construct the corresponding analysis of variance table using `proc glm`, including an additional column for expected mean square, $E(\text{MS})$. You must extract numbers from the SAS output to write down your own table.

b. Using model parameters, express the hypothesis that there is no variability in sodium content among batches, using model parameters. Perform a test of this hypothesis using your analysis of variance table from part (a). Use the p-value for making the decision.

c. If the hypothesis in part (b) is rejected, estimates of the variance components associated with the model in part (a) can be computed. Obtain these estimates depending on the results of part (b).

d. Calculate 95% confidence intervals for the variance components that are found to be nonzero.

6.4 In an experiment to study variability of a blood pH measurements among animals from different dams as well as from different sires described in Sokal and Rohlf (1995), ten female mice (dams) were successfully mated over a period of time to two males (sires). Different sires were employed

Dam	Sire	Blood pH readings			
1	1	48	48	52	54
	2	48	53	43	39
2	1	45	43	49	40
	2	50	45	43	36
3	1	40	45	42	48
	2	45	33	40	46
4	1	44	51	49	51
	2	49	49	49	50
5	1	54	36	36	40
	2	44	47	48	48
6	1	41	42	36	47
	2	47	36	43	38
7	1	40	34	37	45
	2	42	37	46	40
8	1	39	31	30	41
	2	50	44	40	45
9	1	52	54	52	56
	2	56	39	52	49
10	1	50	45	43	44
	2	52	43	38	33

for the ten dams, implying that a total of 20 sires were used in the experiment. Four mice were selected at random from each of the resulting litters, and the blood pH of each mouse was determined. The following data (which have been coded) were extracted from the original data to produce equal sample sizes.

a. Write the two-way nested effects model for these observations, explaining each term and stating any assumptions made. Prepare and run a `proc mixed` program necessary to analyze the data using this model.
b. Construct an analysis of variance table (including expected mean squares) using the output from the program. Test all hypotheses concerning the variance components of interest and interpret the results of these tests.
c. Provide estimates of the variance components depending on the outcome of each of the hypotheses tested in part (b); that is, you need to estimate only those variance components that are determined to be nonzero as a result of the above tests. State results of your analysis in a summary statement.
d. Calculate 95% confidence intervals for the variance components that are found to be nonzero.

6.5 In an experiment described in Dunn and Clark (1987), four brands of airplane tires are compared to assess the differences in the rate of tread wear. The data were collected on eight planes, with two tires used under each wing. The researcher uses each airplane as a block, mounting four test tires, one of each brand, in random order on each airplane. Thus, a randomized complete block design with "airplane" as a blocking factor is the design used. The amount of tread is measured initially and after 6 months and the following wear rates obtained: Note that a larger value

Airplane	Brand			
	A	B	C	D
1	4.02	2.46	2.06	3.49
2	4.50	3.39	2.91	3.18
3	2.73	1.69	2.37	1.48
4	3.74	1.95	3.39	3.09
5	3.21	1.20	1.72	2.65
6	2.53	1.04	2.52	1.23
7	3.07	2.55	2.42	2.07
8	3.10	1.09	2.22	2.57

indicates greater wear. Brand A is currently used by the airline, and Brands B, C, and D from three different competitors are being evaluated to replace A. Thus, the management is interested in

i. Comparing Brand A with the average wear of Brands B C, and D
ii. Comparing Brands B, C, and D

Prepare and run a `proc mixed` program necessary and provide answers to the following questions (on a separate sheet) assuming the model for a randomized complete block design.

a. Construct an analysis of variance table and test the hypothesis H_0 : $\tau_A = \tau_B = \tau_C = \tau_D$. State your conclusion based on the p-value.
b. Use a `contrast` statement for making comparison (i) by testing H_0 : $\tau_A = (\tau_B + \tau_C + \tau_D)/3$.
c. Use a `contrast` statement for making comparison (ii) by testing $H_O : \tau_B = \tau_C = \tau_D$. One way to test this hypothesis is to make the comparisons $\tau_B - \tau_C$ and $\tau_B - \tau_D$ simultaneously in a single `contrast` statement. This results in the computation of a sum of squares with 2 df and an F-test with 2 df for the numerator. Add these results to this ANOVA table as additional lines and summarize conclusions from this analysis.
d. Construct 95% confidence intervals for $\mu_B - \mu_C$ and $\mu_B - \mu_D$, using the output from appropriate `estimate` statements used with `proc mixed`.

6.6 A compound is sent to five randomly selected laboratories in the United States for a routine analysis. At each laboratory, four chemists are chosen at random, and each chemist makes three chemical determinations on the compound using the same method of chemical analysis. The object is to study the variation of this method from laboratory to laboratory and also among chemists within each laboratory. The data obtained are as follows:

Chemist	Laboratory				
	1	2	3	4	5
	2.24	2.44	1.97	2.54	2.44
1	2.51	2.40	2.05	2.49	2.36
	2.37	2.51	2.13	2.42	2.45
	2.65	2.26	2.23	2.46	2.67
2	2.57	2.37	2.20	2.39	2.61
	2.48	2.41	2.27	2.40	2.59
	2.41	2.38	2.25	2.71	2.64
3	2.37	2.19	2.28	2.70	2.58
	2.40	2.35	2.31	2.78	2.55
	2.25	2.75	2.37	2.62	2.38
4	2.38	2.58	2.30	2.55	2.41
	2.40	2.62	2.44	2.59	2.35

Use a SAS program with `proc mixed` to answer the following questions:
a. Write the appropriate model for the analysis of these data, stating the effect each term used in the model represents and the assumptions made about these effects.

b. Construct an appropriate ANOVA table including the required F-statistics, p-values, and expected mean squares.
c. State the two hypotheses of interest for this experiment using the model parameters in part (a). Use the F-statistics and p-values above to make conclusions. Use $\alpha = 0.05$.
d. Estimate parameters of interest in this experiment. Note that these estimates depend on the outcomes of the hypotheses tested above.

6.7 The objective of a case study discussed in Ott and Longnecker (2001) was to determine whether the pressure drop across the expansion joint in electric turbines was related to gas temperature. Also, the researchers wanted to assess the variation in readings from various types of pressure gauge and whether they were consistent across different gas temperatures. Three levels of gas temperatures that cover the operational range of the turbine were selected 15 °C, 25 °C, and 35 °C. Four types of gauge were randomly chosen for use in the study from the hundreds of different types pressure gauges used to monitor pressure in the lines. Six replications of each of the 12 temperature-gauge factorial combinations were run and the pressure measured.

Temperature (°C)	Gauge	Pressure					
15	G1	40	40	37	47	42	41
	G2	43	34	38	42	39	35
	G3	42	35	35	41	43	36
	G4	47	47	40	36	41	47
25	G1	57	57	65	67	63	59
	G2	49	43	51	49	45	43
	G3	44	45	49	45	46	43
	G4	36	49	38	45	38	42
35	G1	35	35	35	46	41	42
	G2	41	43	44	36	42	41
	G3	42	41	34	35	39	36
	G4	41	44	35	46	44	46

Use a SAS program with `proc mixed` to answer the following questions:
a. Construct an appropriate ANOVA table including appropriate F-statistics, p-values, and expected mean squares.
b. Use the expected mean squares in the ANOVA table to determine which ratios of sums of squares are to be used to test the two hypotheses of interest regarding variance components. Check your answers with those provided by `proc glm`
c. Construct an interaction plot appropriate for studying any significant interaction.

d. If there is significant variation among the gauges, suggest some BLUPs that might be useful for comparing the performance of gauges at each temperature. Use `estimate` statements to make appropriate comparisons.

6.8 A manufacturing company wishes to study the variation in tensile strength of yarns produced on four different looms (Sahai and Ageel 2000). In an experiment designed for this purpose, 12 machinists were selected, and each loom was assigned to three different machinists at random. Samples from two different runs by each machinist were obtained and tested. The data in standard units are given as follows:

Loom											
1			2			3			4		
Machinist			Machinist			Machinist			Machinist		
1	2	3	1	2	3	1	2	3	1	2	3
38.2	53.5	15.3	61.3	41.5	35.3	47.1	22.5	14.7	15.5	19.3	21.6
21.6	51.5	26.7	58.3	38.5	27.3	34.3	25.7	26.3	32.3	35.7	26.5

a. Write the appropriate model for the analysis of these data, stating the effect each term used in the model represents and the assumptions made about these effects.
b. Construct an analysis of variance table needed to analyze this data. Include a column of expected mean squares.
c. Use the ANOVA table to test whether the mean tensile strength of yarn produced by the four looms is significantly different using $\alpha = 0.05$.
d. Test an appropriate hypothesis about the variability of tensile strength of yarn among the machinists within each loom using $\alpha = 0.05$.
e. Estimate the variance components of relevant effects of this model by constructing 95% confidence intervals for them.
f. Carry out comparisons of pairwise differences among the mean tensile strength of yarn produced by the looms adjusted for multiple comparisons using the Bonferroni adjustment.

6.9 In a study reported in Dunn and Clark (1987), each of three different sprays were applied to four trees selected at random. After 1 week, the concentration of nitrogen was measured on each of six leaves obtained in a random way from each tree. The data are given in the following table.
a. Write the appropriate model for the analysis of these data, stating the effect each term used in the model represents and the assumptions made about these effects.
b. Construct an analysis of variance table needed to analyze these data. Include a column of expected mean squares.

Spray	Leaf	Tree			
		1	2	3	4
1	1	4.50	5.78	13.32	11.59
	2	7.04	7.69	15.05	8.96
	3	4.98	12.68	12.67	10.95
	4	5.48	5.89	12.42	9.87
	5	6.54	4.07	10.03	10.48
	6	7.20	4.08	13.50	12.79
2	1	15.32	14.53	10.89	15.12
	2	14.97	14.51	10.27	14.79
	3	14.81	12.61	12.21	15.32
	4	14.26	16.13	12.77	11.95
	5	15.88	13.65	10.45	12.56
	6	16.01	14.78	11.44	15.31
3	1	7.18	6.70	5.94	4.08
	2	7.98	8.28	5.78	5.46
	3	5.51	6.99	7.59	5.40
	4	7.48	6.40	7.21	6.85
	5	7.55	4.96	6.12	7.74
	6	5.64	7.03	7.13	6.81

c. Use the ANOVA table to test whether the mean nitrogen content resulting from the three sprays is significantly different using $\alpha = 0.05$.

d. Test an appropriate hypothesis about the variability of nitrogen content among the trees within each spray using $\alpha = 0.05$.

e. Estimate the variance components of relevant effects of this model by constructing 95% confidence intervals for them.

f. Carry out comparisons of pairwise differences among the mean nitrogen content resulting from the three sprays adjusted for multiple comparisons using the Tukey adjustment.

6.10 In a health awareness study (Kutner et al. 2005), each of three states independently devised a health awareness program. Three cities within each state of similar demographics were selected at random, and five households within each city were randomly selected to evaluate the effectiveness of the program. A composite index based on responses of members of the selected households who were interviewed before and after participation in the program was used for measuring the impact of the health awareness program. The data on health awareness are given as follows (the larger the index, the greater the awareness):

a. Write the appropriate model for the analysis of these data, stating the effect each term used in the model represents and the assumptions made about these effects.

b. Construct an analysis of variance table needed to analyze these data. Include a column of expected mean squares.

Household	State								
	1			2			3		
	City			City			City		
	1	2	3	1	1	3	1	2	3
1	42	26	34	47	56	68	19	18	16
2	56	38	51	58	43	51	36	40	28
3	35	42	60	39	65	49	24	27	45
4	40	35	29	62	70	71	12	31	30
5	28	53	44	65	59	57	33	23	21

c. Use the ANOVA table to test whether the mean awareness is significantly different among the three states using $\alpha = 0.1$.
d. Test an appropriate hypothesis about the variability of awareness among cities within states using $\alpha = 0.1$.
e. Construct 90% confidence intervals for pairwise comparisons between state means, using the Tukey procedure.
f. Construct 90% confidence interval for the variance component measuring variability of awareness among cities within states.

6.11 Consider an experiment to examine the variation in the effects of different analysts on chemical analyses for the DNA content of plaque (Montgomery 1991). Three female subjects (ages 18–20 years) were chosen for the study. Each subject was allowed to maintain her usual diet, supplemented with 30 mg (15 tablets) of sucrose per day. No toothbrushing or mouthwashing was allowed during the study. At the end of the week, plaque was scraped from the entire dentition of each subject and divided into three samples. The three samples of plaque from each of the subjects were randomly assigned to three analysts chosen at random. They performed an analysis for the DNA content (in micrograms). The data obtained are as follows:

Analyst	Subject		
	1	2	3
1	13.2	10.6	8.5
2	12.5	9.6	7.9
3	13.0	9.9	8.3

a. Write the appropriate model for the analysis of these data, stating the effect each term used in the model represents and the assumptions made about these effects.
b. Construct an appropriate ANOVA table including appropriate F-statistics, p-values, and expected mean squares.

c. State the hypothesis of interest for this experiment using the model parameters in part (a). Use the F-statistic and p-value above to make conclusions. Use $\alpha = 0.05$.

6.12 An engineer is designing a battery for use in a device that will be subjected to extreme variations in temperature (Montgomery 1991). He considers a two-way factorial with plate material and temperature as the two factors but has a large number of feasible choices for plate material and temperatures. Suppose that three plate materials and three temperatures were chosen, both at random, for use in the study. Four batteries were tested at each combination of plate material and temperature, and the resulting 36 tests are run in a random sequence. The data, observed battery life, are in the following table.

Material	Temperature					
type	15 °F		70 °F		125 °F	
1	130	155	34	40	20	70
	74	180	80	75	82	58
2	150	188	136	122	25	70
	159	126	106	115	58	45
3	138	110	174	120	96	104
	168	160	150	139	82	60

Use a SAS `proc mixed` program to analyze these data. Provide answers to the following questions:

a. Write the appropriate model for the analysis of these data, explaining each term in the model and the assumptions made about these.
b. Construct an analysis of variance table needed to analyze this data. Include a column of expected mean squares.
c. Use the ANOVA table to test whether temperature, type of material, or their interaction contributes significantly to the variation in battery life using $\alpha = 0.1$.
d. Estimate significant variance components by providing point estimates and by calculating 95% confidence intervals.

6.13 In a lab experiment carried out in a completely randomized design, a soil scientist studied the growth of barley plants under three different levels of salinity (control, 6 bars, 12 bars) in a controlled growth medium (Kuehl 2000). Two replications of each treatment were obtained, and three plants were measured in each replication. The data on the dry weight of plants in grams are as follows:

Use a SAS `proc mixed` program to analyze this data and provide answers to the following questions:

a. Write the appropriate model for the analysis of these data, explaining each term in the model and the assumptions made about these.
b. Construct an analysis of variance table needed to analyze these data. Include a column of expected mean squares.

Salinity	Container	Weight(g)		
Control	1	11.29	11.08	11.10
	2	7.37	6.55	8.50
6 Bars	1	5.64	5.98	5.69
	2	4.20	3.34	4.21
12 Bars	1	4.83	4.77	5.66
	2	3.28	2.61	2.69

c. Use the ANOVA table to test whether the mean dry weights are significantly different among the three salinity levels using $\alpha = 0.05$. What is the standard error of a salinity level mean?

d. Test an appropriate hypothesis about the variability of weight among containers within salinity levels using $\alpha = 0.1$.

e. Partition the sum of squares for salinity effect using two orthogonal polynomials corresponding to linear and quadratic effects, each with one degree of freedom. Interpret the results of the F-tests.

6.14 An experiment, conducted in a split-plot design to determine the effect of three bacterial inoculation treatments applied to two cultivars of grasses on dry weight yields, is discussed in Littell et al. (1991). The cultivar is the whole-plot factor and inoculi, the subplot factor. Each of the two cultivars is replicated four times. The data are as follows:

	Replication							
	I		II		III		IV	
Inoculi	Cultivar		Cultivar		Cultivar		Cultivar	
	A	B	A	B	A	B	A	B
Control	27.4	29.4	28.9	28.7	28.6	27.2	26.7	26.8
Live	29.7	32.5	28.7	32.4	29.7	29.1	28.9	28.6
Dead	34.5	34.4	33.4	36.4	32.9	32.6	31.8	30.7

Use a SAS `proc mixed` program to analyze this data and provide answers to the following questions:

a. Write the appropriate model for the analysis of these data, explaining each term in the model and the assumptions made about these.

b. Construct an analysis of variance table needed to analyze these data. Include a column of expected mean squares.

c. Use the ANOVA table to test whether the mean dry weights are significantly different among the three inoculation levels using $\alpha =$ 0.05. What is the standard error of an inoculation level mean?

d. Use the ANOVA table to test whether the mean dry weights are significantly different among the three two cultivars using $\alpha = 0.05$. What is the standard error of a cultivar mean?

e. Construct confidence intervals for comparing the cultivar means of the two inoculi with the control, adjusted for multiple testing.

6.15 Soy protein isolates (SPI), widely used in the food industry, are usually stored in dry powder form produced via spray drying or freeze drying, to enhance shelf life and make them easier to distribute. A study was conducted at Iowa State University (Deak and Johnson 2007) to determine how various properties of SPI are affected by the method used to dry them and to compare those of dried SPI to fresh (undried) or frozen-thawed SPI. Another factor that may affect the properties of SPI is the temperature used in the extraction process to create SPI. Thus, a two-factor experiment was conducted in which the two factors and their levels are temperature at levels 25, 40, 60, and 80 °C and Method with levels 1 = fresh, 2 = frozen and then thawed, 3 = freeze dried, and 4 = spray dried.

Twelve batches of SPI were created so that the four temperature levels were assigned to three batches at random. Each batch was split into four parts, and the four methods were assigned to the four parts of each SPI completely at random. Many response variables were measured for each part of each SPI, but data for emulsion capacity (EC, grams of oil emulsified by 1 gram of product) are reported in the following table (data were graciously provided by the first author of the above reference; the EC values were rounded to the nearest whole number).

Temperature											
25 °C			40 °C			60 °C			80 °C		
Batch	Method	EC	Batch	Method	EC	Batch	Method	EC	Batch	Method	EC
1	1	549	4	1	568	7	1	478	10	1	442
	2	531		2	595		2	503		2	433
	3	573		3	557		3	501		3	473
	4	600		4	591		4	512		4	480
2	1	551	5	1	584	8	1	481	11	1	449
	2	640		2	632		2	526		2	496
	3	559		3	608		3	458		3	448
	4	587		4	602		4	485		4	475
3	1	538	6	1	582	9	1	485	12	1	473
	2	591		2	606		2	524		2	503
	3	557		3	591		3	469		3	458
	4	584		4	583		4	497		4	471

Use `proc mixed` (and other procedures if needed) in a SAS program to perform the following analyses:

a. Write the appropriate model for the analysis of these data, explaining each term in the model and the assumptions made about these.

b. Construct an analysis of variance table needed to analyze these data. Include a column of expected mean squares.
c. Is there a significant interaction between temperature and method? Conduct an appropriate test to answer this question. Provide a test statistic, its degrees of freedom, a p-value, and a brief conclusion.
d. Construct an interaction plot to study the interaction between temperature and method. Use the levels of temperature on the x-axis drawn to scale. Comment.
e. Use the ANOVA table to test whether the mean EC are significantly different among the four temperature levels using $\alpha = 0.05$. What is the standard error of a temperature level mean?
f. Use the ANOVA table to test whether the mean EC are significantly different among the four methods using $\alpha = 0.05$. What is the standard error of a method mean?
g. Construct 95% confidence intervals for pairwise differences among the four temperature means adjusted for multiple testing using the Tukey adjustment.
h. Calculate a t-statistic for testing whether there is a difference between the effects of spray-dry and freeze-dry methods. Perform the test using $\alpha = 0.05$.
i. Calculate an F-statistic for testing whether there is a difference between the effects of temperatures 25 °C and 40 °C when the freeze-dry method is used. Perform the test using $\alpha = 0.05$.

6.16 An experiment is designed to study pigment dispersion in paint. Four different mixes of a particular pigment are studied. The procedure consists of preparing a particular mix and then applying that mix to a panel by three application methods (brushing, spraying, and rolling). The response measured is the percentage reflectance of the pigment. Three days are required to run the experiment, and the data obtained follow. Analyze the data and draw conclusions, assuming that mixes and application methods are fixed.

Day	App method	Mix			
		1	2	3	4
1	1	64.5	66.3	74.1	66.5
	2	68.3	69.5	73.8	70.0
	3	70.3	73.1	78.0	72.3
2	1	65.2	65.0	73.8	64.8
	2	69.2	70.3	74.5	68.3
	3	71.2	72.8	79.1	71.5
3	1	66.2	66.5	72.3	67.7
	2	69.0	69.0	75.4	68.6
	3	70.8	74.2	80.1	72.4

7

Beyond Regression and Analysis of Variance

7.1 Introduction

This chapter applies SAS software to the analysis of nonlinear models and generalized linear models. These models have important uses and cannot be analyzed with SAS procedures for linear models that are covered in earlier chapters. Some basic nonlinear models are introduced in Sect. 7.2, and the NLIN procedure is illustrated with applications to growth curves and data from pharmacokinetic and toxicology studies. Generalized linear models, a special class of nonlinear models, are introduced in Sect. 7.3 and illustrated with applications of the LOGISTIC and GENMOD procedures to popular logistic regression and Poisson regression models. Sect. 7.4 shows how overdispersion in observed counts can be accommodated with the LOGISTIC and GENMOD procedures. Extensions to the analysis of rates and logistic regression with multi-category responses are discussed in Sect. 7.5. Each section is presented like a mini chapter with its own brief introduction.

7.2 Nonlinear Models

7.2.1 Introduction

Model

Nonlinear models are used to model curved relationships between variables that would be difficult to approximate with the linear models discussed in previous chapters. It is useful to think about nonlinear models as consisting of two components: a *systematic* component that describes how the mean response changes with changes in the explanatory variables and a *random*

M. G. Marasinghe, K. J. Koehler, *Statistical Data Analysis Using SAS*,
Springer Texts in Statistics, https://doi.org/10.1007/978-3-319-69239-5_7

component that describes how potential responses vary about the systematic component. Using y to denote a value of the response variable and x to denote a set of values for a collection of explanatory variables, a nonlinear model can be expressed as

$$y = \mu(x;\beta) + \epsilon. \tag{7.1}$$

In this expression $\mu(x;\beta)$ is the systematic part of the model that specifies the relationship between the mean response and the values of the explanatory variables represented by x. This relationship is shaped by a set of population parameters denoted by β. Unlike regression models, and other linear models, $\mu(x;\beta)$ is not a linear function of the parameters.

The random part of the model is represented by ϵ, a random error that describes how potential observations vary about the mean response for a particular set of x values. The distribution of the random errors is assumed to have mean zero. A homogeneous variance condition is often assumed that restricts the variance of the random errors, denoted by σ^2, to be the same for all values of x. Least squares estimation, the default estimation method used by the NLIN procedure, is based on this homogeneous variance condition. The NLIN procedure also has an option for performing weighted least squares estimation when the level of variation in the random errors changes with one or more of the variables in x, but we will not illustrate this option in this section.

Estimation and Inference

To apply a nonlinear model, a formula must be provided for $\mu(x,\beta)$. The formula is developed from expert knowledge of the relationship that is being modeled. For example, if a researcher believes that the growth rate of an organism is slow when the organism is small, becomes faster as the organism grows, and eventually slows down as the size of the organism approaches an upper limit, a logistic growth curve model may be considered. As a function of a single explanatory variable, time, the systematic part of the logistic growth curve model can be written as

$$\mu(t;\alpha,\beta,\gamma) = \frac{\alpha}{1+e^{-(t-\beta)/\gamma}}. \tag{7.2}$$

This describes the average size of organisms in the population at any time t. Adding a random component ϵ, the actual size of a randomly selected organism at time t is represented as

$$y = \frac{\alpha}{1+e^{-(t-\beta)/\gamma}} + \epsilon. \tag{7.3}$$

There are four parameters to estimate, α, β, γ, and σ^2, the variance of ϵ. There are many other nonlinear models that could be used, and a major challenge in applying nonlinear models is the determination of the form of the systematic component. The following discussion, however, is focused on applying SAS procedures to fit nonlinear models to data after the form of the model has been selected.

To estimate parameters in a nonlinear model, the NLIN procedure in SAS uses least squares estimation, minimizing the sum of the squared residuals. This is an iterative process that requires starting values for the parameter estimates. Convergence to the global minimum of the sum of squared residuals is not completely guaranteed. The closer the starting values are to the parameter values that provide the global minimum for the sum of squared residuals, however, the greater the chance that the optimization algorithm will converge to the global minimum. The NLIN procedure provides an option of specifying a grid of possible starting values that may increase the chance of arriving at a global minimum. Under that option, the NLIN procedure computes the sum of squared residuals at each point on the grid and starts the iterative process from the point on the grid that yields the lowest sum of squared residuals. Alternatively, good starting values for parameter estimates may be determined by looking at graphs, summary statistics, and other preliminary examinations of the data.

Inferential procedures for nonlinear models, such as confidence intervals and tests of hypotheses, may be based on large sample approximations to the sampling distributions of the estimators. These approximations generally provide good approximations for large samples, but they may not adequately account for all of the variability in the estimators for small samples. Consequently, standard errors of estimators may be underestimated for small samples leading to confidence intervals that are too short to provide the desired level of confidence and tests of hypotheses with artificially small p-values. The NLIN procedure provides alternative inferential procedures based on bootstrap resampling methods that tend to provide more accurate standard errors and confidence intervals for small samples.

7.2.2 Growth Curve Models

Model

As mentioned above, the logistic growth curve model is one of many models that has been used to model growth of humans, animals, plants, and other organisms. The average growth rate for a logistic growth curve is initially small when the organisms are small, but the growth rate increases as the organisms become larger. After a certain time point, corresponding to the inflection point of the logistic growth curve, the growth rate begins to decline and becomes slower as the organisms approach an upper size limit. As a function of time, t, the systematic part of the logistic growth curve model can be written as

$$\mu(t;\alpha,\beta,\gamma) = \frac{\alpha}{1+e^{-(t-\beta)/\gamma}} \text{ for } t>0. \tag{7.4}$$

This curve has a sigmoidal shape that is symmetric about its inflection point. Adding a random component allows the actual sizes of individual organisms to vary about the population curve described by Eq. (7.4). The resulting model is

$$y = \frac{\alpha}{1 + e^{-(t-\beta)/\gamma}} + \epsilon. \tag{7.5}$$

We will assume that random errors, denoted by ϵ, are uncorrelated, each with mean zero and variance σ^2. Consequently, there are four parameters to estimate, α, β, γ from the systematic component, and σ^2 from the random component. For this model, α represents the average size of mature organisms, β is the time at which the average size of organisms reaches one-half of α, and γ is a scale parameter that controls the rate of growth.

SAS Example G1

The following example examines the growth of a cedar tree in Hollis, Alaska, during one summer. The data are taken from Bliss (1970). Beginning on May 14, weekly radial measurements of the tree trunk, in units of 0.01 inches, were recorded for 19 consecutive weeks. The measurements exhibit a sigmoidal growth pattern that is symmetrical about an inflection point near 7.5 weeks. In this study, variation about the logistic growth curve was partly due to variation in local weather factors. Tree trunk swelling from hydration tended to occur on rainy days that followed cloudy days with rainfall, and tree trunk shrinkage tended to occur on clear days. Consequently, weekly measurements of cumulative radial growth vary about the actual growth curve.

The SAS Example G1 program (see Fig. 7.1) illustrates how the SGPLOT procedure may be used to provide a scatter plot of the cumulative radial growth data. The 19 weeks of data are included in the `data step`, with the variable `week` containing the number of weeks since the beginning of the study and the variable `growth` containing the radial measurement of the tree trunk (0.01 inch) at the end of each week.

In addition to providing information about the pattern of growth, the scatter plot displayed in Fig. 7.2 is useful for determining starting values for parameter estimation. This plot indicates that the inflection point of the growth curve is near 7.5 weeks. Consequently, 7.5 is a reasonable starting value for the estimation of β. This plot also suggests that 8 is a reasonable starting value for the estimation of α, because the points on the plot appear to approach an upper limit near 8. Deriving a good starting value for the estimation of γ requires a bit more work. Equation (7.4) implies that γ is the slope in the following relationship:

$$t = \beta + \gamma \log\left(\frac{\mu(t;\alpha,\beta,\gamma)}{\alpha - \mu(t;\alpha,\beta,\gamma)}\right). \tag{7.6}$$

A value of γ is obtained by evaluating $\mu(t;\alpha,\beta,\gamma)$ at two time points, say $t_2 > t_1$, and computing

```
data cedar;
  input week growth;
  label week = "Week from May 14"
         growth = "Cumulative Radial Growth (.01 in)";
datalines;
 1 0.73
 2 0.99
 3 1.32
 4 1.66
 5 2.45
 6 3.10
 7 4.56
 8 4.95
 9 5.76
10 6.13
11 6.80
12 7.18
13 7.38
14 8.17
15 7.87
16 7.51
17 7.89
18 8.05
19 7.82
  ;

proc sgplot data=cedar;
 scatter x=week y=growth /
    markerattrs=(size=12 symbol=Circle color=black);
 yaxis label="Cumulative Radial Growth (.01 in.)"
          labelattrs=(size=14) valueattrs=(size=12);
 xaxis label="Time(weeks after May 14)"
          labelattrs=(size=14) valueattrs=(size=12);
 run;
```

Fig. 7.1. SAS Example G1: scatter plot program

$$\frac{t_2 - t_1}{\log\left(\frac{\mu(t_2;\alpha,\beta,\gamma)}{\alpha-\mu(t_2;\alpha,\beta,\gamma)}\right) - \log\left(\frac{\mu(t_1;\alpha,\beta,\gamma)}{\alpha-\mu(t_1;\alpha,\beta,\gamma)}\right)}. \tag{7.7}$$

Values of $\mu(t;\alpha,\beta,\gamma)$ are unknown, but they can be approximated from the pattern of data points in Fig. 7.2. For example, at $t_1 = 5$ weeks, it appears that $\mu(5;\alpha,\beta,\gamma)$ is close to 2. At $t_2 = 10$ weeks, it appears that $\mu(10;\alpha,\beta,\gamma)$ is close to 6. These time points are near the lower and upper ends of the time interval in which the logistic growth curve is nearly a straight line. Consequently, the starting value for γ is an estimate of the slope of that line segment. Substituting the starting value for α into (7.7), a starting value for γ is computed as

$$\frac{10-5}{\log\left(\frac{6}{8-6}\right) - \log\left(\frac{2}{8-2}\right)} = 2.28. \tag{7.8}$$

A residual is computed for each time point in the data set as the difference between the measured and predicted cumulative radial trunk values. Initially, predictions and residuals are based on the starting values for the estimates of α, β, and γ. From this starting point, an iterative procedure is used to find the

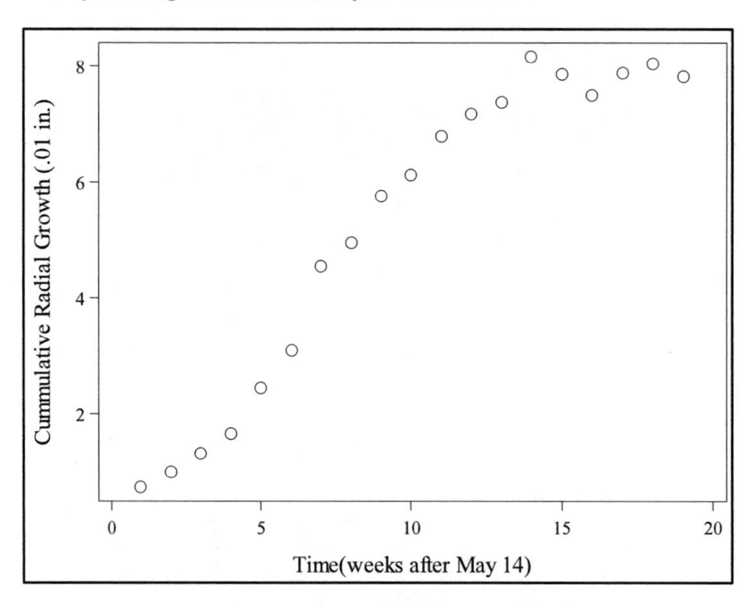

Fig. 7.2. SAS Example G1: scatter plot of growth data

estimates of α, β, and γ that minimize the sum of the squared residuals. In each step of the iteration, the predicted values and residuals are reevaluated with the updated parameter estimates. After the iterative process converges to provide the final set of parameter estimates, $\hat{\alpha}$, $\hat{\beta}$, and $\hat{\gamma}$, the error variance, σ^2, is estimated as the sum of the squared residuals divided by the error degrees of freedom, which are the number of observations minus the number of parameters in the systematic part of the model. For the cedar tree growth data, the error degrees of freedom are $19 - 3 = 16$.

Building on the program in Fig. 7.1, the SAS code displayed in Fig. 7.3 shows how the NLIN procedure is used to fit a logistic growth curve to the data. The `model` statement contains the formula for the systematic component of the growth curve model shown in Eq. (7.4). Initial values for the model parameters are specified in the `parms` statement. Selected parts of the output that provide information about parameter estimates and confidence intervals are shown in Fig. 7.4. The `plots=fit` option included in the `proc nlin` statement produces a plot of the estimated growth curve (see Fig. 7.5). The `bootci` option included in the optional `bootstrap` statement requests bias-corrected bootstrapped confidence intervals and bootstrapped standard errors for the parameter estimates.

The first table displayed in Fig. 7.4 shows that the Gauss–Newton optimization algorithm for finding the least squares parameter estimates converged in three iterations. The first row of this table shows the starting values for the parameter estimates that were specified in the program. The second table in Fig. 7.4 partitions variation in the observed growth values into a model

```
proc nlin data=cedar plots=fit;
parms alpha=8 beta=7.5 gamma=2.28;
model growth = alpha/(1+exp(-(week-beta)/gamma ));
bootstrap / bootci;
run;
```

Fig. 7.3. SAS Example G1: program code for `proc nlin`

The NLIN Procedure

Iterative Phase				
Iter	alpha	beta	gamma	Sum of Squares
0	8.0000	7.5000	2.2800	2.8157
1	8.0147	6.8547	2.3812	0.7738
2	8.0030	6.8346	2.3382	0.7663
3	8.0032	6.8350	2.3384	0.7663

NOTE: Convergence criterion met.

Source	DF	Sum of Squares	Mean Square	F Value	Approx Pr > F
Model	3	662.5	220.8	4610.48	<.0001
Error	16	0.7663	0.0479		
Uncorrected Total	19	663.2			

Parameter	Estimate	Approx Std Error	Approximate 95% Confidence Limits		Bootstrap Std Dev	Bootstrap Bias-Corrected 95% Confidence Limits	
alpha	8.0032	0.1051	7.7805	8.2260	0.1088	7.8198	8.2405
beta	6.8350	0.1364	6.5459	7.1242	0.1398	6.5317	7.0768
gamma	2.3384	0.1194	2.0852	2.5916	0.1161	2.1211	2.5807

Fig. 7.4. SAS Example G1: output from `proc nlin`

sum of squares, which reflects changes across time in the estimated growth curve, and an error sum of squares, which reflects variation in the cumulative radial tree trunk measurements about the estimated growth curve. The error mean square provides an estimate of σ^2, the error variance. The F-test in the `model` row of this table provides a test of the null hypothesis that α, β, and γ are all zero. Because the p-value, shown in the last column of this row, is very small, this null hypothesis may be rejected. The least squares estimates of the parameters are shown in the third table in Fig. 7.4 along with two types of standard errors and two sets of confidence intervals. The estimate of α is $\hat{\alpha} = 8.0032$, indicating that the maximum expected growth approaches 0.080032 inches over the 19-week period. The large sample approximation to

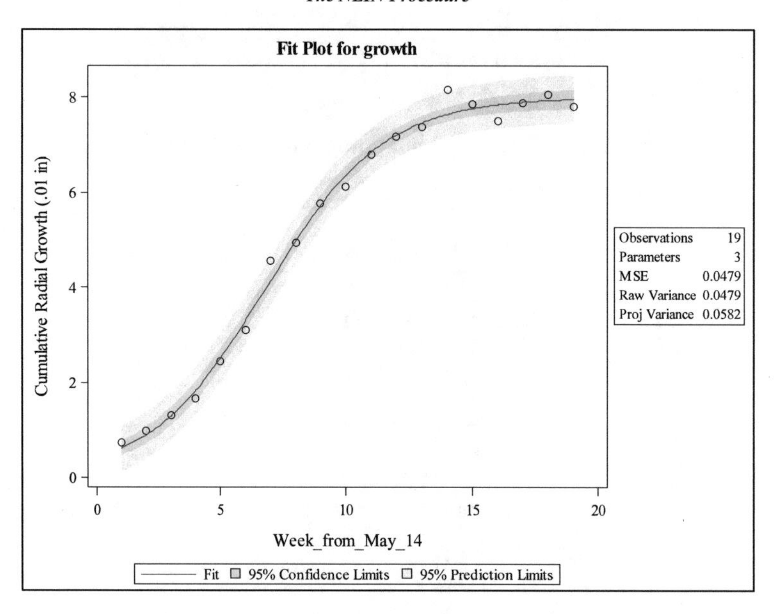

Fig. 7.5. SAS Example G1: growth curve plot from `proc nlin`

the standard error of $\hat{\alpha}$, shown in the first row of the column labeled `Approx Std Error`, is 0.1051. An approximate large sample 95% confidence interval for α runs from 7.7805 to 8.2660 in units of 0.01 inches. The `bootci` option in the `bootstrap` statement produces the bootstrapped standard errors for the parameter estimates and the bias-corrected bootstrap confidence intervals. These are shown in the last three columns of the third table. The bootstrapped confidence interval for α runs from 7.8198 to 8.2405. The estimate of β is $\hat{\beta} = 6.8350$, which indicates that it takes about 6.835 weeks to accomplish half of the maximum expected growth. The 95% bias-corrected bootstrapped confidence interval, (6.5317, 7.0768), indicates that the estimate for β is accurate to about 0.27 weeks, a little less than 2 days. The estimate of the growth rate parameter is $\hat{\gamma} = 2.3384$, and the 95% bias-corrected bootstrapped confidence interval is $(2.1211, 2.5087)$ in units of 0.01 inches per week. For this particular data set, the bias-corrected bootstrapped confidence intervals are similar to the corresponding confidence intervals based on large sample approximations.

Note that bootstrapped standard errors and confidence intervals are computed by randomly selecting new samples from the original data, using simple random sampling with replacement. The value that starts the random number generator to select the samples is obtained from the computer clock. Consequently, running the same code at two different times will result in slightly different values of bootstrapped standard errors and bootstrapped confidence intervals.

The option `plots=fit` produces a plot of the estimated growth curve with confidence bans and prediction bans as displayed in Fig. 7.5. This plot shows that the sinusoidal growth pattern of cedar trees is well represented by a logistic growth curve. The dark-blue shading corresponds to 95% confidence limits for the average growth curve at each time point. The light-blue shading corresponds to 95% prediction bans for potential measurements at each time point. The prediction bans are wider than the confidence bans to account for variation in observed measurements about the estimate of the mean growth curve.

7.2.3 Pharmacokinetic Application of a Nonlinear Model

Model

Compartment models are often used to describe the movement of a substance through blood or tissue in pharmacokinetic studies. Let $\mu(t)$ represent the concentration of a substance in the compartment (the blood) at t time units after it is orally ingested and let $\mu_a(t)$ represent the amount of the substance at the absorption site at time t. Let α represent the absorption rate into the compartment (the blood) and let β represent the elimination rate from the compartment. A one-compartment model for describing changes in the concentration of the substance in the compartment across time is defined by a pair of differential equations:

$$\frac{\partial \mu_a(t)}{\partial t} = -\alpha \mu_a(t) \tag{7.9}$$

and

$$\frac{\partial \mu(t)}{\partial t} = \alpha \mu_a(t) - \beta \mu(t). \tag{7.10}$$

Equation (7.9) dictates that the amount of substance moving from the absorption site into the compartment at time t is proportional to the amount of substance at the absorption site at that point in time. Equation (7.10) specifies the change in the amount of substance in the compartment at time t as the difference between the amount of substance entering the compartment, $\alpha\mu_a(t)$, and the amount of substance leaving the compartment, $\beta\mu(t)$, at that time point. The amount of substance entering the compartment at time t is proportional to the amount of substance at the absorption site at that time point, and the amount of substance eliminated from the compartment is proportional to the amount of substance in the compartment at that time point. The resulting formula for the expected amount of substance in the compartment at time t is

$$\mu(t; \alpha, \beta, \gamma) = \frac{\alpha}{\gamma(\alpha - \beta)} \left(e^{-\beta t} - e^{-\alpha t}\right), \tag{7.11}$$

where t is time since administration of the substance and γ is a proportionality parameter. The model for y_t, the observed amount of the substance in the compartment at time t, is completed by adding a random error to obtain

$$y_t = \frac{\alpha}{\gamma(\alpha - \beta)} \left(e^{-\beta t} - e^{-\alpha t}\right) + \epsilon_t. \tag{7.12}$$

In this model, ϵ_t is a random error with mean zero and variance σ^2 that does not depend on time. The parameters to be estimated are α, β, γ, and σ^2.

In the pharmacokinetics literature, the absorption rate parameter is usually denoted by the symbol κ_a, and the elimination rate parameter is usually denoted by κ_e, instead of α and β. Also, V/x is often used instead of γ, where x is the dose of the substance that is administered and V is a proportionality parameter that relates the amount of substance in the body to serum concentration when blood is the compartment of interest. We have retained the use of Greek letters for parameters, however, to be consistent with the presentation of other models in this chapter.

Other quantities of interest are the maximum concentration that is achieved, C_{max}, and the time at which the maximum concentration is achieved, T_{max}. For the one-compartment model given by Eq. (7.11), the maximum concentration is achieved at time

$$T_{\text{max}} = \frac{1}{\alpha - \beta} \log\left(\frac{\alpha}{\beta}\right) \tag{7.13}$$

and C_{max} is obtained by substituting T_{max} into Eq. (7.11). These relationships can be useful in the determination of initial values of parameter estimates from a plot of the observed data.

If the absorption and elimination rates are the same, Eq. (7.12) simplifies to

$$y_t = \left(\frac{\alpha t}{\gamma}\right) e^{-\alpha t} + \epsilon_t \tag{7.14}$$

where α is the common absorption and elimination rate. This simplified version of the model should be considered when the absorption and elimination rates are expected to be similar.

Compartment models may contain more than one compartment, and they may be extended to situations in which the variation in the random errors changes with time. In the following illustration, however, we will restrict our attention to an application of the one-compartment model corresponding to Eq. (7.12).

SAS Example G2

This example analyzes the rise and fall of plasma concentrations (νg/ml) of the steroid, prednisolone, during the first 24 hours after ingestion of a 5 mg tablet. These data are taken from row 18 of Table C.12 in Lindsey (2001).

Blood samples were taken at 0.25, 0.5, 1, 2, 3, 4, 6, 8, 12, and 24 hours after oral ingestion of the tablet. The data are shown in the listing of the SAS Example G2 program (see Fig. 7.6). This program uses the SGPLOT procedure to produce the scatter plot shown in Fig. 7.7 that is subsequently used to obtain initial values of parameter estimates required to use the NLIN procedure.

```
data pplasma;
   input time conc;
   label time = "Hours after Ingestion"
         conc = "Plasma Concentration (ng/ml)";
datalines;
  0.25  76.6
  0.50 253
    1  267
    2  242
    3  211
    4  164
    6   98.4
    8   57.2
   12   22.5
   24    2.3
    ;

title "Plasma Prednisolone Concentrations";

proc sgplot data=pplasma;
  scatter x=time y=conc /
     markerattrs=(size=12 symbol=Circle color=black);
  yaxis label="Plasma Concentration (ng/ml)"
           labelattrs=(size=14) valueattrs=(size=12);
  xaxis label="Hours after Ingestion"
           labelattrs=(size=14) valueattrs=(size=12);
  run;
```

Fig. 7.6. SAS Example G2: scatter plot program

The scatter plot in Fig. 7.7 shows that prednisolone is quickly absorbed into the blood with the peak concentration occurring near 1 hour. The elimination process appears to be slower, taking about 4 hours to drop halfway down from the peak concentration. Approximating the absorption rate as five times greater than the elimination rate, Eq. (7.13) becomes

$$T_{\max} = \frac{1}{(5-1)\beta} \log(5) \tag{7.15}$$

which implies that $\beta \approx (\log(5))((5-1)T_{\max})$. Because the scatter plot indicates that $T_{\max}$ is close to 1 hour, a good initial value for the estimate of β is $(0.25)\log(5) = 0.4$. It follows that a good initial value for the estimate of α is $(5)(0.4) = 2$. Using these initial values for α and β to evaluate equation (7.11) at time $T_{\max} = 1$ hour, and using 267 as the corresponding maximum concentration, produces an initial value for γ of 0.00075. The addition to the SAS Example G2 program shown in Fig. 7.8 uses these initial estimates to apply `proc nlin` to find the nonlinear least squares estimates of the parameters in the one-compartment model. Selected parts of the output are shown in Fig. 7.9.

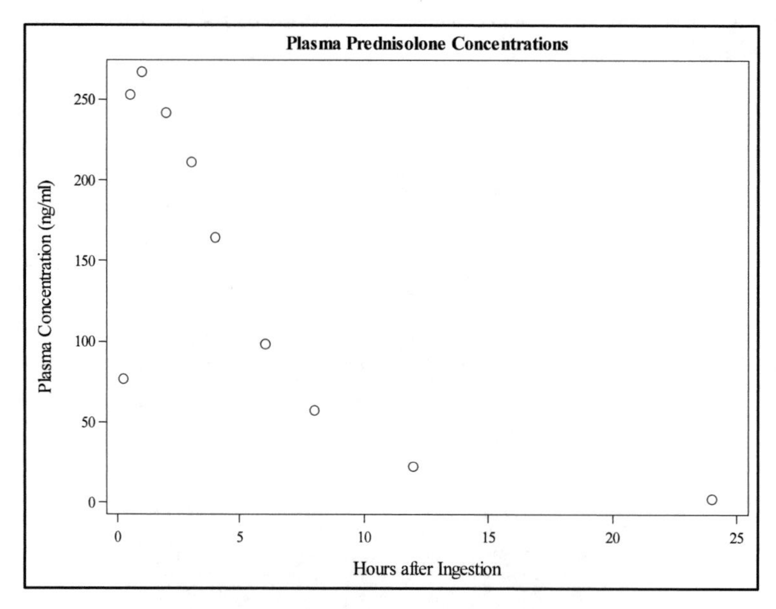

Fig. 7.7. SAS Example G2: scatterplot

```
proc nlin data=pplasma plots=fit;
  parms alpha=2 beta=0.4 gamma=0.00075;
  model conc = alpha/(gamma*(alpha-beta))*(exp(-beta*time)-exp(-alpha*time));
  bootstrap / bootci;
        output out=predictions predicted=prediction stdp=stdp
           lclm=lower95 uclm=upper95 residual=residual;
run;

proc print data=predictions;
  var conc velocity prediction stdp lower95 upper95 residual;
run;
```

Fig. 7.8. SAS Example G2: program code for `proc nlin`

The first table displayed in Fig. 7.9 shows that the Gauss–Newton optimization algorithm for finding the nonlinear least squares estimates of the parameter estimates converged in seven iterations. The first row of this table shows the initial values of the parameter estimates specified in the `parms` statement in Fig. 7.8. The second table in Fig. 7.9 partitions variation in the observed plasma prednisolone concentrations into a model sum of squares and an error sum of squares. The error mean square, 750.1, is an estimate of σ^2,

Plasma Prednisolone Concentrations

The NLIN Procedure
Method: Gauss-Newton

Iterative Phase				
Iter	alpha	beta	gamma	Sum of Squares
0	2.0000	0.4000	0.000750	1157232
1	1.9889	0.3322	0.00131	217936
2	1.8832	0.2812	0.00198	30156.7
3	1.7639	0.2593	0.00246	6190.9
4	1.7373	0.2544	0.00261	5253.6
5	1.7392	0.2538	0.00262	5250.8
6	1.7390	0.2538	0.00262	5250.8
7	1.7390	0.2538	0.00262	5250.8

NOTE: Convergence criterion met.

Source	DF	Sum of Squares	Mean Square	F Value	Approx Pr > F
Model	3	279362	93120.6	124.14	<.0001
Error	7	5250.8	750.1		
Uncorrected Total	10	284613			

Parameter	Estimate	Approx Std Error	Approximate 95% Confidence Limits		Bootstrap Std Dev	Bootstrap Bias-Corrected 95% Confidence Limits	
alpha	1.7390	0.4619	0.6468	2.8312	0.4948	0.9781	2.9622
beta	0.2538	0.0569	0.1193	0.3883	0.0808	0.1682	0.4468
gamma	0.00262	0.000366	0.00175	0.00348	0.000413	0.00182	0.00344

Fig. 7.9. SAS Example G2: output from `proc nlin`

the error variance. The F-test in this table provides an approximate test of the null hypothesis that the systematic component is a horizontal line at zero. Because the p-value is very small, this null hypothesis may be rejected. The nonlinear least squares estimates of the parameters are displayed in the third table in Fig. 7.9 along with asymptotic and bootstrap standard errors and two sets of confidence intervals. The estimate of the absorption rate, $\hat{\alpha} = 1.739$, is about seven times larger than the estimate of the elimination rate, $\hat{\beta} = 0.2538$. The asymptotic standard error for $\hat{\alpha}$, 0.4619, is a bit smaller than the corresponding bootstrap standard error. Relative to the large sample confidence interval for α, which is centered at $\hat{\alpha}$, the end points of the bootstrap confidence interval are both shifted to the right to account for right skewness in

Plasma Prednisolone Concentrations

The NLIN Procedure

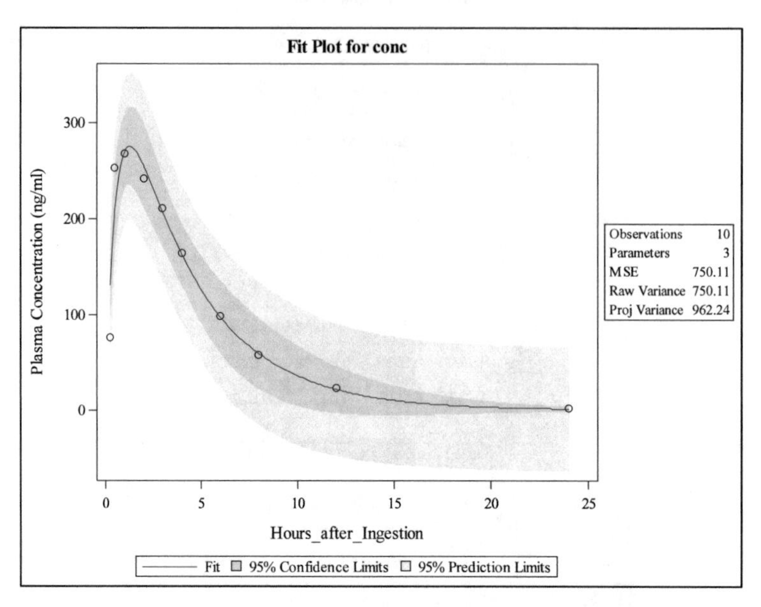

Fig. 7.10. SAS Example G2: estimated plasma prednisolone concentration curve

the sampling distribution for $\hat{\alpha}$. The bootstrap estimate of the standard error of $\hat{\beta}$ is also larger than the asymptotic estimate, and the bootstrap confidence interval for β is also shifted to the right of the asymptotic confidence interval to account for right skewness in the small sample sampling distribution for $\hat{\beta}$. Because the small sample distribution for $\hat{\gamma}$ is more nearly symmetric, the asymptotic and bootstrap results are similar.

The option `plots=fit` produces a plot of the estimated curve with 95% confidence bans and prediction bans as displayed in Fig. 7.10. The plasma prednisolone concentration quickly increases during the first hour after ingestion of the tablet, and then it more gradually declines to zero as absorption slows, and the steroid is eliminated from the blood. This is a consequence of the estimated absorption rate being about seven times larger than the estimated elimination rate. The dark-blue shading corresponds to 95% confidence limits for the mean concentration at individual time points. The lighter-blue shading corresponds to 95% prediction intervals for potential observations. These bans illustrate potential inaccuracy of applying asymptotic methods to small samples as the asymptotic confidence intervals include unrealistic negative concentrations.

7.2.4 A Model for Biochemical Reactions

Model

In biochemistry, the Michaelis–Menten model is one of the best-known models for enzyme kinetics. The Michaelis–Menten equation relates the reaction rate of enzymatic reactions to the concentration of a substrate. The reaction rate, commonly called the *velocity*, corresponds to the number of reactions per second that are catalyzed by an enzyme. As a nonlinear regression model, the Michaelis–Menten model has the form

$$y = \frac{\alpha x}{\beta + x} + \epsilon \tag{7.16}$$

where

$$\begin{aligned} y &= \text{ observed reaction rate} \\ \alpha &= \text{ maximum reaction rate} \\ \beta &= \text{ Michaelis constant} \\ x &= \text{ substrate concentration} \\ \epsilon &= \text{ a random error with mean zero and variance } \sigma^2. \end{aligned}$$

At low substrate concentrations, the reaction rate varies almost linearly with the substrate concentration, but at higher substrate concentrations, the reaction becomes nearly independent of the substrate concentration and asymptotically approaches its maximum rate, represented by the α parameter. This rate is attained when all enzyme is bound to substrate and further addition of substrate does not affect the reaction. The Michaelis constant, β, is the substrate concentration at which the reaction rate is at half maximum. Reactions with smaller β values approach the maximum reaction rate at lower substrate concentrations than reactions with larger β values.

The typical method for estimating α and β involves running a series of enzyme assays at varying substrate concentrations and measuring the reaction rate for each substrate concentration at an early stage of the assay. By plotting the reaction rate (y) against concentration (x), nonlinear regression can be used to estimate parameters and fit a curve. Starting values for parameter estimation can be determined by fitting a line to the plot of x/y against x. The reciprocal of the slope of a line fit to the plot provides a reasonable starting value for $\hat{\alpha}$, and a starting value for $\hat{\beta}$ is the intercept divided by the slope of the line.

The Michaelis–Menten model is used in a variety of biochemical situations other than enzyme–substrate interaction, including antigen-antibody binding, DNA–DNA hybridization, and protein–protein interaction. It has also been used in studies of microbial growth, species richness, geosciences, and various manufacturing processes. More information on Michaelis–Menten kinetics and the design and analysis of enzyme and pharmacokinetic experiments can be found in Chen et al. (2010), Leskovac (2003), and Endrenyi (1981).

SAS Example G3

The data for the following example is taken from Table A1.3 in Bates and Watts (1988). At each substrate concentration (ppm), the initial reaction rate (counts/min^2) was determined from changes in counts per minute of a radioactive product produced from the reaction. The data are included in the `data step` of the SAS Example G3 program shown in Fig. 7.11. This program uses `sgplot` to create the scatter plots shown in Fig. 7.12. In the left panel, the velocity measurements (y) are plotted against the substrate concentrations (x). In the right panel, x/y is plotted against x and a least squares regression line is fit to the points on the plot. The REG procedure was used to obtain estimates of the intercept and slope of that line, and the estimates are displayed in Fig. 7.13. The reciprocal of the estimated slope of the regression line, $1/0.00599 = 166.9$, is used as the starting value for $\hat{\alpha}$. The ratio of the intercept and slope, $0.00033878/0.00599 = 0.0566$, is used as the initial value for $\hat{\beta}$.

```
data enzyme;
   input conc velocity;
   cv = conc/velocity;
   label conc = "Substrate Concentration (ppm)"
         velocity = "Velocity"
         cv = "Concentration / Velocity";
datalines;
  0.02  67
  0.02  51
  0.06  84
  0.06  86
  0.11  98
  0.11 115
  0.22 131
  0.22 124
  0.56 144
  0.56 158
  1.10 160
  0.04  .
  0.80  .
  1.20  .
    ;

proc sgplot data=enzyme;
  scatter x=conc y=velocity /
     markerattrs=(size=12 symbol=Circle color=black);
  yaxis labelattrs=(size=14) valueattrs=(size=12);
  xaxis label="Substrate Concentration (ppm)"
          labelattrs=(size=14) valueattrs=(size=12);
  title "Enzyme Reaction Velocity";
  run;

symbol1 cv=black v=dot i=none h=1.5 w=2;
proc reg data=enzyme;
  model cv = conc;
  plot cv*conc / nostat nomodel lline=1 cline=red;
  title "Concentration/Velocity vs Concentration";
  run;
```

Fig. 7.11. SAS Example G3: plotting and regression program

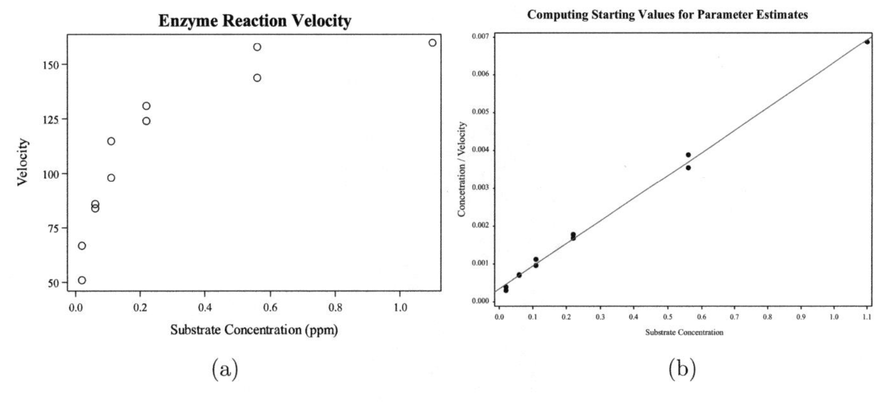

Fig. 7.12. (**a**) Velocity measurements (counts/min^2) versus substrate concentration (ppm). (**b**) Concentration/velocity ratios versus substrate concentration

Computing Starting Values for Parameter Estimates

Parameter Estimates						
Variable	**Label**	**DF**	**Parameter Estimate**	**Standard Error**	**t Value**	**Pr > |t|**
Intercept	Intercept	1	0.00033878	0.00004698	7.21	<.0001
conc	Substrate Concentration	1	0.00599	0.00011113	53.92	<.0001

Fig. 7.13. SAS Example G3: output from `proc reg`

```
proc nlin data=enzyme plots=fit;
  parms alpha=166.9 beta=0.0566;
  model velocity = (alpha * conc)/(beta + conc);
  bootstrap / bootci;
  output out=predictions predicted=prediction stdp=stdp
         lclm=lower95 uclm=upper95 residual=residual;
run;

proc print data=predictions;
  var conc velocity prediction stdp lower95 upper95 residual;
run;
```

Fig. 7.14. SAS Example G3: program code for `proc nlin`

Additional program code for SAS Example G3 is shown in Fig. 7.14. This code applies `proc nlin` to find the nonlinear least squares estimates of the parameters in the Michaelis–Menten model that is specified in the `model` statement. The `parms` statement specifies values for the initial parameter estimates. Selected parts of the output are shown in Fig. 7.15 and the estimated curve is shown in Fig. 7.16.

The first table in Fig. 7.15 indicates that the Gauss–Newton optimization algorithm that `proc nlin` uses to evaluate parameter estimates converged in six iterations. The first row of this table shows the initial values of the param-

Nonlinear Regression for the Michaelis-Menten Model

The NLIN Procedure
Method: Gauss-Newton

Iterative Phase			
Iter	alpha	beta	Sum of Squares
0	166.9	0.0566	959.0
1	160.8	0.0485	860.5
2	160.4	0.0478	859.6
3	160.3	0.0477	859.6
4	160.3	0.0477	859.6
5	160.3	0.0477	859.6
6	160.3	0.0477	859.6

NOTE: Convergence criterion met.

Source	DF	Sum of Squares	Mean Square	F Value	Approx Pr > F
Model	2	147348	73674.2	771.36	<.0001
Error	9	859.6	95.5116		
Uncorrected Total	11	148208			

Parameter	Estimate	Approx Std Error	Approximate 95% Confidence Limits		Bootstrap Std Dev	Bootstrap Bias-Corrected 95% Confidence Limits	
alpha	160.3	6.4802	145.6	174.9	6.5905	148.4	174.7
beta	0.0477	0.00778	0.0301	0.0653	0.00771	0.0347	0.0651

Fig. 7.15. SAS Example G3: output from `proc nlin`

eter estimates specified in the `parms` statement in Fig. 7.14. The second table in Fig. 7.15 partitions variation in the observed enzyme reaction rates into a model sum of squares, corresponding to variation explained by the estimate of the Michaelis–Menten curve, and an error sum of squares corresponding to variation in observed reaction rates about the estimated curve. The error mean square, 95.51, is an estimate of σ^2, the error variance. The F-test in the first row of this table provides a test of the null hypothesis that α and β are both zero, which implies that no reaction occurred. Because the p-value is very small, this null hypothesis may be rejected.

The nonlinear least squares estimates of the parameters are displayed in the third table in Fig. 7.15. The estimate of the maximum reaction rate is $\hat{\alpha} = 160.3$ counts/min^2. The asymptotic standard error for $\hat{\alpha}$ is 6.4802, and an approximate 95% confidence interval for α extends from 145.6 to 174.9

counts/min^2. The `bootstrap` statement in Fig. 7.14 produces the last three columns in the third table. The bootstrapped standard error for $\hat{\alpha}$ is almost identical to the large sample standard error. The bias-corrected bootstrap confidence interval for α, shown in the last two columns of the table, is also similar to the large sample confidence interval shown in columns four and five. Although the sample size is small, the similarity between the asymptotic and bootstrap results occurs because the small sample distribution of possible values for $\hat{\alpha}$ is close to a normal distribution in this case.

The estimate of the substrate concentration at which the reaction achieves half of the maximum reaction rate is $\hat{\beta} = 0.0477$ ppm. The approximate large sample standard error for $\hat{\beta}$ is 0.00778 ppm, very close to the corresponding bootstrapped standard error. The 95% confidence interval for β based on a large sample normal approximation is similar to the bias-corrected bootstrapped confidence interval. Using the bias-corrected bootstrap interval, the data provide enough information to be 95% confident that the substrate concentration that achieves half of the maximum reaction rate is between 0.0347 ppm and 0.0651 ppm.

The least squares estimate of the Michaelis–Menten curve is displayed in Fig. 7.16. This plot is produced by the `plots=fit` option in the `proc nlin` statement shown in Fig. 7.14. To prevent information on summary statistics from being displayed on the right side of the plot, change this option to `plots=fit(stats=none)`. The dark-blue shaded region on the plot corresponds to asymptotic 95% confidence limits for the mean reaction rate at

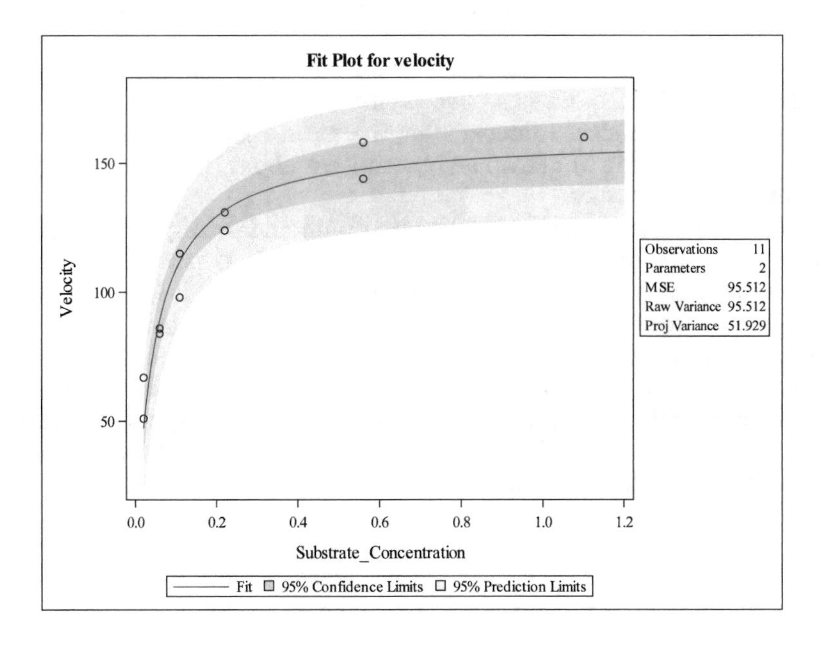

Fig. 7.16. SAS Example G3: estimated curve

Nonlinear Regression for the Michaelis-Menten Model

Obs	conc	velocity	prediction	stdp	lower95	upper95	residual
1	0.02	67	47.344	4.13406	37.992	56.696	19.6557
2	0.02	51	47.344	4.13406	37.992	56.696	3.6557
3	0.06	84	89.286	4.29711	79.565	99.006	-5.2857
4	0.06	86	89.286	4.29711	79.565	99.006	-3.2857
5	0.11	98	111.794	3.48161	103.918	119.670	-13.7938
6	0.11	115	111.794	3.48161	103.918	119.670	3.2062
7	0.22	131	131.717	3.36761	124.099	139.335	-0.7166
8	0.22	124	131.717	3.36761	124.099	139.335	-7.7166
9	0.56	144	147.697	4.65716	137.162	158.232	-3.6973
10	0.56	158	147.697	4.65716	137.162	158.232	10.3027
11	1.10	160	153.618	5.44142	141.308	165.927	6.3825
12	0.04	.	73.097	4.58443	62.726	83.468	.
13	0.80	.	151.260	5.11221	139.695	162.824	.
14	1.20	.	154.151	5.51891	141.667	166.636	.

Fig. 7.17. SAS Example G3: estimated mean velocities and 95% confidence limits

individual substrate concentrations. The widths of the confidence intervals increase as the substrate concentration increases, eventually converging to a constant width as the saturation condition is approached. The lighter-blue shaded region in Fig. 7.16 corresponds to 95% prediction intervals for future observations of reaction rates. The prediction intervals are wider than the corresponding confidence intervals for the mean reaction rates in order to account for variation in individual observations about the curve representing the mean reaction rates.

Values of the estimated mean reaction rates and corresponding 95% confidence intervals are shown in Fig. 7.17. This table is produced by including an `output` statement in the program code as shown in Fig. 7.14. The `out=predictions` option creates a new SAS data set named `predictions` containing columns specified by the remaining options included in the statement. Using `predicted=prediction` creates a column of estimated mean reaction rates under the variable name `prediction`. The asymptotic standard errors for the estimated means are output to a column with variable name `stdp` with the `stdp=stdp` option. The lower and upper limits of approximate 95% confidence intervals for mean reaction rates are output into columns with variable names `lower95` and `upper95`, respectively, with the options `lclm=lower95` and `uclm=upper95`. The residuals, the differences between the observed reaction rates and the corresponding mean reaction rates estimated from the model, are output with the option `residual=residual`. One line is created in the new data set for each line in the original data set. The output data are shown in Fig. 7.17. The last three lines correspond to cases that were not part of the data collected in the experiment. These lines of data are included in the `data step` in Fig. 7.11 with periods inserted as placeholders for the missing velocity values. These three cases are not used to fit the model because the values of the velocity variable are missing. This illustrates how estimates of mean responses and corresponding standard errors and confidence intervals can be evaluated for substrate concentrations not included in the study. At a substrate concentration of 0.04 ppm, for example, the estimate of the mean reaction rate is 73.097 counts/min^2, with a standard error of 4.58443 counts/min^2 and a 95% confidence interval that extends from 62.726 to 83.468 counts/min^2. Note how the standard error of the prediction and the width of the confidence interval depend on the substrate concentration.

7.3 Generalized Linear Models

7.3.1 Introduction

The family of generalized linear models, introduced by Nelder and Wedderburn (1972), contains models for which a function of the mean response is linked to a linear combination of explanatory variables. Examples include linear regression models, logistic regression models, and Poisson regression models. An advantage of this formulation is that a single algorithm can be developed to evaluate parameter estimates for the members of this large family of models. Maximum likelihood estimation is the most popular estimation procedure, and it is the default estimation method in the SAS GENMOD and LOGISTIC procedures. Large sample normal approximations to the sampling distributions of the maximum likelihood estimators are typically used to obtain approximate standard errors, confidence intervals, and tests of hypotheses.

Model

There are three basic components to a generalized linear model:

- **Linear component**: a linear combination of explanatory variables used to describe how a function of the mean response changes as values of the explanatory variables change, e.g., $\beta_0 + \sum_{j=1}^{k} \beta_j x_j$.
- **Link function**: a function of the mean response, $g(\mu)$, that equates to the linear component, i.e.,

$$g(\mu) = \beta_0 + \sum_{j=1}^{k} \beta_j x_j. \tag{7.17}$$

- **Probability distribution**: identifies the conditional distribution of the response variable y given the values of the explanatory variables, x_1, x_2, ..., x_k.

The linear component and link function define the systematic component of the model which describes how the mean response varies with changes in the values of the explanatory variables. The inverse of the link function provides the mean response, i.e.,

$$\mu = g^{-1}(\beta_0 + \beta_1 x_1 + \beta_2 x_2 + \beta_k x_k). \tag{7.18}$$

The probability distribution must be a member of the overdispersed exponential family of distributions, a large class of probability distributions that includes the normal, binomial, Poisson, multinomial, and gamma distributions, among others. The variance of the response variable may be a function of the mean and possibly other parameters.

Estimation and Inference

Some members of the family of generalized linear can be analyzed with the SAS GENMOD and LOGISTIC procedures. The LOGISTIC procedure analyzes logistic regression models. The GENMOD procedure handles a much broader class of models, but it also may be used to analyze logistic regression models. Both procedures use maximum likelihood estimation to evaluate parameter estimates. For a specific model, maximum likelihood estimation finds values of the model parameters that maximize the likelihood of obtaining the data that were actually observed. Because the natural logarithm is a strictly increasing function, maximizing the natural logarithm of the likelihood function, called the log-likelihood function, produces the same parameter estimates as maximizing the likelihood function.

The GENMOD procedure uses a ridge-stabilized Newton–Raphson algorithm to maximize the log-likelihood function with respect to the regression parameters. On the r-th iteration, the algorithm updates the current estimate of the parameter vector $\hat{\beta}^r$ with

$$\hat{\beta}^{r+1} = \hat{\beta}^r - \mathbf{H}^{-1}\mathbf{S} \tag{7.19}$$

where $\mathbf{H}$ is the Hessian matrix, the matrix of second partial derivatives of the log-likelihood function, and $\mathbf{S}$ is the gradient vector, the vector of first partial derivatives of the log-likelihood function. Both $\mathbf{H}$ and $\mathbf{S}$ are evaluated at $\hat{\beta}^r$, the value of the parameter estimates from the previous iteration. This algorithm is determined to have converged to the maximum likelihood estimate, $\hat{\beta}$, when changes in the parameter estimates between successive iterations are small enough to satisfy some convergence criteria. The GENMOD and LOGISTIC procedures have a number of different convergence criteria that users may select, but the default criteria are sufficient for most applications. At convergence $\mathbf{H}$ should be a negative definite matrix and S should be close to a vector of zeros. The covariance matrix for the large sample normal approximation to the sampling distribution of the maximum likelihood estimator for β is estimated with the inverse of $-\mathbf{H}$ evaluated at $\hat{\beta}$, the final vector of values for the maximum likelihood estimates. This large sample approximation is used to evaluate standard errors and approximate confidence intervals and to perform tests of hypotheses.

There is no guarantee that the iterative algorithm for finding $\hat{\beta}$ will converge to the set of parameter values that yield the global maximum of the log-likelihood function. If the log-likelihood is multimodal, the algorithm could converge to a local mode instead of the global mode. In other cases, the algorithm may wander off to a place where $\mathbf{H}$ is not negative definite, an inappropriate solution, and still satisfies the convergence criteria. This will produce either a warning or an error message that should not be ignored. At the very least, users should examine the set of parameter estimates produced at each iteration to determine if the sequence of estimates appears to converge to reasonable values. Sometimes convergence problems arise because the model has too many parameters. This can be avoided by fitting models with fewer parameters and fewer explanatory variables. Success will depend on how close the initial parameter estimates used to start the algorithm are to the actual maximum likelihood estimates and how much information is in the data relative to the complexity of the proposed model.

Goodness of Fit and Overdispersion

Two statistics that are helpful in assessing the fit of a generalized linear model are the scaled deviance and the Pearson goodness-of-fit statistic. Formulas for these statistics depend on the three components of the model, especially the probability distribution specified for the response variable. These statistics do not always have chi-square distributions, so formal chi-square tests of the fit of the proposed model may be misleading and are generally not done. In the examples considered in this section, we will simply compare the value of the Pearson statistic to its degrees of freedom to help judge if the proposed model provides an adequate description of the data. See McCullagh and Nelder (1989) for more advice.

The Akaike information criterion (AIC) is a measure of goodness of model fit that balances model fit against model simplicity (see Akaike (1981)). AIC has the form:

$$\text{AIC} = -2(\text{log-likelihood}) + 2p \tag{7.20}$$

where p is the number of parameters in the model, and the log-likelihood is evaluated with the values of the maximum likelihood estimates of the parameters. An alternative form is the corrected AIC given by

$$\text{AICC} = -2(\text{log-likelihood}) + 2p\frac{n}{n-p-1} \tag{7.21}$$

where n is the total number of cases in the data set. The Bayesian information criterion (BIC) is a similar measure that is defined by

$$\text{BIC} = -2(\text{log-likelihood}) + p\log(n) \tag{7.22}$$

`Proc genmod` uses the full log-likelihood for computing the AIC, AICC, and BIC criteria. Simonoff (2003) discusses applications of AIC, AICC, and BIC to generalized linear models. Smaller values of these criteria indicate better compromises between the accuracy and simplicity of the models. As a model is made more complex by including more explanatory variables and more parameters, biases of predicted responses for the cases in the data set tend to become smaller, but the variances of the predicted responses tend to increase. If the values of the AIC, BIC, or AICC criteria do not become smaller when the model is made more complex, then the reduction in bias gained by the added model complexity is not enough to offset the increased variability in the predictions. Models with smaller values of these criteria are preferred to models with larger values. The AICC and BIC criteria tend to indicate less complex models than the AIC criteria. Keep in mind that these are relative comparisons with respect to the specific set of models under consideration. These criteria can help to select the best models in a set of proposed models, but that does not necessarily imply that any of the models in the set actually provide a good description of the data.

We will illustrate applications of the GENMOD and LOGISTIC procedures to generalized linear models by considering applications to logistic regression and Poisson regression. Additional information on the theory and application of generalized linear models can be found in McCullagh and Nelder (1989), Madsen and Thyregod (2011), and Agresti (2013).

7.3.2 Logistic Regression

Model

Logistic regression models are often used in situations in which there are only two possible responses, such as a patient surviving or not surviving a medical procedure, a seed germinating or not germinating, or someone voting or not voting for a particular candidate. One of the two possible outcomes can be labeled as a `success` and the other outcome labeled as a `failure`. Logistic regression models are used to describe how the probability of a successful outcome changes as the values of one or more of the explanatory variables change. The link function is the natural logarithm of the odds of success, called the logit function. If the probability of a successful outcome is π, then the odds for success are $\frac{\pi}{1-\pi}$. Note that the odds are a monotone increasing function of the probability. As the probability of success increases from zero to one, the odds of success increase from zero to infinity. The logit is the natural logarithm of the odds of success, $\log\left(\frac{\pi}{1-\pi}\right)$, and it is also a monotone increasing function of the probability of success. As the probability of success increases from 0 to 1, the logit increases from minus infinity to plus infinity. When the probability of success is 0.5, the odds of success is 1.0 and the logit is 0. To allow the probability of success to depend on the values of some explanatory variables, say $x_1, x_2, \ldots, x_k$, a logistic regression model of the form

$$\log\left(\frac{\pi}{1-\pi}\right) = \beta_0 + \beta_1 x_1 + \beta_2 x_2 + \cdots + \beta_k x_k \tag{7.23}$$

may be used. Inverting this relationship, the formula for the success probability under conditions corresponding to the values of the explanatory variables is

$$\pi = \frac{\exp(\beta_0 + \beta_1 x_1 + \beta_2 x_2 + \cdots + \beta_k x_k)}{1 + \exp(\beta_0 + \beta_1 x_1 + \beta_2 x_2 + \cdots + \beta_k x_k)}. \tag{7.24}$$

The regression coefficients may be interpreted in the context of odds ratios. To interpret β_2, consider increasing x_2 by one unit in Eq. (7.23) without changing the values of any of the other explanatory variables. The log-odds of success under those conditions are

$$\log\left(\frac{\pi_{x_2+1}}{1-\pi_{x_2+1}}\right) = \beta_0 + \beta_1 x_1 + \beta_2(x_2+1) + \cdots + \beta_k x_k \tag{7.25}$$

The difference in the log-odds for success given by Eqs. (7.25) and (7.23) is

$$\log\left(\frac{\pi_{x_2+1}}{1-\pi_{x_2+1}}\right) - \log\left(\frac{\pi_{x_2}}{1-\pi_{x_2}}\right) = \beta_2. \tag{7.26}$$

Then

$$\beta_2 = \log\left(\frac{\pi_{x_2+1}}{1-\pi_{x_2+1}}\right) - \log\left(\frac{\pi_{x_2}}{1-\pi_{x_2}}\right) = \log\left(\frac{\frac{\pi_{x_2+1}}{1-\pi_{x_2+1}}}{\frac{\pi_{x_2}}{1-\pi_{x_2}}}\right) \tag{7.27}$$

is the logarithm of the ratio of two odds, the odds of success when the second explanatory variable has value $x_2 + 1$ divided by the odds of success when the second explanatory variable has value x_2. Consequently,

$$\exp(\beta_2) = \frac{\text{odds of success at } x_2 + 1}{\text{odds of success at } x_2} \tag{7.28}$$

is a conditional odds ratio, the odds for success at $x_2 + 1$ divided by the odds of success at x_2 when the other explanatory variables in the model are held constant. If $\exp(\beta_2) = 3$, for example, then a one unit change in x_2 triples the odds of success when the other explanatory variables are held constant. Estimates of the exponential functions of the coefficients may be requested in the output from the LOGISTIC procedure. When "success" corresponds to contracting a disease, for example, these quantities are used as approximate measures of relative risk for contracting the disease.

The description of the model is completed by specifying the distribution for the observed counts. When the data contain only two possible outcomes for the response variable, the LOGISTIC procedure assumes that the counts have independent binomial distributions. In `proc genmod` the logit is the default link function when the binomial distribution is specified for the observed counts, but other link functions may be requested.

Estimation and Hypothesis Testing

A logistic regression model can be fit to data using either the GENMOD or LOGISTIC procedure in SAS. The LOGISTIC procedure is preferred because it offers more options for displaying results and assessing the fit of logistic regression models and it provides search procedures for selecting explanatory variables to include in the model that are not available in GENMOD. The LOGISTIC procedure computes maximum likelihood estimates of the regression parameters, $\beta_0, \beta_1, \ldots, \beta_k$. The maximization procedure is initiated with $\hat{\beta}_1, \hat{\beta}_2, \ldots, \hat{\beta}_k$ all set to zero and $\hat{\beta}_0 = \log(\hat{\pi}/(1 - \hat{\pi}))$, where $\hat{\pi}$ is the overall proportion of successes, the total number of successes in the data divided by the total number of trials. Because the log-likelihood function for logistic regression models is unimodal, the estimation algorithm will converge from any starting value, and it is not necessary to search for better starting values. The `clparm=` option is used to specify the method for constructing confidence intervals for regression parameters. By default, the LOGISTIC procedure uses a profile-likelihood (PL) method to construct confidence intervals for parameters and other quantities. The PL method (`clparm=pl`) is based on an asymptotic chi-square approximation to the distribution of a likelihood ratio test. An alternate is the Wald method (`clparm=wald`) which is based on the asymptotic normal approximation to the distribution of the parameter estimates. The PL method is usually more accurate than the Wald method for small samples, but differences between the two methods are inconsequential for sufficiently large sample sizes. The `clparm=both` option produces both sets of confidence intervals. The `clodds=` option is used to specify the method for constructing confidence intervals for odds ratios. The options are `pl`, `wald`, or `both`. The `alpha=` statement is used to specify the confidence level. The default, `alpha=.05`, corresponds to a 95% confidence level.

The default algorithm for optimizing the log-likelihood is the Fisher scoring method, which is equivalent to fitting by iteratively reweighted least squares. The alternative algorithm is a modified Newton–Raphson method. Both algorithms produce the same parameter estimates, but the estimated covariance matrices for the parameter estimators may differ slightly. If it is available and it converges, the results of the Fisher scoring method are preferred. When multi-category logit models are used, however, only the modified Newton–Raphson technique is available.

There are four convergence criteria. The `fconv=` option in the `model` statement terminates the algorithm when the absolute relative change in the values of the log-likelihood function on successive iterations is smaller than the specified `value`. The `absconv=` option terminates the algorithm when the absolute change in the log-likelihood function is smaller than the specified `value`. The `xconx=` option terminates when there are sufficiently small relative changes in all of the regression parameter estimates. The default is the `gconv=` option which converges when the change in a "standardized" gradient vector between two successive iterations is smaller than the specified `value`. For most applications the default criterion is adequate. Because computational problems may occur, it is good practice to examine how the values of the parameter estimates change across iterations to make sure that they actually converge to reasonable values. The sequence of parameter estimates is displayed by including the `itprint=` option in the `model` statement for `proc logistic` or `proc genmod`.

SAS Example G4

The following example is taken from Okada et al. (2010). It examines the effects of egg incubation temperature on sex determination of Japanese pond turtles. In this experiment, groups of eggs were incubated at different temperatures, and the numbers of male and female turtles that hatched under each incubation temperature were recorded. We only use the results for incubation temperatures between 26 °C and 30 °C. No female turtles were observed to hatch below 26 °C, and no male turtles were observed to hatch above 30 °C. The data are embedded in the SAS Example G4 program code (see Fig. 7.18). Each line in the data file contains the incubation temperature (x), number of females (y_1), number of males (y_2), and the total number of eggs ($n = y_1 + y_2$) that hatched for that incubation temperature. The LOGISTIC procedure can estimate success probabilities for cases in the study and also for values of the explanatory variables that were not included in the study. To illustrate how this is achieved, two lines are included in the data file that contain periods as placeholders for missing values for $y1$, $y2$, and n. These optional data lines are used to estimate the proportions of female turtles that would hatch for two incubation temperatures, 27.5 °C and 28.8 °C, that were not used in the experiment. Any data line that has a missing value for any variable used in the model will not be used in the estimation of the model parameters, but estimates of success probabilities are produced for any data line that contains a complete set of values for the explanatory variables in the model.

The LOGISTIC procedure may be used to fit a logistic regression model

$$\log\left(\frac{\pi}{1-\pi}\right) = \beta_0 + \beta_1 x. \tag{7.29}$$

that relates π, the probability that a female turtle hatches from an egg, to the egg incubation temperature (x). In the `proc logistic` statement in Fig. 7.18, the `data=turtles` option identifies the input data set, and the `plots(only)=effect` option produces a graph of the estimated logistic curve (see Fig. 7.19). This curve shows how the estimated proportion of female turtles increases as temperature increases. Because $y1$ contains the number of females and n contains the number of eggs that hatch at each temperature, the `y1/n` notation is used on the left side of the equal sign in the `model` statement. This informs the LOGISTIC procedure that each line in the data file represents the number of eggs specified by the n variable and that $y1$ females hatched from those eggs. The single explanatory variable x is entered on the right side of the equal sign in the `model` statement. Selected parts of the

```
data turtles;
  input x y1 y2 n;
  label  x = 'Incubation Temperature (C)'
        y1 = 'Number of Females'
        y2 = 'Number of Males'
         n = 'Number of Eggs';
  datalines;
   26.0   0   8   8
   26.5   0  24  24
   28.0   0  19  19
   28.5   8  18  26
   29.0  20   7  27
   29.5  12   6  18
   30.0  30   0  30
   27.5   .   .   .
   28.8   .   .   .
    ;

title "Temperature Dependent Sex Determination for Japanese Turtles";

proc logistic data=turtles plots(only)=effect;
  model y1/n = x  / itprint covb clparm=both clodds=both;
  output out=setp p=phat lower=cl_lower upper=cl_upper;
  run;

proc print data=setp; run;
```

Fig. 7.18. SAS Example G4: program for `proc logistic`

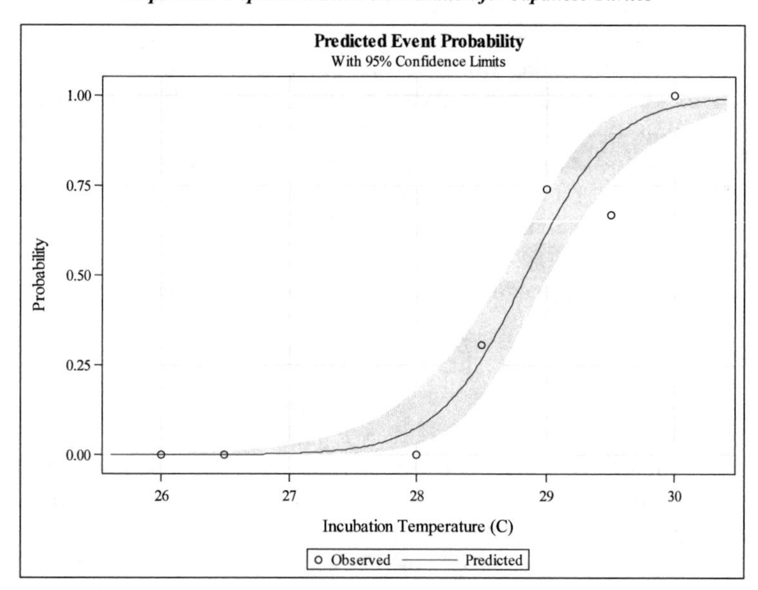

Fig. 7.19. SAS Example G4: output from the `plots(only)=effect` option

Temperature Dependent Sex Determination for Japanese Turtles

The LOGISTIC Procedure

Response Profile		
Ordered Value	**Binary Outcome**	**Total Frequency**
1	Event	70
2	Nonevent	82

Maximum Likelihood Iteration History				
Iter	**Ridge**	**-2 Log L**	**Intercept**	**x**
0	0	209.768388	-0.158224	0
1	0	123.186535	-31.998373	1.117456
2	0	104.932673	-53.670660	1.865577
3	0	99.010662	-74.007605	2.567373
4	0	98.296639	-84.216393	2.920364
5	0	98.283543	-85.852849	2.977048
6	0	98.283538	-85.887416	2.978247

Last Change in -2 Log L	5.569454E-6

Last Evaluation of Gradient	
Intercept	**x**
4.4730151E-7	0.0000149306

Convergence criterion (GCONV=1E-8) satisfied.

Fig. 7.20. SAS Example G4: output from the `itprint` option

output are shown in Fig. 7.20. The first table indicates that 70 female turtles hatched (event) and 82 male turtles hatched (nonevent) from the eggs, adding across all incubation temperatures used in the study. The `itprint` option in the model statement causes the values of the estimates of the regression parameters to be displayed for each iteration of the process used to evaluate the maximum likelihood estimates. Output from this option is displayed in the second table in Fig. 7.20. The first line of that table shows that the starting value for estimating the temperature coefficient is zero and the starting value for estimating the intercept is $209.763388 = \log(70/(70 + 82))$. The Fisher scoring algorithm converges in six steps to an estimate of -85.887416 for the intercept and an estimate of 2.978247 for the temperature coefficient. The third table shows that both entries in the gradient vector, derivatives of the log-likelihood function with respect to the intercept and temperature coefficient, are close to zero. This is what should happen when the algorithm converges. The note under the third table also indicates that the algorithm converged.

The maximum likelihood estimates for the parameters are displayed in the second table in Fig. 7.21. The estimated intercept is an estimately small of the log-odds that a female hatches when the incubation temperature is 0 °C. This corresponds to very small odds of 5.0069×10^{-38} and an extreme probability of 5.0069×10^{-38}. This is an extrapolation to a temperature at which no eggs would actually hatch, but it extends the trend that hatchlings are less likely to be female at lower incubation temperatures. Applying the exponential function to the estimated temperature coefficient indicates that the odds of a female turtle increase by a factor of about $\exp(2.978247) = 19.65$ for each 1 degree increase in incubation temperature. Standard errors for the estimates are displayed in the fourth column of the table. The values of the Wald chi-square test statistics shown in the fifth column of the table are computed by dividing each parameter estimate by its standard error and then squaring the ratio. The null hypothesis is that the corresponding population parameter is zero, and the alternative is that it is not zero. The p-values shown in column six are obtained by comparing the value of the Wald chi-square statistic to percentiles of the chi-square distribution with one degree of freedom. The p-values are approximate and the approximation is more accurate for larger sample sizes.

Temperature Dependent Sex Determination for Japanese Turtles

The LOGISTIC Procedure

Testing Global Null Hypothesis: BETA=0			
Test	Chi-Square	DF	Pr > ChiSq
Likelihood Ratio	111.4849	1	<.0001
Score	75.3842	1	<.0001
Wald	34.2018	1	<.0001

Analysis of Maximum Likelihood Estimates					
Parameter	DF	Estimate	Standard Error	Wald Chi-Square	Pr > ChiSq
Intercept	1	-85.8874	14.7025	34.1254	<.0001
x	1	2.9782	0.5093	34.2018	<.0001

Estimated Covariance Matrix		
Parameter	Intercept	x
Intercept	216.163	-7.48622
x	-7.48622	0.259341

Fig. 7.21. SAS Example G4: output from the `model` statement

The last line of the first table in Fig. 7.21 shows the value of a Wald chi-square test statistic for the null hypothesis that the coefficients are zero for all of the explanatory variables in the model. Because temperature is the only explanatory variable in this model, the results of this test are identical to the Wald chi-square test for the temperature coefficient in the second table. The likelihood ratio and score tests provide alternative tests of the same null hypothesis that all explanatory variables have zero coefficients. All three tests yield similar results when the sample size is large relative to the number of explanatory variables in the model, but they may differ substantially and may all be unreliable for small sample sizes. For the turtle data, all three tests indicate that the true temperature coefficient is different from zero. The bottom table in Fig. 7.21 displays the estimated covariance matrix for the parameter estimates. Estimated variances of the parameter estimates are on the main diagonal of the matrix. These values are the squares of the standard errors reported in the middle table. The estimated covariance between the estimates of the intercept and the temperature coefficient is -7.48622, and the estimated correlation is $-0.9998534 = -7.48622/\sqrt{216.163 \times 0.259341}$.

The `clparm=both` option in the `model` statement produced the 95% confidence intervals for the regression parameters displayed in the first two tables in Fig. 7.22. The two sets of confidence intervals are similar, but the profile-likelihood intervals are not centered at the parameter estimates and are slightly wider than the Wald intervals. The profile-likelihood intervals better reflect the shape of the small sample distributions of the parameter estimates and more nearly provide 95% coverage. The `clodds=both` option in th model statement produced the other two tables in Fig. 7.22. Those tables display profile-likelihood and Wald confidence intervals, respectively, for the ratio of the odds that a female hatches at a temperature $x + 1$ relative to the odds that a female hatches at temperature x. The profile-likelihood interval indicates with 95% confidence that a 1 degree increase in a viable incubation temperature will increase the odds of a female by a factor between 8 and 60. The Wald interval indicates a similar result, although the confidence interval is more narrow than the profile-likelihood interval and it may provide a lower level of confidence.

Figure 7.23 displays the file created by the `output` statement in the `proc logistic` procedure step. The option `p=` computes the maximum likelihood estimates of the proportion of eggs that produce females at each incubation temperature in the data set. In this case `p=phat` outputs those estimated proportions to a column labeled `phat` in the data set created and named by the `out=` option. The options `lower=cl_lower` and `upper=cl_upper` output the lower and upper limits of corresponding approximate 95% confidence intervals for the proportion of eggs that produce females at each incubation temperature. The output data file was printed by the `print` procedure shown in Fig. 7.18. By augmenting the data with two lines containing incubation temperatures that were not used in the experiment and using periods to indicate missing values for y_1, y_2, and n, estimates of the proportion of eggs that produce females at those temperatures are obtained along with corresponding

Temperature Dependent Sex Determination for Japanese Turtles

The LOGISTIC Procedure

Parameter Estimates and Profile-Likelihood Confidence Intervals			
Parameter	Estimate	95% Confidence Limits	
Intercept	-85.8874	-118.4	-60.2855
x	2.9782	2.0918	4.1052

Parameter Estimates and Wald Confidence Intervals			
Parameter	Estimate	95% Confidence Limits	
Intercept	-85.8874	-114.7	-57.0711
x	2.9782	1.9801	3.9764

Odds Ratio Estimates and Profile-Likelihood Confidence Intervals				
Effect	Unit	Estimate	95% Confidence Limits	
x	1.0000	19.653	8.099	60.653

Odds Ratio Estimates and Wald Confidence Intervals				
Effect	Unit	Estimate	95% Confidence Limits	
x	1.0000	19.653	7.244	53.323

Fig. 7.22. SAS Example G4: confidence intervals produced by the `clparm` and `clodds` options in the `model` statement

Temperature Dependent Sex Determination for Japanese Turtles

Obs	x	y1	y2	n	phat	cl_lower	cl_upper
1	26.0	0	8	8	0.00021	0.00001	0.00387
2	26.5	0	24	24	0.00094	0.00008	0.01045
3	28.0	0	19	19	0.07610	0.02951	0.18243
4	28.5	8	18	26	0.26749	0.16477	0.40333
5	29.0	20	7	27	0.61816	0.49215	0.73006
6	29.5	12	6	18	0.87770	0.76267	0.94128
7	30.0	30	0	30	0.96953	0.90244	0.99094
8	27.5	.	.	.	0.01824	0.00434	0.07346
9	28.8	.	.	.	0.47155	0.35119	0.59532

Fig. 7.23. SAS Example G4: output file created by the `output` statement

approximate 95% confidence intervals. With 95% confidence, for example, an incubation temperature of 28.8 °C is expected to produce between 35.1 and 59.5% females.

```
data turtles;
  input x y1 y2 n;
  label  x = 'Incubation Temperature (C)'
        y1 = 'Number of Females'
        y2 = 'Number of Males'
         n = 'Number of Eggs';
  datalines;
   26.0   0   8   8
   26.5   0  24  24
   28.0   0  19  19
   28.5   8  18  26
   29.0  20   7  27
   29.5  12   6  18
   30.0  30   0  30
   27.5   .   .   .
   28.8   .   .   .
    ;

title "Temperature Dependent Sex Determination for Japanese Turtles";

proc genmod data=turtles plots=predicted;
  model y1/n = x  / link=logit dist=binomial itprint covb waldci lrci;
  output out=setp p=phat lower=cl_lower  upper=cl_upper;
  run;

proc print data=setp; run;
```

Fig. 7.24. SAS Example G4: program for `proc genmod`

The same logistic regression model can be fit with the GENMOD procedure. The code is shown in Fig. 7.24. The option `link=logit` is included in the `model` statement to specify the logit link, and the option `dist=binomial` specifies independent binomial distributions for the numbers of female turtles produced by the various incubation temperatures. Wald confidence intervals for the regression parameters are obtained with the `waldci` option, and profile-likelihood confidence intervals are obtained with the `lrci` option. Because the output is similar to the output from `proc logistic`, it is not displayed here. The `plots=predicted` option in the `proc genmod` statement produces a plot of the estimated curve for the probabilities that females hatch at various incubation temperatures, but it is of lower quality than the plot displayed in Fig. 7.19.

SAS Example G5

Logistic regression models may incorporate both quantitative and categorical explanatory variables. Categorical variables are also called classification variables. The LOGISTIC procedure assumes that each explanatory variable is quantitative unless the variable is designated as a classification variable by including it in a `class` statement. Quantitative variables are modeled by simply adding a term to the logistic regression model consisting of the variable multiplied by a regression coefficient. Classification variables are modeled in a different way. A separate parameter is added to the model for each unique value of a classification variable, allowing the logit to be different for each unique level of the classification variable. Consider data from a study reported by Graubard and Korn (1987). This study examined the association between the incidence of congenital sex organ malformations in newborn babies and the level of alcohol consumption by the mothers during the first trimester of their pregnancy. At the end of the first trimester, each woman reported her average alcohol consumption as belonging to one of five categories: no alcohol use, less than 1 drink per day, 1–2 drinks per day, 3–5 drinks per day, or more than 5 drinks per day. Following childbirth, the presence or absence of congenital sex organ malformations was recorded for each baby. The counts are displayed in Table 7.1.

Table 7.1. SAS Example G5: maternal alcohol use and congenital malformations

	Alcohol Consumption (average number of drinks per day)				
Malformation	0	<1	1-2	3-5	>5
Present	48	38	5	1	1
Absent	17,066	14,464	788	126	37

The counts are entered directly into the `data step` of the program code for SAS Example G5 shown in Fig. 7.25. The `freq` statement in the `proc logistic` step indicates that each line in the data set represents the number of cases given by the value of the `y` variable. In the `model` statement, the variable that designates the response categories, `malformation`, is set equal to the explanatory variable. This is a second way of establishing the logit link. By default `proc logistic` uses the alphabetic ordering of the values of the response variables (numerical ordering if the response variable is quantitative) to define the log-odds. Because `absent` precedes `present`, for example, `proc logistic` would relate the log-odds that a malformation is absent to the levels of the alcohol consumption variable unless instructed to do otherwise. In this case it is desirable to relate the log-odds that a malformation is present to the categories of the alcohol consumption variable. This is done by including the `descending` option in the `proc logistic` statement to reverse the ordering of the response variable categories in the construction of the logit. The ordering of the categories for the response variable is reported in a table near the beginning of the output from `proc logistic`. This is the first table displayed in Fig. 7.26. In this case `present` is identified as the first category

```
data set1;
  input malformation \$ drinks y;
  datalines;
     present 1     48
     absent  1 17066
     present 2     38
     absent  2 14464
     present 3      5
     absent  3    788
     present 4      1
     absent  4    126
     present 5      1
     absent  5     37
           ;

 title 'Congenital Malformation and Alcohol Use';

 proc logistic data=set1 descending covout outest=parms;
   class drinks / param=glm ref=first;
   model malformation =  drinks / itprint covb
            clparm=wald expb clodds=wald  ;
   freq y;
   contrast '< 1 vs  none' drinks 1 0 0 0 -1 / alpha=.05 estimate=all;
   contrast '> 5 vs none' drinks  0 0 0 1 -1/ estimate=all;
   contrast 'none' intercept 1 drinks 0 0 0 0 1 / estimate=all;
   contrast '> 5 drinks' intercept 1 drinks 0 0 0 1 0/ estimate=all;
   output out=pred l=lower95 p=phat u=upper95 predprob=i;
   run;

  proc print data=parms;
    title 'Parameter Estimates and Covariance Matrix';
        run;

  proc print data=pred;
    title 'Estimates of Malformation Probabilities';
        run;
```

Fig. 7.25. SAS Example G5: program for `proc logistic`

and `absent` is identified as the second category. The systematic part of the logistic regression model is set up as

$$\log\left(\frac{\pi_i}{1-\pi_i}\right) = \gamma + \tau_i \quad \text{for } i = 1, 2, 3, 4, \text{ or } 5, \tag{7.30}$$

where π_i is the probability of delivering a baby with a congenital malformation for pregnant women in the i-th alcohol consumption category. In forming the logit, the probability for the category of the response variable that is designated as first in the table (presence) is divided by the probability of the response appearing in the second category (absence).

Because the `drinks` variable is included in a `class` statement, a different parameter is included in the model for each category of the `drinks` variable, in this case τ_1, τ_2, τ_3, τ_4, and τ_5. Because six parameters, γ, τ_1, τ_2, τ_3, τ_4, and τ_5, are used to model logits for just five categories, one constraint must be placed on the parameters in order to obtain unique parameter estimates. To be consistent with parameter constraints employed by the `glm` procedure for linear models, the `param=glm` option is included in the `class` statement. Then

Congenital Malformation and Alcohol Use

The LOGISTIC Procedure

Response Profile		
Ordered Value	malformation	Total Frequency
1	present	93
2	absent	32481

Probability modeled is malformation='present'.

Class Level Information						
Class	Value	Design Variables				
drinks	2	1	0	0	0	0
	3	0	1	0	0	0
	4	0	0	1	0	0
	5	0	0	0	1	0
	1	0	0	0	0	1

Fig. 7.26. SAS Example G5: output from `proc logistic`

the highest category of the `drinks` variable is designated as the baseline category by setting the corresponding parameter equal to zero. If not instructed otherwise, the category with the label that comes last alphabetically (or last numerically if the category labels are numbers) is selected as the baseline category. In this case the levels of the `drinks` are labeled 1, 2, 3, 4, and 5, and the category for more than 5 drinks per day would be the baseline category. Because it is more interesting to compare malformation rates for the last four categories for which alcohol is consumed to the first category for which no alcohol is consumed, the baseline category is changed to the first category with the `ref=first` option in the `class` statement. The result of this designation is reflected in the design table shown as the second table in Fig. 7.26. This table shows that level 1 of the `drinks` variable has been designated as the last category and consequently it serves as the baseline category. Consequently, τ_1 is set equal to zero, and τ_2, τ_3, τ_4, and τ_5 are interpreted with respect to the first level of the `drinks` variable (no alcohol use). For example, τ_3 is the log-odds that pregnant women who consume one to two alcohol drinks per day give birth to babies with malformations minus the log-odds that women who consume no alcohol during pregnancy give birth to babies with malformations. Similarly, $\exp(\tau_3)$ is the corresponding ratio of odds. Other choices for the `param=` and `ref=` options will create parameter estimates with different interpretations, although the model produces the same estimates of the malformation probabilities.

Congenital Malformation and Alcohol Use

The LOGISTIC P rocedure

Maximum Likelihood Iteration History								
Iter	**Ridge**	**-2 Log L**	**Intercept**	**drinks2**	**drinks3**	**drinks4**	**drinks5**	**drinks1**
0	0	1275.446887	-5.855811	0	0	0	0	0
1	1.6384	1270.991698	-5.857243	-0.030705	0.459873	0.668740	3.123968	0
2	0	1269.348522	-5.873509	-0.066904	0.876588	1.106242	2.534391	0
3	0	1269.245911	-5.873642	-0.068188	0.815509	1.039651	2.294575	0
4	0	1269.244890	-5.873642	-0.068189	0.813584	1.037363	2.263200	0

Last Change in -2 Log L	0.0010212126

Last Evaluation of Gradient					
Intercept	**drinks2**	**drinks3**	**drinks4**	**drinks5**	**drinks1**
-0.000474447	-1.88436E-11	-9.100914E-6	-2.558799E-6	-0.000462787	7.105427E-15

Convergence criterion (GCONV=1E-8) satisfied.

Fig. 7.27. SAS Example G5: convergence of maximum likelihood estimates

The output displayed in Fig. 7.27 shows that the search for the maximum likelihood estimates of the regression coefficients begins with each $\hat{\tau}_i = 0$ for $i = 1, 2, 3, 4, 5$, and $\hat{\gamma} = \log(93/32481) = -5.855811$. The search converges in four iterations. This output was requested with the `itprint` option in the `model` statement. Maximum likelihood estimates of regression coefficients are shown in Fig. 7.28 along with their standard errors. The estimates of τ_2, τ_3, τ_4, and τ_5 are all positive, suggesting that the incidence of congenital malformations increases with any level of alcohol consumption during pregnancy, but only $\hat{\tau}_5$ is significantly different from zero. Approximate 95% confidence intervals for the alcohol consumption parameters are requested with the `clparm=wald` statement in the `model` statement. Estimates of odds ratios, shown in the last column of the table, are requested with the `expb` option in the `model` statement. Confidence intervals for odds ratios are displayed in the third table in Fig. 7.28. The confidence interval in the last row of the table, for example, provides 95% confidence that the odds that women who consume at least five drinks per day give birth to a baby with malformations are between 1.29 and 71.46 times greater than the odds for women who consume no alcohol during the first trimester of pregnancy. This odds ratio is significantly different from one, but it is not well estimated. Also keep in mind that this is an observational study and pregnant women who heavily use

Congenital Malformation and Alcohol Use

The LOGISTIC Procedure

Testing Global Null Hypothesis: BETA=0			
Test	Chi-Square	DF	Pr > ChiSq
Likelihood Ratio	6.2020	4	0.1846
Score	12.0821	4	0.0168
Wald	9.2847	4	0.0544

Analysis of Maximum Likelihood Estimates							
Parameter		DF	Estimate	Standard Error	Wald Chi-Square	Pr > ChiSq	Exp(Est)
Intercept		1	-5.8736	0.1445	1651.3399	<.0001	0.003
drinks	2	1	-0.0682	0.2174	0.0984	0.7538	0.934
drinks	3	1	0.8136	0.4713	2.9795	0.0843	2.256
drinks	4	1	1.0374	1.0143	1.0460	0.3064	2.822
drinks	5	1	2.2632	1.0235	4.8900	0.0270	9.614
drinks	1	0	0	.	.	.	.

Odds Ratio Estimates and Wald Confidence Intervals				
Effect	Unit	Estimate	95% Confidence Limits	
drinks 2 vs 1	1.0000	0.934	0.610	1.430
drinks 3 vs 1	1.0000	2.256	0.896	5.683
drinks 4 vs 1	1.0000	2.822	0.386	20.602
drinks 5 vs 1	1.0000	9.614	1.293	71.460

Fig. 7.28. SAS Example G5: maximum likelihood estimates of regression coefficients and odds ratios

alcohol may also tend to engage in other activities that could contribute to the incidence of malformations. Consequently, a cause and effect conclusion may not be justified.

Estimates, standard errors, confidence intervals, and tests for linear combinations of model parameters may be requested with `contrast` statements. The program code in Fig. 7.25 contains four contrast statements. The first `contrast` statement requests an estimate of $\tau_2 - \tau_1$ the difference between the log-odds of giving birth to child with a congenital malformation for women who consume less than one drink per week versus women who consume no alcohol. Note that earlier in the program code, the first category of the `drinks` variable was designated as the reference category making it the last category in the design table in Fig. 7.26. Consequently, the request for an estimate of $\tau_2 - \tau_1$ is specified in the `contrast` statement with `drinks 1 0 0 0 -1`. The last value in this list is the coefficient for τ_1, and the first value in this list is the coefficient for τ_2. The `alpha=.05` option specifies a 95% confidence level, and the `estimate=all` option includes the value of the maximum likelihood estimate of $\tau_2 - \tau_1$ in the output. The characters in single quotes provide a label for the contrast in the output i.e., $<$ 1 vs. none. The second contrast statement requests output for the estimation of $\tau_5 - \tau_1$, the difference between the log-odds of giving birth to a child with congenital malformations for women who consume more than five drinks per day relative to women who consume no alcohol. It labels the output `>5 versus none`. The third contrast statement requests output for the estimation of $\mu + \tau_1$, the log-odds that a pregnant women who consumes no alcohol will give birth to a child with a congenital malformation. The output for this contrast is labeled `none`. The fourth contrast statement requests output for the estimation of $\mu + \tau_5$, the log-odds that a pregnant woman who has more than five alcoholic drinks per day will give birth to a child with a congenital malformation. This output is labeled `>5 drinks`.

The output from the four contrast statements is displayed in Fig. 7.29. There are three lines in the table for each of the four `contrast` statements. The line labeled `PARM` presents the estimate of the linear combination of model parameters, its standard error, and an approximate 95% confidence interval. This is an estimate of the natural logarithm of the ratio of odds of giving birth to a child with congenital malformations for women who consume alcohol during the first 3 months of pregnancy but have fewer than one drink per day relative to women who consume no alcohol. The line labeled `EXP` presents the

Congenital Malformation and Alcohol Use

The LOGISTIC Procedure

Contrast Estimation and Testing Results by Row

Contrast	Type	Row	Estimate	Standard Error	Alpha	Confidence Limits		Wald Chi-Square	Pr > ChiSq
< 1 vs none	PARM	1	-0.0682	0.2174	0.05	-0.4943	0.3580	0.0984	0.7538
< 1 vs none	EXP	1	0.9341	0.2031	0.05	0.6100	1.4304	0.0984	0.7538
< 1 vs none	PROB	1	0.4830	0.0543	0.05	0.3789	0.5885	0.0984	0.7538
> 5 vs none	PARM	1	2.2632	1.0235	0.05	0.2573	4.2691	4.8900	0.0270
> 5 vs none	EXP	1	9.6138	9.8393	0.05	1.2934	71.4595	4.8900	0.0270
> 5 vs none	PROB	1	0.9058	0.0873	0.05	0.5640	0.9862	4.8900	0.0270
none	PARM	1	-5.8736	0.1445	0.05	-6.1569	-5.5903	1651.3399	<.0001
none	EXP	1	0.00281	0.000407	0.05	0.00212	0.00373	1651.3399	<.0001
none	PROB	1	0.00280	0.000404	0.05	0.00211	0.00372	1651.3399	<.0001
> 5 drinks	PARM	1	-3.6104	1.0132	0.05	-5.5963	-1.6246	12.6980	0.0004
> 5 drinks	EXP	1	0.0270	0.0274	0.05	0.00371	0.1970	12.6980	0.0004
> 5 drinks	PROB	1	0.0263	0.0260	0.05	0.00370	0.1646	12.6980	0.0004

Fig. 7.29. SAS Example G5: contrast estimates and tests

exponential function of the estimate in the previous line along with a standard error and an approximate 95% confidence interval. This is an estimate of the odds ratio, and the 95% confidence interval extends from 0.61 to 1.43 indicating that the odds that women who consume alcohol but have less than one drink per day give birth to a child with congenital malformations are likely to be between 61% and 143% of the odds for women who consume no alcohol. The data do not provide convincing evidence that the odds of a malformation for this low level of alcohol consumption differ from the odds of a malformation for no alcohol consumption. The line labeled `PROB` does not produce a useful estimate for the $\tau_2 - \tau_1$ contrast: it gives an estimate of $1/(1 + \exp(\tau_1 - \tau_2))$. The 95% confidence interval in the `EXP` line for the $\tau_5 - \tau_1$ contrast indicates that the odds that pregnant women who consume more than five alcoholic drinks per day deliver a child with a congenital malformation are likely to be between 129% and 7146% greater than the corresponding odds for pregnant women who consume no alcohol. The last two columns of the table present values of Wald chi-square test statistics and the corresponding p-value for testing the null hypothesis that the linear combination of the model parameters is zero against the alternative that the linear combination is not zero. The `PARM` lines for the third and fourth contrasts give estimates and confidence intervals of the log-odds of delivering a baby with a congenital malformation for pregnant women who consume no alcohol and pregnant women who have more than five drinks per day, respectively. The `EXP` lines give estimates and confidence intervals for the corresponding odds of delivering a baby with a congenital malformation for these two categories of women. The `PARM` lines display estimates and confidence intervals for the corresponding probabilities of delivering a baby with a congenital malformation for women in those two categories. The 95% confidence interval for women who do not consume any alcohol indicates that the probability of delivering a child with a congenital malformation is likely to be between 0.00211 and 0.00372, but for women who have at least five alcohol drinks per day, the probability is likely to be between 0.0037 and 0.1646.

A file containing estimates of malformation probabilities is created with the `output` statement in the program code shown in Fig. 7.25. This file is displayed in Fig. 7.30. The output file is named with the `out=` option. Estimates of malformation probabilities are inserted into a column called `phat` with the `p=phat` option, and lower and upper limits of approximate 95% confidence intervals are requested and named with the `l=` and `u=` options, respectively. This output is displayed in the last three columns of Fig. 7.30. The `pregprob=i` option outputs estimates of the probability of a malformation and the probability of no malformation displayed in columns 6 and 7 of Fig. 7.30.

Estimates of Malformation Probabilities

Obs	malformation	drinks	y	_FROM_	_INTO_	IP_present	IP_absent	_LEVEL_	phat	lower95	upper95
1	present	1	48	present	absent	0.002805	0.99720	present	0.002805	.002114255	0.00372
2	absent	1	17066	absent	absent	0.002805	0.99720	present	0.002805	.002114255	0.00372
3	present	2	38	present	absent	0.002620	0.99738	present	0.002620	.001907224	0.00360
4	absent	2	14464	absent	absent	0.002620	0.99738	present	0.002620	.001907224	0.00360
5	present	3	5	present	absent	0.006305	0.99369	present	0.006305	.002626806	0.01506
6	absent	3	788	absent	absent	0.006305	0.99369	present	0.006305	.002626806	0.01506
7	present	4	1	present	absent	0.007874	0.99213	present	0.007874	.001108096	0.05373
8	absent	4	126	absent	absent	0.007874	0.99213	present	0.007874	.001108096	0.05373
9	present	5	1	present	absent	0.026328	0.97367	present	0.026328	.003697960	0.16457
10	absent	5	37	absent	absent	0.026328	0.97367	present	0.026328	.003697960	0.16457

Fig. 7.30. SAS Example G5: estimates of malformation probabilities

7.3.3 Poisson Regression

Model

Poisson regression models are used to link expected numbers of occurrences of a specific type of event to values of a set of explanatory variables. For example, Poisson regression models may be used to relate expected numbers of skin cancer tumors in mice to exposure to different levels of toxic substances in their diet. Poisson regression models are used in ecological studies to relate the expected numbers of particular plant or animal species per unit area to features of the environment at different locations. They could be used by credit card companies to analyze associations between numbers of late payments during the past 5 years and explanatory variables such as annual salary, age, gender, and marital status of the card holders. Here, the number of late credit card payments is the response variable, whereas "marital status" and "gender" are categorical explanatory variables, and "age" and "annual salary" are quantitative explanatory variables.

In Poisson regression models, the natural logarithm of the expected count is related to a linear combination of values of the explanatory variables. Suppose $y_1, y_2, \ldots, y_n$ are observed counts provided by n different subjects under different sets of conditions. Let μ denote the expected count under the set of conditions corresponding to a particular set of values for k explanatory variables x_1, x_2, ..., x_k. Then, the systematic part of the Poisson regression model is

$$\log(\mu) = \beta_0 + \beta_1 x_1 + \beta_2 x_2 + \cdots + \beta_k x_k. \tag{7.31}$$

For each set of values for the explanatory variables, the observed count is assumed to have a Poisson distribution with mean or expected count:

$$\mu = e^{\beta_0 + \beta_1 x_1 + \cdots + \beta_k x_k}. \tag{7.32}$$

The parameter β_j may be interpreted as the change in the natural logarithm of the mean count when x_j is increased by one unit while the other explanatory variables are held constant. Holding the other explanatory variables constant, the expected count at $x_j + 1$ is the expected count at x_j multiplied by $\exp(\beta_j)$.

Estimation and Hypothesis Testing

The GENMOD procedure may be used to fit a Poisson regression model to count data. The natural logarithm of the mean is the default link function when the Poisson distribution is specified for the observed counts. Maximum likelihood estimates are computed for the regression coefficients $\beta_0, \beta_1, \ldots, \beta_k$. The large sample normal approximation to the distribution of the parameter estimates can be used to construct Wald confidence intervals and tests of hypothesis. Wald confidence intervals are requested with the `waldci` option in the `model` statement in the `proc genmod` step. A $(1-\alpha)100\%$ Wald confidence interval has the form:

$$\hat{\beta}_j \pm z_{1-\alpha/2} \hat{\sigma}_{\hat{\beta}_j} \tag{7.33}$$

where z_p is the $100p$ percentile of the standard normal distribution, $\hat{\beta}_j$ is the parameter estimate, and $\hat{\sigma}_{\hat{\beta}_j}$ is the large sample estimate of the standard error of $\hat{\beta}_j$. Profile-likelihood confidence intervals for the individual regression parameters are requested with the `LRCI` option in the `model` statement. These confidence intervals tend to provide more accurate coverage levels for smaller samples, but they are more computationally intensive than the Wald intervals.

`Proc genmod` computes likelihood ratio chi-square tests of null hypotheses involving individual parameters or linear combinations of parameters. Wald chi-square tests are computed if the `wald` option is specified.

SAS Example G6

The data for this application of the GENMOD procedure to Poisson regression analysis are reported by Margolin et al. (1981) from an Ames Salmonella reverse mutagenicity assay. Figure 7.31 displays the number of revertant colonies of TA95 Salmonella observed on each of three replicate plates tested at each of six dose levels of quinoline. Margolin et al. (1981) consider several models, but we will consider an approximation to one of those models that was proposed by Breslow (1984). Using μ to represent the mean number of revertant colonies and x to represent the dose of quinoline (μg per plate), this model is expressed as

$$\log(\mu) = \beta_0 + \beta_1 z + \beta_2 x \tag{7.34}$$

where $z = \log(x + 10)$.

The SAS Example G6 program (see Fig. 7.32) illustrates how `proc genmod` can be used to fit this model. The data are included in the `data step` with x representing the level of quinoline and $c1$, $c2$, and $c3$ representing the counts

Dose of Quinoline (μg per plate)					
0	10	33	100	333	1000
15	16	16	27	33	20
21	18	26	41	38	27
29	21	33	60	41	42

Fig. 7.31. SAS Example G6: numbers of revertant colonies of TA98 Salmonella

```
data set1;
input x c1 c2 c3;
  z = log(x+10);
  array cc {3} c1-c3;
  drop c1-c3;
  do i = 1 to 3;
    y = cc(i); output;
        end;
datalines;
   0 15 21 29
  10 16 18 21
  33 16 26 33
 100 27 41 60
 333 33 38 41
1000 20 27 42
;

proc genmod data=set1 plots=(stdreschi);
  model y = z x  / dist=poisson link=log itprint;
  output out=setp p=mean;
 run;

data setnew;
  do i = 1 to 100;
    w=10**(3*i/100);
        m=exp(2.1728 + 0.3198*log(w+10) - 0.0010*w);
        output;
  end;
run;

data setp; set setp setnew; run;

 proc sgplot data=setp noautolegend;
  scatter x=x y=y / markerattrs=(size=10 symbol=CircleFilled color=black);
  loess x=w y=m / interpolation=cubic nomarkers;
  xaxis label="Quinoline Level (micrograms per plate)"
              labelattrs=(size=14) valueattrs=(size=13);
  yaxis label="Number of Revertant TA98 Salmonella Colonies"
              labelattrs=(size=14) valueattrs=(size=13);
  run;
```

Fig. 7.32. SAS Example G6: program for `proc genmod`

for the three plates at each level of quinoline. As described in SAS Example A6 in Chap. 1, the `array`, `do`, and `output` statements are used to convert each of the original lines of data to three lines of data with the count of each plate on a separate line and denoted by the variable y. The values of $z = \log(x+10)$ are also computed in the data step, and the resulting file has values of x, y, and z on each line.

The Poisson regression model is specified in the `model statement` by including the options `dist=poisson` and `link=log`. Actually, the `link=log` option is not needed because it is the default link when the Poisson distribution is specified. The variable `y` that contains the observed counts is on the left side of the equal sign in the `model statement`, and the explanatory variables are on the right side of the equal sign. An intercept is included in the model by default. The `itprint` option in the `model statement` prints the estimates of the regression coefficients at each step of the Fisher scoring procedure. This output, not shown here, shows that the estimation procedure converges in four iterations. Maximum likelihood estimates of the regression coefficients are displayed in Fig. 7.33 along with standard errors and 95% confidence intervals. These results indicate that all of the regression coefficients are significantly different from zero.

Analysis Of Maximum Likelihood Parameter Estimates

Parameter	DF	Estimate	Standard Error	Wald 95% Confidence Limits		Wald Chi-Square	Pr > ChiSq
Intercept	1	2.1728	0.2184	1.7447	2.6009	98.95	<.0001
z	1	0.3198	0.0570	0.2081	0.4315	31.48	<.0001
x	1	-0.0010	0.0002	-0.0015	-0.0005	17.07	<.0001
Scale	0	1.0000	0.0000	1.0000	1.0000		

Note: The scale parameter was held fixed.

Fig. 7.33. SAS Example G6: estimates of regression coefficients

The `plots=(stdreschi)` option in the `proc genmod` statement produces the plot of standardized chi-square residuals displayed in Fig. 7.34. When the proposed model provides a good description of changes in expected counts as values of the explanatory variables change, this plot should exhibit no obvious pattern and appear to be randomly scattered about the horizontal reference line at zero.

The estimated curve for the expected number of revertant TA98 colonies is shown in Fig. 7.35 along with the observed counts. The curve appears to provide a good representation for changes in the expected number of colonies as the level of quinoline changes with the possible exception of one large count for a plate with 100 μg of quinoline. The curve was constructed by evaluating

$$\hat{\mu} = \exp(2.1728 + 0.3198 \log(x + 10) - 0.0010x) \qquad (7.35)$$

at 100 values of x. This was accomplished by using `do` and `output` statements in the data step to output 100 values of x to the `setnew` data set. The subsequent data step attaches the `setnew` data set to the bottom of the `setp` data set created with the `output` statement in the `proc genmod` step. The `setp`

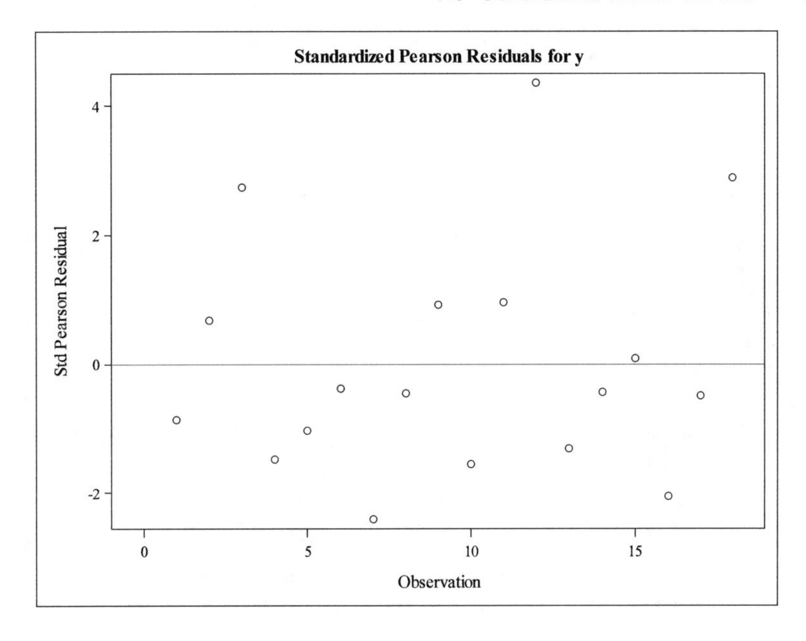

Fig. 7.34. SAS Example G6: output from the `plots=(stdreschi)` option

data set contains the estimated mean counts for the 18 plates used in the experiment. The `scatter` statement in the subsequent `proc sgplot` step plots the original counts against the six quinoline levels used in the study as filled in circles. The `loess` statement passes a smooth curve through the estimated means for the 100 new quinoline values in the `setnew` data set to form the curve displayed in Fig. 7.35.

The table in Fig. 7.36 displays statistics for assessing the fit of the model. The AIC, AICC, and BIC values are useful for comparing models within a specific set of models, but they provide no useful information for assessing the fit of a single model. Consequently, we will examine the deviance and Pearson chi-square statistics to assess the fit of the model. The last entry in the `deviance` line of the table shows that the value of the deviance statistic is almost 3 times larger than the corresponding degrees of freedom. The last entry in the Pearson chi-square line shows that the value of the Pearson chi-square statistic is slightly more than 3 times larger than the degrees of freedom. This is an indication that there is more variation in the counts about the fitted model than a Poisson model can support. One feature of any Poisson distribution is that the variance is equal to the mean, or expected count. This restriction is often violated in real-world situations. In this case, it appears that the variance in the counts is about 3 times the mean. This extra-Poisson variation can be taken into account by including an additional scale parameter in the model. The note under the table in Fig. 7.33 indicates that this was not done and the scale parameter was held fixed at 1.0 in this analysis.

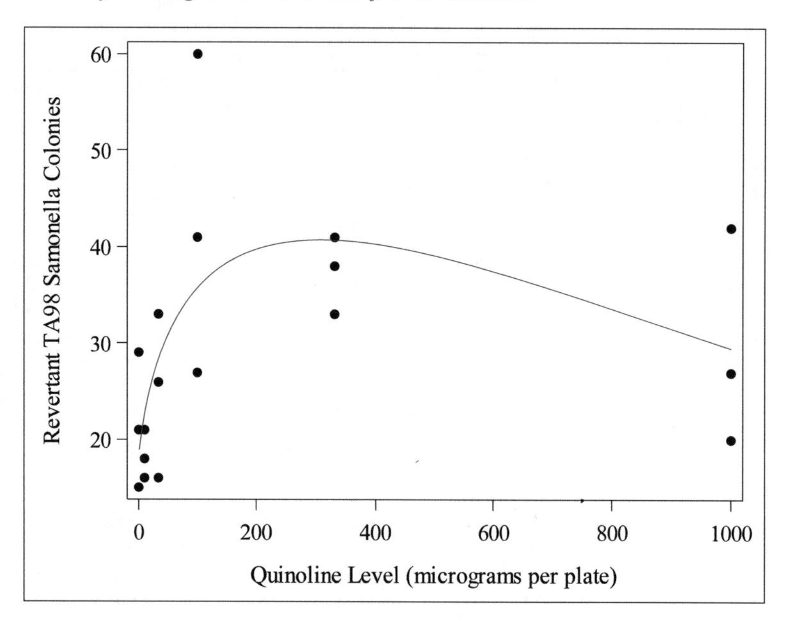

Fig. 7.35. SAS Example G6: plot of the estimated mean response curve and the observed counts

Consequently, the standard errors of the estimated regression coefficients are too small by a factor of about $\sqrt{3} = 1.73$, and the corresponding confidence intervals are too short to provide 95% coverage. Methods for accounting for overdispersion are considered in Sect. 7.4.

7.4 Generalized Linear Models with Overdispersion

7.4.1 Introduction

Overdispersion occurs when variation in observed counts about a logistic regression model is larger than the level of variation that can be accommodated by the binomial probability model. This is often called extra-binomial variation. Overdispersion also occurs when variation in observed counts about a Poisson regression model is larger than the level of variation that can be accommodated by the Poisson probability model. This is often called extra-Poisson variation.

Consider a logistic regression model:

$$\log\left(\frac{\pi}{1-\pi}\right) = \beta_0 + \beta_1 x_1 + \cdots + \beta_k x_k \qquad (7.36)$$

The GENMOD Procedure

Criteria For Assessing Goodness Of Fit			
Criterion	**DF**	**Value**	**Value/DF**
Deviance	15	43.7157	2.9144
Scaled Deviance	15	43.7157	2.9144
Pearson Chi-Square	15	46.2706	3.0847
Scaled Pearson X2	15	46.2706	3.0847
Log Likelihood		1259.7878	
Full Log Likelihood		-68.1260	
AIC (smaller is better)		142.2520	
AICC (smaller is better)		143.9663	
BIC (smaller is better)		144.9231	

Fig. 7.36. SAS Example G6: goodness-of-fit output from `proc genmod`

and

$$\pi = \frac{\exp(\beta_0 + \beta_1 x_1 + \cdots + \beta_k x_k)}{1 + \exp(\beta_0 + \beta_1 x_1 + \cdots + \beta_k x_k)}. \tag{7.37}$$

If y represents the number of successes observed for n independent trials run under conditions corresponding to a particular set of values for $x_1, x_2, \ldots, x_k$, then imposing a binomial distribution for y results in $n\pi$ as the mean value for y and $n\pi(1-\pi)$ as the variance of y. The observed counts are overdispersed if variances tend to be larger than $n\pi(1-\pi)$ for the sets of $x_1, x_2, \ldots, x_k$ values used in the study. This is often called extra-binomial variation. The observed counts are underdispersed if variances tend to be smaller than $n\pi(1-\pi)$ for the sets of $x_1, x_2, \ldots, x_k$ values used in the study. Underdispersion occurs much less frequently than overdispersion.

Overdispersion in Poisson regression occurs when variation in observed counts about a Poisson regression model is larger than the level of variation association with Poisson distributions. Consider a Poisson regression model:

$$\log(\mu) = \beta_0 + \beta_1 x_1 + \cdots + \beta_k x_k \tag{7.38}$$

with

$$\mu = \exp(\beta_0 + \beta_1 x_1 + \cdots + \beta_k x_k). \tag{7.39}$$

Imposing a Poisson distribution for the observed count at some set of $x_1, x_2, \ldots, x_k$ values results in a variance of μ to describe how the much observed counts should vary about the mean, which is also μ. Overdispersion occurs when variances of observed counts tend to be larger than the means, and this is often called extra-Poisson variation. Underdispersion occurs when the variances of the counts tend to be smaller than their means.

Overdispersion or underdispersion can be detected by dividing the value of the Pearson chi-square goodness-of-fit statistic by its degrees of freedom. If there is no problem with overdispersion or underdispersion, then the value of the Pearson chi-square statistic should be close to the degrees of freedom, and the ratio should be close to 1. Overdispersion is indicated by a ratio that is substantially larger than 1, and underdispersion is indicated by a ratio that is much smaller than 1. There is no fixed guideline that applies to all situations, but a ratio larger than 1.25 suggests that the analysis should be adjusted to account for overdispersion, and a ratio smaller than 0.75 may indicate that the analysis should be adjusted for underdispersion. Failure to make adjustments for overdispersion generally leads to standard errors for parameter estimates that are too small, confidence intervals that are too narrow to provide the stated level of confidence, and p-values of tests of hypotheses that are too small. The deviance divided by its degrees of freedom may be used in place of the Pearson chi-square statistic divided by its degrees of freedom, but the latter will be used in the examples presented here.

A large value of the ratio of the Pearson chi-square statistic to its degrees of freedom may also arise from a model that does not adequately describe how the mean counts change as value of the explanatory variables changes. To distinguish this situation from overdispersion, residual plots such as the plot of standardized chi-square residuals should be examined. Any deficiencies in the model should be addressed, and the improved model should be refit to the data before considering the use of an additional scale parameter to model overdispersion.

7.4.2 Binomial and Poisson Models with Overdispersion

Introduction

The binomial and Poisson distributions are members of the exponential family of distributions that do not have scale parameters that can be used to accommodate overdispersion or underdispersion. At the request of the user, however, the GENMOD procedure is able to include a scale parameter while still optimizing the log-likelihood function corresponding to the relevant binomial or Poisson distribution to estimate the regression parameters. The resulting estimates of the regression parameters are the same as when the scale parameter is set equal to one, but they are no longer maximum likelihood estimates because the distribution of the counts is altered by introducing the extra scale parameter. This is sometimes called quasi-likelihood estimation (see Wedderburn (1974) for details).

The GENMOD and LOGISTIC procedures introduce an additional scale parameter ϕ as a multiple of the formula for the standard deviations of the counts for the specified distribution. For the binomial distribution, the adjusted standard deviation of an observed count is $\phi\sqrt{n\pi(1-\pi)}$, and for the

Poisson distribution, the adjusted standard deviation of an observed count is $\phi\sqrt{\mu}$. Then ϕ is estimated as

$$\hat{\phi} = \sqrt{\frac{\text{Pearson chi-square statistic}}{\text{degrees of freedom}}} \tag{7.40}$$

SAS Example G7: Poisson Regression

The SAS Example G7 program (see Fig. 7.37) illustrates how `proc genmod` can be used to fit a Poisson regression model to the data on the effects of quinoline on counts of TA98 Salmonella colonies displayed in Fig. 7.31. The data are entered as described for SAS Example G6. The `scale=Pearson` option has been added to the `model` statement to indicate that the additional scale parameter should be estimated as shown in (7.40).

```
data set1;
input x c1 c2 c3;
  z = log(x+10);
  array cc {3} c1-c3;
  drop c1-c3;
  do i = 1 to 3;
    y = cc(i); output;
        end;
datalines;
   0 15 21 29
  10 16 18 21
  33 16 26 33
 100 27 41 60
 333 33 38 41
1000 20 27 42
;

proc genmod data=set1 plots=(stdreschi);
  model y = z x  / dist=poisson link=log itprint scale=Pearson;
  output out=setp p=mean;
 run;
```

Fig. 7.37. SAS Example G7: program for `proc genmod`

Estimated parameters are shown in Fig. 7.38. The parameter estimates are identical to those in Fig. 7.33 when the scale parameter was set equal to one, but the standard errors have all been increased by a factor of $\hat{\phi} = \sqrt{43.7157/15} = 1.7563$. The value of the Pearson chi-square statistic, 43.7157, is shown in Fig. 7.36. Note that 1.7563 is reported in the `scale` line of the table.

The `plots=(stdreschi)` option in the `proc genmod` statement produces the plot of standardized Pearson residuals displayed in Fig. 7.39. The pattern is the same as the pattern in Fig. 7.34 when the scale parameter was held fixed at 1, but the standardized residuals are closer to zero in Fig. 7.39 because the standard deviations used to compute them have been inflated by a factor of

Analysis Of Maximum Likelihood Parameter Estimates							
Parameter	DF	Estimate	Standard Error	Wald 95% Confidence Limits		Wald Chi-Square	Pr > ChiSq
Intercept	1	2.1728	0.3836	1.4209	2.9247	32.08	<.0001
z	1	0.3198	0.1001	0.1236	0.5160	10.21	0.0014
x	1	-0.0010	0.0004	-0.0019	-0.0002	5.53	0.0187
Scale	0	1.7563	0.0000	1.7563	1.7563		

Note: The scale parameter was estimated by the square root of Pearson's Chi-Square/DOF.

Fig. 7.38. SAS Example G7: estimates of regression coefficients

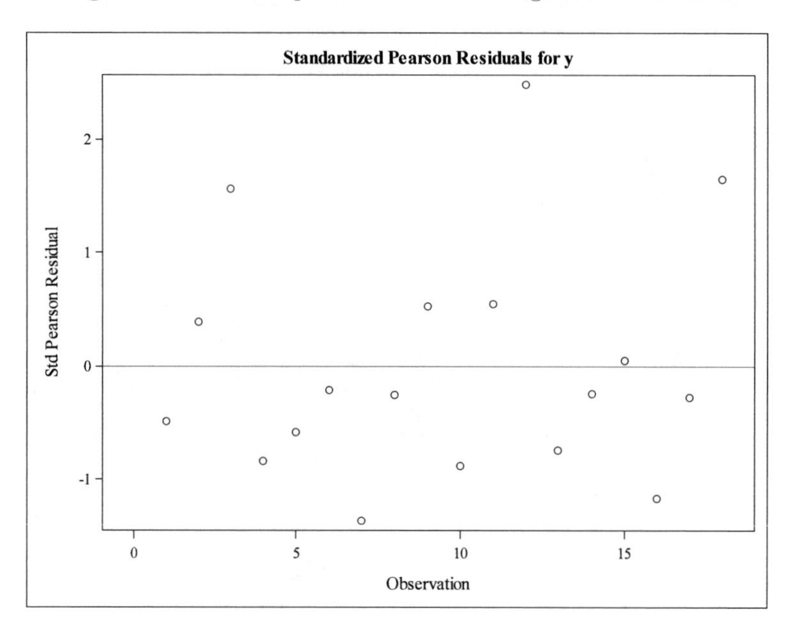

Fig. 7.39. SAS Example G7: output from the `plots=(stdreschi)` option

$\hat{\phi} = 1.7563$. The largest standardized Pearson residual is now smaller than 2.5, and it appears that the proposed model provides a good description of how the expected number of revertant TA98 Salmonella colonies changes as the quinoline concentration changes.

SAS Example G8: Logistic Regression

A scale parameter may be used to account for extra-binomial variation in logistic regression analysis. This example uses seed germination data reported by Crowder (1978). In this study *Orobanche cernua* seeds were brushed onto plates that were covered with different dilutions of a bean root extract. Three different dilutions were used: $1/1 = 1.0$, $1/24 = 0.04$, and $1/625 = 0.0016$. Six plates were prepared for the first dilution and five plates were prepared for the other two dilutions. The data set contains the number of seeds on each plate and the number of seeds that germinated on each plate. A logistic regression model is used to relate the germination rate to the dilution level of the bean root extract. The variation in the observed proportions of germinating seeds among plates with the same dilution of bean root extract is more than can be attributed to independent binomial distributions with the success probability. The extra-binomial variation may arise from plates with the same dilution of bean extract being exposed to slightly different levels of temperature, humidity, and other environmental conditions during the course of the study. This would cause variation in germination rates among plates covered with the same dilution of the extract, resulting in extra-binomial variation. The data are included in the SAS Example G8 program displayed in Fig. 7.40.

```
data seeds;
  input dilution n y;
        p = y/n;
  label n = 'Number of Seeds'
        y = 'Number that Germinate'
        p = 'Germination Rate'
        dilution = 'Bean Root Extract Dilution';
  datalines;
         1.0      43  2
         1.0      51  9
         1.0      44  5
         1.0      71 16
         1.0      24  2
         1.0       7  0
         0.04     19 17
         0.04     56 43
         0.04     87 79
         0.04     55 50
         0.04     10  9
         0.0016   13 11
         0.0016   62 47
         0.0016  104 90
         0.0016   51 46
         0.0016   11  9
           ;

title 'Seed Germination Rates';

proc logistic data=seeds plots(only)=effect;
  model y/n = dilution / itprint scale=Pearson;
  output out=setp p=phat lower=cl_lower upper=cl_upper;
  run;

proc print data=setp; run;
```

Fig. 7.40. SAS Example G8: program for `proc logistic`

The data step enters the information on the extract dilution, the number of seeds (n), and the number of seeds that germinated (y) for each plate. The proportion of seeds that germinated on each plate (p) is computed on the line after the `input` statement in the `data step`. This variable is not used in the analysis, but it is displayed in the output. The `successes/trials` option is used to specify the logit response on the left side of the equal sign in the `model` statement. The explanatory variable `dilution` is on the right side of the equal sign. The resulting model is

$$\log\left(\frac{\pi}{1-\pi}\right) = \beta_0 + \beta_1(\text{dilution}), \tag{7.41}$$

where

$$\pi = \frac{\exp(\beta_0 + \beta_1(\text{dilution}))}{1 + \exp(\beta_0 + \beta_1(\text{dilution}))} \tag{7.42}$$

represents the true germination probability for the specified dilution level. The variance of each observed count is inflated by including the `scale=Pearson` option in the `model` statement. For a particular dilution level, the variance of the observed number of germinating seeds out of the n seeds on the plate is $\phi^2 n\pi(1-\pi)$, where ϕ is the scale parameter.

The `itprint` option in the `model` statement requests the first table of output shown in Fig. 7.41. The Fisher scoring algorithm converged in four iterations. The final estimates of the regression parameters are shown in the third table in Fig. 7.41. The inclusion of the `scale=Pearson` option in the `model` statement has no effect on the estimates of the regression parameters. They are estimated in the same way as for a logistic regression model in which the scale parameter is set to 1. The inclusion of the `scale=Pearson` option does affect the standard errors of the estimates of the regression parameters.

The second table in Fig. 7.41 shows that the value of the Pearson goodness-of-fit statistic is 24.1127 with 14 degrees of freedom. The extra-binomial variation parameter is estimated as $\hat{\phi} = \sqrt{24.1127/14} = \sqrt{1.72234} = 1.31238$. Because the `scale=Pearson` option is used, `proc logistic` prints a note under the second table that indicates that the covariance matrix for the estimates of the regression parameters has been multiplied by 1.72234. Consequently, the standard errors of the regression coefficients are inflated by a factor of 1.31238 relative to what is reported when the `scale=Pearson` is not used. When the `scale=Pearson` option is not used and scale parameter is set equal to 1.0, the standard errors for $\hat{\beta}_0$ and $\hat{\beta}_1$ are reported as 0.1347 and 0.2316, respectively. By including the `scale=Pearson` option, the standard errors become 0.1768 and 0.3040, respectively. For these data, adjusting standard errors for extra-binomial variation does not affect the outcome of the test that the $\beta_1 = 0$, but it does affect the width of a confidence interval for β_1, and it could change the conclusion for tests of regression coefficients for other studies. The inflation of standard errors is carried through to standard errors for estimates of germination probabilities and estimates of contrasts of model parameters and related confidence intervals.

Seed Germination Rates

The LOGISTIC Procedure

Maximum Likelihood Iteration History				
Iter	Ridge	-2 Log L	Intercept	dilution
0	0	944.098189	0.465874	0
1	0	589.305145	1.548952	-3.073846
2	0	581.537153	1.832211	-3.603469
3	0	581.476648	1.860909	-3.654849
4	0	581.476643	1.861193	-3.655319

Convergence criterion (GCONV=1E-8) satisfied.

Deviance and Pearson Goodness -of-Fit Statistics				
Criterion	Value	DF	Value/DF	Pr > ChiSq
Deviance	25.0676	14	1.7905	0.0339
Pearson	24.1127	14	1.7223	0.0444

Note: The covariance matrix has been multiplied by the heterogeneity factor (Pearson Chi-Square / DF) 1.72234.

Analysis of Maximum Likelihood Estimates					
Parameter	DF	Estimate	Standard Error	Wald Chi-Square	Pr > ChiSq
Intercept	1	1.8612	0.1768	110.7999	<.0001
dilution	1	-3.6553	0.3040	144.5729	<.0001

Fig. 7.41. SAS Example G8: output from the `itprint` option

The `plots(only)=effect` option in the `proc logistic` statement creates the plot shown in Fig. 7.42. It appears that the model is reasonable, but it would be easier to judge this if plates had been made with additional dilutions, say 1:2, 1:4, and 1:8.

The `output` statement in the `proc logistic` step creates a data file called `setp` that is printed with the `print` statement in the last line of Fig. 7.40. The output is displayed in Fig. 7.43. The inclusion of the `scale=Pearson` option has no effect on the estimates of mean germination rates displayed in the `phat` column of the table, but it does result in wider confidence intervals that reflect the extra-binomial variation in the observed counts. In this case, the confidence intervals for the germination rates are 1.31238 times wider than what is reported when the `scale=Pearson` option is not used. Because the counts

Seed Germination Rates

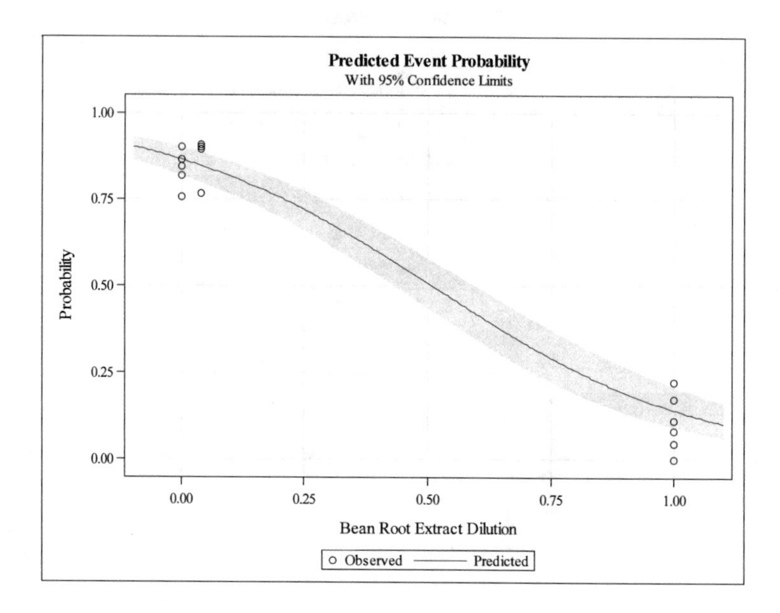

Fig. 7.42. SAS Example G8: output from the `plots(only)=effect` option

exhibit extra-binomial variation, using the `scale=Pearson` option provides a more honest indication of how well the germination rates are estimated with the data from this study.

7.4.3 Negative Binomial Models

The negative binomial distribution may be used to analyze data with extra-Poisson variation. A restrictive feature of the Poisson distribution is that the variances of the potential counts must be equal to their means. The negative binomial distribution has an additional scale parameter that allows the variances of potential counts to be larger than the means. Consequently, applying the GENMOD procedure with the negative binomial distribution instead of the Poisson distribution provides a method of fitting models similar to Poisson regression models that allow for more variation in the observed counts than can be accommodated by the Poisson distribution.

Model

As with Poisson regression models, we will consider models that link the natural logarithm of the mean count μ to a linear combination of explanatory variables, i.e., the systematic part of the model is

Seed Germination Rates

Obs	dilution	n	y	p	phat	cl_lower	cl_upper
1	1.0000	43	2	0.04651	0.14257	0.09374	0.21092
2	1.0000	51	9	0.17647	0.14257	0.09374	0.21092
3	1.0000	44	5	0.11364	0.14257	0.09374	0.21092
4	1.0000	71	16	0.22535	0.14257	0.09374	0.21092
5	1.0000	24	2	0.08333	0.14257	0.09374	0.21092
6	1.0000	7	0	0.00000	0.14257	0.09374	0.21092
7	0.0400	19	17	0.89474	0.84748	0.79936	0.88571
8	0.0400	56	43	0.76786	0.84748	0.79936	0.88571
9	0.0400	87	79	0.90805	0.84748	0.79936	0.88571
10	0.0400	55	50	0.90909	0.84748	0.79936	0.88571
11	0.0400	10	9	0.90000	0.84748	0.79936	0.88571
12	0.0016	13	11	0.84615	0.86475	0.81897	0.90037
13	0.0016	62	47	0.75806	0.86475	0.81897	0.90037
14	0.0016	104	90	0.86538	0.86475	0.81897	0.90037
15	0.0016	51	46	0.90196	0.86475	0.81897	0.90037
16	0.0016	11	9	0.81818	0.86475	0.81897	0.90037

Fig. 7.43. SAS Example G8: output from the `model` statement

$$\log(\mu) = \beta_0 + \beta_1 x_1 + \beta_2 x_2 + \cdots + \beta_k x_k. \tag{7.43}$$

For a specific set of the explanatory variables, the distribution of potential observed counts has a negative binomial distribution with mean μ and variance:

$$\text{Var}(Y) = \mu + \theta\mu^2, \tag{7.44}$$

where $\theta > 0$ is a dispersion parameter. When $\theta = 0$ the variance is equal to the mean as with the Poisson distribution. Otherwise, the variance is larger than the mean, but the relationship between the variance and the mean is not the same as for the extra-Poisson variation model for which $\text{Var}(Y) = \phi^2\mu$. Therefore, those two approaches for handling overdispersion are not identical, and they generally yield slightly different estimates of the regression parameters.

Estimation and Hypothesis Testing

The GENMOD procedure uses maximum likelihood estimation to estimate the regression parameters and the dispersion parameter θ for the negative binomial model. Wald chi-square tests for testing hypotheses about parameters and contrasts are based on the approximate large sample normal distribution for the parameter estimates. Profile-likelihood methods are used to construct confidence intervals.

The negative binomial and quasi-Poisson models cannot be compared with AIC or BIC values because the quasi-likelihood method used to estimate parameters for the quasi-Poisson model is not the same as maximum likelihood method used to estimate parameters for the negative binomial model. The models can be compared less formally with graphical examination of the relationship between the mean and variance. In this study three plates were prepared for each level of quinoline. This replication allows both a sample mean and a sample variance to be directly computed from the three counts observed at each level of quinoline. A plot of the sample variances versus the sample means can be constructed. If the plot exhibits a straight line pattern, then the quasi-Poisson model is appropriate. If the plot exhibits an increasing quadratic pattern, then the negative binomial model may be more appropriate. If there is no replication, a smoothed plot of $(y_i - \hat{\mu}_i)^2$ against $\hat{\mu}_i$ can be used. See Ver Hoef and Boveng (2007) for more details.

SAS Example G9

The SAS Example G9 program (shown in two parts: Figs. 7.44 and 7.45) illustrates how `proc genmod` can be used for a negative binomial model with the logarithm of the means linked to a linear combination of explanatory variables as shown in (7.43). We will use the data for the study of the effect of quinoline on the numbers of revertant TA98 Salmonella colonies that are examined in SAS Example G7. The program shown in Fig. 7.44 is similar to the program used to fit the quasi-Poisson model in Fig. 7.37. The data are entered in the same way. To request the negative binomial model in the `proc genmod`

```
data set1;
input x c1 c2 c3;
  z = log(x+10);
  array cc {3} c1-c3;
  drop c1-c3;
  do i = 1 to 3;
    y = cc(i); output;
        end;
datalines;
   0 15 21 29
  10 16 18 21
  33 16 26 33
 100 27 41 60
 333 33 38 41
1000 20 27 42
;

proc genmod data=set1 plots=(stdreschi);
  model y = z x  / dist=negbin link=log itprint;
  output out=setp p=mean lower=cl_lower upper=cl_upper;
 run;

proc print data=setp; run;
```

Fig. 7.44. SAS Example G9: Program

step, however, the `dist=Poisson` option in the `model` statement is changed to `dist=negbin`. The `link=log` option requests the natural logarithm of the mean count as the link function.

The output in Fig. 7.46 shows that the optimization algorithm converges in four iterations. A column is included for the estimation of the dispersion parameter. Final parameter estimates are shown in the second table in Fig. 7.46. The estimates of the regression parameters are similar to those for the quasi-Poisson model shown in Fig. 7.38 but the sets of estimates that are not identical. The standard errors are a bit smaller and the confidence intervals are shorter for the negative binomial model. *p*-Values for the Wald chi-square tests of significance for individual regression parameters are a bit smaller for the negative binomial model, but the results for both models indicate that all of the regression parameters are significantly different from zero. The plot of the standardized Pearson residuals (not shown here) is almost identical to the plot displayed in Fig. 7.39 for the quasi-Poisson model.

```
data setnew;
  do i = 1 to 100;
    w=10**(3*i/100);
        m=exp(2.1728 + 0.3198*log(w+10) - 0.0010*w);
        output;
  end;
run;

proc sort data=set1; by x; run;

data setnew; set setp setnew; run;

proc sgplot data=setnew noautolegend;
  scatter x=x y=y / markerattrs=(size=10 symbol=CircleFilled color=black);
  loess x=w y=m / interpolation=cubic nomarkers;
  xaxis label="Quinoline Level (micrograms per plate)" labelattrs=(size=14) valueattrs=(size=13);
  yaxis label="Revertant TA98 Salmonella Colonies" labelattrs=(size=14) valueattrs=(size=13);
  run;

proc means data=set1 noprint; by x;
  var y;
  output out=setv mean= mean var=var;
  run;

proc sgplot data=setv noautolegend;
  scatter x=mean y=var / markerattrs=(size=10 symbol=CircleFilled color=black);
  reg x=mean y=var / degree=1;
  reg x=mean y=var / degree=2 lineattrs=(pattern=MediumDash);
  xaxis label="Estimate of Mean Count" labelattrs=(size=14) valueattrs=(size=13);
  yaxis label="Variance Estimate" labelattrs=(size=14) valueattrs=(size=13);
  run;
```

Fig. 7.45. SAS Example G9: part of the program for obtaining plots

The plot of the estimated curve shown in Fig. 7.47 is very similar to the plot of the estimated curve for the quasi-Poisson model shown in Fig. 7.35. The negative binomial and quasi-Poisson models both provide good descriptions of these data. Estimates of mean numbers of Salmonella colonies are displayed in the `mean` column of Fig. 7.48, and approximate 95% confidence intervals

The GENMOD Procedure

Iteration History For Parameter Estimates						
Iter	Ridge	LogLikelihood	Prm1	Prm2	Prm3	Dispersion
0	0	1264.44336	2.1739641	0.3080112	-0.000959	0.0821862
1	0	1264.99986	2.2029631	0.3152694	-0.000992	0.0489248
2	0	1265.02414	2.1980081	0.3124097	-0.00098	0.048543
3	0	1265.02417	2.1976286	0.3125094	-0.00098	0.0487687
4	0	1265.02417	2.1976286	0.3125094	-0.00098	0.0487687

Analysis Of Maximum Likelihood Parameter Estimates							
Parameter	DF	Estimate	Standard Error	Wald 95% Confidence Limits		Wald Chi-Square	Pr > ChiSq
Intercept	1	2.1976	0.3213	1.5679	2.8274	46.78	<.0001
z	1	0.3125	0.0868	0.1424	0.4826	12.97	0.0003
x	1	-0.0010	0.0004	-0.0017	-0.0002	6.62	0.0101
Dispersion	1	0.0488	0.0275	0.0161	0.1473		

Note: The negative binomial dispersion parameter was estimated by maximum likelihood.

Fig. 7.46. SAS Example G9: parameter estimates

are displayed in the least two columns of the table. The confidence intervals reflect the extra variation associated with the negative binomial model.

The `means` procedure is used to compute the sample mean and sample variance for the three counts observed for each of the six quinoline levels. The `by x` statement causes the mean and variance of the counts to be computed for each level of the quinoline variable (x). Note that the `means` procedure is preceded by a `sort` procedure that sorts the lines in the input data file (`set1`) according to the value of the variable x that is used in the `by` statement of the `means` procedure. The presorting is needed to insure that the observed counts are properly grouped by the values of the x variable when the `means` procedure operates on the subsets of data defined by the levels of the x variables. The `noprint` option is specified in the `proc means` statement to suppress the display of the results in the program output. The `out=setv` command in the `output` statement creates a temporary SAS data set with the name `setv` that contains one line for each value of the quinoline variable (x), with the sample mean and sample variance along with the value of x. Next the `proc sgplot` step creates the plot displayed in Fig. 7.49. The `scatter` statement uses `x=mean` to indicate that the values of the `mean` variable are plotted on the horizontal axis and `y=var` to indicate that the values of the

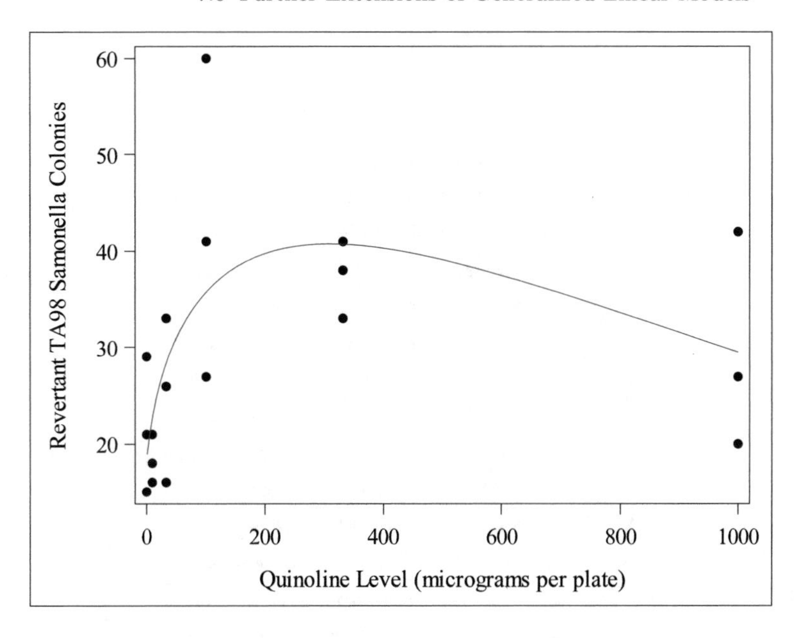

Fig. 7.47. SAS Example G9: plot of the estimated mean response curve and the observed counts

`var` variable are on the vertical axis. The options after the backslash in the `scatter` statement specify the plotting symbols as black-filled circles. The first `reg` statement draws a straight line on the plot, and the second `reg` statement draws the best-fitting quadratic curve on the plot. There are not enough levels of quinoline in this study to provide a clear choice between the quasi-Poisson and negative binomial models. The pattern in the plot does not exhibit an obvious curved trend, indicating that the quasi-Poisson model may be adequate. For these data, both the quasi-Poisson and the negative binomial models provide essentially the same results.

7.5 Further Extensions of Generalized Linear Models

7.5.1 Introduction

In section we explore some extensions of models examined in Sects. 7.3 and 7.4. Poisson and negative binomial regression models are modified to analyze rates of occurrence of events instead of mean numbers of events when event counts are obtained from inspection intervals of different lengths or from examination of areas of different sizes. This is accomplished by including an offset variable in the `proc genmod` step to adjust for different levels of exposure. Logistic regression models are extended to accommodate response variables

Obs	x	z	i	y	mean	cl_lower	cl_upper
1	0	2.30259	1	15	18.4896	14.0812	24.2781
2	0	2.30259	2	21	18.4896	14.0812	24.2781
3	0	2.30259	3	29	18.4896	14.0812	24.2781
4	10	2.99573	1	16	22.7376	18.7220	27.6146
5	10	2.99573	2	18	22.7376	18.7220	27.6146
6	10	2.99573	3	21	22.7376	18.7220	27.6146
7	33	3.76120	1	16	28.2386	24.0774	33.1189
8	33	3.76120	2	26	28.2386	24.0774	33.1189
9	33	3.76120	3	33	28.2386	24.0774	33.1189
10	100	4.70048	1	27	35.4649	29.1934	43.0837
11	100	4.70048	2	41	35.4649	29.1934	43.0837
12	100	4.70048	3	60	35.4649	29.1934	43.0837
13	333	5.83773	1	33	40.2671	32.3475	50.1257
14	333	5.83773	2	38	40.2671	32.3475	50.1257
15	333	5.83773	3	41	40.2671	32.3475	50.1257
16	1000	6.91771	1	20	29.3467	21.3228	40.3899
17	1000	6.91771	2	27	29.3467	21.3228	40.3899
18	1000	6.91771	3	42	29.3467	21.3228	40.3899

Fig. 7.48. SAS Example G9: plot of the estimated mean response curve and the observed counts

with more than two response categories. The LOGISTIC procedure offers several link functions and other options for accommodating multi-category response variables.

7.5.2 Poisson Regression with Rates

In analyzing count data, it is sometimes necessary to adjust for varying levels of exposure, such as differences in lengths of exposure or inspection times or differences in sizes of inspection areas. In a comparative study of occurrence of side effects from several medical treatments in which patients are not all treated for the same amount of time, it may be more meaningful to analyze rates, the number of adverse events per unit time, than the numbers of adverse events. Similarly, in ecological studies of species abundance, it is generally more meaningful to analyze species counts per unit area when counts are obtained from inspecting areas of different sizes at different locations. If a quantitative measure of exposure is available, such as the lengths of inspection intervals, then rates can be analyzed by including the values of that measure of exposure as an `offset` variable in a Poisson or negative binomial regression analysis.

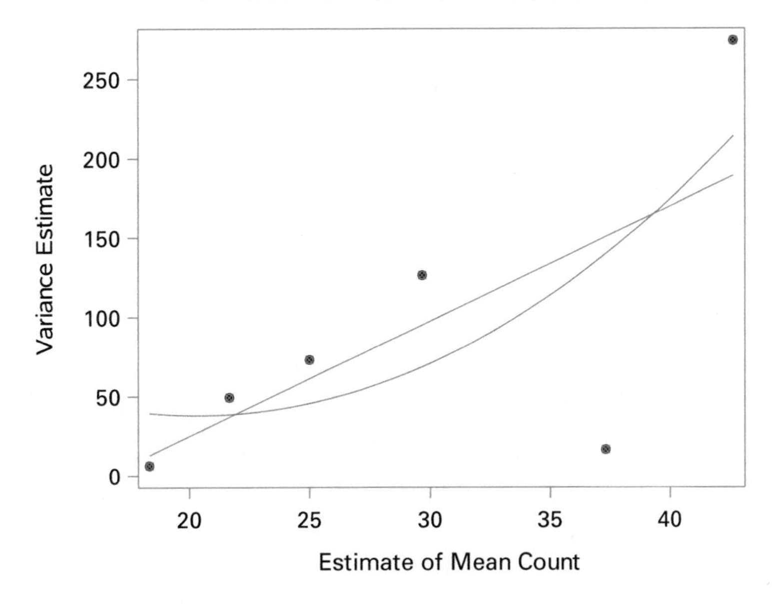

Fig. 7.49. SAS Example G9: estimated variances plotted against estimated means

Model

Suppose the response variable has either a Poisson or negative binomial distributions with mean count μ that changes with the values of some set of explanatory variables, $x_1, x_2, \ldots, x_k$. Using z to represent the corresponding measure of exposure, the expected rate, expected number of events per unit of exposure, is μ/z. Using a log link function, the expected rate is related to the values of the explanatory variables as

$$\log\left(\frac{\mu}{z}\right) = \beta_0 + \beta_1 x_1 + \beta_2 x_2 + \cdots + \beta_k x_k. \tag{7.45}$$

Note that this model can be reexpressed as

$$\log(\mu) = \log(z) + \beta_0 + \beta_1 x_1 + \beta_2 x_2 + \cdots + \beta_k x_k. \tag{7.46}$$

This model is fit with the `proc genmod` code used to fit Poisson or negative binomial regression models with an `offset=` option added to the `model` statement. As indicated on the right side of Eq. (7.46), the `offset=` option must identify the variable in the data file that contains values of the natural logarithm of the exposure variable.

For a quantitative explanatory variable x_j, the coefficient β_j represents the change in the expected incident rate when the x_j variable is increased by one unit while the values of the other explanatory variables are held constant. Then $\exp(\beta_j)$ represents the ratio of expected incidence rates at $x_j + 1$ versus x_j when the values of the other explanatory variables are held constant.

Effects of categorical explanatory variables are interpreted in the application considered below.

Estimation and Hypothesis Testing

The GENMOD procedure uses maximum likelihood estimation to obtain estimates of model parameters. Confidence intervals and test of hypotheses are computed as described in Sects. 7.3.3 and 7.4.3.

Using PROC GENMOD *to Analyze Rates*

Moore and Beckman (1988) analyze data from a study of failures for 90 valves from one pressurized nuclear reactor. The number of failures (y) was recorded for each valve along with the operating time (z) as a measure of exposure. Operating times were recorded in 100 hour units. Five explanatory variables which may be associated with differences in failure rates were also recorded:

System:	1 = containment, 2 = nuclear, 3 = power conversion,
	4 = safety, 5 = process auxiliary.
Operator type:	1 = air, 2 = solenoid, 3 = motor driven, 4 = manual.
Valve type:	1 = ball, 2 = butterfly, 3 = diaphram, 4 = gate,
	5 = globe, 6 = directional control.
Head size:	1 =$\leq$ 2 inches, 2 = 2 − 10 inches, 3 = 10 − 30inches.
Operation mode:	0 = normally open, 1 = normally closed.

SAS Example G10

The SAS Example G10 program (see Fig. 7.50) illustrates how `proc genmod` can be used to perform the necessary computations. The data are read from an external file identified by the `infile` statement. There is one line of data for each valve. The information for each valve includes values of the system, operation type (otype), valve type (vtype), head size (hsize), and operation mode (mode) variables in addition to the number of valve failures (failures) and the operation time (time) in hundreds of hours. These data are shown in Table B.16.

To indicate that the analysis will pertain to rates (numbers of failures per 100 hours of operation time) rather than counts, the `offset=logtime` option is included in the `model` statement options. The variable `logtime`, the natural logarithm of the operation time, is created in the data step with the `logtime=log(time)` statement. Note that the natural logarithm of the exposure variable must be used in the `offset` option.

The form of the model for the expected failure rate, mean number of valve failures per 100 hours of operation, for valves of the j-th valve type, k-th operator type, and ℓ-th head size operating in a specific mode in the i-th system is

$$\log\left(\frac{\mu_{ijk\ell m}}{time}\right) = \beta_0 + \gamma_i + \delta_j + \tau_k + \lambda_\ell + \beta_1(mode), \qquad (7.47)$$

```
data set1;
input system otype vtype hsize mode failures time @@;
logtime = log(time);
infile "C:\Users\user_name\Documents\...\valve.txt";
run;

proc genmod data=set1;
  class system otype vtype hsize ;
  model failures = system otype vtype hsize mode /
        dist=poisson link=log offset=logtime itprint;
  estimate 'System: safety vs. process aux.' system 0 0 0 1 -1 ;
  estimate 'System: power conv. vs. containment' system -1 0 1 0 0 ;
  estimate 'Mode: closed vs. open' mode 1 ;
  estimate 'Specific case' intercept 1 system 1 0 0 0 0 otype 0 0 1 0
                           vtype 0 0 0 1 0 0 hsize 0 0 1 mode 1 / exp;
  output out=setp p=mean;
 run;

proc print data=setp; run;
```

Fig. 7.50. SAS Example G10: program for `proc genmod` to analyze rates

where $\mu_{ijk\ell m}$ is the mean number of failures for the corresponding Poisson distribution, with the value of the time variable providing the length of the period of operation. Because the system otype, vtype, and hsize variables are included in a `class` statement, they are modeled as categorical explanatory variables using a separate parameter for each level of each of these variables. By default, the set of parameters for each of these variables is constrained by setting the parameter for the highest level of each variable equal to zero, making those levels the baseline levels for the corresponding variables. In this case, the constraints are $\gamma_5 = \delta_4 = \tau_6 = \lambda_3 = 0$, and zero values are reported in corresponding positions in the table of estimated coefficients shown in Fig. 7.51. Because the mode variable is not included in the `class` statement, it is treated as a quantitative explanatory variable and multiplied by a coefficient parameter (β_1). The `model` statement specifies that the natural logarithms of the true failure rates are related to the additive effects of the five explanatory variables. The log link function is specified with the `link=log` option in the `model` statement, and the `dist=poisson` option is used to specify Poisson distributions for the failure counts.

Maximum likelihood estimates of the model parameters produced by `proc genmod` are displayed in Fig. 7.51. For the system variable with levels coded 1, 2, 3, 4, and 5, the model parameters and associated systems are containment (γ_1), nuclear (γ_2), power conversion (γ_3), safety (γ_4), and process auxiliary (γ_5), respectively. The process auxiliary system is the baseline category, and expected failure rates of the other four systems are compared to it. For example, $\hat{\gamma}_4 = 0.8902$ is the estimate of the parameter for the safety system. This indicates that the estimate of the logarithm of the failure rate for the safety system is 0.8902 units larger than the estimate of the logarithm of the failure rate for the process auxiliary system. This is an estimate of the logarithm of the ratio of expected failure rates for the two systems, the safety

The GENMOD Procedure

Analysis Of Maximum Likelihood Parameter Estimates								
Parameter		DF	Estimate	Standard Error	Wald 95% Confidence Limits		Wald Chi-Square	Pr > ChiSq
Intercept		1	-9.5762	0.9311	-11.4010	-7.7513	105.79	<.0001
system	1	1	-0.3329	0.5841	-1.4777	0.8119	0.32	0.5687
system	2	1	0.5826	0.4145	-0.2298	1.3951	1.98	0.1599
system	3	1	0.6859	0.3167	0.0652	1.3066	4.69	0.0303
system	4	1	0.8902	0.4496	0.0090	1.7713	3.92	0.0477
system	5	0	0.0000	0.0000	0.0000	0.0000	.	.
otype	1	1	2.4723	0.4766	1.5382	3.4064	26.91	<.0001
otype	2	1	3.1767	0.7325	1.7411	4.6123	18.81	<.0001
otype	3	1	1.2797	0.5009	0.2979	2.2615	6.53	0.0106
otype	4	0	0.0000	0.0000	0.0000	0.0000	.	.
vtype	1	1	-1.0089	0.9301	-2.8319	0.8140	1.18	0.2780
vtype	2	1	-0.8236	0.8788	-2.5461	0.8989	0.88	0.3487
vtype	3	1	-0.4022	0.8728	-2.1128	1.3085	0.21	0.6450
vtype	4	1	1.9500	0.7365	0.5065	3.3936	7.01	0.0081
vtype	5	1	0.7843	0.7479	-0.6817	2.2502	1.10	0.2944
vtype	6	0	0.0000	0.0000	0.0000	0.0000	.	.
hsize	1	1	-1.6146	0.3210	-2.2438	-0.9853	25.29	<.0001
hsize	2	1	-1.6268	0.2341	-2.0857	-1.1679	48.27	<.0001
hsize	3	0	0.0000	0.0000	0.0000	0.0000	.	.
mode		1	0.2093	0.1903	-0.1637	0.5824	1.21	0.2714
Scale		0	1.0000	0.0000	1.0000	1.0000		

Note: The scale parameter was held fixed.

Fig. 7.51. SAS Example G10: parameter estimates

system relative to the process auxiliary system. The values of the chi-square test (3.92) and corresponding p-value (0.0477) indicate that the failure rates for these two systems are significantly different at the 0.05 level of significance. The ratio of the expected failure rates for these two systems is estimated as $\exp(\hat{\gamma}_4) = \exp(0.8902) = 2.4356$. This estimate is shown in the first row of the `Mean Estimate` column in Fig. 7.52. The 95% confidence interval (1.0091, 5.8787) shown in the next two columns of the first row indicates that the expected failure rate of valves in the safety system is likely to be between 1.0091 and 5.8787 times higher than the expected failure rate of valves in the process auxiliary system. This output was produced by the first `estimate` statement in the `proc genmod` step. The labels given in quotes in the `estimate` statement are used to label the rows in Fig. 7.52. The comparison between the safety and process auxiliary systems is specified by inserting the name of the system variable to the right of the label and then inserting a value for each level of the system variable. To obtain the difference between the logarithms of the expected valve failure rates for the two systems, a one is inserted in the fourth position corresponding to the safety system, and a minus one is inserted in the fifth position corresponding to the process auxiliary system. Zeros are inserted in the first three positions because the first three systems are not included in this comparison. This produces

$$(0)(\hat{\gamma_1}) + (0)(\hat{\gamma_2}) + (0)(\hat{\gamma_3}) + (1)(\hat{\gamma_4}) + (-1)(\hat{\gamma_5}) = \hat{\gamma_4} - \hat{\gamma_5}.$$

Note that this is $\hat{\gamma_4}$ because $\hat{\gamma_5}$ is zero under the model constraints.

Comparisons do not have to be made with the baseline category. The second `estimate` statement in the `proc genmod` step compares the estimates of the logarithms of expected failure rates for the power conversion ($i = 3$) and containment ($i = 1$) systems. Using the estimates reported in Fig. 7.51, the estimated difference is

$$\hat{\gamma_3} - \hat{\gamma_1} = 0.6859 - (-0.3329) = 1.0188.$$

The GENMOD Procedure

Contrast Estimate Results										
		Mean					L'Beta			
Label	Mean Estimate	Confidence Limits		L'Beta Estimate	Standard Error	Alpha	Confidence Limits		Chi-Square	Pr > ChiSq
System: safety vs. process aux.	2.4356	1.0091	5.8787	0.8902	0.4496	0.05	0.0090	1.7713	3.92	0.0477
System: power conv. vs. containment	2.7699	1.0285	7.4598	1.0188	0.5055	0.05	0.0281	2.0095	4.06	0.0439
Mode: closed vs. open	1.2329	0.8490	1.7903	0.2093	0.1903	0.05	-0.1637	0.5824	1.21	0.2714
Specific case	0.0015	0.0006	0.0038	-6.4700	0.4552	0.05	-7.3622	-5.5778	202.02	<.0001
Exp(Specific case)				0.0015	0.0007	0.05	0.0006	0.0038		

Fig. 7.52. SAS Example G10: estimates of linear combinations of parameters

This estimate is shown in the second row of the `L'Beta Estimate` column in Fig. 7.52. The standard error is 0.5055, and a 95% confidence interval extends from 0.0281 to 2.0095. Because the confidence interval is shifted to the right of zero, it indicates that the expected failure rates are significantly different for the two systems at the 0.05 level of significance. The value of the chi-square test statistic is 4.06 for test of the null hypothesis that the expected failure rates of valves are the same for the two systems. The corresponding p-value is 0.0439, indicating a significant difference at the 0.05 level of significance. The value under the `Mean Estimate` column in this row, $2.7699 = \exp(1.0188)$, is an estimate of the ratio of the expected failure rates for the two systems, power conversion over containment. The corresponding 95% confidence interval indicates that the ratio of expected failure rates is likely to be between 1.0285 and 7.4598.

The coefficient, β_1, for the `mode` variable represents the difference between the logarithms of the expected failure rates for valves in the closed and open operation modes for any particular combination of the levels of the system, valve type, operator type, and head size variables. From the `mode` row in Fig. 7.51, the estimate is $\hat{\beta}_1 = 0.2093$ with a standard error of 0.1930. The 95% confidence interval extends from -0.1637 to 0.5824. The value of a chi-square test of the null hypothesis that $\beta_1 = 0$ is 1.21 with a p-value of 0.2714. This analysis does not establish a significant difference between the expected failure rates for the two operation modes. The parameter is also estimated with the third `estimate` statement in the `proc genmod` step, and the same results are included in last seven columns of the third row in Fig. 7.52. The `estimate` statement provides additional information in the first three columns. An estimate of $\exp(\beta_1)$, the ratio of expected failure rates for closed operation relative to open operation, is reported in the `Mean Estimate` column as 1.2329. This is followed by a 95% confidence interval (0.8490, 1.7903).

An example of estimation of the failure rate for a specific combination of levels of the explanatory variables is provided by the fourth `estimate` statement in the `proc genmod` step. In this example the estimate of the logarithm of the failure rate is -6.47 for the containment system (level 1), operation type (level 3), and valve type (level 4) with head size between 10 and 30 inches (level 3) running in closed mode. The estimate of the failure rate, shown in the `Mean Estimate` column, is $0.0015 = \exp(-6.47)$ valve failures per 100 hours of operation, with a 95% confidence interval extending from 0.0006 to 0.0038 valve failures per 100 hours of operation. To obtain a standard error for the estimate of the value failure rate under these specific conditions, the `exp` option is included in the `estimate` statement. This produces the last row of values in Fig. 7.52. The standard error is 0.0007 valve failures per 100 hours of operation. The rest of the information on this line also appears on the previous line.

The `output` statement in the `proc genmod` step creates an output file named `outp` and writes the values of the explanatory variables, the exposure variable (time), and the response variable (failures) onto the file. The `p=mean` option adds the estimate of the mean number of failures during the entire exposure (operation time) for each case to this output file under the user-specified variable name `mean`; ninety-fine percent confidence limits are added in the columns labeled `lower` and `upper`, respectively. There is one line in this output file for each of the 90 cases in the data set. The first five and last three lines are displayed in Fig. 7.53.

Note that the estimate of the expected failure rate is obtained by dividing the estimate of the mean count by the exposure time. The first line in Fig. 7.53 corresponds to the case in which valves in the containment system (level 1), operation type (level 3), and valve type (level 4) with head size between 10 and 30 inches (level 3) are running in closed mode. The `mean` column in Fig. 7.53 displays an estimate of 2.7143 for the expected number of valve failures during $1752 \times 100 = 175{,}200$ hours of operation, and a 95% confidence interval is

	system	otype	vtype	hsize	mode	failures	time	logtime	mean	lower	upper
1	1	3	4	3	1	2	1752	7.46851	2.7143	1.1122	6.6240
2	1	3	4	3	0	2	1752	7.46851	2.2016	0.8945	5.4188
3	1	3	5	1	1	1	876	6.77537	0.0842	0.0283	0.2507
4	2	1	2	2	0	0	876	6.77537	0.1112	0.0309	0.4004
5	2	1	3	2	1	0	876	6.77537	0.2090	0.0636	0.6871
	.	.	.	.	.	.	.	.	.	.	.
	.	.	.	.	.	.	.	.	.	.	.
	.	.	.	.	.	.	.	.	.	.	.
88	5	4	4	1	0	0	438	6.08222	0.0425	0.0132	0.1363
89	5	4	4	2	1	0	438	6.08222	0.0517	0.0170	0.1574
90	5	4	5	2	0	0	438	6.08222	0.0131	0.0042	0.0411

Fig. 7.53. SAS Example G10: estimates of mean counts

(1.1122, 6.6240). The estimated failure rate is $2.7143/1752 = 0.001549$ failures per 100 hours of operation. A 95% confidence interval for the expected failure rate is (1.1122/1752, 6.6240/1752) or (0.000635, 0.003781). This matches with the results obtained for this case from the fourth `estimate` statement in the `proc genmod` step.

Criteria For Assessing Goodness Of Fit			
Criterion	DF	Value	Value/DF
Deviance	74	195.6781	2.6443
Scaled Deviance	74	195.6781	2.6443
Pearson Chi-Square	74	334.8645	4.5252
Scaled Pearson X2	74	334.8645	4.5252
Log Likelihood		33.1933	
Full Log Likelihood		-150.0116	
AIC (smaller is better)		332.0232	
AICC (smaller is better)		339.4753	
BIC (smaller is better)		372.0202	

Fig. 7.54. SAS Example G10: goodness-of-fit statistics from `proc genmod`

Estimates of parameters and standard errors and corresponding confidence intervals may have little value if the model specified in the `proc genmod` statement is incorrect. Goodness-of-fit information shown in Fig. 7.54 indicates that there is more variation in the observed failure rates about the estimate of the proposed model than the a Poisson distribution can accommodate. Note that the value of the Pearson chi-square statistic is larger than its degrees of freedom by a factor of 4.52. Either the logarithms of the expected failure rates are not well described by a model with additive effects of the five explanatory variables or the Poisson distribution is incorrect. To investigate the latter possibility, the same model is fit to the data with the Poisson distribution replaced by the negative binomial distribution. The `proc genmod` step is shown in Fig. 7.55. Only the `dist=nb` option in the `model` statement differs from the program code in Fig. 7.50. The data are entered in the same way.

```
proc genmod data=set1;
  class system otype vtype hsize ;
  model failures = system otype vtype hsize mode /
          dist=nb link=log offset=logtime itprint;
  estimate 'System: safety vs. process aux.' system 0 0 0 1 -1 ;
  estimate 'System: power conv. vs. containment' system -1 0 1 0 0 ;
  estimate 'Mode: closed vs. open' mode 1 ;
  estimate 'Specific case' intercept 1 system 1 0 0 0 0 otype 0 0 1 0
                           vtype 0 0 0 1 0 0 hsize 0 0 1 mode 1 / exp;
  output out=setp p=mean;
 run;

proc print data=setp; run;
```

Fig. 7.55. SAS Example G10: program for `proc genmod`: negative binomial model fit

Goodness-of-fit information for this model, displayed in Fig. 7.56, indicates that the negative binomial distribution is able to account for the variability in the observed failure rates relative to the fitted model. As shown in Fig. 7.57, standard errors of parameter estimates are larger than when the Poisson dis-

The GENMOD Procedure

Criteria For Assessing Goodness Of Fit			
Criterion	DF	Value	Value/DF
Deviance	74	70.5502	0.9534
Scaled Deviance	74	70.5502	0.9534
Pearson Chi-Square	74	82.2224	1.1111
Scaled Pearson X2	74	82.2224	1.1111
Log Likelihood		65.3058	
Full Log Likelihood		-117.8991	
AIC (smaller is better)		269.7983	
AICC (smaller is better)		278.2983	
BIC (smaller is better)		312.2951	

Fig. 7.56. SAS Example G10: goodness of fit for negative binomial model

tribution was used. These larger standard errors better reflect the level of variability in the observed failure rates. Note that some of the parameter estimates in Fig. 7.57 are quite different from the corresponding estimates in Fig. 7.51 when the Poisson distribution was imposed. The Poisson model restriction that the mean is equal to the variance of each count affects the estimates of the parameters in the systematic part of the model. The relationship between the means and variances of the counts is less of an issue with the negative binomial model because it has an extra scale parameter that allows the variances to be inflated relative to the means. For the negative binomial model, the variances are $\text{Var}(Y) = \mu + \theta\mu^2$. The maximum likelihood estimate of the negative binomial dispersion parameter is $\hat{\theta} = 1.3131$, as shown in the last line of table in Fig. 7.57.

Output from the `estimate` statements in the program code for the negative binomial model is displayed in Fig. 7.58. These estimates are interpreted in the same way as for the Poisson model. Note that the standard errors of the estimates are larger and the confidence intervals are wider than for the Poisson model. By accounting for the extra-Poisson variation in the valve failure counts, the negative binomial model provides standard errors and confidence intervals that more accurately reflect the variation in the processes that generated the data.

As described above for the Poisson model, estimates of mean counts for valve failures are written to a user-specified file `outp` by the `output` statement in the `proc genmod` step. The `p=mean` option adds the estimate of the mean number of failures during the entire exposure period (operation time) for each case to this output file under the user-specified variable name `mean`; ninety-five percent confidence limits are added in the columns labeled `lower` and `upper`, respectively. Results displayed in Fig. 7.59 reveal much wider confidence intervals than those reported for the Poisson model in Fig. 7.53, and the estimates of the mean counts differ as well. The first line in Fig. 7.59 shows an estimate of 5.4510 for the mean number of failures in 175,200 hours of operation for the negative binomial model, while the estimated mean number of failures is 2.7143 under the Poisson model. The confidence interval under the negative binomial model is (0.7982, 27.2238), much wider than the 95% confidence interval under the Poisson model. Remember that the Poisson model does not adequately account for the variation in the counts and produces confidence intervals that are too short to actually provide 95% confidence. The estimated failure rate is $5.4510/1752 = 0.00311$ failures per 100 hours of operation, and a 95% confidence interval for the expected failure rate is (0.79823/1752, 37.2238/1752) or (0.000456, 0.02125). This matches with the results for this case obtained from the fourth `estimate` statement in the `proc genmod` step and displayed in the last two rows of Fig. 7.58.

The GENMOD Procedure

Analysis Of Maximum Likelihood Parameter Estimates

Parameter		DF	Estimate	Standard Error	Wald 95% Confidence Limits		Wald Chi-Square	Pr > ChiSq
Intercept		1	-8.1760	1.3706	-10.8624	-5.4897	35.58	<.0001
system	1	1	0.0207	1.0925	-2.1205	2.1619	0.00	0.9849
system	2	1	0.4450	0.6322	-0.7941	1.6842	0.50	0.4815
system	3	1	0.4323	0.5472	-0.6401	1.5047	0.62	0.4295
system	4	1	1.4278	0.9950	-0.5225	3.3780	2.06	0.1513
system	5	0	0.0000	0.0000	0.0000	0.0000	.	.
otype	1	1	1.1931	0.7601	-0.2966	2.6828	2.46	0.1165
otype	2	1	1.8574	1.1258	-0.3492	4.0640	2.72	0.0990
otype	3	1	-0.1772	0.8036	-1.7522	1.3979	0.05	0.8255
otype	4	0	0.0000	0.0000	0.0000	0.0000	.	.
vtype	1	1	-0.0770	1.4076	-2.8359	2.6820	0.00	0.9564
vtype	2	1	-1.1566	1.2562	-3.6187	1.3056	0.85	0.3572
vtype	3	1	-0.3407	1.1727	-2.6391	1.9576	0.08	0.7714
vtype	4	1	1.6891	0.9773	-0.2264	3.6045	2.99	0.0839
vtype	5	1	1.2005	1.0357	-0.8294	3.2304	1.34	0.2464
vtype	6	0	0.0000	0.0000	0.0000	0.0000	.	.
hsize	1	1	-1.6942	0.6909	-3.0483	-0.3401	6.01	0.0142
hsize	2	1	-1.7955	0.5069	-2.7890	-0.8019	12.55	0.0004
hsize	3	0	0.0000	0.0000	0.0000	0.0000	.	.
mode		1	0.8707	0.4261	0.0357	1.7058	4.18	0.0410
Dispersion		1	1.3131	0.4171	0.7046	2.4472		

Note: The negative binomial dispersion parameter was estimated by maximum likelihood.

Fig. 7.57. SAS Example G10: parameter estimates

7.5.3 Logistic Regression with Multiple Response Categories

Logistic regression models may be used when the categorical response variable has more than two categories. If there are J response categories, a set of $J-1$ binary logistic regression models is required, but there are many ways to define a set $J-1$ of logistic regression models. Different sets of logistic regression models may lead to different estimates of the response probabilities.

For a particular set of values for the explanatory variables, $x_1, x_2, \ldots, x_k$, let π_j denote the true probability of observing a response in the j-th category for $j = 1, 2, \ldots, J$. The response categories must be defined so that any possible response must fall into one of the categories and it cannot fall into

The GENMOD Procedure

Contrast Estimate Results										
		Mean					L'Beta			
Label	Mean Estimate	Confidence Limits		L'Beta Estimate	Standard Error	Alpha	Confidence Limits		Chi-Square	Pr > ChiSq
System: safety vs. process aux.	4.1693	0.5931	29.3112	1.4278	0.9950	0.05	-0.5225	3.3780	2.06	0.1513
System: power conv. vs. containment	1.5093	0.2001	11.3807	0.4116	1.0308	0.05	-1.6087	2.4319	0.16	0.6897
Mode: closed vs. open	2.3886	1.0363	5.5057	0.8707	0.4261	0.05	0.0357	1.7058	4.18	0.0410
Specific case	0.0031	0.0005	0.0212	-5.7727	0.9802	0.05	-7.6939	-3.8516	34.68	<.0001
Exp(Specific case)				0.0031	0.0030	0.05	0.0005	0.0212		

Fig. 7.58. SAS Example G10: estimates of linear combinations of parameters

Obs	system	otype	vtype	hsize	mode	failures	time	logtime	mean	lower	upper
1	1	3	4	3	1	2	1752	7.46851	5.4510	0.79823	37.2238
2	1	3	4	3	0	2	1752	7.46851	2.2821	0.34002	15.3161
3	1	3	5	1	1	1	876	6.77537	0.3073	0.03767	2.5062
4	2	1	2	2	0	0	876	6.77537	0.0662	0.00831	0.5281
5	2	1	3	2	1	0	876	6.77537	0.3577	0.05639	2.2692
	.		.	.	.	.	.	.	.	.	.
	.		.	.	.	.	.	.	.	.	.
	.		.	.	.	.	.	.	.	.	.
88	5	4	4	1	0	0	438	6.08222	0.1226	0.02187	0.6872
89	5	4	4	2	1	0	438	6.08222	0.2646	0.04955	1.4130
90	5	4	5	2	0	0	438	6.08222	0.0680	0.00936	0.4937

Fig. 7.59. SAS Example G10: estimates of mean counts

more than one category. Then $\Sigma_{j=1}^{J} \pi_j = 1$. It is also assumed that $\pi_j > 0$ for every category. By default, `proc logistic` designates the response variable category with the highest alphanumeric value, or the highest numeric value when category labels are numbers, as the baseline category.

Model

There are a number of ways to construct a set of $J - 1$ logistic regression models when the response variable has J categories. One possibility is to use the log-odds of observing a response in the j-th category relative to the J-th category for each of the first $j = 1, 2, \ldots, J - 1$ categories. Then the J-th category becomes the baseline category to which each of the other categories is compared. This is called a *baseline category* model, and the $J - 1$ logistic regression models used to describe how the log-odds change with changes in the explanatory variables are

$$\begin{aligned}
\log\left(\frac{\pi_1}{\pi_J}\right) &= \beta_{01} + \beta_{11}x_1 + \beta_{21}x_2 + \cdots + \beta_{k1}x_k \\
\log\left(\frac{\pi_2}{\pi_J}\right) &= \beta_{02} + \beta_{12}x_1 + \beta_{22}x_2 + \cdots + \beta_{k2}x_k \\
&\;\;\vdots \\
\log\left(\frac{\pi_{J-1}}{\pi_J}\right) &= \beta_{0,J-1} + \beta_{1,J-1}x_1 + \cdots + \beta_{k,J-1}x_k
\end{aligned} \tag{7.48}$$

For this set of models the probability that the response falls into the j-th category under conditions given by a particular set of values for $x_1, x_2, \ldots, x_k$ is

$$\pi_j = \frac{\exp(\beta_{0j} + \beta_{1j}x_1 + \cdots + \beta_{kj}x_k)}{1 + \Sigma_{\ell=1}^{J-1}\exp(\beta_{0\ell} + \beta_{1\ell}x_1 + \cdots + \beta_{k\ell}x_k)} \tag{7.49}$$

for $j = 1, 2, \ldots, J-1$, and

$$\pi_J = \frac{1}{1 + \Sigma_{\ell=1}^{J-1}\exp(\beta_{0\ell} + \beta_{1\ell}x_1 + \cdots + \beta_{k\ell}x_k)}. \tag{7.50}$$

This model is invoked in the `proc logistic` step by including the option `link=glogit` in the `model` statement. It can be used for either nominal or ordinal categorical response variables. Although the set of $J-1$ logistic regression models changes when a different category is designated as the baseline category for the response variable, the estimates of the response category probabilities are unchanged. In this sense, the choice of the baseline category for the response variable does not matter.

In some applications the user may want to restrict the model by using the same regression coefficient for a particular explanatory variable in all of the $J-1$ logistic regression models. This can be done by including that name of the explanatory variable in the `equalslopes` option in the `model` statement.

Although the $J-1$ logistic regression models define the log-odds of observing a response in each of the first $J-1$ categories relative to a baseline category, a regression model for the log-odds can be obtained for any pair of response categories. The log-odds of the j-th category relative to the ℓ-th category, for example, are

$$\begin{aligned}\log\left(\frac{\pi_j}{\pi_\ell}\right) &= \log\left(\frac{\pi_j}{\pi_J}\cdot\frac{\pi_J}{\pi_\ell}\right) = \log\left(\frac{\pi_j}{\pi_J}\right) - \log\left(\frac{\pi_\ell}{\pi_J}\right)\\ &= [\beta_{0j} + \beta_{1j}x_1 + \cdots + \beta_{kj}x_k]\\ &\quad - [\beta_{0\ell} + \beta_{1\ell}x_1 + \cdots + \beta_{k\ell}x_\ell]\\ &= (\beta_{0j} - \beta_{0\ell}) + (\beta_{1j} - \beta_{1\ell})x_1 + \cdots + (\beta_{kj} - \beta_{k\ell})x_k.\end{aligned} \tag{7.51}$$

An equivalent model is obtained from the set of $J-1$ logistic regression models corresponding to successive pairs of adjacent response categories. This set of $J-1$ logistic regression models is

$$\begin{aligned}\log\left(\frac{\pi_1}{\pi_2}\right) &= \gamma_{01} + \gamma_{11}x_1 + \cdots + \gamma_{k1}x_k\\ \log\left(\frac{\pi_2}{\pi_3}\right) &= \gamma_{02} + \gamma_{12}x_1 + \cdots + \gamma_{k2}x_k\\ &\ \ \vdots\\ \log\left(\frac{\pi_{J-1}}{\pi_J}\right) &= \gamma_{0,J-1} + \gamma_{1,J-1}x_1 + \cdots + \gamma_{k,J-1}x_k\end{aligned} \tag{7.52}$$

The *adjacent categories* model is invoked with the `proc logistic` step by including the `link=alogit` option in the `model` statement. It can be used for either nominal or ordinal categorical response variables. This set of $J-1$ logistic equations produces the same estimates for the response category probabilities as the *baseline category* logistic regression model. With the exception of the intercepts, `proc logistic` uses the same set of regression parameters for all $J-1$ logistic regression models by default when the `link=alogit` option is specified. To remove this restriction and allow the sets of regression parameters to vary across the different logistic regression models, as shown in Eq. (7.52), the names of all of the explanatory variables must be included in the `unequalslopes` option in the `model` statement. The equal slopes constraint can be relaxed for a subset of the explanatory variables by including only the names of the specific subset of explanatory variables in the `unequalslopes` option.

A set of $J-1$ logistic regression models based on cumulative logits may be used for an ordinal categorical response variable with J categories. The set of logistic regression models is

$$
\begin{aligned}
\log\left(\frac{\pi_1}{\pi_2+\pi_3+\cdots+\pi_J}\right) &= \beta_{01}+\beta_{11}x_1+\cdots+\beta_{k1}x_k \\
\log\left(\frac{\pi_1+\pi_2}{\pi_3+\cdots+\pi_J}\right) &= \beta_{02}+\beta_{12}x_1+\cdots+\beta_{k2}x_k \\
&\ \ \vdots \\
\log\left(\frac{\pi_1+\pi_2+\cdots+\pi_{J-1}}{\pi_J}\right) &= \beta_{0,J-1}+\beta_{1,J-1}x_1+\cdots+\beta_{k,J-1}x_k
\end{aligned}
\qquad (7.53)
$$

This model is invoked with the `proc logistic` step by including the `link=clogit` option in the `model` statement. It is the default model for the LOGISTIC procedure when the response variable has more than two categories. It is not equivalent to the *baseline category* and *adjacent categories* models as it produces different estimates of the response category probabilities. The log-odds for adjacent categories are not linear functions of the explanatory variables for this model. With exception of the intercepts, `proc logistic` uses the same set of regression parameters for all $J-1$ logistic regression models by default when the `link=clogit` option is specified. To remove this restriction and allow the sets of regression parameters to vary across the different logistic regression models, as shown in Eq. (7.53), the names of all of the explanatory variables must be included in the `unequalslopes` option in the `model` statement. The equal slopes constraint can be relaxed for a subset of the explanatory variables by including the names of the specific subset of explanatory variables in the `unequalslopes` option.

Estimation and Hypothesis Testing

The LOGISTIC procedure uses maximum likelihood estimation to estimate regression coefficients and response category probabilities. Approximate tests of hypotheses and confidence intervals are based on the large sample properties of maximum likelihood estimators. These results are more accurate for larger sample sizes.

Using PROC LOGISTIC *to Fit Logistic Regression Models with Multi-Category Response Variables*

The application of multi-category logistic regression is illustrated with the analysis of data from a study of the toxic effects of diethylene glycol dimethyl ether (DIEGdiMe) on pregnant mice reported by Price (1987) (also see Agresti, 2013, pp. 312–313). Early in its pregnancy, each female mouse was exposed to exactly one of five randomly assigned concentrations of DIEGdiMe for exactly 10 days. Subsequently, each fetus was classified as either non-live

($j = 1$), malformed ($j = 2$), or normal ($j = 3$). The percentages of fetuses in the three response categories are displayed in Table 7.2 for each of the five DIEGdiMe concentrations. The counts are included with the program code shown in Fig. 7.60.

Table 7.2. SAS Example G11: observed percentages for three response categories

Concentration	Response rates (percentages)			Number of
(mg/kg/day)	Non-live	Malformed	Normal	fetuses exposed
0	5.05	0.34	94.61	297
62.5	7.02	0.00	92.98	242
125	7.05	2.24	90.71	312
250	12.71	19.73	67.56	299
500	50.53	46.32	3.16	285

There are five lines of data shown in the data step in Fig. 7.60, one line for each concentration of DIEGdiME used in the study. On each line, the concentration of DIEGdiME is followed by the counts for the three possible outcomes, non-live, malformed, and normal, respectively. This data step also creates a new variable, `conc2`, containing the squares of the concentrations. The second data step in Fig. 7.60 puts the counts for the three response categories on three different lines. The counts are now stored in the variable y, and information on the corresponding outcome categories is stored in the `outcome` variable. The `outcome` variable uses 1 to designate a `non-live` outcome, 2 to designate a `malformed` outcome, and 3 to designate a `normal` outcome. The `keep` statement in this data step retains only the values of the concentration (`conc`), square of the concentration (`conc2`), outcome category (`outcome`), and corresponding count (`y`) variables.

The third data step creates a second data file, called `set2`, that contains 101 concentration values starting at 0 and running up to 500 in increments of 5. It stores the concentrations in the `conc` variable, and it stores the squares of those concentrations in the `conc2` variable. It creates three lines in the data file for each concentration, one for each of the three outcome categories, and the category labels are stored in the `outcome` variable. The corresponding counts, stored in the y variable, are all set equal to zero. These data lines are used to obtain maximum likelihood estimates of the three outcome probabilities for the 101 concentrations, for the purpose of plotting probability curves to show how those probabilities change as the concentration of DIEGdiME increases. The names of the variables in the `set2` data set match the corresponding variable names in the `set1` data set. This enables the two data sets to be combined into a single data set, called `set1` in the fourth data step. Because the values of y are all zero in `set2`, the cases in the combined data set that come from `set2` are not used to estimate the coefficients in the multi-category logistic regression model, but `proc logistic` does compute maximum likelihood estimates of the three response category probabilities for those cases.

```
data set1;
  input conc y1 y2 y3;
   conc2 = conc*conc;
 datalines;
   0   15 1 281
62.5   17 0 225
125    22 7 283
250    38 59 202
500   144 132 9
run;

data set1; set set1;
  outcome=1; y=y1; output;
  outcome=2; y=y2; output;
  outcome=3; y=y3; output;
        keep conc conc2 outcome y;
run;

data set2;
  do conc = 0 to 500 by 5;
  do outcome = 1 to 3;
  conc2=conc*conc;
  y=0;
  output; end; end;
  run;

data set1; set set1 set2; run;

proc logistic data=set1;
  model outcome(ref="3") = conc conc2 / itprint covb maxiter=50 link=glogit;
  weight y;
  output out=setp p=phat;
  run;
```

Fig. 7.60. SAS Example G11: program for a baseline log-odds model

Because both conc and conc2 appear on the right side of the equal side in the model statement shown in Fig. 7.60, t proc logistic fits a pair of logistic regression models in which log-odds are related to quadratic functions of the concentration of DIEGdiME. The baseline logistic regression model is constructed with the third category (normal $= 3$) designated as the baseline category by including the (ref="3") option on the left side of the equal sign in the model statement. The weight statement indicates that the counts are provided by the y variable.

Maximum likelihood estimates of the regression coefficients are displayed in Fig. 7.61. One set of parameter estimates is reported for each of the logistic regression models that comprise the baseline model. The estimated models are

The LOGISTIC Procedure

Analysis of Maximum Likelihood Estimates						
Parameter	outcome	DF	Estimate	Standard Error	Wald Chi-Square	Pr > ChiSq
Intercept	1	1	-2.7824	0.2099	175.6874	<.0001
Intercept	2	1	-6.7154	0.7725	75.5763	<.0001
conc	1	1	-0.00168	0.00208	0.6513	0.4196
conc	2	1	0.0252	0.00512	24.1908	<.0001
conc2	1	1	0.000025	4.203E-6	36.6010	<.0001
conc2	2	1	-0.00001	7.732E-6	2.7826	0.0953

Fig. 7.61. SAS Example G11: estimates of model parameters

$$\begin{aligned}\log\left(\frac{\hat{\pi}_{\text{non-live}}}{\hat{\pi}_{\text{normal}}}\right) &= -2.7824 - 0.00168(\text{conc}) + 0.000025(\text{conc})^2 \\ \log\left(\frac{\hat{\pi}_{\text{malformed}}}{\hat{\pi}_{\text{normal}}}\right) &= -6.7156 + 0.0252(\text{conc}) - 0.00001(\text{conc})^2\end{aligned} \tag{7.54}$$

Including the squared concentration as an explanatory variable allows the estimated probability of malformed fetuses to initially increase as the DIEGdiME concentration increases, reach a peak, and then decline with further increases in the DIEGdiME concentration, while the probability of non-live fetuses continues to increase. The estimated probability of normal fetuses monotonically declines as the DIEGdiME concentration increases.

Corresponding formulas for the estimates of the probabilities of the three response categories are

$$\hat{\pi}_{\text{non-live}} = \frac{\exp(-2.7824 - (0.00168)\text{conc} + (0.000025)\text{conc}^2)}{\delta}$$

$$\hat{\pi}_{\text{malformed}} = \frac{\exp(-6.7154 + (0.0252)\text{conc} - (0.00001)\text{conc}^2)}{\delta}$$

and

$$\hat{\pi}_{\text{normal}} = \frac{1}{\delta}$$

where

$$\begin{aligned}\delta = 1 &+ \exp(-2.7824 - (0.00168)\text{conc} + (0.000025)\text{conc}^2) \\ &+ \exp(-6.7154 + (0.0252)\text{conc} - (0.00001)\text{conc}^2)\end{aligned}$$

The probability curves are displayed in Fig. 7.62. The probability of a normal fetus declines monotonically, and the probability of a non-live fetus increases monotonically as the concentration of DIEGdiME increases. The probability of a malformed fetus initially increases as the DIEGdiME concentration increases, but it reaches a peak and then declines with further increase in the DIEGdiME concentration, while the probability of a non-live fetus continues to increase.

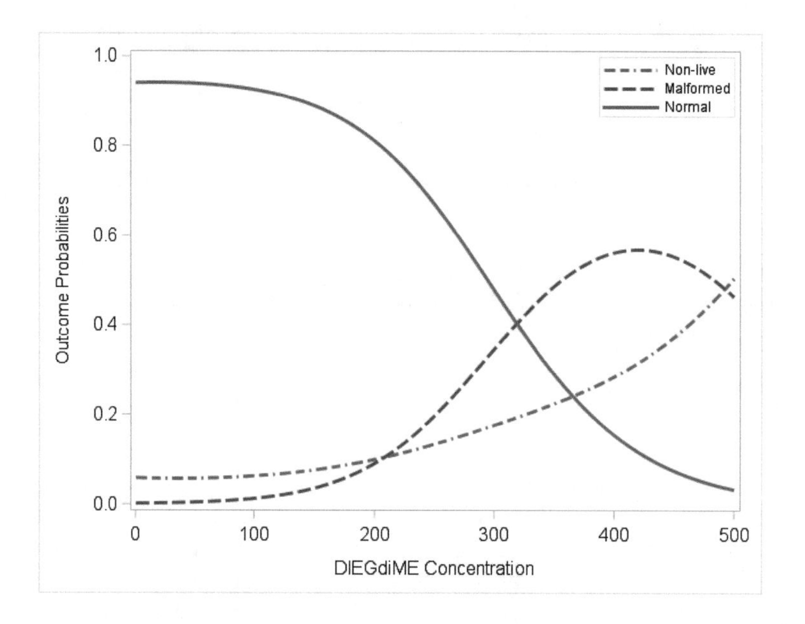

Fig. 7.62. SAS Example G11: estimated probability curves

Additional code for creating probability curves is displayed in Fig. 7.63. Maximum likelihood estimates of the outcome category probabilities are produced with the `p=` option in the `output` statement, as shown in the code for `proc logistic` in Fig. 7.60. The `out=setp` option creates an output file named `setp` that contains the estimated probabilities in a variable named `phat`, created by the `p=phat` option. The `setp` output file also contains a `SAS` generated variable `_LEVEL_` that uses the values of the `outcome` variable to define the response categories, in this case 1, 2, or 3. The first line of code in Fig. 7.63 sorts the file with respect to the values of the explanatory variable `conc` that defines the horizontal axis of the plot of probability curves. The next three data steps create separate files of estimated probabilities for each of the three outcome categories. The fourth data step combines the three data files side by side so that the estimated outcome category probabilities are in three different columns of the same file and matched with the respective concentration values. A `proc sgplot` step is used to draw smooth curves to show how the outcome category probability changes with changes in the DIEGdiME concentration for each of the three outcome categories.

```
proc sort data=setp; by conc;  run;

data setp1; set setp;
  if(_LEVEL_=1);  p1=phat; run;

data setp2; set setp;
  if(_LEVEL_=2); p2=phat; run;

data setp3; set setp;
  if(_LEVEL_=3); p3=phat; run;

data setpall; merge setp1 setp2 setp3;
    by conc; run;

proc sgplot data=setpall;
   yaxis label="Outcome Probabilities" valueattrs=(size=12)
                     labelattrs=(size=11);
   xaxis label="DIEGdiME Concentration" valueattrs=(size=12)
                     labelattrs=(size=12);
   pbspline x=conc y=p1 / lineattrs = (pattern=14 thickness=3)
          markerattrs=(size=0 symbol=none) legendlabel="Non-live";
   pbspline x=conc y=p2 / lineattrs = (pattern=4 thickness=3)
          markerattrs=(size=0 symbol=none) legendlabel="Malformed";
   pbspline x=conc y=p3 / lineattrs = (pattern=1 thickness=3)
          markerattrs=(size=0 symbol=none)  legendlabel="Normal";
   keylegend / location=inside position=topright  down=3;
  run;
```

Fig. 7.63. SAS Example G11: program for a baseline log-odds model (continued)

`SAS` code for fitting a cumulative logit model to the same data is displayed in two parts in Figs. 7.64 and 7.65. The data are entered in the same manner as shown in Fig. 7.60. The first logistic regression model in this set models changes in the log-odds of fetus mortality as DIEGdiME concentration changes. The odds are the probability of a non-live fetus divided by the sum of the probabilities for the other two possible outcomes, malformed or normal. The second logistic regression model in this set models changes in the conditional log-odds of fetus malformation versus a normal fetus given that the fetus is alive. This is the default set of links for `proc logistic`, and no `link=` option is needed in the `model` statement. This model uses only the DIEGdiME concentration as the single explanatory variable in this set of logistic regression models; only the `conc` variable appears on the right side of the equal sign in the `model` statement. The cumulative odds model does not need to include the square of the concentration as an explanatory variable in order to provide a good description of the data from this study. Also by default, `proc logistic` will force the regression coefficient for the `conc` variable to be the same for both logistic regression models, creating a special proportional odds model. To remove this restriction, the `unequalslopes` option is included in the `model` statement options.

```
data set1;
  input conc y1 y2 y3;
   conc2 = conc*conc;
 datalines;
   0   15 1 281
62.5   17 0 225
125    22 7 283
250    38 59 202
500   144 132 9
run;

/*  Now put the result for each
    response on a separate line */

data set1; set set1;
  outcome=1; y=y1; output;
  outcome=2; y=y2; output;
  outcome=3; y=y3; output;
        keep conc conc2 outcome y;
run;

/*  Add some more concentration levels to be used
    to obtain estimated proportions for plotting */

data set2;
  do conc = 0 to 500 by 5;
  do outcome = 1 to 3;
  conc2=conc*conc;
  y=0;
  output; end; end;
  run;

data set1; set set1 set2; run;
```

Fig. 7.64. SAS Example G11: program for a cumulative log-odds model

Maximum likelihood estimates of regression coefficients for the cumulative odds model are shown in Fig. 7.66 along with approximate standard errors. The estimated logistic regression models are

$$\log\left(\frac{\hat{\pi}_{\text{non-live}}}{\hat{\pi}_{\text{malformed}} + \hat{\pi}_{\text{normal}}}\right) = -3.5988 + 0.00720(\text{conc}) \qquad (7.55)$$

and conditional on a live fetus

$$\log\left(\frac{\hat{\pi}_{\text{malformed}}}{\hat{\pi}_{\text{normal}}}\right) = -3.5733 + 0.0122(\text{conc}) \qquad (7.56)$$

The estimated coefficient for the `conc` variable in Eq. (7.55) indicates that the log-odds for mortality increase by about 0.00720 for each unit increase in the DIEGdiME concentration. The estimated odds ratio is $\exp(0.00720) = 1.0072$, which indicates that the odds of fetus mortality increase by about 0.72% for each unit increase in the DIEGdiME concentration to which pregnant females are exposed. The value of 0.0122 for the estimated coefficient for the `conc` variables in Eq. (7.56) indicates that if the fetus does not die, then the log-odds of a malformed fetus increase by about 0.0122 for each unit increase in the DIEGdiME concentration. The estimated odds ratio is $\exp(0.0122) = 1.0123$,

```
proc logistic data=set1;
  model outcome = conc / itprint unequalslopes;
  output out=setp p=phat;
  weight y;
        conc: test conc_1 = conc_2;
  run;

proc sort data=setp; by conc;  run;

data setp1; set setp;
  if(_LEVEL_=1);  p1=phat; run;

data setp2; set setp;
  if(_LEVEL_=2); p2=phat; run;

data setpall; merge setp1 setp2; by conc;
  pnonlive = p1;
  pmalform = p2*(1-p1);
  pnormal = 1-pnonlive-pmalform;
run;

proc sgplot data=setpall;
   yaxis label="Outcome Probabilities" valueattrs=(size=12)
                  labelattrs=(size=11);
   xaxis label="DIEGdiME Concentration" valueattrs=(size=12)
                  labelattrs=(size=12);
   pbspline x=conc y=pnonlive / lineattrs = (pattern=14 thickness=3)
         markerattrs=(size=0) legendlabel="Non-live";
   pbspline x=conc y=pmalform / lineattrs = (pattern=4 thickness=3)
         markerattrs=(size=0) legendlabel="Malformed";
   pbspline x=conc y=pnormal / lineattrs = (pattern=1 thickness=3)
         markerattrs=(size=0)  legendlabel="Normal";
   keylegend / location=inside position=topright  down=3;
  run;
```

Fig. 7.65. SAS Example G11: program for a cumulative log-odds model (continued)

which indicates that the conditional odds of malformation increase by about 1.23% for each unit increase in the DIEGdiME concentration, among fetuses that do not die.

Equation (7.55) can be converted into a formula for estimating how the probability of a non-live fetus changes as the DIEGdiME concentration changes. The formula is

$$\hat{\pi}_{\text{non-live}} = \frac{\exp(-3.5988 + (0.00720)\text{conc})}{1 + \exp(-3.5988 + (0.00720)\text{conc})}. \tag{7.57}$$

Equation (7.56), however, provides formulas for conditional probabilities of malformed and normal fetuses among fetuses that survive. These equations are

$$\hat{\pi}_{\text{malformed}|\text{survival}} = \frac{\exp(-3.5733 + (0.0122)\text{conc})}{1 + \exp(-3.5733 + (0.0122)\text{conc})} \tag{7.58}$$

and

$$\hat{\pi}_{\text{normal}|\text{survival}} = \frac{1}{1 + \exp(-3.5733 + (0.0122)\text{conc})} \tag{7.59}$$

The LOGISTIC Procedure

Analysis of Maximum Likelihood Estimates						
Parameter	outcome	DF	Estimate	Standard Error	Wald Chi-Square	Pr > ChiSq
Intercept	1	1	-3.5988	0.1571	524.9261	<.0001
Intercept	2	1	-3.5733	0.1559	525.6605	<.0001
conc	1	1	0.00720	0.000423	288.8285	<.0001
conc	2	1	0.0122	0.000562	467.6829	<.0001

Odds Ratio Estimates				
Effect	outcome	Point Estimate	95% Wald Confidence Limits	
conc	1	1.007	1.006	1.008
conc	2	1.012	1.011	1.013

Linear Hypotheses Testing Results			
Label	Wald Chi-Square	DF	Pr > ChiSq
conc	216.0784	1	<.0001

Fig. 7.66. SAS Example G11: estimates of model parameters for the cumulative log-odds model

The probabilities in (7.57) and (7.58), $\hat{\pi}_{\text{non-live}}$ and $\hat{\pi}_{\text{malformed}|\text{survival}}$, are computed and written to an output file in rows corresponding to `_LEVEL_=1` and `_LEVEL_=2`, respectively, by the `p=` option in the `output` statement in the `proc logistic` step. Estimates of probabilities of malformed and normal fetuses are computed by multiplying the estimates of the conditional probabilities by the corresponding estimate of the probability of survival, i.e.,

$$\hat{\pi}_{\text{malformed}} = \hat{\pi}_{\text{malformed}|\text{survival}}(1 - \hat{\pi}_{\text{non-live}}) \tag{7.60}$$

and

$$\hat{\pi}_{\text{normal}} = (1 - \hat{\pi}_{\text{non-live}} - \hat{\pi}_{\text{malformed}}) \tag{7.61}$$

The calculations are performed in the step that creates the data set called `setpall` in Fig. 7.64.

The `proc sgplot` step in Fig. 7.64 produced the plot of smooth curves for these probabilities shown in Fig. 7.67. These probability curves are similar to those displayed in Fig. 7.62 for the baseline logit model. The probability of a normal fetus declines monotonically, and the probability of a non-live fetus increases monotonically as the concentration of DIEGdiME increases. The probability of a malformed fetus initially increases as the DIEGdiME concentration increases, but it reaches a peak and then declines with further increase in the DIEGdiME concentration, while the probability of a non-live

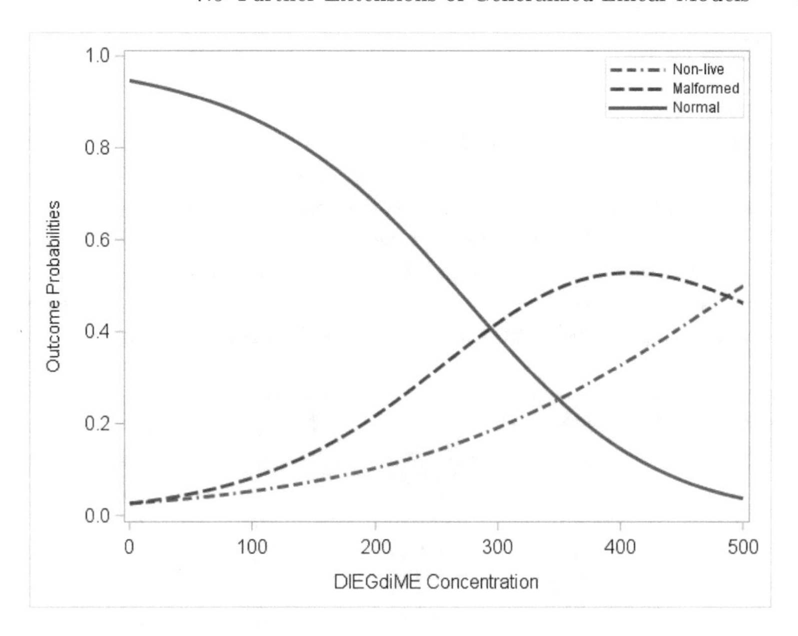

Fig. 7.67. SAS Example G11: estimated probability curves for a cumulative log-odds model

fetus continues to increase. Both models provide similar descriptions of how the outcome category probabilities change with changes in the DIEGdiME concentrations to which pregnant female mice are exposed.

The `conc: test conc_1 = conc_2;` statement in the code for the `proc logistic` step in Fig. 7.64 is used to test null hypotheses that the coefficients for the `conc` variable are the same for both logistic regression models that comprise the cumulative odds model. The name on the left side of the colon is the name of one of the explanatory variables in the model statement. The `conc_1` and `conc_2` notation refers to the coefficients on the `conc` variable in the first and second logistic regression models needed to specify the cumulative logit model for the three outcome categories. You will need $J-1$ labels when there are J outcome categories. The numbering must use the values used in the table of parameter estimates to distinguish the regression models. This will vary according to how the outcome categories are coded in the data set. The results of a chi-square test are shown in the third table in Fig. 7.66. In this case that value of the chi-square statistic is 216.0784, and the corresponding p-value is smaller than 0.001 which suggests that the proportional odds model with equal regression coefficients does not provide a good description of the data. The degrees of freedom for this test are one less than the number of outcome categories. When the model contains more than one explanatory variable, a similar test can be requested for each of the explanatory variables.

7.6 Exercises

7.1 Chester Bliss (1935) examined data from a study of the toxic effects of exposing a certain species of beetle to various levels of concentrations of gaseous carbon disulfide for a period of 5 hours. The first column in the following table shows the eight concentrations of carbon disulfide (mg/liter) used in the experiment, and the second column shows the natural logarithms of those concentrations. The fourth column shows the number of beetles exposed to each concentration, and the third column shows the number of the beetles that survived after 5 hours of exposure.

Concentration (mg/l)	Log concentration	Number of survivors	Number exposed
49.057	3.893	53	59
52.991	3.970	47	60
56.911	4.041	44	62
60.842	4.108	28	56
64.759	4.171	11	63
68.691	4.230	6	59
72.611	4.258	1	52
76.542	4.338	0	60

a. Using the concentration of carbon disulfide as the explanatory variable, write the *logistic regression model* you will use to analyze this data. Explain what each parameter in your model represents.
b. Using `proc logistic`, compute the maximum likelihood estimates of the model parameters. Construct a table with one column for the parameter estimates, one column for their standard errors, and two more columns for the lower and upper limits of 95% confidence intervals. Interpret the confidence intervals.
c. Express the null hypothesis that the concentration of carbon disulfide has no effect on the probability of survival in terms of a model parameter. Perform a test of this hypothesis using the analysis in part (b). Use a significance level of $\alpha = 0.05$.
d. Plot the estimated logistic curve. Looking at the plot, determine the concentration corresponding to a survival probability of 0.5.
e. Use a `data step` to add 141 lines with new concentration values to the data file. Start at 49 and go up to 77 by increments of 0.2. Designate the counts for the number of survivors and the number exposed as missing values for the new data lines. Use `proc logistic` to fit the logistic regression model to the combined data and to compute and output estimates of the survival probabilities at each concentration to a new data set. Use those results to plot the estimated survival probability curve against the concentration of carbon disulfide. Include

the observed survival probabilities on the same plot. Does it appear that the logistic regression model provides a good description of the decreasing trend in the survival probabilities as the concentration of carbon disulfide increases?

7.2 This problem explores an alternative model for the beetle survival data from problem (7.1) in which the log-odds for survival are related to the natural logarithm of the carbon disulfide concentration to which the beetles are exposed. The log concentration appears in the second column of the data table.

a. Using the log concentration of carbon disulfide as the explanatory variable, fit a logistic regression model to the beetle survival data. Construct a table with one column for the parameter estimates, one column for their standard errors, and two more columns for the lower and upper limits of 95% confidence intervals. Interpret the confidence intervals.
b. Plot the estimated logistic regression curve and use that curve to determine the concentration corresponding to a survival probability of 0.5.
c. Use a `data step` to add 141 lines with new concentration values to the data file. Start at 49 and go up to 77 by increments of 0.2. Then compute the natural logarithm of each of those concentrations. Designate the counts for the number of survivors and the number exposed as missing values for the new data lines. Use `proc logistic` to fit the logistic regression model to the combined data and with the log concentration as the explanatory variable. Output estimates of the survival probabilities at each concentration. Use those results to plot the estimated survival probability curve against the concentration of carbon disulfide. Include the observed survival probabilities on the same plot. Does it appear that this logistic regression model provides a good description of the decreasing trend in the survival probabilities as the concentration of carbon disulfide increases?
d. Plot the estimates of the survival probability curves from problems 1 and 2 on the same plot. How do these curves differ?
e. Compare the AIC values for those two models. Comment on the results.

7.3 Lloyd (1999, Chapter 6) examines the following data on the relationship of age and marital status from a survey of 185 people in Denmark over the age of 16. A respondent is classified as divorced if the person had been divorced at any time prior to the survey, regardless of whether or not that person is currently remarried. A respondent is classified as single if that person has never married, and a respondent is classified as married if that person has married and never been divorced. At the time of the survey, the legal age of marriage in Denmark was 16.

		Observed counts			
Age group	X	Single	Married	Divorced	Total
17–21	19	17	1	0	18
21–25	23	16	8	0	24
25–30	27.5	8	17	1	26
30–40	35	6	22	4	32
40–50	45	5	21	6	32
50–60	55	3	17	8	28
60–70	65	2	8	6	16
70+	75	1	3	5	9

Use `proc logistic` to fit logistic models to these data, with a multi-category response variable corresponding to the three marital status categories. Use the X variable in the second column of the table as a quantitative explanatory variable representing age.

a. Using marriage as the baseline category, consider the following logistic regression model:

$$\log\left(\frac{\pi_{\text{single}}}{\pi_{\text{married}}}\right) = \alpha_1 + \beta_1(x - 16)$$

$$\log\left(\frac{\pi_{\text{divorced}}}{\pi_{\text{married}}}\right) = \alpha_2 + \beta_2(x - 16)$$

Interpret the parameters α_1, α_2, β_1, and β_2 in this model. Interpret $\exp(\beta_1)$ and $\exp(\beta_2)$ as odds ratios.

b. Compute the maximum likelihood estimates of the model parameters and corresponding standard errors.

c. Compute 95% confidence intervals for β_1 and β_2 and interpret the intervals.

d. Plot the estimated probability curves for the three response categories against age on the same plot. Describe what this plot implies about the association between marital status and age.

7.4 Collett (2003) discusses data from a study of the specific gravity of two species of rotifers. Rotifers are microscopic organisms that make up a substantial proportion of freshwater zooplankton. When rotifers are centrifuged in a solution that has lower specific gravity than their own, they will settle to the bottom of the tube. When they are centrifuged in a solution of specific gravity equal or greater than their own, they will remain in suspension. By using a series of solutions with different specific gravities, it is possible to estimate the proportions of the population with various specific densities. The data for the *Polyarthra major* species of rotifers is displayed in the following table: The first column shows the specific

Specific gravity of solution	Number in suspension	Number in the tube
1.019	11	58
1.020	7	86
1.021	10	76
1.030	19	83
1.030	9	56
1.030	21	73
1.031	13	29
1.040	34	44
1.040	10	31
1.041	36	56
1.048	20	27
1.049	54	59
1.050	20	22
1.050	9	14
1.060	14	17
1.061	10	22
1.063	64	66
1.070	68	86
1.070	488	492
1.070	88	89

gravity of the solution in the tube, and the third column shows the number of rotifers placed in the tube. The second column shows the number of rotifers remaining in suspension after centrifugation.

a. Use `proc logistic` to fit the following logistic regression model:

$$\log\left(\frac{\pi}{1-\pi}\right) = \beta_0 + \beta_1 x$$

where π represents the proportion of the population of rotifers that would remain in suspension in a solution with specific gravity x. Compute estimates of the model parameters and corresponding standard errors.

b. Construct and interpret 95% confidence intervals for the model parameters.

c. Test the null hypothesis that β_1 is zero. State your conclusion.

d. Construct a plot of the estimated logistic curve. Also compute the empirical logit, log(number of survivors/number that settle out), for each tube centrifuged in this study, and plot the empirical logits against the specific densities of the solution on the same plot. Does the logistic regression model appear to provide a good description of the data? Explain.

e. Compute the value of the Pearson goodness-of-fit statistic divided by its degrees of freedom. A value substantially greater than 1.0 would indicate that there is extra-binomial variation, i.e., more variation in the sample proportions from the different tubes than can be explained by binomial distributions. Does there appear to be extra-binomial variation?

f. Use `proc logistic` to fit the same logistic regression model with the `scale=Pearson` option. This will adjust standard error of the parameter estimates for extra-binomial variation. Did the estimates of the model parameters change? How did the values of the standard errors change?

g. Compare 95% confidence intervals for the model parameters to those computed in part(b) where the `scale` option was not used. How do the centers of the confidence intervals change when the `scale` option is used? How do the widths of the confidence intervals change?

7.5 Myers (1990) presents data from a study of the growth behavior for protozoa colonies in a particular lake. Fifteen sponges were placed in the lake, and three of the sponges were removed from the lake at each of five time points, 1, 3, 6, 15, and 21 days after the sponges were put into the lake. The number of protozoa was counted on each sponge. The counts are displayed in the following table.

Day	Number of protozoa
1	17
1	21
1	16
3	30
3	25
3	25
6	33
6	31
6	32
15	34
15	33
15	33
21	39
21	35
21	36

Use the MacArthur–Wilson model to describe the growth pattern. This model is given by

$$y = \alpha(1 - e^{-\beta t}) + \epsilon$$

where

y is the number of protozoa on a sponge.
α is a species equalization parameter.
β is a parameter related to how quickly growth occurs.
t is time in days.
ϵ is a random error with mean zero and variance σ^2.

a. Use `proc nlin` to estimate α and β. Present nonlinear least squares estimates and corresponding standard errors.
b. Construct and interpret 95% bootstrapped confidence intervals for α and β.
c. Obtain an estimate of the error variance σ^2.
d. Test the null hypothesis that β is zero, i.e., no growth occurs. State your conclusion.
e. Construct a plot of the estimated growth curve. What happens to the expected number of protozoa as time increases? Use the estimated model to estimate the expected number of protozoa at 50 days. Give a standard error for your estimate.
f. Construct and interpret a 95% confidence interval for the expected number of protozoa on sponges that are submerged in the lake for 50 days.

7.6 The data for this exercise are from a study of the frequencies of urinary tract infections in $n = 98$ men infected with the HIV virus. CD4+ cell counts were also measured. They are used as an indication of how well the immune system is working in people infected with HIV. CD4+ counts are reported as the number of cells per cubic millimeter of blood. Normal levels of CD4+ counts typically range from 500 to 1500 cells per cubic millimeter of blood. In general, lower CD4+ counts are an indication of progression of HIV and a weakening immune system. As a result of a weakening immune system, people are less likely to resist other infections. The data shown in the following table were collected at Utrecht University Hospital in the Netherlands and reported by Morel and Neerchal (2012). Note that different subjects were exposed to different follow-up times. Consequently, you will need to use the square root of the follow-up time as an offset. Use `proc genmod` to complete the following exercises.

a. Write an appropriate Poisson regression model to relate the expected number of urinary tract infections per month of follow-up time to the CD4+ cell counts. In the context of this study, explain what each parameter in your model represents.
b. Use `genmod` to compute estimates of the model parameters and corresponding standard errors. Be sure to include the square root of the follow-up time in the offset option to adjust for variation in follow-up times.

Urinary tract infections in men infected with HIV								
Number of episodes	Follow-up time (months)	CD4+ cell count	Number of episodes	Follow-up time (months)	CD4+ cell count	Number of episodes	Follow-up time (months)	CD4+ cell count
0	24	125	0	5	10	0	11	290
0	12	50	0	10	5	0	9	380
1	6	30	0	6	50	0	9	420
0	6	80	0	11	15	0	9	240
0	3	170	0	11	55	0	9	470
0	6	95	0	11	80	0	3	310
0	4	35	0	9	140	0	3	460
0	3	50	0	9	60	2	3	345
2	6	25	2	4	5	0	5	670
1	13	15	0	22	45	0	8	1280
0	10	80	0	5	30	0	6	780
0	24	130	0	5	110	0	19	1585
0	5	70	0	5	40	0	6	615
0	3	40	1	3	104	0	15	880
2	12	70	0	6	410	0	6	645
0	16	30	0	10	280	0	13	560
1	13	65	0	5	480	0	6	710
0	24	40	0	7	300	0	13	640
1	3	55	0	8	230	0	12	1150
1	16	25	0	3	210	0	11	530
0	18	70	2	23	380	0	10	620
0	15	5	0	18	310	0	12	980
3	23	20	0	5	275	0	6	600
2	11	105	0	5	390	0	7	1240
0	12	60	0	18	440	0	7	530
1	6	85	0	18	360	0	5	590
0	17	5	0	12	300	0	7	735
0	17	50	0	5	290	0	3	1075
1	9	15	0	5	370	0	5	840
2	16	50	0	12	460	0	4	520
0	8	10	0	11	275	0	3	540
0	4	175	0	11	290	0	4	860
0	12	20	1	11	270			

c. Test the null hypothesis that the coefficient for the CD4+ count variable is zero, against a one-sided alternative that the coefficient is positive. Use a 0.05 significance level. State your conclusion.
d. Estimate the increase in risk of a urinary tract infection, as measured by the odds ratio, for a decrease of 100 in the CD4+ cell count. Report a standard error for your estimate.

e. Check for overdispersion by looking at the value of the Pearson goodness-of-fit statistic divided by its degrees of freedom. Discuss what you see.
f. Fit a negative binomial regression model to these data, and report the estimates of the regression coefficients and their standard errors. Also report the estimates of the variance inflation parameter for the negative binomial distribution.
g. Based on the negative binomial model, is there a significant relationship between the expected rates of urinary tract infections per month of follow-up time and levels of CD4+ counts? Provide a statistical test or confidence interval to support your conclusion.

Appendix A

SAS Templates

A.1 Introduction

From references relating to the SAS Output Delivery System (ODS) appearing throughout this book, it is evident that at least an abbreviated discussion on this topic will be useful to many SAS users. As has been previously observed, SAS procedures that support ODS produce output objects consisting of tables and graphs that are deliverable to various destinations such as HTML, RTF, PDF, SAS Listing, and others. These output objects are created using the results of a procedure and SAS supplied *templates* that describe how each object will be formatted to be displayed. For example, the UNIVARIATE procedure used with a *plots* option will produce five tables of statistics and one graphics panel containing three plots. The appearance of these tables and graphs are controlled by the *table and graphical templates* that are in effect and a *style template* that dictate the overall appearance of the entire output.

A.1.1 What Are Templates?

Templates describe characteristics of various parts of the output tables, such as headers and footers, cell contents, and colors and symbols used in graphs. Each object in the output has an associated template, and all such templates supplied by SAS are stored in the SASHELP library. For most SAS users, these templates facilitate tables and graphs that are satisfactory in appearance of the output from the analysis performed.

Lists of procedure-dependent templates associated with a particular output object can be obtained either by using the ODS TRACE statement or the SAS Results window. A modified procedure step from SAS Example B6 (see Fig. 2.18) is used here to illustrate the use of ODS TRACE:

M. G. Marasinghe, K. J. Koehler, *Statistical Data Analysis Using SAS*,
Springer Texts in Statistics, https://doi.org/10.1007/978-3-319-69239-5

```
ods trace on;
proc univariate data=biology normaltest;
   var Height;
   probplot Height;
   title 'Biology class: Analysis of Height Distribution';
   run;
ods trace off;
```

This proc step produced seven output objects: the six tables named `Moments`, `BasicMeasures`, `TestsForLocation`, `Quantiles`, `ExtremeObs` created by default, plus the table named `TestsForNormality` additionally requested, and the graphical object named `ProbPlot`, resulting from the `probplot` statement requesting a normal probability plot of the variable named `Height`. Recall that the Results window (when the folder is expanded) displays all output objects created by the `proc univariate` step. The results of the trace appears in the LOG window and is reproduced in Fig. A.1. Note that the names of templates corresponding to every output object that appears in the Results window are listed in the trace.

On the other hand, using the Results window is simpler to use if the user only needs to obtain the template name of a specific object. From the results window, right click on the object of interest (say, Basic Measures table from the UNIVARIATE output), and select the item Properties to open a properties window. In the above example, the properties window for the BasicMeasures table is displayed below:

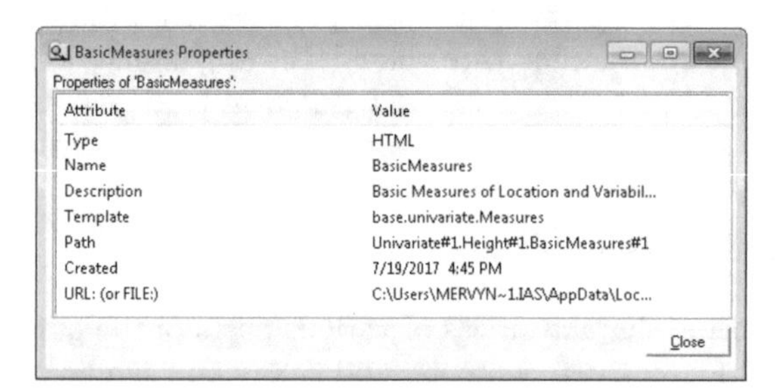

An entirely new template can be created from scratch or an existing template may be modified to obtain a new template using the TEMPLATE procedure. For example, different column headings may be specified or columns in a table may be reordered. Colors and fonts for text in various parts of the output can be changed by altering contents of an existing table template that may be then renamed and saved as a new template for future use. In fact, all SAS default templates have been created using the TEMPLATE procedure.

There are a number of different types of templates: those discussed in this section are style templates, table templates, and graphical templates. A table template applies to a specific table that references the template. Graphical templates are used for producing template-based statistical graphics with

```
Output Added:
-------------
Name:       Moments
Label:      Moments
Template:   base.univariate.Moments
Path:       Univariate.Height.Moments
-------------

Output Added:
-------------
Name:       BasicMeasures
Label:      Basic Measures of Location and Variability
Template:   base.univariate.Measures
Path:       Univariate.Height.BasicMeasures
-------------

Output Added:
-------------
Name:       TestsForLocation
Label:      Tests For Location
Template:   base.univariate.Location
Path:       Univariate.Height.TestsForLocation
-------------

Output Added:
-------------
Name:       TestsForNormality
Label:      Tests For Normality
Template:   base.univariate.Normal
Path:       Univariate.Height.TestsForNormality
-------------

Output Added:
-------------
Name:       Quantiles
Label:      Quantiles
Template:   base.univariate.Quantiles
Path:       Univariate.Height.Quantiles
-------------

Output Added:
-------------
Name:       ExtremeObs
Label:      Extreme Observations
Template:   base.univariate.ExtObs
Path:       Univariate.Height.ExtremeObs
-------------

Output Added:
-------------
Name:       ProbPlot
Label:      Panel 1
Template:   base.univariate.Graphics.ProbPlot
Path:       Univariate.Height.ProbPlot.ProbPlot
-------------
```

Fig. A.1. Log window resulting from ODS trace

ODS as discussed in Chap. 3. A style template controls the overall appearance of the output produced by a SAS program, including all tables and graphs, and can be specified with the `style=` option in an ODS statement and directed to a valid ODS destination, such as HTML, RTF, or PDF. One can request a style in a SAS program by including a statement such as

```
ods html style=HTMLBlue;
```

However, as discussed in Chap. 1, under the SAS windowing environment, the default style for HTML output has been set as the HTMLBlue style. This setting can be modified by selecting the `Results` Tab from the dialog that is opened by selecting the `Tools➡Options➡Preferences` from the menu on top of the main SAS window. It is to be noted that the HTMLBlue style template produces output that is supported by the HTML destination. In the SAS windowing environment, this HTML output file will be opened using an internal browser by default, unless a different browser has been selected from the preferences dialog above.

A.1.2 Where Are the SAS Default Templates Located?

As mentioned above, SAS system supplies users with a large number of all types of templates that are easily accessible using a two-level naming system that is similar to the method used for saving and accessing SAS data sets stored in SAS libraries. Recall that in that system, the first level is a name called *libref* which is associated with the physical location of the library (synonymous with the folder or the directory under Windows operating environment) where the SAS data set is stored. The name of the SAS data set is the second level of the two-level name. For example, see SAS Examples B2 and B3 in Sect. 2.1.3 to review this notation, where SAS data sets are saved/retrieved using two-level names such as *mylib1.first.*

In the case of templates, the first level or the libref refers to a name associated with a SAS library. Familiar examples of SAS libraries are Sashelp, Sasuser, and Work, as can be found by selecting the Libraries icon in the Explorer window under the SAS windowing environment. The second level references a member of SAS library called an *item store.* The directories (or folders) where the templates are stored form the *items* in the item store (though physically they are organized as elements of a single file).

As stated previously, the SAS default templates are stored in the Sashelp library. SAS users may conveniently browse the SAS template library under the SAS windowing environment by selecting the `Results` folder, right clicking on the `Results` icon seen in this folder, and selecting `Templates`. This will open a templates window as displayed below:

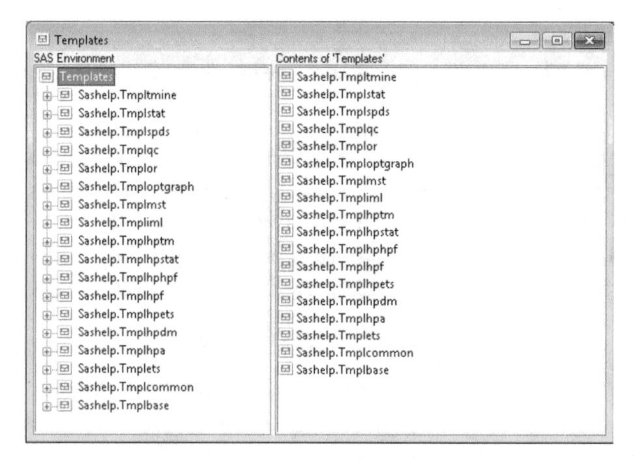

A list of item stores is displayed in the templates window. For example, the item store named *Sashelp.Tmplmst* to be found here contains the default templates that SAS provides. Note carefully that a template store contains many folders (or directories), the contents of which may be examined by double-clicking on the name of the item store of interest to open the folder. For example, double-clicking on *Sashelp.Tmplmst* displays the list of folders shown below on the right:

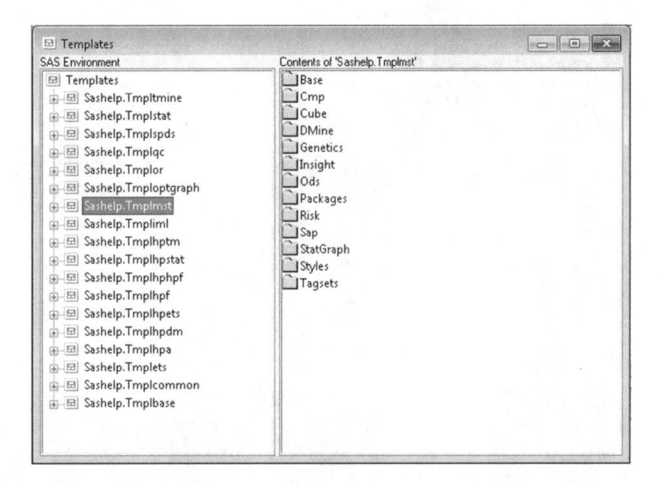

One of these folders of interest is the folder named `Styles` that contains the listings of all the style templates that SAS provides. By opening this folder, its contents may be displayed as usual:

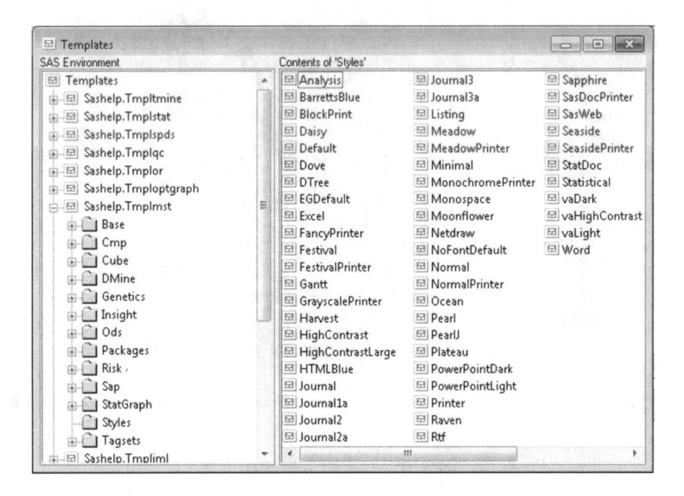

It is observed that the contents of the Styles folder are files containing various *template definitions*. These consist of listings of the actual SAS code for the

TEMPLATE procedure steps that generated the respective styles. One may select any of these to view its contents as a listing. Screenshot of the SAS program for the HTMLBlue style is shown in Fig. A.2.

```
proc template;
   define style Styles.HTMLBlue;
      parent = styles.statistical;
      class GraphColors /
         'gndata12' = cxECE8C4
         'gndata11' = cxDBD8F8
         'gndata10' = cxC6E4BF
         'gndata9' = cxE6CEAD
         'gndata8' = cxE5C1D4
         'gndata7' = cxCCDFF0
         'gndata6' = cxDDDEB5
         'gndata5' = cxDBC7E7
         'gndata4' = cxD5C6B4
         'gndata3' = cxB7D4D3
         'gndata2' =
         'gndata1' =
         'gndata' =
         'gofill' =
         'gblockhead
         'gcphasebox
         'gphasebox'
         'gczonec' =
         'gzonec' =
         'gczoneb' =
         'gzoneb' =
         'gzonea' =
         'gconramp3c
         'gconramp3c
         'gconramp3c
         'gramp3cend
         'gramp3cneu
         'gramp3csta
         'gcontrolli
         'gccontroll
         'gruntest'
         'gcruntest'
         'gclipping'
         'gcclipping
         'gaxis' = c
         'greference
      class colors /
         'link2' = c
         'link1' = c
         'docbg' = c
         'contentbg'
         'systitlebg
         'titlebg' =
         'proctitleb
         'headerbg'
         'captionbg'
         'captionfg'
         'bylinebg'
         'notebg' =
         'tablebg' =
         'batchbg' =
         'systitlefg
```

```
      class colors /
         'link2' = cx0000FF
         'link1' = cx800080
         'docbg' = cxFAFBFE
         'contentbg' = cxFAFBFE
         'systitlebg' = cxFAFBFE
         'titlebg' = cxFAFBFE
         'proctitlebg' = cxFAFBFE
         'headerbg' = cxEDF2F9
         'captionbg' = cxFAFBFE
         'captionfg' = cx112277
         'bylinebg' = cxFAFBFE
         'notebg' = cxFAFBFE
         'tablebg' = cxFAFBFE
         'batchbg' = cxFAFBFE
         'systitlefg' = cx112277
         'titlefg' = cx112277
         'proctitlefg' = cx112277
         'bylinefg' = cx112277
         'notefg' = cx112277;
      class Header /
         bordercolor = cxB0B7BB
         backgroundcolor = cxEDF2F9
         color = cx112277;
      class Footer /
         bordercolor = cxB0B7BB
         backgroundcolor = cxEDF2F9
         color = cx112277;
      class RowHeader /
         bordercolor = cxB0B7BB
         backgroundcolor = cxEDF2F9
         color = cx112277;
      class RowFooter /
         bordercolor = cxB0B7BB
         backgroundcolor = cxEDF2F9
         color = cx112277;
      class Table /
         cellpadding = 5;
      class Graph /
         attrpriority = "Color";
      class GraphFit2 /
         linestyle = 1;
      class GraphClipping /
         markersymbol = "circlefilled";
   end;
run;
```

Fig. A.2. Screen shot of HTMLblue style template

Alternatively, if the name of the style is known, the TEMPLATE procedure may be used to view the contents of the style file:

```
proc template;
source styles.htmlblue;
run;
```

Although a detailed discussion of the TEMPLATE procedure is omitted from this book, it is helpful to recognize a few statements and options in a `proc template` step. For example, in Fig. A.2, the statement `define style` begins a *define style block* that is used to create the style named HTMLBlue. The define style statement can be used to create or modify existing styles for destinations used in ODS statements that support the `style=` option. The `parent=styles.statistical` statement in this block specifies that HTMLBlue style inherit from the STATISTICAL style; that is, a parent-child *inheritance* relationship is established between these two styles. As will be observed later, the STATISTICAL style itself also inherits from another style named DEFAULT. A more elaborate discussion of inheritance between styles will follow in the next section after the terms *style element* and *style attribute* are introduced below.

A.1.3 More on Template Stores

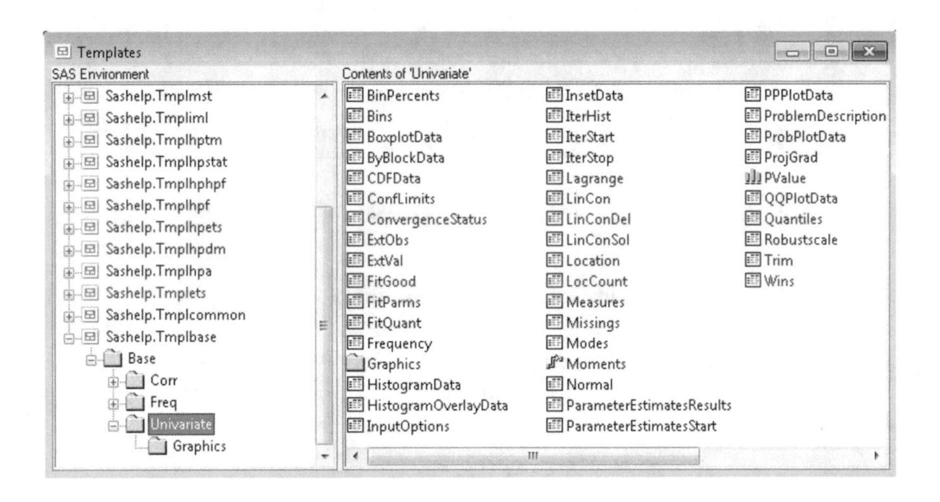

Fig. A.3. Screen shot of the Univariate tables folder in the base item store

In Sect. A.1.1 it was demonstrated how ODS TRACE is used to obtain a list of objects produced by the `proc univariate` step (see Fig. A.1).

At this point, it is important to point out that these objects were actually formatted using various default table and graphical templates specific to the UNIVARIATE procedure. Templates specific to SAS procedures are stored in the Sashelp library item stores reserved for those procedures. For example, since UNIVARIATE is a SAS/BASE procedure, the folder where the templates for base procedures will be found in the item store named *Sashelp.Tmplbase.* By double-clicking on this and then expanding the `Base` folder, the sub-folder `Univariate` where the table templates used by UNIVARIATE are stored can be located (see Fig. A.3).

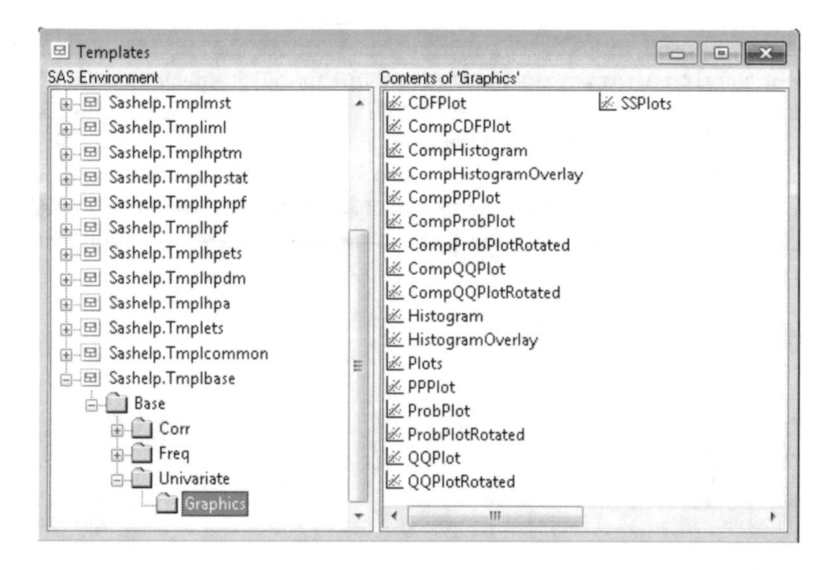

Fig. A.4. Screen shot of the Univariate Graphics folder in the base item store

Select a file, say `Measures`, to view a listing of the SAS program for producing the template for the `BasicMeasures` table (not shown here). The SAS code for the template for the graphical object `ProbPlot` can be similarly found in the sub-folder `Graphics` in the expanded Base/Univariate folder as seen in Fig. A.4.

A.2 Templates and Their Composition

In the introduction (see Sect. A.1.1), templates were defined as a description of how the output should appear when displayed in the intended destination. A template consists of *style elements*, each of which is a named collection of *style attributes*. A style element name identifies a specific area of the output and is associated with a group of style attributes that describe how the material in that area is to be formatted. Each style attribute specifies a value for a single characteristic of a style element. For example, in the HTMLBlue style template displayed in Fig. A.2, the statement

```
class Header /
          bordercolor = cxB0B7BB
          backgroundcolor = cxEDF2F9
          color = cx112277;
```

references the style element named `Header` that controls the header area of the output tables and `backgroundcolor = cxEDF2F9` is the attribute that specifies the background color of this area and `color = cx112277`specifies the color of the textual material of the header in an ODS output table formatted using the HTMLBlue style. Recall that HTMLBlue style *inherits* from the STATISTICAL style; thus, the above assignments either replace or add to the attributes already defined in the parent template. This is because the CLASS statement in `proc template` is used for the creation of a style element from an existing (like-named) style element in a parent style. It is to be noted that users are more likely to modify an existing style than create a new style from scratch. Figure A.5 highlights the difference in the formatting of the Basic Measures table in the `proc univariate` output when using the two styles STATISTICAL and HTMLBlue, respectively. Note the differences

Basic Statistical Measures			
Location		Variability	
Mean	67.89545	Std Deviation	4.58595
Median	67.95000	Variance	21.03093
Mode	67.30000	Range	18.00000
		Interquartile Range	4.90000

Basic Statistical Measures			
Location		Variability	
Mean	67.89545	Std Deviation	4.58595
Median	67.95000	Variance	21.03093
Mode	67.30000	Range	18.00000
		Interquartile Range	4.90000

Fig. A.5. Basic measures table from UNIVARIATE using STATISTICAL and HTMLBlue styles in effect, respectively

in the background and foreground colors of both the table header and the column headers. As a simple illustration of the use of the TEMPLATE procedure, the SAS program shown in Fig. A.6 is used to create a new style named

MYSTYLE using the STATISTICAL style as a parent in a `proc template` step. Notice that several attributes of the HEADER style elements are changed in the newly created style. Then the new style is invoked in an ODS statement to create the output from the same UNIVARIATE step that created the previous tables. The Basic Measures table output from this program is shown in Fig. A.7, showing the changes expected from the use of the new style.

```
data biology;
infile "C:\users\user_name\Documents\...\Class\biology.txt";
input Id Sex $ Age Year Height Weight;
run;
proc template;
define style styles.mystyle;
   parent=styles.statistical;
   style header from header/
        background=lightcyan
        fontstyle=italic
        fontfamily="arial"
        fontweight=bold;
 end;
run;
ods html path="C:\users\user_name\Documents\...\Class"
         file="table_new.htm"style=mystyle;
proc univariate data=biology normaltest;
  var Height;
  probplot Height;
  title 'Biology class: Analysis of Height Distribution';
run;
ods html close;
```

Fig. A.6. Illustrating TEMPLATE procedure for creating a new style

Basic Statistical Measures			
Location		Variability	
Mean	67.89545	Std Deviation	4.58595
Median	67.95000	Variance	21.03093
Mode	67.30000	Range	18.00000
		Interquartile Range	4.90000

Fig. A.7. Basic measures table from UNIVARIATE using MYSTYLE style

A.2.1 Style Templates

A style template (usually just called a style) controls the general visual *attributes* such as text colors and fonts, line styles, marker symbols, etc. of the entire SAS ODS output for which the style is in effect. Examine examples of output produced from the HTMLBlue style in the previous chapters (such as Fig. 3.5) to get an idea of the default characteristics of this style. A screenshot of the HTMLBlue style template is displayed in Fig. A.2.

While the user can choose the style template to be in effect, not all destinations may be compatible with the selected template. For example, when the HTMLBlue style is in effect, output directed to PRINTER destinations will not generate an output with the expected appearance. That is, the style templates are compatible only with destinations that provide support for the intended attributes such as color, font, size, etc. However, the table and column templates (discussed below) are supported by all destinations because these templates are internal to the output object.

The HTMLBlue style inherits most of its attributes from the STATISTICAL style, which in turn inherits from the DEFAULT style. In the HTMLBlue style, for example, the dominant color is blue; in the DEFAULT style, the dominant color is gray. This implies that *style elements* that control these visual attributes specified in the DEFAULT style may have been overridden in the HTMLBlue style, which is a child style. Thus, styles are organized in a hierarchical manner where the lower level ("child" template) may inherit from or override attributes of a higher level (or "parent") template.

The advantage (as found in many object-oriented systems) is that a style definition can be created or modified using already defined attributes of similar style elements. A change in the attribute of a parent style element will also affect all related child style elements. As previously mentioned, by referencing a style element already present in a parent definition, the user can make sure that a child style element defined with the same name carries forward attributes from the parent style element, supplemented by any new attributes. The TEMPLATE procedure incorporates the `STYLE` statement that enables the user to either create entirely new styles or modify existing styles by adding new style elements or modifying existing style elements.

A.2.2 Style Elements and Attributes

Each style consists of style elements. Style element is a collection of *style attributes* that apply to a particular feature of the output. For example, a style element can contain instructions for the presentation of column headings or for the presentation of the data inside cells in a table. Style elements control default colors and fonts for the entire output that uses the style. Each style attribute specifies a value for one aspect of presentation. For example, the `background=` attribute specifies the color for the background of an HTML table, and the `fontstyle=` attribute specifies whether to use a roman or an italic font styles. In the STATISTICAL style template (mentioned earlier but was not shown), the following statements extracted from the `proc template` step:

```
style fonts /
        'TitleFont2' = ("<sans-serif>, <MTsans-serif>, Helvetica, Helv",2,bold)
        'TitleFont' = ("<sans-serif>, <MTsans-serif>, Helvetica, Helv",3,bold)
        'StrongFont' = ("<sans-serif>, <MTsans-serif>, Helvetica, Helv",2,bold)
        'EmphasisFont' = ("<sans-serif>, <MTsans-serif>, Helvetica, Helv",2,italic)
        'FixedFont' = ("<monospace>, Courier",2)
        'BatchFixedFont' = ("SAS Monospace, <monospace>, Courier, monospace",2)
        'FixedHeadingFont' = ("<monospace>, Courier, monospace",2)
        'FixedStrongFont' = ("<monospace>, Courier, monospace",2,bold)
        'FixedEmphasisFont' = ("<monospace>, Courier, monospace",2,italic)
        'headingEmphasisFont' = ("<sans-serif>, <MTsans-serif>,
                                Helvetica, Helv",2,bold italic)
        'headingFont' = ("<sans-serif>, <MTsans-serif>, Helvetica, Helv",2,bold)
        'docFont' = ("<sans-serif>, <MTsans-serif>, Helvetica, Helv",2);
```

and

```
    style Header /
       font = fonts('HeadingFont')
       backgroundcolor = cxF5F7F1
       bordercolor = cxC1C1C1
       bordertopwidth = 0px
       borderleftwidth = 0px
       borderbottomwidth = 1px
       borderrightwidth = 1px
       borderstyle = solid;
```

are used to create the FONTS and HEADER style elements. The value assigned to the `font=` option in defining the HEADER style `fonts('HeadingFont')` is an example of a technique known as *referencing.* This particular example is a style reference that asks for the value of `font=` to be the value for the style attribute 'HeadingFont' from the definition of the style element named `fonts`. Definition of FONTS is shown immediately above the definition of HEADER.

The STYLE statement in a DEFINE STYLE block in the TEMPLATE procedure can be used to create new styles or modify style elements in an existing style as was illustrated in the SAS example displayed in Fig. A.6. In this example, a new style named MYSTYLE was created as a child of the STATISTICAL style. Thus, the new style inherited every style element from its parent style. This is called *style definition inheritance* and is controlled by the `parent=` statement in the DEFINE STYLE block.

In contrast, the style statement controls *style element inheritance.* The user can use the FROM option in the style statement to inherit attributes from a style element within the current style. One might want to do this when defining a new style element drawing from attributes already defined for a specific style element and perhaps add to or modify some of those attributes. Alternately, one may create a new style element that inherits from a style element in the parent style identified in the `parent=` statement.

Thus, if the new style element being created has the same name as one in a parent style specified in the FROM option, and then all attributes for the new style element are copied from the parent style except those attributes that will be defined new (see Fig. A.6). If a FROM option is not used, then the new style element will not inherit any attributes from the parent; that is, the new style element will have ALL new attributes as defined in the current style.

A.2.3 Tabular Templates

All SAS procedures (except PRINT, REPORT, and TABULATE) use tabular templates to determine how to present the results generated by each procedure in table form. A table template determines the order of table headers and footers, the order of columns, and the overall appearance of the output object. Each table template contains or references table elements: columns, headers, and footers. Each table element can specify the use of one or more style elements for various parts of the table and thus is a collection of attributes that apply to a particular column, header, or footer. Thus, table templates are similar to the style templates discussed previously and are customized styles for various ODS destinations. As already seen in style templates table, elements such as columns, headers, and footers can also be defined outside of table templates, usually in a style template. Any table template can then directly reference these table elements, as well.

As a simple example of a table template, the SAS code that output the Basic Measures table for the variable `Height` in the Biology data set and produced from executing SAS Example B6 in Sect. A.1.1 will be examined. The actual table output is reproduced in Fig. A.5 (lower table). Recall that the style in use is the default style for HTML output HTMLBlue. The table template that created this table can be found in the item store *Sashelp.Tmplbase* in the folder named Univariate as described in Sect. A.1.2. Also see Fig. A.3. By double-clicking on the file named `Measures`, a listing of the code for the TEMPLATE step that produced the Basic Measures table template can be viewed. This is reproduced in Fig. A.8.

This program is similar to those used for the creation of a style template except that it makes use of several other statements available with the TEMPLATE procedure. The DEFINE TABLE statement creates a new table template named `Base.Univariate.Measures`. It also begins a DEFINE TABLE block that contains other DEFINE statements that create header, column, or footer templates. The COLUMN statement names the four columns in this table (in the order of occurrence). These are named LocMeasure, LocValue, VarMeasure, and VarValue. These names are used in define statements to create *column templates* for the corresponding columns that describe how the contents of these columns are to be formatted. The HEADER statement similarly names the three headers (named h1, h2, and h3) that are used to identify the headers in the actual order they appear in the table. These are also used in define statements to create *header templates* that define the location and the appearance of the headers.

It is important to understand that a table template only provides the *attributes* for presentation of the output data provided by the procedure. Thus, one may change the content, placement, and attributes of headers, footers, and columns and but only edit the attributes of the contents. This can be done either by creating a new table template or customizing an existing table template. In the next subsection, an example is provided to illustrate

```
proc template;
   define table Base.Univariate.Measures;
      notes "Basic measures of location and variability";
      dynamic MeasHdr varname varlabel;
      column LocMeasure LocValue VarMeasure VarValue;
      header h1 h2 h3;
      translate _val_=._ into "";

      define h1;
         text MeasHdr;
         space = 1;
         spill_margin;
      end;

      define h2;
         text "Location";
         end = LocValue;
         start = LocMeasure;
      end;

      define h3;
         text "Variability";
         end = VarValue;
         start = VarMeasure;
      end;

      define LocMeasure;
         space = 3;
         glue = 2;
         style = Rowheader;
         print_headers = OFF;
      end;

      define LocValue;
         space = 5;
         print_headers = OFF;
      end;

      define VarMeasure;
         space = 3;
         glue = 2;
         style = Rowheader;
         print_headers = OFF;
      end;

      define VarValue;
         format = D10.;
         print_headers = OFF;
      end;
   end;
run;
```

Fig. A.8. Template for basic measures table from UNIVARIATE

how the TEMPLATE procedure may be used to customize the appearance of the Basic Measures table from UNIVARIATE.

It is to be pointed out that tables can also be produced in a DATA step using data from a SAS data set and table and/or style templates designed to format the entire table. In this case, the output is produced with a `file print` statement in a data step that does not create a SAS data set (using `data _null_`) using the `ods=` option in the file statement to a specify a table template to be used. A `put _ods_` statement in this data step creates the ODS output using a default style or a style specified in the ODS HTML statement. Examples can be found in the section dealing with ODS Tabular Templates in the TEMPLATE procedure chapter of the ODS Procedures guide.

A.2.4 Simple Table Template Modification

There are many statements available in the TEMPLATE procedure for modifying an existing table without defining a new table template. For simple customization of an existing table template, the EDIT statement in the TEMPLATE procedure is useful. Although TEMPLATE procedure is sufficiently flexible to perform a wide array of modifications, this example does not attempt to illustrate all of its capabilities. The SAS program displayed in Fig. A.9 uses the EDIT statement to modify the attributes of several elements of the table template and several elements of its composite sub-templates. Recall that the HTMLBlue style is in effect when this program is executed; thus, the default attributes displayed in the Basic Measures table are a result of that.

```
data biology;
infile "C:\users\user_name\Documents\...\Class\biology.txt";
input Id Sex $ Age Year Height Weight;
run;
ods path sasuser.templat(update) sashelp.tmplmst(read);
proc  template;
edit Base.Univariate.Measures;
  edit h1;
    style=header{color=red fontstyle=italic};
    text 'Sample Distribution Statistics';
    just=left;
  end;
  edit LocMeasure;
     cellstyle _val_='Mean' as data{background=lightcyan color=blue fontweight=bold};
  end;
 edit VarMeasure;
     cellstyle _val_='Variance' as data{background=lightcyan color=blue fontweight=bold};
  end;
end;
run;

ods select basicmeasures;
proc univariate data=biology normaltest;
  var Height;
  probplot Height;
  title 'Biology class: Analysis of Height Distribution';
run;

proc template;
delete Base.Univariate.Measures;
run;
```

Fig. A.9. Illustrating TEMPLATE procedure for creating a new style

This program again makes use of the SAS code for SAS Example B6 used in Sect. A.1.1; thus, the Basic Measures table that results will be the effectively the same as the table that appears in Fig. A.5, with the modifications resulting from the Base.Univariate.Measures table template modified in this program. For the sake of brevity and clarity, the more important statements or groups of statements in this program are described below:

```
ods path sasuser.templat(update) sashelp.tmplmst(read);
```

This makes sure that the original table template for Basic Measures is taken from *sashelp.tmplmst* and the modified template stored in the default location *sasuser.templat.* (However, this statement may be not needed as this is the default path.)

```
edit Base.Univariate.Measures;
```

This begins an EDIT block for editing the Base.Univariate.Measures table template, and ends with an END statement when the editing statements are complete. Statements that comprise an EDIT block are the same as those in a DEFINE block (in this case a DEFINE HEADER block).

```
edit h1;
   style=header{color=red fontstyle=italic};
   text 'Statistics for Sample Distribution';
   just=left;
end;
```

The `style=` attribute specifies some new attributes for the style element of the `h1` table template, that is, the first table header. Recall that the original attributes for Header element are specified in the style template in use, i.e., the HTMLBlue style template. Here, text color is changed to red, and the font is changed to italics. Note that the `style=` attribute only affects HTML output. The `text` statement replaces the original header text, and `just=` attribute is changed to left-justified.

```
edit LocMeasure;
   cellstyle _val_='Mean' as
     data{background=lightcyan color=blue fontweight=bold};
end;
edit VarMeasure;
   cellstyle _val_='Variance' as
      data{background=lightcyan color=blue fontweight=bold};
end;
```

The above two EDIT blocks are used for editing the two *column templates* named LocMeasure and VarMeasure. They illustrate how attributes of the contents of a table (identified by `data`) may be conditionally changed using the `cellstyle as` statement. *_val_* is a SAS name for the data value of a cell in the column. The statement

```
cellstyle _val_='Mean' as
     data{background=lightcyan color=blue fontweight=bold} ;
```

specifies that when the logical expression *_val_* =`'Mean'` is true for the values in the column LocMeasure, the attributes of the `data` element be changed to those indicated, for example, background color is changed to lightcyan etc.

Here the Mean and Variance fields are highlighted using a different background color, etc. A more complex statement could be constructed to modify the attributes of the actual values for the mean and variance instead. The reader may verify that the `data` element was originally defined in the Default style and then altered in the Statistical style and that HTMLBlue style inherits it from both.

```
ods select basicmeasures;
```

The above ODS statement sends the output table produced in the `proc univariate` step to the default HTML destination. Finally, the last `proc template` step removes the Base.Univariate.Measures template from the *Sasuser.Templat* item store. The modified table template produced the Basic Measures table displayed in Fig. A.10

Statistics for Sample Distribution			
Location		Variability	
Mean	67.89545	**Std Deviation**	4.58595
Median	67.95000	**Variance**	21.03093
Mode	67.30000	**Range**	18.00000
		Interquartile Range	4.90000

Fig. A.10. Modified basic measures table from UNIVARIATE

A.2.5 Other Types of Templates

The SAS system incorporates several other types of templates that exist for a variety of output. The contingency or cross-tabulation tables produced by the FREQ procedure (discussed in Sect. 2.2.2) require a separate template named Base.Freq.CrossTabsFreqs which is found in the `Freq` sub-folder in the `Base` folder in *Sashelp.Tmplbase* item store. One may also find other table templates associated with FREQ procedure such as Base.Freq.OneWayFreqs or Base.Freq.Measures at this location. Customization of Base.Freq.CrossTabsFreqs table requires the DEFINE CROSSTABS statement available with the TEMPLATE procedure.

Templates for markup language tagsets are in *Sashelp.Tmplmst* in the `Tagsets` folder. For example, Tagsets.Chtml, the template for CHTML, is found here and can be customized using the TEMPLATE procedure, where DEFINE TAGSET blocks are used to create tagset templates.

Finally, STATGRAPH templates are used to create template-based or ODS Graphics. They are found in *Sashelp.Tmplmst* in the `StatGraph` folder and use Graph Template Language (GTL) statements within the `proc` `template` step. STATGRAPH templates are also found in Graphics folders, within folders named for SAS procs that produce ODS graphics. In the next section, methods for customizing graphics templates are briefly introduced via an example.

A.3 Customizing Graphs by Editing Graphical Templates

A brief introduction is provided in this section primarily by discussing an example. Consider the SAS program used in Sect. 4.1.1 to produce a simple linear regression analysis. A modified version of this program is displayed in Fig. A.11. Executing this program produces the usual tabular output from the

```
data d1;
input Traffic Lead;
label Traffic="Traffic Flow (Thousands/24 hours)" Lead="Lead Content (mcg/gm)";
datalines;
 8.3  227
 8.3  312
12.1  362
12.1  521
17.0  640
17.0  539
17.0  728
24.3  945
24.3  738
24.3  759
33.6 1263
;

proc reg data=d1 plots(only)=qqplot;
  model Lead=Traffic/r p;
  title "Simple Linear Regression of Lead Content Data";
run;
```

Fig. A.11. Obtaining a standard QQ plot of residuals from proc reg

regression analysis of the data and the Q-Q plot of the residuals as shown in Fig. A.12.

In the SAS program shown in Fig. A.14 colors, line attributes, and marker symbol attributes of the previous QQ plot are customized by modifying the graph template that produced the graphical output from REG procedure. This template is found using the template browser from the Results window under the SAS windowing environment as usual. In particular, the item store named *Sashelp.Tmplstat* contains templates for all SAS/Stat procedures, where the folder `Reg` contains the table templates for the REG procedure and a sub-folder named `Graphics` contains the statistical graphics templates for the graphs produced by REG procedure. (These are called STATGRAPH templates and are created using DEFINE STATGRAPH blocks in `proc template` using SAS Graph Template Language (GTL) syntax.) The template for the QQ plot named *Stat.Reg.Graphics.QQPlot* can be found in the above folder and is reproduced in Fig. A.13.

The modification of this graph is done by first copying the template into a SAS program, modifying it as discussed below, and then executing it. This will create a modified template by the same name and store it temporarily in the item store *Sasuser.Templat.* Then the `proc reg` step can be executed so

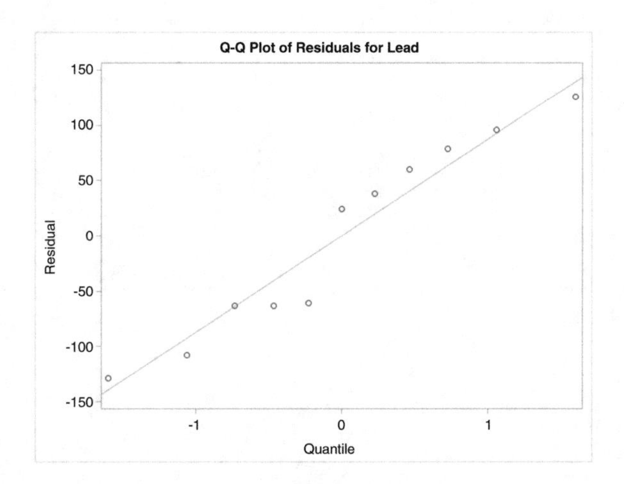

Fig. A.12. Original QQ plot of the residuals from the REG procedure

Template Browser

```
proc template;
   define statgraph Stat.Reg.Graphics.QQPlot;
      notes "QQ Plot";
      dynamic _DEPNAME _MODELLABEL _byline_ _bytitle_ _byfootnote_;
      BeginGraph;
         entrytitle halign=left textattrs=GRAPHVALUETEXT _MODELLABEL halign=
            center textattrs=GRAPHTITLETEXT "Q-Q Plot of Residuals" " for "
            _DEPNAME;
         layout overlay / yaxisopts=(label="Residual" shortlabel="Resid")
            xaxisopts=(label="Quantile");
            lineparm slope=eval (STDDEV(RESIDUAL)) y=eval (MEAN(RESIDUAL)) x=0
               / clip=false extend=true lineattrs=GRAPHREFERENCE;
            scatterplot y=eval (SORT(DROPMISSING(RESIDUAL))) x=eval (
               PROBIT((NUMERATE(SORT(DROPMISSING(RESIDUAL))) -0.375)/(0.25 + N
(RESIDUAL)))) / markerattrs=GRAPHDATADEFAULT primary=true rolename=(s=eval (
               SORT(DROPMISSING(RESIDUAL))) nq=eval (
               PROBIT((NUMERATE(SORT(DROPMISSING(RESIDUAL))) -0.375)/(0.25 + N
(RESIDUAL))))) tiplabel=(nq="Quantile" s="Residual") tip=(nq s);
         endlayout;
         if (_BYTITLE_)
            entrytitle _BYLINE_ / textattrs=GRAPHVALUETEXT;
         else
            if (_BYFOOTNOTE_)
               entryfootnote halign=left _BYLINE_;
            endif;
         endif;
      EndGraph;
   end;
run;
```

Fig. A.13. Graph template for the QQ plot from the REG procedure

that the modified template will be used (using the ODS PATH statement as before). Instead, both proc steps (`proc template` and `proc reg` steps) may be executed together in a single program as shown in Fig. A.14.

```
*****Data step as in original program ****

proc template;
   define statgraph Stat.Reg.Graphics.QQPlot;
      notes "QQ Plot";
      dynamic _DEPNAME _MODELLABEL _byline_ _bytitle_ _byfootnote_;
      BeginGraph;
         entrytitle halign=left textattrs=GRAPHVALUETEXT _MODELLABEL halign=
            center textattrs=GRAPHTITLETEXT "Q-Q Plot of Residuals" " for "
            _DEPNAME;
         layout overlay / yaxisopts=(label="Residual" shortlabel="Resid")
            xaxisopts=(label="Quantile");
            lineparm slope=eval (STDDEV(RESIDUAL)) y=eval (MEAN(RESIDUAL)) x=0
               / clip=false extend=true lineattrs=GRAPHREFERENCE(color=red pattern=shortdash);
            scatterplot y=eval (SORT(DROPMISSING(RESIDUAL))) x=eval (
               PROBIT((NUMERATE(SORT(DROPMISSING(RESIDUAL))) -0.375)/(0.25 + N
(RESIDUAL)))) / markerattrs=GRAPHDATADEFAULT(symbol=CircleFilled) primary=true rolename=(s=eval (
               SORT(DROPMISSING(RESIDUAL))) nq=eval (
               PROBIT((NUMERATE(SORT(DROPMISSING(RESIDUAL))) -0.375)/(0.25 + N
(RESIDUAL))))) tiplabel=(nq="Quantile" s="Residual") tip=(nq s);
         endlayout;
         if (_BYTITLE_)
            entrytitle _BYLINE_ / textattrs=GRAPHVALUETEXT;
         else
            if (_BYFOOTNOTE_)
               entryfootnote halign=left _BYLINE_;
            endif;
         endif;
      EndGraph;
   end;
run;

**** proc reg step here ****

proc template;
   delete Stat.Reg.Graphics.QQPlot/store=Sasuser.Templat;
run;
```

Fig. A.14. Illustrating graphic template modification using proc reg

The attributes of the line color and pattern are specified by the `lineattrs=` option in the `lineparm` statement: lineattrs=GRAPHREFERENCE. This references the style element GraphDataDefault in the HTMLBlue style. The LINEPARM statement results in a straight line specified by slope and intercept parameters. The color of the line is changed to red and the line pattern to dashes, by overriding those attributes of the style specification:

```
lineattrs=GRAPHREFERENCE(color=red pattern=shortdash)
```

The attributes of the marker symbol in the scatter plot are specified by MarkerAttrs=GraphDataDefault. As above this is a reference to the style element GraphDataDefault in the HTMLBlue style where the marker symbol is an empty circle. This can be changed to a filled circle as the marker symbol by overriding the symbol attribute of the style specification as follows:

```
markerattrs=GRAPHDATADEFAULT(symbol=CircleFilled)
```

The `proc reg` step that follows produces the graph (in addition to the tabular output) that is displayed in Fig. A.15. The modified graph template is then deleted from the *Sasuser.Templat* store.

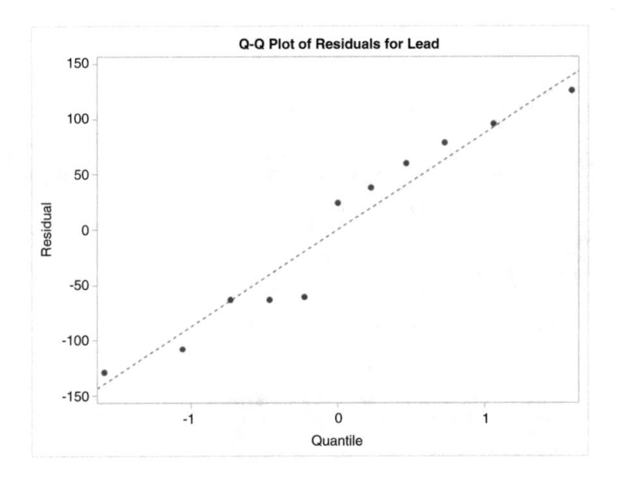

Fig. A.15. Original QQ plot of the residuals from the REG procedure

A.4 Creating Customized Graphs by Extracting Code from Standard Graphical Templates

A user may not be able to use statistical graphics (SG) procedures alone to produce a customized graph such as that shown in Fig. 5.18, which is a plot consisting of four panels, each containing a different graphs produced from REG procedure output. One might immediately think of creating a graphical template from scratch to produce this plot. As mentioned earlier, graphical templates are based on statements and syntax from the Graphic Template Language (GTL). Unless one is an experienced SAS programmer, it will be fairly cumbersome for the standard SAS user to sufficiently master the syntax of GTL to be able to design an entirely new templates for the production of complex statistical graphics. However, if the user is able to peruse an existing graphical template and recognize portions of code that create relevant parts of the graphical output produced by the template, it will be possible to construct a new template using code segments extracted from the original template. This technique is perhaps best suited for putting together several graphs in panels to form a composite plot such the Fit Diagnostics plot produced by the REG procedure.

This technique is illustrated in this section by building a graphical template for producing Fig. 5.18. Since each of the graphs shown in this figure is extracted from the Fit Diagnostics panel output from `proc reg`, first examine the graphical template *Stat.Reg.Graphics.DiagnosticsPanel* using the template browser. This is found in the `Graphics` folder expanding the `Stat` folder `Stat ➡Reg ➡Graphics` as in previous discussions. A portion of this template is reproduced in Fig. A.16.

```
proc template;
  define statgraph Stat.Reg.Graphics.DiagnosticsPanel;
    notes "Diagnostics Panel";
    dynamic _DEPLABEL _DEPNAME _MODELLABEL _OUTLEVLABEL _TOTFREQ _NPARM
      _NOBS _OUTCOOKSDLABEL _SHOWSTATS _NSTATSCOLS _DATALABEL _SHOWNObs
      _SHOWTOTFREQ _SHOWNParm _SHOWEDF _SHOWMSE _SHOWRSquare _SHOWAdjRSq
      _SHOWSSE _SHOWDepMean _SHOWCV _SHOWAIC _SHOWBIC _SHOWCP _SHOWGMSEP
      _SHOWJP _SHOWPC _SHOWSBC _SHOWSP _EDF _MSE _RSquare _AdjRSq _SSE
      _DepMean _CV _AIC _BIC _CP _GMSEP _JP _PC _SBC _SP _byline_ _bytitle_
      _byfootnote_;
    BeginGraph / designheight=defaultDesignWidth;
      entrytitle halign=left textattrs=GRAPHVALUETEXT _MODELLABEL halign=
        center textattrs=GRAPHTITLETEXT "Fit Diagnostics" " for " _DEPNAME
;
      layout lattice / columns=3 rowgutter=10 columngutter=10 shrinkfonts=
        true rows=3;
        layout overlay / xaxisopts=(shortlabel='Predicted');
          referenceline y=0;
          scatterplot y=RESIDUAL x=PREDICTEDVALUE / primary=true
            datalabel=_OUTLEVLABEL rolename=(_tip1=OBSERVATION _id1=ID1
            _id2=ID2 _id3=ID3 _id4=ID4 _id5=ID5) tip=(y x _tip1 _id1
            _id2 _id3 _id4 _id5);
        endlayout;

        layout overlay / xaxisopts=(shortlabel='Predicted');
          referenceline y=-2;
          referenceline y=2;
          scatterplot y=RSTUDENT x=PREDICTEDVALUE / primary=true
            datalabel=_OUTLEVLABEL rolename=(_tip1=OBSERVATION _id1=ID1
            _id2=ID2 _id3=ID3 _id4=ID4 _id5=ID5) tip=(y x _tip1 _id1
            _id2 _id3 _id4 _id5);
        endlayout;
        layout overlay / xaxisopts=(label='Leverage' offsetmax=0.05)
          yaxisopts=(offsetmin=0.05 offsetmax=0.05);
          referenceline y=2;
          referenceline y=-2;
          referenceline x=eval (MIN(1,2*_NPARM/_TOTFREQ));
          scatterplot y=RSTUDENT x=HATDIAGONAL / primary=true datalabel=
            _OUTLEVLABEL rolename=(_tip1=OBSERVATION _id1=ID1 _id2=ID2
            _id3=ID3 _id4=ID4 _id5=ID5) tip=(y x _tip1 _id1 _id2 _id3
            _id4 _id5);
        endlayout;
        layout overlay / yaxisopts=(label="Residual" shortlabel="Resid")
          xaxisopts=(label="Quantile");
          lineparm slope=eval (STDDEV(RESIDUAL)) y=eval (MEAN(RESIDUAL))
            x=0 / clip=false extend=true lineattrs=GRAPHREFERENCE;
          scatterplot y=eval (SORT(DROPMISSING(RESIDUAL))) x=eval (
            PROBIT((NUMERATE(SORT(DROPMISSING(RESIDUAL))) -0.375)/(0.25
+ N(RESIDUAL)))) / markerattrs=GRAPHDATADEFAULT primary=true rolename=(s=eval
            (SORT(DROPMISSING(RESIDUAL))) nq=eval (
            PROBIT((NUMERATE(SORT(DROPMISSING(RESIDUAL))) -0.375)/(0.25
+ N(RESIDUAL))))) tiplabel=(nq="Quantile" s="Residual") tip=(nq s);
        endlayout;
.....................
.....................
.....................
    EndGraph;
  end;
run;
```

Fig. A.16. Diagnostic panel graphical template

The highlighted box is the block of code in this template that represent the Residual versus Predicted plot in the diagnostic panel of plots. Similarly, it is possible to identify blocks of code that produce each of the other plots. The DYNAMIC statement seen in the template contains names of variables passed on from the procedure output; thus, they may be used in the code to represent those variables for plotting purposes. By combining similar blocks of code, the proc template step in the SAS program shown in Fig. A.17 and continued in Fig. A.18, was constructed.

```
data mileage;
input Size $ @;
  do  i=1 to 4;
  input  MPG @;
  output;
  end;
drop i;
datalines;
300 16.6 16.9 15.8 15.5
350 14.4 14.9 14.2 14.1
400 12.4 12.7 13.3 13.6
450 11.5 12.8 12.1 12.0
;

proc glm data=mileage plots=Diagnostics;
 class Size;
 model MPG = Size/p;
 output out=stats1 p=Fitted r=Residual;
 title 'Analysis of Gas Mileage Data';
 run;

proc template;
  define statgraph ResidPanel;
    notes "Fit Summary Plots";
    dynamic Residual Predicted Dependent Xvar
      Observation _DEPLABEL _DEPNAME _OutLevLabel _TOTFREQ _NPARM
      _RsByLevGroup _NOBS _OutCooksDLabel _SHOWTOTFREQ _EDF _MSE _RSquare
      _AdjRSq _byline_ _bytitle_ _byfootnote_;
    BeginGraph / designHeight=defaultDesignWidth;
      entrytitle "Fit Diagnostics";
      layout lattice / columns=2 rowgutter=10 columngutter=10 shrinkfonts=
        true rows=2;
      layout overlay / xaxisopts=(shortlabel='Size');
         referenceline y=0/lineattrs=(Color=magenta Pattern=2);
         scatterplot y=RESIDUAL x=Xvar / primary=true datalabel=
           _OUTLEVLABEL rolename=(_tip1=OBSERVATION _id1=_ID1 _id2=_ID2
           _id3=_ID3 _id4=_ID4 _id5=_ID5) tip=(y x _tip1 _id1 _id2 _id3
           _id4 _id5);
      endlayout;
```

Fig. A.17. SAS program for producing the graph Fig. 5.18

```
      layout overlay / xaxisopts=(shortlabel='Predicted');
        referenceline y=0/lineattrs=(Color=magenta Pattern=2);
        scatterplot y=RESIDUAL x=PREDICTED / primary=true datalabel=
          _OUTLEVLABEL rolename=(_tip1=OBSERVATION _id1=_ID1 _id2=_ID2
          _id3=_ID3 _id4=_ID4 _id5=_ID5) tip=(y x _tip1 _id1 _id2 _id3
          _id4 _id5);
      endlayout;
      layout overlay / xaxisopts=(shortlabel='X');
        scatterplot y=DEPENDENT x=Xvar / primary=true datalabel=
          _OUTLEVLABEL rolename=(_tip1=OBSERVATION _id1=_ID1 _id2=_ID2
          _id3=_ID3 _id4=_ID4 _id5=_ID5) tip=(y x _tip1 _id1 _id2 _id3
          _id4 _id5);
        seriesplot y=PREDICTED x=Xvar/lineattrs=(Color=magenta Pattern=2);
      endlayout;
      layout overlay / yaxisopts=(label="Residual" shortlabel="Resid")
        xaxisopts=(label="Quantile");
        lineparm slope=eval (STDDEV(RESIDUAL)) y=eval (MEAN(RESIDUAL))
          x=0 / clip=false extend=true lineattrs=GRAPHREFERENCE;
        scatterplot y=eval (SORT(DROPMISSING(RESIDUAL))) x=eval (
          PROBIT((NUMERATE(SORT(DROPMISSING(RESIDUAL))) -0.375)/(0.25
+ N(RESIDUAL)))) / markerattrs=GRAPHDATADEFAULT primary=true rolename=(s=eval
          (SORT(DROPMISSING(RESIDUAL))) nq=eval (
          PROBIT((NUMERATE(SORT(DROPMISSING(RESIDUAL))) -0.375)/(0.25
+ N(RESIDUAL))))) tiplabel=(nq="Quantile" s="Residual") tip=(nq s);
      endlayout;

    endlayout;
    if (_BYTITLE_)
      entrytitle _BYLINE_ / textattrs=GRAPHVALUETEXT;
    else
      if (_BYFOOTNOTE_)
        entryfootnote halign=left _BYLINE_;
      endif;
    endif;
  EndGraph;
 end;
run;

proc sgrender data=stats1 template=ResidPanel;
dynamic Residual="Residual" Predicted="Fitted" Dependent="MPG" Xvar="Size";
run;

proc template;
  delete ResidPanel/store=Sasuser.Templat;
run;
quit;
```

Fig. A.18. SAS program for producing the graph Fig. 5.18 (continued)

The reader can examine the four LAYOUT segments of code and figure out how these code blocks produce the four plots appearing in Fig. 5.18. Note that the new template was named ResidPanel and is a STATGRAPH template and will be automatically saved in *Sasuser.Templat*. After the template has been designed, SGRENDER procedure is used to associate the template with the `stats1` data set and the output variables from the REG procedure and produce the graph. Then the template is removed from the item store.

Appendix B

Tables

See Tables B.1, B.2, B.3, B.4, B.5, B.6, B.7, B.8, B.9, B.10, B.11, B.12, B.13, B.14, B.15, B.16, B.17, and B.18.

Table B.1. Fuel consumption data for the 48 contiguous states (Weisberg 1985)

(1) STATE:	State FIPS code
(2) POP:	1771 Population, in thousands
(3) TAX:	1772 Motor Fuel Tax Rate, in cents per gallon
(4) NUMLIC:	1771 Number of Licensed Drivers, in thousands
(5) INCOME:	1772 Per Capita income, in thousands of dollars
(6) ROADS:	1771 Miles of Federal-Aid Primary Highways, in thousands
(7) FUELC:	1772 Fuel Consumption, in millions of gallons

STATE	POP	TAX	NUMLIC	INCOME	ROADS	FUELC
ME	1027	7.0	540	3.571	1.776	557
NH	771	7.0	441	4.072	1.250	404
VT	462	7.0	268	3.865	1.586	257
MA	5787	7.5	3060	4.870	2.351	2376
RI	768	8.0	527	4.377	0.431	377
CT	3082	7.0	1760	5.342	1.333	1408
NY	18,366	8.0	8278	5.317	11.868	6312
NJ	7367	8.0	4074	5.126	2.138	3437
PA	11,726	8.0	6312	4.447	8.577	5528
OH	10,783	7.0	5748	4.512	8.507	5375
IN	5271	8.0	2804	4.371	5.737	3068
IL	11,251	7.5	5703	5.126	14.186	5301
MI	7082	7.0	5213	4.817	6.730	4768
WI	4520	7.0	2465	4.207	6.580	2274
MN	3876	7.0	2368	4.332	8.157	2204
IA	2883	7.0	1,687	4.318	10.340	1830
MO	4753	7.0	2717	4.206	8.508	2865
ND	632	7.0	341	3.718	4.725	451
SD	577	7.0	417	4.716	5.715	501
NE	1525	8.5	1033	4.341	6.010	776
KS	2258	7.0	1476	4.573	7.834	1466
DE	565	8.0	340	4.783	0.602	305
MD	4056	7.0	2073	4.877	2.447	1883
VA	4764	7.0	2463	4.258	4.686	2604
WV	1781	8.5	782	4.574	2.617	817
NC	5214	7.0	2835	3.721	4.746	2753

(continued)

M. G. Marasinghe, K. J. Koehler, *Statistical Data Analysis Using SAS*, Springer Texts in Statistics, https://doi.org/10.1007/978-3-319-69239-5

Table B.1. (continued)

(1) STATE:	State FIPS code
(2) POP:	1771 Population, in thousands
(3) TAX:	1772 Motor Fuel Tax Rate, in cents per gallon
(4) NUMLIC:	1771 Number of Licensed Drivers, in thousands
(5) INCOME:	1772 Per Capita income, in thousands of dollars
(6) ROADS:	1771 Miles of Federal-Aid Primary Highways, in thousands
(7) FUELC:	1772 Fuel Consumption, in millions of gallons

STATE	POP	TAX	NUMLIC	INCOME	ROADS	FUELC
SC	2665	8.0	1460	3.448	5.377	1537
GA	4720	7.5	2731	3.846	7.061	2777
FL	7257	8.0	4084	4.188	5.775	4167
KY	3277	7.0	1626	3.601	4.650	1761
TN	4031	7.0	2088	3.640	6.705	2301
AL	3510	7.0	1801	3.333	6.574	1746
MS	2263	8.0	1307	3.063	6.524	1306
AR	1778	7.5	1081	3.357	4.121	1242
LA	3720	8.0	1813	3.528	3.475	1812
OK	2634	7.0	1657	3.802	7.834	1675
TX	11,647	5.0	6575	4.045	17.782	7451
MT	717	7.0	421	3.877	6.385	506
ID	756	8.5	501	3.635	3.274	470
WY	345	7.0	232	4.345	3.705	334
CO	2357	7.0	1475	4.447	4.637	1384
NM	1065	7.0	600	3.656	3.785	744
AZ	1745	7.0	1173	4.300	3.635	1230
UT	1126	7.0	572	3.745	2.611	666
NV	527	6.0	354	5.215	2.302	412
WA	3443	7.0	1766	4.476	3.742	1757
OR	2182	7.0	1360	4.276	4.083	1331
CA	20,468	7.0	12,130	5.002	7.774	10,730

Table B.2. Daily maximum ozone concentrations at Stamford, Connecticut (Stmf) and Yonkers, New York (Ykrs), during the period May 1, 1774 to September 30, 1774, recorded in parts per billion (ppb), reproduced from Chambers et al. (1983)

May		June		July		August		September	
Stmf	Ykrs	Stmf	Ykrs	Stmf	Ykrs	Stmf	Ykrs	Stmf	Ykrs
66	47	61	36	152	76	80	66	113	66
52	37	47	24	201	108	68	82	38	18
–	27	–	52	134	85	24	47	38	25
–	37	176	88	206	76	24	28	28	14
–	38	131	111	72	48	82	44	52	27
–	–	173	117	101	60	100	55	14	7
47	45	37	31	117	54	55	34	38	16
64	52	47	37	124	71	71	60	74	67
68	51	215	73	133	–	87	70	87	74
26	22	230	106	83	50	64	41	77	74
86	27	–	47	–	27	–	67	150	75
52	25	67	64	60	37	–	127	146	74
43	–	78	83	124	47	170	76	113	42
75	55	125	77	142	71	–	56	38	–
87	72	74	77	124	46	86	54	66	38
188	132	72	36	64	41	202	100	38	23
118	–	72	51	76	47	71	44	80	50
103	106	125	75	103	57	85	44	80	34
82	42	143	104	–	53	122	75	77	58
71	45	172	107	46	25	155	86	71	35
103	80	–	56	68	45	80	70	42	24
240	107	122	68	–	78	71	53	52	27
31	21	32	17	87	40	28	36	33	17
40	50	114	67	27	13	212	117	38	21
47	31	32	20	–	25	80	43	24	14
51	37	23	35	73	46	24	27	61	32
31	17	71	30	57	62	80	77	108	51
47	33	38	31	117	80	167	75	38	15
14	22	136	81	64	37	174	87	28	21
–	67	167	117	–	70	141	47	–	18
71	45			111	74	202	114		

Note: Missing observations are shown as "–"

Source: Stamford, Connecticut Department of Environmental Protection
Yonkers, Boyce Thompson Institute

Table B.3. Rainfall in acre-feet from 52 clouds, of which 26 were chosen randomly and seeded with silver oxide, reproduced here from Chambers et al. (1983)

Rainfall from control clouds	Rainfall from seeded clouds
1202.6	2745.6
830.1	1677.8
372.4	1656.0
345.5	778.0
321.2	703.4
244.3	487.1
163.0	430.0
147.8	334.1
75.0	302.8
87.0	274.7
81.2	274.7
68.5	255.0
47.3	242.5
41.1	200.7
36.6	178.6
27.0	127.6
28.6	117.0
26.3	118.3
26.1	115.3
24.4	72.4
21.7	40.6
17.3	32.7
11.5	31.4
4.7	17.5
4.7	7.7
1.0	4.1

Source: Simpson et al. (1975)

Table B.4. Measurements from an aerobic fitness study

Id:	Subject Identification No.:
WtLoss:	Monthly weight loss (in lbs.) (assume 1 decimal place)
Height:	Height (in inches)
Weight:	Weight (in lbs.)
Intake:	Food intake (calories/day)
Aero:	Aerobic points
BodyFat:	Body fat (assume 1 decimal place)
RunTime:	Time to run 1.5 miles (min) (assume 1 decimal place
RstPulse:	Heart rate while resting
Oxygen:	Oxygen uptake rate (in ml/kg body weight/min (assume 1 decimal place)
Age:	Age (Yrs.)
Gender:	Gender (alphabetic character)

Id	WtLoss	Height	Weight	Intake	Aero	BodyFat	Run Time	RstPulse	Oxygen	Age	Gender
1	3.5	64	140	1420	8	16.7	7.2	55	47.8	38	F
2	2.1	68	165	1510	5	11.2	14	56	37.4	45	M
3	1.6	71	177	1840	30	8.4	8.5	55	56.8	27	M
4	5.3	66	175	750	5	25.6	11.6	58	44.8	47	F
5	4.8	74	253	1560	25	22.3	8.6	48	60.1	36	M
6	4.2	67	186	1580	40	14.2	8.2	40	57.6	42	M
7	5.7	60	156	1170	20	21.7	8.6	48	46.7	30	F
8	2.1	67	183	1550	35	16.4	7.2	55	44.6	26	M
7	2.7	66	151	1300	32	17.1	11	60	40.2	21	F
10	3.6	63	120	1420	50	13.7	10.1	54	45.6	23	F
11	4	73	221	1080	18	20	8.1	40	57.5	32	M
12	3.1	70	211	1000	15	21.2	7.6	48	45.4	52	M
13	2.5	71	176	1270	36	15.8	10.5	64	47.8	27	M
14	2.8	67	170	1340	26	.	11.2	53	44.8	31	M
15	2.7	68	162	1270	10	15.7	11.7	70	45.7	40	F
16	4.3	65	164	1110	22	18.7	13.1	63	37.5	44	M
17	1.2	71	188	1680	15	15.8	10	48	46.8	31	M
18	3.8	74	233	1200	18	21.3	8.7	51	60.2	25	M
17	1.6	64	138	1570	5	18.6	10.3	48	46.8	47	F
20	1.5	65	147	1670	0	14.7	10.5	53	47.5	52	F
21	2.1	64	155	1770	.	10.5	10.1	67	50.4	50	F
22	1.7	67	167	2150	38	18.1	11.5	57	47.7	35	M
23	1.1	65	176	2450	25	18.3	12.8	44	37.2	54	M
24	1.3	67	167	2310	20	17.6	11.7	47	50.1	57	M
25	0.8	71	185	2580	16	18.7	12.7	65	41.7	55	F
26	0.7	61	125	1620	38	12.7	8.7	55	47.5	55	F
27	0.5	67	135	1550	48	11.5	10.5	57	51.5	56	M
28	1.5	70	130	1350	28	13	11	61	45.5	52	M
27	1	72	128	1450	18	12	11.2	64	48.5	47	M
30	1.6	64	130	1610	25	12.1	8.7	65	48.2	22	F
31	1	67	132	1710	20	8.4	8.5	53	50.7	24	M
32	1.1	68	125	1810	30	10.5	7.5	60	47.8	17	F
33	1.3	74	131	1510	21	8.7	10.5	57	40.2	21	M
34	2	70	118	1450	28	17.1	8.8	60	43.5	31	M
35	2.5	68	128	1400	32	20.1	8.5	65	45.2	31	M
36	1.1	65	127	1350	36	7.5	7.5	53	43.5	31	F
37	1.6	67	135	1850	48	17.5	7	50	47	35	F
38	1	68	145	1700	27	17.5	7.5	64	33	25	M
39	2.1	73	142	1680	37	17.5	7.8	66	45.2	25	M

(continued)

Table B.4. (continued)

`Id:`	Subject Identification No.:
`WtLoss:`	Monthly weight loss (in lbs.) (assume 1 decimal place)
`Height:`	Height (in inches)
`Weight:`	Weight (in lbs.)
`Intake:`	Food intake (calories/day)
`Aero:`	Aerobic points
`BodyFat:`	Body fat (assume 1 decimal place)
`RunTime:`	Time to run 1.5 miles (min) (assume 1 decimal place
`RstPulse:`	Heart rate while resting
`Oxygen:`	Oxygen uptake rate (in ml/kg body weight/min (assume 1 decimal place)
`Age:`	Age (Yrs.)
`Gender:`	Gender (alphabetic character)

Id	WtLoss	Height	Weight	Intake	Aero	BodyFat	RunTime	RstPulse	Oxygen	Age	Gender
40	2.2	71	165	2050	37	20.5	8.7	50	55.7	20	F
41	1.4	75	157	1700	44	10.5	10.3	62	53.2	24	F
42	3	76	150	1770	42	16	10.8	56	35.5	22	M
43	1.2	72	175	1850	22	10.5	10.4	75	43.2	22	F
44	1.6	70	185	1710	12	13.6	7.6	60	45.4	20	M
45	2.5	68	175	1450	23	7.5	10.8	53	40.7	18	M
46	1.8	77	167	1730	35	11.3	7.3	74	47.6	23	F
47	1.5	67	145	1651	58	17.8	10.6	57	54.5	56	M
48	2	75	147	1680	20	18.5	11.5	65	60	54	M
47	1.5	77	145	1650	17	16.5	11.5	67	41.5	56	M
50	1	75	150	1450	23	13.7	12	61	45.5	52	F
51	1.2	70	158	1750	18	17	10.2	64	38.5	47	M
52	1.3	68	177	2210	17	23.6	11.7	56	60.1	58	M
53	0.8	67	128	1820	30	15.7	7.7	57	75.7	55	F
54	3.4	74	160	1080	18	20	7.1	70	57.5	38	M
55	2.2	78	175	1750	24	17.5	7.4	65	53.2	22	F
56	1.5	76	225	1810	15	15.6	7.2	50	65.4	20	M
57	2.3	67	175	1410	20	17.5	11.8	60	47.5	45	F

Table B.5. Demographic data for 60 countries reproduced from Ott et al. (1987)

(1) COUNTRY:	Country name (20 characters max.)
(2) BIRTHRAT:	Crude birth rate
(3) DEATHRAT:	Crude death rate
(4) INF_MORT:	Infant mortality rate
(5) LIFE_EXP:	Life expectancy in years
(6) POPURBAN:	Percent population in urban areas
(7) PERC_GNP:	Per capita GNP in US dollars
(8) LEV_TECH:	Level of technology (100 is maximum)
(7) CIVILLIB:	Degree of civil liberties (1 = minimal denial of civil liberties, 7 = maximal denial)

(1)	(2)	(3)	(4)	(5)	(6)	(7)	(8)	(7)
Algeria	45	12	107	60	52	2400	17	6
Argentina	24	8	35	70	83	2030	23	3
Australia	16	7	10	75	86	7210	71	1
Austria	12	12	12	73	56	7210	50	1
Bolivia	42	16	124	51	46	510	10	3
Brazil	31	8	71	63	68	1870	15	3
Bulgaria	14	11	17	72	65	3700	44	7
Canada	15	7	7	75	76	12,000	75	1
Chile	24	6	24	70	83	1870	22	5
Colombia	28	7	53	64	67	1410	15	3
Czech	15	12	16	71	74	5800	72	6
Denmark	10	11	8	74	83	11,470	71	1
Egypt	37	10	80	57	44	700	13	5
Finland	14	7	6	74	60	10,440	57	2
France	14	10	7	75	73	11,370	62	2
Ghana	47	15	107	52	40	320	10	5
Greece	14	7	15	74	70	3770	23	2
Hungary	12	14	17	70	54	2150	47	5
Italy	11	10	12	74	72	6350	41	2
India	34	13	118	53	23	260	6	3
Iraq	46	13	72	57	68	3400	12	7
Ireland	17	7	11	73	56	4810	48	1
Israel	24	6	14	74	87	5360	33	2
Ivory Coast	46	18	122	47	42	720	7	5
Japan	13	6	6	77	76	10,100	53	1
Kenya	54	13	80	53	16	340	11	5
Madagascar	45	17	67	50	22	270	12	6
Malawi	52	20	165	45	12	210	12	7
Malaysia	27	7	27	67	32	1870	14	4
Morocco	41	12	77	58	42	750	12	5

(continued)

Table B.5. (continued)

(1) COUNTRY:	Country name (20 characters max.)
(2) BIRTHRAT:	Crude birth rate
(3) DEATHRAT:	Crude death rate
(4) INF_MORT:	Infant mortality rate
(5) LIFE_EXP:	Life expectancy in years
(6) POPURBAN:	Percent population in urban areas
(7) PERC_GNP:	Per capita GNP in US dollars
(8) LEV_TECH:	Level of technology (100 is maximum)
(7) CIVILLIB:	Degree of civil liberties (1 = minimal denial of civil liberties, 7 = maximal denial)

(1)	(2)	(3)	(4)	(5)	(6)	(7)	(8)	(7)
Netherlands	12	8	8	76	88	7710	68	1
New Zealand	16	8	13	74	83	7410	66	1
Nigeria	48	17	105	50	28	760	8	3
Norway	12	10	8	76	71	13,820	63	1
Pakistan	43	15	120	50	17	370	8	5
Peru	35	10	77	57	65	1040	12	3
Philippines	32	7	50	64	37	760	15	5
Poland	20	10	17	71	57	4200	53	5
Portugal	14	7	20	71	30	2170	22	2
Romania	15	10	28	71	47	2200	33	6
Senegal	50	17	141	43	42	440	11	4
South Africa	35	14	72	54	56	2450	33	6
Spain	13	7	10	74	71	4800	28	2
Sri Lanka	27	6	34	68	22	330	7	4
Sweden	11	11	7	76	83	12,400	81	1
Switzerland	11	7	8	76	58	16,370	57	1
Syria	47	7	57	64	47	1680	16	7
Thailand	25	6	51	63	17	810	12	4
Togo	45	17	113	47	20	280	15	6
Tunisia	33	10	85	61	52	1270	15	5
Turkey	35	10	110	63	45	1230	14	5
U.S.S.R.	20	10	32	67	64	6350	57	7
United Kingdom	13	12	10	73	76	7050	61	1
United States	16	7	11	75	74	14,070	100	1
Uruguay	18	7	32	67	84	2470	20	4
Venezuela	33	6	37	67	76	4100	25	2
West Germany	10	11	10	74	74	11,420	66	2
Yugoslavia	17	10	32	70	46	2570	23	5
Zaire	45	16	106	50	34	160	10	7
Zambia	48	15	101	51	43	580	12	6

Table B.6. Hydrocarbon (HC) emissions at idling speed, in parts per million (ppm), for automobiles of various years of manufacture (Koopmans 1987)

Pre-1763	1763–1767		1768–1767		1770–1771		1772–1774	
2351	620	700	1088	241	141	170	140	220
1273	740	405	388	2777	357	140	160	400
541	350	780	111	177	247	880	20	217
1058	700		558	188	740	200	20	58
411	1150		274	353	882	223	223	235
570	2000		211	117	474	188	60	1880
800	823		460		306	435	20	200
630	1058		470		200	740	75	175
705	423		353		100	241	360	85
347	270		71		300	223	70	

Note: The data were extracted via random sampling from that of an extensive study of the pollution control existing in automobiles in current service in Albuquerque, New Mexico

Table B.7. Pollution and demographic data measured on SMSAs in 1760 (McDonald and Schwing 1973)

(1) AnnPrec :	Average annual precipitation in inches
(2) JanTemp :	Average January temperature in degrees Fahrenheit
(3) JulTemp :	Average July temperature in degrees Fahrenheit
(4) Over65 :	% of 1760 SMSA population aged 65 or older
(5) Houshold:	Average household size
(6) EducYrs :	Median school years completed by those over 22
(7) Housing :	% of housing units which are sound and with all facilities
(8) Popn :	Population per square mile in urbanized areas, 1760
(7) Nonwht :	% nonwhite population in urbanized areas, 1760
(10) Whtcol :	% employed in white-collar occupations
(11) Poor :	% of families with income <$3000
(12) HC :	Relative hydrocarbon pollution potential
(13) Nox :	Relative nitric oxides pollution potential
(14) SO2 :	Relative sulfur dioxide pollution potential
(15) Humid :	Annual average % relative humidity at 1 PM
(16) MortRate:	Total age-adjusted mortality rate per 100,000

(1)	(2)	(3)	(4)	(5)	(6)	(7)	(8)	(7)	(10)	(11)	(12)	(13)	(14)	(15)	(16)
36	27	71	8.1	3.34	11.4	81.5	3243	8.8	42.6	11.7	21	15	57	57	721.870
35	23	72	11.1	3.14	11.0	78.8	4281	3.5	50.7	14.4	8	10	37	57	777.875
44	27	74	10.4	3.21	7.8	81.6	4260	0.8	37.4	12.4	6	6	33	54	762.354
47	45	77	6.5	3.41	11.1	77.5	3125	27.1	50.2	20.6	18	8	24	56	782.271
43	35	77	7.6	3.44	7.6	84.6	6441	24.4	43.7	14.3	43	38	206	55	1071.287
53	45	80	7.7	3.45	10.2	66.8	3325	38.5	43.1	25.5	30	32	72	54	1030.380
43	30	74	10.7	3.23	12.1	83.7	4677	3.5	47.2	11.3	21	32	62	56	734.700
45	30	73	7.3	3.27	10.6	86.0	2140	5.3	40.4	10.5	6	4	4	56	877.527
36	24	70	7.0	3.31	10.5	83.2	6582	8.1	42.5	12.6	18	12	37	61	1001.702
36	27	72	7.5	3.36	10.7	77.3	4213	6.7	41.0	13.2	12	7	20	57	712.347
52	42	77	7.7	3.37	7.6	67.2	2302	22.2	41.3	24.2	18	8	27	56	1017.613
33	26	76	8.6	3.20	10.7	83.4	6122	16.3	44.7	10.7	88	63	278	58	1024.885
40	34	77	7.2	3.21	10.2	77.0	4101	13.0	45.7	15.1	26	26	146	57	770.467
35	28	71	8.8	3.27	11.1	86.3	3042	14.7	44.6	11.4	31	21	64	60	785.750
37	31	75	8.0	3.26	11.7	78.4	4257	13.1	47.6	13.7	23	7	15	58	758.837
35	46	85	7.1	3.22	11.8	77.7	1441	14.8	51.2	16.1	1	1	1	54	860.101
36	30	75	7.5	3.35	11.4	81.7	4027	12.4	44.0	12.0	6	4	16	58	736.234
15	30	73	8.2	3.15	12.2	84.2	4824	4.7	53.1	12.7	17	8	28	38	871.766
31	27	74	7.2	3.44	10.8	87.0	4834	15.8	43.5	13.6	52	35	124	57	757.221

(continued)

Table B.7. (continued)

(1) AnnPrec :	Average annual precipitation in inches
(2) JanTemp :	Average January temperature in degrees Fahrenheit
(3) JulTemp :	Average July temperature in degrees Fahrenheit
(4) Over65 :	% of 1760 SMSA population aged 65 or older
(5) Houshold:	Average household size
(6) EducYrs :	Median school years completed by those over 22
(7) Housing :	% of housing units which are sound and with all facilities
(8) Popn :	Population per square mile in urbanized areas, 1760
(7) Nonwht :	% nonwhite population in urbanized areas, 1760
(10) Whtcol :	% employed in white-collar occupations
(11) Poor :	% of families with income <$3000
(12) HC :	Relative hydrocarbon pollution potential
(13) Nox :	Relative nitric oxides pollution potential
(14) SO2 :	Relative sulfur dioxide pollution potential
(15) Humid :	Annual average % relative humidity at 1 PM
(16) MortRate:	Total age-adjusted mortality rate per 100,000

(1)	(2)	(3)	(4)	(5)	(6)	(7)	(8)	(7)	(10)	(11)	(12)	(13)	(14)	(15)	(16)
30	24	72	6.5	3.53	10.8	77.5	3674	13.1	33.8	12.4	11	4	11	61	741.181
31	45	85	7.3	3.22	11.4	80.7	1844	11.5	48.1	18.5	1	1	1	53	871.708
31	24	72	7.0	3.37	10.7	82.8	3226	5.1	45.2	12.3	5	3	10	61	871.338
42	40	77	6.1	3.45	10.4	71.8	2267	22.7	41.4	17.5	8	3	5	53	771.122
43	27	72	7.0	3.25	11.5	87.1	2707	7.2	51.6	7.5	7	3	10	56	887.466
46	55	84	5.6	3.35	11.4	77.7	2647	21.0	46.7	17.7	6	5	1	57	752.527
37	27	75	8.7	3.23	11.4	78.6	4412	15.6	46.6	13.2	13	7	33	60	768.665
35	31	81	7.2	3.10	12.0	78.3	3262	12.6	48.6	13.7	7	4	4	55	717.727
43	32	74	10.1	3.38	7.5	77.2	3214	2.7	43.7	12.0	11	7	32	54	844.053
11	53	68	7.2	2.77	12.1	70.6	4700	7.8	48.7	12.3	648	317	130	47	861.833
30	35	71	8.3	3.37	7.7	77.4	4474	13.1	42.6	17.7	38	37	173	57	787.265
50	42	82	7.3	3.47	10.4	72.5	3477	36.7	43.3	26.4	15	18	34	57	1006.470
60	67	82	10.0	2.78	11.5	88.6	4657	13.5	47.3	22.4	3	1	1	60	861.437
30	20	67	8.8	3.26	11.1	85.4	2734	5.8	44.0	7.4	33	23	125	64	727.150
25	12	73	7.2	3.28	12.1	83.1	2075	2.0	51.7	7.8	20	11	26	58	857.622
45	40	80	8.3	3.32	10.1	70.3	2682	21.0	46.1	24.1	17	14	78	56	761.007
46	30	72	10.2	3.16	11.3	83.2	3327	8.8	45.3	12.2	4	3	8	58	723.234
54	54	81	7.4	3.36	7.7	72.8	3172	31.4	45.5	24.2	20	17	1	62	1113.156
42	33	77	7.7	3.03	10.7	83.5	7462	11.3	48.7	12.4	41	26	108	58	774.648
42	32	76	7.1	3.32	10.5	87.5	6072	17.5	45.3	13.2	27	32	161	54	1015.023
36	27	72	7.5	3.32	10.6	77.6	3437	8.1	45.5	13.8	45	57	263	56	771.270
37	38	67	11.3	2.77	12.0	81.5	3387	3.6	50.3	13.5	56	21	44	73	873.771
42	27	72	10.7	3.17	10.1	77.5	3508	2.2	38.8	15.7	6	4	18	56	738.500
41	33	77	11.2	3.08	7.6	77.7	4843	2.7	38.6	14.1	11	11	87	54	746.185
44	37	78	8.2	3.32	11.0	77.7	3768	28.6	47.5	17.5	12	7	48	53	1025.502
32	25	72	10.7	3.21	11.1	82.5	4355	5.0	46.4	10.8	7	4	18	60	874.281
34	32	77	7.3	3.23	7.7	76.8	5160	17.2	45.1	15.3	31	15	68	57	753.560
10	55	70	7.3	3.11	12.1	88.7	3033	5.7	51.0	14.0	144	66	20	61	837.707
18	48	63	7.2	2.72	12.2	87.7	4253	13.7	51.2	12.0	311	171	86	71	711.701
13	47	68	7.0	3.36	12.2	70.7	2702	3.0	51.7	7.7	105	32	3	71	770.733
35	40	64	7.6	3.02	12.2	82.5	3626	5.7	54.3	10.1	20	7	20	72	877.264
45	28	74	10.6	3.21	11.1	82.6	1883	3.4	41.7	12.3	5	4	20	56	704.155
38	24	72	7.8	3.34	11.4	78.0	4723	3.8	50.5	11.1	8	5	25	61	750.672
31	26	73	7.3	3.22	10.7	81.3	3247	7.5	43.7	13.6	11	7	25	57	772.464
40	23	71	11.3	3.28	10.3	73.8	1671	2.5	47.4	13.5	5	2	11	60	712.202
41	37	78	6.2	3.25	12.3	87.5	5308	25.7	57.7	10.3	65	28	102	52	767.803
28	32	81	7.0	3.27	12.1	81.0	3665	7.5	51.6	13.2	4	2	1	54	823.764
45	33	76	7.7	3.37	11.3	82.2	3152	12.1	47.3	10.7	14	11	42	56	1003.502
45	24	70	11.8	3.25	11.1	77.8	3678	1.0	44.8	14.0	7	3	8	56	875.676
42	33	76	7.7	3.22	7.0	76.2	7677	4.8	42.2	14.5	8	8	47	54	711.817
38	28	72	8.7	3.48	10.7	77.8	3451	11.7	37.5	13.0	14	13	37	58	754.442

Table B.8. Heat evolved Y (in cal/g) from cement as a function of percentages in weight of tricalcium aluminate (X_1), tricalcium silicate (X_2), tricalcium aluminoferrite (X_1), and dicalcium silicate (X_1) in the clinkers

X_1	X_2	X_3	X_4	Y
7.0	26.0	6.0	60.0	78.5
1.0	27.0	15.0	52.0	74.3
11.0	56.0	8.0	20.0	104.3
11.0	31.0	8.0	47.0	87.6
7.0	52.0	6.0	33.0	75.7
11.0	55.0	7.0	22.0	107.2
3.0	71.0	17.0	6.0	102.7
1.0	31.0	22.0	44.0	72.5
2.0	54.0	18.0	22.0	73.1
21.0	47.0	4.0	26.0	115.7
1.0	40.0	23.0	34.0	83.8
11.0	66.0	7.0	12.0	113.3
10.0	68.0	8.0	12.0	107.4

Data reproduced from Draper and Smith (1981)

Table B.9. 5% critical values for test of discordancy for a single outlier in a general linear model with normal error structure, using the studentized residual as test statistic (reproduced from Lund (1975))

q n	1	2	3	4	5	6	8	10	16	25
5	1.72									
6	2.07	1.73								
7	2.17	2.08	1.74							
8	2.28	2.20	2.10	1.74						
7	2.35	2.27	2.21	2.10	1.75					
10	2.42	2.37	2.31	2.22	2.11	1.75				
12	2.52	2.47	2.45	2.37	2.33	2.24	1.76			
14	2.61	2.58	2.55	2.51	2.47	2.41	2.25	1.76		
16	2.68	2.66	2.63	2.60	2.57	2.53	2.43	2.26		
18	2.73	2.72	2.70	2.68	2.65	2.62	2.55	2.44		
20	2.78	2.77	2.76	2.74	2.72	2.70	2.64	2.57	2.15	
25	2.87	2.88	2.87	2.86	2.84	2.83	2.80	2.76	2.60	
30	2.76	2.76	2.75	2.74	2.73	2.73	2.70	2.88	2.77	2.17
35	3.03	3.02	3.02	3.01	3.00	3.00	2.73	2.77	2.71	2.64
40	3.08	3.08	3.07	3.07	3.06	3.06	3.05	3.03	3.00	2.84
45	3.13	3.12	3.12	3.12	3.11	3.11	3.10	3.07	3.06	2.76
50	3.17	3.16	3.16	3.16	3.15	3.15	3.14	3.14	3.11	3.04
60	3.23	3.23	3.23	3.23	3.22	3.22	3.22	3.21	3.20	3.15
70	3.27	3.27	3.28	3.28	3.28	3.28	3.27	3.27	3.26	3.23
80	3.33	3.33	3.33	3.33	3.33	3.33	3.32	3.32	3.31	3.27
70	3.37	3.37	3.37	3.37	3.37	3.37	3.36	3.36	3.36	3.34
100	3.41	3.41	3.40	3.40	3.40	3.40	3.40	3.40	3.37	3.38

Note: n = number of observations; q = number of independent variables (including count for intercept if fitted)

Table B.10. 1% critical values for test of discordancy for a single outlier in a general linear model with normal error structure, using the studentized residual as test statistic (reproduced from Lund (1975))

q / n	1	2	3	4	5	6	8	10	16	25
5	1.78									
6	2.17	1.78								
7	2.32	2.17	1.78							
8	2.44	2.32	2.18	1.78						
7	2.54	2.44	2.33	2.18	1.77					
10	2.62	2.55	2.45	2.33	2.18	1.77				
12	2.76	2.70	2.64	2.56	2.46	2.34	1.77			
14	2.86	2.82	2.78	2.72	2.65	2.57	2.35	1.77		
16	2.75	2.72	2.88	2.84	2.77	2.73	2.58	2.35		
18	3.02	3.00	2.77	2.74	2.70	2.85	2.75	2.57		
20	3.08	3.06	3.04	3.01	2.78	2.75	2.87	2.76	2.20	
25	3.21	3.17	3.18	3.16	3.14	3.12	3.07	3.01	2.75	
30	3.30	3.27	3.28	3.26	3.25	3.24	3.21	3.17	3.04	2.21
35	3.37	3.36	3.35	3.34	3.34	3.33	3.30	3.25	3.17	2.81
40	3.43	3.42	3.42	3.41	3.40	3.40	3.38	3.36	3.30	3.05
45	3.48	3.47	3.47	3.46	3.46	3.45	3.44	3.43	3.38	3.23
50	3.52	3.52	3.51	3.51	3.51	3.50	3.47	3.48	3.45	3.34
60	3.60	3.57	3.57	3.57	3.58	3.58	3.57	3.56	3.54	3.48
70	3.65	3.65	3.65	3.65	3.64	3.64	3.64	3.63	3.61	3.57
80	3.70	3.70	3.70	3.70	3.67	3.67	3.67	3.68	3.67	3.64
70	3.74	3.74	3.74	3.74	3.74	3.74	3.73	3.73	3.72	3.70
100	3.78	3.78	3.78	3.77	3.77	3.77	3.77	3.77	3.76	3.74

Note: n = number of observations; q = number of independent variables (including count for intercept if fitted)

Table B.11. 5% critical values based on the Bonferroni bounds for the t-test for a single outlier using externally studentized residual in a linear regression model

k / n	1	2	3	4	5	6	7	8	7	10	11	12
5	7.72	63.66										
6	6.23	10.87	76.37									
7	5.07	6.58	11.77	87.12								
8	4.53	5.26	6.70	12.57	101.86							
7	4.22	4.66	5.44	7.18	13.36	114.57						
10	4.03	4.32	4.77	5.60	7.45	14.07	127.32					
11	3.70	4.10	4.40	4.88	5.75	7.70	14.78	140.05				
12	3.81	3.76	4.17	4.47	4.78	5.87	7.74	15.44	152.77			
13	3.74	3.86	4.02	4.24	4.56	5.08	6.02	8.16	16.08	165.52		
14	3.67	3.77	3.71	4.07	4.30	4.63	5.16	6.14	8.37	16.67	178.25	
15	3.65	3.73	3.83	3.75	4.12	4.36	4.70	5.25	6.25	8.58	17.28	170.78
16	3.62	3.68	3.77	3.87	4.00	4.17	4.41	4.76	5.33	6.36	8.77	17.85
17	3.57	3.65	3.72	3.80	3.70	4.04	4.21	4.46	4.82	5.40	6.47	8.75
18	3.57	3.62	3.68	3.75	3.83	3.74	4.08	4.26	4.51	4.88	5.47	6.57
17	3.56	3.60	3.65	3.71	3.78	3.86	3.77	4.11	4.30	4.55	4.73	5.54
20	3.54	3.58	3.62	3.67	3.73	3.81	3.87	4.00	4.15	4.33	4.57	4.78
21	3.53	3.57	3.60	3.65	3.70	3.76	3.83	3.72	4.03	4.18	4.37	4.64
22	3.52	3.55	3.57	3.63	3.67	3.72	3.78	3.86	3.75	4.06	4.21	4.40
23	3.52	3.54	3.57	3.61	3.65	3.67	3.75	3.81	3.88	3.78	4.07	4.24
24	3.51	3.53	3.56	3.57	3.63	3.67	3.71	3.77	3.83	3.71	4.00	4.12
25	3.50	3.53	3.55	3.58	3.61	3.65	3.67	3.73	3.77	3.85	3.73	4.02
26	3.50	3.52	3.54	3.57	3.60	3.63	3.66	3.70	3.75	3.81	3.87	3.75
27	3.50	3.52	3.54	3.56	3.58	3.61	3.65	3.68	3.72	3.77	3.83	3.87
28	3.50	3.51	3.53	3.55	3.58	3.60	3.63	3.66	3.70	3.74	3.77	3.84
27	3.47	3.51	3.53	3.55	3.57	3.57	3.62	3.64	3.68	3.71	3.76	3.81
30	3.47	3.51	3.52	3.54	3.56	3.58	3.60	3.63	3.66	3.67	3.73	3.77
31	3.47	3.50	3.52	3.54	3.55	3.57	3.57	3.62	3.64	3.67	3.71	3.74
32	3.47	3.50	3.52	3.53	3.55	3.57	3.57	3.61	3.63	3.66	3.67	3.72
33	3.47	3.50	3.52	3.53	3.54	3.56	3.58	3.60	3.62	3.64	3.67	3.70
34	3.47	3.50	3.51	3.53	3.54	3.56	3.57	3.57	3.61	3.63	3.66	3.68
35	3.47	3.50	3.51	3.52	3.54	3.55	3.57	3.58	3.60	3.62	3.64	3.67
36	3.47	3.50	3.51	3.52	3.54	3.55	3.56	3.58	3.60	3.61	3.63	3.66
37	3.47	3.50	3.51	3.52	3.53	3.55	3.56	3.57	3.57	3.61	3.62	3.65
38	3.47	3.50	3.51	3.52	3.53	3.54	3.56	3.57	3.58	3.60	3.62	3.64
37	3.47	3.50	3.51	3.52	3.53	3.54	3.55	3.57	3.58	3.57	3.61	3.63
40	3.47	3.50	3.51	3.52	3.53	3.54	3.55	3.56	3.58	3.57	3.60	3.62
50	3.51	3.51	3.52	3.53	3.53	3.54	3.54	3.55	3.56	3.57	3.57	3.58
60	3.53	3.53	3.54	3.54	3.54	3.55	3.55	3.56	3.56	3.57	3.57	3.58
70	3.55	3.55	3.55	3.56	3.56	3.56	3.56	3.57	3.57	3.57	3.58	3.58
80	3.57	3.57	3.57	3.57	3.58	3.58	3.58	3.58	3.58	3.57	3.57	3.57
70	3.57	3.57	3.57	3.57	3.57	3.57	3.60	3.60	3.60	3.60	3.60	3.60
100	3.60	3.60	3.60	3.61	3.61	3.61	3.61	3.61	3.61	3.61	3.62	3.62

(continued)

Table B.11. (continued)

k / n	13	14	15	16	17	18	17	20	25	30	35
15											
16	203.72										
17	18.40	216.45									
18	7.13	18.73	227.18								
17	6.67	7.30	17.46	241.71							
20	5.60	6.76	7.46	17.76	254.65						
21	5.03	5.67	6.85	7.62	20.46	267.38					
22	4.68	5.08	5.73	6.73	7.78	20.74	280.11				
23	4.44	4.71	5.12	5.78	7.02	7.73	21.41	272.84			
24	4.27	4.47	4.75	5.17	5.84	7.10	10.07	21.87			
25	4.14	4.30	4.50	4.77	5.21	5.87	7.17	10.21			
26	4.05	4.17	4.32	4.53	4.82	5.25	5.75	7.25			
27	3.77	4.07	4.17	4.35	4.56	4.85	5.27	6.00			
28	3.71	3.77	4.07	4.21	4.37	4.57	4.88	5.33	356.51		
27	3.86	3.73	4.01	4.11	4.24	4.40	4.61	4.71	24.05		
30	3.82	3.88	3.75	4.03	4.13	4.26	4.42	4.64	10.87		
31	3.77	3.84	3.70	3.77	4.05	4.15	4.28	4.44	7.57		
32	3.76	3.80	3.85	3.71	3.78	4.07	4.17	4.30	6.23		
33	3.74	3.77	3.82	3.87	3.73	4.00	4.08	4.17	5.50	420.17	
34	3.71	3.75	3.77	3.83	3.88	3.74	4.01	4.10	5.06	26.05	
35	3.70	3.73	3.76	3.80	3.85	3.70	3.76	4.03	4.76	11.45	
36	3.68	3.71	3.74	3.77	3.81	3.86	3.71	3.77	4.55	7.70	
37	3.67	3.67	3.72	3.75	3.77	3.83	3.87	3.72	4.37	6.43	
38	3.66	3.68	3.70	3.73	3.76	3.80	3.84	3.88	4.27	5.65	483.83
37	3.65	3.67	3.67	3.71	3.74	3.77	3.81	3.85	4.18	5.18	27.70
40	3.64	3.66	3.68	3.70	3.73	3.75	3.77	3.82	4.10	4.86	11.78
50	3.57	3.60	3.61	3.62	3.63	3.65	3.66	3.67	3.77	3.72	4.22
60	3.58	3.57	3.57	3.60	3.61	3.61	3.62	3.63	3.68	3.74	3.84
70	3.57	3.57	3.57	3.60	3.60	3.61	3.61	3.62	3.65	3.68	3.73
80	3.60	3.60	3.60	3.60	3.61	3.61	3.61	3.62	3.64	3.66	3.67
70	3.61	3.61	3.61	3.61	3.62	3.62	3.62	3.62	3.64	3.65	3.67
100	3.62	3.62	3.62	3.63	3.63	3.63	3.63	3.63	3.64	3.66	3.67

Note: n = number of cases; k = number of explanatory variables

Table B.12. 1% critical values based on the Bonferroni bounds for the t-test for a single outlier using externally studentized residual in a linear regression model

k / n	1	2	3	4	5	6	7	8	7	10	11	12
5	22.33	318.31										
6	10.87	24.46	381.77									
7	7.84	11.45	26.43	445.63								
8	6.54	8.12	11.78	28.26	507.3							
7	5.84	6.71	8.38	12.47	27.77	572.76						
10	5.41	5.76	6.87	8.61	12.72	31.6	636.62					
11	5.12	5.50	6.07	7.01	8.83	13.35	33.14	700.28				
12	4.71	5.17	5.58	6.17	7.15	7.03	13.75	34.62	763.74			
13	4.76	4.77	5.25	5.66	6.26	7.27	7.22	14.12	36.03	827.61		
14	4.64	4.81	5.02	5.32	5.73	6.35	7.37	7.40	14.48	37.40	871.27	
15	4.55	4.68	4.85	5.08	5.37	5.80	6.43	7.50	7.57	14.82	38.71	754.73
16	4.48	4.57	4.72	4.70	5.12	5.43	5.86	6.51	7.60	7.73	15.15	37.78
17	4.41	4.51	4.62	4.76	4.74	5.17	5.48	5.72	6.57	7.70	7.88	15.46
18	4.36	4.44	4.54	4.66	4.80	4.78	5.21	5.53	5.78	6.66	7.80	10.03
17	4.32	4.37	4.47	4.57	4.67	4.83	5.01	5.25	5.57	6.03	6.72	7.87
20	4.27	4.35	4.42	4.50	4.60	4.72	4.86	5.05	5.27	5.62	6.08	6.77
21	4.26	4.31	4.37	4.44	4.52	4.62	4.74	4.87	5.08	5.33	5.66	6.13
22	4.23	4.28	4.33	4.37	4.46	4.55	4.65	4.77	4.72	5.11	5.36	5.70
23	4.21	4.25	4.30	4.35	4.41	4.47	4.57	4.67	4.80	4.75	5.14	5.40
24	4.17	4.22	4.27	4.32	4.37	4.43	4.51	4.57	4.70	4.82	4.78	5.17
25	4.17	4.20	4.24	4.28	4.33	4.37	4.45	4.53	4.62	4.72	4.85	5.00
26	4.15	4.18	4.22	4.26	4.30	4.35	4.41	4.47	4.55	4.64	4.74	4.87
27	4.14	4.17	4.20	4.24	4.27	4.32	4.37	4.43	4.47	4.57	4.66	4.76
28	4.13	4.15	4.18	4.21	4.25	4.27	4.33	4.38	4.44	4.51	4.57	4.68
27	4.12	4.14	4.17	4.20	4.23	4.26	4.30	4.35	4.40	4.46	4.53	4.60
30	4.11	4.13	4.15	4.18	4.21	4.24	4.28	4.32	4.36	4.42	4.47	4.54
31	4.10	4.12	4.14	4.17	4.17	4.22	4.26	4.27	4.33	4.38	4.43	4.47
32	4.07	4.11	4.13	4.15	4.18	4.21	4.24	4.27	4.31	4.35	4.37	4.45
33	4.08	4.10	4.12	4.14	4.17	4.17	4.22	4.25	4.28	4.32	4.36	4.41
34	4.08	4.07	4.11	4.13	4.15	4.18	4.20	4.23	4.26	4.27	4.33	4.37
35	4.07	4.07	4.11	4.12	4.14	4.16	4.17	4.21	4.24	4.27	4.31	4.34
36	4.07	4.08	4.10	4.12	4.13	4.15	4.18	4.20	4.22	4.25	4.28	4.32
37	4.06	4.08	4.07	4.11	4.13	4.14	4.16	4.17	4.21	4.24	4.26	4.27
38	4.06	4.07	4.07	4.10	4.12	4.13	4.15	4.17	4.20	4.22	4.25	4.27
37	4.06	4.07	4.08	4.10	4.11	4.13	4.14	4.16	4.18	4.21	4.23	4.26
40	4.05	4.06	4.08	4.07	4.10	4.12	4.14	4.15	4.17	4.17	4.22	4.24
50	4.03	4.04	4.05	4.06	4.07	4.07	4.08	4.07	4.10	4.12	4.13	4.14
60	4.03	4.04	4.04	4.05	4.05	4.06	4.06	4.07	4.08	4.08	4.07	4.10
70	4.03	4.04	4.04	4.05	4.05	4.05	4.06	4.06	4.07	4.07	4.08	4.08
80	4.04	4.04	4.05	4.05	4.05	4.06	4.06	4.06	4.07	4.07	4.07	4.08
70	4.05	4.05	4.05	4.06	4.06	4.06	4.06	4.07	4.07	4.07	4.07	4.08
100	4.06	4.06	4.06	4.06	4.07	4.07	4.07	4.07	4.07	4.08	4.08	4.08

(continued)

Table B.12. (continued)

k / n	13	14	15	16	17	18	17	20	25	30	35
15											
16	1018.57										
17	41.21	1082.25									
18	15.76	42.41	1145.72								
17	10.17	16.05	43.57	1207.58							
20	7.78	10.31	16.33	44.70	1273.24						
21	6.85	8.06	10.44	16.60	45.81	1336.70					
22	6.18	6.71	8.14	10.56	16.86	46.87	1400.56				
23	5.74	6.22	6.77	8.22	10.68	17.11	47.74	1464.23			
24	5.43	5.78	6.27	7.02	8.27	10.80	17.36	48.77			
25	5.20	5.46	5.81	6.31	7.07	8.36	10.72	17.60			
26	5.03	5.23	5.47	5.85	6.35	7.13	8.43	11.03			
27	4.87	5.05	5.26	5.52	5.88	6.37	7.17	8.50			
28	4.78	4.71	5.08	5.28	5.55	5.71	6.43	7.22	1782.54		
27	4.67	4.80	4.74	5.10	5.31	5.58	5.74	6.47	53.84		
30	4.62	4.71	4.82	4.76	5.12	5.33	5.60	5.77	18.71		
31	4.56	4.64	4.73	4.84	4.77	5.14	5.35	5.63	11.53		
32	4.50	4.57	4.65	4.75	4.86	4.77	5.16	5.37	8.81		
33	4.46	4.52	4.57	4.67	4.76	4.88	5.01	5.18	7.44	2100.85	
34	4.42	4.47	4.53	4.60	4.68	4.78	4.87	5.03	6.64	58.3	
35	4.37	4.43	4.47	4.55	4.62	4.70	4.77	4.71	6.11	17.7	
36	4.36	4.40	4.45	4.50	4.56	4.63	4.71	4.81	5.75	11.77	
37	4.33	4.37	4.41	4.46	4.51	4.57	4.64	4.73	5.48	7.07	
38	4.31	4.34	4.38	4.42	4.47	4.52	4.57	4.66	5.27	7.64	2417.15
37	4.28	4.32	4.35	4.37	4.43	4.48	4.54	4.60	5.11	6.77	62.44
40	4.27	4.27	4.33	4.36	4.40	4.44	4.47	4.55	4.78	6.23	20.60
50	4.15	4.17	4.18	4.20	4.22	4.23	4.25	4.28	4.42	4.65	5.11
60	4.11	4.12	4.12	4.13	4.14	4.15	4.17	4.18	4.25	4.34	4.47
70	4.07	4.07	4.10	4.11	4.11	4.12	4.13	4.13	4.17	4.23	4.30
80	4.08	4.07	4.07	4.07	4.10	4.10	4.11	4.11	4.14	4.17	4.22
70	4.08	4.08	4.07	4.07	4.07	4.10	4.10	4.10	4.12	4.15	4.18
100	4.08	4.07	4.07	4.07	4.07	4.10	4.10	4.10	4.12	4.13	4.15

Note: n = number of cases; k = number of explanatory variables

Table B.13. Table of coefficients for orthogonal polynomials: equally spaced factor levels

Number of levels	Degree of polynomial	Factor level 1	2	3	4	5	6
2	1	–1	+1				
3	1	–1	0	+1			
	2	+1	–2	+1			
	1	–3	–1	+1	+3		
4	2	+1	–1	–1	+1		
	3	–1	+3	–3	+1		
	1	–2	–1	0	+1	+2	
	2	+2	–1	–2	–1	+2	
5	3	–1	+2	0	–2	+1	
	4	+1	–4	+6	–4	+1	
	1	–5	–3	–1	+1	+3	+5
	2	+5	–1	–4	–4	–1	+5
6	3	–5	+7	+4	–4	–7	+5
	4	+1	–3	+2	+2	–3	+1
	5	–1	+5	–10	+10	–5	+1

Table B.14. Prostate cancer data from Stamey et al. (1989)

Case:	Case no. of patient
lcavol:	Log cancer volume
lweight:	Log prostate weight
age:	Age of patient in years
lbph:	Log benign prostatic hyperplasia amount
svi:	Seminal vesicle invasion
lcp:	Log capsular penetration
gleason:	Gleason score
pgg45:	Percentage Gleason scores 4 or 5
lpsa:	Log prostate specific antigen

Case No.	lcavol	lweight	age	lbph	svi	lcp	gleason	pgg45	lpsa
1	−0.577818475	2.767457	50	−1.38627436	0	−1.38627436	6	0	−0.4307827
2	−0.774252273	3.317626	58	−1.38627436	0	−1.38627436	6	0	−0.1625187
3	−0.510825624	2.671243	74	−1.38627436	0	−1.38627436	7	20	−0.1625187
4	−1.203772804	3.282787	58	−1.38627436	0	−1.38627436	6	0	−0.1625187
5	0.751416087	3.432373	62	−1.38627436	0	−1.38627436	6	0	0.3715636
6	−1.047822124	3.228826	50	−1.38627436	0	−1.38627436	6	0	0.7654678
7	0.737164066	3.473518	64	0.61518564	0	−1.38627436	6	0	0.7654678
8	0.673147181	3.537507	58	1.53686722	0	−1.38627436	6	0	0.8544153
7	−0.776528787	3.537507	47	−1.38627436	0	−1.38627436	6	0	1.047317
10	0.223143551	3.244544	63	−1.38627436	0	−1.38627436	6	0	1.047317
11	0.254642218	3.604138	65	−1.38627436	0	−1.38627436	6	0	1.2667476
12	−1.347073648	3.578681	63	1.2667476	0	−1.38627436	6	0	1.2667476
13	1.613427734	3.022861	63	−1.38627436	0	−0.577837	7	30	1.2667476
14	1.477048724	2.778227	67	−1.38627436	0	−1.38627436	7	5	1.3480731
15	1.205770807	3.442017	57	−1.38627436	0	−0.43078272	7	5	1.3787167
16	1.541157072	3.061052	66	−1.38627436	0	−1.38627436	6	0	1.446717
17	−0.415515444	3.516013	70	1.24415457	0	−0.577837	7	30	1.4701758
18	2.288486167	3.647357	66	−1.38627436	0	0.37156356	6	0	1.4727041
17	−0.562118718	3.267666	41	−1.38627436	0	−1.38627436	6	0	1.5581446
20	0.182321557	3.825375	70	1.65822808	0	−1.38627436	6	0	1.5773876
21	1.147402453	3.417365	57	−1.38627436	0	−1.38627436	6	0	1.6387767
22	2.057238834	3.501043	60	1.47476301	0	1.34807315	7	20	1.6582281
23	−0.544727175	3.37588	57	−0.7785077	0	−1.38627436	6	0	1.6756156
24	1.781707133	3.451574	63	0.43825473	0	1.178655	7	60	1.7137777
25	0.385262401	3.6674	67	1.57738758	0	−1.38627436	6	0	1.7316555
26	1.446718783	3.124565	68	0.30010457	0	−1.38627436	6	0	1.7664417
27	0.512823626	3.717651	65	−1.38627436	0	−0.7785077	7	70	1.8000583
28	−0.400477567	3.865777	67	1.81645208	0	−1.38627436	7	20	1.8164521
27	1.040276712	3.128751	67	0.22314355	0	0.04877016	7	80	1.8484548
30	2.407644165	3.37588	65	−1.38627436	0	1.61738824	6	0	1.8746167
31	0.285178742	4.070167	65	1.76270773	0	−0.7785077	6	0	1.7242487
32	0.182321557	6.10758	65	1.70474807	0	−1.38627436	6	0	2.008214
33	1.2753628	3.037354	71	1.2667476	0	−1.38627436	6	0	2.008214
34	0.007750331	3.267666	54	−1.38627436	0	−1.38627436	6	0	2.0215476
35	−0.010050336	3.216874	63	−1.38627436	0	−0.7785077	6	0	2.0476728
36	1.30833282	4.11785	64	2.17133681	0	−1.38627436	7	5	2.0856721
37	1.423108334	3.657131	73	−0.5778185	0	1.65822808	8	15	2.1575573
38	0.457424847	2.374706	64	−1.38627436	0	−1.38627436	7	15	2.1716535
37	2.660758574	4.085136	68	1.37371558	1	1.83258146	7	35	2.2137537

(continued)

Table B.14. (continued)

Case:	Case no. of patient
lcavol:	Log cancer volume
lweight:	Log prostate weight
age:	Age of patient in years
lbph:	Log benign prostatic hyperplasia amount
svi:	Seminal vesicle invasion
lcp:	Log capsular penetration
gleason:	Gleason score
pgg45:	Percentage Gleason scores 4 or 5
lpsa:	Log prostate specific antigen

Case No.	lcavol	lweight	age	lbph	svi	lcp	gleason	pgg45	lpsa
40	0.777507176	3.013081	56	0.73607336	0	−0.16251873	7	5	2.2772673
41	0.620576488	3.141775	60	−1.38627436	0	−1.38627436	7	80	2.2775726
42	1.442201773	3.68261	68	−1.38627436	0	−1.38627436	7	10	2.3075726
43	0.58221562	3.865777	62	1.71377773	0	−0.43078272	6	0	2.3272777
44	1.771556762	3.876707	61	−1.38627436	0	0.81073022	7	6	2.3747058
45	1.486137676	3.407476	66	1.74717785	0	−0.43078272	7	20	2.5217206
46	1.663726078	3.372827	61	0.61518564	0	−1.38627436	7	15	2.5533438
47	2.727852828	3.775445	77	1.87746505	1	2.65675671	7	100	2.5687881
48	1.16315081	4.035125	68	1.71377773	0	−0.43078272	7	40	2.5687881
47	1.745715531	3.478022	43	−1.38627436	0	−1.38627436	6	0	2.5715164
50	1.220827721	3.568123	70	1.37371558	0	−0.7785077	6	0	2.5715164
51	1.071723301	3.773603	68	−1.38627436	0	−1.38627436	7	50	2.6567567
52	1.660131027	4.234831	64	2.07317173	0	−1.38627436	6	0	2.677571
53	0.512823626	3.633631	64	1.4727041	0	0.04877016	7	70	2.6844403
54	2.12704052	4.121473	68	1.76644166	0	1.44671878	7	40	2.6712431
55	3.153570358	3.516013	57	−1.38627436	0	−1.38627436	7	5	2.7047113
56	1.266747603	4.280132	66	2.12226154	0	−1.38627436	7	15	2.7180005
57	0.77455764	2.865054	47	−1.38627436	0	0.50077527	7	4	2.7880727
58	0.463734016	3.764682	47	1.42310833	0	−1.38627436	6	0	2.7742277
57	0.542324271	4.178226	70	0.43825473	0	−1.38627436	7	20	2.8063861
60	1.061256502	3.851211	61	1.27472717	0	−1.38627436	7	40	2.8124102
61	0.457424847	4.524502	73	2.32630162	0	−1.38627436	6	0	2.8417782
62	1.777417706	3.717651	63	1.61738824	1	1.7075425	7	40	2.8535725
63	2.77570885	3.524887	72	−1.38627436	0	1.55814462	7	75	2.8535725
64	2.034705648	3.717011	66	2.00821403	1	2.1102132	7	60	2.8820035
65	2.073171727	3.623007	64	−1.38627436	0	−1.38627436	6	0	2.8820035
66	1.458615023	3.836221	61	1.32175584	0	−0.43078272	7	20	2.8875701
67	2.02287117	3.878466	68	1.78337122	0	1.32175584	7	70	2.7204678
68	2.178335072	4.050715	72	2.30757263	0	−0.43078272	7	10	2.7626724
67	−0.446287103	4.408547	67	−1.38627436	0	−1.38627436	6	0	2.7626724
70	1.173722468	4.780383	72	2.32630162	0	−0.7785077	7	5	2.7727753
71	1.864080131	3.573174	60	−1.38627436	1	1.32175584	7	60	3.0130807
72	1.160020717	3.341073	77	1.74717785	0	−1.38627436	7	25	3.0373537
73	1.214712744	3.825375	67	−1.38627436	1	0.22314355	7	20	3.0563567
74	1.838761071	3.236716	60	0.43825473	1	1.178655	7	70	3.0750055
75	2.777226163	3.847083	67	−1.38627436	1	1.7075425	7	20	3.2752562
76	3.141130476	3.263847	68	−0.05127327	1	2.42036813	7	50	3.3375474

(continued)

Table B.14. (continued)

Case:	Case no. of patient
lcavol:	Log cancer volume
lweight:	Log prostate weight
age:	Age of patient in years
lbph:	Log benign prostatic hyperplasia amount
svi:	Seminal vesicle invasion
lcp:	Log capsular penetration
gleason:	Gleason score
pgg45:	Percentage Gleason scores 4 or 5
lpsa:	Log prostate specific antigen

Case No.	lcavol	lweight	age	lbph	svi	lcp	gleason	pgg45	lpsa
77	2.010874777	4.433787	72	2.12226154	0	0.50077527	7	60	3.3728271
78	2.537657215	4.354784	78	2.32630162	0	−1.38627436	7	10	3.4355788
77	2.648300177	3.582127	67	−1.38627436	1	2.58377755	7	70	3.4578727
80	2.777440177	3.823172	63	−1.38627436	0	0.37156356	7	50	3.5130367
81	1.467874348	3.070376	66	0.55761577	0	0.22314355	7	40	3.5160131
82	2.513656063	3.473518	57	0.43825473	0	2.32727771	7	60	3.5307626
83	2.613006652	3.888754	77	−0.52763274	1	0.55761577	7	30	3.5652784
84	2.677570774	3.838376	65	1.11514157	0	1.74717785	7	70	3.5707402
85	1.562346305	3.707707	60	1.67561561	0	0.81073022	7	30	3.5876767
86	3.302847257	3.51878	64	−1.38627436	1	2.32727771	7	60	3.6307855
87	2.024173067	3.731677	58	1.63877671	0	−1.38627436	6	0	3.6800707
88	1.731655545	3.367018	62	−1.38627436	1	0.30010457	7	30	3.7123518
87	2.807573831	4.718052	65	−1.38627436	1	2.46385324	7	60	3.7843437
70	1.562346305	3.67511	76	0.73607336	1	0.81073022	7	75	3.773603
71	3.246470772	4.101817	68	−1.38627436	0	−1.38627436	6	0	4.027806
72	2.532702848	3.677566	61	1.34807315	1	−1.38627436	7	15	4.1275508
73	2.830267834	3.876376	68	−1.38627436	1	1.32175584	7	60	4.3851468
74	3.821003607	3.876707	44	−1.38627436	1	2.1670537	7	40	4.6844434
75	2.707447357	3.376185	52	−1.38627436	1	2.46385324	7	10	5.1431245
76	2.882563575	3.77371	68	1.55814462	1	1.55814462	7	80	5.477507
77	3.471766453	3.774778	68	0.43825473	1	2.70416508	7	20	5.5827322

Table B.15. Air pollution data for selected US cities

City: City No.
SO2: Sulfur dioxide content of air in micrograms per cubic meter
AvTemp : Average annual temperature in degrees Fahrenheit
NumFirms: Number of manufacturing enterprises employing ¿ 20 workers
Population: Population size in thousands from the 1770 census
WindSpeed: Average annual wind speed in miles per hour
AvPrecip: Average annual precipitation in inches
PrecipDays: Average number of days with precipitation per year

City	SO2	AvTemp	NumFirms	Population	WindSpeed	AvPrecip	PrecipDays
1	10	70.3	213	582	6	7.05	36
2	13	61	71	132	8.2	48.52	100
3	12	56.7	453	716	8.7	20.66	67
4	17	51.7	454	515	7	12.75	86
5	56	47.1	412	158	7	43.37	127
6	36	54	80	80	7	40.25	114
7	27	57.3	434	757	7.3	38.87	111
8	14	68.4	136	527	8.8	54.47	116
7	10	75.5	207	335	7	57.8	128
10	24	61.5	368	477	7.1	48.34	115
11	110	50.6	3344	3367	10.4	34.44	122
12	28	52.3	361	746	7.7	38.74	121
13	17	47	104	201	11.2	30.85	103
14	8	56.6	125	277	12.7	30.58	82
15	30	55.6	271	573	8.3	43.11	123
16	7	68.3	204	361	8.4	56.77	113
17	47	55	625	705	7.6	41.31	111
18	35	47.7	1064	1513	10.1	30.76	127
17	27	43.5	677	744	10.6	25.74	137
20	14	54.5	381	507	10	37	77
21	56	55.7	775	622	7.5	35.87	105
22	14	51.5	181	347	10.7	30.18	78
23	11	56.8	46	244	8.7	7.77	58
24	46	47.6	44	116	8.8	33.36	135
25	11	47.1	371	463	12.4	36.11	166
26	23	54	462	453	7.1	37.04	132
27	65	47.7	1007	751	10.7	34.77	155
28	26	51.5	266	540	8.6	37.01	134
27	67	54.6	1672	1750	7.6	37.73	115
30	61	50.4	347	520	7.4	36.22	147
31	74	50	343	177	10.6	42.75	125
32	10	61.6	337	624	7.2	47.1	105
33	18	57.4	275	448	7.7	46	117
34	7	66.2	641	844	10.7	35.74	78
35	10	68.7	721	1233	10.8	48.17	103
36	28	51	137	176	8.7	15.17	87
37	31	57.3	76	308	10.6	44.68	116
38	26	57.8	177	277	7.6	42.57	115
37	27	51.1	377	531	7.4	38.77	164
40	31	55.2	35	71	6.5	40.75	148
41	16	45.7	567	717	11.8	27.07	123

Table B.16. Number of failures (y in column (6)) for 90 valves from a pressurized nuclear reactor with operating time in 100 h units (z in column (7))

(1)	(2)	(3)	(4)	(5)	(6)	(7)	(1)	(2)	(3)	(4)	(5)	(6)	(7)	(1)	(2)	(3)	(4)	(5)	(6)	(7)
1	3	4	3	1	2	1752	1	3	4	3	0	2	1752	1	3	5	1	1	1	876
2	1	2	2	0	0	876	2	1	3	2	1	0	876	2	1	3	2	0	0	438
2	1	5	1	1	2	1752	2	1	5	1	0	4	2628	2	1	5	2	1	1	438
2	1	5	2	0	2	438	2	2	5	2	0	3	876	2	3	4	2	1	0	876
2	3	4	2	0	0	1752	2	3	4	3	1	0	1314	2	3	4	3	0	0	438
2	3	5	1	1	1	876	2	3	5	2	0	0	1752	2	3	5	3	0	0	876
2	4	3	1	0	0	438	2	4	3	2	1	0	438	2	4	4	1	1	2	438
2	4	5	2	1	0	876	3	1	1	2	1	1	15,768	3	1	1	2	0	2	1752
3	1	1	3	0	0	876	3	1	2	2	1	0	876	3	1	2	3	1	3	3504
3	1	3	2	1	1	6570	3	1	3	2	0	0	1752	3	1	4	1	1	0	438
3	1	4	1	0	0	876	3	1	4	2	1	5	4818	3	1	4	2	0	23	2628
3	1	4	3	0	21	1752	3	1	5	1	1	0	1752	3	1	5	1	0	0	1752
3	1	5	2	1	11	13,578	3	1	5	2	0	3	13,578	3	1	5	3	0	2	438
3	1	6	2	1	1	876	3	1	6	2	0	0	438	3	1	6	3	0	0	438
3	2	6	2	0	1	876	3	3	2	2	1	0	438	3	3	2	3	0	0	438
3	3	4	1	1	0	3066	3	3	4	1	0	0	1752	3	3	4	2	1	8	3504
3	3	4	2	0	0	1314	3	3	4	3	1	13	876	3	3	4	3	0	3	1314
3	3	5	1	0	0	1314	3	3	5	2	0	0	2190	3	4	4	2	0	1	1752
3	4	4	3	0	1	4380	3	4	5	2	0	0	1752	4	3	3	3	0	2	438
4	3	4	2	1	2	3504	4	3	4	2	0	0	1752	4	3	4	3	0	7	1314
4	3	5	1	0	0	438	5	1	2	2	1	0	1314	5	1	2	2	0	0	876
5	1	2	3	1	0	438	5	1	2	3	0	0	2190	5	1	3	1	1	0	438
5	1	3	1	0	0	1314	5	1	3	2	0	0	876	5	1	4	2	1	3	1752
5	1	4	2	0	0	1752	5	1	5	1	1	3	438	5	1	5	1	0	2	1314
5	1	5	2	0	0	3504	5	1	6	1	1	0	438	5	1	6	2	0	0	876
5	2	3	2	0	0	4818	5	2	4	1	1	0	438	5	3	2	2	1	0	438
5	3	2	2	0	0	876	5	3	2	3	1	2	1752	5	3	2	3	0	0	876
5	3	4	2	1	2	2190	5	3	4	2	0	1	6132	5	3	5	2	0	0	876
5	4	3	1	1	1	2190	5	4	3	1	0	0	876	5	4	3	2	1	0	1314
5	4	4	1	0	0	438	5	4	4	2	1	0	438	5	4	5	2	0	0	438

The five explanatory variables are System, Operator type, Valve type, Head size, and Operation mode (in columns (1)–(5), respectively), taking values as described in the text

Table B.17. ODS graphics plots options for the REG procedure

Plot description	PLOTS option	ODS graph name
Adjusted R-square statistic for models	ADJRSQ	AdjrsqPlot
examined doing variable selection AIC statistic for models examined doing variable selection	AIC	AICPlot
BIC statistic for models examined doing variable selection	BIC	BICPlot
Cooks D statistic versus observation number	COOKSD	CooksDPlot
Cp statistic for models examined doing variable selection	CP	CPPlot
Panel of fit diagnostics	DIAGNOSTICS	DiagnosticsPanel
Regression line, confidence limits, and	FIT	FitPlot
prediction limits overlaid on scatter plot of data Dependent variable versus predicted values	OBSERVEDBYPREDICTED	ObservedByPredicted
Partial regression plot	PARTIAL	PartialPlot
Normal quantile plot of residuals	QQ	QQPlot
Residuals versus predicted values	RESIDUALBYPREDICTED	ResidualByPredicted
Plot of residuals versus regressor	RESIDUALS	ResidualPlot
R-square statistic for models examined doing variable selection	RSQUARE	RSquarePlot
Studentized residuals versus leverage	RSTUDENTBYLEVERAGE	RStudentByLeverage
Studentized residuals versus predicted values	RSTUDENTBYPREDICTED	RStudentByPredicted
SBC statistic for models examined doing variable selection	SBC	SBCPlot
Panel of fit statistics for models examined doing variable selection	CRITERIA	SelectionCriterionPanel

Table B.18. ODS tables produced by the REG procedure

ODS table name	Description	Statement	Option
ANOVA	Model ANOVA table	MODEL	Default
CollinDiag	Collinearity Diagnostics table	MODEL	COLLIN
Corr	Correlation matrix for analysis variables	PROC	ALL, CORR
CorrB	Correlation of estimates	MODEL	CORRB
CovB	Covariance of estimates	MODEL	COVB
CrossProducts	Bordered model XX matrix	MODEL	ALL, XPX
FitStatistics	Model fit statistics	MODEL	Default
InvXPX	Bordered XX inverse matrix	MODEL	I
OutputStatistics	Output statistics table	MODEL	ALL, CLI, CLM, INFLUENCE, P, R
ParameterEstimates	Model parameter estimates	MODEL	Default if SELECTION= is not specified
ResidualStatistics	Residual statistics and PRESS statistic	MODEL	ALL, CLI, CLM, INFLUENCE, P, R
SelParmEst	Parameter estimates for selection methods	MODEL	SELECTION=BACKWARD | FORWARD | STEPWISE | MAXR | MINR
SelectionSummary	Selection summary for FORWARD, BACKWARD, and STEPWISE methods	MODEL	SELECTION=BACKWARD | FORWARD | STEPWISE
SimpleStatistics	Simple statistics for analysis variables	PROC	ALL, SIMPLE
SubsetSelSummary	Selection summary for R-square, Adj-RSq, and Cp methods	MODEL	SELECTION=RSQUARE | ADJRSQ | CP
USSCP	Uncorrected SSCP matrix for analysis variables	PROC	ALL, USSCP

References

Agresti, A. (2013). *Categorical data analysis* (3rd ed.). Hoboken, NJ: Wiley.

Akaike, H. (1981). Likelihood of a model and information criteria. *Journal of Econometrics, 16*, 3–14.

Armitage, P., & Berry, G. (1994). *Statistical methods in medical research* (3rd ed.). Malden, MA: Blackwell.

Bates, D.M., & Watts, D.G. (1988). *Nonlinear regression analysis and its applications*. New York, NY: Wiley.

Bliss, C. I. (1935). The calculation of the dosage-mortality curve. *Annals of Applied Biology, 22*(1), 134–167.

Bliss, C. I. (1970). *Statistics in biology* (Vol. 2). New York, NY: McGraw-Hill.

Bowerman, B. L., & O'Connell, R. T. (2004). *Business statistics in practice* (4th ed.). Chicago, IL: McGraw-Hill/Irwin.

Box, G. E. P., Hunter, W. G., & Hunter, J. S. (1978). *Statistics for experimenters*. New York, NY: Wiley.

Breslow, N. E. (1984). Extra-Poisson variation in Log-linear models. *Applied Statistics, 33*(1), 38–44.

Chambers, J. M., Cleveland, W. S., Kleiner, B., & Tukey, P. A. (1983). *Graphical methods in data analysis*. Belmont, CA: Wadsworth.

Chen, W. W., Neipel, M., & Sorger, P. K. (2010). Classic and contemporary approaches to modeling biochemical reactions. *Genes & Development, 24*(17), 1861–1875.

Collett, D. (2003). *Modelling binary data*. London: Chapman & Hall.

Crowder, M. J. (1978). Beta-binomial Anova for proportions. *Applied Statistics, 27*(1), 34–37.

Deak, N. A., & Johnson, L. A. (2007). Effects of extraction temperature and preservation method on functionality of soy protein. *Journal of the American Oil Chemists' Society, 84*, 259–268.

Devore, J. L. (1982). *Probability and statistics for engineering and the sciences*. Monterey, CA: Brooks/Cole.

M. G. Marasinghe, K. J. Koehler, *Statistical Data Analysis Using SAS*, Springer Texts in Statistics, https://doi.org/10.1007/978-3-319-69239-5

Draper, N. R., & Smith, H. (1981) *Applied regression analysis* (2nd ed.). New York, NY: Wiley.

Draper, N. R., & Smith, H. (1998). *Applied regression analysis* (3rd ed.). New York, NY: Wiley.

Dunn, O. J., & Clark, V. A. (1987). *Applied statistics: Analysis of variance and regression analysis* (2nd ed.). New York, NY: Wiley.

Efron, B., Hastie, T. J., Johnstone, I. M., & Tibshirani, R. (2004). Least angle regression (with discussion). *Annals of Statistics, 32*, 407–499.

Endrenyi, L. (1981). *Kinetic data analysis.* New York: Springer.

Graubard, B. I., & Korn, E. L. (1987). Choice of column scores for testing independence in ordered 2xk contingency tables (with discussion). *Biometrics, 43*, 471–476.

Henderson, C. R., Kempthorne, O., Searle, S. R., & von Krosigk, C. N. (1959). Estimation of environmental and genetic trends from records subject to culling. *Biometrics, 15*, 192–218.

Kenward, M. G., & Roger, J. H. (1997). Small sample inference for fixed effects from restricted maximum likelihood. *Biometrics, 53*, 983–997.

Kirk, R. E. (1982). *Experimental design* (2nd ed.). Monterey, CA: Brooks/Cole.

Koopmans, L. H. (1987). *Introduction to contemporary statistical methods* (2nd ed.). Boston, MA: Duxbury.

Kuehl, R. O. (2000). *Design of experiments: Statistical principles of research design and analysis.* Pacific Grove, CA: Brooks/Cole.

Kutner, M. H., Nachtsheim, C. J., & Neter, J. (2004). *Applied linear regression models* (4th ed.). Chicago, IL: McGraw-Hill/Irwin.

Kutner, M. H., Nachtsheim, C. J., Neter, J., & Li, W. (2005). *Applied linear statistical models* (5th ed.). Chicago, IL: McGraw-Hill/Irwin.

Leskovac, V. (2003). *Comprehensive enzyme kinetics.* New York: Kluwer Academic/Plenum.

Lindsey, J. K. (2001). *Nonlinear models in medical sciences.* New York, NY: Oxford University Press.

Littell, R. C., Freund, R. J., & Spector, P. C. (1991). *SAS system for linear models* (3rd ed.). Cary, NC: SAS Institute Inc.

Lloyd, C. J. (1999). *Statistical analysis of categorical data.* New York, NY: Wiley.

Lund, R. E. (1975). Tables for an approximate test for outliers in linear models. *Technometrics, 17*, 473–476.

Madsen, H., & Thyregod, P. (2011). *Introduction to general and generalized linear models.* Boca Raton, FL: Chapman & Hall/CRC.

Margolin, B. H., Kaplan, N., & Zeiger, E. (1981). Statistical analysis of the Ames Salmonella test. *Proceedings of the National Academy of Sciences of the United States of America, 78*(6), 3779–3783.

Mason, R. L., Gunst, R. F., & Hess, J. L. (1989). *Statistical design & analysis of experiments.* New York, NY: Wiley.

McClave, J. T., Benson, G. P., & Sincich T. L. (2000). *Statistics for business and economics* (8th ed.). Englewood Cliffs, NJ: Prentice Hall Inc.

McCullagh, P., & Nelder, J. (1989). *Generalized linear models* (2nd ed.). Boca Raton, FL: Chapman & Hall/CRC.

McDonald, G. C., & Schwing, R. C. (1973). Instabilities of regression estimates relating air pollution to mortality. *Technometrics, 15*, 463–482.

Milliken, G. A., & Johnson, D. E. (2001). *Analysis of Messy data, volume III: Analysis of covariance.* Boca Raton, FL: Chapman & Hall/CRC.

Montgomery, D. C. (1991). *The design and analysis of experiments* (3rd ed.). New York, NY: Wiley.

Montgomery, D. C. (2013). *The design and analysis of experiments* (8th ed.). New York, NY: Wiley.

Moore, L. M., & Beckman, R. J. (1988). Approximate one-sided tolerance bounds on the number of failures using Poisson regression. *Technometrics, 30*, 283–290.

Morel, J. G., & Neerchal, N. K. (2012). *Overdispersion models in SAS.* Cary, NC: SAS Institute Inc.

Morrison, D. F. (1983). *Applied linear statistical methods.* Englewood Cliffs, NJ: Prentice Hall Inc.

Myers, R. H. (1990). *Classical and modern regression with applications* (2nd ed.). Boston, MA: PWS-KENT Publishing.

Nelder, J., & Wedderburn, R. (1972). Generalized linear models. *Journal of the Royal Statistical Society, Series A, 135*, 370–384.

Okada, Y., Yabe, T., & Oda, S. (2010). Temperature-dependent sex determination in Japanese pond turtles, *Mauremys japonica* (Reptilia: Geoemydidea). *Current Herpetology, 29*(1), 1–10.

Ostle, B. (1963). *Statistics in research* (2nd ed.). Ames, IA: Iowa State University Press.

Ott, R. L., Larson, R. F., & Mendenhall, W. (1987). *Statistics: A tool for the social sciences* (4th ed.). Boston, MA: Duxbury.

Ott, R. L., & Longnecker, M. (2001). *An introduction to statistical methods and data analysis* (5th ed.). Pacific Grove, CA: Duxbury.

Price, C. J., Kimmel, C. A., George, J. D., & Marr, M.C. (1987). The developmental toxicity of diethylene glycol dimethyl ether in mice. *Fundamental and Applied Toxicology, 8*, 115–126.

Rice, J. A. (1988). *Mathematical statistics and data analysis.* Pacific Grove, CA: Wadsworth & Brooks/Cole.

Sahai, H., & Ageel M. I. (2000). *The analysis of variance.* Boston, MA: Birkhäuser.

Schlotzhauer, S. D., & Littell, R. C. (1997). *SAS system for elementary statistical analysis* (2nd ed.). Cary, NC: SAS Institute Inc.

Searle, S. R. (1971). *Linear models.* New York, NY: Wiley.

Searle, S. R., Casella, G., & McCulloch, C. E. (1992). *Variance components.* New York, NY: Wiley.

Simonoff, J. S. (2003). *Analyzing categorical data.* New York: Springer-Verlag.

Simpson, J., Olsen, A., & Eden, J. (1975). A Bayesian analysis of a multiplicative treatment effect in weather modification. *Technometrics, 17*, 161–166.

Snedecor, G. W., & Cochran, W. G. (1989). *Statistical methods* (8th ed.). Ames, IA: Iowa State University Press.

Sokal, R. R., & Rohlf, J. F. (1995). *Biometry: The principles and practice of statistics in biological research* (3rd ed.). New York, NY: Freeman.

Stamey, T., Kabalin, J., McNeal, J., Johnstone, I., Freiha, F., Redwine, E., & Yang, N. (1989). Prostate specific antigen in the diagnosis and treatment of adenocarcinoma of the prostate II radical prostatectomy treated patients. *Journal of Urology, 16*, 1076–1083.

Tibshirani, R. (1996). Regression shrinkage and selection via the lasso. *Journal of the Royal Statistical Society, Series B, 58*, 267–288.

Tukey, J. W. (1949). One degree of freedom for nonadditivity. *Biometrics, 5*, 232–242.

Ver Hoef, J., & Boveng, P. (2007). Quasi-Poisson vs. negative binomial regression: How should we model overdispersed count data? *Ecology, 11*, 2766–2772.

Wedderburn, R. W. M. (1974). Quasi-likelihood functions, generalized linear models, and the Gauss-Newton method. *Biometrika, 61*, 439–447.

Weisberg, S. (1985). *Applied linear regression analysis* (2nd ed.). New York, NY: Wiley.

Index

M. G. Marasinghe, K. J. Koehler, *Statistical Data Analysis Using SAS*, Springer Texts in Statistics, https://doi.org/10.1007/978-3-319-69239-5

Printed in the United States
By Bookmasters